WASTEWATER TREATMENT

WASTEWATER TREATMENT

Edited by

David H.F. Liu
Béla G. Lipták

Paul A. Bouis
Special Consultant

CRC Press
Taylor & Francis Group
Boca Raton London New York

CRC Press is an imprint of the
Taylor & Francis Group, an **informa** business

CRC Press
Taylor & Francis Group
6000 Broken Sound Parkway NW, Suite 300
Boca Raton, FL 33487-2742

First issued in paperback 2019

ISBN-13: 978-1-56670-515-8 (hbk)
ISBN-13: 978-0-367-39912-2 (pbk)

Library of Congress Card Number 99-052047

Library of Congress Cataloging-in-Publication Data

Wastewater treatment / edited by David H. F. Liu, Béla G. Lipták
p. cm.
Includes bibliographical references and index.
ISBN 1-56670-515-0 (alk. paper)
1. Sewage—Purification. 2. Factory and trade waste—Management. I. Liu, David H. F.
II. Lipták, Béla G.
TD745 .W367 1999
628.3—dc21 99-052047

Visit the Taylor & Francis Web site at
http://www.taylorandfrancis.com

and the CRC Press Web site at
http://www.crcpress.com

Preface

Dr. David H.F. Liu passed away prior to the preparation of this book.
He will be long remembered by his coworkers,
and the readers of this book will carry his memory into the 21st Century.

Engineers respond to the needs of society with technical innovations. Their tools are the basic sciences. Some engineers might end up working *on* these tools instead of working *with* them. Environmental engineers are in a privileged and challenging position, because their tools are the totality of man's scientific knowledge, and their target is nothing less than human survival through making man's peace with nature.

In the natural life cycle of the water bodies (Figure 1), the sun provides the energy source for plant life (algae), which produces oxygen while converting the inorganic molecules into larger organic ones. The animal life obtains its muscle energy (heat) by consuming these molecules and by also consuming the dissolved oxygen content of the water.

When a town or industry discharges additional organic material into the waters (which nature intended to be disposed of as fertilizer on land), the natural balance is upset. The organic effluent acts as a fertilizer, therefore the algae overpopulates and eventually blocks the transparency of the water. When the water becomes opaque, the ultraviolet rays of the sun can no longer penetrate it. This cuts off the algae from its energy source and it dies. The bacteria try to protect the life cycle in the water by attempting to break down the excess organic material (including the dead body cells of the algae), but the bacteria require oxygen for the digestion process. As the algae is no longer producing fresh oxygen, the dissolved oxygen content of the water drops, and when it reaches zero, all animals suffocate. At that point the living water body has been converted into an open sewer.

In the United States, the setting of water quality standards and the regulation of discharges have been based on the "assimilative capacity" of the receiving waters (a kind of pollution dilution approach), which allows discharges into as yet unpolluted waterways. The Water Pollution Act of 1972 would have temporarily required industry to apply the "best practicable" and "best available" treatments

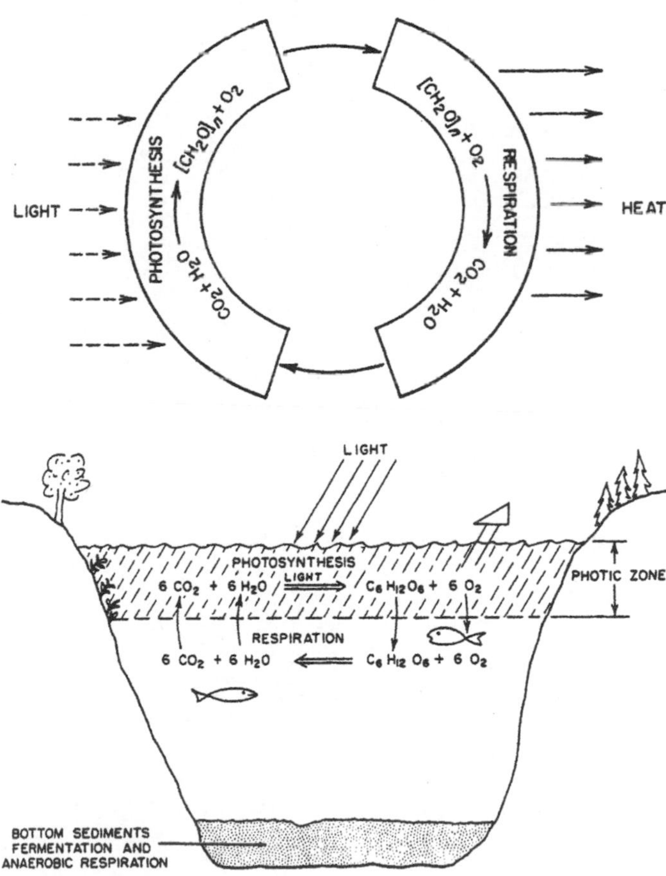

FIG. 1 The natural life cycle.

of waste emissions and aimed for zero discharge by 1985. While this last goal has not been reached, the condition of American waterways generally improved during the last decades, while on the global scale water quality has deteriorated.

Water availability has worsened since the first edition of this handbook. In the United States the daily withdrawal rate is about 2,000 gallons per person, which represents roughly one-third of the total daily runoff. The bulk of this water is used by agriculture and industry. The average daily water consumption per household is about 1000 gallons and, on the East Coast, the daily cost of that water is \$2–\$3. As some 60% of the discharged pollutants (sewage, industrial waste, fertilizers, pesticides, leachings from landfills and mines) reenter the water supplies, there is a direct relationship between the quality and cost of supply water and the degree of waste treatment in the upstream regions.

There seems to be some evidence that the residual chlorine from an upstream wastewater treatment plant can combine in the receiving waters with industrial wastes to form carcinogenic chlorinated hydrocarbons, which can enter the drinking water supplies downstream. Toxic chemicals from the water can be further concentrated through the food chain. Some believe that the gradual poisoning of the environment is responsible for cancer, AIDS, and other forms of immune deficiency and self-destructive diseases.

While the overall quality of the waterways has improved in the United States, worldwide the opposite occurred. This is caused not only by overpopulation, but also by ocean dumping of sludge, toxins, and nuclear waste, as well as by oil leaks from off-shore oil platforms. We do not yet fully understand the likely consequences, but we can be certain that the ability of the oceans to withstand and absorb pollutants is not unlimited and, therefore, international regulation of these discharges is essential. In terms of international regulations, we are just beginning to develop the required new body of law.

Protecting the global environment, protecting life on this planet, must become a single-minded, unifying goal for all of us. The struggle will overshadow our differences, will give meaning and purpose to our lives and, if we succeed, it will mean survival for our children and the generations to come.

Béla G. Lipták

Contributors

Carl E. Adams, Jr.
BSCE, MSSE, PhDCE, PE; Technical Director, Associated Water & Air Resources Engineers, Inc.

Donald B. Aulenbach
BSCh, MS, PhDS; Associate Professor, Bio-Environmental Engineering, Rensselaer Polytechnic Institute

Joseph L. Bollyky

Jerry L. Boyd

Thomas F. Brown, Jr.
BSAE, EIT; Assistant Director, Environmental Engineering, Commercial Solvents Corp.

Robert D. Buchanan
BSCE, MSCE, PE; Chief Sanitary Engineer, Bureau of Indian Affairs

Don E. Burns
BSCE, MSCE, PhD-SanE; Senior Research Engineer, Eimco Corp.

Larry W. Canter
BE, MS, PhD, PE;
Sun Company Chair of Ground Water Hydrology, University of Oklahoma

George J. Crits
BSChE, MSChE, PE; Technical Director, Cochrane Division, Crane Company

Donald Dahlstrom
PhDChE; Vice President and Director of Research & Development, Eimco Corp.

Stacy L. Daniels
BSChE, MSSE, MSChE, PhD; Development Engineer, The Dow Chemical Company

Frank W. Dittman
BSChE, MSChE, PhD, PE;
Professor of Chemical Engineering, Rutgers University

Wayne F. Echelberger, Jr.
BSCE, MSE, MPH, PhD; Associate Professor of Civil Engineering, University of Notre Dame

Ronald G. Gantz
BSChE; Senior Process Engineer, Continental Oil Company

Louis C. Gilde, Jr.
BSSE; Director, Environmental Engineering, Campbell Soup Company

Brian L. Goodman
BS, MS, PhD; Director, Technical Services, Smith & Loveless Division, Ecodyne Corp.

Negib Harfouche
PhD; President, NH Environmental Consultants

R. David Holbrook
BSCE, MSCE; Senior Process Engineer, I. Krüger, Inc.

Sun-Nan Hong
BSChE, MSChE, PhD; Vice President, Engineering, I. Krüger, Inc.

Derk T.A. Huibers
BSChE, MSChE, PhDChE, FAIChE; Manager, Chemical Processes Group, Union Camp Corp.

Frederick W. Keith, Jr.
BSChE, PhDChE, PE; Manager, Applications Research, Pennwalt Corp.

Mark K. Lee
BSChE, MEChE; Project Manager, Westlake Polymers Corp.

Béla G. Lipták
ME, MME, PE; Process Control and Safety Consultant, President, Liptak Associates, P.C.

Janos Lipták
CE, PE; Senior Partner, Janos Liptak & Associates

David H.F. Liu
PhD, ChE; Principal Scientist, J.T. Baker, Inc. a division of Procter & Gamble

Francis X. McGarvey
BSChE, MSChE; Manager, Technical Center, Sybron Chemical Company

Thomas J. Myron, Jr.
BSChE; Senior Systems Design Engineer, The Foxboro Company

Van T. Nguyen
BSE, MSE, PhD; Department of Civil Engineering, California State University, Long Beach

Joseph G. Rabosky
BSChE, MSE, PE; Senior Project Engineer, Calgon Corp.

LeRoy H. Reuter
MS, PhD, PE; Consultant

Bernardo Rico-Ortega
BSCh, MSSE; Product Specialist, Pollution Control Department, Nalco Chemical Company

Chakra J. Santhanam
BSChE, MSChE, ChE, PE; Senior Environmental Engineer, Crawford & Russell, Inc.

E. Stuart Savage
BSChE, PE; Manager, Research and Development, Water & Waste Treatment, Dravco Corp.

Frank P. Sebastian
MBA, BSME; Senior Vice President, Envirotech Corp.

Gerald L. Shell
MSCE, PE; Director of Sanitary Engineering, Eimco Corp.

Wen K. Shieh
PhD; Department of Systems Engineering, University of Pennsylvania

John R. Snell
BECE, MSSE, DSSE, PE; President, John R. Snell Engineers

Paul L. Stavenger
BSChE, MSChE; Director of Technology, Process Equipment Division, Dorr-Oliver, Inc.

Michael S. Switzenbaum
BA, MS, PhD; Professor, Environmental Engineering Program, Department of Civil and Environmental Engineering, University of Massachusetts, Amherst

Contents

11 Sludge Disposal 405

1

Sources and Characteristics

Larry W. Canter | Negib Harfouche

1.1 NATURE OF WASTEWATER

The nature of wastewater is described by its flow and quality characteristics. In addition, wastewater discharges are classified based on whether they are from municipalities or industries. Flow rates and quality characteristics of industrial wastewater are more variable than those for municipal wastewater.

Flow Rates

Municipal wastewater is comprised of domestic (or sanitary) wastewater, industrial wastewater, infiltration and inflow into sewer lines, and stormwater runoff. Domestic wastewater refers to wastewater discharged from residences and from commercial and institutional facilities (Metcalf and Eddy, Inc. 1991). Domestic water usage, and the resultant wastewater, is affected by climate, community size, density of development, community affluence, dependability and quality of water supply, water conservation requirements or practices, and the extent of metered services. Metcalf and Eddy, Inc. (1991) provide details on the influence of these factors. Additional factors influencing water use include the degree of industrialization, cost of water, and supply pressure (Qasim 1985). One result of the combined influence of these factors is water use fluctuations. Table 1.1.1 summarizes such fluctuations (Metcalf and Eddy, Inc. 1991). About 60 to 85% of water usage becomes wastewater, with the lower percentages applicable to the semiarid region of the southwestern United States (Metcalf and Eddy, Inc. 1991).

Environmental engineers can use unit flow rate data to develop estimates for wastewater flow rates from residential areas, commercial districts, and institutional facilities. Tables 1.1.2 through 1.1.4 depict data for these use categories, respectively. Industrial wastewater flow rates vary and are a function of the type and size of industry. For estimation purposes, typical design flows from industrial areas that have little or no wet-process-type industries are 1000 to 1500 gal/acre per day (9 to 14 $m^3/ha \cdot d$) for light industrial developments and 1500 to 3000 gal/acre per day (14 to 28 $m^3/ha \cdot d$) for medium industrial developments (Metcalf and Eddy, Inc. 1991). Better estimates for industries can be developed with industry-specific information.

Wastewater volume generated in a municipality depends on the population served, the per capita contribution, and other nondomestic sources such as industrial wastewater discharges. Environmental engineers may need to use population forecasting to project future rates of wastewater generation in the service area of a wastewater

TABLE 1.1.1 TYPICAL FLUCTUATIONS IN WATER USE IN COMMUNITY SYSTEMS

	Percentage of Average for Year	
Water Use	*Range*	*Typical*
Daily average in maximum month	110–140	120
Daily average in maximum week	120–170	140
Maximum day	160–220	180
Maximum hr	225–320	270[a]

Source: Metcalf and Eddy, Inc., 1991, *Wastewater engineering,* 3d ed. (New York: McGraw-Hill).
Note: [a]1.5 × maximum day value.

TABLE 1.1.2 TYPICAL WASTEWATER FLOW RATES FROM RESIDENTIAL SOURCES

		Flow, gal/unit·d	
Source	*Unit*	*Range*	*Typical*
Apartment:			
High-rise	Person	35–75	50
Low-rise	Person	50–80	65
Hotel	Guest	30–55	45
Individual residence:			
Typical home	Person	45–90	70
Better home	Person	60–100	80
Luxury home	Person	75–150	95
Older home	Person	30–60	45
Summer cottage	Person	25–50	40
Motel:			
With kitchen	Unit	90–180	100
Without kitchen	Unit	75–150	95
Trailer park	Person	30–50	40

Source: Metcalf and Eddy, Inc., 1991.
Note: l = gal × 3.7854.

TABLE 1.1.3 TYPICAL WASTEWATER FLOW RATES FROM COMMERCIAL SOURCES

Source	Unit	Flow, gal/unit·d Range	Typical
Airport	Passenger	2–4	3
Automobile service station	Vehicle served	7–13	10
	Employee	9–15	12
Bar	Customer	1–5	3
	Employee	10–16	13
Department store	Toilet room	400–600	500
	Employee	8–12	10
Hotel	Guest	40–56	48
	Employee	7–13	10
Industrial building (sanitary waste only)	Employee	7–16	13
Laundry (self-service)	Machine	450–650	550
	Wash	45–55	50
Office	Employee	7–16	13
Restaurant	Meal	2–4	3
Shopping center	Employee	7–13	10
	Parking space	1–2	2

Source: Metcalf and Eddy, 1991.
Note: l = gal × 3.7854.

TABLE 1.1.4 TYPICAL WASTEWATER FLOW RATES FROM INSTITUTIONAL SOURCES

Source	Unit	Flow, gal/unit·d Range	Typical
Hospital, medical	Bed	125–240	165
	Employee	5–15	10
Hospital, mental	Bed	75–140	100
	Employee	5–15	10
Prison	Inmate	75–150	115
	Employee	5–15	10
Rest home	Resident	50–120	85
School, day			
With cafeteria, gym, and showers	Student	15–30	25
With cafeteria only	Student	10–20	15
Without cafeteria and gym	Student	5–17	11
School, boarding	Student	50–100	75

Source: Metcalf and Eddy, Inc., 1991.
Note: l = gal × 3.7854.

treatment plant. Some mathematical or graphical methods used to project population data to a design year include (Qasim 1985):

- arithmetic growth
- geometric growth
- decreasing rate of increase
- mathematical or logistic curve fitting
- graphical comparison with similar cities
- ratio method
- employment forecast
- birth cohort

Water usage exhibits daily, weekly, and seasonal patterns; and wastewater flow rates can also exhibit such patterns. Figures 1.1.1 and 1.1.2 show typical hourly, daily,

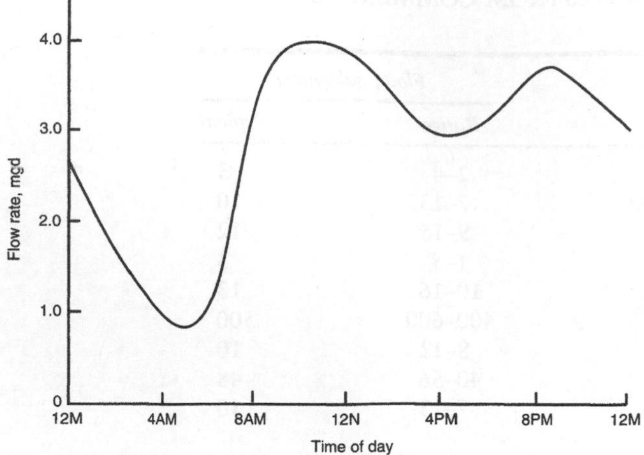

FIG. 1.1.1 Typical pattern of hourly variations in domestic wastewater flow rates. (Reprinted, with permission, from Metcalf and Eddy, Inc., 1991.)

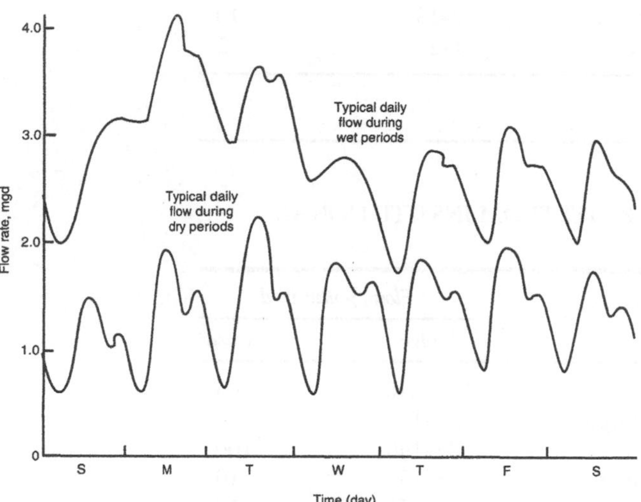

FIG. 1.1.2 Typical patterns of daily and weekly variations in domestic wastewater flow rates. (Reprinted, with permission, from Metcalf and Eddy, Inc., 1991.)

and weekly wastewater flow rates, respectively (Metcalf and Eddy, Inc. 1991).

Wide variations of wastewater flow rates can occur within a municipality. For example, minimum to maximum flow rates range from 20 to 400% of the average daily rate for small communities with less than 1000 people, from 50 to 300% for communities with populations between 1000 and 10,000, and up to 200% for communities up to 100,000 in population (Water Pollution Control Federation and American Society of Civil Engineers 1977). Large municipalities have variations from 1.25 to 1.5 average flow. When storm water runoff goes into municipal sewerage systems, the maximum flow rate is often two to four times the average dry-weather flow (Water Pollution Control Federation and American Society of Civil Engineers 1977).

Flow rate information needed in designing a wastewater treatment plant includes (Metcalf and Eddy, Inc. 1991):

AVERAGE DAILY FLOW—The average flow rate occurring over a 24-hr period based on total annual flow rate data. Environmental engineers use average flow rate in evaluating treatment plant capacity and in developing flow rate ratios.

MAXIMUM DAILY FLOW—The maximum flow rate occurring over a 24-hr period based on annual operating data. The maximum daily flowrate is important in the design of facilities involving retention time, such as equalization basins and chlorine-contact tanks.

PEAK HOURLY FLOW—The peak sustained hourly flow rate occurring during a 24-hr period based on annual operating data. Data on peak hourly flows are needed for the design of collection and interceptor sewers, wastewater pumping stations, wastewater flowmeters, grit chambers, sedimentation tanks, chlorine-contact tanks, and conduits or channels in the treatment plant.

MINIMUM DAILY FLOW—The minimum flow rate that occurs over a 24-hr period based on annual operating data. Minimum flow rates are important in sizing conduits where solids deposition might occur at low flow rates.

MINIMUM HOURLY FLOW—The minimum sustained hourly flow rate occurring over a 24-hr period based on annual operating data. Environmental engineers need data

TABLE 1.1.5 PHYSICAL, CHEMICAL, AND BIOLOGICAL WASTEWATER CHARACTERISTICS CONSIDERED FOR DESIGN

Physical	*Chemical*	*Biological*
Solids	Organics	Plants
Temperature	Proteins	Animals
Color	Carbohydrates	Viruses
Odor	Lipids	
	Surfactants	
	Phenols	
	Pesticides	
	Inorganics	
	pH	
	Chloride	
	Alkalinity	
	Nitrogen	
	Phosphorus	
	Heavy metals	
	Toxic materials	
	Gases	
	Oxygen	
	Hydrogen sulfide	
	Methane	

Source: Water Pollution Control Federation and American Society of Civil Engineers, 1977, *Wastewater treatment plant design* (Washington, D.C.), 10.

on the minimum hourly flow rate to determine possible process effects and size wastewater flow meters, particularly those that pace chemical-feed systems.

SUSTAINED FLOW—The flow rate value sustained or exceeded for a specified number of consecutive days based on annual operating data. Data on sustained flow rates may be used in sizing equalization basins and other plant hydraulic components.

Wastewater Characteristics

Wastewater quality can be defined by physical, chemical, and biological characteristics. Physical parameters include color, odor, temperature, solids (residues), turbidity, oil, and grease. Solids can be further classified into suspended and dissolved solids (size and settleability) as well as organic (volatile) and inorganic (fixed) fractions. Chemical parameters associated with the organic content of wastewater include the biochemical oxygen demand (BOD), chemical oxygen demand (COD), total organic carbon (TOC), and total oxygen demand (TOD). BOD is a measure of the organics present in the water, determined by measuring the oxygen necessary to biostabilize the organics (the oxygen equivalent of the biodegradable organics present). Inorganic chemical parameters include salinity, hardness, pH, acidity, alkalinity, iron, manganese, chlorides, sulfates, sulfides, heavy metals (mercury, lead, chromium, copper, and zinc), nitrogen (organic, ammonia, nitrite, and nitrate), and phosphorus. Bacteriological parameters include coliforms, fecal coliforms, specific pathogens, and viruses.

Design considerations for wastewater treatment facilities are based in part on the characteristics of the wastewater; Table 1.1.5 lists some key characteristics of concern.

Table 1.1.6 shows the typical concentration range of various constituents in untreated domestic wastewater. Depending on the concentrations, wastewater is classified as strong, medium, or weak. Table 1.1.7 shows typical mineral increases resulting from domestic water use. The types and numbers of microorganisms in untreated domestic wastewater vary widely; examples of such variations are shown in Table 1.1.8.

TABLE 1.1.6 TYPICAL COMPOSITION OF UNTREATED DOMESTIC WASTEWATER

		Concentration		
Contaminants	*Unit*	*Weak*	*Medium*	*Strong*
Total solids (TS)	mg/l	350	720	1200
total dissolved solids (TDS)	mg/l	250	500	850
fixed	mg/l	145	300	525
volatile	mg/l	105	200	325
suspended solids (SS)	mg/l	100	220	350
fixed	mg/l	20	55	75
volatile	mg/l	80	165	275
Settleable solids	mL/l	5	10	20
BOD, mg/l:				
5-day, 20°C (BOD_5, 20°C)	mg/l	110	220	400
TOC	mg/l	80	160	290
COD	mg/l	250	500	1000
Nitrogen (total as N)	mg/l	20	40	85
organic	mg/l	8	15	35
free ammonia	mg/l	12	25	50
nitrites	mg/l	0	0	0
nitrates	mg/l	0	0	0
Phosphorus (total as P)	mg/l	4	8	15
organic	mg/l	1	3	5
inorganic	mg/l	3	5	10
Chlorides[a]	mg/l	30	50	100
Sulfate[a]	mg/l	20	30	50
Alkalinity (as $CaCO_3$)	mg/l	50	100	200
Grease	mg/l	50	100	150
Total coliform	no/100 ml	10^6–10^7	10^7–10^8	10^7–10^9
Volatile organic compounds (VOCs)	μg/L	<100	100–400	>400

Source: Metcalf and Eddy, Inc., 1991, *Wastewater engineering,* 3d ed. (New York: McGraw-Hill).
Notes: °F = 1.8(°C) + 32.
[a]Values should be increased by the amount present in the domestic water supply; see Table 1.1.7.

TABLE 1.1.7 TYPICAL MINERAL INCREASE FROM DOMESTIC WATER

Constituent	*Increment Range,[a] mg/l*
Anions	
Bicarbonate (HCO_3)	50–100
Carbonate (CO_3)	0–10
Chloride (Cl)	20–50[b]
Nitrate (NO_3)	20–40
Phosphate (PO_4)	5–15
Sulfate (SO_4)	15–30
Cations	
Calcium (Ca)	6–16
Magnesium (Mg)	4–10
Potassium (K)	7–15
Sodium (Na)	40–70
Other constituents	
Aluminum (Al)	0.1–0.2
Boron (B)	0.1–0.4
Fluoride (F)	0.2–0.4
Manganese (Mn)	0.2–0.4
Silica (SiO_2)	2–10
Total alkalinity (as $CaCO_3$)	60–120
TDS	150–380

Source: Metcalf and Eddy, Inc., 1991.
Notes: [a]Reported values do not include commercial and industrial additions.
[b]Excluding the addition from domestic water softeners.

TABLE 1.1.8 TYPES AND NUMBER OF MICROORGANISMS TYPICALLY FOUND IN UNTREATED DOMESTIC WASTEWATER

Organism	*Concentration, number/ml*
Total coliform	10^5–10^6
Fecal coliform	10^4–10^5
Fecal streptococci	10^3–10^4
Enterococci	10^2–10^3
Shigella	Present[a]
Salmonella	10^0–10^2
Pseudomonas aeroginosa	10^1–10^2
Clostridium perfringens	10^1–10^3
Mycobacterium tuberculosis	Present[a]
Protozoan cysts	10^1–10^3
Giardia cysts	10^{-1}–10^2
Cryptosporidium cysts	10^{-1}–10^1
Helminth ova	10^{-2}–10^1
Enteric virus	10^1–10^2

Source: Metcalf and Eddy, Inc., 1991.
Note: [a]Results for these tests are usually reported as positive or negative rather than quantified.

The remainder of this subsection focuses on selected quality characteristics of wastewater. Details related to the analytical procedures for quality characteristics are available in *Standard methods for the examination of water and wastewater* (Clesceri, Greenberg, and Trussell 1989).

The TS content of wastewater can be defined as the matter that remains as residue upon evaporation at 103 to 105°C; Figure 1.1.3 shows the categories of solids in wastewater. Table 1.1.9 shows particle sizes related to different categories. Clesceri, Greenberg, and Trussell (1989) provide detailed information related to testing procedures. Figure 1.1.4 shows typical values for solids found in medium strength wastewater.

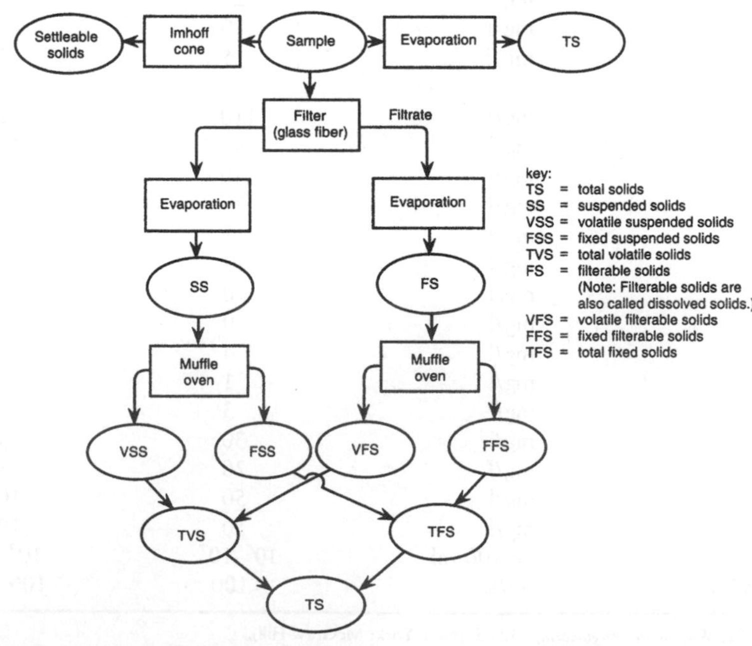

FIG. 1.1.3 Interrelationships of wastewater solids (also called dissolved solids). (Reprinted, with permission, from Metcalf and Eddy, Inc., 1991.)

TABLE 1.1.9 GENERAL CLASSIFICATION OF WASTEWATER SOLIDS

Particle Classification	*Particle Size, mm*
Dissolved	Less than 10^{-6}
Colloidal	10^{-6} to 10^{-3}
Suspended	Greater than 10^{-3}
Settleable	Greater than 10^{-2}

Source: R.A. Corbitt, 1990, Wastewater disposal, Chap. 6 in *Standard handbook of environmental engineering,* edited by R.A. Corbitt (New York: McGraw-Hill Publishing Company).

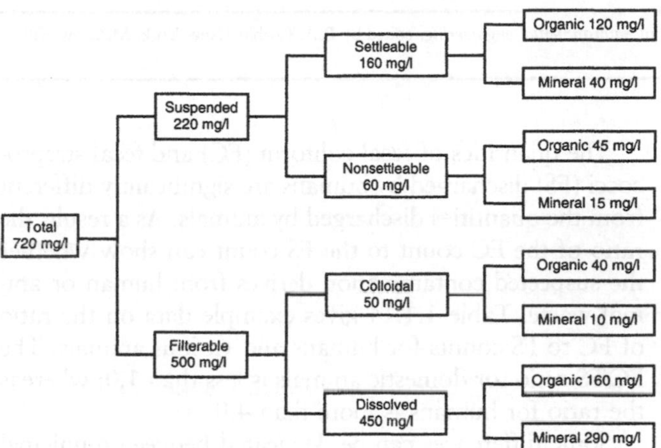

FIG. 1.1.4 Classification of solids found in medium-strength wastewater. (Metcalf and Eddy, Inc., 1991.)

Odors in wastewater are caused by gases from decomposition or by odorous substances within the wastewater. Table 1.1.10 lists examples of odorous compounds associated with untreated wastewater.

Organic matter in wastewater can be related to oxygen demand and organic carbon measurements. Table 1.1.11 shows several measures of organic matter. BOD refers to the amount of oxygen used by a mixed population of microorganisms under aerobic conditions that stabilize organic matter in the wastewater. The 5-day BOD is primarily composed of carbonaceous oxygen demand, while the 20-day BOD includes both carbonaceous and nitrogenous oxygen demands. Figure 1.1.5 is a typical BOD curve, developed via laboratory measurements, for domestic wastewater. The following relationships and definitions are associated with Figure 1.1.5 (Qasim 1985):

$$\begin{aligned} UOD &= L_o + L_n \\ y &= L_o(1 - e^{-Kt}) \\ L_{oT} &= L_o(0.02T + 0.60) \\ K_T &= K(1.047)^{T-20} \end{aligned} \qquad 1.1(1)$$

where:

UOD = ultimate oxygen demand, mg/l

L_n = nitrogenous oxygen demand or second stage BOD, mg/l

L_o = ultimate carbonaceous BOD, or first stage BOD at 20°C, mg/l (for domestic wastewater BOD_5 is approximately equal to $\frac{2}{3}$ L_o)

y = carbonaceous BOD at any time t, mg/l

L_{oT} = ultimate carbonaceous BOD at any temperature T°C, mg/l

K = reaction rate constants at 20°C, d^{-1} (for domestic wastewater K = 0.2 to 0.3 per d)

K_T = reaction rate constant to any temperature T°C, d^{-1}

For medium-strength domestic wastewater, about 75% of SS and 40% of FS are classified as organic. The main groups of organic substances in wastewater include proteins (40–60%), carbohydrates (25–50%), and fats and oils (10%) (Metcalf and Eddy, Inc. 1991). Urea, a constituent of urine, is also found in wastewater along with numerous synthetic organic compounds. Synthetic compounds include surfactants, organic priority pollutants, VOCs, and agricultural pesticides.

The U.S. Environmental Protection Agency (EPA) has identified 129 priority pollutants in wastewater; these pol-

TABLE 1.1.10 ODOROUS COMPOUNDS ASSOCIATED WITH UNTREATED WASTEWATER

Odorous Compound	*Chemical Formula*	*Odor, Quality*
Amines	CH_3NH_2, $(CH_3)_3H$	Fishy
Ammonia	NH_3	Ammoniacal
Diamines	$NH_2(CH_2)_4NH_2$, $NH_2(CH_2)_5NH_2$	Decayed flesh
Hydrogen sulfide	H_2S	Rotten eggs
Mercaptans (e.g., methyl and ethyl)	CH_3SH, $CH_3(CH_2)SH$	Decayed cabbage
Mercaptans (e.g., *T* = butyl and crotyl)	$(CH_3)_3CSH$, $CH_3(CH_2)_3SH$	Skunk
Organic sulfides	$(CH_3)_2S$, $(C_6H_5)_2S$	Rotten cabbage
Skatole	C_9H_9N	Fecal matter

Source: Metcalf and Eddy, Inc., 1991, *Wastewater engineering,* 3d ed. (New York: McGraw-Hill).

TABLE 1.1.11 OXYGEN DEMAND AND ORGANIC CARBON PARAMETERS

Parameter	*Description*
BOD_5	Biochemical or biological oxygen demand exerted in 5 days; the oxygen consumed by a waste through bacterial action; generally about 45–55% of THOD.
COD	Chemical oxygen demand. The amount of strong chemical oxidant (chromic acid) reduced by a waste; results are expressed in terms of an equivalent amount of oxygen; generally about 80% of THOD.
THOD	Theoretical oxygen demand. The amount of oxygen theoretically required to completely oxidize a compound to CO_2, H_2O, PO_4^{-3}, SO_4^{-2}, and NO_3.
TOC	Total organic carbon; generally about 30% of THOD.
BOD_L	The ultimate BOD exerted by a waste in an infinite time.
IOD	Immediate oxygen demand. The amount of oxygen consumed by a waste within 15 min (chemical oxidizers and bacteria not used).

Source: R.A. Corbitt, 1990, Wastewater disposal, Chap. 6 in *Standard handbook of environmental engineering*, edited by R.A. Corbitt (New York: McGraw-Hill Publishing Company).

lutants are subject to discharge standards. Examples of priority pollutants and their health-related concerns are listed in Table 1.1.12.

Pathogenic organisms in wastewater can be categorized as bacteria, viruses, protozoa, and helminths; Table 1.1.13 lists examples of such organisms present in raw domestic wastewater. Because of the many types of pathogenic organisms and the associated measurement difficulties, coliform organisms are frequently used as indicators of human pollution. On a daily basis, each person discharges from 100 to 400 billion coliform organisms, in addition to other kinds of bacteria (Metcalf and Eddy, Inc. 1991). In terms of the indicator concept, the presence of coliform organisms indicates that pathogenic organisms may also be present, and their absence indicates that the water is free from disease-producing organisms (Metcalf and Eddy, Inc. 1991). Total coliform and fecal coliform are often used as indicators of wastewater effluent disinfection.

The quantities of fecal coliform (FC) and fecal streptococci (FS) discharged by humans are significantly different from the quantities discharged by animals. As a result, the ratio of the FC count to the FS count can show whether the suspected contamination derives from human or animal waste. Table 1.1.14 gives example data on the ratio of FC to FS counts for humans and various animals. The FC/FS ratio for domestic animals is less than 1.0; whereas the ratio for humans is more than 4.0.

Some differences can be delineated between municipal and industrial wastewater discharges. For example, fluctuations in industrial wastewater flow rates typically exceed those for municipal wastewater. Industrial plants may not operate continuously; there may be daily, weekly, or seasonal variations in operations, reflected by flow rate variations. In addition, the number and types of contaminants in industrial wastewater can vary widely, and the concentrations can range from near zero to 100,000 mg/l

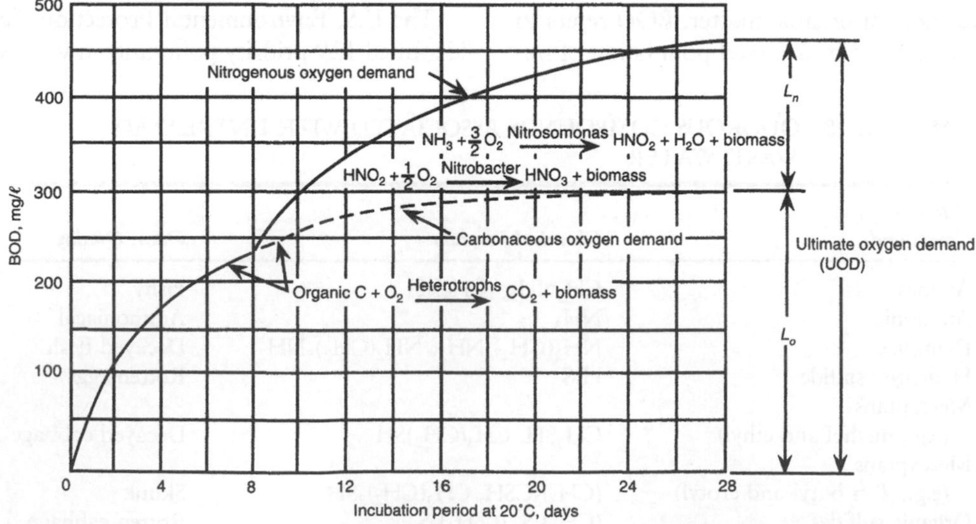

FIG. 1.1.5 Typical BOD curve for domestic wastewater showing carbonaceous and nitrogenous oxygen demands. (Reprinted, with permission, from S.R. Qasim, 1985, *Wastewater treatment plants—Planning, design, and operation,* New York: Holt, Rinehart and Winston.)

TABLE 1.1.12 TYPICAL WASTE COMPOUNDS PRODUCED BY COMMERCIAL, INDUSTRIAL, AND AGRICULTURAL ACTIVITIES AND CLASSIFIED AS PRIORITY POLLUTANTS

Name (Formula)	*Use*	*Concern*
Nonmetals		
Arsenic (As)	Alloying additive for metals, especially lead and copper as shot, battery grids, cable sheaths, and boiler tubes. High purity (semiconductor) grade	Carcinogen and mutagen. *Long term*—can cause fatigue, loss of energy and dermatitis.
Selenium (Se)	Electronics, xerographic plates, TV cameras, photocells, magnetic computer cores, solar batteries, rectifiers, relays, ceramics (colorant for glass) steel and copper, rubber accelerator, catalyst, and trace element in animal feeds	*Long term*—red staining of fingers, teeth, and hair; general weakness; depression; and irritation of the nose and mouth.
Metals		
Barium (Ba)	Getter alloys in vacuum tubes, deoxidizer for copper, Frary's metal, lubricant for anode rotors in X-ray tubes, and spark-plug alloys	Flammable at room temperature in powder form. *Long term*—increased blood pressure and nerve block.
Cadmium (Cd)	Electrodeposited and dipped coatings on metals, bearing and low-melting alloys, brazing alloys, fire protection systems, nickel-cadmium storage batteries, power transmission wire, TV phosphors, basis of pigments used in ceramic glazes, machinery enamels, fungicide photography and lithography, selenium rectifiers, electrodes for cadmium-vapor lamps, and photoelectric cells	Flammable in powder form. Toxic by inhalation of dust or fume. A carcinogen. Soluble compounds of cadmium highly toxic. *Long term*—concentrates in the liver, kidneys, pancreas, and thyroid; hypertension suspected effect.
Chromium (Cr)	Alloying and plating element on metal and plastic substrates for corrosion resistance, chromium-containing and stainless steels, protective coating for automotive and equipment accessories, nuclear and high-temperature research and constituent of inorganic pigments	Hexavalent chromium compounds are carcinogenic and corrosive to tissue. *Long term*—skin sensitization and kidney damage.
Lead (Pb)	Storage batteries, gasoline additive, cable covering, ammunition, piping, tank linings, solder and fusible alloys, vibration damping in heavy construction, foil, babbit, and other bearing alloys	Toxic by ingestion or inhalation of dust or fumes. *Long term*—brain and kidney damage and birth defects.
Mercury (Hg)	Amalgams, catalyst electrical apparatus, cathodes for production of chlorine and caustic soda, instruments, mercury vapor lamps, mirror coating, arc lamps, and boilers	Highly toxic by skin absorption and inhalation of fume or vapor. *Long term*—toxic to central nervous system, and can cause birth defects.
Silver (Ag)	Manufacture of silver nitrate, silver bromide, photo chemicals; lining vats and other equipment for chemical reaction vessels, water distillation, etc.; mirrors, electric conductors, silver plating electronic equipment; sterilant; water purification; surgical cements; hydration and oxidation catalyst special batteries, solar cells, reflectors for solar towers; low-temperature brazing alloys; table cutlery; jewelry; dental, medical, and scientific equipment; electrical contacts; bearing metal; magnet windings; and dental amalgams. Colloidal silver used as a nucleating agent in photography and medicine, often combined with protein	Toxic metal. *Long term*—permanent grey discoloration of skin, eyes, and mucus membranes.

Continued on next page

TABLE 1.1.12 *Continued*

Name (Formula)	*Use*	*Concern*
Organic Compounds		
Benzene (C_6H_6)	Manufacturing of ethylbenzene (for styrene monomer); dodecylbenzene (for detergents); cyclohexane (for nylon); phenol; nitrobenzene (for aniline); maleic anhydride; chlorobenzene hexachloride; benzene sulfonic acid; and as a solvent	A carcinogen. Highly toxic. Flammable, and a dangerous fire risk
Ethylbenzene ($C_6H_5C_2H_5$)	Intermediate in the production of styrene and as a solvent	Toxic by ingestion, inhalation, and skin absorption; irritant to skin and eyes. Flammable and a dangerous fire risk
Toluene ($C_6HC_5H_3$)	Aviation gasoline and high-octane blending stock; benzene, phenol, and caprolactam; solvent for paints and coatings, gums, resins, most oils, rubber, vinyl organosols; diluent and thinner in nitrocellulose lacquers; adhesive solvent in plastic toys and model airplanes; chemicals (benzoic acid, benzyl and bezoyl derivatives, saccharine, medicines, dyes, perfumes); source of toluenediisocyanates (polyurethane resins); explosives (TNT); toluene sulfonates (detergents); and scintillation counters	Flammable and a dangerous fire risk. Toxic by ingestion, inhalation, and skin absorption
Halogenated Compounds		
Chlorobenzene (C_6H_5Cl)	Phenol, chloronitrobenzene, aniline, solvent carrier for methylene diisocyanate, solvent, pesticide intermediate, heat transfer	Moderate fire risk. Recommend avoiding inhalation and skin contact
Chloroethene (CH_2CHCl)	Polyvinyl chloride and copolymers, organic synthesis, and adhesives for plastics	An extremely toxic and hazardous material by all avenues of exposure. A carcinogen
Dichloromethane (CH_2Cl_2)	Paint removers, solvent degreasing, plastics processing, blowing agent in foams, solvent extraction, solvent for cellulose acetate, and aerosol propellant	Toxic. A carcinogen and a narcotic
Tetrachloroethene (CCl_2CCl_2)	Dry cleaning solvent, vapor-degreasing solvent, drying agent for metals and certain other solids, vermifuge, heat transfer medium, and manufacture of fluorocarbons	Irritant to eyes and skin
Pesticides, Herbicides, Insecticides[a]		
Endrin ($C_{12}H_6OCl_6$)	Insecticide and fumigant	Toxic by inhalation and skin absorption and a carcinogen
Lindane ($C_6H_6Cl_6$)	Pesticide	Toxic by inhalation, ingestion, and skin absorption
Methoxychlor ($Cl_3CCH(C_6H_4OCH_3)_2$)	Insecticide	Toxic material
Toxaphene ($C_{10}H_{10}Cl_6$)	Insecticide and fumigant	Toxic by ingestion, inhalation, and skin absorption
Silvex ($Cl_3C_6H_2OCH(CH_3)COOH$)	Herbicides and plant growth regulator	Toxic material; use restricted

Source: Metcalf and Eddy, Inc., 1991, *Wastewater engineering,* 3d ed. (New York: McGraw-Hill).
Note: [a]Pesticides, herbicides, and insecticides are listed by trade name. The compounds listed are also halogenated organic compounds.

TABLE 1.1.13 INFECTIOUS AGENTS POTENTIALLY PRESENT IN RAW DOMESTIC WASTEWATER

Organism	*Disease*	*Remarks*
Bacteria		
Escherichia coli (enteropathogenic)	Gastroenteritis	Diarrhea
Legionella pneumophila	Legionellosis	Acute respiratory illness
Leptospira (150 spp.)	Leptospirosis	Jaundice, and fever (Weil's disease)
Salmonella typhi	Typhoid fever	High fever, diarrhea, and ulceration of small intestine
Salmonella (~1700 spp.)	Salmonellosis	Food poisoning
Shigella (4 spp.)	Shigellosis	Bacillary dysentery
Vibrio cholerae	Cholera	Extremely heavy diarrhea and dehydration
Yersinia enterolitica	Yersinosis	Diarrhea
Viruses		
Adenovirus (31 types)	Respiratory disease	
Enteroviruses (67 types, e.g., polio, echo, and Coxsackie viruses)	Gastroenteritis, heart anomalies, and meningitis	
Hepatitis A	Infectious hepatitis	Jaundice and fever
Norwalk agent	Gastroenteritis	Vomiting
Reovirus	Gastroenteritis	
Rotavirus	Gastroenteritis	
Protozoa		
Balantidium coli	Balantidiasis	Diarrhea and dysentery
Cryptosporidium	Cryptosporidiosis	Diarrhea
Entamoeba histolytica	Amebiasis (amoebic dysentery)	Prolonged diarrhea with bleeding and abscesses of the liver and small intestine
Giardia lamblia	Giardiasis	Mild to severe diarrhea, nausea, and indigestion
Helminths[a]		
Ascaris lumbricoides	Ascariasis	Roundworm infestation
Enterobius vericularis	Enterobiasis	Pinworm
Fasciola hepatica	Fascioliasis	Sheep liver fluke
Hymenolepis nana	Hymenolepiasis	Dwarf tapeworm
Taenia saginata	Taeniasis	Beef tapeworm
T. solium	Taeniasis	Pork tapeworm
Trichuris trichiura	Trichuriasis	Whipworm

Source: Metcalf and Eddy, Inc., 1991.
Note: [a]The helminths listed are those with a worldwide distribution.

(Nemerow and Dasgupta 1991). Some industrial wastewater contaminants are toxic; while others exhibit deoxygenation rates about five times greater than that for municipal wastewater (Nemerow and Dasgupta 1991).

Permits and Effluent Limitations

The *Federal Water Pollution Control Act Amendments of 1972* (PL 92-500) established basic water quality goals and policies for the United States (U.S. Congress 1972). The objective of the *Clean Water Act of 1987* (also known as the *Water Quality Act of 1987*) was to restore and maintain the chemical, physical, and biological integrity of the nation's waters; this objective was also in the precursor laws, including PL 92-500. Three key components of the *Clean Water Act of 1987* relevant to wastewater discharges include water quality standards and planning, discharge permits, and effluent limitations.

New point sources of wastewater discharge must apply for National Pollutant Discharge Elimination System (NPDES) permits under the auspices of Section 402 of the *Clean Water Act of 1987* and its precursors back to 1972. Permits typically address pertinent effluent limitations (discharge standards) for conventional and toxic pollutants, monitoring and reporting requirements, and schedules of

TABLE 1.1.14 ESTIMATED PER CAPITA CONTRIBUTION OF INDICATOR MICROORGANISMS FROM HUMANS AND SOME ANIMALS

	Average Indicator Density/g of Feces		*Average Contribution/capita · 24h*		
Animal	*Fecal Coliform, 10^6*	*Fecal Streptococci, 10^6*	*Fecal Coliform, 10^6*	*Fecal Streptococci, 10^6*	*Ratio FC/FS*
Chicken	1.3	3.4	240	620	0.4
Cow	0.23	1.3	5400	31,000	0.2
Duck	33.0	54.0	11,000	18,000	0.6
Human	13.0	3.0	2000	450	4.4
Pig	3.3	84.0	8900	230,000	0.04
Sheep	16.0	38.0	18,000	43,000	0.4
Turkey	0.29	2.8	130	1300	0.1

Source: Metcalf and Eddy, Inc., 1991.
Note: lb = g × 0.0022.

compliance (Miller, Taylor, and Monk 1991). Effluent limitations can be based on control technologies or required removal (treatment) efficiencies, or they can be based on achieving water quality standards. Water quality-based effluent limitations can require greater treatment levels as determined via a waste load allocation study for the relevant stream segment.

Section 402(p) of the *Clean Water Act of 1987* requires NPDES permits for storm water (runoff water) discharges associated with industrial activity, discharges from large municipal separate storm water systems (systems serving a population of 250,000 or more), and discharges from medium municipal separate storm water systems (systems serving a population of 100,000 or more, but less than 250,000). NPDES permits for storm water from industrial areas require the development of a pollution prevention plan to reduce pollution at the source (U.S. EPA 1992b). Detailed information on developing pollution prevention plans for construction activities in urban or industrial areas is also available (U.S. EPA 1992a).

—Larry W. Canter

References

Clesceri, L.S., A.E. Greenberg, and R.R. Trussell, eds. 1989. *Standard methods for the examination of water and wastewater.* 17th ed. Washington, D.C.: American Public Health Association, American Water Works Association, and Water Pollution Control Federation.

Metcalf and Eddy, Inc. 1991. *Wastewater engineering.* 3d ed. 36, 50–57, 65–70, 93–96, 100–101, and 108–112. New York: McGraw-Hill, Inc.

Miller, L.A., R.S. Taylor, and L.A. Monk. 1991. *NPDES permit handbook.* 3d Printing. Rockville, Md.: Government Institutes, Inc.

Nemerow, N.L. and A. Dasgupta. 1991. *Industrial and hazardous waste treatment.* 10. New York: Van Nostrand Reinhold.

Qasim, S.R. 1985. *Wastewater treatment plants—Planning, design, and operation.* 9, 40–41. New York: Holt, Rinehart and Winston.

U.S. Congress. 1972. *Federal Water Pollution Control Act Amendments of 1972.* PL 92-500, 92nd Congress, S. 2770, 18 October 1972.

U.S. Environmental Protection Agency (EPA). 1992a. *Storm water management for construction activities—Developing pollution prevention plans and best management practices.* EPA 832-R-92-005. Washington, D.C.: Office of Water. (September).

———. 1992b. *Storm water management for industrial activities—Developing pollution prevention plans and best management practices.* EPA 832-R-92-006. Washington, D.C.: Office of Water. (September).

Water Pollution Control Federation and American Society of Civil Engineers. 1977. *Wastewater treatment plant design.* 10. Washington, D.C.

1.2 SOURCES AND EFFECTS OF CONTAMINANTS

Sources of Contaminants

Contaminants in municipal wastewater are introduced as a result of water usage for domestic, commercial, or institutional purposes; water usage for product processing or cooling purposes within industries discharging liquid effluents into municipal sewerage systems; and infiltration/inflow and/or stormwater runoff. Table 1.2.1 shows examples of typical physical, chemical, and biological characteristics of municipal wastewater, along with potential sources. Table 1.2.2 summarizes selected sources and effects of industrial wastewater constituents.

POINT AND NONPOINT SOURCES

Two main sources of water pollutants are point and nonpoint. Nonpoint sources are also referred to as area or diffuse sources. Nonpoint pollutants are substances introduced into receiving waters as a result of urban area, industrial area, or rural runoff; e.g., sediment and pesticides or nitrates entering surface water due to surface runoff from agricultural farms. Point sources are specific discharges from municipalities or industrial complexes; e.g., organics or metals entering surface water due to wastewater discharge from a manufacturing plant. In a surface water body, nonpoint pollution can contribute significantly to total pollutant loading, particularly with regard to nutrients and pesticides. To illustrate the relative contributions, Figure 1.2.1 shows the estimated nationwide loadings of four key water pollutants.

Municipal and industrial wastewater discharges are primary contributors to point source discharges in the United States. Table 1.2.3 provides quantitative information on BOD, total suspended sediments, and phosphorus discharges from municipal treatment facilities and several industrial categories.

Over the past two decades, more than $75 billion in federal, state, and local funds were used to construct municipal sewage treatment facilities, and the private sector has spent additional billions to limit discharges of conventional pollutants (Council on Environmental Quality 1992). Between 1972 and 1988, the number of people served by municipal treatment plants with secondary treatment or better increased from 85 million to 144 million; Figure 1.2.2 depicts these changes in population served. These trends suggest that further reductions in pollutant loadings from point sources will occur.

VIOLATIONS OF WATER QUALITY STANDARDS

From a holistic perspective, a major source of wastewater discharges is associated with increases in the violations of receiving water quality standards. The 1990 National Water Quality Inventory in the United States assessed, in relation to applicable water quality standards and designated beneficial uses of the water, about one-third of the total river miles, half of the acreages of lakes, and three-quarters of the estuarine square miles. Table 1.2.4 summarizes these results relative to supporting designated uses. About one-third of the three assessed water resources did not fully meet their respective designated uses (Council on Environmental Quality 1993). Table 1.2.5 indicates the causes and sources of pollution for the three types of water resources.

Figure 1.2.3 shows pollution sources for impaired river miles; Figure 1.2.4 shows pollution in estuarine waters. While they are not the only sources of pollution, municipal and industrial wastewater discharges contribute significantly to impaired water resources.

Table 1.2.6 summarizes violation rates of water quality criteria in U.S. rivers and streams from 1975–1989. FC bacteria violations are well above the rates for other constituents for each year. In addition, the percent of violations has declined for all listed parameters since 1975.

Effects of Contaminants

The effects of pollution sources on receiving water quality are manifold and depend on the type and concentration of pollutants (Nemerow and Dasgupta, 1991). Soluble organics, as represented by high BOD waste, deplete oxygen in surface water. This results in fish kills, the growth of undesirable aquatic life, and undesirable odors. Trace quantities of certain organics cause undesirable tastes and odors, and certain organics can be biomagnified in the aquatic food chain.

SSs decrease water clarity and hinder photosynthetic processes; if solids settle and form sludge deposits, changes in benthic ecosystems result. Color, turbidity, oils, and floating materials influence water clarity and photosynthetic processes and are aesthetically undesirable.

TABLE 1.2.1 PHYSICAL, CHEMICAL, AND BIOLOGICAL CHARACTERISTICS OF WASTEWATER AND THEIR SOURCES

Characteristic	*Sources*
Physical Properties	
Color	Domestic and industrial wastes and natural decay of organic materials
Odor	Decomposing wastewater and industrial wastes
Solids	Domestic water supply, domestic and industrial wastes, soil erosion, and inflow/infiltration
Temperature	Domestic and industrial wastes
Chemical Constituents	
ORGANIC	
Carbohydrates	Domestic, commercial, and industrial wastes
Fats, oils, and grease	Domestic, commercial, and industrial wastes
Pesticides	Agricultural wastes
Phenols	Industrial wastes
Proteins	Domestic, commercial, and industrial wastes
Priority pollutants	Domestic, commercial, and industrial wastes
Surfactants	Domestic, commercial, and industrial wastes
VOCs	Domestic, commercial, and industrial wastes
Other	Natural decay of organic materials
INORGANIC	
Alkalinity	Domestic wastes, domestic water supply, and groundwater infiltration
Chlorides	Domestic wastes, domestic water supply, and groundwater infiltration
Heavy metals	Industrial wastes
Nitrogen	Domestic and agricultural wastes
pH	Domestic, commercial, and industrial wastes
Phosphorus	Domestic, commercial, and industrial wastes and natural runoff
Priority pollutants	Domestic, commercial, and industrial wastes
Sulfur	Domestic water supply and domestic, commercial, and industrial wastes
GASES	
Hydrogen sulfide	Decomposition of domestic wastes
Methane	Decomposition of domestic wastes
Oxygen	Domestic water supply and surface-water infiltration
Biological Constituents	
Animals	Open watercourses and treatment plants
Plants	Open watercourses and treatment plants
Protists:	
Eubacteria	Domestic wastes, surface-water infiltration, and treatment plants
Archaebacteria	Domestic wastes, surface-water infiltration, and treatment plants
Viruses	Domestic wastes

Source: Metcalf and Eddy, Inc., 1991, *Wastewater engineering,* 3d ed. (New York: McGraw-Hill, Inc.).

Excessive nitrogen and phosphorus lead to algal overgrowth with concomitant water treatment processes. Chlorides cause a salty taste in water; and in sufficient concentration, water usage must be limited. Acids, alkalies, and toxic substances can cause fish kills and create other imbalances in stream ecosystems.

Thermal discharges can also cause imbalances and reduce the stream waste assimilative capacity. Stratified flows from thermal discharges minimize normal mixing patterns in receiving streams and reservoirs. Table 1.2.7 provides an overview of certain contaminants and their impact in surface waters. Table 1.2.8 summarizes the impact of certain pollutants with regard to use impairment of the water.

ECOLOGICAL EFFECTS

The effects of pollutant discharges in municipal or industrial wastewaters can be considered from an ecological perspective. For example, Welch (1980) discusses the effects

TABLE 1.2.2 EXAMPLES OF SOURCES AND EFFECTS OF WASTEWATER CONSTITUENTS

Component Group	*Effects*	*Typical Sources*
Biooxidizables expressed as BOD_5	Deoxygenation, anaerobic conditions, fish kills, odors	Large amounts of soluble carbohydrates: sugar refining, canning, distilleries, breweries, milk processing, pulping, and paper making
Primary toxicants: As, CN, Cr, Cd, Cu, F, Hg, Pb, and Zn	Fish kills, cattle poisoning, plankton kills, and accumulations in flesh of fish and mollusks	Metal cleaning, plating, and pickling; phosphate and bauxite refining; chlorine generation; battery making; and tanning
Acids and alkalines	Disruption of pH buffer systems and disordering previous ecological system	Coal-mine drainage, steel pickling, textiles, chemical manufacture, wool scouring, and laundries
Disinfectants: Cl_2, H_2O_2, formalin, and phenol	Selective kills of microorganisms, taste, and odors	Bleaching of paper and textiles; rocketry; resin synthesis; penicillin preparation; gas, coke, and coal-tar making; and dye and chemical manufacture
Ionic forms: Fe, Ca, Mg, Mn, Cl, and SO_4	Changed water characteristics: staining, hardness, salinity, and encrustations	Metallurgy, cement making, ceramics, and oil-well pumpage
Oxidizing and reducing agents: NH_3, NO_2^-, NO_3^-, S^-, and SO_3^{-2}	Altered chemical balances ranging from rapid oxygen depletion to overnutrition, odors, and selective microbial growths	Gas and coke making, fertilizer plants, explosive manufacture, dyeing and synthetic fiber making, wood pulping, and bleaching
Evident to sight and smell	Foaming, floating, and setteable solids; odors; anaerobic bottom deposits; oils, fats, and grease; and waterfowl and fish injuries	Detergent wastes, tanning, food and meat processing, beet sugar mills, woolen mills, poultry dressing, and petroleum refining
Pathogenic organisms: *B. anthracis, Leptospira,* fungi, and viruses	Infections in humans, reinfection of livestock, plant diseases from fungi-contaminated irrigation water and risks to humans slight	Abattoir wastes, wool processing, fungi growths in waste treatment works, and poultry-processing waste waters

Source: R.A. Corbitt, 1990, Wastewater disposal, Chap. 6 in *Standard handbook of environmental engineering,* edited by R.A. Corbitt (New York: McGraw-Hill Publishing Company).

of wastewater discharges on the ecological characteristics of the receiving water environment; his topics include the specific effects on phytoplankton, zooplankton, periphyton, macrophytic rooted plants, benthic macroinvertebrates, and fish.

Material and energy flow diagrams demonstrate biogeochemical cycles and system interrelationships. For example, Figure 1.2.5 shows the material and energy flow in an aquatic ecosystem. Food web (or food chain) relationships and energy flow considerations indicate the dynamic aspects of the biological environment. They are also used to develop qualitative and quantitative models of aquatic or terrestrial systems, useful in predicting aquatic impacts of wastewater discharges. Wetland loss or degradation from municipal or industrial wastewater discharges illustrates an ecosystem effect. Such loss or degradation also occurs as a consequence of other human activities or natural occurrences (Mannion and Bowlby 1992).

To analyze the potential effects of wastewater discharges, the engineer may consider environmental cycling

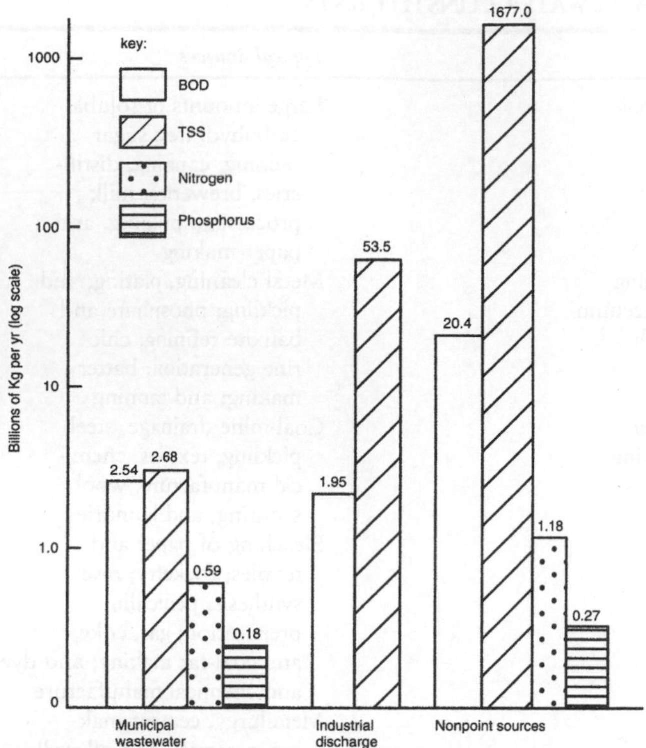

FIG. 1.2.1 Estimated nationwide loadings of selected water pollutants. (Reprinted, with permission, from Corbitt, 1990, *Wastewater disposal,* Chap. 6 in Standard handbook of environmental engineering, edited by R.A. Corbitt, New York: McGraw-Hill Publishing Company.)

	1960	*1978*	*1988*
Not served	70.0*	66.0	69.9
Raw discharge	na†	na	1.5
Secondary treatment	na	56.0	78.0
No discharge	na	na	6.1
Less than secondary treatment	36.0	na	26.5
Greater than secondary treatment	4.0	49.0	65.7

*Numbers represent millions of people
†na = not available.

FIG. 1.2.2 Population served by municipal wastewater treatment systems. (Reprinted from Council on Environmental Quality 1992, *Environmental trends,* Washington, D.C.).

of specific pollutants. Figure 1.2.6 shows various transfer routes and processes related to pollutant movement within the hydrosphere and biosphere. Examples of specific biogeochemical cycles for one nutrient (nitrogen) and one metal (mercury) are shown in Figures 1.2.7 and 1.2.8, respectively. The information in these examples indicates that the biological environment is a dynamic system that can be stressed as a result of various wastewater discharges.

Environmental engineers can use chromium to illustrate changes within aqueous systems since it is a transition metal that exhibits various oxidation states and behavior patterns (Canter and Gloyna 1968). Trivalent chromium is generally present as a cation, $Cr(OH)^+$, and is chemically reactive, tending to sorb on suspended materials and subsequently settle from the liquid phase. Hexavalent chromium is anionic (CrO_4^-) and chemically unreactive,

TABLE 1.2.3 QUANTITATIVE POLLUTANT LOADINGS INTO SURFACE WATERS

	Point Sources[a]		*Nonpoint Sources*[a]
Pollutant	*Total (10⁶ tn/yr)*	*Source Contribution*[b] *(% of total)*	*Total (10⁶ tn/yr)*
BOD	1.87	MTF = 72.3 AFI = 21.6 CMI = 5.8 MMI = 0.3	14
TSS	2.13	MTF = 61.5 AFI = 13.0 CMI = 8.9 MMI = 16.7	3130
Phosphorus	2.66	MTF = 81.8 AFI = 18.0 CMI = 0.05 MMI = 0.05	515

Source: Council on Environmental Quality, 1989, *Environmental trends* (Washington, D.C.).
Notes: [a]Point source information is from mid-1980s; nonpoint source information is from 1980.
[b]MTF = municipal treatment facilities, AFI = agriculture and fisheries industry, CMI = chemical and manufacturing industry, and MMI = minerals and metals industry.

TABLE 1.2.4 DESIGNATED USE SUPPORT IN SURFACE WATERS OF THE UNITED STATES, 1990

Designated Use Support	*Rivers*	*Lakes & Reservoirs*	*Estuaries*
	mi	*acres*	*sq mi*
Fully supporting	407,162	8,173,917	15,004
Threatened	43,214	2,902,809	3052
Partially supporting	134,472	3,471,633	6573
Not supporting	62,218	3,940,277	2064
Not assessed	1,153,000	20,910,000	8931
Total	1,800,000	39,400,000	35,624

Source: U.S. Environmental Protection Agency (EPA), 1992, *National water quality inventory: 1990,* Report to Congress (Washington, D.C.).

TABLE 1.2.5 CAUSES AND SOURCES OF SURFACE WATER POLLUTION IN THE UNITED STATES, 1990

Causes of Pollution	*Rivers*	*Lakes and Reservoirs*	*Estuaries*
	impaired mi	*impaired acres*	*impaired sq mi*
Siltation	67,059	702,857	312
Nutrients	51,747	1,793,022	3279
Organic enrichment	47,545	1,072,184	1876
Pathogens	35,151	129,286	1781
Metals	28,287	2,672,427	431
Salinity	21,914	243,482	nr
Habitat modification	20,258	nr	nr
Pesticides	20,701	nr	105
Priority organics	nr	247,317	917
SS	20,819	731,993	467
Flow alteration	15,565	677,413	nr
pH	9368	192,153	36
Sources of Pollution	*Rivers*	*Lakes and Reservoirs*	*Estuaries*
	impaired mi	*impaired acres*	*impaired sq mi*
Agriculture	103,439	1,996,772	1074
Municipal	27,994	593,518	2038
Habitat modification	24,884	1,407,827	307
Resource extraction	24,015	301,398	91
Storm sewers and runoff	18,129	973,077	1790
Industrial	15,568	318,446	615
Silviculture	15,459	106,502	89
Construction	9810	106,398	640
Land disposal	7188	846,892	1137
Combined sewers	2836	3015	359
Unknown	9266	403,080	189

Source: U.S. EPA, 1992.

Notes: nr = not reported. In addition to the causes and sources listed, thermal modifications impair 9970 river mi. Taste and odor impairments affect 105,288 acres of lakes and reservoirs, and noxious aquatic plants impair 711,323 acres. Additional causes of pollution in estuaries are ammonia (50 sq mi), oil and grease (36 sq mi), and unknown (109 sq mi). Estimates of impairment are based on the sums of partially and not supporting designated uses in Table 1.2.4 which represent 9.5% of total U.S. river mi, 8.9% of total U.S. lake acres, and 16.7% of total U.S. estuary square mi.

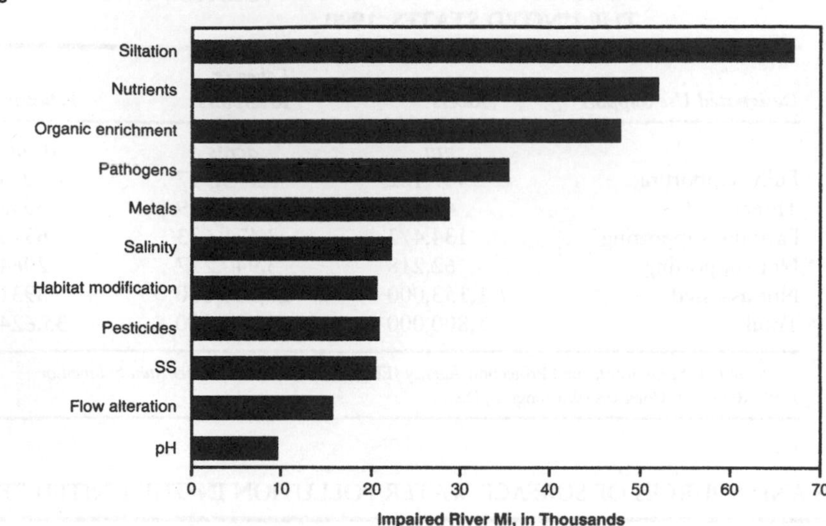

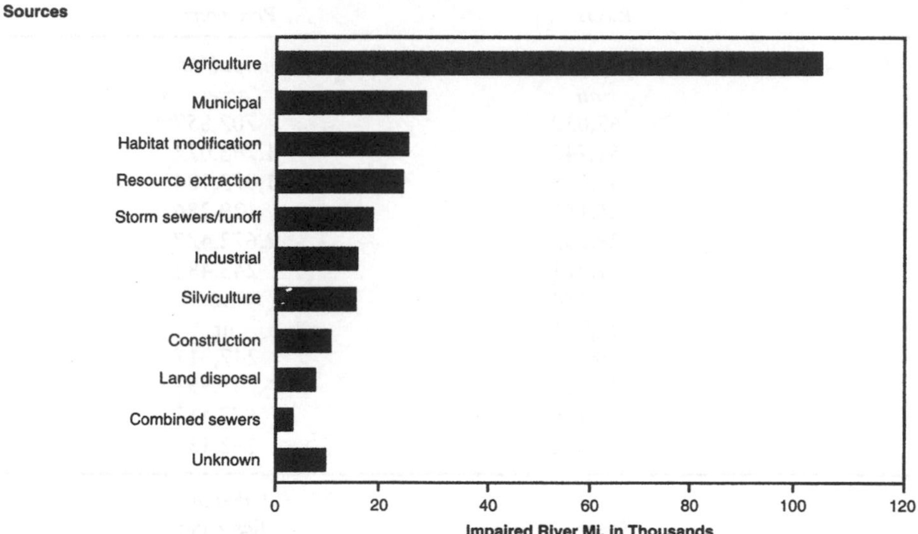

FIG. 1.2.3 Types and sources of pollution in rivers of the United States. Figure based on river mi monitored in 1990, which represent 9.5% of total U.S. river mi. (Reprinted from U.S. Environmental Protection Agency (EPA), 1992, *National water quality inventory:* 1990, Report to Congress, Washington, D.C.)

thus tending to remain in solution. Changes can occur in the chromium oxidation state due to stream water quality. For example, hexavalent chromium can be chemically reduced to trivalent chromium under anaerobic conditions, whereas trivalent chromium can be oxidized to hexavalent chromium under aerobic conditions. This information qualitatively predicts the impact of chromium discharges into river systems.

An additional concern is the potential reconcentration of pollutant materials (e.g., heavy metals and pesticides) into aquatic organisms and their subsequent harvesting and consumption by man. One example is the study from 1974–1990 of pesticide contaminant levels in herring gull eggs from the five Great Lakes (Council on Environmental Quality 1992). Another example is the closure of shellfish beds due to bacterial contamination of the water and shellfish (Council on Environmental Quality 1993).

TOXICITY EFFECTS

Depending on their constituents, some municipal or industrial wastewater discharges exhibit toxicity effects on aquatic organisms. As a result, biomonitoring can be required for effluent discharges. In biomonitoring, indicator organisms are chosen to represent all segments of the aquatic community of the water body under study (U.S. Department of Energy 1985). Five groups of organisms have traditionally been studied as indicators of water quality (U.S. Department of Energy 1985): bacteria, fish, plankton, periphyton, and macroinvertebrates. The choice of an

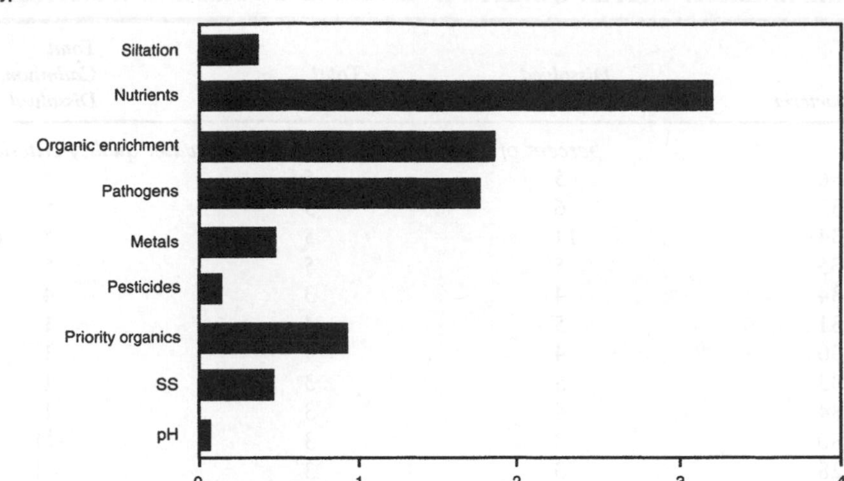

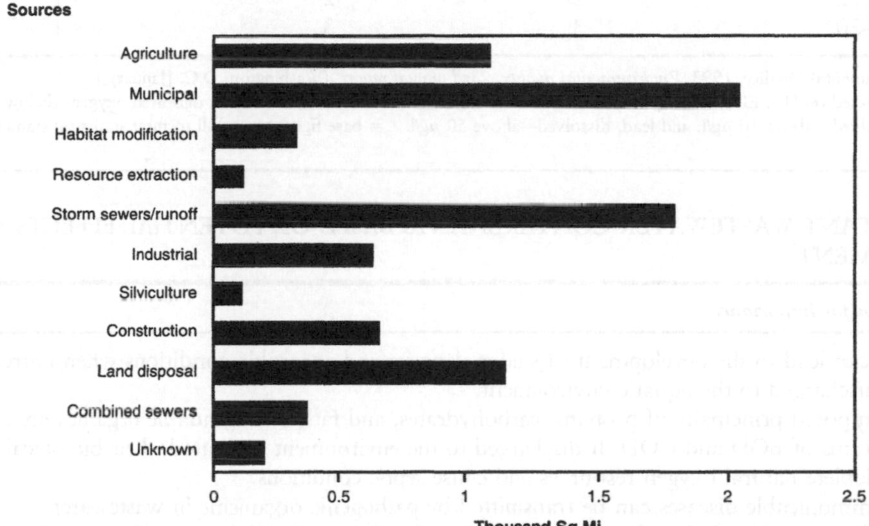

FIG. 1.2.4 Types and sources of pollution in estuaries of the United States. Figure based on sq mi of estuaries monitored in 1990, which represent 16.7% of total U.S. estuaries. An estuary is a tidal body of water, formed where a river meets the sea. Estuaries, such as bays, have a measurable quantity of salt. (Reprinted from U.S. EPA 1992.)

indicator species is important in a biomonitoring study because it can affect the results of the study phase and toxicity characterization procedures.

Bacteria are indicators of fecal contamination and the presence of pathogenic organisms (U.S. Department of Energy 1985). Fish are useful in assessing water quality because in aquatic communities, they often represent the highest trophic level. If fish are lacking or exist only in small numbers, decreased water quality may be affecting the fish directly or indirectly by affecting those organisms on which the fish feed.

Plankton are good indicators of the water quality because they float passively with the currents (U.S. Department of Energy 1985). Estimating the plankton population indicates, in part, the nutrient-supplying capability of that water body (U.S. Department of Energy 1985). Periphyton, because of their stationary or sessile existence, also indicate nutrient availability in an aquatic system; furthermore, they represent the integration of the physical and chemical conditions of water passing through their locations (U.S. Department of Energy 1985).

Benthic species are excellent indicators of water quality because they are generally restricted to the area where they are found (U.S. Department of Energy 1985). Benthic macroinvertebrates, because of their varying environmental requirements, form communities characteristic or associated with physical and chemical conditions. For example, the presence of immature insects, certain mollusks,

TABLE 1.2.6 NATIONAL AMBIENT WATER QUALITY IN RIVERS AND STREAMS: VIOLATION RATES, 1975–1989

Year	*FC Bacteria*	*Dissolved Oxygen*	*Total Phosphorus*	*Total Cadmium, Dissolved*	*Total Lead, Dissolved*
		percent of all measurements exceeding water quality criteria			
1975	36	5	5	*	*
1976	32	6	5	*	*
1977	34	11	5	*	*
1978	35	5	5	*	*
1979	34	4	3	4	13
1980	31	5	4	1	5
1981	30	4	4	1	3
1982	33	5	3	1	2
1983	34	4	3	1	5
1984	30	3	3	<1	<1
1985	28	3	3	<1	<1
1986	24	3	3	<1	<1
1987	23	2	2	<1	<1
1988	22	2	2	<1	<1
1989	20	3	3	<1	<1

Source: Council on Environmental Quality, 1993, *Environmental quality, 23rd annual report,* (Washington, D.C. [January]).

Note: Violation levels are based on U.S. EPA water quality criteria: FC bacteria—above 200 cells per 100 mi; dissolved oxygen—below 5 mg/l; total phosphorus—above 1.0 mg/l; cadmium, dissolved—above 10 μg/l; and lead, dissolved—above 50 μg/l. * = base figure too small to meet statistical standards for reliability of derived figures.

TABLE 1.2.7 IMPORTANT WASTEWATER CONTAMINANTS BASED ON POTENTIAL EFFECTS AND CONCERNS IN TREATMENT

Contaminants	*Reason for Importance*
SS	SS can lead to the development of sludge deposits and anaerobic conditions when untreated wastewater is discharged to the aquatic environment.
Biodegradable organics	Composed principally of proteins, carbohydrates, and fats, biodegradable organics are commonly measured in terms of BOD and COD. If discharged to the environment untreated, their biological stabilization can deplete natural oxygen resources and cause septic conditions.
Pathogens	Communicable diseases can be transmitted by pathogenic organisms in wastewater.
Nutrients	Both nitrogen and phosphorus, along with carbon, are essential nutrients for growth. When discharged to the aquatic environment, these nutrients can lead to the growth of undesirable aquatic life. When discharged in excessive amounts on land, they can also lead to groundwater pollution.
Priority pollutants	These pollutants include organic and inorganic compounds selected on the basis of their known or suspected carcinogenicity, mutagenicity, teratogenicity, or high acute toxicity. Many of these compounds are found in wastewater.
Refractory organics	These organics tend to resist conventional wastewater treatment. Typical examples include surfactants, phenols, and agricultural pesticides.
Heavy metals	Heavy metals are usually added to wastewater from commercial and industrial activities and may have to be removed if the wastewater is to be reused.
Dissolved inorganics	Inorganic constituents such as calcium, sodium, and sulfate are added to the original domestic water supply as a result of water use and may have to be removed if the wastewater is to be reused.

Source: Metcalf and Eddy, Inc., 1991, *Wastewater engineering,* 3d ed. (New York: McGraw-Hill, Inc.).

and crayfish usually indicates relatively clean water; while communities of sludge worms, air breathing snails, midges, and aquatic earthworms indicate the presence of oxygen-consuming materials and deteriorating conditions.

Toxicity tests, the second category of biomonitoring, are less expensive and labor intensive than ecological surveys. Toxicity tests use indicator organisms in a controlled situation to examine water and effluent toxicity. The toxic effects of concern are death, immobilization, serious incapacitation, reduced fecundity, or reduced growth. Toxicity tests can be used to assess acute or chronic toxicity. Acute toxicity is a severe toxic effect resulting from a brief exposure, while chronic toxicity results from prolonged exposure. Tests for acute toxicity can be conducted within

TABLE 1.2.8 WATER USE LIMITS DUE TO WATER QUALITY DEGRADATION

	Use						
Pollutant	*Drinking Water*	*Aquatic Wildlife, Fisheries*	*Recreation*	*Irrigation*	*Industrial Uses*	*Power and Cooling*	*Transport*
Pathogens	xx	0	xx	x	xx[1]	na	na
SS	xx	xx	xx	x	x	x[2]	xx[3]
Organic matter	xx	x	xx	+	xx[4]	x[5]	na
Algae	x[5,6]	x[7]	xx	+	xx[4]	x[5]	x[8]
Nitrate	xx	x	na	+	xx[1]	na	na
Salts[9]	xx	xx	na	xx	xx[10]	na	na
Trace elements	xx	xx	x	x	x	na	na
Organic micropollutants	xx	xx	x	x	?	na	na
Acidification	x	xx	x	?	x	x	na

Source: D. Chapman, ed., 1992, *Water quality assessments—A guide to the use of biota, sediments, and water in environmental monitoring,* 9 (London: Chapman and Hall, Ltd.).

Notes:

xx Marked impairment causing major treatment or excluding desired use
x Minor impairment
0 No impairment
na Not applicable
+ Degraded water quality can be beneficial for this specific use.
? Effects not fully realized
[1] Food industries
[2] Abrasion
[3] Sediment settling in channels
[4] Electronic industries
[5] Filter clogging
[6] Odor and taste
[7] In fish ponds, higher algal biomass can be accepted.
[8] Development of water hyacinth (*Eichhomia crassiodes*)
[9] Also includes boron and fluoride
[10] Ca, Fe, and Mn in textile industries

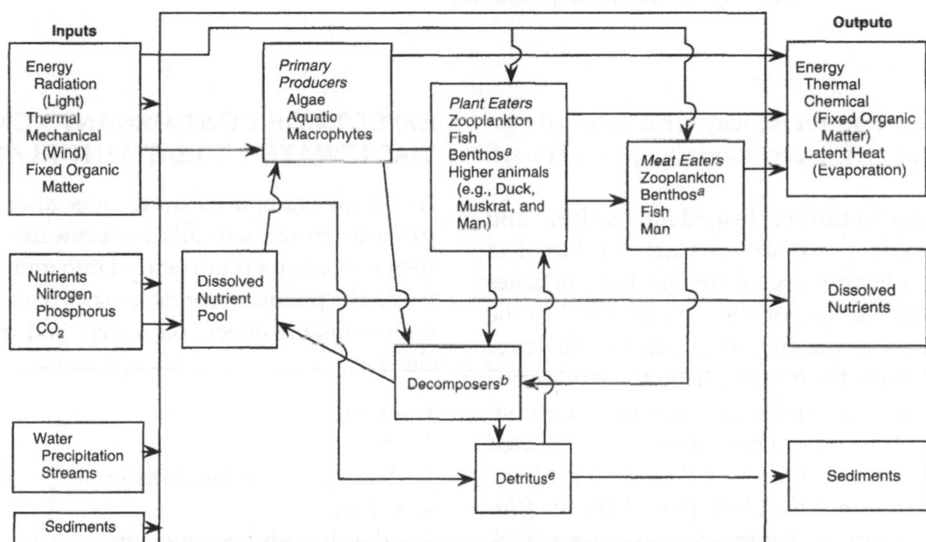

FIG. 1.2.5 Material and energy flow in an aquatic ecosystem: [a]Organisms living at or on the bottom of bodies of water, [b]Fungi and bacteria, [c]Small particles of organic matter. (Reprinted, with permission, from D.C. Watts and D.L. Loucks, 1969, *Models for describing exchange within ecosystems,* Madison, Wis.: Institute for Environmental Studies, University of Wisconsin.)

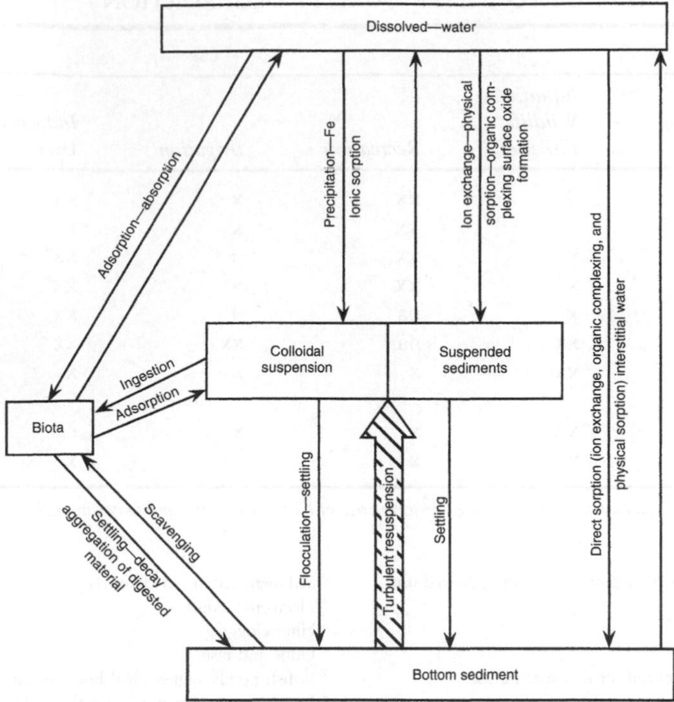

FIG. 1.2.6 Pollutant transfer routes and processes within the hydrosphere and biosphere. (Reprinted, with permission, from R.A. Corbitt, 1990, Wastewater disposal, Chap. 6 in *Standard handbook of environmental engineering,* edited by R.A. Corbitt, New York: McGraw-Hill Publishing Company.)

24–96 hr, and chronic tests for toxicity are conducted usually in four or more days (Clesceri, Greenberg, and Trussell 1989).

Toxicity testing can be further divided into ambient and laboratory tests. Ambient toxicity tests are conducted in situ. In such tests, indicator organisms are kept in cages within the water under study, or they are exposed to the test water in chambers at the site. The water in the latter test is renewed daily with fresh water from the study sites.

Laboratory tests are dissimilar because they are conducted offsite in a laboratory. These tests are conducted within a set of conditions such as those described in *Short-term methods for estimating the chronic toxicity of effluents and receiving waters to freshwater organisms* (U.S. EPA 1989). In such tests, the test water is fractionally diluted with synthetic water. Replicate groups of indicator species are exposed to different concentrations of a dilution series, and toxic effects are recorded for each dilution. The recorded data are then analyzed using a series of statistical tests to determine the effective or lethal concentrations. Effective and lethal concentrations are those point estimates of the toxicant concentration that cause an observable adverse effect (such as death, immobilization, serious incapacitation, reduced fecundity, or reduced growth) in a percentage of the test organisms (U.S. EPA 1989).

EFFECTS OF CONTAMINANTS ON WASTEWATER TREATMENT PLANTS

As a final example of wastewater discharge effects, industrial wastewater can affect wastewater treatment plants if such wastewater is introduced into municipal sewerage systems. The pollution characteristics of industrial wastewater with definable effects on sewers and treatment plants include

1. BOD
2. SS
3. floating and colored material
4. volume
5. other harmful constituents

Excessive BOD and volume can cause organic and hydraulic overloads, respectively. SS can create operational problems due to excess solids and sludge production. Floating and colored material are related to visible pollution. Examples of other harmful industrial wastewater constituents include (Nemerow and Dasgupta 1991):

Toxic metal ions (Cu^{++}, Cr^{+6}, Zn^{++}, and Cn^{-}), which interfere with biological oxidation by tying up the enzymes required to oxidize organic matter

Acids and alkalis, which can corrode pipes, pumps, and treatment units, interfere with settling, upset the bio-

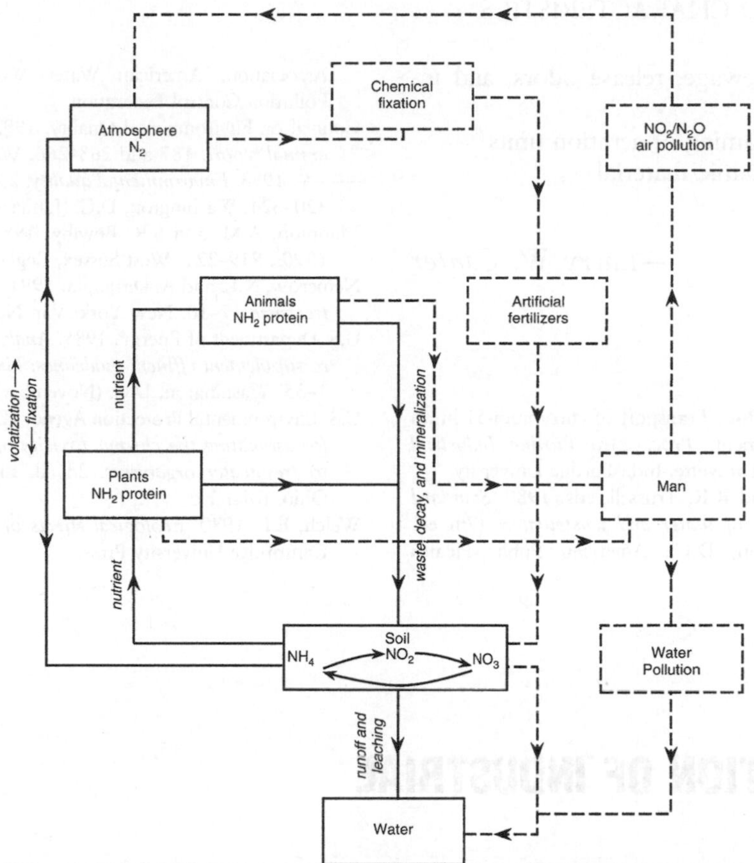

FIG. 1.2.7 The biogeochemical cycle for nitrogen. Dotted lines denote human intervention. (Reprinted, with permission, from D. Drew, 1983, *Man–environment processes*, London: George Allen and Unwin, Ltd.)

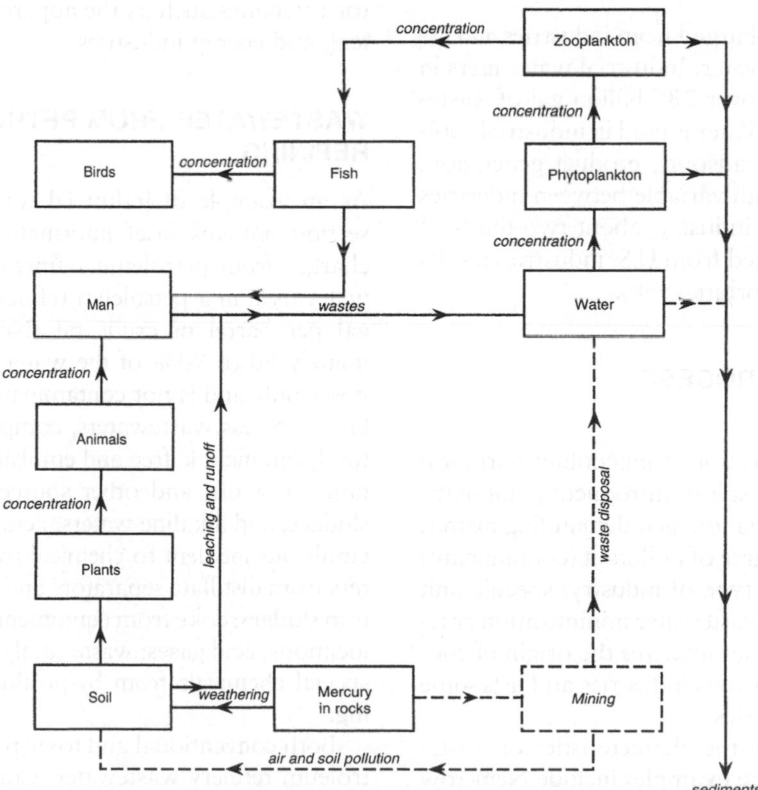

FIG. 1.2.8 The biogeochemical cycle for mercury. Special reference is given to the biological concentration in the aquatic environment. Broken lines show areas of human intervention. (Reprinted, with permission, from Drew, 1983.)

logical purification of sewage, release odors, and intensify color

Detergents, which cause foaming of aeration units

Phenols and other toxic organic material

—Larry W. Canter

References

Canter, L.W. and E.F. Gloyna. 1968. Transport of chromium-51 in an organically polluted environment. *Proc. 23rd Purdue Industrial Waste Conference.* 374–387. Lafayette, Ind.: Purdue University.

Clesceri, L.S., A.E. Greenberg, and R.R. Trussell, eds. 1989. *Standard methods for the examination of water and wastewater.* 17th ed. 10-1 to 10-201. Washington, D.C.: American Public Health Association, American Water Works Association, and Water Pollution Control Federation.

Council on Environmental Quality. 1992. *Environmental quality, 22nd annual report.* 187 and 263–266. Washington, D.C. (March).

———. 1993. *Environmental quality, 23rd annual report.* 225–227 and 320–324. Washington, D.C. (January).

Mannion, A.M. and S.R. Bowlby. 1992. *Environmental issues in the 1990s.* 219–221. West Sussex, England: John Wiley and Sons, Ltd.

Nemerow, N.L. and A. Dasgupta. 1991. *Industrial and hazardous waste treatment.* 7–10. New York: Van Nostrand Reinhold.

U.S. Department of Energy. 1985. *Analysis of biomonitoring techniques to supplement effluent guidelines: Final report.* DOE/PE/16036—TI. 1–35. Washington, D.C. (November).

U.S. Environmental Protection Agency (EPA). 1989. *Short-term methods for estimating the chronic toxicity of effluents and receiving waters to freshwater organisms.* 2d ed. EPA 600–4–89–001. Cincinnati, Ohio. (March).

Welch, E.B. 1980. *Ecological effects of waste water.* 1–2. Cambridge: Cambridge University Press.

1.3 CHARACTERIZATION OF INDUSTRIAL WASTEWATER

Origins of Industrial Wastewater

Industrial wastewater is discharged from industries and associated processes utilizing water. Industrial water users in the United States discharge over 285 billion gal of wastewater daily (Corbitt 1990). Water is used in industrial cooling, product washing and transport, product generation, and other purposes. Although variable between industries and plants within the same industry, about two-thirds of the total wastewater generated from U.S. industries results from cooling operations (Corbitt 1990).

WASTEWATER FROM PROCESS OPERATIONS

Water used for process operations (noncooling purposes) can become degraded as a result of introducting nutrients, suspended sediments, bacteria, oxygen-demanding matter, and toxic chemicals. The degree of pollutant (contaminant) loading is a function of the type of industry, specific unit processes, and the extent of wastewater minimization practices employed. Table 1.3.1 summarizes the origin of contaminant introductions in various industries and lists some key contaminant characteristics.

Several books summarize the characteristics of wastewater from various industries; examples include Nemerow (1978) and Nemerow and Dasgupta (1991). Nemerow and Dasgupta (1991) provide information on the types and quantities of water pollutants from major private sector categories such as the apparel, food, materials, chemical, and energy industries.

WASTEWATER FROM PETROLEUM REFINING

As an example of industrial wastewater discharges, this section presents brief information on the pollutant discharges from petroleum refineries. The total amount of water used in a petroleum refinery is estimated to be 770 gal per barrel of crude oil (Nemerow 1978). Approximately 80 to 90% of the water is used for cooling purposes only and is not contaminated except by leaks in the lines. Process wastewaters, comprising 10 to 20% of the total, can include free and emulsified oil from leaks, spills, tank draw-off, and other sources; waste caustic, caustic sludges, and alkaline waters; acid sludges and acid waters; emulsions incident to chemical treatment; condensate waters from distillate separators and tank draw-off; tank-bottom sludges; coke from equipment tubes, towers, and other locations; acid gases; waste catalyst and filtering clays; and special chemicals from by-product chemical manufacturing.

Both conventional and toxic pollutants are found in petroleum refinery wastewater. Conventional pollutants are those that have received historical attention, while toxic pollutants relate to parameters receiving increasing atten-

TABLE 1.3.1 SUMMARY OF THE ORIGIN AND CHARACTERISTICS OF SELECTED INDUSTRIAL WASTEWATER

Industries Producing Wastes	*Origin of Major Wastes*	*Major Characteristics*
Apparel		
Textiles	Cooking of fibers and desizing of fabric	Highly alkaline, colored, high BOD and temperature, and high SS
Leather goods	Unhairing, soaking, deliming, and bating of hides	High total solids, hardness, salt, sulfides, chromium, pH, precipitated lime, and BOD
Laundry trades	Washing of fabrics	High turbidity, alkalinity, and organic solids
Dry cleaning	Solvent cleaning of clothes	Condensed, toxic, and organic vapors
Food and Drugs		
Canned goods	Trimming, culling, juicing, and blanching of fruits and vegetables	High in SS, colloidal, and dissolved organic matter
Dairy products	Dilutions of whole milk, separated milk, buttermilk, and whey	High in dissolved organic matter, mainly protein, fat, and lactose
Brewed and distilled beverages	Steeping and pressing of grain; residue from distillation of alcohol; and condensate from stillage evaporation	High in dissolved organic solids containing nitrogen and fermented starches or their products
Meat and poultry products	Stockyards; slaughtering of animals; rendering of bones and fats; residues in condensates; grease and wash water; and picking of chickens	High in dissolved and suspended organic matter, blood, other proteins, and fats
Beet sugar	Transfer, screening, and juicing waters; drainings from lime sludge; condensates after evaporator; and juice and extracted sugar	High in dissolved and suspended organic matter containing sugar and protein
Pharmaceutical products	Mycelium, spent filtrate, and wash waters	High in suspended and dissolved organic matter, including vitamins
Yeast	Residue from yeast filtration	High in solids (mainly organic) and BOD
Pickles	Lime water; brine, alum and turmeric, syrup, seeds, and pieces of cucumber	Variable pH, high SS, color, and organic matter
Coffee	Pulping and fermenting of coffee beans	High BOD and SS
Fish	Rejects from centrifuge; pressed fish, and evaporator and other wash water wastes	Very high BOD, total organic solids, and odor
Rice	Soaking, cooking, and washing of rice	High BOD, total and suspended solids (mainly starch)
Soft drinks	Bottle washing; floor and equipment cleaning; and syrup storage tank drains	High pH, suspended solids, and BOD
Bakeries	Washing and greasing of pans and floor washing	High BOD, grease, floor washing, sugars, flour, and detergents
Materials		
Pulp and paper	Cooking, refining, washing of fibers, and screening of paper pulp	High or low pH, color, high suspended, colloidal, and dissolved solids, and inorganic fillers
Photographic products	Spent solutions of developer and fixer	Alkaline containing various organic and inorganic reducing agents
Steel	Coking of coal, washing of blast-furnace flue gases, and pickling of steel	Low pH, acids, cyanogen, phenol, ore, coke, limestone, alkali, oils, mill scale, and fine SS
Metal-plated products	Stripping of oxides and cleaning and plating of metals	Acid, metals, toxic, low volume, and mainly mineral matter
Iron-foundry products	Wasting of used sand by hydraulic discharge	High SS, mainly sand; and some clay and coal
Oil fields and refineries	Drilling muds, salt, oil, and some natural gas; acid sludges; and miscellaneous oils from refining	High dissolved salts from field and high BOD, odor, phenol, and sulfur compounds from refinery

Continued on next page

TABLE 1.3.1 *Continued*

Industries Producing Wastes	*Origin of Major Wastes*	*Major Characteristics*
Petrochemicals	Contaminated water from chemical production and transportation of second generation oil compounds	High COD, TDS, metals, COD/BOD ratio, and cpds. inhibitory to biologic action
Cement	Fine and finish grinding of cement, dust leaching collection, and dust control	Heated cooling water, SS, and some inorganic salts
Chemicals		
Acids	Dilute wash waters and many varied dilute acids	Low pH and low organic content
Detergents	Washing and purifying soaps and detergents	High in BOD and saponified soaps
Cornstarch	Evaporator condensate or bottoms when not reused or recovered, syrup from final washes, and wastes from bottling-up process	High BOD and dissolved organic matter; mainly starch and related material
Explosives	Washing trinitrotoluene (TNT) and guncotton for purification and washing and pickling of cartridges	TNT, colored, acid, odorous, and containing organic acids and alcohol from powder and cotton, metals, acid, oils, and soaps
Pesticides	Washing and purification products such as 2,4-D and dichlorodiphenyl trichloroethane (DDT)	High organic matter, benzenering structure, toxic to bacteria and fish, and acid
Phosphate and phosphorus	Washing, screening, floating rock, and condenser bleedoff from phosphate reduction plant	Clays, slimes and tall oils, low pH, high SS, phosphorus, silica, and fluoride
Plastic and resins	Unit operations from polymer preparation and use and spills and equipment washdowns	Acids, caustic, and dissolved organic matter such as phenols and formaldehyde
Fertilizer	Chemical reactions of basic elements and spills, cooling waters, washing of products, and boiler blowdowns	Sulfuric, phosphorous, and nitric acids; mineral elements, P, S, N, K, Al, NH_3, and NO_3; Fl; and some SS
Toxic chemicals	Leaks, accidental spills, and refining of chemicals	Various toxic dissolved elements and compounds such as Hg and polychlorinated biphenyls (PCBs)
Mortuary	Body fluids, washwaters, and spills	Blood salt, formaldehydes, high BOD, and infectious diseases
Hospital and Research Laboratories	Washing, sterilizing of facilities, used solutions, and spills	Bacteria and various chemicals and radioactive materials
Chloralkali wastes	Electrolytic cells and making chlorine and caustic soda	Mercury and dissolved metals
Organic chemicals	Various chemical productive processes	Varied types of organic chemicals
Energy		
Steam power	Cooling water, boiler blowdown, and coal drainage	Hot, high volume, and high inorganic and dissolved solids
Scrubber power plant wastes	Scrubbing of gaseous combustion products by liquid water	Particulates, SO_2, impure absorbents or NH_3 and NaOH
Coal processing	Cleaning and classification of coal and leaching of sulfur strata with water	High SS (mainly coal), low pH, high H_2SO, and $FeSO_4$

Source: N.L. Nemerow and A. Dasgupta, 1991, *Industrial and hazardous waste treatment* (New York: Van Nostrand Reinhold).

TABLE 1.3.2 SUBCATEGORIES OF THE PETROLEUM REFINING INDUSTRY REFLECTING SIGNIFICANT DIFFERENCES IN WASTEWATER CHARACTERISTICS

Topping: Topping and catalytic reforming whether the facility includes any other process in addition to topping and catalytic process. This subcategory is not applicable to facilities that include thermal processes (e.g., coking and visbreaking) or catalytic cracking.

Cracking: Topping and cracking, whether the facility includes any processes in addition to topping and cracking, unless specified in one of the following subcategories listed.

Petrochemical: Topping, cracking, and petrochemical operations, whether the facility includes any process in addition to topping, cracking, and petrochemical operations[a], except lube oil manufacturing operations.

Lube: Topping, cracking, and lube oil manufacturing processes, whether the facility includes any process in addition to topping, cracking, and lube oil manufacturing processes, except petrochemical operations.[a]

Integrated: Topping, cracking, lube oil manufacturing processes, and petrochemical operations, whether the facility includes any processes in addition to topping, cracking, lube oil manufacturing processes, and petrochemical operations.[a]

Source: U.S. Environmental Protection Agency (EPA), 1980, *Treatability manual,* Vol. II—Industrial descriptions, Sec. II.14—Petroleum refining, EPA-600/8-80-042b (Washington, D.C. [July]).

Notes: [a]The term petrochemical operations means the production of second generation petrochemicals (i.e., alcohols, ketones, cumene, and styrene) or first generation petrochemical and isomerization products (i.e., benzene/toluene/xylene [BTX], olefins, and cyclohexane) when 15% or more of refinery production is as first generation petrochemicals and isomerization products.

tion due to their potential environmental toxicity (U.S. EPA 1980). Five refinery subcategories, based on throughputs and process capacities, delineate information on wastewater characteristics as defined in Table 1.3.2. Table 1.3.3 presents ranges and median loadings in raw wastewater of conventional pollutants for the petroleum refining industry subcategories. Raw wastewater is the effluent from the oil separator, which is an integral part of refinery process operations for product and raw material recovery prior to wastewater treatment. Table 1.3.4 lists the toxic pollutants that have been measured in wastewater generated at petroleum refineries; no concentration data are included.

Wastewater Discharge Standards

The flow rates and quality characteristics of wastewater within and between types of industries vary widely. Therefore, wastewater discharge standards are related to industry type. For example, Table 1.3.5 summarizes the effluent limitations based on best practicable treatment (BPT) for point sources associated with the cracking subcategory of petroleum refining.

Effluent limitations can be technology-based or water quality-based, with the former considering BPT, best available technology economically achievable (BAT), best conventional pollutant control technology (BCT), or new source performance standards (NSPSs). BPT emphasizes end-of-pipe controls and reflects the average of the best for the industry category; it deals primarily with conventional pollutants such as BOD, oil and grease, solids, pH, and some metals. BAT can include pollution prevention through process control and end-of-pipe technology; it deals primarily with toxics such as organics and heavy metals. BCT is used with BAT. NSPSs are based on the best available demonstrated control technology (BADCT); it is typically similar to BAT with BCT.

Water quality-based effluent limitations can require greater levels of treatment as dictated by a waste load allocation scheme based on the total maximum daily load for the stream segment. Adjustment factors for the effluent limitations, which account for facility size and specific processes, are in the regulations (40 *CFR* Chap. 1, part 419). Effluent limitation information is also available for BAT, BCT, and NSPS for the cracking subcategory and for BPT, BAT, BCT, and NSPS for the topping subcategory, petrochemical subcategory, lube subcategory, and integrated subcategory (40 *CFR* Chap. 1, part 419)

Effluent discharge standards have been developed for wastewater from many industry types. This section does not address multiple types of industries; however, as an illustration of variability, Table 1.3.6 lists the water quality parameters for fourteen types of industries (Corbitt 1990). As the table shows the parameters are industry-specific.

Wastewater Characterization Surveys

Environmental engineers use wastewater characterization surveys, also referred to as industrial waste surveys, to establish flows, quality characteristics, and pollutant loadings at an individual industrial plant. They use the results of such surveys in (1) determining the treatment level necessary to meet effluent discharge standards, (2) selecting treatment processes, (3) making the discharge permit application for the facility, (4) establishing pretreatment requirements for the facility prior to discharge into municipal sewerage systems, and (5) developing a wastewater flow and loading minimization program. They also use survey information in establishing industrial user charges for discharges into municipal systems (Water Pollution Control Federation and American Society of Civil Engineers 1977).

TABLE 1.3.3 RAW WASTEWATER[a] LOADINGS IN NET KILOGRAMS/1000 M^3 OF FEEDSTOCK THROUGHPUT BY SUBCATEGORY IN PETROLEUM REFINING

	Topping Subcategory		*Cracking Subcategory*		*Petrochemical Subcategory*		*Lube Subcategory*		*Integrated Subcategory*	
Characteristics	*Range*[b]	*Median*	*Range*[b]	*Median*	*Range*[b]	*Median*	*Range*[b]	*Median*	*Range*[b]	*Median*
Flow[c]	8.00–558	66.6	3.29–2,750	93.0	26.6–443	109	68.6–772	117	40.0–1.370	235
BOD_5	1.29–217	3.43	14.3–466	72.9	40.9–715	172	62.9–758	217	63.5–615	197
COD	3.43–486	37.2	27.7–2,520	217	200–1,090	463	166–2.290	543	72.9–1,490	329
TOC	1.09–65.8	8.01	5.43–320	41.5	48.6–458	149	31.5–306	109	28.6–678	139
TSS	0.74–286	11.7	0.94–360	18.2	6.29–372	48.6	17.2–312	71.5	15.2–226	59.1
Sulfides	0.002–1.52	0.054	0.01–39.5[d]	0.94[d]	0.009–91.5	0.86	0.00001–20.0	0.014	0.52–7.87[d]	2.00[d]
Oil and grease	1.03–88.7	8.29	2.86–365	31.2	12.0–235	52.9	23.7–601	120	20.9–269	74.9
Phenols	0.001–1.06	0.034	0.19–80.1	4.00	2.55–23.7	7.72	4.58–52.9	8.29	0.61–22.6	3.78
Ammonia	0.077–19.5	1.20	2.35–174	28.3	5.43–206	34.3	6.5–96.2	24.1		
Chromium	0.0002–0.29	0.007	0.0008–4.15	0.25	0.014–3.86	0.234	0.002–1.23	0.046	0.12–1.92	0.49

Source: U.S. EPA, 1980.

Notes: [a]After refinery oil separator.

[b]Probability of occurrence less than or equal to 10 or 90% respectively.

[c]1000 m^3/1000 m^3 of feedstock throughput.

[d]Sulfur.

TABLE 1.3.4 QUALITATIVE LISTING OF TOXIC WATER POLLUTANTS POTENTIALLY IN PETROLEUM REFINERY WASTEWATERS

Metals and inorganics	Polycrylic aromatic hydrocarbons
Antimony	Acenaphthene
Arsenic	Acenaphthylene
Asbestos	Anthracene
Beryllium	Benzo(a)pyrene
Cadmium	Chrysene
Chromium	Fluoranthene
Copper	Flourene
Cyanide	Naphthalene
Lead	Phenanthrene
Mercury	Pyrene
Nickel	Polychlorinate biphenyls and related compounds
Selenium	Aroclor 1016
Silver	Aroclor 1221
Thallium	Aroclor 1232
Zinc	Aroclor 1242
Phthalates	Aroclor 1248
Bis(2-ethylhexyl) phthalate	Aroclor 1254
Di-n-butyl phthalate	Aroclor 1260
Diethyl phthalate	Halogenated aliphatics
Dimethyl phthalate	Carbon tetrachloride
Phenols	Chloroform
2-Chlorophenol	Dichlorobromomethane
2,4-Dichlorophenol	1,2-Dichloroethane
2,4-Dinitrophenol	1,2 *Trans* dichloroethylene
2,4-Dimethylphenol	Methylene chloride
2-Nitrophenol	1,1,2,2-Tetrachloroethane
4-Nitrophenol	Tetrachloroethylene
Pentachlorophenol	1,1,1-Trichloroethane
Phenol	Trichloroethylene
4,6-Dinitro-*o*-cresol	Pesticides and metabolites
Parachlorometa cresol	Aldrin
Aromatics	α-BHC
Benzene	β-BHC
1,2-Dichlorobenzene	δ-BHC
1,4-Dichlorobenzene	γ-BHC
Ethylbenzene	Chlordane
Toluene	4,4′-DDE
	4,4′-DDD
	α-Endosulfan
	β-Endosulfan
	Endosulfan sulfate
	Heptachlor
	Isophorone

Source: U.S. EPA, 1980.

TABLE 1.3.5 BPT EFFLUENT LIMITATION GUIDELINES FOR CRACKING SUBCATEGORY POINT SOURCES FROM PETROLEUM REFINING

Pollutant or pollutant property	*BPT Effluent Limitations* *Maximum for any 1 day*	*Average of daily values for 30 consecutive days shall not exceed*
	Metric Units (kg per 1000 m³ of feedstock)[3]	
BOD5	28.2	15.6
TSS	19.5	12.6
COD[1]	210	109
Oil and grease	8.4	4.5
Phenolic compounds	0.21	0.10
Ammonia as N	18.8	8.5
Sulfide	0.18	0.082
Total chromium	0.43	0.25
Hexavalent chromium	0.035	0.016
pH	([2])	([2])
	English Units (lb per 1000 bbl feedstock)	
BOD5	9.9	5.5
TSS	6.9	4.4
COD[1]	74.0	38.4
Oil and grease	3.0	1.6
Phenolic compounds	0.074	0.036
Ammonia as N	6.6	3.0
Sulfide	0.065	0.029
Total chromium	0.15	0.088
Hexavalent chromium	0.012	0.0056
pH	([2])	([2])

Source: Code of Federal Regulations, Title 40, Chap. 1, part 419—Petroleum refining point source category, 419–457 (1 July 1991).

Notes: [1]In any case where the applicant can demonstrate that chloride ion concentration in the effluent exceeds 1000 mg/l (1000 ppm), the Regional Administrator can substitute TOC as a parameter in lieu of COD. Effluent limitations for TOC shall be based on effluent data from the plant correlating TOC to BOD5. If the Regional Administrator judges that adequate correlation data are not available, the effluent limitations for TOC shall be established at a ratio of 2.2 to 1 to the applicable effluent limitations on BOD5.

[2]Within the range of 6.0 to 9.0.

[3]Feedstock denotes the crude oil and natural gas liquids fed to the topping units.

Figure 1.3.1 shows the components of a comprehensive industrial wastewater survey. The two main elements in a survey are (1) definition of the physical characteristics of the plant's sewer systems and (2) development of individual wastewater stream profiles. Defining the physical systems is required before information on flow characteristics and composition of individual wastewater streams can be obtained.

A survey of aircraft paint stripping wastewater generated at U.S. Navy facilities is an example of industrial wastewater characterization. The survey was conducted at six Naval Air Rework Facilities (NARFs). The field survey at each NARF was conducted for periods ranging from 5 to 7 days (Law, Olah, and Torres 1985). The waste-

TABLE 1.3.6 PARAMETERS ADDRESSED IN DISCHARGE STANDARDS FOR SELECTED INDUSTRIAL WASTEWATERS

	Industry													
Parameter	*Auto-mobile*	*Bever-age*	*Can-ning*	*Fertil-izer*	*Inor-ganic Chemi-cals*	*Or-ganic Chemi-cals*	*Meat Prod-ucts*	*Metal Finish-ing*	*Plas-tics & Syn-thetics*	*Pulp & Paper*	*Petro-leum Refin-ing*	*Steel*	*Tex-tiles*	*Dairy*
BOD_5	x	x	x		x	x	x		x	x	x		x	x
COD	x		x	x	x	x		x	x	x	x		x	x
TOC			x			x				x	x			x
TOD					x									
pH		x	x	x	x	x	x		x	x	x	x	x	x
Total solids	x				x									
SS	x	x	x	x	x	x	x	x	x	x	x	x	x	x
Settleable solids		x					x							
TDS		x	x	x	x	x	x		x	x	x		x	
Volatile SS											x			
Oil and grease	x	x		x		x	x	x	x	x	x	x	x	
Heavy metals, general						x		x		x			x	
Chromium	x			x	x						x	x	x	
Copper											x			
Nickel	x													
Iron	x			x	x						x	x		
Zinc	x			x					x		x	x		
Arsenic					x									
Mercury				x	x									
Lead	x				x						x			
Tin	x											x		
Cadmium	x													
Calcium				x										
Fluoride				x	x									
Cyanide					x	x		x	x		x	x		

Parameter	Industry													
	Auto-mobile	*Bever-age*	*Can-ning*	*Fertil-izer*	*Inor-ganic Chemi-cals*	*Or-ganic Chemi-cals*	*Meat Prod-ucts*	*Metal Finish-ing*	*Plas-tics & Syn-thetics*	*Pulp & Paper*	*Petro-leum Refin-ing*	*Steel*	*Tex-tiles*	*Dairy*
Cyanide					x	x		x	x		x	x		
Chloride	x			x	x	x					x	x		x
Sulfate	x			x	x				x		x	x		
Ammonia	x			x		x	x		x	x	x	x		
Sodium				x										
Silicates					x									
Sulfite										x				
Nitrate	x			x		x			x	x	x			x
Phosphorus			x	x	x	x	x		x	x	x			x
Urea or organic nitrogen				x										
Color		x	x				x			x	x		x	x
Total coliform		x					x			x				
FC			x							x				
Toxic materials		x					x		x	x	x		x	x
Temperature		x	x	x	x						x	x	x	x
Turbidity		x					x			x	x			x
Foam		x												
Odor											x			
Phenols	x				x	x			x	x	x	x	x	
Chlorinated benezoids & polynuclear aromatics					x				x					
Mercaptan sulfide									x		x		x	

Source: R.A. Corbitt, 1990, Wastewater disposal, Chap. 6 in *Standard handbook of environmental engineering,* edited by R.A. Corbitt (New York: McGraw-Hill, Inc.).

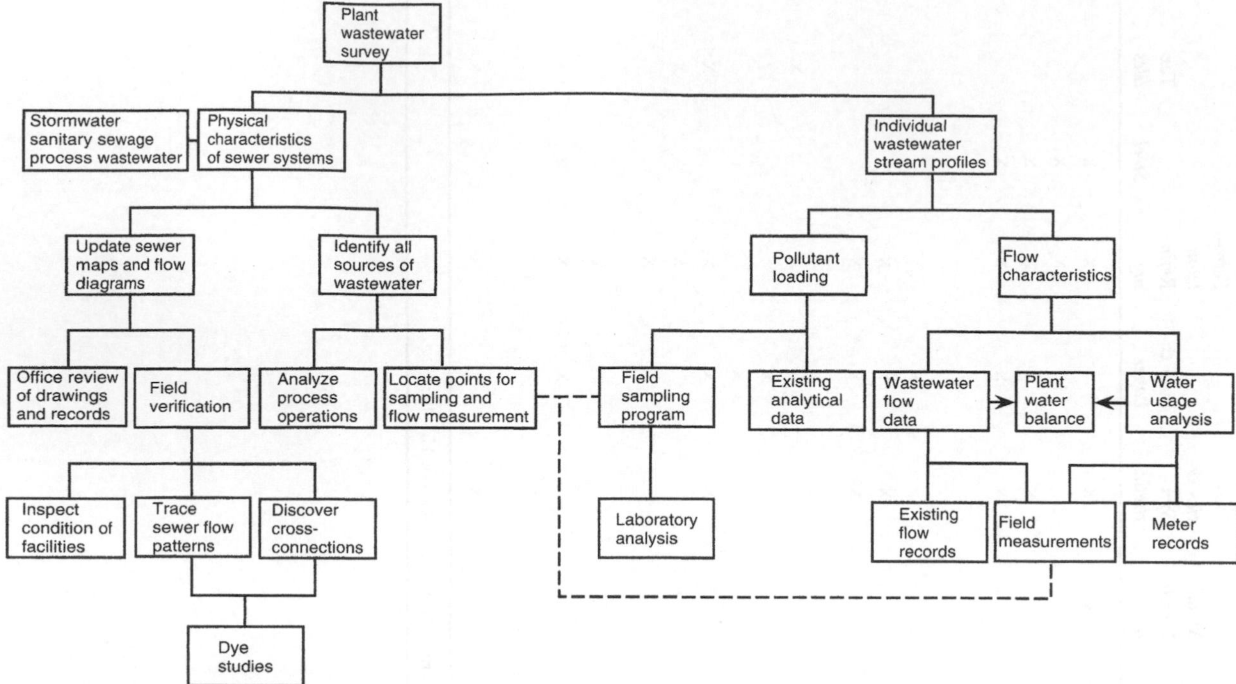

FIG. 1.3.1 Components of an industrial wastewater survey. (Reprinted, with permission, from R.A. Corbitt, 1990, Wastewater disposal, Chap. 6 in *Standard handbook of environmental engineering,* edited by R.A. Corbitt (New York: McGraw-Hill, Inc.)

water characteristics at these NARFs varied due to differences in missions and operations. Thus, the survey required characterization of wastewaters from all six NARFs to adequately represent aircraft paint stripping wastewater generated by the U.S. Navy.

The survey team selected 24 test parameters to characterize the quality of the aircraft paint stripping wastewater. Some parameters were important in monitoring chemical or biological treatment, while others were related to monitoring requirements for discharge regulations. The four parameters identified as most important, either because of high concentrations or limitations imposed by regulatory agencies, were oil and grease, phenol, chromium, and total toxic organics (TTO). The TTO parameter includes up to 110 toxic organics identified by the U.S. EPA. It is being incorporated into discharge permits by various regulatory agencies. The EPA set a value of 2.13 mg/l for TTO as the wastewater pretreatment standard for metal finishing industries, and aircraft paint stripping wastewater is in this category. The 2.13 mg/l value is based on the summation of all quantifiable concentrations greater than 0.01 mg/l for the 110 listed toxic organics.

To reduce sample analysis costs, the TTO analyses conducted during the survey excluded organics not present in paint stripping wastewater, including PCBs and pesticides. Thus, the analyses for TTO included only volatile organics (EPA Methods 601 and 602), acid extractables, and base neutral extractable organics (EPA Method 625). A total of eighty-four toxic organic compounds were analyzed from the list of 110 compounds; the twenty-six compounds not analyzed were PCBs and pesticides.

In nearly two months of site visits to the six NARFs, eighty-three aircraft paint stripping wastewater samples were collected, including nineteen composite samples (collected by automatic samplers) and sixty-four grab samples. For each sample, twenty-three analytical tests were performed. In addition, the TTO tests were performed on seventeen grab samples.

Table 1.3.7 shows the results of the analyses. Naval aircraft paint stripping wastewater is characterized by high organic pollutant contents and, except for chromium, low concentrations of heavy metals. The concentration level of oil and grease is generally lower than 500 mg/l; however, for one NARF, it averaged as high as 1215 mg/l. The average pH values varied from 5.2 to 9.4. Averages for phenol concentrations were typically in the hundreds range, but it was as high as 800 to 1300 mg/l for two NARFs. Due to the use of a nonphenolic paint stripper, one NARF (North Island) was able to keep its wastewater phenol concentration to around 1 mg/l; however, at this NARF, both stripper and labor usage increased due to the reduced effectiveness of the phenol-free stripper. The average TDS and SS values are typical for many wastewaters. Total chromium levels varied from 1.6 to 76 mg/l, while hexavalent chromium levels ranged from below 0.002 to about 13 mg/l. For the TTO parameter, values ranged from 124 to 2765 mg/l.

TABLE 1.3.7 COMPARISON OF COMPOSITE SAMPLE AVERAGES FROM THE NARFs

	Concentrations (mg/l, except for pH)					
	Norfolk	*Cherry Pt*	*Jacksonville*	*Pensacola*	*Norls*	*Alameda*
	3-Day Time	*3-Day Time*	*4-Day Flow*	*3-Day Time*	*3-Day Flow*	*3-Day Time*
BOD_5	3564	904	2537	2492	1540	311[a]
COD	8767	1823	4760	3957	4730	32032[a]
Cyanide, CN^-	0.1	0.1	<0.003	0.052	0.127	<0.1
Nitrate, NO_3^-	2.28	0.91	0.05	0.60	8.6	0.067
Oil and grease	1215	375	73	53.3	167.5	119
pH	5.6	5.2	8.7	8.54	9.44	7.3
Phenol	509	123	838	513	0.96	1346
Phosphorus, Total	1.26	0.81	4.4	0.95	3.02	<0.025[a]
Sulfate, SO_4	1533	150	94	113	147	10.7
TDS	1042	343	572	696	994	378
Total SS	214	51	260	54	184	88
Aluminum, Al	1.6	1.0	0.14	0.28	0.87	0.456
Arsenic, As	<0.003	<0.003	0.006	<0.001	0.003	0.0059
Cadmium, Cd	0.5	0.22	0.156	0.093	0.05	0.0437
Chromium, Hexavalent	<0.002	<0.002	13.1	1.45	7.71	2.95
Chromium, Total	13	1.60	15.6	27.3	18.70	76.0
Copper, Cu	0.23	0.03	0.054	0.36	0.18	0.119
Lead, Pb	0.51	0.12	0.04	0.13	0.12	0.099
Mercury, Hg	50×10^{-5}	48×10^{-5}	20×10^{-5}	0.0001	<0.0002	0.0006
Nickel, Ni	0.77	0.04	<0.03	0.022	<0.05	0.039
Selenium, Se	<0.003	<0.003	<0.005	<0.001	<0.003	0.0053
Silver, Ag	<0.01	0.01	0.005	0.001	0.01	0.009
Zinc, Zn	1.22	0.26	0.28	0.88	0.277	0.365
TTO[b]						
1. 601/602						
2. 625	124	312	1328	490	2765	986

Source: A. Law, N.J. Olah, and T. Torres, 1985, *Navy aircraft paint stripping waste characterization,* Technical memorandum 71-85-30 (Port Hueneme, Calif.: Naval Civil Engineering Laboratory).

Notes: [a]Alameda's data are suspected.

[b]Grab sample analysis averages.

TABLE 1.3.8 U.S. AIR FORCE AND U.S. NAVY PAINT STRIPPING WASTEWATERS

	Concentrations (mg/l)	
Parameter	*Air Force*[a]	*Navy*[b]
1. COD	9200–36,400	1800–8800
2. Oil and grease	8.4–66.3	50–1300
3. pH	8–8.6	5–9.5
4. Phenol	1040–4060	<1–1300
5. Total phosphorus	10–28	0.8–4.5
6. TSS	107–303	50–300
7. Total chromium	17.5–59.5	1.5–80
8. TTO	NM	124–2765
9. Methylene chloride	75–2000	70–2760[c]
10. TOC	1710–14,400	NM

Source: Law, Olah, and Torres, 1985.

Notes: NM = Not measured.

[a]From Perrotti (1975).

[b]Approximate range from Table 1.3.7.

[c]Approximate range from TTO tests.

Table 1.3.8 compares the data on paint stripping wastewater characteristics reported by Perrotti (1975) for the U.S. Air Force and the data procured by Law, Olah, and Torres (1985). The significant differences between the two are related to the concentration ranges of COD, phenol, oil and grease, total chromium, and pH. The Navy and Air Force paint stripping wastewaters are comparable with regard to the concentration ranges of total SS and methylene chloride. The TTO concentration was not measured in the Air Force study. These data show that the Navy's aircraft paint stripping wastewater is different from that generated at the corresponding Air Force facilities. These differences have implications in terms of potential treatment technologies.

As noted throughout this section, quality characteristics of industrial wastes vary considerably depending on the type of industry. A useful parameter in describing industrial wastes is an organic matter-based population equivalent defined as:

$$PE = \frac{A \times B \times 8.34}{0.17} \quad 1.3(1)$$

where:

PE = population equivalent based on organic constituents in the industrial waste
A = industrial waste flow, mg/day
B = industrial waste BOD, mg/l
8.34 = number of lb/gal
0.17 = number of lb BOD/person–day

Similar population equivalent calculations can be made for SS, nutrients, and other constituents. Expressing all waste loadings on a similar basis requires population equivalent calculations to be made for various pollutants from both point and nonpoint sources in a geographical area.

—*Larry W. Canter*

References

Code of Federal Regulations, Title 40, Chap. 1, Part 419—Petroleum refining point source category. 419–457. (1 July 1991).

Corbitt, R.A. 1990. Wastewater disposal. Chap. 6 in *Standard handbook of environmental engineering,* edited by R.A. Corbitt. 6.26–6.33. New York: McGraw-Hill, Inc.

Law, A., N.J. Olah, and T. Torres. 1985. *Navy aircraft paint stripping waste characterization.* Technical Memorandum 71-85-30. Port Hueneme, Calif.: Naval Civil Engineering Laboratory.

Nemerow, N.L. 1978. *Industrial water pollution: Origin, characteristics and treatment.* 529–549. Reading, Mass.: Addison-Wesley Publishing Company.

Nemerow, N.L. and A. Dasgupta. 1991. *Industrial and hazardous waste treatment.* 387–391. New York: Van Nostrand Reinhold.

Perrotti, A.E. 1975. *Activated carbon treatment of phenolic paint stripping wastewater.* AFCEC-TR-75-14. Tyndall AFB, Fla.: U.S. Air Force Civil Engineering Center. (May).

U.S. Environmental Protection Agency (EPA). 1980. *Treatability manual.* Vol. II, Industrial descriptions. Section II.14, Petroleum refining. EPA-600/8-80-042b. Washington, D.C. (July).

Water Pollution Control Federation and American Society of Civil Engineers. 1977. *Wastewater treatment plant design.* 9–10. Washington, D.C.

1.4 WASTEWATER MINIMIZATION

Wastewater minimization can be considered in terms of volume (or flow) reduction, strength (or pollutant concentration) reduction, or combinations of these reductions. Emphases on flow rate reductions are generally associated with both municipal and industrial wastewater, while pollutant concentration reductions are the focus of attention within industries. Municipalities also emphasize decreases in pollutant concentration via pretreatment ordinances for industries discharging wastewater into municipal sewerage systems.

Benefits resulting from wastewater minimization include the following:

- Reductions of flow and wastewater loadings on existing treatment plants with finite available capacities
- Minimization of hydraulic or organic overloads and toxicity effects on existing treatment facilities
- Reductions in unit process sizes, and possibly the types of processes, for new treatment facilities
- Reductions in initial and operation and maintenance costs for wastewater treatment
- Reductions in water costs for domestic and industrial users and in chemical or manufacturing costs for industries
- Reductions in energy requirements within municipalities and industries
- Reductions in water usage demands on limited water supplies in specific areas
- Facilitation of compliance with treatment plant effluent discharge standards
- Reductions in undesirable impacts on receiving water quality and aquatic ecology

Municipal Wastewater Flow Reduction

Flow reductions related to municipal wastewater include: (1) water conservation; (2) reuse of water in homes; (3) reduction of infiltration and inflow; and (4) reduction in stormwater runoff via best management practices. Water savings of up to 20 to 30% can be accomplished in homes and businesses using flow reduction devices and practicing simple water conservation measures (Qasim 1985). Table 1.4.1 gives examples of home water savings devices and their potential for water use reduction. Table 1.4.2 provides brief descriptions of flow reduction devices and water-efficient appliances. Municipal ordinances can require water saving devices for new homes and businesses.

Qasim (1985) suggests that home water reuse can lead to a 30 to 40% reduction in water consumption and a 40 to 50% reduction in wastewater volume. Wastewater from sinks, bathtubs, showers, and laundry can be treated on-site and reused for toilet flushing and lawn sprinkling.

Reducing infiltration and inflow into sewer lines also reduces wastewater flow rates arriving at a treatment facility. Infiltration refers to the volume of groundwater entering sewers and building sewer connections from the soil and through defective joints, broken or cracked pipes, improper connections, and manhole walls. Inflow denotes the volume of water discharged into sewer lines from sources such as roof leaders, cellar and yard area drains, foundation drains, commercial and industrial clean water discharges, and drains from springs and swampy areas (Sullivan et al. 1977). Wet weather flows containing large volumes of infiltration and inflow can create hydraulic overload difficulties at treatment plants (Qasim 1985). For existing sewers, a program involving cleaning, inspection, testing, and rehabilitation measures may be necessary; rehabilitation includes root control, grouting, and pipe lining (Sullivan et al. 1977). New, smaller diameter sewers can also be placed inside existing sewers that have major infiltration and inflow problems. New sewers should be designed, constructed, and inspected to remain within an infiltration limit of 20 gal/in diameter/mi/day (185.2 l/cm diameter/km/day) (Sullivan et al. 1977).

The 1987 *Clean Water Act* requires the use of best management practice (BMP) for minimizing the flow and pollutional characteristics of storm water runoff from industrial areas and municipalities with populations of 100,000 or more. The act emphasizes minimizing nonpoint pollution from the introduction of storm water into sanitary sewer systems or from direct discharge into receiving bodies of water.

BMP refers to a combination of practices that are most effective in preventing or reducing pollution generated by nonpoint sources to a level compatible with water quality goals (Novotny and Chesters 1981). After problem assessment, examination of alternative practices, and public participation, the state (or designated area-wide planning agency) determines the BMP based on technological, economic, and institutional considerations. Examples of BMPs include spill prevention and response, sediment and erosion control measures, and runoff management measures (U.S. EPA 1992a,b). Applying these measures can reduce infiltration in areas with separate (storm water and sanitary) sewer systems and reduce water flows and pollutional characteristics in areas with combined sewer systems.

Industrial Wastewater Flow Reduction

Industrial plants can achieve wastewater volume reductions by using

1. classification (segregation) of wastewater according to quality characteristics

TABLE 1.4.1 COMPARISON OF HOME WATER USE WITH AND WITHOUT CONSERVATION DEVICES

	Flow, gal/capita · day		
		With Conservation Devices	
Use	*Without Conservation Devices*	*Level 1*[a]	*Level 2*[a]
Baths	7	7	7
Dishwashers	2	1	1
Faucets	9	9	8
Showers	16	12	8
Toilets	22	19	14
Toilet leakage	4	4	8
Washing machines	16	14	13
Total	76	66	59

Source: Metcalf and Eddy, Inc., 1991, *Wastewater engineering,* 3d ed. (New York: McGraw-Hill, Inc.).
Note: [a]Level 1 uses retrofit devices such as flow restrictors and toilet dams. Level 2 uses water-conserving devices and appliances such as low-flush toilets and low-water-use washing machines.

TABLE 1.4.2 SUMMARY OF FLOW REDUCTION DEVICES AND APPLIANCES

Device/Appliance	*Description and Application*
Faucet aerators	Increases rinsing power of water by adding air and concentrating flow, reducing the amount of wash water used
Limiting-flow shower heads	Restricts and concentrates water passage by means of orifices that limit and divert shower flow for optimum use by the bather
Low-flush toilets	Reduces the discharge of water per flush
Pressure-reducing valve	Maintains home water pressure at a lower level than the water distribution system; decreases the probability of leaks and dripping faucets
Retrofit kits for bathroom fixtures	Consists of shower-flow restrictors, toilet dams, or displacement bags, and toilet leak detector tablets
Toilet dam	A partition in the water closet that reduces the amount of water per flush
Toilet leak detectors	Tablets that dissolve in the water closet and release dye to indicate leakage of the flush valve
Water-efficient dishwasher	Reduces water used
Water-efficient clothes washer	Reduces water used

Source: Metcalf and Eddy, Inc., 1991.

2. conservation of water use within industrial processes
3. production decreases
4. reuse of municipal and industrial treatment plant effluents as a water supply, and
5. elimination of batch or slug discharges of process wastewater (Nemerow and Dasgupta 1991).

Plants can achieve wastewater classification by considering the flow rates, quality characteristics, and treatment needs of wastewater used for manufacturing processes, cooling purposes, and sanitary uses.

Water conservation within industrial processes occurs when the industry changes from open to closed systems (Nemerow and Dasgupta 1991). For example, the sequential reuse of water within canning plants can achieve savings up to 20 to 25% and the use of shutoff valves on hoses and water lines can lead to another 20 to 25% reduction in water use. Nemerow and Dasgupta (1991) give additional examples of production decreases, effluent reuse, and elimination of slug discharges.

Pollutant Concentration Reduction

Clean technology, pollution prevention, and waste minimization are technical and managerial activities that can reduce the pollution emissions from industrial operations (Freeman et al. 1992; Hirschhorn and Oldenburg 1991; and Office of Technology Assessment 1986). Clean technology refers to applying technical processes to minimize waste material from the processes themselves (Johansson 1992). Pollution prevention relates to approaches that prevent pollution from occurring, including the incorporation of clean technology. Other housekeeping and conservation practices can be included in pollution prevention. Waste minimization tries to minimize negative impacts on the environment by reducing the amount of waste material from operations. Such waste reductions include applying pollution control technologies, chemical substitutions, clean technologies, and other activities that minimize the waste generated.

Chapter 3 provides additional information on pollution prevention.

—*Larry W. Canter*

References

Freeman, H.M., T. Harten, J. Springer, P. Randall, M.A. Curran, and K. Stone. 1992. Industrial pollution prevention: A critical review. *Journal of Air and Waste Management Association* 42, no. 5 (May): 618–656.

Hirschhorn, J.S. and K.U. Oldenburg. 1991. *Prosperity without pollution: The prevention strategy for industry and consumers.* New York: Van Nostrand Reinhold.

Johansson, A. 1992. *Clean technology.* Boca Raton, Fla.: Lewis Publishers, Inc.

Nemerow, N.L. and A. Dasgupta. 1991. *Industrial and hazardous waste treatment.* 101–105. New York: Van Nostrand Reinhold.

Novotny, V. and G. Chesters. 1981. *Handbook of nonpoint pollution.* New York: Van Nostrand Reinhold.

Office of Technology Assessment. 1986. *Serious reduction of hazardous waste: For pollution prevention and industrial efficiency.* OTA-ITE-317. 3–20. Washington, D.C. (September).

Qasim, S.R. 1983. *Wastewater treatment plants—Planning, design, and operation.* 32–37. New York: Holt, Rinehart and Winston.

Sullivan, R.H., M.M. Cohn, T.J. Clark, W. Thompson, and J. Zaffle. 1977. *Sewer system evaluation, rehabilitation, and new construction.* EPA 600-2-77-017d. iv-vii and 1. Cincinnati, Ohio: U.S. EPA. (December).

U.S. Environmental Protection Agency (EPA). 1992a. *Storm water management for construction activities—Developing pollution prevention plans and best management practices.* EPA 832-R-92-005. Washington, D.C.: Office of Water. (September).

———. 1992b. *Storm water management for industrial activities—Developing pollution prevention plans and best management practices.* EPA 832-R-92-006. Washington, D.C.: Office of Water. (September).

1.5 DEVELOPING A TREATMENT STRATEGY

In-plant treatments are a set of cost-effective onsite unit operations and processes installed in an industrial facility to remedy the production or discharge of hazardous or conventional waste to the environment. These remedial techniques consist of preliminary and primary process equipment, instrumentation, and control units related to the industry type and wastewater characteristics.

Factors affecting unit process selection are influent water characteristics, effluent quality required, reliability, sludge handling, and costs. Figure 1.5.1 shows the proposed strategy for wastewater management at a manufacturing complex.

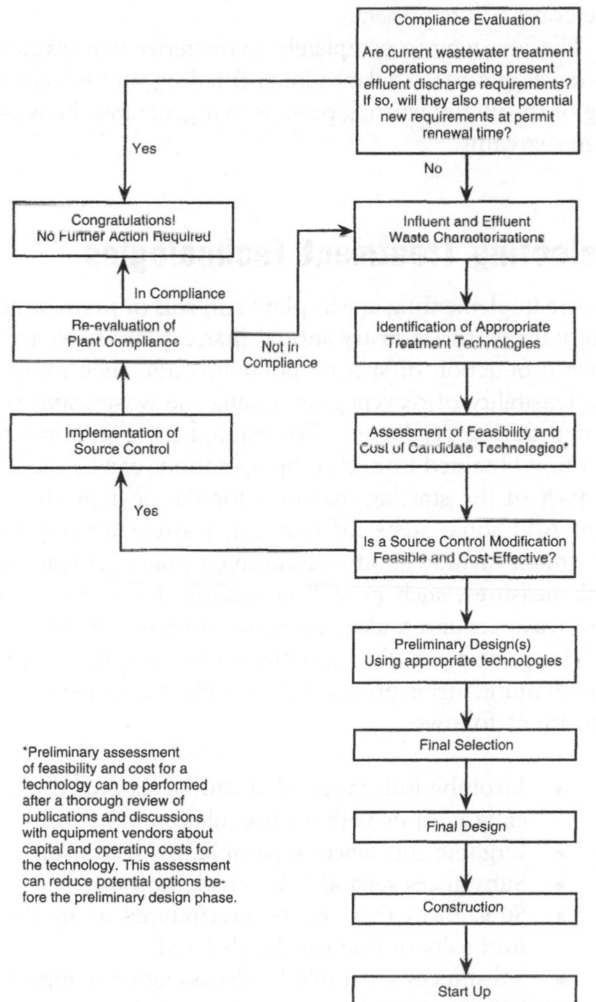

FIG. 1.5.1 Developing a wastewater treatment strategy. (Reprinted, with permission, from L.A. McLaughlin, H.S. McLaughlin, and K.A. Groff, 1992, Develop an effective wastewater treatment strategy, Chem. Eng. Prog. [September].)

Evaluating Compliance

The ultimate goal of any wastewater treatment system is to comply with regulations in a cost-effective manner. Common outlets for wastewater discharges are as follows:

Discharge to Surface Water. Effluent from wastewater treatment operations is piped directly to a surface water body and is subject to NPDES regulations. Effluent limitations depend on the ambient water quality criteria, the condition of the receiving stream, and the amount of mixing available. Discharge to surface water is usually a viable outlet for effluents containing benign contaminants or being treated to a level guaranteeing that the receiving stream is not impacted.

Discharge to the Sewer. Effluent from wastewater treatment operations is sent to the sewer, which is connected to a POTW. The wastewater is subject to municipal pretreatment regulations. Typically this outlet is good for effluents containing constituents that the POTW can effectively degrade, principally biodegradable organics of moderate strength. The capacity of the POTW to accept the waste must be considered.

Offsite Disposal. Effluents and other residues (sludge) from wastewater treatment operations are transported to an offsite treatment facility. The handler determines the level of pretreatment required for off-site disposal. This method is appropriate for low-volume, high-toxicity effluents and residuals. Effluents and residuals in this category are usually prohibited from discharge through other outlets (NPDES outfalls or municipal sewers).

Compliance evaluations can have either of the following forms:

Assessment of whether a plant's wastewater treatment operations are meeting effluent discharge limitations. If a facility is consistently not in compliance on a critical NPDES permit parameter, such as a primary pollutant concentration, this noncompliance is more urgent than an occasional minor deviation from a composite parameter such as BOD. The former situation requires immediate revision of a facility's wastewater management strategy, involving significant modifications or even complete replacement of existing treatment units.

Assessment of whether the facility can meet newer, more restrictive discharge limitations to be imposed when NPDES permits are renewed. These facilities should examine their current permits and allow adequate time before renewal to determine whether they can meet the anticipated discharge limitations.

Characterizing Wastewater

Figure 1.5.1 is a guide to the decisions involved in developing an appropriate waste management strategy.

The most important step in developing a wastewater management strategy is to completely characterize the wastewater. Although compliance monitoring indicates current compliance status, it is not an adequate starting point for the cost-effective design of a wastewater treatment system. Environmental engineers must characterize both the source to and effluent from current wastewater treatment operations. Understanding how the wastewater is produced is as important as knowing what contaminants are present (McLaughlin, McLaughlin, and Groff 1992). A review of manufacturing processes provides the knowledge base needed to evaluate the best place to reduce, recover, or treat individual waste streams.

The data should include the following information:

- All production activities within the facility, i.e., raw materials used and production records
- Detailed drawings of the plant showing the locations of processing units, their water distribution, and wastewater production and collection systems
- The quantity, analysis, frequency, and flow rate of the waste stream discharge from each unit process
- The frequency, extent, and type of monitoring and sampling used in accordance with the nature and variability of each waste stream
- The flow measurement and location of sample collection points within the facility indicating the type of monitoring stations (permanent or temporary) used

The constituents to be assayed and quantified in the influent and effluent depend on manufacturing process characteristics and should be determined on a case-by-case basis (McLaughlin, McLaughlin, and Groff 1992). In general, environmental engineers should analyze the constituents to assess compliance with current and future regulatory requirements and consider:

- Options for treating individual wastewater resources
- Potentials for modifying the manufacturing process to reduce, eliminate, or modify contaminants.

Table 1.5.1 lists the constituents and parameters that should be analyzed.

To assess whether current treatment systems require modification or replacement, environmental engineers must be aware of the target compounds for treatment and the additional constraints of individual treatment processes. For example, surfactants in metal waste streams must be analyzed because these compounds can chelate with metals and deteriorate conventional metal-removal technologies. Another example is evaluating the presence of salts when recycling is considered since equipment restrictions (such as preventing scaling) can limit the level of salt during recycling. Thus, environmental engineers should develop the list of constituents to analyze with both current and potential treatment technologies in mind.

Environmental engineers should also characterize wastewater in terms of flow rate. A comprehensive understanding of flow rates and patterns of flow to wastewater treatment operations is critical in the design of either a new system or system modifications. A good waste characterization accounts for all components of the final discharge including all resources and losses of water and the constituents present.

The best way to completely characterize wastewater is to develop a mass balance augmented by an understanding of the manufacturing process that generates the wastewater streams.

Selecting Treatment Technologies

Before implementing any in-plant controls or pretreatment alternatives, the industry should first explore ways to reduce production of specific pollutants and then examine the feasibility of recycling or reusing the wastewater generated during production. For example, the concentrated solution obtained from cleanup operations can be recycled as part of the starting materials for the next production run. Additional steps for reducing wastewater requiring treatment include good housekeeping practices; spill control measures, such as spill containment enclosures and drip trays around tanks; and eliminating wet floor areas.

The principal pollutants affected by modifying industrial manufacturing processes and in-plant treatment methods are as follows:

- Insoluble substances that can be separated physically with or without flocculation
- Organic substances separable by adsorption
- Substances separable by precipitation
- Substances that can be precipitated as insoluble iron salts or that can be chelated
- Substances separable by degassing or stripping
- Substances requiring a redox reaction
- Acids and bases
- Substances that can be concentrated by ion exchange or reverse osmosis
- Substances treatable by biological methods

TABLE 1.5.1 EXAMPLES OF COMMON POLLUTANTS AND OTHER PARAMETERS FOR WHICH EFFLUENTS SHOULD BE CHARACTERIZED

Constituent/Parameter	*Description*
VOCs Acid-extractable organics Base- and Neutral-extractable organics Metals, total and metals, soluble	Priority pollutants. Concentrations of these compounds are typically regulated on both sewer and NPDES permits.
BOD COD TOC TSS Temperature pH	Conventional pollutants. Permissible levels and values are also typically regulated on both sewer and NPDES permits.
Whole-effluent toxicity (LC_{50})	A relatively new parameter, it is usually only evaluated for NPDES permits.
Surfactants	Potential interfering agents
Ammonia Nitrate, nitrite Phosphorus	Nutrients. Determination is needed to adequately evaluate the potential for biological treatment.
Sulfate Chloride Sodium	Inorganic salts. Potential interfering agents
Flow rates	Necessary to perform a mass balance on the facility

Source: McLaughlin, McLaughlin, and Groff, 1992.

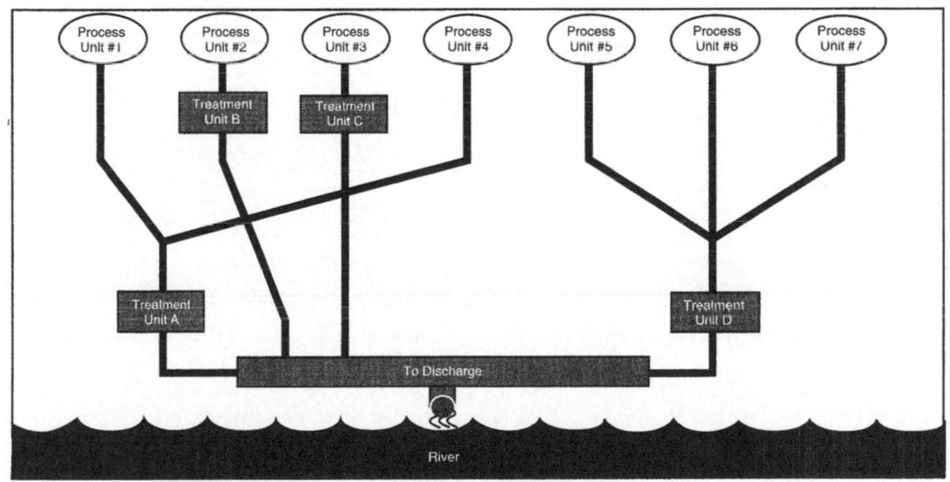

FIG. 1.5.2 Combining streams that use the same treatment technology and treating other streams at the source. (Reprinted, with permission, from McLaughlin, McLaughlin, and Groff, 1992.)

Environmental engineers should identify viable technologies for individual wastewater streams. Then, they can combine, on paper, the streams using the same technologies to create composite waste treatment trains. They then compare the resulting wastewater treatment trains to the current manufacturing and waste treatment practices to identify possible candidates for waste segregation and independent treatment (see Figure 1.5.2).

Treatability testing may be needed, especially when a plant that already has a physical-chemical or biological treatment facility is confronted with new wastes. For example, at a large chemicals complex, wastewater is screened for treatability as follows. The stream is pretreated to remove heavy metals and SS, and pH is adjusted. It is then fed to a batch-activated sludge reactor, and primed with biomass from the plant treatment facility. If the wastewater degrades quickly, as it should, it can be fed into the plant's main flow. If it does not, the choices are in-plant pretreatments, PAC addition to the bioreactor, or granular actuated carbon (GAC) treatment of the effluent.

The problem associated with combining two streams that require different technologies is that the cost of treating the combined stream is almost always more than individual treatment of the separate streams. This is because the capital cost of most treatment operations is proportional to the total flow of the wastewater, and the operating cost for treatment increases with a decreasing concentration for a given mass of contaminant.

Thus, if two waste streams use the same treatment, combining them improves the economics of scale for capital investment and similar operating costs. In contrast, if two treatment operations are required, combining the two streams increases capital costs for both treatment operations. In addition, if the streams are combined before the treatment, both treatments have lower contaminant concentrations for the same net contaminant mass, resulting in higher operating costs per lb of contaminant removed (McLaughlin et al. 1992).

—*Negib Harfouche*

Reference

McLaughlin, L.A., H.S. McLaughlin, and K.A. Groff. 1992. Develop an effective wastewater treatment strategy. *Chem. Eng. Prog.* (September).

2

Monitoring and Analysis

Béla G. Lipták

2.1 FLOW AND LEVEL MONITORING

The control and monitoring of flows and levels in the wastewater treatment industry involve the measurement of water, biological sludge, solid and liquid additives, and reagent flows. This section discusses methods of flow detection followed by a summary of wastewater-related level detection techniques.

Flow Sensors for the Wastewater Industry

Flow detection applications in the wastewater treatment industry include the measurement of large flows in partially filled pipes using weirs, flumes, or ultrasonic sensors. When water is flowing in regular pipelines, magnetic flowmeters, venturi tubes, flow nozzles, and pitot tubes are the usual sensors. In smaller pipelines, orifice plates, vortex flowmeters, or variable area flowmeters are used. For sludge services, doppler-type ultrasonic and magnetic flowmeter (provided with electrode cleaners), V-cone detector, and segmental wedge-type detector can be used. Gas, liquid, or solid additives can be charged by Coriolis mass flowmeters (gas or liquid), metering pumps, turbine or positive displacement meters (liquids), variable-area flowmeters (gas or liquid), or gravimetric feeders (solids). Table 2.1.1 summarizes flowmeter features and capabilities. The following sections provide a brief summary of the features and capabilities of the flowmeters used in the wastewater treatment industry.

Magnetic Flowmeters

DESIGN PRESSURE

Varies with pipe size. For a 4 in (100 mm) unit, the maximum pressure is 285 psig (20 bars); special units are available with pressure ratings up to 2500 psig (172 bars).

DESIGN TEMPERATURE

Up to 250°F (120°C) with Teflon liners and up to 360°F (180°C) with ceramic liners.

MATERIALS OF CONSTRUCTION

Liners: ceramics, fiberglass, neoprene, polyurethene, rubber, Teflon, vitreous enamel, and Kynar; Electrodes: platinum, Alloy 20, Hastelloy C, stainless steel, tantalum, titanium, tungsten carbide, Monel, nickel, and platinum-alumina cermet.

TYPE OF FLOW DETECTED

Volumetric flow of conductive liquids, including slurries and corrosive or abrasive materials.

MINIMUM CONDUCTIVITY REQUIRED

The majority of designs require 1 to 5 μS/cm. Some probe types require more. Special designs can operate at 0.05 or 0.1 μS/cm.

FLOW RANGES

From 0.01 to 100,000 gpm (0.04 to 378,000 liters per minute (lpm)).

SIZE RANGES

From 0.1 to 96 in (2.5 mm to 2.4 m) in diameter.

VELOCITY RANGES

0–0.3 to 0–30 ft/sec (0–0.1 to 0–10 m/sec).

ERROR (INACCURACY)

±1% of actual flow with pulsed direct current (dc) units within a range of up to 10:1 if flow velocity exceeds 0.5 ft/sec (0.15 m/sec). ±1% to ±2% full-scale with alternating current (ac) excitation.

COST

The probe designs are least expensive, at a cost of about $1500. A 1-in (25-mm) ceramic tube unit can be obtained for under $2000. A 1-in (25-mm) metallic wafer unit can be obtained for under $3000. An 8-in (200-mm) flanged meter that has a Teflon liner and stainless electrodes and is provided with 4 to 20 mA dc output, grounding ring, and calibrator costs about $8000. The scanning magmeter probe used in open-channel flow scanning costs about $10,000.

PARTIAL LIST OF SUPPLIERS

ABB Kent-Taylor Inc.; AccuDyne Systems Inc.; Accurate Metering Systems Inc.; ADE-Applied Digital Electronics; Badger Meter Inc.; Baily Controls Co.; Brooks Instrument Div. of Rosemount; Colorado Engineering Experimental Station; Dantec Electronics; H.R. Dulin Co.; Dynasonics Inc. (probe-type); Edinboro Computer Instruments Corp.; Electromagnetic Controls Corp.; Endress + Hauser Instruments; Engineering Measurements Co.; Fischer & Porter Co.; Foxboro Co.; Harwil Corp.; Honeywell, Industrial Controls Div.; Instrumark International Inc.; Johnson Yokogawa Corp.; K & L Research Co. (probe-type); Krone-America Inc.; Marsh-McBirney Inc. (probe-type); Meter Equipment Mfg.; Mine Safety Appliances Co.; Monitek Tech. Inc.; Montedoro Whitney; MSR Magmeter Manufacturing Ltd. (probe-type); Omega Engineering; Rosemount Inc.; Sarasota Measurements & Controls; Schlumberger Industries Inc.; Signet Industrial (probe-type); Sparling Instruments Co.; Toshiba International; Turbo Instruments Inc.; Vortab Corp.; Wallace & Tiernan Inc.; Wilkerson Instrument Co.; XO Technologies Inc.; Yokogawa Electric Corp.

Magnetic flowmeters use Faraday's Law of electromagnetic induction for measuring flow. Faraday's Law states that when a conductor moves through a magnetic field of given strength, a voltage level is produced in the conductor that depends on the relative velocity between the conductor and the field. This concept is used in electric gen-

TABLE 2.1.1 ORIENTATION TABLE FOR FLOW SENSORS

Flow Range scales:

- Solids Flow Units: 0.1, 1.0, 10, 10^2, 10^3, 10^4 kgm/hr; 0.1, 1.0, 10, 10^2, 10^3, 10^4, 10^5, 10^6 lbm/hr
- Gas Flow Units: 0.05, 0.3, 2.8, 28.3 cc/min; 10^{-6}, 10^{-5}, 10^{-4}, 10^{-3}, 10^{-2}, 0.1, 1.0, 10, 10^2, 10^3, 10^4 m³/hr or Am³/hr; 10^{-6}, 10^{-5}, 10^{-4}, 10^{-3}, 10^{-2}, 0.1, 1.0, 10, 10^2, 10^3, 10^4, 10^5 scfm or acfm
- Liquid Flow Units: .004, 0.04, 0.4, 3.8, 38, 379 cc/min; 10^{-6}, 10^{-5}, 10^{-4}, 10^{-3}, 10^{-2}, 0.1, 1.0, 10, 10^2, 10^3, 10^4 m³/hr; 10^{-6}, 10^{-5}, 10^{-4}, 10^{-3}, 10^{-2}, 0.1, 1.0, 10, 10^2, 10^3, 10^4, 10^5, 10^6 gpm

Type of Design	Applicable to Detect the Flow of: Clean Liquids	Viscous Liquids	Slurry	Gas	Solids	Direct Mass—Flow Sensor	Volumetric Flow Detector	Flow Rate Sensor	Inherent Totalizer	Direct Indicator	Transmitter Available	Linear Output	Rangeability	Pressure Loss Thru Sensor	Approximate Straight Pipe-Run Requirement[1] (Upstream Diameter/Downstream Diameter)	Accuracy * ±% Full Scale ** ±% Rate *** ±% Registration	Flow Range
Elbow taps	√	L	L	√			√	√			√	SR	3:1⊕	N	25/10⊕	5–10*	gpm—m³/hr; scfm—Sm³/hr
Jet deflection				√			√	√			√	√	25:1	M	20/5⊕	2*	scfm—Sm³/hr
Laminar flowmeters	√	√		√			√	√			√	√	10:1	H	15/5	½–5*⊕	gpm—m³/hr; scfm—Sm³/hr
Magnetic flowmeters	√⊕	√⊕	√⊕				√	√			√	√	10:1⊕	N	5/3	½**–2*	gpm—m³/hr
Mass flowmeters and miscellaneous coriolis	√ √	√ √	√ L	√ √	SD	√ √	√ √	√ √	SD SD	SD SD	√ √	√ √	100:1 20:1	A H	N N	½** 0.15–½**	lbm/hr–kgm/hr; scfm—Sm³/hr
Metering pumps	√	√	√				√		√		SD	√	20:1	—	N	1/10–1*	gpm—m³/hr
Orifice (plate or integral cell)	√	L	L	√			√	√			√	SR	3:1⊕	H	20/5⊕	½**–2*	gpm—m³/hr; scfm—Sm³/hr
Pitot tubes	√		L	√			√	√			√	SR	3:1⊕	M	30/5⊕	0.5–5*	gpm—m³/hr; scfm—Sm³/hr
Positive displacement gas meters				√			√		√	√	SD	√	10:1 to 200:1	M	N	½–1***	scfm—Sm³/hr

Positive displacement liquid meters											SD		10:1↗	H	N	0.1–2**	gpm—m^3/hr
Segmental wedge												SR	3:1	M	15/5	3**	gpm—m^3/hr
Solids flowmeters		SD	SD			SD	SD			SD			20:1	—	5/3	1/2**–4*	lbm/hr-kgm/hr
Target meters			L							SD		SR	4:1	H	20/5	0.5*–5*	----- gpm—m^3/hr ----- SCFM—Sm^3/hr
Thermal meters (mass flow)		L	L									L	20:1↗	A	5/3	1–2*	gpm—m^3/hr SCFM—Sm^3/hr
Turbine flowmeters		L		SD									10:1⊕	H	15/5[1]	1/4**	gpm—m^3/hr SCFM—Sm^3/hr
V-cone flowmeter		L	L									SR	3:1↓	M	2/5	1/2–2**	gpm—m^3/hr ACFM—Sm^3/hr
Ultrasonic flowmeters																	
Transit		L											20:1	N	15/5→	1**–2*	gpm—m^3/hr^3
Doppler		L		L	L								10:1	N	15/5→	2–3*	SCFM—Sm^3/hr
Variable-area flowmeters		L	L										5:1	A	N	1/2*–10**	gpm—m^3/hr SCFM—Sm^3/hr
Venturi tubes		L	L									SR	3:1↓	M	15/5→	1/2**–1*	gpm—m^3/hr^3
Flow nozzles			L									SR	3:1↓	H	20/5→	1**–2*	SCFM—Sm^3/hr
Vortex shedding													10:1⌀	H	20/5	0.5–1.5**	gpm—m^3/hr
Fluidic													20:1⌀	H	20/5	1–2**	ACFM—Sm^3/hr
Oscillating													10:1⌀	H	20/5	0.5*	
Weirs and flumes		L	L									SD	100:1	M	See Text	2–5*	gpm—m^3/hr^3

- - - - - = Nonstandard Range
L = Limited
SD = Some Designs
H = High
A = Average
M = Minimal
N = None
SR = Square Root

⊙ = The data in this column is for general guidance only.
↓ = The inherent rangeability of the primary device is substantially greater than shown. The value used reflects the limitation of the differential pressure sensing device when 1% of the actual flow accuracy is needed. With multiple-range intelligent transmitters, the rangeability can reach 10:1.
↳ = The pipe size establishes the upper limit.
↖ = Practically unlimited with the probe-type design.
⚲ = Must be conductive.
· = Can be reranged over 100:1.
→ = Varies with upstream disturbance.
⌀ = Can be more at high Re. No. services.
↗ = Up to 100:1 with high precision design.
▣ = Commercially available gas flow elements can be ±1% of rate.
⊕ = More for gas turbine meters.

erators. Faraday foresaw the practical application of the principle to flow measurement because many liquids are adequate electrical conductors. In fact, he attempted to measure the flow velocity of the Thames River using this principle. He failed because his instrumentation was not adequate, but 150 years later, the principle is successfully applied in magnetic flowmeters.

THEORY

Figure 2.1.1 shows how Faraday's Law is applied in the electromagnetic flowmeter. The liquid is the conductor that has a length equivalent to the inside diameter of the flowmeter D. The liquid conductor moves with an average velocity V through the magnetic field of strength B. The induced voltage is E. The mathematical relationship is:

$$E = BDV/C \qquad 2.1(1)$$

where:

C is a constant to take care of the proper units

When the pair of magnetic coils is energized, a magnetic field is generated in a plane mutually perpendicular to the axis of the liquid conductor and the plane of the electrodes. The velocity of the liquid is along the longitudinal axis of the flowmeter body; therefore, the voltage induced within the liquid is mutually perpendicular to the velocity of the liquid and the magnetic field.

The liquid should be considered as an infinite number of conductors moving through the magnetic field with each element contributing to the voltage that is generated. An increase in the flow rate of the liquid conductors moving through the field increases the instantaneous value of the voltage generated. Also, each of the individual generators contributes to the instantaneously generated voltage.

Whether the profile is essentially square (characteristic of a turbulent velocity profile), parabolic (characteristic of a laminar velocity profile), or distorted (characteristic of poor upstream piping), the magnetic flowmeter is excellent at averaging the voltage contribution across the metering cross section. The sum of the instantaneous voltages generated represents the average liquid velocity because each increment of liquid velocity within the plane of the electrode develops a voltage proportional to its local velocity. The signal voltage generated is equal to the average velocity almost regardless of the flow profile. The magnetic flowmeter detects the volumetric flow rate by sensing the linear velocity of the liquid.

The equation of continuity (Q = VA) is the relationship that converts the velocity measurement to volumetric flow rate if the area is constant. The area must be known and constant and the pipe must be full for a correct measurement.

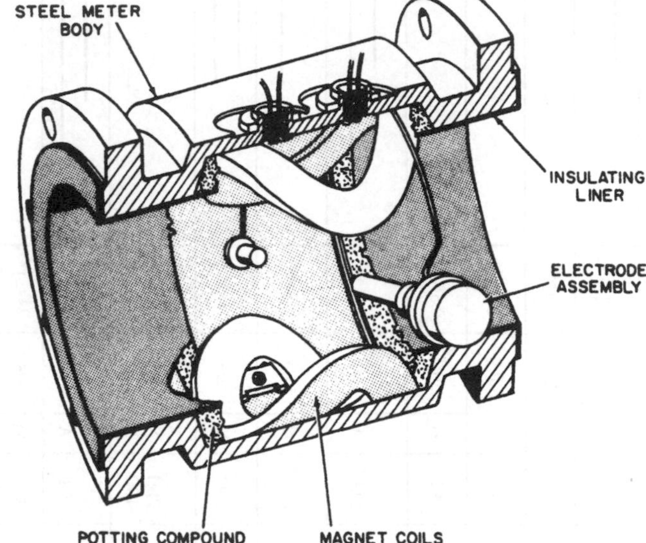

FIG. 2.1.2 The short-form magnetic flowmeter.

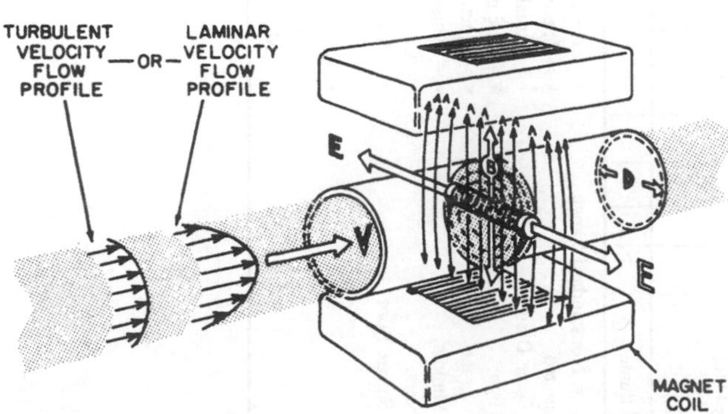

FIG. 2.1.1 Schematic representation of the magnetic flowmeter.

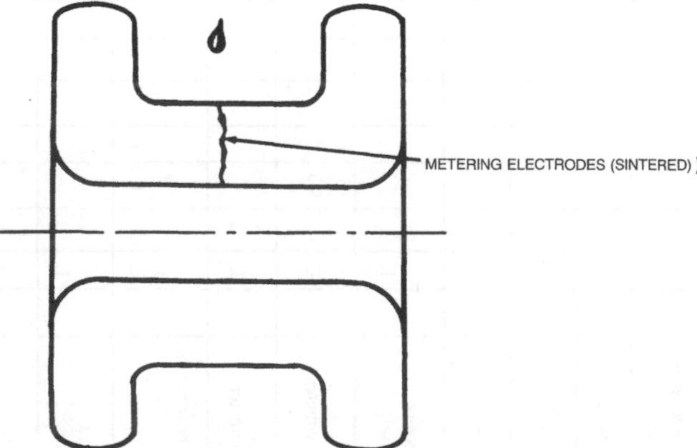

FIG. 2.1.3 The ceramic insert-type magnetic flowmeter.

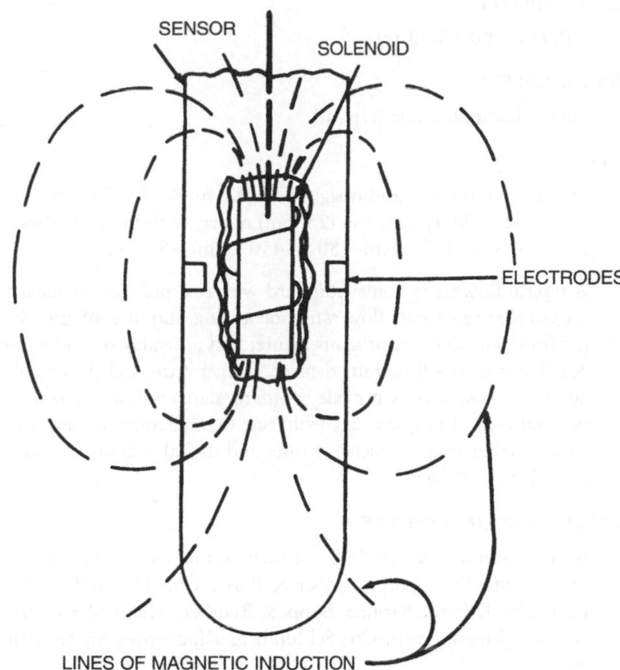

FIG. 2.1.4 The probe-type magnetic flowmeter.

DESIGNS AND APPLICATIONS

Magnetic flowmeters are available in conventional (see Figure 2.1.2), ceramic (see Figure 2.1.3), and probe (see Figure 2.1.4) constructions.

Most liquids or slurries are adequate electrical conductors to be measured by electromagnetic flowmeters. If the liquid conductivity is equal to 20 μS per cm or greater, most conventional magnetic flowmeters can be used. Special designs are available to measure the flow of liquids with threshold conductivities as low as 0.1 μS.

Magnetic flowmeters are not affected by viscosity or consistency (referring to Newtonian and nonNewtonian fluids, respectively). Changes in the flow profile due to changes in Reynolds numbers or upstream piping do not greatly affect the performance of magnetic flowmeters. The voltage generated is the sum of the incremental voltages across the entire area between the electrodes, resulting in a measure of the average fluid velocity. Nevertheless, the meter should be installed with five diameters of straight pipe before and three diameters of straight pipe following the meter.

Magnetic flowmeters are bidirectional. Manufacturers offer converters with output signals for both direct and reverse flows.

The magnetic flowmeter must be full to assume accurate measurement. If the pipe is only partially full, the electrode voltage, which is proportional to the fluid velocity, is still multiplied with the full cross section, and the reading will be high. Similarly, if the liquid contains entrained gases, the meter measures them as liquid, the reading will be high.

The meter's electrodes must remain in electrical contact with the fluid being measured and should be installed in the horizontal plane. In applications where a buildup or coating occurs on the inside wall of the flowmeter, periodic flushing or cleaning is recommended.

Special meters for measuring sewage sludge flow are designed to prevent the buildup and carbonizing of sludge on the meter electrodes. They use self-heating to elevate the metering body temperature to prevent sludge and grease accumulation.

ADVANTAGES

Magnetic flowmeters have the following advantages:

1. The magnetic flowmeter has no obstructions or moving parts. Flowmeter pressure loss is no greater than that of the same length of pipe. Pumping costs are thereby minimized.
2. Electric power requirements can be low, particularly with the pulsed dc types. Electric power requirements as low as 15 or 20 W are common.
3. The meters are suitable for most acids, bases, waters, and aqueous solutions because the lining materials are not only good electrical insulators but are also corrosion-resistant. Only a small amount of electrode metal is required, and stainless steel, Alloy 20, the Hastelloys, nickel, Monel, titanium, tantalum, tungsten carbide, and even platinum are all available.
4. The meters are widely used for slurry services not only because they are obstructionless but also because some of the liners, such as polyurethane, neoprene, and rubber, have good abrasion or erosion resistance.
5. The meters are capable of handling extremely low flows. Their minimum size is less than $\frac{1}{8}$ in (3.175 mm) inside diameter. The meters are also suitable for high volume flow rates with sizes as large as 10 ft (3.04 m).
6. The meters can be used as bidirectional meters.

LIMITATIONS

Magnetic flowmeters do have some specific application limitations:

1. The meters work only with conductive fluids. Pure substances, hydrocarbons, and gases cannot be measured. Most acids, bases, water, and aqueous solutions can be measured.
2. The conventional meters are relatively heavy, especially in larger sizes. Ceramic and probe-type units are lighter.
3. Electrical installation care is essential.
4. The price of magnetic flowmeters ranges from moderate to expensive. Their corrosion resistance, abrasion resistance, and accurate performance over wide turndown ratios can justify the cost. Ceramic and probe-type units are less expensive.

5. Periodically checking the zero on ac-type magnetic flowmeters requires block valves on either side to bring the flow to zero and keep the meter full. Cycled dc units do not have this requirement.

Coriolis Mass Flowmeters

SIZES

$\frac{1}{16}$ to 6 in (1.5 to 150 mm).

FLOW RANGE

0 to 25,000 lb/m (0 to 11,340 kg/m).

FLUIDS

Liquids, slurries, compressed gases, and liquified gases; not gas-liquid mixtures or gases at below 150 psig (10.3 bars).

OUTPUT SIGNAL

Linear frequency, analog, digital, scaled pulse, and display.

DETECTOR TYPES

Electromagnetic, optical, and capacitive.

OPERATING PRESSURE

Depends upon tube size and flange rating: 1800 psig (124 bars) typical standard; 5000 psig (345 bars) typical high-pressure.

PRESSURE DROP REQUIRED

From under 10 psig (0.7 bars) to over 100 psig (6.9 bars) as a function of viscosity and design.

OPERATING TEMPERATURE

Depends on the design: −100 to 400°F (−73 to 204°C) typical standard; 32 to 800°F (0 to 426°C) high-temperature.

MATERIALS OF CONSTRUCTION

Stainless steel, Hastelloy, titanium, and NiSpan C as standard; tantalum and Tefzel-lined as special.

INACCURACY

±0.15 to 0.5% of rate

$$\pm 0.15\% \text{ of rate} \pm \frac{\text{zero offset}}{\text{mass flow rate}} \times 100\%$$

Zero offset depends on flowmeter size and design; for a 1 in (25 mm) meter with a typical maximum flow rate of 400 to 1000 lb/m (180 to 450 kg/m), the zero offset typically ranges from 0.03 to 0.1 lb/m (0.014 to 0.045 kg/m), which is under 0.01%.

REPEATABILITY

±0.05 to ±0.2% of rate.

RANGEABILITY

20:1 calibration range (typical).

COST

Depends on the size and design: $\frac{1}{16}$ in (1.5 mm)—$3950; 6 in (150 mm)—$21,000; typical 1 in (25 mm) meter, with full-scale flow rate of 400 to 1000 lb/m (180 to 450 kg/m)—$5300.

A typical flowmeter comes standard with one pulse or frequency output that represents flow rate; one analog output configurable for flow rate, density, or temperature; and a display or digital output that provides flow rate, density, temperature, and flow total. In addition, most devices provide standard alarm outputs. The number and type of outputs vary from one manufacturer to another. Additional analog, frequency, pulse, and digital outputs are often provided as options.

PARTIAL LIST OF SUPPLIERS

Bailey Controls; Danfoss A/S (Denmark); Endress & Hauser Instruments; Exac Corp.; Fischer & Porter Co.; The Foxboro Co.; Heinrichs, K-Flow; Krohne, Bopp & Reuther; Micro Motion Inc.; Neptune Measurement Co.; Schlumberger Industries; Smith Meter Inc.

Coriolis flowmeters are not often used in wastewater applications. They are used on additive charging applications where the chemical is added on a weight basis or where their capability to detect both the mass flow and density of slurry streams is an advantage.

Since the appearance of the first commercial meters in the late 1970s, Coriolis flowmeters (see Figure 2.1.5) have become widely used. Their ability to measure mass flow directly with high accuracy and rangeability and to measure a variety of fluids makes Coriolis flowmeters the preferred flow measurement instrument for many applications. Coriolis flowmeters are also capable of measuring process fluid density and temperature. Since Coriolis flow measurement is a relatively new technology, many of the subtleties of its operation are still being investigated.

ADVANTAGES

Coriolis flowmeters have the following advantages:

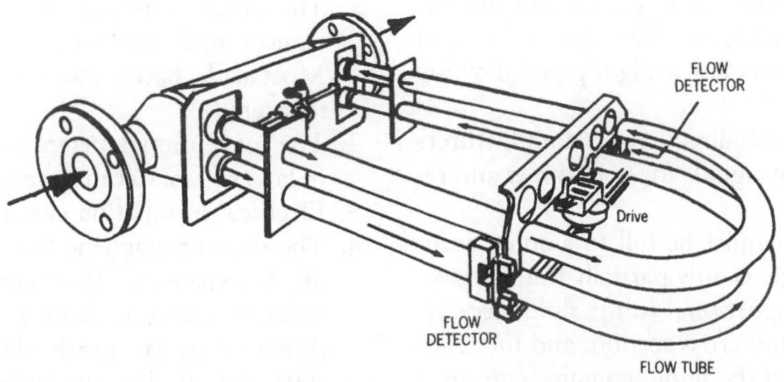

FIG. 2.1.5 Coriolis mass flowmeter.

1. They are capable of measuring a range of fluids that are often incompatible with other flow measurement devices. The operation of the flowmeter is independent of the Reynolds number; therefore, extremely viscous fluids can also be measured. A Coriolis flowmeter can measure the flow rate of Newtonian fluids, all types of nonNewtonian fluids, and slurries. Compressed gases and cryogenic liquids can also be measured by some designs.
2. Coriolis flowmeters provide a direct mass-flow measurement without the addition of external measurement instruments. While the volumetric flow rate of the fluid varies with changes in density, the mass-flow rate of the fluid is independent of density changes.
3. Coriolis flowmeters have outstanding accuracy. The base inaccuracy is commonly 0.2%. In addition, the flowmeters are extremely linear over their entire flow range.
4. The rangeability of the flowmeters is usually 20:1 or greater. Coriolis flowmeters have been successfully applied at flow rates 100 times lower than their rated full-scale flow rate.
5. A Coriolis flowmeter is capable of measuring mass-flow rate, volumetric flow rate, fluid density, and temperature—all from one instrument.
6. The operation of the flowmeter is independent of flow characteristics such as turbulence and flow profile. Therefore, upstream and downstream straight run requirements and flow conditioning are not necessary. They can also be used in installations that have pulsating flow.
7. Coriolis flowmeters do not have internal obstructions that can be damaged or plugged by slurries or other types of particulate matter in the flow stream. Entrained gas or slugs of gas in the liquid do not damage the flowmeter. The flowmeter has no moving parts that wear out and require replacement. These design features reduce the need for routine maintenance.
8. The flowmeter can be configured to measure flow in either the forward or reverse direction. In reverse flow, a time or phase difference occurs between the flow detector signals, but the relative difference between the two detector signals is reversed.
9. Coriolis flowmeter designs are available for use in sanitary applications and for the measurement of shear sensitive fluids. Materials are available that permit the measurement of corrosive fluids.

LIMITATIONS

Coriolis flowmeters have the following limitations:

1. They are not available for large pipelines. The largest Coriolis flowmeter has a maximum flow rating of 25,000 lb/min (11,340 kg/min) and is equipped with 6-in. (15-cm) flanges. Larger flow rates require more than one flowmeter mounted in parallel.
2. Some flowmeter designs require high fluid velocities to achieve significant time or phase difference between the flow detector signals. These high velocities can result in high pressure drops across the flowmeter.
3. Coriolis flowmeters are expensive. However, the cost of a Coriolis meter is often comparable to (or below) the cost of a volumetric meter plus a densitometer used together to determine the mass-flow rate.
4. Coriolis flowmeters have difficulty measuring the flow rate of low-pressure gas. Applications with pressures less than 150 psig are marginal with the flowmeter designs currently available. Low-pressure gases have low density, and their mass-flow rate is usually low. Generating sufficient Coriolis force requires a high gas velocity. This high velocity can lead to prohibitively high pressure drops across the meter.

Metering Pumps

TYPES

A. Peristaltic
B. Piston or plunger types (provided with packing glands)
C. Diaphragm or glandless types (mechanical, hydraulic, double-diaphragm, and pulsator designs)

CAPACITY

A. 0.0005 cc/min to 20 gpm (90 lpm)
B. 0.001 gph to 280 gpm (0.005 lph to 1250 lpm)
C. Mechanical diaphragms: from 0.01 to 50 gallons per hour (gph) (0.05 to 3.7 lpm); mechanical bellows: from 0.01 to 250 gph (0.05 to 18 lpm); and others: from 0.01 to 800 gph (0.05 liters per hour (lph) to 60 lpm); pulsator pumps: from 30 to 1800 gph (2 to 130 lpm)

ERROR (INACCURACY)

A. ±0.1 to ±0.5% of full scale over a 10:1 range
B & C. ±0.25 to ±1% of full scale over a 10:1 range; can be as good as ±0.1% full scale at 100% stroke and tends to drop as stroke is reduced

MAXIMUM DISCHARGE PRESSURE

A. 50 psig (3.5 bars)
B. 50,000 psig (3450 bars)
C. Mechanical bellows: up to 75 psig (5 bars); mechanical diaphragm: up to 125 psig (8.5 bars); hydraulic Teflon diaphragm: 1500 psig (104 bars); pulsator pumps: up to 5000 psig (345 bars); and hydraulic metallic diaphragms: up to 40,000 psig (2750 bars)

MAXIMUM OPERATING TEMPERATURE

A. 70 to 600°F (−57 to 315°C)
B. Jacketed designs: up to about 500°F (260°C)
C. Units containing hydraulic fluids can handle from −95 to 360°F (−71 to 182°C), Teflon and Viton diaphrams are limited to 300°F (150°C), and neoprene and Buna N are limited to 200°F (92°C). The metal bellows and the remote head designs can operate from cryogenic to 1600°F (870°C).

MATERIALS OF CONSTRUCTION

A. Neoprene, Tygon, Viton, and silicone
B. Cast iron, steel, stainless steel, Hastelloy C, Alloy 20, Carpenter 20, Monel, nickel, titanium, glass, ceramics, Teflon, polyvinyl chloride (PVC), Kel-F, Penton, polyethylene, and other plastics

C. Polyethylene, Teflon, PVC, Kel-F, Penton, steel, stainless steel, Carpenter 20, Monel, Hastelloy B & C

COST

A. \$200 to \$800
B. \$1000 to \$6000
C. \$1000 to \$12,000

PARTIAL LIST OF SUPPLIERS

American LEWA Inc. (A,B,C); Barnant Co. (A); Blue White Industries; Bran & Luebbe Inc.; Clark-Cooper Corp. (B,C); Cole-Parmer Instrument Co.; Flo-Tron Inc. (B); Fluorocarbon Co.; Gerber Industries; Hydroflow Corporation; LDC Analytical; Leeds & Northrup, Unit of General Signal; Liquid Metronics Inc. Milton Roy Div. (B); Plast-O-Matic Valves Inc.; Ruska Instrument Corp.; S J Controls Inc.; Valcor Scientific; Wallace & Tiernan Inc. (B,C)

In the wastewater treatment industry, metering pumps are often used to charge reagents, coagulants, or other additives. While they require periodic recalibration, their advantages include high accuracy (similar to turbine or positive displacement flowmeters), high rangeability, suitability for slurry service, and the ability to both pump and meter the fluid.

Orifices

DESIGN PRESSURE

For plates, limited by the readout device only; integral orifice transmitter to 1500 psig (10.3 MPa)

DESIGN TEMPERATURE

Function of the associated readout system when the differential pressure unit must operate at the elevated temperature. For the integral orifice transmitter, the standard range is −20 to 250°F (−29 to 121°C).

SIZES

Maximum size is the pipe size.

FLUIDS

Liquids, vapors, and gases

FLOW RANGE

From a few cc/min using integral orifice transmitters to any maximum flow; limited only by pipe size

MATERIALS OF CONSTRUCTION

No limitation on plate materials. Integral orifice transmitter wetted parts can be obtained in steel, stainless steel, Monel, nickel, and Hastelloy.

INACCURACY

The orifice plate, if the bore diameter is correctly calculated and prepared, can be accurate to ±0.25 to ±0.5% of the actual flow. When a conventional d/p cell is used to detect the orifice differential, that adds a ±0.1 to ±0.3% of the full-scale error. The error contribution of smart d/p cells is only 0.1% of the actual span.

INTELLIGENT D/P CELLS

Inaccuracy of ±0.1%, rangeability of 40:1, the built-in proportional integral and derivative (PID) algorithm

RANGEABILITY

If rangeability is defined as the flow range within which the combined flow measurement error does not exceed ±1% of the actual flow, then the rangeability of conventional orifice installations is 3:1. When intelligent transmitters with automatic switching capability between the high and low spans are used, the rangeability can approach 10:1.

COST

A plate only is \$50 to \$300, depending on size and materials. For steel orifice flanges from 2 to 12 in (50 to 300 mm), the cost ranges from \$200 to \$1000. For flanged meter runs in the same size range, the cost ranges from \$400 to \$3000. The cost of electronic or pneumatic integral orifice transmitters is between \$1500 and \$2000. The cost of d/p transmitters ranges from \$900 to \$2000, depending on type and intelligence.

PARTIAL LIST OF SUPPLIERS

ABB Kent-Taylor Inc. (includes integral orifices); Crane Manufacturing Inc.; Daniel Flow Products Inc. (orifice plates and plate changers); Fischer & Porter Co. (includes integral orifices); Fluidic Techniques, a Div. of FTI Industries; Foxboro Co. (includes integral orifices); Honeywell Industrial Div.; Lambda Square Inc.; Meriam Instrument, Div. Scott & Fetzer (orifice plates); Rosemount Inc.; Vickery-Simms, a Div. of FTI Industries. In addition, orifice plates, flanges, and accessories can be obtained from most major instrument manufacturers.

The orifice plate, when installed in a pipeline, causes an increase in flow velocity and a corresponding decrease in pressure. The flow pattern shows an effective decrease in the cross-section beyond the orifice plate, with a maximum velocity and minimum pressure at the vena contracta (see Figure 2.1.6). This location can be from .35 to .85 pipe diameters downstream from the orifice plate depending on the β ratio and the Reynolds number.

This flow pattern and the sharp leading edge of the orifice plate (see Figure 2.1.6) that produces it are important. The sharp edge results in an almost pure line contact between the plate and the effective flow, with negligible fluid-to-metal friction drag at this boundary. Any nicks, burrs, or rounding of the sharp edge can result in large measurement errors.

When differential pressure is measured at a location close to the orifice plate, friction effects between the fluid and the pipe wall upstream and downstream from the orifice are minimized so that pipe roughness has a minimum effect. Fluid viscosity, as reflected in the Reynolds number, has a considerable influence, particularly at low Reynolds numbers. Since the formation of the vena contracta is an inertial effect, a decrease in the ratio of inertial to frictional forces (decrease in Reynolds number), and the corresponding change in flow profile, results in less constriction of flow at the vena contracta and an increase of the flow coefficient. In general, the sharp edge orifice plate should not be used at pipe Reynolds numbers under 10,000. The minimum recommended Reynolds number varies from 10,000 to 15,000 for 2-in (50-mm) through 4-in (102-mm) pipe sizes for β ratios up to 0.5 and from 20,000 to 45,000 for higher β ratios. The Reynolds number requirement increases with pipe size and β ratio and can range up to 200,000 for pipes 14 in (355 mm) and

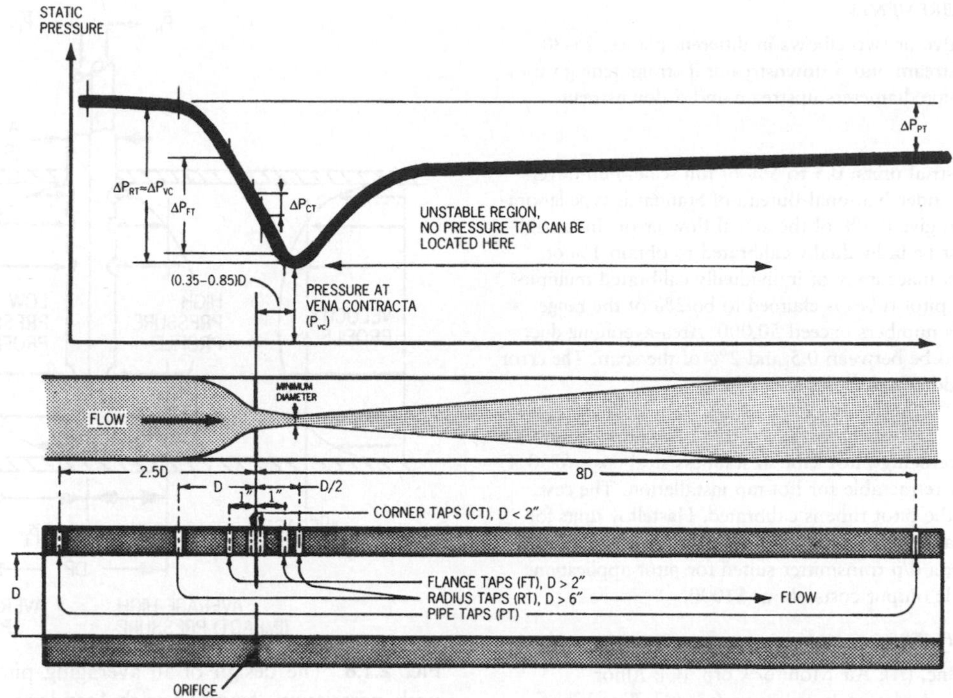

FIG. 2.1.6 Pressure profile through an orifice plate and the different methods of detecting the pressure drop.

larger. Maximum Reynolds numbers can be 10^6 for 4-in (102-mm) pipe and 10^7 for larger sizes.

WASTEWATER APPLICATIONS

If the water is dirty, containing solids or sludge, the pressure taps must be protected by clean water purging or by use of chemical seals and the orifice plates should be the segmental or eccentric orifice type (see Figure 2.1.7). Annular orifices and V-cone meters are also applicable to dirty services. Because the pressure recovery of orifices is low, they are not recommended to measure larger flows due to the excessive pumping costs. In these applications, venturi-type, high-recovery flow elements should be used.

The main advantages of orifices are their familiarity, simplicity, and the fact that they do not need calibration.

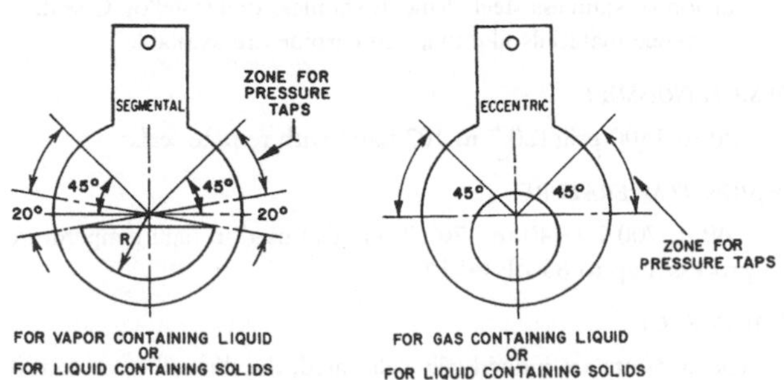

FIG. 2.1.7 Segmental and eccentric orifice plates.

The disadvantages include their low rangeability, low accuracy, high pressure drop, and potential plugging.

Pitot Tubes

TYPES

A. Standard, single-port
B. Multiple-opening, averaging
C. Area averaging for ducts

APPLICATIONS

Liquids, gases, and steam

OPERATING PRESSURE

Permanently installed carbon or stainless steel units can operate at up to 1400 psig (97 bars) at 100°F (38°C) or 800 psig (55 bars) at approximately 700°F (371°C). The pressure rating of retractable units is a function of the isolating valve.

OPERATING TEMPERATURE

Up to 750°F (399°C) in steel and 850°F (454°C) in stainless steel construction when permanently installed

FLOW RANGES

Generally 2-in (50-mm) pipes or larger; no upper limit

MATERIALS OF CONSTRUCTION

Brass, steel, and stainless steel

MINIMUM REYNOLDS NUMBER

Range from 20,000 to 50,000

RANGEABILITY

Same as orifice plates

STRAIGHT-RUN REQUIREMENTS

Downstream of valve or two elbows in different planes, 25–30 pipe diameters upstream and 5 downstream; if straightening vanes are provided, 10 pipe diameters upstream and 5 downstream

INACCURACY

For standard industrial units: 0.5 to 5% of full scale. Full-traversing Pitot Venturis under National-Bureau-of-Standards-type laboratory conditions can give 0.5% of the actual flow error. Industrial Pitot Venturis must be individually calibrated to obtain 1% of range performance. Inaccuracy of individually calibrated multiple-opening averaging pitot tubes is claimed to be 2% of the range when the Reynolds numbers exceed 50,000. Area-averaging duct units are claimed to be between 0.5 and 2% of the span. The error of the d/p cell is additional to the errors listed.

COSTS

A 1-in-diameter averaging pitot tube in stainless steel costs $750 if fixed and $1400 if retractable for hot-tap installation. The cost usually doubles if the pitot tube is calibrated. Hastelloy units for smokestack applications can cost $2000 or more. A local pitot indicator costs $400; a d/p transmitter suited for pitot applications with 4 to 20 mA dc output costs about $1000.

PARTIAL LIST OF SUPPLIERS

ABB Kent-Taylor Inc. (A); Air Monitor Corp. (C); Alnor Instrument Co. (A); Andersen Instruments Inc. (A); Blue White Industries (A); Brandt Instruments (C); Davis Instrument Mfg. Co. (A); Dietrich Standard, a Dover Industries Company (Annubar—B); Dwyer Instruments Inc. (B); Fischer & Porter Inc. (A); Foxboro Co. (Pitot Venturi—A); Land Combustion Inc. (A); Meriam Instrument, a Scott Fetzer Company (B); Mid-West Instrument (Delta Tube—B); Preso Industries (Elliptical—B); Sirco Industries Ltd. (A); Ultratech Industries Inc. (A); United Electric Controls Co. (A)

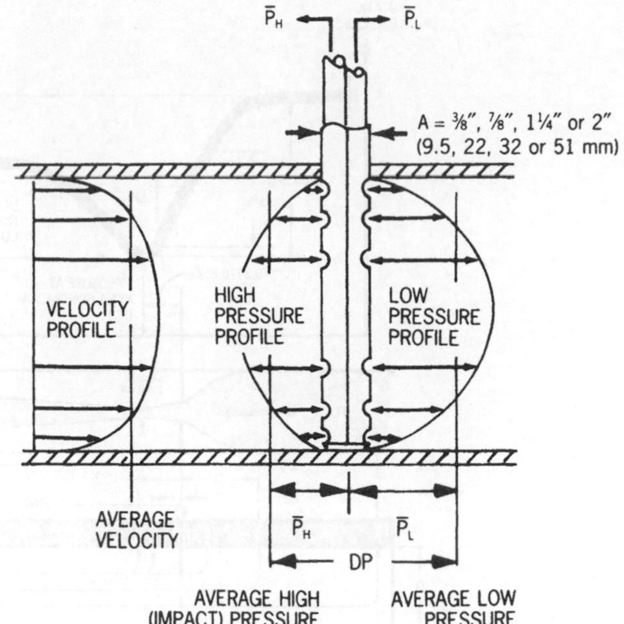

FIG. 2.1.8 The design of an averaging pitot tube. (Reprinted, with permission, from Dietrich Standard, a Dover Industries Company.)

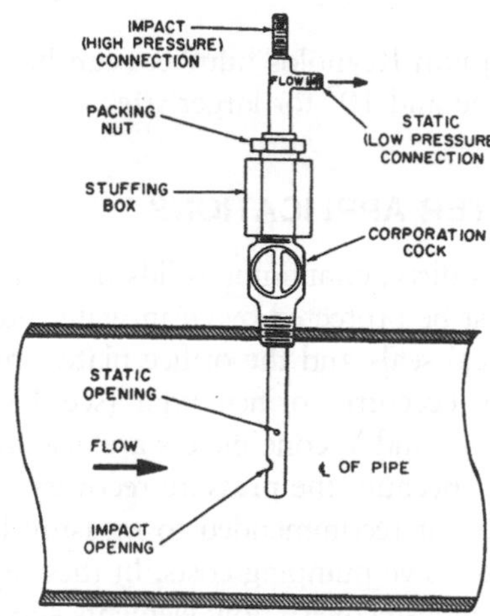

FIG. 2.1.9 Schematic diagram of an industrial device for sensing static and dynamic pressures in a flowing fluid.

While pitot sensors are low-accuracy and low-rangeability detectors, they do have a place in wastewater treatment-related flow measurement. Pitot tubes should be used when the measurement is not critical, the water is reasonably clean, and a low cost measurement is needed. These sensors can be inserted in the pipe without shutdown and can also be removed for periodic cleaning while the pipe is in use. Multiple-opening pitot tubes (see Figure 2.1.8) are less sensitive to flow velocity profile variations than single-opening (see Figure 2.1.9) tubes. In some dirtier applications, purged pitot tubes are also used.

Segmental Wedge Flowmeters

APPLICATIONS

Clean, viscous liquids or slurries and fluids with solids

SIZES

1- to 12-in (25.4- to 305-mm) diameter pipes

DESIGNS

For smaller sizes (1 and 1.5 in), the wedge can be integral; for larger pipes, remote seal wedges are used with calibrated elements.

WEDGE OPENING HEIGHT

From 0.2 to 0.5 of the inside pipe diameter

PRESSURE DROPS

25 to 200 in H_2O (6.2 to 49.8 kPa)

MATERIALS OF CONSTRUCTION

Carbon or stainless steel element; stainless or Hastelloy C seal; special wedge materials like tungsten carbide are available.

DESIGN PRESSURE

300 to 1500 psig (20.7 to 103 bars) with remote seals

DESIGN TEMPERATURE

−40 to 700°F (−40 to 370°C) but also used in high-temperature processes up to 850°F (454°C)

INACCURACY

The elements are individually calibrated; the d/p cell error contribution to the total measurement inaccuracy is 0.25% of full scale. The error over a 3:1 flow range is usually not more than 3% of the actual flow.

COST

A 3-in (75-mm) calibrated stainless steel element with two stainless steel chemical tees and an electronic d/p transmitter provided with remote seals is about $3500.

PARTIAL LIST OF SUPPLIERS

ABB Kent-Taylor Inc.

The segmental wedge flow element provides a flow opening similar to that of a segmental orifice, but flow obstruction is less abrupt (more gradual), and its sloping entrance makes the design similar to the flow tube family. It is primarily used on slurries. Its main advantage is its ability to operate at low Reynolds numbers. While the square root relationship between the flow and pressure drop in sharp-edged orifices, venturis, or flow nozzles requires a Reynolds number above 10,000, segmental wedge flowmeters require a Reynolds number of only 500 or 1000. For this reason the segmental wedge flowmeter can measure flows at low flow velocities and when process fluids are viscous. In that respect, it is similar to conical or quadrant edge orifices.

For pipe sizes under 2 in (50 mm), the segmental wedge flow element is made by a V-notch cut into the pipe and a solid wedge welded accurately in place (see Figure 2.1.10). In sizes over 2 in, the wedge is fabricated from two flat plates that are welded together before insertion into the spool piece. On clean services, regular pressure taps are located equidistant from the wedge (see Figure 2.1.10), while on applications where the process fluid contains solids in suspension, chemical tees are added upstream and downstream of the wedge flow element. The chemical seal element is flush with the pipe, eliminating pockets and making the assembly self-cleaning. The seals are made of corrosion-resistant materials and are also suited for high-temperature services. Some users have reported applications on processes at 3000 psig (210 bars) and 850°F (454°C).

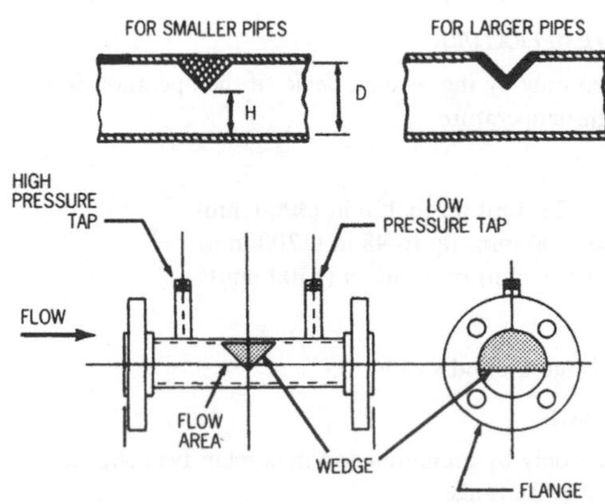

FIG. 2.1.10 The segmental wedge flowmeter designed for clean fluid service.

Variable-Area Flowmeters

Variable-area flowmeters are used to regulate purge flow and as flow indicators or transmitters.

PURGE FLOWMETER

One variety of variable-area flowmeters is the purge flowmeter (see Figure 2.1.11). The features and characteristics of these instruments are summarized next.

APPLICATIONS

Low flow regulation for air bubblers, for purge protection of instruments, for purging electrical housings in explosion-proof areas, and for purging the optical windows of smokestack analyzers

PURGE FLUIDS

Air, nitrogen, and liquids

OPERATING PRESSURE

Up to 450 psig (3 MPa)

OPERATING TEMPERATURE

For glass tubes up to 200°F (93°C)

RANGES

From 0.01 cc/min for liquids and from 0.5 cc/min and higher for gases. A ¼-in (6-mm) glass tube rotameter can handle 0.05 to 0.5 gpm (0.2 to 2 lpm) of water or 0.2 to 2 scfm (0.3 to 3 cmph) of air

INACCURACY

Generally 2 to 5% of the range (laboratory units are more accurate)

COSTS

A 150-mm glass-tube unit with ⅛-in (3-mm) threaded connection, 316 stainless steel frame, and 16-turn high-precision valve is $260; the same with aluminum frame and standard valve is $100. Adding a differential pressure regulator of brass or aluminum construction costs about $150 (of stainless steel, about $500). For highly corrosive services, all-Teflon, all-PTFE, all-PFA, and all-

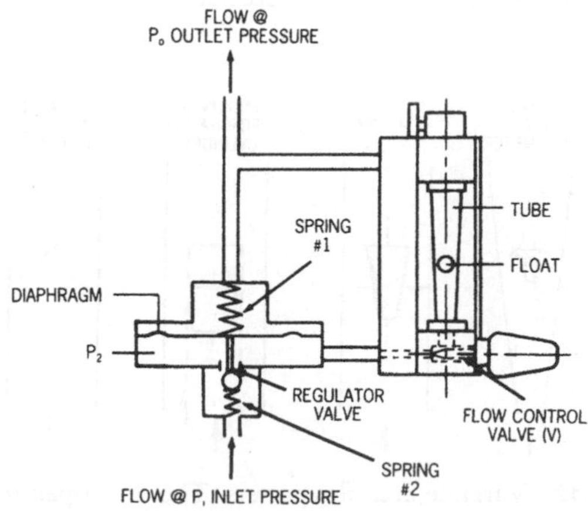

FIG. 2.1.11 A purge flow regulator consisting of a glass tube rotameter, an inlet needle valve, and a differential pressure regulator. (Reprinted, with permission, from Krone America Inc.)

CTFA units are available which, when provided with valves, cost \$550 with ¼-in (6-mm) and \$1300 with ¾-in (19-mm) connections.

PARTIAL LIST OF SUPPLIERS

Aaborg Instruments & Controls Inc.; Blue White Industries; Brooks Instrument, Div. of Rosemount; Fischer & Porter Co.; Fisher Scientific; Flowmetrics Inc.; ICC Federated Inc.; Ketema Inc. Schutte and Koerting Div.; Key Instruments; King Instrument Co.; Krone America Inc.; Matheson Gas Products Inc.; Omega Engineering Inc.; Porter Instrument Co. Inc.; Scott Specialty; Wallace & Tiernan Inc.

VARIABLE-AREA FLOWMETERS

In the wastewater treatment industry, variable-area flowmeters are also used as flow indicators or transmitters if the process fluid is clean. Figure 2.1.12 shows their operating principles, and their features and capabilities are listed next.

TYPES

A. Rotameter (float in tapered tube)
B. Orifice/rotameter combination
C. Open-channel variable gate
D. Spring and vane or piston

STANDARD DESIGN PRESSURE

A. 350 psig (2.4 MPa) average maximum for glass metering tubes, dependent on size. Up to 720 psig (5 MPa) for metal tubes and special designs to 6000 psig (41 MPa)

STANDARD DESIGN TEMPERATURE

A. Up to 400°F (204°C) for glass tubes and up to 1000°F (538°C) for some models of metal tube meters

END CONNECTIONS

Female pipe thread or flanged

FLUIDS

Liquids, gases, and vapors

FLOW RANGE

A. 0.01 cc/min to 4000 gpm (920 m^3/hr) of liquid
0.3 cc/min to 1300 scfm (2210 m^3/hr) of gas

INACCURACY

A. Laboratory rotameters can be accurate to ±½% of actual flow; most industrial rotameters perform within ±1 to 2% of full scale over a 10:1 range, and purge or bypass meters perform within ±5 to 10% of full range.
B and D. ±2 to ±10% of full range
C. ±7.5% of actual flow

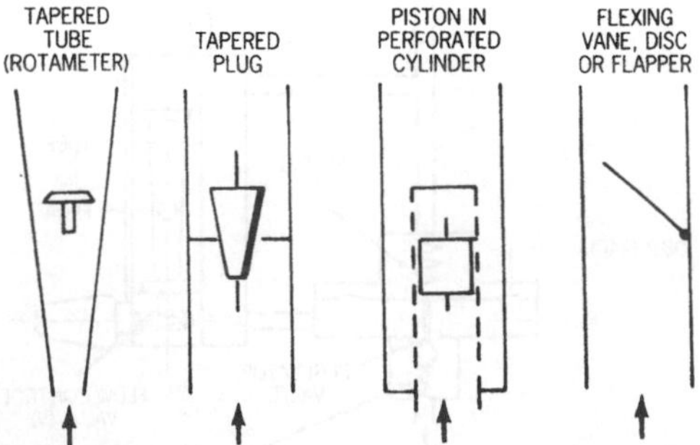

FIG. 2.1.12 Variable-area flowmeters. The area open to flow is changed by the flow itself in a variable-area flowmeter. Either gravity or spring action can be used to return the float or vane as flow drops.

MATERIALS OF CONSTRUCTION

A. Tube: Borosilicate glass, stainless steel, Hastelloy, Monel, and Alloy 20. Float: *Conventional type*—brass, stainless steel, Hastelloy, Monel, Alloy 20, nickel, titanium, or tantalum, and special plastic floats. *Ball type*—glass, stainless steel, tungsten carbide, sapphire, or tantalum. End Fittings: Brass, stainless steel, or alloys for corrosive fluids. Packing: The generally available elastomers are used and O-rings of commercially available materials; Teflon is also available.

COST

A ¼-in (6-mm) glass tube purge meter starts at \$100. A ¼-in stainless steel meter is about \$300. Transmitting rotameters start at about \$1000; with 0.5% of rate accuracy, their costs are over \$2000. A 3-in (75-mm) standard bypass rotameter is about \$500; a 3-in stainless steel tube standard rotameter is about \$2000. A 3-in tapered-plug variable-area meter in aluminum construction is about \$1000; the same unit in spring and vane design is around \$750.

PARTIAL LIST OF SUPPLIERS

Aaborg Instruments & Controls Inc. (A); Aquamatic Inc. (B); Blue White Industries (A); Brooks Instrument Div. of Rosemount (A); Dwyer Instruments Inc. (A); ERDCO Engineering Corp. (D); ESKO Industries Ltd. (A); Fischer & Porter Co. (A); Flowmetrics Inc. (A); Gilflo Metering & Instrumentation Inc. (D); Gilmont Instruments Div. of Barnant Co. (B); Headland Div. of Racine Federated Inc. (D); ICC Federated Inc. (A); ISCO Environmental Div. (C); Ketema Inc. Schutte and Koerting Div. (A); Key Instruments (A); King Instrument Co. (A); Kobold Instruments Inc.; Krone America Inc. (A); Lake Monitors Inc.; Matheson Gas Products Inc. (A); McMillan Co.; Meter Equipment Mfg. Inc. (D); Metron Technology (A); Omega Engineering Inc. (A); G. A. Planton Ltd. (D); Porter Instrument Co. Inc. (A); Turbo Instruments Inc. (D); Universal Flow Monitors Inc. (D); Wallace & Tiernan Inc. (A); Webster Instruments (D)

Venturi and Flow Tubes

DESIGN TYPES

A. Venturi tubes; B. Flow tubes; C. Flow nozzles

DESIGN PRESSURE

Usually limited only by the readout device or pipe pressure ratings

DESIGN TEMPERATURE

Limited only by the readout device if the operation is at very low or high temperature

SIZES

A. 1 in (25 mm) up to 120 in (3000 mm)
B. 4 in (100 mm) up to 48 in (1200 mm)
C. 1 in (25 mm) up to 60 in (1500 mm)

FLUIDS

Liquids, gases, and steam

FLOW RANGE

Limited only by minimum and maximum beta (β) ratio and available pipe size range

INACCURACY

Values given are for flow elements only; d/p cell and readout errors are additional.
A. ±0.25% of rate if calibrated in a flow laboratory and ±0.75% of rate if uncalibrated
B. Can range from ±0.5 to ±3% of rate depending upon the design and variations in fluid operating conditions
C. ±1% of rate when uncalibrated to ±0.25% when calibrated

MATERIALS OF CONSTRUCTION

Virtually unlimited. Cast venturi tubes are usually cast iron, but fabricated venturi tubes can be made from carbon steel, stainless steel, most available alloys, and fiberglass plastic composites. Flow nozzles are commonly made from alloy steel and stainless steel.

PRESSURE RECOVERY

90% of the pressure loss is recovered by a low-loss venturi when the beta (β) ratio is 0.3, while an orifice plate recovers only 12%. (The corresponding energy savings in a 24-in (600-mm) waterline is about 20 hp.)

REYNOLDS NUMBERS

Venturi and flow tube discharge coefficients are constant at Re > 100,000. Flow nozzles are used at high pipeline velocities (100 ft/sec or 30.5 m/sec), usually corresponding to Re > 5 million. Critical-flow venturi nozzles operate under choked conditions at sonic velocity.

COSTS

Flow nozzles are less expensive than venturi or flow tubes but cost more than orifices. American Society of Mechanical Engineers (ASME) gas flow nozzles in aluminum for 3- to 8-in (75- to 200-mm) lines cost from $200 to $750. Epoxy–fiberglass nozzles for 12- to 32-in (300- to 812-mm) lines cost from $750 to $2500. The relative costs of Herschel venturis and flow tubes in different sizes and materials are as follows:

	6-in Stainless Steel	*8-in Cast Iron*	*12-in Steel*
Herschel venturi	$8000	$5500	$6000
Flow tube	$3600	$2100	$2900

PARTIAL LIST OF SUPPLIERS

ABB Kent Taylor (B); Badger Meter Inc. (A,B); Bethlehem Corp. (B); BIF Products of Leeds & Northrup (A,B,C); Daniel Flow Products Inc. (A,C); Delta-T Co. (C); Digital Valve Co. (critical-flow venturi nozzles); Fielding Crossman Div. of Lisle Metrix Ltd. (A,C); Fischer & Porter Co. (B); Flow Systems Inc. (B); Fluidic Techniques Inc. (A); Fox Valve Development Corp. (A); F.B. Leopold Co. (A,B); Permutit Co. Inc. (A,C); Perry Equipment Corp. (B); Henry Pratt Co. (A,B); Preso Industries (A,B); Primary Flow Signal Inc. (A,C); STI Manufacturing Inc.; Tri-Flow Inc. (A); Vickery-Simms Div. of FTI Industries (A); West Coast Research Corp.

In applications where the flows of large volumes of water are measured, considerations of the measurement pumping costs often outweigh the initial cost of the sensor. Because the venturi flowmeters (see Figure 2.1.13) require less pressure drop than any other d/p-type flow sensor, their designs (see Figure 2.1.14) are frequently used in the wastewater treatment industry.

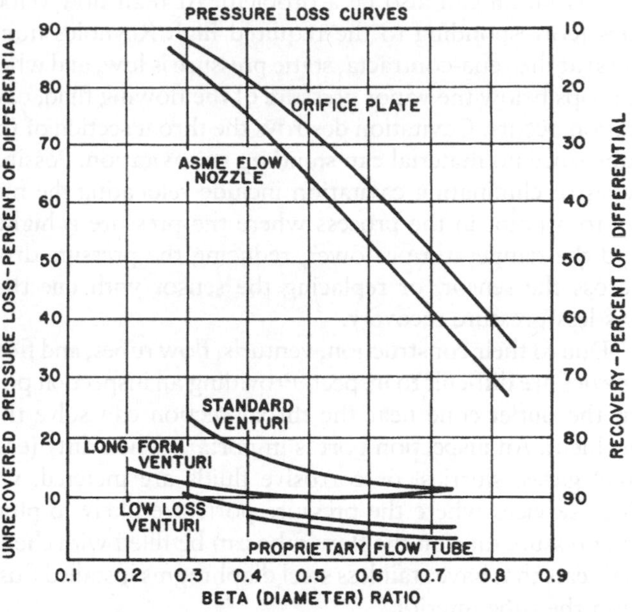

FIG. 2.1.13 Pressure loss curves.

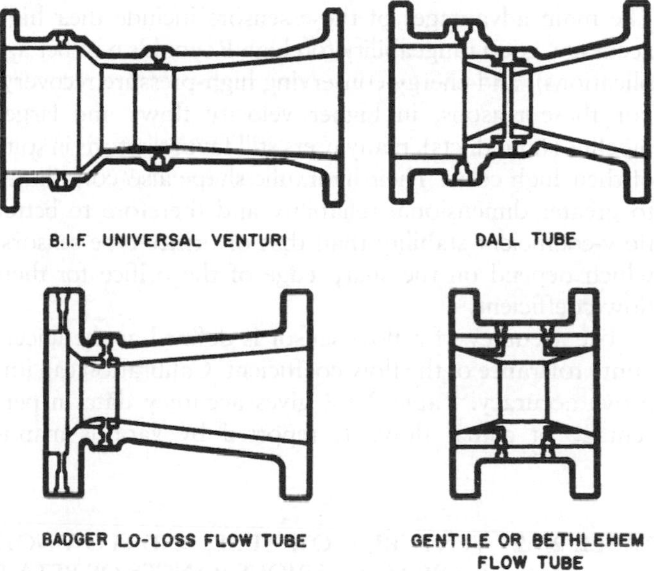

FIG. 2.1.14 Proprietary flow tubes.

LIMITATIONS

The main limitation of venturi tubes is cost, both for the tube itself and often for the long piping required for the larger sizes. However, the energy cost savings attributable to their higher pressure recovery and reduced pressure loss usually justify the use of venturi tubes in larger pipes.

Another limitation is the high minimum Reynolds number required to maintain accuracy. For venturis and flow tubes, this minimum is around 100,000; while for flow nozzles, it is over 1 million. Correction factors are available for Reynolds numbers below these limits, and measurement performance also suffers.

Cavitation can also be a problem. At high flow velocities (corresponding to the required high Reynolds numbers) at the vena-contracta, static pressure is low, and when it drops below the vapor pressure of the flowing fluid, cavitation occurs. Cavitation destroys the throat section of the tube since no material can stand up to cavitation. Possible ways of eliminating cavitation include relocating the meter to a point in the process where the pressure is higher and the temperature is lower, reducing the pressure drop across the sensor, or replacing the sensor with one that has less pressure recovery.

Due to their construction, venturis, flow tubes, and flow nozzles are difficult to inspect. Providing an inspection port on the outlet cone near the throat section can solve this problem. An inspection port is important when dirty (erosive) gases, slurries, or corrosive fluids are metered. On dirty services where the pressure ports are likely to plug, the pressure taps on the flow tube can be filled with chemical seals that have stainless steel diaphragms installed flush with the tube interior.

ADVANTAGES

The main advantages of these sensors include their high accuracy, good rangeability (on high Reynolds number applications), and energy-conserving high-pressure recovery. For these reasons, in higher velocity flows and larger pipelines (and ducts), many users still favor venturis in spite of their high costs. Their hydraulic shape also contributes to greater dimensional reliability and therefore to better flow-coefficient stability than that of orifice-type sensors, which depend on the sharp edge of the orifice for their flow coefficient.

The accuracy of a flow sensor is defined as the uncertainty tolerance of the flow coefficient. Calibration can improve accuracy. Table 2.1.2 gives accuracy data in percentage of actual flow, as reported by various manufacturers. These values are likely to hold true only for the stated ranges of beta ratios and Reynolds numbers, and they do not include the added error of the readout device or d/p transmitter.

Vortex Flowmeters

TYPES

A. Vortex
B. Fluidic shedding coanda effect
C. Oscillating vane in orifice bypass

SERVICES

A. Gas, steam, and clean liquids
B and C. Clean liquids

SIZE RANGES AVAILABLE

A. 0.5 to 12 in (13 to 300 mm), also probes
B. 1 to 4 in (25 to 100 mm)
C. 1 to 4 in (25 to 100 mm)

DETECTABLE FLOWS

A. Water—2 to 10,000 gpm (8 lpm to 40 m^3/hr)
Air—3 to 12,000 scfm (0.3 to 1100 scmm)
Steam (D&S at 150 psig [10.4 bars])—25 to 250,000 lbm/hr (11 to 113,600 kg/hr)
B. Water—1 to 1000 gpm (4 to 4000 lpm)
C. Water—5 to 800 gpm (20 to 3024 lpm)

FLOW VELOCITY RANGE

A. Liquids—1 to 33 ft/sec (0.3 to 10 m/sec)
Gas and steam—20 to 262 ft/sec (6 to 80 m/sec)

MINIMUM REYNOLDS NUMBERS

A. Under Reynolds number of 8000 to 10,000, meters do not function at all; for best performance, Reynolds number should exceed 20,000 in sizes under 4 in (100 mm) and 40,000 in sizes above 4 in.
B. Reynolds number = 3000

OUTPUT SIGNALS

A, B, and C. Linear pulses or analog

TABLE 2.1.2 VENTURI, FLOW TUBE, AND FLOW NOZZLE INACCURACIES (ERRORS) IN PERCENT OF ACTUAL FLOW FOR VARIOUS RANGES OF BETA RATIOS AND REYNOLDS NUMBERS

Flow Sensor		*Line Size in Inches (1 in = 25.4 mm)*	*Beta Ratio*	*Pipe Reynolds Number Range for Stated Accuracy*	*Inaccuracy in % of Actual Flow*
Herschel standard	Cast[1]	4–32	.30–.75	2×10^5 to 1×10^6	±0.75%
	Welded	8–48	.40–.70	2×10^5 to 2×10^6	±1.5%
Proprietary true venturi	Cast[2]	2–96	.30–.75	8×10^4 to 8×10^6	±0.5%
	Welded	1–120	.25–.80	8×10^4 to 8×10^6	±1.0%
Proprietary flow tube	Cast[3]	3–48	.35–.85	8×10^4 to 1×10^6	±1.0%
ASME flow nozzles[4]		1–48	.20–.80	7×10^6 to 4×10^7	±1.0%

[1]No longer manufactured because of long laying length and high cost.
[2]Badger Meter Inc.; BIF Products of Leeds & Northrup; Fluidic Techniques, Inc.; F.B. Leopold Co.; Permutit Co., Inc.; Henry Pratt Co.; Primary Flow Signal, Inc.; Tri-Flow Inc.
[3]Badger Meter Inc.; Bethlehem Corp.; BIF Products of Leeds & Northrup; Fischer & Porter Co.; F.B. Leopold Co.; Henry Pratt Co.; Preso Industries.
[4]BIF Products of Leeds & Northrup; Daniel Flow Products, Inc.; Permutit Co., Inc.; Primary Flow Signal, Inc.

DESIGN PRESSURE

A. 2000 psig (138 bars)
B. 600 psig (41 bars) below 2 in (50 mm); 150 psig (10.3 bars) above 2 in
C. 300 psig (30.6 bars)

DESIGN TEMPERATURE

A. −330 to 750°F (−201 to 400°C)
B. 0 to 250°F (−18 to 120°C)
C. −14 to 212°F (−25 to 100°C)

MATERIALS OF CONSTRUCTION

A. Mostly stainless steel; some in plastic
B. 316 stainless steel with Viton A O-rings
C. Wetted body is Kynar, sensor is Hastelloy C

RANGEABILITY

A. Reynolds number at maximum flow divided by minimum Reynolds number of 20,000 or more
B. Reynolds number at maximum flow divided by minimum Reynolds number of 3000
C. 10:1 for Reynolds number at maximum flow divided by minimum Reynolds numbers of 14,000 for 1 in, 28,000 for 2 in, 33,000 for 3 in, and 56,000 for 4 in

INACCURACY

A. 0.5 to 1% of rate for liquids and 1 to 1.5% of rate for gases and steam with pulse outputs; for analog outputs, add 0.1% of full scale.
B. 1 to 2% of actual flow
C. 0.5% of full scale over 10:1 range

COST

A. Plastic and probe units cost about $1500; stainless steel units in small sizes cost about $2500; insertion-types cost about $3000.
C. The sensor with only unscaled pulse output in 1-, 2-, 3-, and 4-in sizes costs $535, $625, $875, and $1295, respectively. The additional cost of a scaler is $250 and of a 4–20 mA transmitter is $350.

PARTIAL LIST OF SUPPLIERS

ABB Kent (A); Alphasonics Inc. (A); Badger Meter Inc. (C—proximity switch sensor); Brooks Div. of Rosemount (A—ultrasonic); EMC Co. (A—dual piezoelectric sensor); Endress + Hauser Instruments (A—capacitance sensor); Fischer & Porter Co. (A—internal strain gauge sensor); Fisher Controls (A—dual piezoelectric sensor); Flowtec AG of Switzerland (A); Foxboro Co. (A—piezoelectric sensor); Johnson Yokogawa Corp. (A—dual piezoelectric sensor); J-Tec Associates Inc. (A—retractable design available, ultrasonic sensor); MCO/Eastech (A—including insertion-type, mechanical, thermal, or piezoelectric sensors); Moore Products Co. (B); Nice Instrumentation Inc. (A—dual piezoelectric sensor); Oilgear/Ball Products (A—vortex velocity); Sarasota Automation Inc. (A); Schlumberger Industries Inc. (A—dual piezometric sensor); Turbo Instruments Inc. (A); Universal Flow Monitors Inc. (A—plastic body, piezoelectric sensor); Universal Vortex (A—piezoelectric sensor)

Weirs and Flumes

TYPES

These devices measure open-channel flow by causing level variations in front of primaries. Bubblers, capacitance, float and hydrostatic and ultrasonic devices are used as level sensors. These devices can also measure open-channel flows without primaries by calculating the flow from depth and velocity data obtained from ultrasonic and magnetic sensors.

OPERATING CONDITIONS

Atmospheric

APPLICATIONS

Waste or irrigation water flows in open channels

FLOW RANGE

From 1 gpm (3.78 lpm)—no upper limit

RANGEABILITY

Most devices provide 75:1, V-notch weirs can reach 500:1.

INACCURACY

2 to 5%

COSTS

Primaries used as pipe inserts cost under $1000. A 6-in (150-mm) Parshall flume costs about $1500, and a 48-in (1.22-m) Parshall flume costs about $5000. Primaries for irrigation applications are usually field-fabricated. Manual depth sensors can be obtained for $200; local bubbler or float indicators for $750 to $1500; and programmable transmitting capacitance, ultrasonic, or bubbler units from $1800 to $3000. Open-channel flowmeters, when calculating flow based on depth and velocity, range from $5000 to over $10,000.

PARTIAL LIST OF SUPPLIERS

ABB Kent Taylor Inc. (primaries); American Sigma Inc. (bubbler); Badger Meter Inc. (Parshall or manhole flume, ultrasonic and open channel computing); Bernhar Inc. (ultrasonic for partially filled pipes); Bestobell/Mobrey (ultrasonic); BIF Unit of Leeds & Northrup (primary and detector); Drexelbrook Engineering Co. (capacitance for flumes); Endress + Hauser Inc. (ultrasonic and capacitance); Fischer & Porter Co. (ultrasonic); Free Flow Inc. (primaries); Greyline Instruments Inc. (ultrasonic); Inventron Inc. (ultrasonic); ISCO Inc. (bubbler, hydrostatic, and ultrasonic); Key-Ray/Sensall Inc. (ultrasonic); Leeds & Northrup BIF (flow nozzles); Leupold & Stevens Inc. (float); Manning Environmental Corp. (primaries); Marsh-Mcbirney Inc. (electromagnetic); Mead Instruments Corp. (velocity probe); Milltronics Inc. (ultrasonic); Minitek Technologies Inc. (open-channel magmeter and ultrasonic); Montedoro-Whitney Corp. (open-channel flow by ultrasonics); MSR Magmeter Mfg. Ltd. (robotic magmeter probe for open channel); N.B. Instruments Inc. (computer monitoring of sewers); Plasti-Fab Inc. (primaries); Princo Instruments Inc. (capacitance); J.L. Rochester Co. (manual depth sensor); Sparling Instruments Co. (primaries); TN Technologies Inc. (ultrasonic)

In the wastewater treatment industry, the flow in large, open pipes or channels must be measured. The weir and flume designs, particularly the Parshall flume (see Figure 2.1.15), make such measurements. The common feature of all these flow sensors is that they detect the level rise in front of a restriction in the flow channel.

DETECTORS FOR OPEN-CHANNEL SENSORS

The level rise generated by flumes or weirs can be measured by any level detector including simple devices such as air bubblers.

The flow in open channels can also be detected without using flumes, weirs, or any other primary devices. One such design computes flow in round pipes or open channels by ultrasonically measuring the depth, calculating the flowing cross-sectional area on that basis, and multiplying the area by the velocity to obtain volumetric flow (see Figure 2.1.16).

Another open-channel flowmeter that does not need a primary element uses a robotically operated magnetic flowmeter probe to scan the velocity profile in the open channel (see Figure 2.1.17). In this design, the computer algorithm separately calculates and adds the flow segments through each slice of the velocity profile as the velocity sensor moves down to the bottom of the channel.

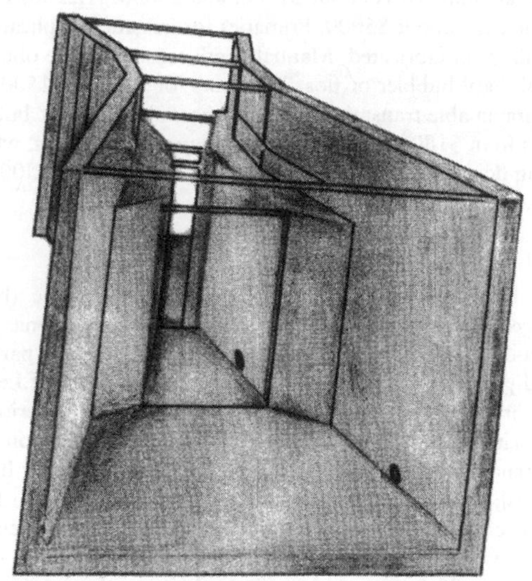

FIG. 2.1.15 Dual-range Parshall flume. (Reprinted, with permission, from Fischer & Porter Co.)

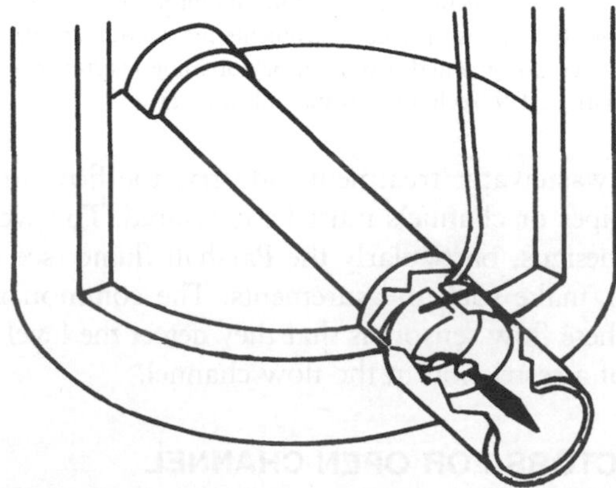

FIG. 2.1.16 Volumetric flow computer measuring depth and velocity in an open channel without a primary device. (Reprinted, with permission, from Montedoro-Whitney Corp.)

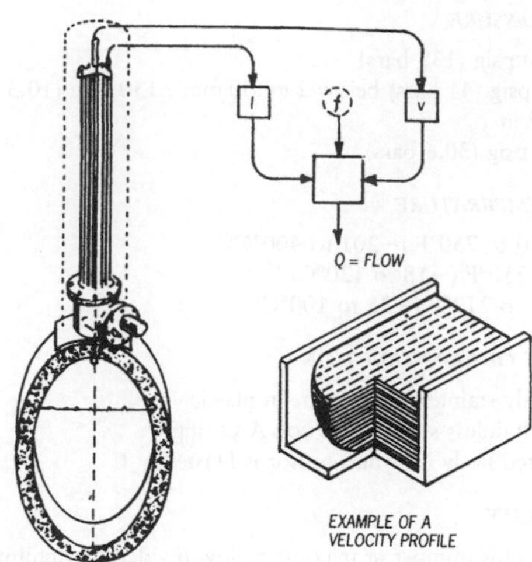

FIG. 2.1.17 Robotically operated magmeter probe sensor used to compute channel flow. (Reprinted, with permission, from MSR Magmeter Mfg. Ltd.)

Level Sensors

Most level sensors used in the wastewater industry do not need to be very accurate; reliable operation, rugged design, and low maintenance are more important. For these reasons, the newer level detector designs (laser, microwave, radar, gamma radiation, and time-domain reflectometry types) are seldom used. Similarly, the designs that use mechanical motion (float, displacer, or tape designs) are used infrequently since the solid-state or force-balance designs are more maintenance free.

On clean water level applications for local level indication, reflex-type level gauges, resistance tapes, and bubbler gauges are used most often. For high- and low-level switches, conductivity, capacitance, vibrational, ultrasonic and thermal level switches are used. For level transmitter applications, d/p and ultrasonic designs are often used.

For dirty or sludge-type level measurement, extended-diaphragm-type or purged d/p sensors, capacitance probes, and ultrasonic detectors are usually used. Lately, electronic load cells have also been used to detect the level on the basis of weight measurement in some larger tanks. For sludge or oil interface detection, ultrasonic, optical, vibrational, thermal, and microwave level switches work well. Table 2.1.3 provides an overall summary of the features and capabilities of all level measuring devices.

INTERFACE MEASUREMENT

When detecting the interface between two liquids, the measurement can be based on the difference of densities, dielectric constants, electric or thermal conductivities, opacity, or the sonic and ultrasonic transmittance of the two fluids. Environmental engineers should base their measurement on the process property with the largest step change between the upper and lower fluids. If, instead of a clean interface, a rag layer (a mix of the two fluids) exists between the two fluids, the interface detector cannot change that fact (it cannot eliminate the rag layer); but if properly selected, the interface detector can signal its beginning and end and thereby measure its thickness.

Interface level switches are usually ultrasonic, optical (Figure 2.1.18), capacitance, float, conductivity, thermal, microwave, or radiation designs. The ultrasonic switch described in Figure 2.1.19 uses a gap-type probe installed at a 10-degree angle from the horizontal. At one end of the gap is the ultrasonic source, at the other end is the receiver. As long as the probe is in the upper or lower liquid, the detector receives the ultrasonic pulse.

When the interface enters the gap, the pulse is deflected, and the switch is actuated. This switch can detect the interface between water and oil or other hydrocarbons, such as vinyl-acetate. If the thickness of the light layer rather than the location of the interface in the tank is of interest, the ultrasonic gap sensor can be attached to a float as shown in Figure 2.1.20.

Continuous measurement of the interface between two liquids can be detected by d/p transmitters if P_1 is detected in the heavy liquid and P_2 in the light liquid. In atmospheric vessels, three bubbler tubes can achieve the same interface measurement. The configuration shown in Figure 2.1.21 is appropriate for applications where the density of the light layer is constant and the density of the heavy liquid is variable. In these differential pressure-type systems, the movement of the interface level must be large enough to cause a change that satisfies the minimum span of the d/p transmitters. If the difference between the dielectric constants is substantial, such as in crude oil desalting, capacitance probes can also serve as continuous interface detectors.

On clean services, float- and displacer-type sensors can also be used as interface level detectors. For float-type units a float density heavier than the light layer but lighter than the heavy layer must be selected. In displacer-type sensors, the displacer must always be flooded, the upper connection of the chamber must be in the light liquid layer, and the lower connection must be in the heavy liquid layer. In this arrangement, the displacer becomes a density sensor. Therefore, the smaller the difference between the densities of the fluids and the smaller the range within which the interface can move, the larger displacer diameter will be required. Displacer density can be the same or more than that of the heavy layer.

Bubblers

APPLICATIONS

Usually local indicator on open tanks containing corrosive, slurry, or viscous process liquids. Can also be used on pressurized tanks but only up to the pressure of the air supply.

OPERATING PRESSURE

Usually atmospheric.

OPERATING TEMPERATURE

Limited only by pipe material; purging has also been used on high-temperature, fluidized-bed combustion processes to detect levels.

MATERIALS

Any pipe material available.

COSTS

\$100 to \$500 depending on accessories.

INACCURACY

Depends on readibility of pressure indicator, usually ±0.5% to 2% of full scale.

RANGE

Unlimited.

PARTIAL LIST OF SUPPLIERS

Automatic Switch Co.; Computer Instruments Corp.; Davis Instrument Mfg.; Dwyer Instruments Inc.; Fischer & Porter Co.; King Engineering Corp.; Meriam Instrument Div. of Scott & Fetzer; Petrometer Corp.; Scannivalve Corp.; Time Mark Corp.; Trimount Div. of Custom Instrument Components; Uehling Instrument Co.; Wallace & Tiernan Inc.

Capacitance Probes

SERVICE

Point and continuous level measurement of solids and liquids (both conductive and nonconductive) using both the wetted probe and the noncontacting proximity designs.

DESIGN PRESSURE

Up to 4000 psig (28 MPa)

DESIGN TEMPERATURE

PTFE insulation can be used from −300 to 500°F (−185 to 296°C). Uncoated bare probes can be used up to 1800°F (982°C). Alumina insulation can be used up to 2000°F (1128°C). Proximity designs can also be used to measure the level of molten metals.

EXCITATION

A few MHz

MATERIALS OF CONSTRUCTION

Generally stainless steel for nonconductive and Teflon-coated stainless steel for both conductive and nonconductive services, but higher alloys, ceramics, PVC, Kynar, and other plastic coatings are also available.

SPANS

From 0.25 to 4000 picofarad (pf). Because of sensitivity limitations, a minimum span of 10 pf is preferred.

INACCURACY

On–off point sensors usually actuate within $\frac{1}{4}$ in (6 mm) of their setpoints. For continuous level detection, dividing the sensitivity by the span calculates the minimum percentage error of 1 to 2% of full scale.

SENSITIVITY AND DRIFT

Depending on design, sensitivities vary from 0.1 to 0.5 pf, while

TABLE 2.1.3 ORIENTATION TABLE FOR LEVEL DETECTORS

Type	Level Range (In feet: 1, 3, 6, 12, 24, 48, 96, 100, 150, 200; In meters: 0.3, 1, 2, 4, 8, 16, 32, 34, 50, 67)	Maximum Temperature (°F) °C = (°F − 32)/1.8	Available as Noncontact	Inaccuracy (1 in = 25.4 mm)	Cost			Available Designs			Applications: Liquids					Applications: Solids			Limitations
					Under $1000	$1000–$5000	Over $5000	Switch	Local Indicator	Transmitter	Clean	Viscous	Slurry/Sludge	Interface	Foam	Powder	Chunky	Sticky	
Air bubblers		UL		1–2% FS	√			√	√	√	G	F	P	F					Introduces foreign substance to process; high maintenance
Capacitance		2000	√	1–2% FS	√	√		√		√	G	F–G	F	G–L	P	F	F	P	Problem with interface between conductive layers and detection of foam
Conductivity switch	Point sensor	1800		⅛ in	√			√			F	P	F	L	L	L	L	L	Can detect interface only between conductive and nonconductive liquids; field effect design for solids
Diaphragm		350		0.5% FS	√	√		√		√	G	F	F			F	F	P	Switches only for solids service
Differential pressure		1200		0.1% AS	√	√		√	√	√	E	G–E	G	P					Plugging eliminated by only extended diaphragm seals or repeaters. Purging and sealing legs also used
Displacer		850		0.5% FS	√	√		√	√	√	E	P	P	F–G					Not recommended for sludge or slurry service
Float		500		1% FS	√			√	√	√	G	P	P	F					Most designs limited by moving parts to clean service. Only preset density floats following interfaces
Laser		UL	√	0.5 in			√			√	L	G	G		F	F	F	F	Limited to cloudy liquids or bright solids in tanks with transparent vapor spaces
Level gauges		700		0.25 in	√				√		G	F	P	F					Glass not allowed in some processes
Microwave switch	Point sensor	400	√	0.5 in	√	√		√			G	G	F	G		G	G	F	Thick coating
Optical switches	Point sensor	260	√	0.25 in	√	√		√			G	F	E	F–G	F	F	P	F	Refraction-type for clean liquids only; reflection-type requires clean vapor space
Radar		450	√	0.12 in			√		√	√	G	G	F	P		P	F	P	Interference from coating, agitator blades, spray, or excessive turbulence
Radiation		UL	√	0.25 in		√	√	√		√	G	E	E	G	F	G	E	E	Requires a Nuclear Regulatory Commission (NRC) license

Resistance tape		225		0.5 in		√			√	√	G	G	G						Limited to liquids under near-atmospheric pressure and temperature conditions
Rotating paddle switch	Point sensor	500		1 in	√											G	F	P	Limited to detection of dry, noncorrosive, low-pressure solids
Slip tubes		200		0.5 in	√						F	P	P						An unsafe manual device
Tape-type level sensors		300		0.1 in		√			√	√	E	F	P	G		G	F	F	Only the inductively coupled float suited for interface measurement. Float hangup a potential problem with most designs
Thermal		850		0.5 in	√	√		√		√	G	F	F	P	F				Foam and interface detection limited by the thermal conductivities involved
TDR/PDS		221		3 in		√				√	F	F	F			G	G	F	Limited performance on sticky process materials
Ultrasonic		300	√	1% FS	√	√		√		√	F–G	G	G	F–G	F	F	F	G	Presence of dust, foam, dew in vapor space; performance limited by sloping or fluffy process material
Vibrating switches	Point sensor	300		0.2 in	√			√			F	G	G	F		F	G	G	Operation limited by excessive material buildup can prevent

TDR = Time Domain Reflectometry
PDS = Phase Difference Sensors
AS = in % of actual span
E = Excellent
F = Fair
FS = in % of full scale
G = Good
L = Limited
P = Poor
UL = Unlimited

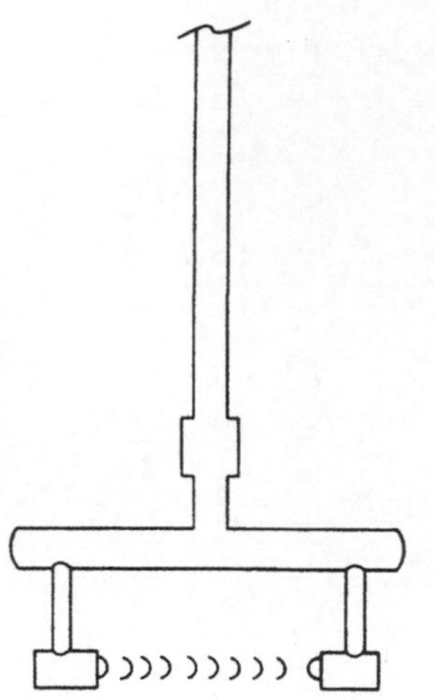

FIG. 2.1.18 Optical or ultrasonic sludge level or interface switch. (Courtesy of Sensall Inc.)

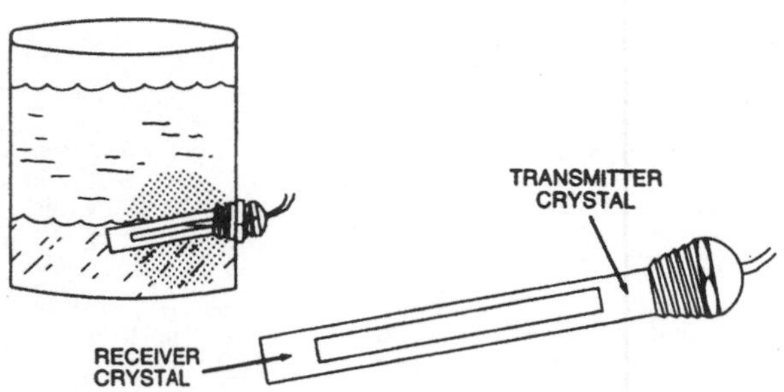

FIG. 2.1.19 Ultrasonic interface level switch. (Courtesy of Sensall Inc.)

the drift per 100°F (56°C) temperature change can vary from 0.2 to 5 pf.

RANGE

Proximity devices can be used from a fraction of an inch to a few feet; probes can be used up to 20 ft (6 m) and cables up to 200 ft (61 m).

DEADBAND AND TIME DELAY

Capacitance-type level switches are usually provided with deadband settings adjustable over the full span of the unit and time delays adjustable over a 0- to 25-sec range.

COST

From $600 for a simple level switch with power supply and output relay, plus $600 for a continuous indicator. Microprocessor-based intelligent units with special probe configurations start at $2000.

PARTIAL LIST OF SUPPLIERS

(* indicates that the supplier also markets proximity probes.) *ADE Corp.; Aeroquip Corp.; Agar Corp. Inc.; Amprodux Corp. Inc.; *Arjay Engineering Ltd.; ASC Computer Systems; ASI Instruments Inc.; Babbitt International Inc.; Bailey Controls Co.; Bedford Control Systems; Bernhard Inc.; Bindicator; Controlotron Corp.; *Custom Control Sensors Inc.; *Delavan Inc.; Delta Controls Corp.; *Drexelbrook Engineering Co.; *Electromatic

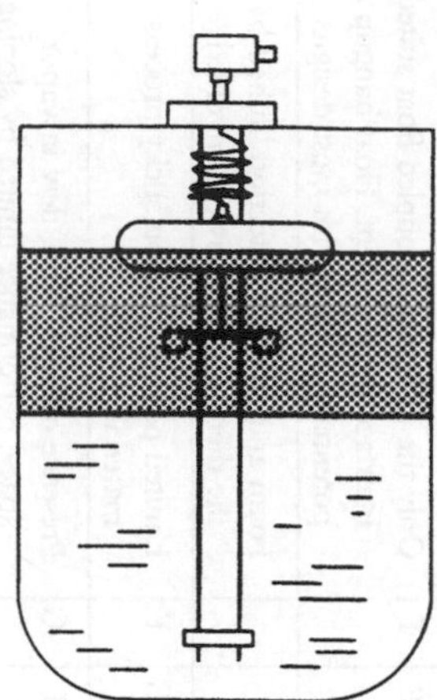

FIG. 2.1.20 Detecting the thickness of the top layer.

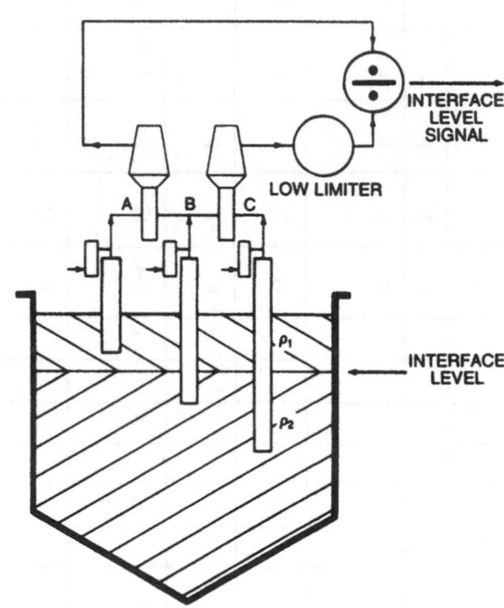

FIG. 2.1.21 Interface detection with bubbler tubes. (Courtesy of Fischer & Porter Co.)

Controls Corp.; Endress + Hauser Instruments; Enraf-Nonius; ETA Control Instruments; Fischer & Porter; Fowler Co.; Free Flow Inc.; *FSI/Fork Standards Inc.; Great Lakes Instruments Inc.; HITech Technologies Inc.; Hyde Park Electronics; Hydril P.T.D.; Invalco; KDG Mobrey Ltd.; Lumenite Electronic Co.; Magne-Sonics; Magnetrol International; Monitor Manufacturing Co.; *MTI Instruments Div.; Omega Engineering; Penberthy Inc.; Princo Instruments Inc.; *Robertshaw Controls Co.; Rosemount Inc.; Systematic Controls; Transducer Technologies Inc.; TVC Instruments Co.; Vega B.V.; Zi-Tech Instrument Corp.

Conductivity Probes

APPLICATIONS

Point or differential level detection of conductive liquids or slurries with dielectric constants of 20 or above. For electric types, the maximum fluid resistivity is 20,000 ohm/cm; electronic types can work on even more resistive fluids. Field effect probes are used on both conductive and nonconductive solids and liquids.

DESIGN PRESSURE

Up to 3000 psig (21 MPa) for conductivity probes and 100 psig (6.9 bars, or 0.69 MPa) for field conductivity probes.

DESIGN TEMPERATURES

From −15°F (−26°C) to 140°F (60°C) for units with integral electronics and from −15°F (−26°C) to 1800°F (982°C) for units with remote electronics when detecting conductivity. Field effect probes can operate up to 212°F (100°C).

MATERIALS OF CONSTRUCTION

Conductivity probes are made of 316 stainless steel, Hastelloy, titanium, or Carpenter 20 rods with Teflon, Kynar, or PVC sleeves. The housing is usually corrosion-resistant plastic or aluminum for NEMA 4 and 12 service. The field effect probe has a Ryton probe and aluminum housing.

PROBE LENGTHS

$\frac{1}{4}$-in (6-mm) solid rods are available in lengths up to 6 ft (1.8 m); $\frac{1}{8}$-in (2-mm) stainless steel cables can be obtained in lengths up to 100 ft (30 m) for conductivity applications. Field effect probes are 8 in (200 mm) long.

SENSITIVITY

Adjustable from 0 to 50,000 ohms for conductivity probes

INACCURACY

$\frac{1}{8}$ in (3 mm)

COST

From $50 to $400. The typical price of an industrial conductivity switch is about $300.

PARTIAL LIST OF SUPPLIERS

BL Tec.; Burt Process Equipment; B/W Controls—Magatek Controls; Conax Buffalo Corp.; Control Engineering Inc.; Delavan Inc. Division Colt Industries; Delta Controls Corp.; Electromatic Controls Corp.; Endress + Hauser Instruments; Great Lakes Instruments Inc.; Invalco Inc.; Lumenite Electronic Co.; Monitor Mfg.; National Controls Corp.; Revere Corp. of America; Vega B.V.; Warrick Controls Inc.; Zi-Tech Instrument Corp.

D/P Cells

DESIGN PRESSURE

To 10,000 psig (69 MPa)

DESIGN TEMPERATURE

To 350°F (175°C) for d/p cells and to 1200°F (650°C) for filled systems; others to 200°F (93°C). Standard electronics are generally limited to 140°F (60°C).

RANGE

d/p cells and indicators are available with full-scale ranges as low as 0 to 5 in (0 to 12 cm) H_2O. The higher ranges are limited only by physical tank size since d/p cells are available with ranges over 433 ft H_2O (7 MPa or 134 m H_2O).

INACCURACY

±0.5 to 2% of full scale for indicators and switches. For d/p transmitters, the basic error is from ±0.1 to 0.5% of the actual span. Added to this error are the temperature and pressure effects on the span and zero. In intelligent transmitters, pressure and temperature correction is automatic, and the overall error is ±0.1 to 0.2% of the span with analog outputs and even less with digital outputs.

MATERIALS OF CONSTRUCTION

Plastics, brass, steel, stainless steel, Monel, and special alloys for the wetted parts. Enclosures and housings are available in aluminum, steel, stainless steel, and fiberglass composites, with aluminum and fiberglass the most readily available.

COST

$200 to $1500 for transmitters in standard construction and $100 to $500 for local indicators. Add $400 to $800 for extended diaphragms and $300 to $600 for smart features such as communications and digital calibration. Expert tank systems cost approximately $1500 for the basic transmitter plus $3500 to $4500 for the interface unit and $1500 to $4000 for software plus a handheld communicator.

PARTIAL LIST OF D/P CELL SUPPLIERS

ABB Kent-Taylor; Dresser Industries; Enraf Nonius; Fischer & Porter Co.; Foxboro Co.; Honeywell, Inc.; ITT Barton; Johnson Yokogawa; King Engineering Corp.; L&J Engineering Inc.; Major Controls, Inc.; Rosemount Inc., Measurement Div., Varec Div.; Schlumberger Industries, Statham Div.; Smar International; Texas Instruments; Uehling Instrument

PARTIAL LIST OF TANKFARM PACKAGE SUPPLIERS

The Foxboro Co.; King Engineering Corp.; L&J Engineering Inc.; Sarasota M&C Inc.; Texas Instruments Inc.; Varec, a Rosemount Div.

The level measurement device used most often on slurry and sludge services is the extended-diaphragm-type differential pressure transmitter (see Figure 2.1.22). The diaphragm extension eliminates the dead-ended cavity in the nozzle, where materials accumulate, and brings the sens-

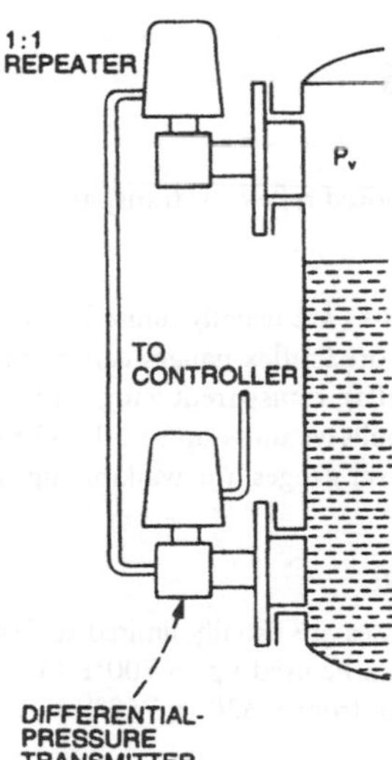

FIG. 2.1.22 Schematic diagram showing the clean and cold air output of the repeater repeating the vapor pressure (P_v) in the tank.

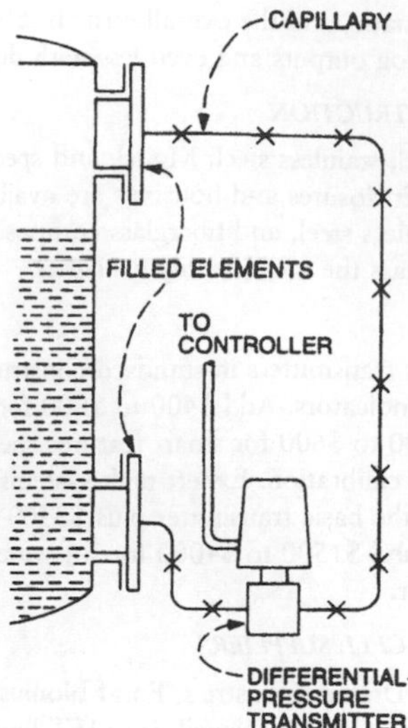

FIG. 2.1.23 Schematic diagram that shows how the temperature compensated, extended-diaphragm-type, chemical seals protect the d/p cell from plugging.

ing diaphragm flush with the inside surface of the tank. The sensing diaphragm is sometimes coated with Teflon to further minimize material buildup. One of the best methods of keeping the low-pressure side of the d/p cell clean is to insert another extended-diaphragm device in the upper nozzle. This device can be a pressure repeater, which can repeat both vacuums and pressures within the range of the available vacuum and plant or instrument air supply pressures. When air or vacuum is unavailable at the process pressures, extended-diaphragm-type chemical seals can be used (see Figure 2.1.23) if properly compensated for ambient temperature variations and sun exposure.

Level Gauges

TYPES

Tubular glass, armored reflex, or transparent and magnetic gauges

DESIGN PRESSURE

Tubular gauge glasses are usually limited to 15 psig (1 bar). At 100°F (38°C), armored-reflex gauges can be rated to 4000 psig (270 bars = 27 MPa); transparent gauges to 3000 psig (200 bars = 20 MPa), and bullseye units up to 10,000 psig (690 bars = 69 MPa). Magnetic level gauges are available up to 3500 psig (230 bars = 23 MPa).

DESIGN TEMPERATURE

Tubular gauge glasses are usually limited to 200°F (93°C). Armored gauges can be used up to 700°F (371°C), and magnetic gauges are available from −320 to 750°F (−196 to 400°C).

MATERIALS OF CONSTRUCTION

The wetted parts of armored gauges are available in steel, stainless steel, and tempered borosilicate glass. Magnetic level gauges are available with steel flanges and stainless steel, K-monel, Hastelloy-B, and solid PVC chambers. Available chamber and float liner materials include Kynar, Teflon, and Kel-F.

RANGE

For armored gauges, the visible length of a section is 10 to 20 in (250 to 500 mm). A maximum of four sections per column is recommended with a maximum total distance between gauge connections of 5 ft (1.5 m).

INACCURACY

Level gauges can be provided with scales. The reading accuracy is limited by visibility (foaming and boiling), and the height of the liquid column in the gauge can also differ from the process level. If the liquid in the gauge is warmer, it is also lighter, and therefore the error is on the high side; if the liquid in the gauge is colder (heavier), the indication is low. Readout wafer size limits magnetic gauge display accuracy to $\frac{1}{4}$ in (6 mm).

COSTS

Excluding the cost of shutoff valves or pipe stands, the per-ft (300 mm) unit cost of tubular glass gauges is about $25; armored-reflex and transparent gauges cost about $150/ft and $200/ft, respectively, while magnetic level gauges in stainless steel construction cost about $500/ft.

PARTIAL LIST OF SUPPLIERS

Daniel Industries Inc.; Essex Brass Co.; Imo Industries Inc. (magnetic); Jerguson Gauge and Valves, Div. of the Clark Reliance Corp. (regular and magnetic); Jogler Inc.; Kenco Engineering Co. (magnetic); Krohne America Inc.; K-Tek Corp. (magnetic); MagTech Div. ISE of Texas Inc. (magnetic); Metron Technology (magnetic); Oil-Rite Corp.; Penberthy Inc. (regular and magnetic)

Optical Sensors

TYPES

Visible or infrared (IR) light reflection (noncontacting type usually for solids and laser type for molten glass applications), light transmission (usually for sludge level), and light refraction in clean liquid level services

APPLICATIONS

Point sensor probes for liquid, sludge, or solids (some continuous detectors also available)

DESIGN PRESSURE

Up to 150 psig (10.3 bars) with polypropylene, polysulfone, PVDF, or Teflon and up to 500 psig (35 bars) with stainless steel probes

DESIGN TEMPERATURE

Between 150 and 200°F (66 to 93°C) with plastic probes and up to 260°F (126°C) with stainless steel probes

MATERIALS OF CONSTRUCTION

Quartz reflectors with Viton-A or Rulon seals, mounted in polypropylene, polysulfone, Teflon, polyvinyl fluoride, phenolic, aluminum, or stainless steel probes

HOUSINGS

Can be integral with the probe or remote. Explosion-proof enclosures and intrinsically safe probes are both available. With remote electronics, the fiber-optic cable can be from 50 to 250 ft (15 to 76 m) long.

DIMENSIONS

Refraction probe lengths vary from 1 to 24 in (25 to 600 mm), and the probe diameter is usually 0.5 to 1 in (12 to 25 mm).

COSTS

Fiberoptic level switches cost from $100 and $300. Portable sludge level detectors cost $900. Continuous transmitters to measure sludge depth or sludge interface cost $4000 and up.

PARTIAL LIST OF SUPPLIERS

Automata Inc. (noncontacting IR); BTG Inc. (IR); Conax Buffalo Corp. (fiber optic); Enraf Nonius Tank Inventory Systems Inc. (IR); Gems Sensors Div. IMO Industries Inc. (fiber optics); Genelco Div. of Bindicator Inc. (IR switch); Kinematics & Controls Corp. (switch); Markland Specialty Engineering Ltd. (IR for sludge); OPW Division of Dover Corp.; 3M Specialty Optical Fibers; Zi-Tech Instrument Corp. (switch)

Resistance Tapes

APPLICATIONS

Liquids including slurries but not solids. Can also measure temperature

RESOLUTION

⅛ in, which is the distance between helix turns

ACTUATION DEPTH (AD)

The depth required to short out the tape varies with the specific gravity (SG) as follows: AD (in inches) = 4/(SG). Therefore, AD at the minimum SG of 0.5 is 8 in (200 mm).

TEMPERATURE EFFECT

A 100°F (55°C) change in temperature changes the resistance of the unshorted tape by 0.1%. Temperature compensation is available.

INACCURACY

0.5 in if the AD is zeroed out and both AD and temperature are constant. If SG varies, a zero shift based on AD = 4/(SG) occurs.

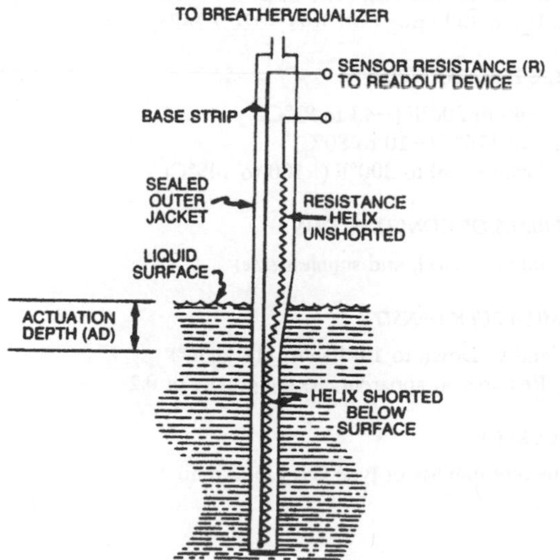

FIG. 2.1.24 Schematic diagram of resistance tape sensor operation.

Cold temperature also raises the AD.

WETTED MATERIAL

Fluorocarbon polymer film

ALLOWABLE OPERATING PRESSURE

From 10 to 30 psia (0.7 to 2.1 bars absolute)

OPERATING TEMPERATURE RANGE

−20 to 225°F (−29 to 107°C)

COSTS

Resistance tape unit cost varies with service and with tape length. A 10 ft (3 m) tape with breather and transmitter for water service costs from $600 to $1000. The added cost for longer tapes is $25 or more per foot, depending on the service.

SUPPLIERS

Metritape Inc.; R-Tape Corp.; Sankyo Pio-Tech

The resistance tape (see Figure 2.1.24) for continuous liquid level measurement was invented in the early 1960s, initially for water well gauging and subsequently for marine and industrial usage. The sensor is a flat, coilable strip (or tape), ranging from 3 to 100 ft (1 to 30 m) in length, suspended from the top of the tank. It is small enough in cross-section to be held within a perforated pipe (diameter of 2 to 3 in), which also supports the transducer and acts as a stilling pipe when the process is turbulent.

While resistance tapes are not widely used in the wastewater treatment industry today, their low cost, low maintenance, and adaptability for multipoint scanning makes them a candidate for use in new plants.

Thermal Switches

TYPES

Switches operate on either thermal difference or thermal dispersion. Transmitters utilize the thermal conductivity difference between liquids and vapors. Metal mold level controllers use direct temperature detection.

APPLICATIONS

Liquid, interface, and foam level detection. Special units are available for molten metal level measurement.

DESIGN PRESSURE

Up to 3000 psig (207 bars = 20.7 MPa)

DESIGN TEMPERATURE

Standard units can be used from −100 to 350°F (−73 to 177°C); high-temperature units operate from −325 to 850°F (−198 to 490°C).

RESPONSE TIME

10 to 300 sec for standard response units and 1 to 150 sec for fast response units. The time constant in molten metal applications is under 1 sec.

AREA CLASSIFICATION

Explosion-proof and intrinsically safe designs are both available.

MATERIALS OF CONSTRUCTION

316 stainless steel, PVC, and Teflon

INACCURACY

The repeatability is 0.25 in (6 mm) for side-mounted and 0.5 in (13 mm) for top-mounted level switches. Transmitters are less accurate. Molten metal level error depends on thermocouple spacing.

COST

The cost of a thermal level switch is about $250. Transmitters cost about $1000. Mold level systems are field-installed.

PARTIAL LIST OF SUPPLIERS

Chromalox Instruments and Control; Delta M Corp. (transmitter); Fluid Components Inc. (switch and monitor); Intek Inc. Rheotherm Div. (switch); Scientific Instruments Inc.; Scully Electronic Systems, Inc. (switch)

Ultrasonic Detectors

APPLICATIONS

Wetted and noncontacting switch and transmitter applications for liquid level or interface and solids level measurement. Also used as open-channel flow monitors.

DESIGN PRESSURE

Probe switches are used up to 3000 psig (207 bars = 20.7 MPa); transmitters are usually used for atmospheric service up to 7 psig (0.5 bar), but some special units are available for use up to 150 psig (10.3 bars).

DESIGN TEMPERATURE

Switches from −100 to 300°F (−73 to 149°C); transmitters from −30 to 150°F (−34 to 66°C)

MATERIALS OF CONSTRUCTION

Aluminum, stainless steel, titan, Monel, Hastelloy B & C, Kynar, PVC, Teflon, polypropylene, PVDF, and epoxy

RANGES

For tanks and silos (pulse usually travels in vapor space), up to 200 ft (60 m) for some special designs and up to 25 ft (7.6 m) for most standard systems. For wells (usually submerged), up to 2000 ft (600 m)

INACCURACY

⅛ in (3 mm) for a horizontal probe switch. For transmitters, the error varies from 0.25 to 2% of full scale depending on the dust and dew in the vapor space and the quality of the surface that reflects the ultrasonic pulse.

COSTS

Level switches cost from $200 to $500; transmitters cost from under $1000 to $2500, with the average cost around $1800.

PARTIAL LIST OF SUPPLIERS

Bindicator Co.; Contaq Technologies Corp.; Controltron Corp.; Crane/Pro-Tech Environmental Instruments; Delavan Inc. Process Instrumentation Operations; Delta Controls Corp.; Electro Corp.; Electronic Sensors Inc.; Endress + Hauser Inc.; Enterra; Fischer and Porter Co.; Genelco Div. Bindicator; Gordon Products Inc.; Greyline Instruments Inc.; HiTech Technologies Inc. (fly ash application); Hyde Park Electronics Inc.; Introkek, Subsidiary of Magnetrol International; Inventron Inc.; Kay Ray/Sensall Inc.; KDG Mobrey Ltd.; Kistler-Morse Corp.; Krone America Inc. (sludge interface); Magnetrol International; Markland Specialty Engineering Ltd. (sludge level); Marsh-McBirney Inc.; Massa Products Corp.; Microswitch/Honeywell, Milltronics Inc.; Monitek Technologies Inc.; Monitor Mfg.; Monitrol Mfg. Co.; Panametrics Inc.; Penberthy; Sirco Industrial Ltd.; SOR Precision Sensors; TN Technologies Inc.; Ultrasonic Arrays Inc. (thickness, texture, surface reflectivity); United Sensors Inc.; Vega B.V.; Zevex Inc.

As is shown in Figures 2.1.18 and 2.1.19, ultrasonic level sensors are used widely on sludge level and sludge interface detection services. Ultrasonic sludge blanket detectors can also be lowered periodically into the tank for transmittance measurements, or they can be permanently positioned for echo detection. In the newer designs, targets or sounding pipe ridges are used for automatic calibration. Even more recently, flexural sensors are installed to measure the transit time or echo in the tank wall instead of through the process liquid.

Vibrating Switches

TYPES

A. Tuning fork
B. Vibrating probe
C. Vibrating reed

APPLICATIONS

Liquid, slurry, and solids level switches

DESIGN PRESSURE

A and B. To 150 psig (10.3 bars = 1 MPa)
C. Up to 3000 psig (207 bars = 20.7 MPa)

DESIGN TEMPERATURE

A. −45 to 200°F (−43 to 93°C)
B. 8 to 176°F (−10 to 80°C)
C. From −150 to 300°F (−100 to 149°C)

MATERIALS OF CONSTRUCTION

Aluminum, steel, and stainless steel

MINIMUM BULK DENSITY

A and C. Down to 1.0 lbm/ft^3 (16 kg/m^3)
B. Requires an apparent specific gravity of 0.2

INACCURACY

The repeatability of type C is ⅛ in (3 mm)

COST

Standard type A, $300; other designs up to $500

PARTIAL LIST OF SUPPLIERS

Automation Products Inc.; Bindicator Co.; Endress + Hauser Inc.; KDG Mobrey Ltd.; Monitor Mfg.; Monitrol Bin Level Manufacturing Co.; Nohken Co. Ltd.; Vega B.V.; Zi Tech Instrument Corp.

—*Béla G. Lipták*

2.2 pH, OXIDATION-REDUCTION PROBES (ORP) AND ION-SELECTIVE SENSORS

Because the goal of the wastewater treatment industry is to purify and neutralize industrial and municipal waste streams, sensors are needed to detect the activity and concentration of various ionic substances. An important water parameter is the pH, which indicates the activity of the hydrogen ion and describes the acidity or alkalinity of the stream. Ion selective electrodes detect the activity of other ions, while ORPs describe the chemical or biological processes in progress. This section describes the features and capabilities of these three sensor types.

Probes and Probe Cleaners

In wastewater applications, environmental engineers use analytical probes to detect concentrations in the sludge layers situated in the lower parts of scraped bottom tanks. These probes are installed on pivoted hinges so that the mechanical scraper assembly can pass (see Figure 2.2.1).

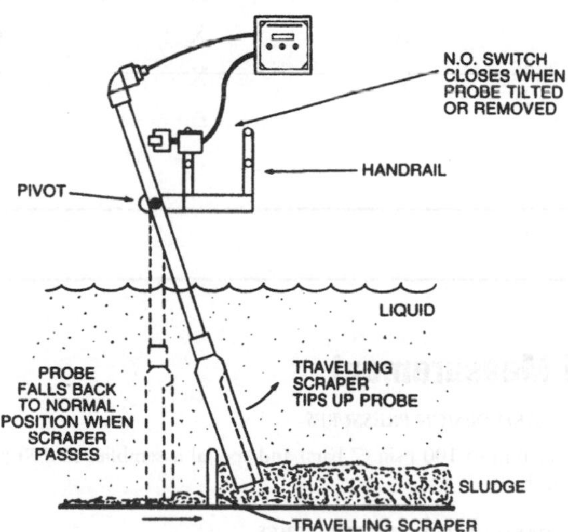

FIG. 2.2.1 Probe-type sensors used to detect the composition of sludge and slurry layers in clarifiers. (Courtesy of Markland Specialty Engineering Ltd.)

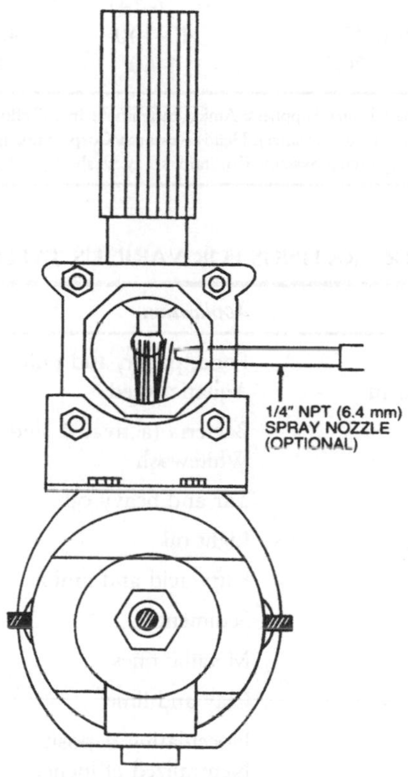

FIG. 2.2.2 Probe cleaner mounted in sight-flow glasses for good visibility. (Courtesy of Aimco Instruments Inc.)

While probe-type in-line analyzers eliminate the transportation lag and sample deterioration problems associated with offline analysis, they illustrate the need for efficient probe cleaners. A probe cleaner should be placed inside a sight glass so that clearer performance can be continuously observed by the operator (see Figure 2.2.2). A variety of probe-cleaning devices are available. Table 2.2.1 lists features and capabilities for the removal of various coatings and Table 2.2.2 lists suppliers.

If no sampling system is used, sample integrity is automatically guaranteed, and sensors that penetrate the

TABLE 2.2.1 SELECTION OF AUTOMATIC PROBE CLEANERS

	Applicable Choice of Probe Cleaner						
	Mechanical		*Chemical*			*Hydro-Dynamic (self-cleaning)*	*Acoustical or Ultrasonic*
	Brush	*Rotary Scraper*	*Acid*	*Base*	*Emulsifier*		
Service							
Oils and fats		√			√		√
Resins (wood and pulp)				√			√
Latex emulsions		√					
Fibers (paper and textile)						√	
SS						√	
Crystalline precipitations (carbonates)	√	√	√				
Amorphous precipitations (hydroxides)	√	√	√				√
Material of construction	Stainless steel (brush pH 7–14)	Stainless steel	PVC	PVC	PVC	Stainless steel	Polypropylene, stainless steel
Temperature °F	40–140	40–140	40–140	40–140	40–140	40–250	40–195
°C	4–60	4–60	4–60	4–60	4–60	4–120	4–90

Note: Probe Cleaner Suppliers; Amico Instrument Inc. (Teflon brush); Branson Cleaning Equipment Co. (ultrasonic cleaners); Fetterolf Corp. (spray rinse valve); Graphic Controls (brush cleaner); Helios Research Corp. (tank spray washers); Polymetron, Div. of Uster Corp. (probe cleaners); Sybron/Gamlen, Gamajet Div. (tank spray washers); Spraying Systems Co. (tank spray washers); Toftejorg Inc. (tank spray washers).

TABLE 2.2.2 RATINGS FOR VARIOUS TYPES OF CLEANERS

	Application	*Ultrasonic*	*Water-jet*	*Brushing*	*Chemical*
Slime Microorganism	Food, paper, and pulp, Aquatic weed	X	O	O	△
	Bacteria (activated sludge) Whitewash	△	O	O	△
Oil	Tar and heavy oil	X	X	X	△
	Light oil	O	△	△	O
	Fatty acid and amine	X	O	X	O
Suspension	Sediment	O	X	X	O
	Metallic fines	△	X	X	O
	Clay and lime	△	O	X	O
Scale	Flocculating deposit Neutralized effluent $CaCO_3$	△	△	△	△

Source: Horiba Instruments.
Notes: O: Recommend, △: Applicable, X: Not applicable.

process pipe with a retractable, cleanable probe are preferred. Probe sensors, either solid or membrane, require periodic cleaning. This can be done manually, by withdrawing the probe through an isolating valve so that the process is not opened when the electrode is cleaned; or automatically, using automatic probe-cleaning devices such as pressurized liquid or gas jets, and thermal, mechanical, or ultrasonic cleaning and scraping devices.

pH Measurement

STANDARD DESIGN PRESSURES

Vacuum to 100 psig (7 bars) and special assemblies to 500 psig (35 bars)

STANDARD DESIGN TEMPERATURES

Generally −5 to 100°C (23 to 212°F); sterilizable, −30 to 130°C (−22 to 266°F); Glasteel, −5 to 140°C (<5 pH) (23 to 284°F)

MATERIALS OF CONSTRUCTION

Electrode hardware: stainless steel, Monel, Hastelloy, titanium, epoxy, Kynar, halar, PVC, chlorinated polyvinyl chloride, polyethylene, polypropylene, polyphenylene sulfide, ryton, Teflon, and various elastomer materials

ASSEMBLIES

Flow-through, submersion, insertion, and retractable

CLEANERS

Ultrasonic, jet washer (chemical and water), and brush

INACCURACY

Electrodes 0.02 pH; lab meters and displays 0.01 pH; transmitters 0.02 mA; and installation effects 0.2 pH

RANGE

0 to 14 pH

COSTS

Electrodes cost $100 to $500 (Glasteel is $2000); lab meters, $200 to $800; transmitters, $500 to $2000; assemblies, $200 to $1000; and cleaners, $400 to $2000. A brush cleaner in 316 SS is $1900, and a retractable cleaner in 316 SS is $1750. A fiber-optic pH assembly is $15,000, with associated electronics costing an additional $10,000.

PARTIAL LIST OF SUPPLIERS

Amico-Instruments; Bailey-TBI; Beckman Instruments (Process Instruments and Control Group); Broadley-James; Custom Sensors & Technology; Electro Chemical Devices; Foxboro Analytical; George Fischer Signet; Great Lakes Instruments; Horiba Instruments; Ingold; Innovative Sensors; Johnson Yokogawa Electrofac; Lakewood Instruments; Leeds & Northrup Instruments; McNab; Monitek; Orion Industrial Division of Orion Research; Pfaudler; Phoenix; Uniloc Division of Rosemount; SensoreX; Van London

An important step in wastewater treatment is neutralization. The neutralization process includes the reagent delivery system, the mixing equipment, the reaction and equalization tanks, and the associated controls. In general, a single stirred-reaction vessel can neutralize influents between 4 and 10 pH. If the influent pH varies from 2 to 12 pH, one stirred and one attenuation tanks are needed. If the influent pH drops below 2 or rises above 12 pH, two stirred and one attenuation tanks are needed. Section 9.3 describes these aspects of the overall pH control system. This section discusses only pH measurement probes.

The measurement of pH covers a wide range of dilute acid and base concentrations (see Figure 2.2.3). For strong acids and bases, these measurements can track changes from one to one millionth percent. Thus, pH is a sensitive indicator of deficient and excess acid and base reactant concentrations for chemical reactors and scrubbers. For example, a few millionths of a percent of excess of sodium hydroxide (a strong base) is needed for chlorine destruction with sodium bisulfite. pH measurement can reduce the addition of sodium hydroxide to a minimum and still ensure complete use of sodium bisulfite.

Biological reactors use acids and bases to supply food or neutralize the waste products of organisms. Cells are

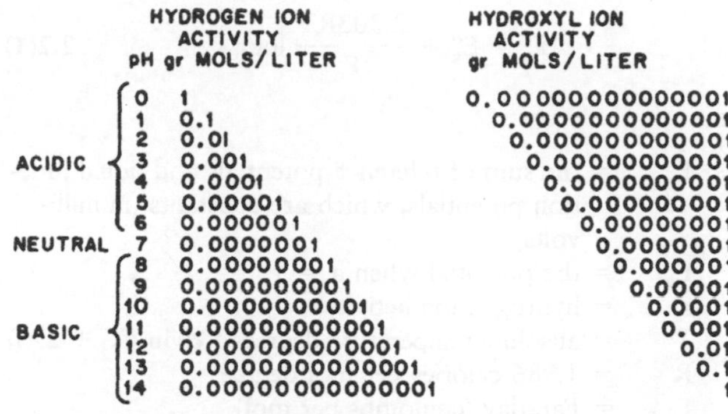

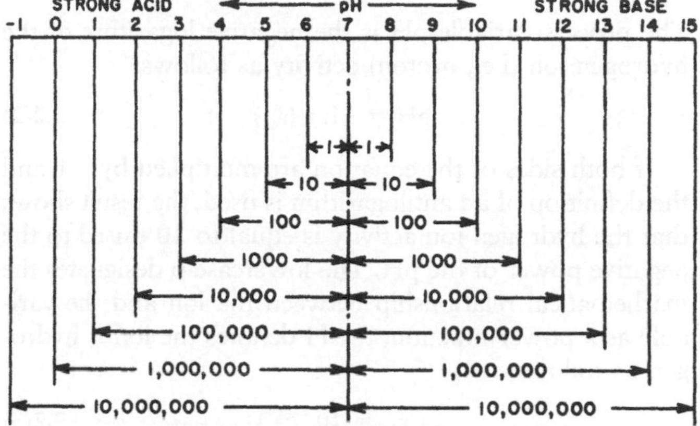

FIG. 2.2.3 The logarithmic nature of pH.

extremely sensitive to pH fluctuations. Genetically engineered bacteria tend to be weak and need tight pH control. Thus, pH is critical to the cell growth rate, enzyme reactions, and the extraction of intercellular products. The sensitivity of cells to pH has even wider significance in that any food, drink, or drug ingested or injected and any waste discharged to the environment must have pH specifications to prevent damage to living matter and ecological systems. Stricter environmental regulations have increased the number and importance of pH measurements.

Some environmental regulations have instantaneous limits on pH. An excursion outside the acceptable range for a fraction of a second can be a recordable violation. The pH measurement system must be designed to prevent violations from spurious readings due to installation effects. For most discharges, the acceptable range lies between 6 and 9 pH.

THEORETICAL REVIEW

In pH measurement systems, a pH responsive glass takes up hydrogen ions and establishes a potential at the glass surface with respect to the solution. This potential is related to the hydrogen ion activity of the solution by the Nernst relationship as follows:

$$E_g = E_g^\circ + \frac{2.303RT}{F} \log_{10^a} \qquad 2.2(1)$$

where:

E_g = the sum of reference potentials and liquid junction potentials, which are constants (in millivolts)
E_g° = the potential when a = 1
a = hydrogen ion activity
T = absolute temperature degrees Kelvin (°C + 273)
R = 1.986 calories per mol degree
F = Faraday (coulombs per mol)
2.303 = logarithm conversion factor

The process variable pH is the negative logarithm of the hydrogen ion (i.e., proton) activity as follows:

$$pH = -\log (a_H) \qquad 2.2(2)$$

If both sides of the equation are multiplied by -1 and the definition of an antilogarithm is used, the result shows that the hydrogen ion activity is equal to 10 raised to the negative power of the pH. The lowercase p designates the mathematical relationship between the ion and the variable as a power function; the H denotes the ion is hydrogen as follows:

$$a_H = 10^{-pH} \qquad 2.2(3)$$

For dilute aqueous (water) solutions, the activity coefficient is approximately unity, and the hydrogen ion concentration is essentially equal to the hydrogen ion activity. As the concentrations of acids, bases, and salts increase, the crowding effect of the ions reduces hydrogen ion activity. Thus, an increase in salt concentration can increase the pH reading even though the hydrogen ion concentration is constant.

An acid is a proton donor, and a base is a proton acceptor. When an acid dissociates (breaks apart into its component ions), it yields a hydrogen ion and a negative acid ion. When a base dissociates, it gives a positive base ion and a hydroxyl ion that is a proton acceptor. When water dissociates, the result is both a hydrogen ion (proton) and hydroxyl ion (proton acceptor). Thus, water acts as both an acid and a base. Neutralization is the association of hydrogen ions from acids and hydroxyl ions from bases to form water.

pH measurement can track fourteen decades of hydrogen ion concentration and detect changes as small as 10^{-14} (at 14 pH). The concentration changes of strong acids and bases also follow the decade change per pH unit within this range. No other concentration measurement has such rangeability and sensitivity. These characteristics have profound implications for pH control.

Concentrated strong acids and bases have a pH that lies outside this range. For example, concentrated sulfuric acid has a pH of -10, and concentrated sodium hydroxide has a pH of 19 as measured by a hydrogen electrode. However, the set point of a pH loop is usually well within the 0 to 14 range. Some feedforward pH loops can require measurements outside this range, but the shortened life expectancy and increased error from the electrode at range extremes make such measurements impractical.

The neutral point is where the hydrogen ion concentration equals the hydroxyl ion concentration. At 25°C, this point occurs at 7 pH (see Table 2.2.3).

MEASUREMENT ELECTRODE

The hydrogen ion does not exist alone in aqueous solutions. It associates with a water molecule to form a hy-

TABLE 2.2.3 CONCENTRATIONS OF ACTIVE HYDROGEN AND HYDROXYL IONS AT 25°C AT DIFFERENT pH VALUES (IN GRAM-MOLES/LITER) AND SOME FLUIDS WITH CORRESPONDING pH VALUES

pH	*Fluid Example*	*Hydrogen Ions*	*Hydroxyl Ions*
0	4% Sulfuric acid	1.0	0.0000000000000
1		0.1	0.0000000000000
2	Lemon juice	0.01	0.000000000001
3		0.001	0.00000000001
4	Orange juice	0.0001	0.0000000001
5	Cottage cheese	0.00001	0.000000001
6	Milk	0.000001	0.00000001
7	Pure water	0.0000001	0.0000001
8	Egg white	0.00000001	0.000001
9	Borax	0.000000001	0.00001
10	Milk of magnesia	0.0000000001	0.0001
11		0.00000000001	0.001
12	Photo developer	0.000000000001	0.01
13	Lime	0.0000000000001	0.1
14	4% sodium hydroxide	0.00000000000001	1.0

dronium ion (H_3O^+). The glass measurement electrode (see Figure 2.2.4) develops a potential when hydronium ions get close enough to the glass surface for hydrogen ions to associate with hydronium ions in an outer layer of the glass surface. This thin hydrated gel layer is essential for electrode response. The input to the pH measurement circuit is a potential difference between the external glass surface exposed to the process (E_1) and the internal glass surface wetted by a 7 pH solution (E_2). If the external glass surface is in exactly the same condition as the internal glass surface, the Nernst equation states that the potential difference in millivolts is proportional to the deviation of the process pH from 7 pH at 25°C.

Flat glass electrodes minimize glass damage and maximize a sweeping action to prevent fouling. A small button flat glass electrode has a range of 0 to 10 pH, and a large flush flat glass electrode has a range of 2 to 12 pH. High sodium ion concentrations and low hydrogen ion activity have a larger effect on flat glasses.

The photometer-type pH sensor shown in Figure 2.2.5 uses a fiber-optic sensor. It should be free of problems due to sodium ions, temperature, coating, and abrasion in a properly designed sample system. The time delay caused by the temperature lag in the sampling system and the higher cost and maintenance are disadvantages. Also, the color dyes are sensitive to oxidants.

CLEANERS

Significant natural self-cleaning by turbulent eddies requires a velocity of 5 or more feet per second (fps) past the electrode. A velocity of greater than 10 fps can cause excessive measurement noise and sensor wear. The area obstructed by the electrode must be subtracted from the total cross-sectional area when estimating the total area open to flow around the electrode. The pressure drop at the restricted cross section should be calculated to ensure

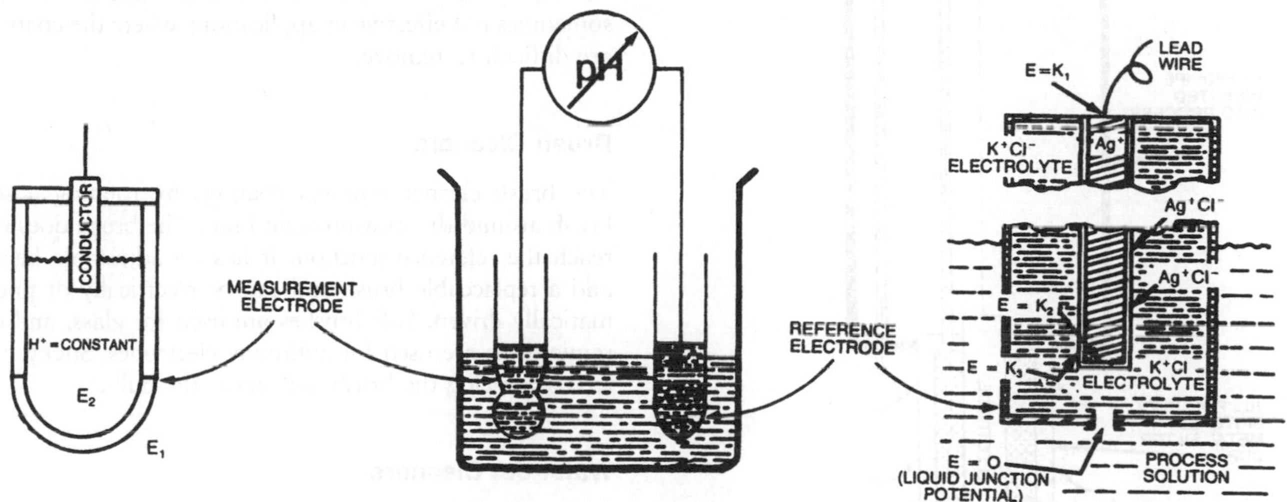

FIG. 2.2.4 The traditional configuration of a glass measurement electrode and a flowing junction reference electrode.

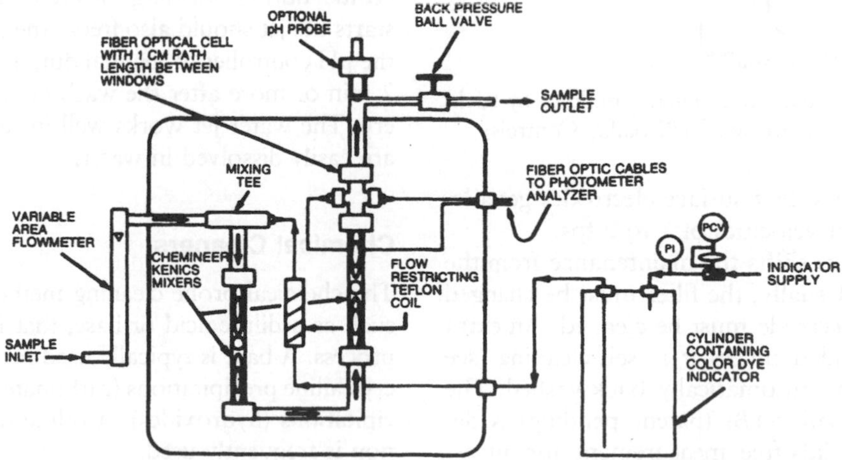

FIG. 2.2.5 Fiber-optic photometer-type pH detector. (Courtesy of Custom Sensors & Technology)

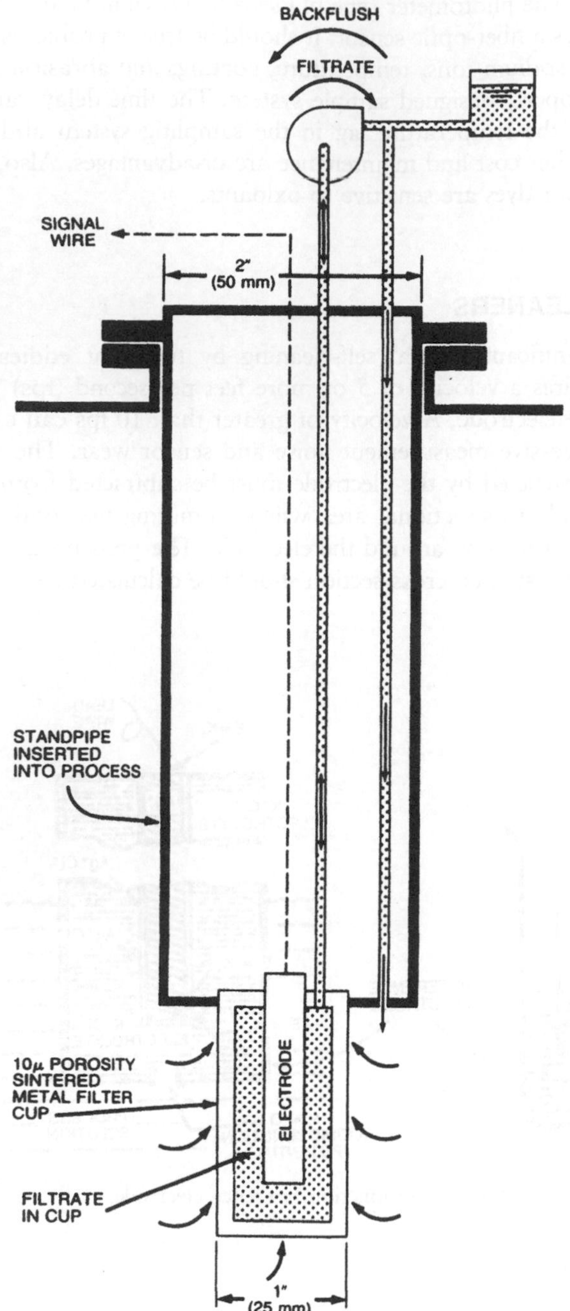

FIG. 2.2.6 Electrode protected from material buildup by backflushed porous filter cup. (Courtesy of TBI-Bailey Controls)

that no cavitation occurs. Flat surface electrodes get adequate cleaning action at velocities of 1 to 2 fps.

The addition of filters shifts the maintenance from the electrode to the filter. Usually, the filter must be changed more often than the electrode must be cleaned. An extra filter is not recommended unless it is self-cleaning (see Figure 2.2.6) or can be automatically backwashed. The Filtrate Master from Bailey-TBI (patent pending) is designed to provide a solids-free measurement for an assembly submerged in a slurry. It reverses flow and pulses loose particles caught in the 10-μm metal filter.

Four types of automatic cleaners are ultrasonic, brush, water-jet, and chemical. Table 2.2.4 shows the performance ratings for various applications, and Figure 2.2.7 shows the components of these assemblies. These methods concentrate on the removal of coatings from the measurement bulb. Particles and material clogged in the porous reference junction are generally difficult to dislodge. The impedance of plugged reference junctions can get so high that it approaches an open circuit, and the pH reading exceeds the scale.

Ultrasonic Cleaners

Ultrasonic cleaners use ultrasonic waves to vibrate the liquid near electrode surfaces. Effectiveness depends on the vibration energy and fluid velocity past the electrodes. Heavy-duty electrodes are needed to withstand the ultrasonic energy. The ultrasonic cleaner works well in processes where fine particles and easily supersaturated sediments are formed or in suspension. It can move loose and light particles and oil deposits. Ultrasonic cleaners are sometimes not effective in applications where the coatings are difficult to remove.

Brush Cleaners

The brush cleaner removes coatings by rotating a soft brush around the measurement bulb. The brush does not reach the reference junction. It has an adjustable height and a replaceable brush and can be electrically or pneumatically driven. Soft brushes are used for glass, and ceramic disks are used for antimony electrodes. Sticky materials can clog the brush and smear the bulb.

Water-Jet Cleaners

The water-jet cleaner directs a high-velocity water jet to the measurement bulb. The reading of the loop becomes erratic during washing. Therefore, the cycle timer that starts the jet should also freeze the pH reading and switch the pH controller to manual during the wash cycle and for 2 min or more after the wash period for electrode recovery. The water jet works well in removing materials that are easily dissolved in water.

Chemical Cleaners

The chemical probe cleaning method uses a chemical jet, such as a dilute acid or base, that is compatible with the process. A base is typically used for resins and an acid for crystalline precipitations (carbonates) and amorphous precipitations (hydroxides). A dilute hydrochloric acid solution is frequently used.

Chemical cleaning tends to be the most effective method, but acid and base cleaners chemically attack the

TABLE 2.2.4 RATINGS FOR VARIOUS CLEANERS

	Application	*Ultrasonic*	*Water-jet*	*Brushing*	*Chemical*
Slime Microorganism	Food, paper, and pulp, Aquatic weed Bacteria (activated sludge)	X	O	O	△
	Whitewash	△	O	O	△
Oil	Tar and heavy oil	X	X	X	△
	Light oil	O	△	△	O
	Fatty acid and amine	X	O	X	O
Suspension	Sediment	O	X	X	O
	Metallic fines	△	X	X	O
	Clay and lime	△	O	X	O
Scale	Flocculating deposit Neutralized effluent $CaCO_3$	△	△	△	△

Source: Horiba Instruments.
Notes: O: Recommend, △: Applicable, X: Not applicable.

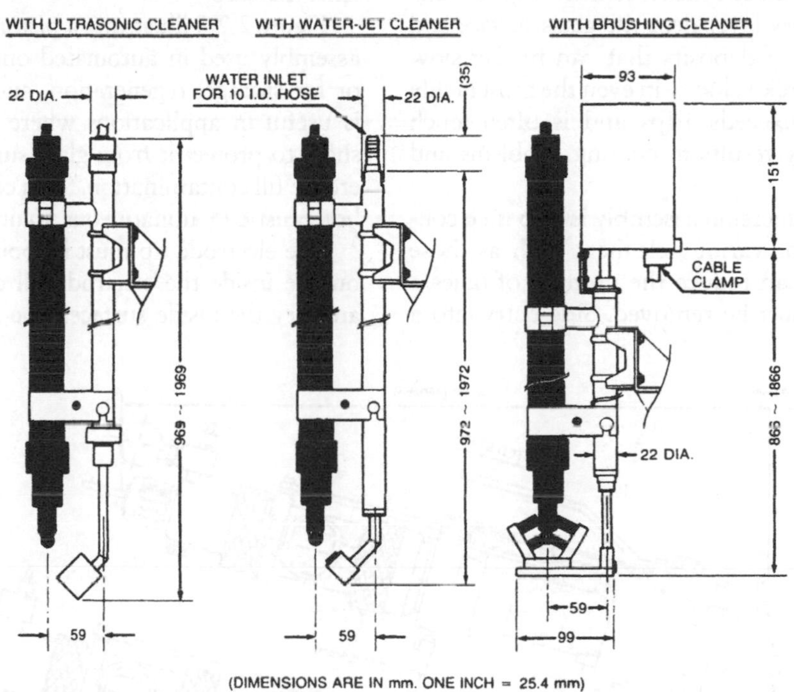

FIG. 2.2.7 Submersion assemblies with various cleaners. (Courtesy of Horiba Instruments)

glass. In addition, cleaning cycles that are too frequent or too long can cause premature failure of the glass electrode. As with the water jet, the cycle timer must hold the last pH reading and suspend control action during the wash cycle. Redundant pH sensors can be installed in parallel so that while one electrode assembly is being reconditioned, the other is in control. In such installations, the reconditioned assembly automatically returns to control after the wash cycle.

Wastewater treatment facilities often manually clean electrodes by soaking them in a dilute hydrochloric acid solution for several hours. Soaking electrodes for 1 min in a dilute solution of hydrofluoric acid in a nonglass container can reactivate electrodes that are sluggish or have too small a span or efficiency. The reactivation occurs by the hydrofluoric acid dissolving part of the aged gel layer. The electrode should then be soaked overnight in its normal storage solution (typically a 4-pH buffer).

INSTALLATION METHODS

Installation design has two main principles. First, for pH control, the sensor location and assembly must minimize transportation delays and sensor time constant. The additional dead time from a delayed and slow measurement increases the loop's period, control error, and sensitivity to nonlinearity. Second, the installation must minimize the number of times the electrodes must be removed for maintenance (e.g., calibration and cleaning). Removal and manual handling increase error and reduce electrode life. The fragile gel layer is altered by handling, and the equilibrium achieved by the reference junction is upset.

Submersion Assemblies

Sample systems are undesirable because they add transportation delay and increase cost, and problems arise with winterization and plugging. Therefore, a submersion assembly is best for control. However, velocities below 1 fps dramatically slow the electrode measurement response due to the increased boundary layer near the glass surface and promote the formation of deposits that can further slow the measurement. The bulk velocity in even the most highly agitated vessels rarely exceeds 1 fps and is often much lower. This low velocity results in coating problems and a slow response.

The removal of a submersion assembly is also time-consuming. The addition of various cleaners such as those shown in Figure 2.2.7 can reduce the number of times a submersion assembly must be removed. Side entry into a vessel with a retractable probe is the standard installation for fermentors, as shown in Figure 2.2.8.

Retractable and Brush-Cleaned Units

The best location for most probes or assemblies, except in the most abrasive services, is in a recirculation line close to the vessel outlet. Installing the probe downstream of the pump is preferred because the strainer blocks and the pump breaks up clumps of material that could damage the electrodes. The retractable electrode is the most straightforward and economical solution. However, accidents caused by removing the restraining strap or omitting tubing ferrules have caused these assemblies to be banned from many plants.

Many wastewater treatment facilities use flow-through assemblies or direct-probe insertions with block, drain, and bypass valves. The flow is returned to the suction of the pump or vessel. If the flow chamber has a cross-sectional area much larger than the process connections, the velocity drops too low, response time slows, and coating problems increase.

Figure 2.2.8 illustrates a piston-actuated, retractable pH assembly used in automated online cleaning applications or for storage, regeneration, and calibration. This design is useful in applications where probe exposure must be short to protect it from glass surface deterioration or reference fill contamination. Such contamination is caused by hot caustic or nonaqueous solutions.

The electrode tip must be pointed down so that the air bubble inside the electrode fill does not reside in the tip and dry the inside surface. The air bubble provides some

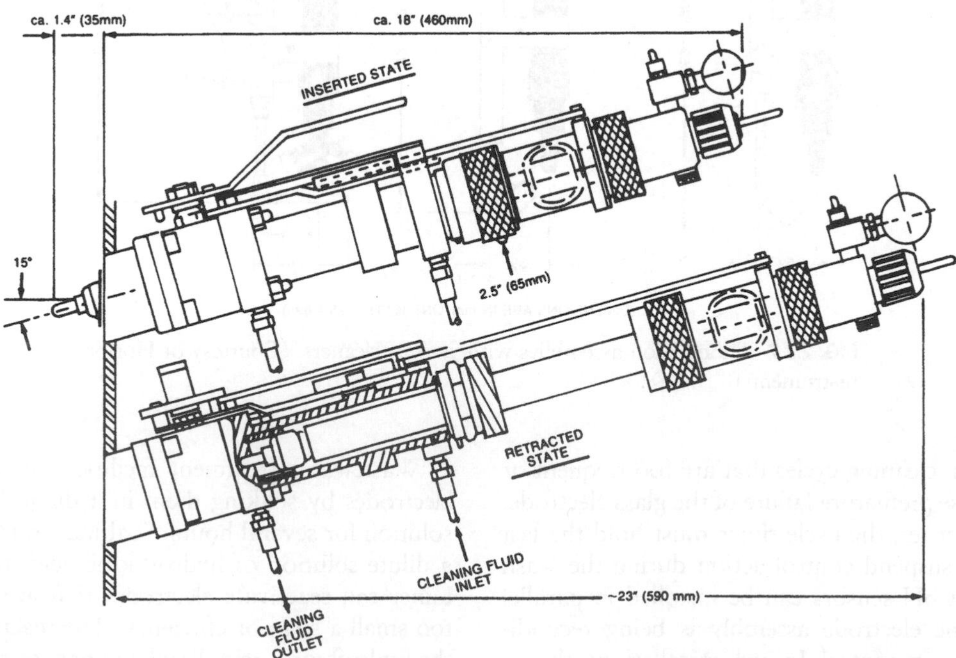

FIG. 2.2.8 Retractable side entry probes often used on fermentors. (Courtesy of Ingold)

compressibility to accommodate thermal expansion. An installation angle of 15 degrees or more from the horizontal is sufficient to keep the bubble out of the tip. Some electrode designs eliminate the bubble and provide a flexible diaphragm for fill contraction and expansion.

Section 9.2 discusses pH control systems.

ORP Probes

DESIGN PRESSURES

Vacuum to 150 psig (10.6 bars) is standard; special assemblies are available for up to 500 psig (35 bars).

DESIGN TEMPERATURES

Generally from 23 to 212°F (−5 to 100°C)

MATERIALS OF CONSTRUCTION

Mounting hardware is available in stainless steel, Hastelloy, titanium, PVC, CPVC, polyethylene, polypropylene, epoxy, polyphenylene sulfide, Teflon, and various elastomer materials; electrodes are available in platinum or gold.

ASSEMBLIES

Submersion, insertion, flow-through, and retractable

CLEANERS

Ultrasonic, water- or chemical-jet washer, and brushes

RANGE

Any span between −2000 mV and +2000 mV

INACCURACY

Typically, inaccuracy is ±10 mV and is a function of the noble metal electrode condition and reference electrode drift; repeatability is about ±3 mV.

COSTS

Electrodes cost from $100 to $500; portable or bench-top laboratory display and control units range from $300 to $1000; transmitters range from $500 to $2000; and cleaners are available from $500 to $2000.

PARTIAL LIST OF SUPPLIERS

Broadley James Corp.; Capital Controls Inc.; Electro-Chemical Devices Inc.; Foxboro Analytical Co.; George Fischer Signet Inc.; Great Lakes Instruments Inc.; Johnson Yokogawa Corp.; Lakewood Instruments; Leeds & Northup, Unit of General Signal; Polymetron; Rosemount Analytical Inc. Uniloc Div.; Sensorex; TBI-Bailey Controls Co.

ORP measurement is important in wastewater treatment applications. Examples of these applications are the removal of heavy metals, such as chromium, from metal finishing wastewater and cooling tower blowdown streams. ORP sensors are also used in cyanide removal, which is often required when heavy metals are removed. In sanitary wastewater treatment, ORP measurement controls the addition of an oxidant for odor control. Also, in most aerobic or anaerobic biological digestion processes, bacterial health can be judged on the basis of ORP measurements. Section 9.4 discusses various ORP control systems; this section describes the ORP detector only.

PRINCIPLES OF ORP MEASUREMENT

In oxidation-reduction reactions, the substances involved gain or lose electrons and show different electron configurations before and after the reaction. Oxidation is the overall process by which a specie in a chemical reaction loses one or more electrons and increases its state of oxidation. An oxidant is a substance capable of oxidizing a chemical specie; it acquires the electron or electrons lost by the specie and is itself reduced in the overall process. Reduction is the overall process in which a specie in a chemical reaction gains one or more electrons and decreases its state of oxidation. A reductant is a substance capable of reducing a chemical specie; it loses the electrons gained by the specie and is itself oxidized in the overall process.

An ORP reaction, therefore, involves an electron exchange capable of doing work. This capability is expressed in terms of potential for a half-cell, or electron, reaction. Table 2.2.5 lists the potentials for standard conditions, that is, where reactants and products are at unit activity. The voltages in this table are referenced to the standard hydrogen electrode (SHE), which is assigned the value of 0.000 V.

Note that the reactions in Table 2.2.5 are written as reductions, which is the almost universally used convention. In this section, the term ox/red indicates the oxidized form on the left side of the equation and the reduced form on the right. For example, the standard potential for ferric iron, Fe^{3+}, reduced to ferrous iron, Fe^{2+}, is written as E°Ox/Red = +0.770 V.

$E_{Red/Ox} = -E_{Ox/Red}$ simply means that the polarity is reversed when the reaction is written as an oxidation reaction. For example, $Fe^{2+} = Fe^{3+} + e^-$. $E^\circ_{Red/Ox} = -0.770$ V.

For example, in a common industrial process, hexavalent chromium is reduced with a ferrous sulfate solution. The half-reactions are:

TABLE 2.2.5 REDUCTION POTENTIALS OF SOLUTION IN ORP MEASUREMENT

Reduction	*E°, Volts*
$O_3 + 2H_3O^+ + 2e^- = O_2 + 3H_2O$	+2.070
$Cr_2O_7^{2-} + 14H_3O^+ + 6e^- = 2Cr^{3+} + 21H_2O$	+1.330
$ClO^- + H_2O + 2e^- = Cl^- + 2OH^-$	+0.890
$Fe^{3+} + e^- = Fe^{2+}$	+0.770
Ag/AgCl electrode 4 M KCl	+0.199
$2H_3O^+ + 2e^- = H_2 + 2H_2O$	0.000
$Zn^{2+} + 2e^- = Zn$	−0.763
$CNO^- + H_2O + 2e^- = CN^- + 2OH^-$	−0.970
$Na^+ + e^- = Na$	−2.711

$$Cr_2O_7^{2-} + 14\ H_3O^+ + 6e^- = 2\ Cr^{3+} + 21\ H_2O$$

$$E^{\circ}_{Ox_2/Red_2} = 1.330$$

$$Fe^{2+} = Fe^{3+} + 1\ e^-$$

$$E_{Red_1/Ox_1} = -0.770 \qquad 2.2(4)$$

The reaction is not complete at pH 4.0 but is substantially complete at pH 2.0, assuming that a concentration of 10^{-6} M represents completion.

In industrial and laboratory work, ORP cell potentials are measured, not against a SHE but against an Ag/AgCl, 4 M KCl reference, $E^{\circ}_{Ox/Red} = +0.199$ V, or a saturated calomel electrode (SCE) whose $E^{\circ} = 0.244$ V. The following equation converts to the potential measured, designated as $E_{AgCl\ ref}$:

$$E_{meas} = E_{cell} - E_{AgCl} \qquad 2.2(5)$$

For example, the E° for the cell is:

$$Fe^{3+} + e^- = Fe^{2+}\ E_{Ox/Red} = 0.770 \text{ vs SHE} \qquad 2.2(6)$$

and the potential measured versus the silver–silver chloride reference is:

$$E_{meas/AgCl\ ref} = +.770\ V - .199\ V$$
$$= 0.571\ V \qquad 2.2(7)$$

Absolute potentials in ORP measurement are not always used. Most equipment manufacturers use slightly modified pH analyzers for ORP measurement. These instruments normally have the standard or zero adjustment of the parent pH meter. Furthermore, in general, the reverse of polarity is achieved merely by reversing the inputs.

Microprocessor-based ORP analyzers generally have wide rangeability with high-resolution digital displays, alarm–control setpoints, and output signal scaling limits.

Most chemical reaction systems involving electron exchange are controlled near the equivalence point with a controlled excess of added reagent so that the reactions are driven to completion. Thus, most ORP reactions are controlled just beyond the steep portion of the titration curve.

EQUIPMENT FOR ORP MEASUREMENT

The basic instrumentation for ORP measurement closely parallels that for pH measurement. In fact, many instrument suppliers use slightly modified pH analyzers, with a changed sensitivity and an mV scale in place of a pH scale. The hardware used to install the electrodes in the process stream is generally the same as that used in pH systems (see Figure 2.2.9).

Two major differences exist between an ORP system and a pH system. One difference is the sensing electrode, which is normally a noble metal, typically platinum or gold, although other metals and carbon have been used.

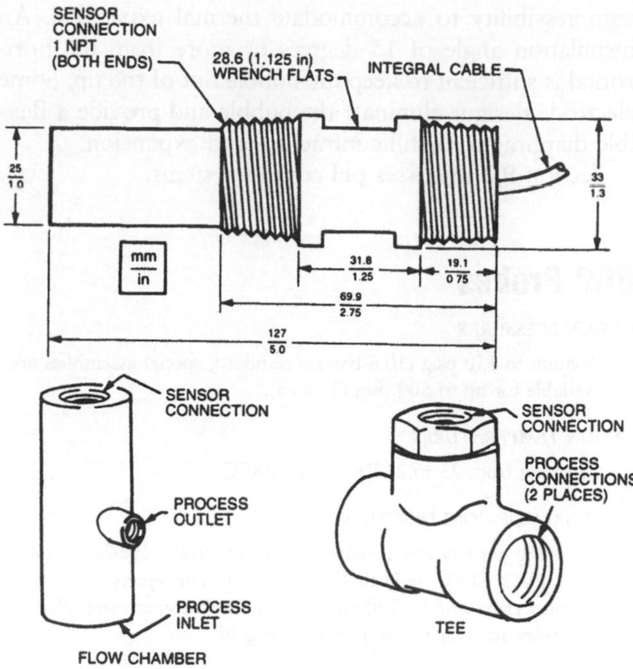

FIG. 2.2.9 ORP and pH probes packaged and mounted in the same way. (Courtesy of The Foxboro Co.)

The second major difference is in temperature compensation. Process pH systems are typically temperature-compensated, whereas ORP systems are almost never temperature-compensated.

Basic thermodynamics apply to pH and ORP as expressed by the classic Nernst equation. For oxidation-reduction half-cell reactions, this equation can be represented as:

$$E_{cell} = E^{\circ}_{Ox/Red} + \frac{2.303\ RT}{nF} \log \left[\frac{Ox}{Red}\right] \qquad 2.2(8)$$

where:

- E° = the potential under standard conditions of unit activity referred to the SHE
- R = the gas constant, 1.986 cal per mol degree
- F = Faraday's constant
- T = temperature in °K
- n = number of electrons exchanged in the reaction

Table 2.2.5 shows how the n values change from reaction to reaction. These changes, plus the fact that a given ORP reaction can encompass side reactions, reveal why it is difficult, if not impossible, to temperature-compensate an ORP reaction. In the Nernstian representation of pH, n always equals 1.

In Equation 2.2(8), the standard potential $E^0_{Ox/Red}$ is found in tables in handbooks and in the value relative to the standard hydrogen electrode. Therefore, the E_{cell} for the prevailing concentration is also relative to the SHE.

Often, ORP and pH electrodes can be mounted in the same tees or flow-through chambers and can have identical appearances. The electrode shown in Figure 2.2.9 is a ruggedized flat-glass electrode with identical dimensions for both pH and ORP services; the only difference is the gold or platinum wire tip in the ORP version.

PRACTICAL APPLICATION OF ORP

The chromium reaction typically takes place at a pH of 2.0 to 2.5. At this pH, the smell of sulfur dioxide is present when the sulfite ion is in slight excess. An experienced operator can adjust his control point potential to attain a slight odor of sulfur dioxide and then make further adjustments based on laboratory analysis for hexavalent chromium. Setting up a system based on calculation is possible when all reactants and products are known. However, this case is rare in industrial processes, and most applications require analytical verification of results.

The ORP responses in two common applications are illustrated by the titration curves in Figures 2.2.10 and 2.2.11. These curves are only examples. The responses can vary considerably from one installation and process composition to another. The actual control points must be finely adjusted after system startup.

CARE OF AN ORP SYSTEM

Maintaining an ORP measuring and control system is comparable to maintaining a similar pH system. However, because standards analogous to pH buffers are less common, maintenance is sometimes cut short. The most pressing maintenance problem with ORP is the noble metal sensing electrode. It is subject not only to coating but also to poisoning, both of which can result in sluggish or inaccurate measurement of the potential. This inaccuracy can result in improper demand for the reagent because a control point not representative of the reagent excess is held. Jones reported on the use of standards using quinhydrone-saturated pH buffers to establish known potentials as a check on the condition and response of the electrode system. The recommended treatment for either a change of span or a shift in potential is cleaning with aqua regia.

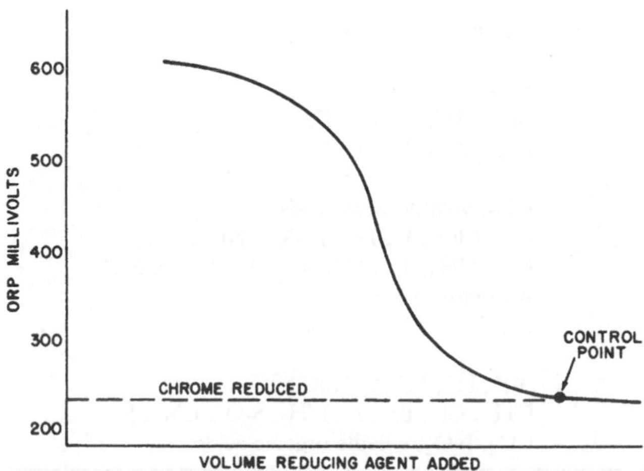

FIG. 2.2.10 Chrome reduction titration curve.

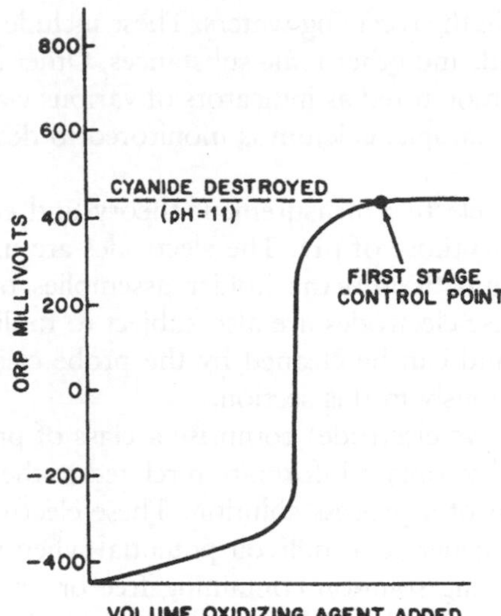

FIG. 2.2.11 Cyanide oxidation titration curve.

Ion-Selective Electrodes

TYPE OF ELECTRODE

Glass, solid-state, solid matrix, liquid-ion exchanger, and gas sensing

STANDARD DESIGN PRESSURE

Generally dictated by the electrode holder; 0 psig for the solid-state and liquid-ion exchanger, 0 to 100 psig (0 to 7 bars) for most electrode types, and over 100 psig (over 7 bars) for solid-state designs

STANDARD DESIGN TEMPERATURE

32 to 122°F (0 to 50°C) for the solid matrix and liquid ion exchanger; 23 to 176°F (−5 to 80°C) for most others, with 212°F (100°C) intermittent exposure permissible

RANGE

From fractional ppm to concentrated solutions

RELATIVE ERROR

For direct measurements, an absolute error of ±1.0 mV is equivalent to a relative error of ±4% for monovalent ions and ±8% for divalent ions; for end-point detection or batch control, ±0.25% or better is possible; and for expanded scale commercial amplifiers, the error is better than ±1% of full scale.

COST

Similar to those of pH installations; electrodes, $300 to $700; systems, $2000 to $10,000

PARTIAL LIST OF SUPPLIERS

Corning; Fisher Scientific; The Foxboro Co.; Great Lakes Instruments, Inc.; HNU Systems, Inc.; Horiba Instruments, Inc.; Ingold Electrodes, Inc.; Leeds and Northrup; Orion Research, Inc., Radiometer; Rosemount Analytical, Inc.

Wastewater treatment plant effluents must be monitored for all ionic substances that are deleterious to humans or animal life in the receiving waters. These include cyanides, sulfides, lead, and other ionic substances. Other ionic substances are monitored as indicators of various water properties. For example, calcium is monitored to detect water hardness.

The ion-selective measurement theory and equipment are similar to those of pH. The electrodes are usually the same size and fit into the holder assemblies of the pH probes. These electrodes are also subject to fouling by oil or slimes, and can be cleaned by the probe cleaners discussed previously in this section.

Ion-selective electrodes comprise a class of primary elements used to obtain information related to the chemical composition of a process solution. These electrochemical transducers generate a millivolt potential when immersed in a conducting solution containing free or unassociated ions to which the electrodes are responsive. The potential magnitude is a function of the logarithm of measured ion activity (*not* the total concentration of that ion) as expressed by the Nernst equation (see Equation 2.2[9]). The familiar pH electrode for measuring hydrogen ion activity is the best known ion selective electrode and was the first to be commercially available. The potential developed across an ion-selective membrane is related to the ionic activity as shown by the Nernst equation as follows:

$$E = \frac{2.3\ RT}{nF} \log \frac{a_1}{a_{int}} \qquad 2.2(9)$$

where:

E = the potential developed across the membrane

a_1 = the activity of the measured ion in the sample or process

a_{int} = the activity of the same ion in the internal solution

$2.3\ RT/nF$ = the Nernst slope, or the slope of the calibration curve, and is a function of the absolute temperature T and the charge on the ion being measured n

R = the gas law constant

With few exceptions—notably the silver-billet electrode for halide measurements and the sodium-glass electrode—the pH electrode was the only satisfactory electrode available to the process industry prior to 1966. Currently, more than two dozen electrodes are suitable for industrial use. Table 2.2.6 lists several electrodes commercially available for

TABLE 2.2.6 ION-SELECTIVE ELECTRODES

Ion/Specie	*Type of Electrode*	*Lower Detectable Limit, ppm*	*Principal Interferences*
Ammonia	Gas-sensing	0.009	Volatile amines
Bromide	Solid-state	0.04	CN^-, I^-, $S^=$
Cadmium	Solid-state	0.01	Ag^+, Hg^{++}, Cu^{++}, Fe^{++}, Pb^{++}
Calcium	Solid matrix/ liquid membrane	0.2	Zn^{++}, Fe^{++}, Pb^{++}, Cu^{++}, Ni^{++}, Sr^{++}, Mg^{++}, Ba^{++}
Carbon dioxide	Gas-sensing	0.4	Volatile weak acids
Chloride	Solid-state	0.2	Br^-, CN^-, $S^=$, SCN^-, I^-
Chloride	Liquid membrane	0.2	ClO_4^-, Br^-, I^-, NO_3^-, OH^-, F^-, OAc^-, $SO_4^=$, HCO_3^-
Copper(II)	Solid-state	0.006	Ag^+, Hg^{++}, Fe^{+++}
Cyanide	Solid-state	0.01	$S^=$, I^-
Divalent cation*	Solid matrix/ liquid membrane	0.001	—
Fluoroborate (BF_4^-), (boron)	Liquid membrane	0.11	I^-, HCO_3^-, NO_3^-, F^-
Iodide	Solid-state	0.006	$S^=$, CN^-
Lead	Solid-state	0.2	Ag^+, Hg^{++}, Cd^{++}, Fe^{++}
Nitrate	Solid matrix/ liquid membrane	0.3	Cl^-, ClO_4^-, I^-, Br^-
Nitrite	Gas-sensing	0.002	CO_2, volatile weak acids
Perchlorate	Liquid membrane	0.7	Cl^-, ClO_3^-, I^-, Br^-, HCO_3^-, NO_3^-, etc.
Potassium	Liquid membrane	0.04	Cs^+, NH_4^+, Tl^+, H^+, Ag^+, $Tris^+$, Li^+, Na^+
Redox (platinum)	Solid-state	Varies	All redox systems
Silver/sulfide	Solid-state	0.01 Ag 0.003 S	Hg^{++}
Sodium	Glass	0.02	Ag^+, H^+, Li^+, Cs^+, K^+, Tl^+
Thiocyanate	Solid-state	0.3	OH^-, Cl^-, Br^-, I^-, NH_3, $S_2O_3^=$, CN^-, $S^=$
Sulfur dioxide	Gas-sensing	0.06	CO_2, NO_2, volatile organic acids

*The water hardness electrode is also known as the divalent cation electrode.

process applications. Other research sensors are also available.

Electrode Types

Electrodes are classified by the sensing membrane used. Environmental engineers use glass electrodes to detect sodium, ammonium, and potassium. They can use pH electrodes for carbon dioxide or ammonia detection by covering the glass membrane with a permeable membrane sac filled with a pH buffer. In this case, the gas diffuses in and out of the permeable membrane, and the resulting pH change is related to gas activity.

Solid-state electrodes are made from crystalline membranes. The fluoride electrode has a single crystal, which is dropped in lanthanum fluoride to form a sensing membrane. Silver and sulfide membranes are silver sulfide pellets. These membranes are sealed in epoxy (see Figure 2.2.12). Table 2.2.7 lists some solid-state electrodes and the composition of their membranes. Metal can be deposited on and an electrical lead can be connected to the surface of some pressed pellets and single crystalline silver-salt membranes (see Figure 2.2.13).

Liquid Ion Exchange Electrodes

A membrane can be saturated with an organic ion exchange material dissolved in an organic solvent. The electrode in Figure 2.2.14 has an internal aqueous filling solution, in which the reference electrode is immersed, and an ion exchange reservoir of nonaqueous water-immiscible solution that wicks into the porous membrane. The bottom layer of the material is the unknown process solution, the center is the nonaqueous liquid ion exchange solution, and the top layer is the internal aqueous solution. The electrodes cannot be used in nonaqueous solutions because the liquid ion exchanger would dissolve. In the solid-state version of this probe, the ion exchanger is permanently embedded in a plastic matrix with a nonporous membrane.

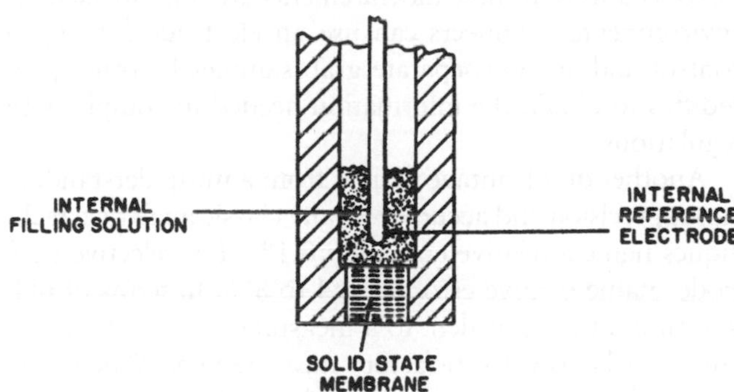

FIG. 2.2.12 Solid-state membrane electrode.

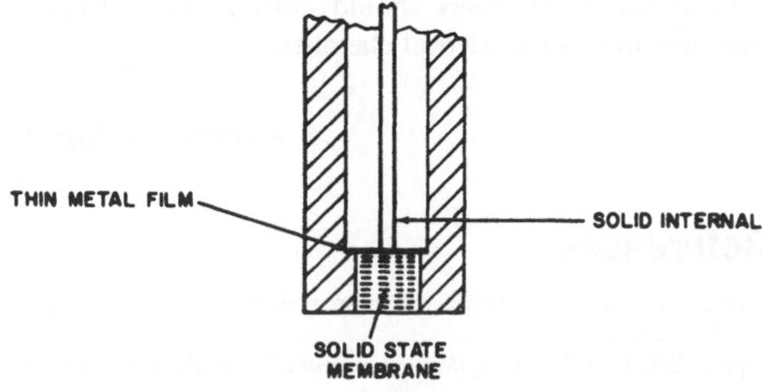

FIG. 2.2.13 Solid-state membrane electrode with solid internals.

TABLE 2.2.7 SOLID-STATE ELECTRODES AND THEIR MEMBRANE COMPOSITION

Electrode	*Membrane*	*Form*
Fluoride	LaF_3	Single crystal
Silver/sulfide	Ag_2S	Pressed pellet
Chloride,	AgX*	Single crystal
bromide or iodide	AgX-Ag_2S	Pressed pellet
Cyanide	AgI-Ag_2S	Pressed pellet

*X = Cl, Br or I.

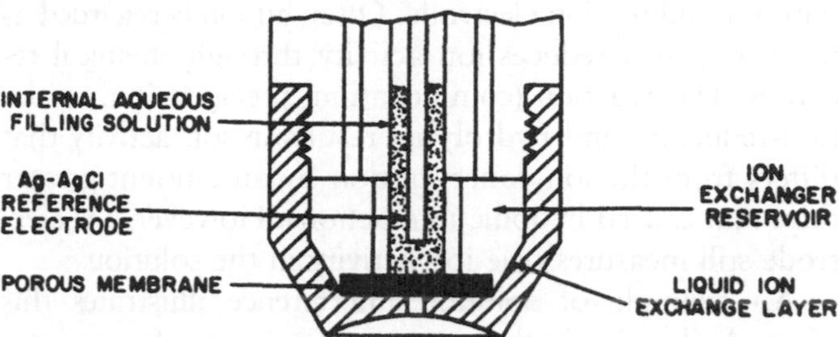

FIG. 2.2.14 Cross section of a divalent cation electrode tip.

INTERFERENCES

All ion selective electrodes are similar in operation and use. They differ only in the process by which the ion to be measured moves across the membrane and by which other ions are kept away. Therefore electrode interferences must be discussed in terms of membrane materials.

Glass electrodes and solid-matrix/liquid-ion exchange electrodes both function by an exchange of mobile ions within the membrane, and ion exchange processes are not specific. Reactions occur among many ions with similar chemical properties, such as alkali metals, alkaline earths, or transition elements. Thus, a number of ions can produce a potential when an ion selective electrode is immersed in a solution. Even the pH glass electrode responds to sodium ions at a high pH (low hydrogen ion activity). Fortunately, an empirical relationship can predict electrode interferences, and a list of selectivity ratios for the interfering ions is available from the manufacturers' specifications or other chemical publications.

Solid-state matrix electrodes are made of crystalline materials. Interferences resulting from ions moving into the solid membrane are not expected. Interference usually occurs from a chemical reaction with the membrane. An interference with the silver-halide membranes (for chloride, bromide, iodide, and cyanide activity measurements) involves a reaction with an ion in the sample solution, such as sulfide, to form a more insoluble silver salt.

A true interference produces an electrode response that can be interpreted as a measure of the ion of interest. For example, the hydroxyl ion, OH^-, causes a response with the fluoride electrode at fluoride levels below 10 ppm. Also, the hydrogen ion, H^+, creates a positive interference with the sodium ion electrode. Often an ion is regarded as interfering if it reduces ion activity through chemical reaction. This reaction (complexation, precipitation, oxidation-reduction and hydrolysis) results in ion activity that differs from the ion concentration by an amount greater than that caused by ionic interactions. However, the electrode still measures true ion activity in the solution.

An example of solution interference illustrates this point. A silver ion in the presence of ammonia forms a stable silver–ammonia complex that is not measured by the silver electrode. Only the free, uncombined silver ion is measured. Environmental engineers can obtain the total silver ion from calculations involving the formation constant of the silver–ammonia complex and the fact that the total silver equals the free silver plus the combined silver. Alternately, they can draw a calibration curve relating the total silver (from analysis or sample preparation) to the measured activity. The ammonia is *not* an electrode interference.

Most confusion stems from the fact that analytical measurements are in terms of concentration without regard to the actual form of the material in solution, and electrode measurements often disagree with the laboratory analyst's results. However, the electrode reflects what is actually taking place in the solution at the time of measurement. This information can be more important in process applications than the classic information. With suggested techniques, environmental engineers can reconcile the two measurements.

ADVANTAGES AND DISADVANTAGES

Compared to other composition-measuring techniques, such as photometric, titrimetric, chromotographic, or automated-classic analysis, ion-selective electrode measurement has several advantages. Electrode measurement is simple, rapid, nondestructive, direct, and continuous. Therefore, it is easily applied to closed-loop process control. In this respect, it is similar to using a thermocouple for temperature control. Electrodes can also be used in opaque solutions and viscous slurries. In addition, the electrodes measure the free or active-ionic species under process conditions and the status of a process reaction.

However, several disadvantages exist. The specificity of ion-selective electrodes is not as good as that of the glass pH electrode. Interferences vary from minor to major; environmental engineers must consult publications and manufacturers' data on limitations for each electrode. Also, the electrodes do not measure total ion concentration, although this parameter is often requested. Prior to the introduction of electrodes, concentration information was the only information available from the chemists' classic measurement techniques. Control laboratory chemists and process engineers do not think in terms of activity, even when making pH measurements. This habit may disappear as the ion-selective technique becomes more popular.

Sometimes concentration is a beneficial measurement, for example, in material-balance calculations or pollution control. Knowledge of material balance allows engineers to predict where a process reaction will occur. This information is necessary if a process is to be controlled by introducing changes that nullify those predicted.

In pollution control, environmental engineers believe that many ions, even in the combined state, are detrimental to life forms. For example, fluorides, cyanides, and sulfides are deleterious to fish and humans in many combined forms. However, they are not detected by ion-selective electrodes in the combined state. Consequently, pollution control agencies usually require concentration information. Electrodes can be used for concentration measurement if they are calibrated with solutions matching the process or ISAB solutions. If these measurements are not satisfactory, environmental engineers can use an electrode for online control and analyze separate grab samples by other procedures to obtain the information needed to comply with regulations.

Another disadvantage derives from a misunderstanding about precision and accuracy. Many classic analytical techniques name a relative error of ±0.1%. Ion-selective electrodes name relative errors of ±4 to 8%. In terms of pH, this amount is equivalent to a measurement of ±0.02 pH units—ordinarily a satisfactory measurement. When used with understanding, ion-selective electrodes can supply satisfactory composition information and afford closed-loop control that was previously unattainable. When in doubt, environmental engineers should consult with electrode manufacturers or analytical chemists.

—*Béla G. Lipták*

References

Jones, R.H. 1966. Oxidation reduction potential measurement. *JISA* (November).

Lipták, Béla G. 1995. *Analytical Instrumentation.* Radnor, Pa.: Chilton.

2.3 OXYGEN ANALYZERS

In the aquatic life cycle, oxygen plays a critical role. If the water is transparent, algae and other plant life generate oxygen as they build their body cells through photosynthesis. The other half of this cycle is respiration, in which bacteria and other animal life forms consume algae and other larger organic molecules while using the dissolved oxygen (DO) in the water and exhaling carbon dioxide. Therefore, two kinds of oxygen measurements are required in the operation of wastewater treatment plants:

The DO concentration of receiving waters must be monitored because the DO amount signals the life-supporting capacity of water. When the DO content drops to zero, the water body can no longer support aerobic life, bacteria and other animals suffocate, and the water becomes an open sewer.

The second oxygen measurement is made on the wastewater effluent discharged into receiving water. Here, the measurement determines the amount of damage that the discharged effluent will do to the receiving water. This damage is measured in the milligrams of DO that a liter of effluent will take from the receiving water, as bacteria decomposes the organic material in the effluent. This measurement is called the BOD of the wastewater effluent.

In addition to measuring the oxygen demand biologically (by bacteria), environmental engineers can measure it chemically; this demand is called COD. Environmental engineers also measure the carbon content of the effluent using total carbon analyzers or total organic carbon analyzers.

This section briefly discusses the use of in-place probes versus sampling and sample filtering versus homogenization. Then it describes DO detection probes and lists BOD and other oxygen demand sensors.

Bypass Filters and Homogenizers

When detecting DO, an environmental engineer can use a probe and insert it directly into the process or take a sample and deliver it to a DO analyzer. The probe eliminates the dead time of sample transportation, which degrades closed-loop control. However, the probe requires effective and unattended cleaning attachments. The advantage of sampling is that loop components are more accessible. Therefore, DO probes are used for online control, and sampling systems are used only for monitoring.

In DO detection, the measured parameter is in the liquid phase and the solids can be filtered out. However, in BOD measurement, oxygen demand occurs mostly in the solids. Therefore, they must be included in the evaluation. Consequently, most BOD and COD analyzers require a sampling system in which homogenization (liquification of solids), is used instead of filters, to prevent plugging and retain sample integrity.

SLIPSTREAM AND BYPASS FILTERS

To minimize transportation lag, the flow rate system takes a large slipstream from the process and tubes it to the analyzer. Because the sample flow to the analyzer is small, the analyzer uses only a small portion of this stream and returns the bulk to the process (see Figure 2.3.1). This arrangement permits the high-flow rate system to continuously sweep the main volume of the filter, minimizing lag time; at the same time, only the low-flow stream to the analyzer is filtered, maximizing filter life.

A slipstream filter requires inlet-to-outlet ports at opposite ends of the filter element to allow the high flow rate of the bypassed material to sweep the surface of the filter element and reservoir. It also requires a third port connected to the low-flow rate line to the analyzer so that filtered samples can be withdrawn from the filter reservoir.

If bubble removal from a liquid is required, this function can be combined with slipstream filtration since the recommended flow direction for bubble removal is outside-to-inside and the separated bubbles are swept out of the housing by the bypass stream. In this case, the liquid feed should enter at the bottom of the housing, and the bypass liquid should exit at the top of the housing.

Some samples can be separated using cyclone separators. In this device (see Figure 2.3.2), the process stream enters tangentially to provide a swirling action, and the cleaned sample is taken near the center. The transportation lag can be kept to less than 1 min, and the unit is applicable to both gas and liquid samples. This type of cen-

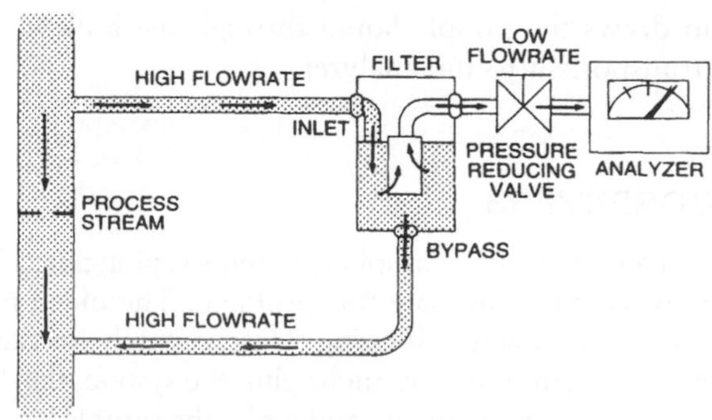

FIG. 2.3.1 Slipstream or bypass filtration.

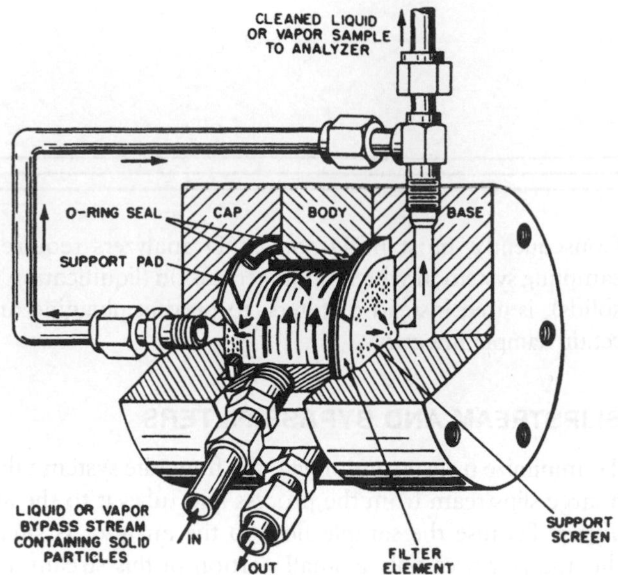

FIG. 2.3.2 Bypass filter with cleaning action amplified by the swirling tangentially entering sample.

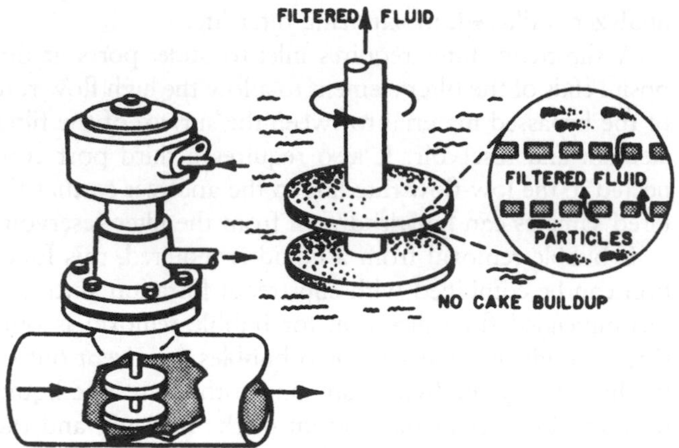

FIG. 2.3.3 Rotary disc filter.

trifuge can also separate sample streams by gravity into their aqueous and organic constituents.

Another good filter design is the rotary disc filter (see Figure 2.3.3). Here, the filtered liquid enters through the small pores in the self-cleaning disc surfaces. The sample pump draws the sample liquid through the hollow shaft and transports it to the analyzer.

HOMOGENIZERS

A frequent problem of sampling systems is plugging. There are two ways to eliminate this problem. The older, more traditional approach is filtering. Unfortunately, as the filters remove materials that might plug the system, they also remove process constituents and make the sample less representative.

The newer approach is to eliminate plugging potential by reducing solid particle size (homogenization) while maintaining sample integrity. Thus, when pulverizers replace filters, analyzer samples become more representative.

Homogenizers disperse, disintegrate, and reduce the solid particle size, reducing agglomerates and liquifying the sample. Homogenizers can be mechanical, using rotor-stator-type disintegrator heads. In this design, the rotor acts as a centrifugal pump to recirculate the slurry, while the shear, impact collision, and cavitation at the disintegrator head provide homogenization.

In ultrasonic homogenizers, high-frequency mechanical vibration is introduced into a probe (horn), which creates pressure waves as it vibrates in front of an orifice (see Figure 2.3.4). As the horn moves away, it creates large numbers of microscopic bubbles (cavities). When it moves forward, these bubbles implode, producing powerful shearing action and agitation due to cavitation. Such homogenizers are available with continuous flow-cells for flow rates up to 4 gph (16 lph) and can homogenize liquids to less than 0.1-μm particle sizes. The flow cell is made of stainless steel and can operate at sample pressures of up to 100 psig (7 bars).

AUTOMATIC LIQUID SAMPLERS

Automatic liquid samplers collect intermittent samples from pressurized pipelines and deposit them in sample containers. Samples are collected on a time-proportional or flow-proportional basis. Figure 2.3.5 shows a sampler that withdraws a predetermined volume of sample every time the actuator piston is stroked. In time-proportional mode,

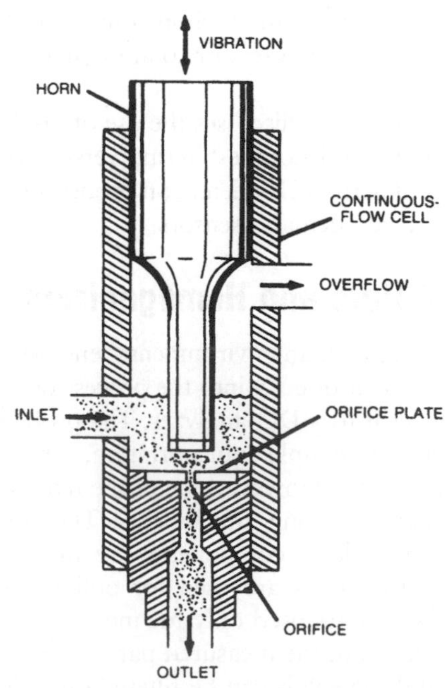

FIG. 2.3.4 Ultrasonic homogenizer. (Reprinted, with permission, from Cole-Parmer Instrument Co.)

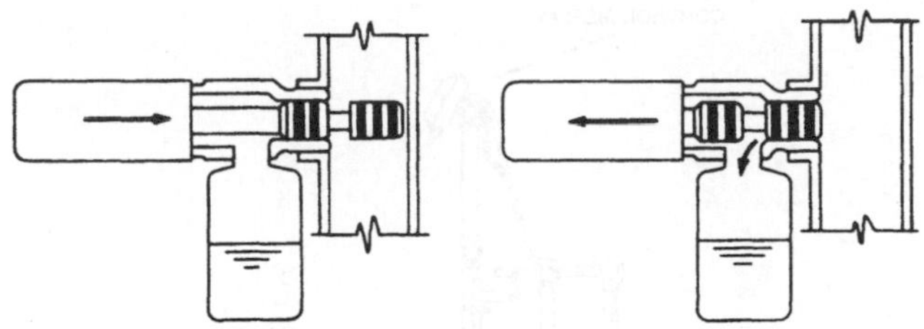

FIG. 2.3.5 Intermittent collection of samples. (Reprinted, with permission, from Bristol Equipment Co.)

this sampling frequency is constant, while in flow-proportional mode, this unit varies sampling frequency as a function of flow.

In some automatic liquid samplers, sampling frequency is adjusted by pneumatic pulse relays or electronic controls. Pulse duration is usually adjustable from 0.25 sec to 1 min, while pulse frequency is adjusted from a few sec up to an hr.

DO Analyzers

TYPES

A. Polarographic; B. Galvanic; C. Coulometric; D. Multiple anode; E. Thallium

OPERATING PRESSURE

Up to 50 psig (3.5 bars) or submersion depths of up to 25 ft (8.3 m)

OPERATING TEMPERATURE RANGE

32 to 122°F (0 to 50°C); special designs up to 175°F (80°C)

FLOW VELOCITY AT SENSING MEMBRANE

Preferably in excess of 1 fps (0.3 m/sec); some can operate down to 0.2 fps (0.06 m/sec).

MATERIALS OF CONSTRUCTION

Typical material for sensor housing is PVC; for electrodes, gold and silver or copper; and for membrane assembly, ABS plastic or stainless steel, mesh-reinforced Teflon membrane.

SPEED OF RESPONSE

90% in 30 sec; 98% in 60 sec

RANGES

Common ranges are 0 to 5, 0 to 10, 0 to 15, and 0 to 20 ppm; special units are available with ranges up to 0 to 150 ppm or down to the 0 to 20 ppb range used on boiler feedwater applications. Systems can also be calibrated in partial pressure units.

INACCURACY

Generally ±1 to ±2% of the span; industrial transmitter errors are generally within 0.02 ppm over a 0 to 20 ppm range. Thallium cells are available with a 0 to 10 ppb range and can read the DO within an error of 0.5 ppb.

COSTS

Portable, battery-operated, 1.5 to 2% FS units that also read temperature cost from \$300 to \$700; replacement probes cost about \$250; 1% FS, microprocessor-based, portable benchtop units for laboratory or plant service cost from \$1000 to \$2000; industrial-quality (0.02 ppm error limit) DO probe and a 4- to 20-mA transmitter cost about \$3500; and cleaning assemblies cost from \$500 to \$2000.

PARTIAL LIST OF SUPPLIERS

ABB Kent Inc.; Ametek Inc.; Cole-Parmer Co.; Delta F Corp.; Electro-Nite Co.; Enterra Instrument Technologies Inc.; Fischer & Porter Co.; Foxboro Co.; Great Lakes Instruments Inc.; Hays Republic Corp.; Honeywell Industrial Controls; Horiba Ltd.; Ingold Electrodes Inc.; Leeds & Northrup Co.; Milton Roy Co.; MTL (B); Ohmart Corp.; Orbisphere Laboratories (Switzerland); Robertshaw Controls Co.; Rosemount Analytical, Uniloc Div.; Royce Instrument Corp.; Teledyne Analytical Instruments; Waltron Ltd.; Yokogawa Corp. of America; YSI Inc.

In wastewater treatment applications, the DO concentration is often measured within the slimy sludge layer. Therefore, the membrane surface must be kept clean. Keeping the membrane surface clean is achieved partly by coating the membrane with a growth-inhibiting chemical, toxic to bacteria, and partly by attaching a vibratory paddle-cleaner, which cleans the membrane by back-and-forth motion close to the surface (see Figure 2.3.6).

POLAROGRAPHIC CELL

The basic polarographic cell shown in Figure 2.3.7 has two noble-metal electrodes and requires a polarizing voltage to reduce oxygen. Sample DO diffuses through the membrane into the electrolyte, which is usually an aqueous KCl solution. If the polarizing voltage is constant (usually 0.8 V supplied by a mercury battery) across the electrodes, the oxygen is reduced at the cathode, and the resulting current flow is directly proportional to the electrolyte oxygen content.

Polarographic cells, like galvanic cells, are affected by temperature. Therefore, they require controlled sample temperature or temperature compensation to attain high-precision measurements of ±1 to 2% in accuracy. If sample temperature varies between 32 and 110°F (0 and 43°C), measurement error rises to approximately ±6% in some designs.

Both galvanic and polarographic cells require a minimum sample flow velocity. This velocity eliminates stagnant layers of sample over the membrane, which would otherwise interfere with continuous oxygen transfer into the cell. Higher sample velocities are also beneficial because of their scrubbing action. Some suppliers provide a combination cell and pump unit where the flow velocity of 5 fps (1.5 m/sec) is directed against the membrane for maximum cleaning effect.

GALVANIC CELL

The ranges of the galvanic-cell DO analyzer are as low as 0 to 20 ppb for applications such as measuring DO content in boiler feedwater.

All galvanic cells consist of an electrolyte and two electrodes (see Figure 2.3.8). Electrolyte oxygen content is

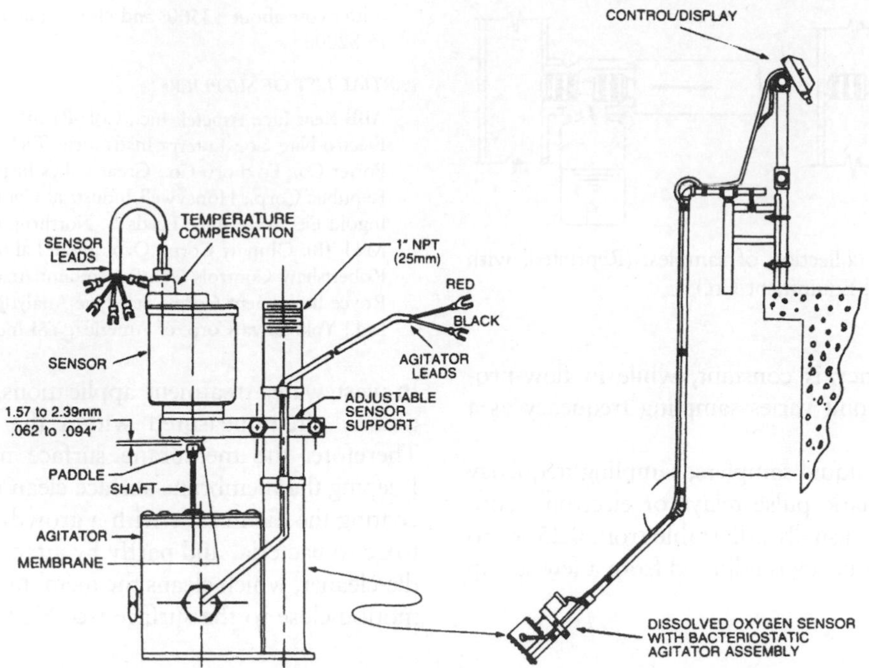

FIG. 2.3.6 Cleaner assembly for freeing the membrane surface of a DO probe of buildup or biological growths. (Reprinted, with permission, from Robertshaw Controls Co.)

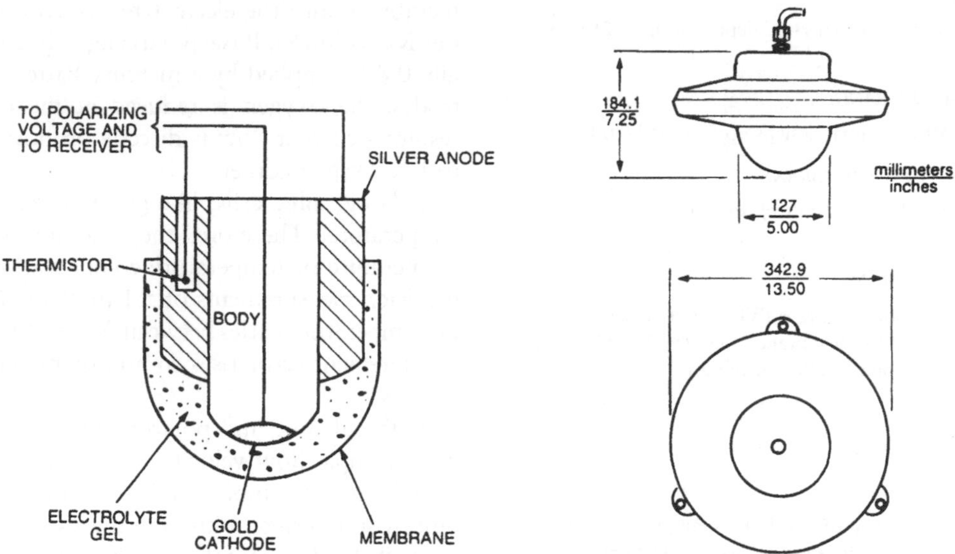

FIG. 2.3.7 Probe-type polarographic cell oxygen detector and flotation collar mount.

brought into equilibrium with that of the sample. The electrodes are polarized by an applied voltage that causes electrochemical reactions when oxygen contacts the electrodes. In this reaction, the cathode reduces oxygen into hydroxide, thus releasing four electrons for each molecule of oxygen. These electrons cause a current flow through the electrolyte, with magnitude in proportion to the electrolyte oxygen concentration.

The following gases are likely to contaminate the cell: chlorine and other halogens, high concentrations of carbon dioxide, hydrogen sulfide, and sulfur dioxide.

Special cells have been developed to minimize the effect of background gases. When an acid gas (such as CO_2) that would neutralize a potassium hydroxide electrolyte solution is present in the background, a potassium bicarbonate electrolyte can be used. Special cells are also available

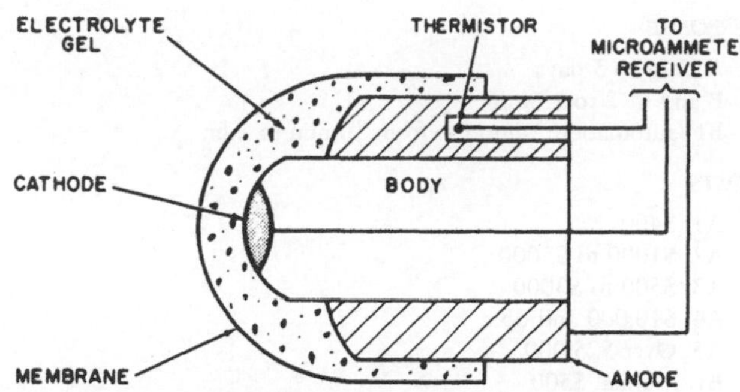

FIG. 2.3.8 Probe-type galvanic cell oxygen detector.

for measuring oxygen in acetylene and fuel gases.

In flow-through cell designs, sampling systems bring the process stream to the analyzer and filter it, scrub it with caustic, or otherwise prepare it for measurement. The probe-type membrane design does not require a sampling system if it can be located in a representative process area where process stream pressure, temperature, and velocity are compatible with the cell's mechanical and chemical design.

Probe Design

In this design (see Figure 2.3.8), the electrodes are wetted by an electrolytic solution retained by a membrane (usually Teflon). This membrane acts as a selective diffusion layer, allowing oxygen to diffuse into the sensor while keeping foreign matter out. The sensor is usually mounted in a thermostatically controlled housing; therefore, the thermistor compensates for minor temperature variations.

Membrane characteristics are critical to performance. The ideal membrane is inert, stable, strong, permeable to oxygen, and impermeable to other ions and water molecules. In most cases, a compromise solution is accepted.

Figure 2.3.9 shows the design of a gold–copper electrode, galvanic (amperometric) cell and its rail mounting installation. The maintenance requirements of this design

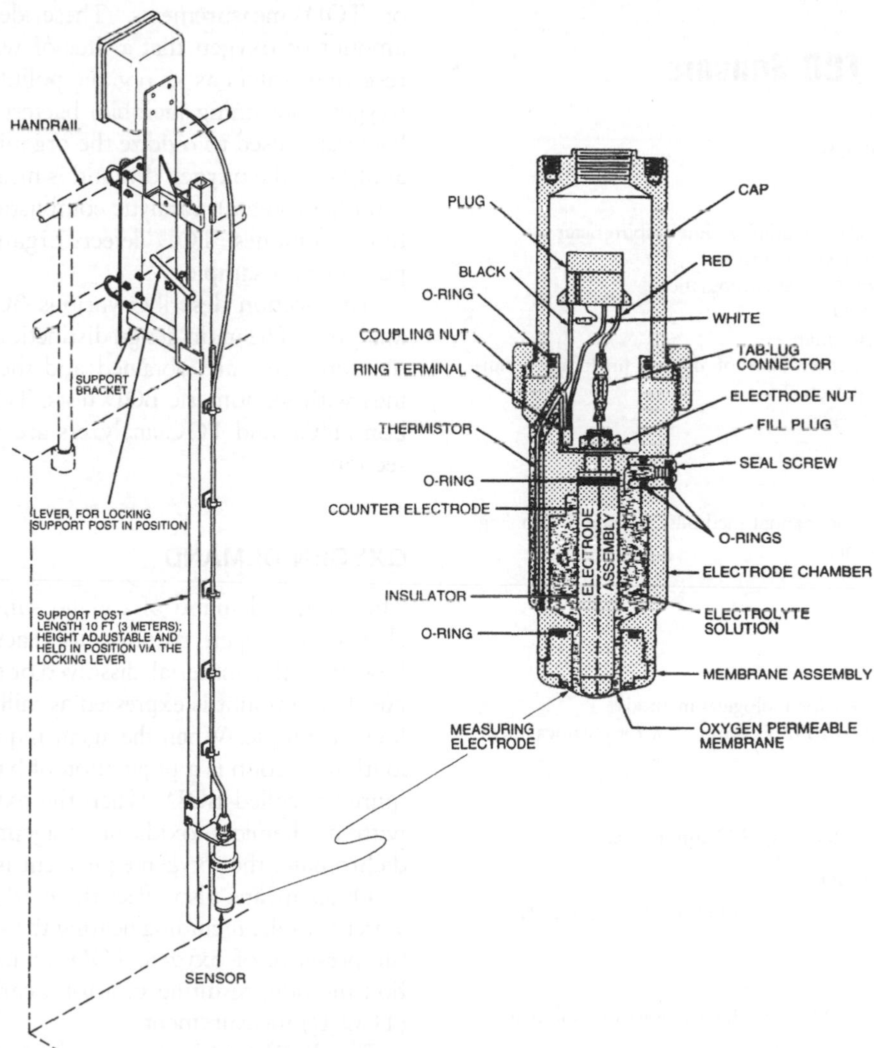

FIG. 2.3.9 Galvanic DO cell and rail mounting installation design. (Reprinted, with permission, from Fischer & Porter Co.)

are reduced with an electrolyte supply that lasts for 2 to 3 years and an easily replaceable membrane assembly. These analyzer systems are available in weatherproof housing, 1% of span inaccuracy, and with 4 to 20 mA of transmitter output.

Flow-Through Design

In these cells, the process sample stream bubbles through the electrolyte. Therefore, the oxygen concentration of the electrolyte is in equilibrium with the sample oxygen content, and the resulting ion current between electrodes represents this concentration.

In some trace analyzer designs, the cathode is made of a porous metal, and the sample gas passes through this electrode, immersed in the electrolyte. Oxygen reduction is usually complete within the pores of this electrode.

Sampling systems are usually provided with these cells, consisting of filtering and scrubbing components and flow, pressure, and temperature regulators.

BOD, COD, and TOD Sensors

TYPES OF MEASUREMENTS

A. Biological agency (BOD)
A1. Winkler titration
A2. DO sensor
A3. Manometric methods (including online respirometer)
A4. Coulometric (electrolysis) methods
A5. BOD for eleven samples, semiautomatic
B. Chemical agency (COD)
B1. Oxidation with dichromate
B2. Combustion (catalytic) with carbon dioxide (including nondispersive IR [NDIR] detector)
B3. Combustion with oxygen
C. TOD

SAMPLING TECHNIQUE

A and B. Grab samples for manual methods; automatic sampling for continuous instruments

SAMPLE PRESSURE

Essentially atmospheric

SAMPLE TEMPERATURE

A. 20°C (68°F) during test for biological methods
B. 150 to 1000°C (302 to 1832°F) during test for chemical methods

SUSPENDED SOLIDS

Sample can contain particles up to 200 μm in size.

MATERIALS OF CONSTRUCTION

A and B. Glass, quartz, Teflon, polyethylene, Tygon, and PVC

RANGES

A and B. 0.1 mg/l and up
C. Standard 0–100 to 0–5000 ppm; higher ranges by dilution

INACCURACY

A and B. 3 to 20% depending on method
C. ±2% of range at the 95% confidence level

RESPONSE

A. 2 hr to 5 days
B and C. 2 to 15 min
B1 (automatic). Adjustable from 10 min to 2 hr

COSTS

A1. $400
A2. $1000 to $5000
A3. $500 to $3000
A4. $10,000 and up
A5. Over $25,000
B1. Manual $500
B2. $18,000
C. Over $15,000

PARTIAL LIST OF SUPPLIERS

Anatel Corp.; Badger Meter Inc. (A); Bran and Luebbe Analyzing, Technicon Industrial Systems (B); Horiba Instruments Inc. (A); Ionics Inc. (B,C); Robertshaw Controls Co. (A); Tech-Line Instruments, Div. of Artech International Inc. (A); Xertex Corp. Delta Analytical (A); YSI Inc. (A)

The total damage caused by discharging wastewater into lakes or rivers is expressed and quantified by BOD, COD, or TOD measurements. These detectors measure the amount of oxygen that a liter of wastewater takes from receiving waters as its organic pollutants are degraded by oxygen-consuming (aerobic) bacteria. In BOD analyzers, bacteria is used to oxidize the organic pollutants; in COD analyzers, the oxygen demand is measured through chemical (dichromate), catalytic combustion, or direct combustion techniques. TOD detects organic and inorganic impurities in a sample.

This section describes various BOD, COD, and TOD analyzers. The main design distinction is the speed at which measurements are obtained and the correlation of readings with manomeric BOD tests. TC, total inorganic carbon (TIC), and TOC analyzers are described later in this section.

OXYGEN DEMAND

The oxygen demand of a water sample is the amount of elemental oxygen required to react with oxidizable or biodegradable material, dissolved or suspended in the sample. This amount is expressed as milligrams of oxygen per liter of sample. When the agent required to effect the oxidation reaction is a population of bacteria, the oxygen required is called BOD. When the oxidation is carried out with a chemical oxidizing reagent such as potassium dichromate, the oxygen equivalent is called the COD.

Other means also effect the oxidation of material in a water sample, including heating the sample in a furnace in the presence of oxygen, TOD, or in the presence of carbon dioxide, resulting in a total carbon dioxide demand (TCO_2D) measurement.

The BOD test is the most important oxygen demand measurement for analyzing effluents and receiving waters (streams, lakes, and rivers). The BOD test measures the amount of oxygen used by microorganisms feeding on organic water pollutants under aerobic conditions.

In this test, a bacterial culture is added to the sample under well-defined conditions, and oxygen utilization is measured. Although test procedures are carefully defined, obtaining reproducible results is difficult and the procedure is subject to the influence of many variables, particularly when the wastewater contains a variety of complex materials.

Factors contributing to variations in BOD results are:

- Biological seed used
- pH if not near neutrality
- Temperature if other than 20°C (68°F) (as shown in Figure 2.3.10)
- Sample toxicity
- Incubation time

When the incubation time and temperature are not stated, the general assumption is that the test was run at 20°C for a period of 5 days. For example, in Figure 2.3.10, the BOD test result can be stated as $BOD_5 = 100$ mg/l. Figure 2.3.10 also shows that bacteria first consume carbonaceous material, and only when the carbonaceous materials are all oxidized (around the 15th day) are nitrogenous material consumed.

Five-Day BOD Procedure

If a water sample BOD at 20°C (68°F) is measured as a function of time, a curve such as the one in Figure 2.3.10 is obtained. For the first 10 to 15 days, the curve is approximately exponential, but at about the fifteenth day, a sharp increase occurs that then falls to a steady BOD rate. Because of the length of time and because the curve does not flatten, environmental engineers have universally adopted a standard test period of 5 days for the BOD procedure. This laboratory procedure requires some skill and training to obtain concordant results.

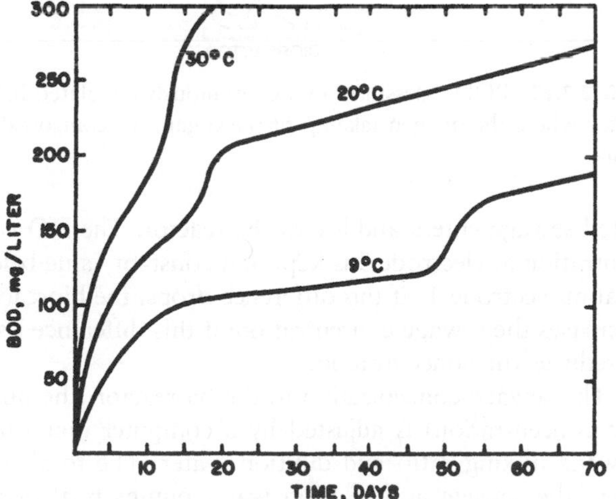

FIG. 2.3.10 Progress of BOD at 9°, 20°, and 30°C (48, 68, and 86°F). The break in each curve corresponds to the onset of nitrification.

In the procedure, the environmental engineer mixes a measured sample portion to be analyzed with seeded dilution water so that after 5 days of incubation, the DO in the mixture is still sufficient for biological oxidation of the material in the sample. Of course, this portion cannot be known beforehand; consequently, the environmental engineer must run several dilutions simultaneously for an unknown sample, or use experience as a guide for well-defined samples. Seeded dilution water contains a phosphate buffer (including ammonium chloride), magnesium sulfate, calcium chloride, and ferric chloride, as well as a portion of seeding material. The former group of inorganic materials is frequently referred to as nutrients. The latter group is a suspension of bacteria in water, usually supernatant liquor from a domestic sewage plant.

Seeds can also be prepared from soil, developed from cultures in the laboratory, or obtained from receiving water 2 to 5 mi downstream of the discharge. The environmental engineer determines the DO content of the mixture at the start of the test and again after 5 days of incubation at 20°C in a special BOD bottle. The DO can be determined by the Winkler titration method or instrumentally with a DO membrane electrode. The difference in DO after 5 days is used to calculate the BOD of the original sample. Corrections must be applied for the immediate oxygen demand (due to inorganic reducing materials) and for the oxygen required by the bacteria for sustaining life (endogenous metabolism).

No standard exists for measuring the accuracy of the BOD test. The precision of the method is also difficult to ascertain because of the many variables. However, environmental engineers have tested the single-operator precision of the method using a standard glucose–glutamic acid solution. Using eight types of seed materials, they found the single-operator precision to be 11 mg/l at a level of 223 mg/l, or about 5%. These results were obtained with highly skilled personnel under well-controlled laboratory conditions.

A semiautomatic instrument can measure the BOD of as many as eleven samples. The samples must be manually placed on the instrument turntable, and the controls must be manually set. The instrument provides for the automatic reaeration of samples in which the DO has fallen to low values. The polarographic DO sensor measures the DO on a preset time schedule. The automatic reaeration capability for low DO eliminates the need for dilution, leading to improved precision in BOD results.

The instrument consists of a measuring unit (DO probe, aerator, water-sealing mechanism, unplugging mechanism, sample bottle, and turntable) and a control unit by which all operations are programmed. The measuring unit is housed in a chamber maintained at 20°C. The instrument can store DO data on each sample and calculate the BOD from the DO values as just described.

Manometric BOD Test

In the standard dilution method, all oxygen required must be inside the BOD bottle since it is sealed in a gas-tight manner at the initiation of the incubation period and air can not enter the sample. In the manometric procedure, the seeded sample is confined in a closed system that includes an appreciable amount of air. As the oxygen in the water is depleted, it is replenished by the gas phase. A potassium hydroxide absorber within the system removes any gaseous carbon dioxide generated by bacterial action. The oxygen removed from the air phase causes a drop in pressure that is measured by a manometer. This drop is then related to sample BOD.

The manometric method lends itself to automatic recording of oxygen utilization since the pressure can be monitored continuously. This monitoring is accomplished in a commercially available automatic respirometer (see Figure 2.3.11). The manometric method introduces the sample, from 1 to 4 liters, into a closed system containing air. The countercurrent circulation of both air and water in the system insures equilibrium between the dissolved and gaseous oxygen. The system provides a carbon dioxide scrubber in the gas-circulation line. A recording manometer detects the utilization of oxygen, and the test is run for several hours. Published data indicate a correlation between 4-hr respirometer BOD and standard BOD. Laboratory and automatic online versions of this instrument are also available.

BOD measurement is inherently a time-consuming process and ill-suited to the requirements of process monitoring or control. The shortest period for the automatic respirometer is 2 hr, much too long for an effective control instrument. However, it is an excellent device for laboratory studies since it can simulate the activated sludge process.

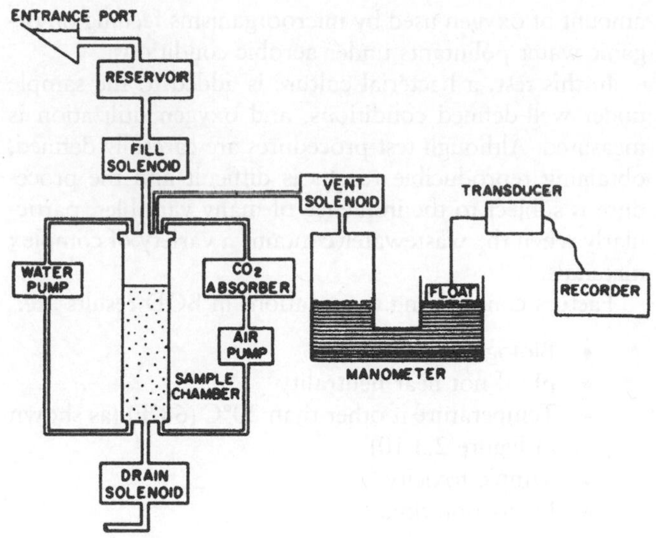

FIG. 2.3.11 BOD determination by an automatic respirometer.

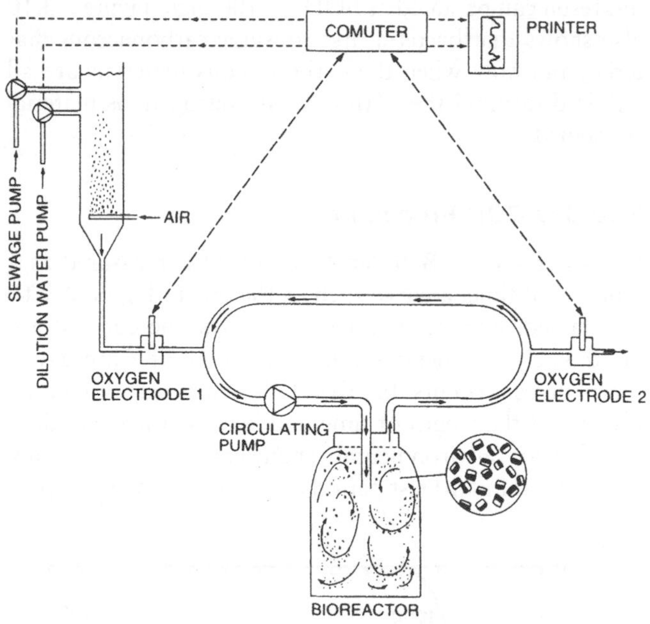

FIG. 2.3.12 BOD assessment in a continuously circulated bioreactor where the oxygen take-up of the organisms controls dilution.

BOD Assessment in Minutes

When BOD concentration is determined in groundwater, waiting 5 days for the results may be acceptable. However, in the control and operation of sewage treatment plants, it is not. The hold-up capacity of industrial and municipal wastewater treatment facilities and the need for closed-loop control necessitates faster sensors. Figure 2.3.12 shows such an analyzer.

In this design, a bioreactor is filled with several plastic rings, the interior of which are protected against mechanical abrasion to provide a growth surface for organisms. A circulation pump quickly distributes the sewage in the bioreactor and keeps the plastic rings in continuous motion. The sewage concentration (nutrient level) in the reactor is at a constant low value, resulting in an oxygen demand of about 3 mg/l. The bioreactor measures and maintains a constant oxygen demand by detecting a decrease in the oxygen concentration at points where the diluted sewage enters and leaves the reactor. The DO concentration at electrode 2 is kept at a constant value below that at electrode 1. If this difference drops, the bioreactor increases the sewage concentration; if this difference rises, it reduces the concentration.

The sewage concentration in the bioreactor (the nutrient concentration) is adjusted by a computer that varies sewage mixing ratio and dilution water. The total flow from the sewage and dilution water pumps is always 1 l/min, and the ratio of the two streams is modulated. Therefore, this pumping ratio indicates sewage sample BOD concentration. Environmental engineers have found the correlation between this fast BOD measurement and the 5-day BOD obtained through conventional methods acceptable. Pipe fouling was minimal, and weekly recalibration of oxygen electrodes was satisfactory.

COD

This laboratory method requires skill and training similar to that required for the BOD test. The environmental engineer heats a sample to its boiling point with known amounts of sulfuric acid and potassium dichromate. Using a reflux condenser minimizes the loss of water. After 2 hr, the environmental engineer cools the solution and determines the dichromate amount that reacted with the oxidizable material in the water sample by titrating the excess potassium dichromate with ferrous sulfate using ferrous 1,10-phenanthraline (ferroin) as the indicator. The environmental engineer calculates the dichromate consumed as to the oxygen equivalent for the sample and reports it as oxygen mg/l of the sample.

Interpretations of COD values are difficult since this method of oxidation is markedly different from the BOD method. Although ultimate BOD values can agree with COD values, a number of factors can prevent this concordance including:

1. Many organic materials are oxidizable by dichromate but not biochemically oxidizable and vice versa. For example, pyridine, benzene, and ammonia are not attacked by the dichromate procedure.
2. A number of inorganic substances such as sulfide, sulfites, thiosulfates, nitrites, and ferrous iron are oxidized by dichromate creating an inorganic COD that is misleading when the organic content of wastewater is estimated. Although the seed acclimation factor gives erroneously low results on BOD tests, COD results do not depend on acclimation.
3. Chlorides interfere with the COD analysis and their effect must be minimized for consistent results. The standard procedure provides for a limited amount of chlorides in the sample. Despite these limitations, the dichromate COD is useful in the control of wastewater effluents containing caustic and chlorine, dyeing and textile effluents, organic and inorganic chemicals, paper, paints, plating, plastics, steel, aluminum, and ammonia.

COD Detector

The COD method usually refers to the laboratory dichromate oxidation procedure. It is also applied to other procedures that differ from the dichromate method but involve chemical reaction. These methods are embodied in instruments both for manual operation in the laboratory and for automatic operation online. They have the advantage of reducing the analysis time from days (5-day BOD) and hours (dichromate and respirometer) to minutes.

Automatic Online Designs

Figure 2.3.13 shows an online analyzer with COD ranges from 0–100 ppm to 0–5000 ppm and adjustable measurement cycle times from 10 min to 5 hr. The sample flow can be continuous at rates to 0.25 gpm (1.0 lpm) and can contain solid particles to 100 μ.

The automatic COD analyzer periodically injects a 5-cc sample from the flowing process stream into the reflux chamber, after mixing it with dilution water (if any) and two reagents: dichromate solution and sulfuric acid. The reagents also contain an oxidation catalyst (silver sulfate) and a chemical that complexes chlorides in the solution (mercuric sulfate). The heater boils the mixture at 302°F (150°C), and the cooling water in the reflux condenser recondenses the vapors. The solution is refluxed for a preset time during which the dichromate ions are reduced to trivalent chromic ions as the oxygen-demanding organics are oxidized in the sample.

The chromic ions give the solution a green color. The environmental engineer measures the COD concentration from the amount of dichromate converted to chromic ions by measuring green color intensity through a fiber-optic detector. The microprocessor-controlled package is available with automatic zeroing, calibration, and flushing features.

In one instrument, a 20-μl water sample is manually injected into a carbon dioxide carrier stream and swept

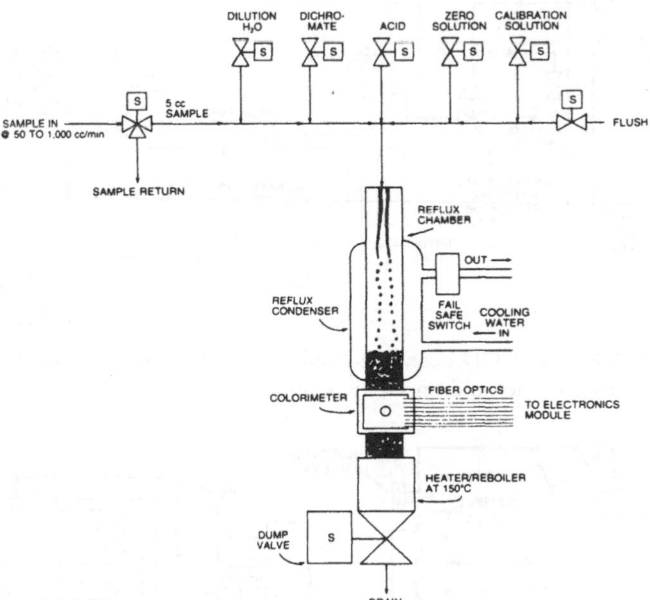

FIG. 2.3.13 Automatic COD analyzer. This analyzer uses a dichromate reagent and fiber-optic colorimeter detector and provides features of adjustable reflux time and autocalibration. (Courtesy of Ionics Inc.)

through a platinum catalyst combustion furnace. In this furnace, pollutants are oxidized to carbon monoxide and water, and the water is removed from the stream by a drying tube. Then, the reaction products receive a second platinum catalytic treatment. An NDIR detector then measures carbon monoxide concentration. The environmental engineer can convert the readings to COD using a calibration chart. An analysis can be completed in 2 min. This instrument is commercially available for manual operation (see Figure 2.3.14). Data obtained on domestic sewage indicate excellent correlation between this method (frequently called CO_2D) and the standard COD method.

TOD

The TOD method is based on quantitative measurement of the oxygen used to burn the impurities in a liquid sample. Thus, it is a direct measure of the oxygen demand of the sample. Measurement is by continuous analysis of the oxygen concentration in the combustion gas effluent (see Figure 2.3.15).

The TOD analyzer converts oxidizable components in a liquid sample in its combustion tube into stable oxides by a reaction that disturbs oxygen equilibrium in the carrier gas stream. An oxygen detector detects momentary depletion in the oxygen concentration in the carrier gas and records it as a negative oxygen peak on a potentiometric recorder. The analyzer obtains sample TOD by comparing recorded peak heights to peak heights of standard TOD calibration solutions, e.g., potassium acid phthalate (KHP).

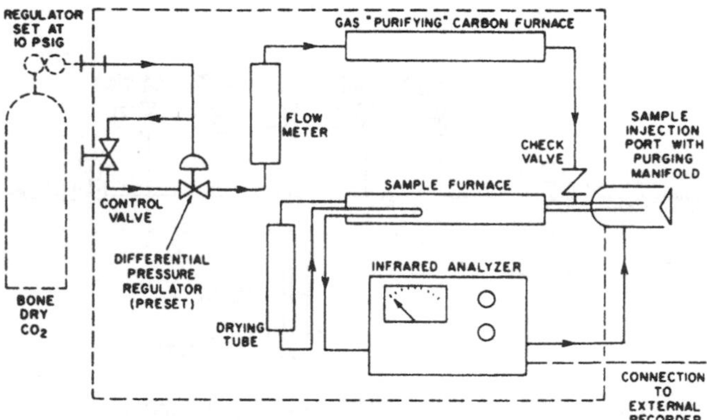

FIG. 2.3.14 COD detection employing combustion in a carbon dioxide carrier and an NDIR sensor.

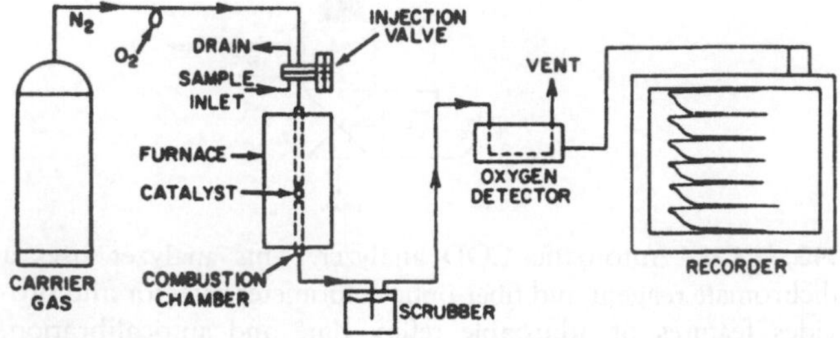

FIG. 2.3.15 Basic components of a TOD analyzer.

Prepurified nitrogen from a cylinder passes through a fixed length of oxygen permeable tube into the combustion chamber, gas scrubber, and oxygen detector. The environmental engineer can vary the baseline oxygen concentration, determined as the nitrogen passes through the temperature-controlled permeation tube, to accommodate different TOD ranges by changing the nitrogen flow rate.

The combustion chamber is a length of Vicor tubing or quartz tube containing a platinum catalyst mounted in an electric furnace and held at a temperature of 900°C (1652°F). The aqueous sample is injected into this chamber and the combustible components are oxidized.

CORRELATION AMONG BOD, COD, AND TOD

Many regulatory agencies recognize only the BOD or COD measurements of the pollution load as the basis for pollution control. They are concerned with the pollution load on the receiving water, which is related to lowering the DO due to bacterial activity. Thus, if environmental engineers use other methods to satisfy the legal requirements of pollution load in effluents or measure BOD removal, they need an established correlation between the other methods and BOD or COD (preferably BOD).

A correlation of the various methods begins with the assumption that the BOD is the standard reference method. The salient features of this method are (1) a property measurement of the sample, i.e., the amount of oxygen required for bacterial oxidation of the bacterial food in the water, the BOD; (2) the dependence of oxygen demand on the nature of the food as well as on its quantity; and (3) the dependence of oxygen demand on the nature and amount of the bacteria.

The variation in OD due to variation in the amount (lb/gal) of food in the wastewater is expected; variation in OD when the amount of food is constant but changes occur in BOD requirements is difficult to predict. The same observations apply to the bacterial seed. Thus, variation in OD due to variation in the number or activity of bacteria, or changes in the nature of bacterial food leads to systematic or bias errors in BOD measurement that cannot be predicted or corrected for.

Therefore, the standard reference method is inherently variable and subject to analytical error. Researchers in an interlaboratory comparative study employing a synthetic waste found standard deviations around the mean of ±20% for BOD and ±10% for COD.

Another extensive study (Ford, Eller, and Gloyna 1971) made the following conclusions:

1. A reliable statistical correlation between wastewater BOD and COD and the corresponding TOC or TOD

can frequently be achieved, particularly when organic strength is high and diversity in dissolved organic constituents is low.
2. The relationship is best described by a least squares regression with the degree of fit expressed by the correlation coefficient—this relationship applies to the characterization of individual chemical-processing and oil-refining wastewaters, not to all types of samples across the board.
3. The observed correspondence COD–TOD was better than COD–BOD for the wastewater mentioned (generally, correlating BOD with TOD was difficult, particularly when the wastewater contained low concentrations of complex organic materials).
4. The BOD–COD or BOD–TOC ratios of untreated wastewater indicate the biological treatment possible with wastewater. As these ratios increase, higher organic removal treatment efficiencies occur by biological methods.

Several papers indicate high correlation between BOD and other methods. This correlation is achieved when the nature of the pollutant is constant and only its amount changes. For complex and varying mixtures, obtaining good correlations is difficult.

An interesting example is given in the work of Nelson, Lysyj, and Nagano (1970), who discuss a pyrolytic method combined with flame ionization detection (FID). Values from the new method agreed with BOD values within ±15% for BOD values greater than 100 ppm on raw sewage and primary effluent. However, they found discrepancies of several hundred percent when the BOD was 20 ppm or less. These poor results can be attributed to a marked variation in biodegradability of carbonaceous products in the secondary effluent compared with the products before treatment as well as to the small amount of total material left.

Total Carbon Analyzers

METHODS OF DETECTION

A. NDIR; B. FID; C. Aqueous conductivity with ultraviolet (UV) irradiation; D. Colorimetry.

SAMPLES

Laboratory samples range from 0.01 to 10 cc. For in-line applications, sample flow rates range from 0.25 to 30 cc/min on continuous units. When a continuous bypass sample is used on cyclically analyzed injection samples, the bypass flow is from 50 to 1000 cc/min.

FLOWING SAMPLE SOLIDS CONTENT

Up to 1000 mg/l; size of particles up to 200 μ in diameter

MATERIALS OF CONSTRUCTION

Glass, quartz, Teflon, stainless steel, Hastelloy, polyethylene, and PVC

MEASUREMENT CYCLE TIME

A and B. 2 to 15 min (for type A, TC requires 2.5 min, and TOC requires 6 min)
C. Continuous with speed of response of 90 sec

UTILITIES OR REAGENTS REQUIRED

Air (10 atmospheric cubic feet per hour [acfh], or 4.6 alpm), oxygen, nitrogen (carrier gas flow is 100 cc/min at 50 psig, or 3.5 bars), hydrogen, mineral acid (1.0 gal per month of sulfuric or phosphoric), oxydizing reagent, buffer, and cooling water

RANGES

A and B. 0–2 ppm to 0–30,000 ppm (0 to 3%)
C. 0–100 ppb to 0–1 ppm

SENSITIVITY

A and B. 0.1 ppm or 0.5% of full-scale, whichever is greater

INACCURACY

2 to 5% full-scale as a function of design, sample size, and range

COSTS

$10,000 to $20,000; NDIR TC is $20,000, and NDIR TOC with differential detectors is $25,000.

PARTIAL LIST OF SUPPLIERS

Anatel Corp., Astro International (A); Bran and Luebbe Analyzing, Technicon Industrial Systems (D); Ionics (A,C); Rosemount Analytical Inc. Dohrmann Div. (A,B); Xertex Corp. Delta Analytical (C)

The damage done by discharged wastewater treatment effluent to the environment is a function of its organic loading because organic molecule decomposition consumes the DO of the receiving waters. BOD analysis measures all molecules that exert an oxygen demand, however it is slow, and readings vary depending on the bioassay used. COD analysis is faster but is affected by variations in chemical oxidation efficiencies. Carbon analyzers (TC and TOC) further increase speed but do not detect the load represented by nitrogen-based molecules. A direct correlation between BOD, COD, and TOC is not possible. Yet, TOC provides a rapid and reasonably accurate indication of pollution levels.

TOC, TC, AND TIC

To determine the TOC content of a sample, an environmental engineer must use one of two techniques to eliminate the TIC usually present but of little or no interest to the TOC analysis. The TIC in a water sample is usually in the form of inorganic bicarbonates and carbonates. One technique analyzes these components independently and then subtracts them from the TC. The TOC is then determined by the difference between TC and TIC (TC − TIC = TOC).

The other technique acidifies the sample to a pH of 2 to 3 followed by a brief gas sparging to drive off the carbon dioxide formed by the acidification. Any carbon remaining after the sparging should be TOC. Thus, the TC in the sparged sample is equal to the TOC content of the sample. A weakness in this technique is the possible loss of some VOCs that may be in the sample. Further techniques can account for these VOCs.

Development of TOC Analyzers

The TOC method was introduced in 1964 as a single-channel TC analyzer using a catalytic oxidation combustion technique followed by analysis of the resulting carbon dioxide. This method removed inorganic carbon (IC) by acid sparging or determined its concentration by titration. A few years later a second channel was added to the method that permitted parallel determination of IC in a second heated-reaction chamber. Several other techniques have since appeared with various changes in methodology and detection.

Because of the rapid acceptance and usefulness of TOC analysis as a laboratory method, online TOC analyzers became available in the late 1960s. Their success was limited by the relative complexity of these continuous analyzers. Today, there are a half dozen distinctly different methodologies and means of detection for TOC analysis.

Automatic Online Design

Figure 2.3.16 shows the catalytic-oxidation-type TOC analyzer for continuous online operation. This design incorporates an acid injection system that converts IC into carbon dioxide and removes it in a sparging chamber before it reaches the analyzer. If this sparging portion of the sampling system is left off, the analyzer becomes a TC analyzer. If a second, low-temperature reaction chamber is added in parallel with the one shown, the analyzer becomes a TOC differential analyzer.

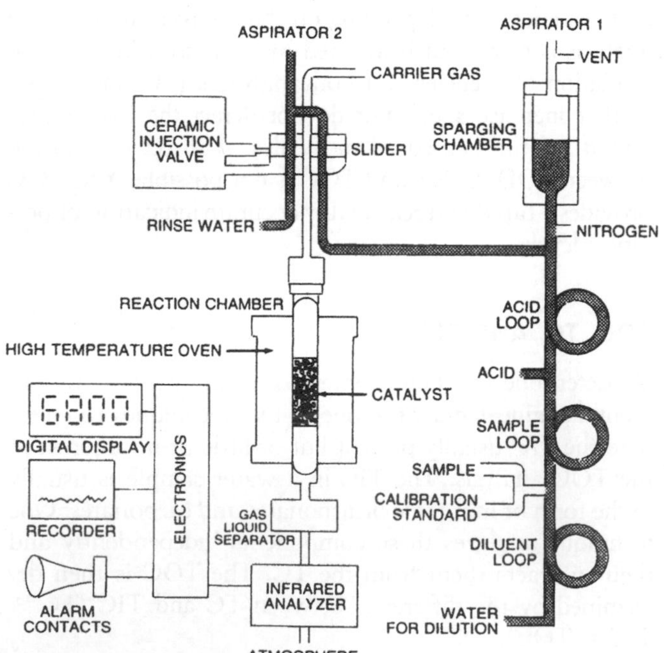

FIG. 2.3.16 Online, NOIR-type TOC analyzer. (Courtesy of Ionics Inc.)

The cycle time of this instrument is 2.5 min in the TC mode and 6 min in the TOC mode. It can operate unattended indoors or outdoors in an analyzer house. It can handle samples with solids content to 1000 ppm and particle sizes to 200 μ because an automatic water rinse is applied after each measurement cycle.

The design of the ceramic injection valve guarantees the accuracy of the sample size. The calibration of the instrument is automatically checked each time a known standard is introduced. During autocalibration, the analyzer runs three consecutive calibration standards, averages the results, and adjusts instrument calibration within preset limits or activates an alarm. The analyzer also has a dilution and automatic range change capability for concentrations that exceed the operating range. Carrier gas (nitrogen) consumption is 100 cc/min at 50 psig (3.5 bars), while acid consumption in the TOC mode is about 1.0 gal (3.8 l) per month.

FID

In the FID analyzer, a small acidified sample is transported in the presence of an oxidizer through a heated vaporization zone. Here, the IC, in the form of CO_2 plus any VOC, is driven off. The residual sample is sent through a pyrolysis zone to convert the remaining TOC to CO_2. The CO_2 is subsequently converted to methane in a nickel-reduction step. An FID detector measures the resulting methane.

The VOC is separated from the CO_2 in a bypass column, reduced to methane, and routed to the same FID for an additional VOC analysis to be added to the dissolved organic carbon value.

Figure 2.3.17 shows another method that uses the FID to analyze the VOC directly after the TIC (CO_2) is removed. In this method, catalytic oxidation combustion is replaced by a wet oxidation method. This method adds persulfate to the sample and exposes the solution to UV radiation to enhance oxidation efficiency. The resulting CO_2 is sparged and converted in the nickel-reduction methanator, and its concentration is measured in the FID analyzer. This wet oxidation technique is available as an online analyzer.

AQUEOUS CONDUCTIVITY

Another method employs wet oxidation and UV irradiation of the sample, which is contained in a recirculating

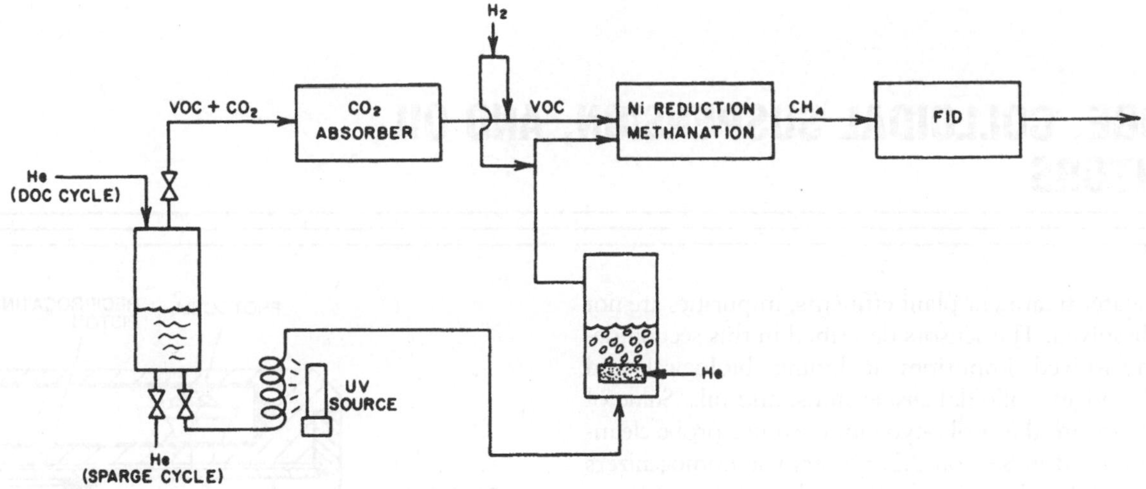

FIG. 2.3.17 FID analyzer with wet oxidation.

stream of demineralized water. A conductivity cell located in this stream measures the increase in conductivity due to the CO_2 resulting from the TOC. A larger sample of water is acidified and sparged with the carrier air. As CO_2 is driven from the sample due to the TIC, it is dissolved in 18.3 mΩ-cm demineralized water. The resulting increase in conductance becomes the new baseline for the next step,

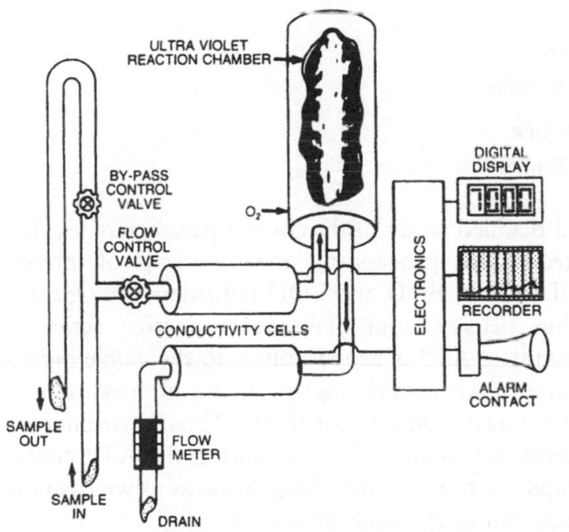

FIG. 2.3.18 Differential-conductivity-type online organic carbon analyzer capable of detecting ppb levels of TOC in high-purity water. (Reprinted, with permission, from Ionics Inc.)

which entails oxidation of the TOC remaining in the water.

This oxidation is accomplished by UV radiation which oxidizes the TOC to CO_2. The added CO_2 raises conductivity to a logarithmically higher level in proportion to the TOC present. The water is then automatically demineralized as it passes through an ion exchange resin bed to prepare it for the next analysis.

This method is best suited for the measurement of low-TOC levels in solids-free samples, such as in drinking water. This method claims sensitivities in the ppb range.

Figure 2.3.18 shows a high-sensitivity, online TOC analyzer used in boiler feedwater; condensate return; or semiconductor, nuclear, or pharmaceutical plant water supply applications where ultrapure water is required. This analyzer takes a 25 cc/min continuous sample from the process water, mixes it with oxygen, and irradiates it with UV light in the reaction chamber. In the presence of oxygen and catalyzed by the UV radiation, the carbon molecules are oxidized into carbon dioxide. As the carbon dioxide is dissolved in the water, its conductivity increases. The analyzer interprets the difference between the conductivities of inlet and outlet water as a measure of the TOC.

This analyzer is continuous, fast (90 sec response time), and sensitive. Its span can be as narrow as 0 to 100 ppb. The sample can contain solids, but their particle size must be under 200 μ.

—Béla G. Lipták

2.4 SLUDGE, COLLOIDAL SUSPENSION, AND OIL MONITORS

In wastewater treatment plant effluents, impurities are not always dissolved. The sensors described in this section detect nondissolved impurities including biological and chemical sludges, colloidal suspensions, and oils. Some of these sensors are the probe-type and require probe cleaners, as discussed in Section 2.2. Others use homogenizers (see Section 2.3), which liquify the solids and simplify sample handling in sample-based analyzers.

SS and Sludge Density Sensors

Figure 2.4.1 shows a probe-type SS detector widely used in biological sludge applications. The probe contains a reciprocating piston. This piston expels a sample (every 15 to 40 sec) during its forward stroke while wiping clean the optical glass of its internal measurement chamber, and pulls in a fresh sample during its return stroke. This device measures the total attenuation of light, which in biological sludge applications is mostly due to SS. Due to its self-cleaning capabilities, cleaning and maintenance are minimal.

This unit is available with a 4- to 20-mA transmitter output that is updated every 15 to 40 sec and has a full-scale inaccuracy of 5% over SS ranges between 0 to 0.1% and 0 to 10%.

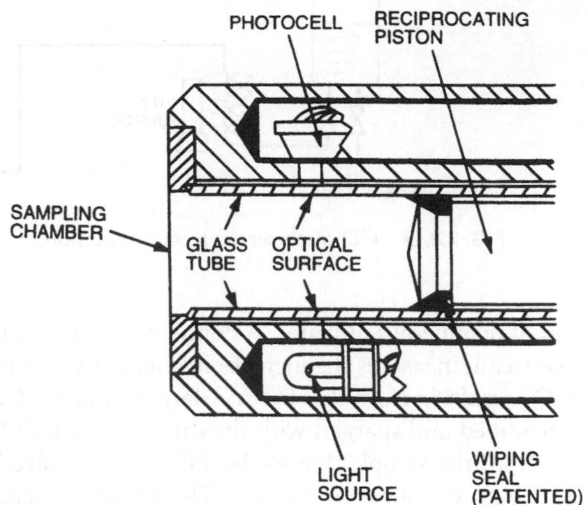

FIG. 2.4.1 Probe-type SS detector used on biological sludge applications. (Courtesy of Monitek Technologies Inc.)

Activated Sludge Monitors

METHOD OF DETECTION

Photometric measurement of light emitted by chemical reaction

SAMPLE PRESSURE

Atmospheric

SAMPLE TEMPERATURE

Ambient

SAMPLE TYPE

Grab sample

MATERIALS OF CONSTRUCTION

Glass

RANGE

10^{-7} to 10^{-2} μg adenosine triphosphate (ATP) per 10 ml sample of bacterial extract. Sensitivity to 10^{-7} μg per 10 μl sample. Calibratable for number of bacteria per μg ATP

RESPONSE

Laboratory method: minutes after starting reaction

COST

$10,000

SUPPLIER

Du Pont Instruments

In a detailed study of the control parameters for the activated sludge process, the measurements of interest are BOD, COD, BOD and COD reduction, biological population density, and biological oxidative activity. The amount of ATP is proportionate to the viable biomass in a sample, whereas changes in ATP concentration measure the biomass oxidative capability. Thus, environmental engineers are interested in measuring the ATP content of samples in the activated sludge process as well as in rivers, lakes, and other receiving waters.

An ATP assay procedure has been developed based on the reactions just described. Briefly, the procedure involves the rapid killing of live bacterial cells and the immediate extraction of ATP into an aqueous solution. The latter is then treated with firefly lantern extract, and a photometer measures the light emission of the resultant solution. The firefly lantern extract and the ATP required for calibration are commercially available. Du Pont is the only supplier who has designed a manually operated instrument specifically for this measurement.

The instrument is supplied with the required reagents. The lab technician dissolves a tablet containing buffer and magnesium sulfate in water and adds a homogeneous powder of luciferin and luciferase. Then, the technician filters the sample through a coarse filter to remove solid matter, which is discarded. The filtrate passes through a bacterial filter to catch the living bacteria. The technician treats the bacteria on the filter with butanol, which ruptures the cell walls and releases the ATP. To take the filtrate up to volume, the technician adds water and a microliter aliquot to the prepared reagent already in a cuvette. He then places the cuvette in the instrument to read its light emission. The instrument automatically converts the light flash to ATP or microorganism concentration per milliliter, depending on how it is calibrated.

Slurry Consistency Monitors

TYPES

Blade, rotary, polarized light, probe, level detector, and flow bridge

ELEMENT MATERIALS

316 stainless steel

NORMAL DESIGN TEMPERATURE

Up to 250°F (120°C)

NORMAL DESIGN PRESSURE

Up to 125 psig (8.6 bars)

RANGE

1.75 to 8% consistency

SENSITIVITY

0.01 to 0.03% consistency

REPEATABILITY

0.5% of reading

INACCURACY

Function of empirical calibration, usually 1% of reading

COST

Laboratory units cost $1000 to $2000; continuous industrial units cost $3000 to $5000. SS transmitters cost $4000 to $5000.

PARTIAL LIST OF SUPPLIERS

Automation Products, Dynatrol Div.; Berthold Systems Inc.; C.W. Brabender; BTG Inc.; DeZurik, a Unit of General Signal; EG&G Chandler Engineering; The Foxboro Co.; Gam Rad West Inc.; IRD Mechanalysis Inc.; Kajaani Electronics Ltd. (Finland); LT Industries Inc.; Markland Specialty Engineering Ltd.; Measurex Corp.; Monitek Technologies Inc.; Ronan Engineering Co.; Schlumberger Industries, Solarotron; TECO, Thompson Equipment Co.; Testing Machines Inc.; Valmet Automation Inc.

While *density* is mass per unit volume, *consistency* is mass per unit mass. It is a percentage obtained by dividing the weight of the solids by the unit weight of the wet sample. The most direct method of measuring consistency is to dry a sample unit weight and measure the weight of the dried solids. Consistency should not be confused with *basis weight,* which is the weight of a unit area of a sheet product. *Freeness* expresses how readily a slurry releases water. Consistency should be measured at a constant velocity because it affects the reading. Consistency increases with freeness or alkalinity (pH) and decreases with temperature and inorganic material content.

INLINE CONSISTENCY MEASUREMENT

Ideally, the complete process stream should be exposed to the sensor, but in large flows this exposure is not practical. Therefore, samples are taken. The sample should be taken from the center of the pipe, preferably from the discharge of a centrifugal pump, so that separation or settling of solids is minimized (see Figure 2.4.2).

Consistency-measuring instruments detect the consistency of the process fluid as shear forces acting on the sensing element. Two basic types of consistency detectors are the fixed and rotary. In the latter, the shear force is reflected as the torque required to maintain a rotary sensor at constant speed, as the imbalance of a strain-gauge resistance bridge, or as a turning moment. The instruments are calibrated inline; thus the output is not in terms of dry consistency but rather some arbitrary, reproducible value.

Fixed sensors depend on the process flow for measurement, and for such instruments, the output is affected by the velocity of the flow. The sensor contour minimizes the flow effects on the output over the operating flow range. On the other hand, rotating sensors do not depend on process flow for measurement. While these units are also sensitive to flow velocity variations, they can be used over wider flow ranges. In addition, the rotary motion of the sensor produces some self-cleaning action while fixed sensors depend solely on a properly designed contour to prevent material obstructions.

The sensing element of this instrument is a blade, specially shaped to minimize the effects of velocity. The instrument can be mounted on any line 4 in (100 mm) or larger. The mounting is through a 2-in (50-mm) flange supplied with the instrument.

A variation of this design uses a shaped float inserted through a pipeline tee. The shear forces acting on the float are transmitted to the force bar of a pneumatic transmitter mounted on top of the tee. The unit can only be installed in a vertical pipe, with 5 pipe diameters of straight

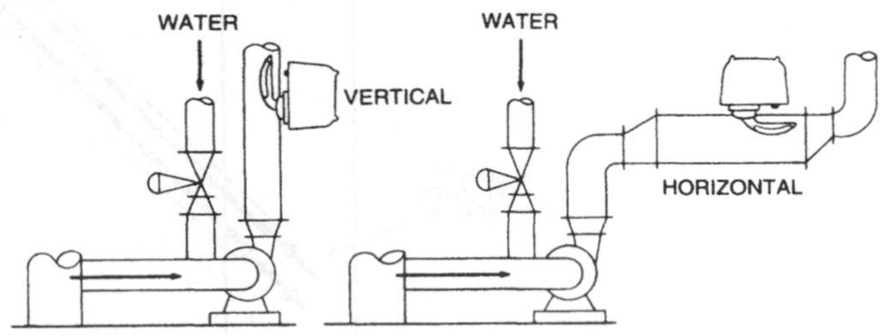

FIG. 2.4.2 Installation of a blade-type consistency transmitter in vertical and horizontal pipelines. (Courtesy of DeZurik, a Unit of General Signal)

run required on the upstream side. The minimum line size is 6 in (150 mm).

PROBE TYPE

This sensor transmitter functions as a resistance-bridge strain gauge. The bridge elements are bonded to the inner wall of a hollow cylinder that is inserted into the process. The shear force acting in the cylinder, due to the consistency of the process fluid, causes an imbalance of the resistance bridge. The amount of imbalance is proportional to the shear force and the consistency of the process fluid. The resistance bridge is powered from a recorder that also contains ac potentiometer electronics.

The sensor is mounted through a threaded bushing furnished with the unit. The flowing velocity must be between 0.5 and 5.0 ft/sec (0.15 and 1.5 m/sec) for repeatability of around 0.1% of bone dry consistency.

OPTICAL SENSOR

Optical consistency meters use a sample cell through which polarized light is passed. Because only solids scatter polarized light, the amount of depolarization is a measure of SS concentration or consistency. Another optical consistency detector operates in the IR region and uses a self-cleaning, multiple fiber-optic probe as its sensor (see Figure 2.4.3). The probes can be inserted into the pipeline to adjustable depths through an isolating ball valve (2 in or 50 mm) and can monitor the consistency in the range of 2 to 6%.

Summary

While convenient from an installation standpoint, inline instruments are sensitive to flow variations. Fixed sensors are often plagued by material buildup, particularly if the sample contains fibers. Rotating sensors are self-cleaning because the sensor motion spins off any material; however, variations in shaft seal friction can be troublesome. The flow-bridge method of consistency measurement is applicable to a range of materials with better accuracy than the other instruments. Since this instrument is not installed inline, the process flow need not be shut down for instrument maintenance.

Sludge and Turbidity Monitors

TYPES

Laboratory units can be manual or flow-through; turbidity transmitters marketed for the process industry are available in probe and flow-through designs.

DESIGN PRESSURES

Up to 250 psig (17 bars)

DESIGN TEMPERATURES

250°F (120°C); 450°F (232°C) special

CONSTRUCTION MATERIALS

Stainless steel, glass, and plastics

RANGES

In ppm silica units, from 0–0.5 to 0–1000; backscattering designs available from 10–5000 ppm to 5–15%. Ranges in Jackson turbidity units (JTU) units from 0–0.1 to 0–10,000; in nephelometric turbidity units (NTU) units from 0–1 to 0–200; in Formazin turbidity units (FTU) units from 0–3 to 0–1100. A sludge density probe with reciprocating piston has a range from 0–0.1% to 0–10% of SS.

INACCURACY

0.5 to 2% of full scale for most and 5% of full scale for reciprocating-piston, probe-type, sludge SS sensor

COSTS

Standard solutions for calibration cost $100 per bottle; laboratory turbidity meters cost from $600 to $1000; laboratory nephelometers with a continuous-flow attachment cost $1500; and process industry transmitters cost from $2000 to $4000 depending on the features and materials of construction. A sludge-density-detecting, self-cleaning probe with an internal reciprocating piston and indicating transmitter costs $6700.

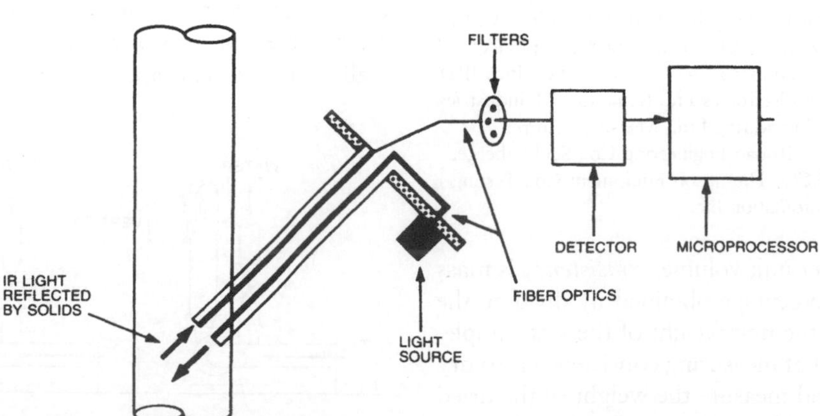

FIG. 2.4.3 Self-cleaning, fiber-optic probe used in consistency measurement. (Courtesy of Kajaani Electronics Ltd., Finland)

PARTIAL LIST OF SUPPLIERS

Bailey Controls Co. Div. Babcock & Wilcox; BTG Inc.; Cole-Parmer Instrument Co.; Custom Sensor & Technology; Du Pont Co. Instrument Systems; Fischer & Porter Co.; Foxboro Co.; Gam Rad West Inc.; Great Lakes Instruments Inc.; Honeywell Industrial Controls; HF Scientific Inc.; Interocean Systems Inc.; Kajaani Electronics Ltd. (Finland); Kernco Instrument Co.; Lisle-Matrix Ltd.; Markland Specialty Engineering Ltd.; Maselli Measurements Inc.; McNab Inc.; Merlab (Hungary); Monitek Technologies Inc.; Ohmart Corp.; Photronic Inc.; Rosemount Analytical Inc.; Sigrist-Photometer AG; Turner Design; Wedgewood Technology Inc.

Turbidity is a measure of water cloudiness caused by finely dispersed SS that scatter visible or IR light. The higher the turbidity (cloudier the fluid), the more scattering occurs. Therefore, transmitted light intensity is reduced while the scattered light intensity (detected at a 90° angle to the light path) increases. Turbidity can also be detected indirectly by colorimeters, activated sludge monitors, or consistency meters.

TURBIDITY UNITS

Different turbidity instruments detect light intensity differently. The three main techniques are *perpendicular* scattering (nephelometry), *backscattering,* and *forward* scattering. Different turbidity units have evolved in connection with different designs. The JTU is a purely optical scale and correlates with forward scattering measurements. The value of one JTU corresponds to the turbidity of a liter of distilled water with 1 mg (1 ppm) of suspended diatomaceous fullers earth (an inert material).

NTUs are based on a U.S. EPA–approved stable polymeric suspension standard and correlate with perpendicular scattering designs.

FTUs use a Formazin polymer standard and also correlate with perpendicular scattering designs. Two Formazin scales are used, and according to some sources, the NTU reference standards are more stable and last longer than the FTU standards. Turbidity measurement error cannot be less than the accuracy at which the standard calibrating solution is available. In Formazin standards, this variation can approach 1%.

All turbidity units measure the amount of solid particles in suspension. Parts per million (ppm) units refer to the weight of the solids in suspension. However, because this measurement requires individual calibration, they usually refer to ppm of silica (silicon dioxide). Therefore, if the cloudiness (turbidity) of the process sample is the same as the turbidity resulting when 1 mg of silica is mixed in a liter of distilled water, the turbidity reading is 1 ppm on the silica scale.

FORWARD SCATTERING OR TRANSMISSION TYPES

In the forward scattering or transmission design (see Figure 2.4.4), the turbidity meter light source is on one side of the process sample, and the detector is on the other. This design determines the total attenuation. When attenuation is due to color absorption, the unit is a colorimeter; when attenuation is caused by light scattered by solid particles, the unit is a turbidity meter.

Dual-Beam Design

When both color and solids are present, the total attenuation is the sum of absorption and scattering effects. Therefore, the single-beam turbidity analyzer can only be used if no color is present or the color is constant and its effect can be zeroed out. When background absorbance or color varies, a dual-beam or split-beam analyzer is needed. Such a unit is described in Figure 2.4.5.

This unit uses two light paths, one passing through the unfiltered process sample cell and the other through a reference sample cell containing filtered process fluid. Analyzer output is proportional to the difference of optical absorbance between the two cells, which corresponds to the solid particles present in the sample but not in the reference cell.

The design shown in Figure 2.4.5 has an oscillating mirror that alternately directs the light beam (alternating 600 times per sec) to the measuring and reference cells. The photocell detector converts the intensity differential of the two beams into a photocurrent that modulates the opening of a mechanical shutter so that the differential is zero.

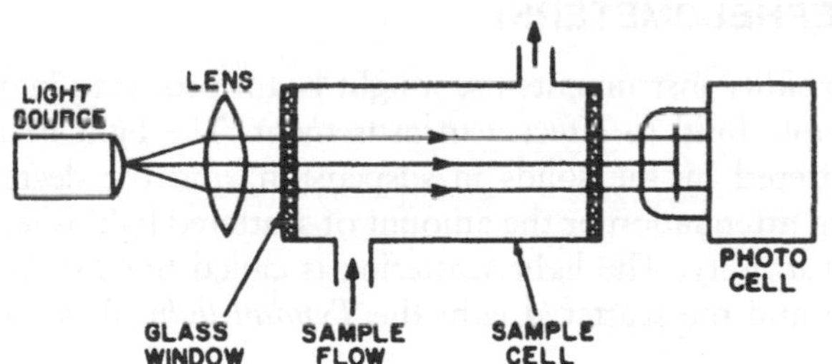

FIG. 2.4.4 Schematic diagram of a transmission-type turbidity meter.

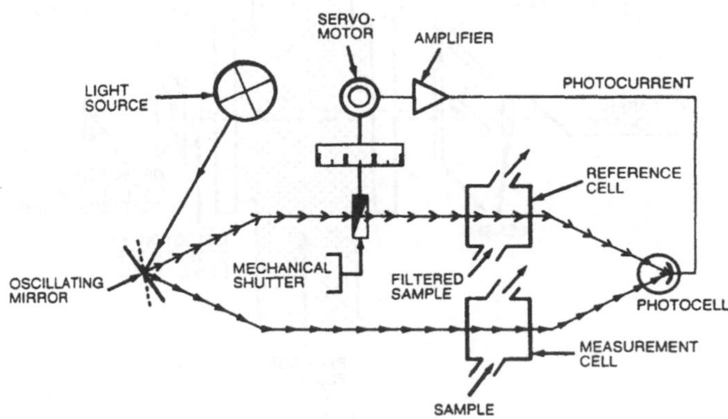

FIG. 2.4.5 Oscillating dual-beam, forward-scattering turbidity analyzer. (Courtesy of Sigrist Photometer AG)

Therefore, the more solid particles in the sample, the more the shutter needs to be closed, and the position of the shutter can be read as turbidity. If the reference cell is filled by other reference materials, the same instrument can measure other properties such as color and fluorescence.

Laser Type

In the laser-type, in-line turbidity meter shown in Figure 2.4.6, a thin ribbon of light is transmitted across the process stream. This light is attenuated by the process fluid and then falls on detector 1. If there are solids in the process fluid, some of the light is scattered. This scattered light is collected and falls on detector 2. The ratio of the two detector signals relates to the amount of solids in the process stream (turbidity); being a ratio signal, it is unaffected by light source aging, line voltage variations, or background light intensity variations. The laser-type detector is less sensitive to interference by gas bubbles than other turbidity meters because the laser-based light ribbon is so thin (about 2 mm). Therefore, when a bubble passes through it, it causes a pulse, which can be filtered out.

In general, in-line turbidity meters are less subject to bubble interference than turbidity analyzers that require sampling. Because in-line units do not lower the operating pressure of the stream, dissolved gases are not encouraged to come out of solution.

SCATTERED LIGHT DETECTORS (NEPHELOMETERS)

Turbidity instruments use a light beam projected into the sample fluid to effect a measurement. The light beam is scattered by the solids in suspension, and the degree of light attenuation or the amount of scattered light is related to turbidity. The light scattering is called the *Tyndall effect* and the scattered light the *Tyndall light.* A constant-candlepower lamp provides a light beam for measurement, and one or more photosensors convert the measured light intensity to an electrical signal for readout.

Usually the photosensor comes with a heater and thermostat to maintain a constant temperature because the device output is temperature sensitive. The supply voltage to the lamp must be regulated to at least $\frac{1}{2}$%. This regulation eliminates errors due to source intensity variations because the measured light is referenced to the source. Because deposits formed on the flow chamber windows by the sample interfere with measurement, the windows require frequent cleaning or automatic compensation.

Transmission-Type Design

Instruments measuring scattered light vary in design. One type uses a flow chamber similar to the one in Figure 2.4.4, except that the window for the measured light is located at 90° to the window for the incident light (see Figure 2.4.7). One window transmits light beams into the measuring chamber and the other, at right angles to the first, transmits scattered light to the photosensor. A light trap is located opposite the incident light window to eliminate reflection.

With this arrangement, dissolved colors do not affect the measurement; however, instrument sensitivity decreases with the presence of color because some light is absorbed. Variations of the basic unit include the use of two source beams and two photosensors in conjunction with two pairs of opposed windows.

Some designs use a separate photosensor to monitor lamp output and adjust the lamp supply voltage through a feedback circuit to maintain constant light intensity.

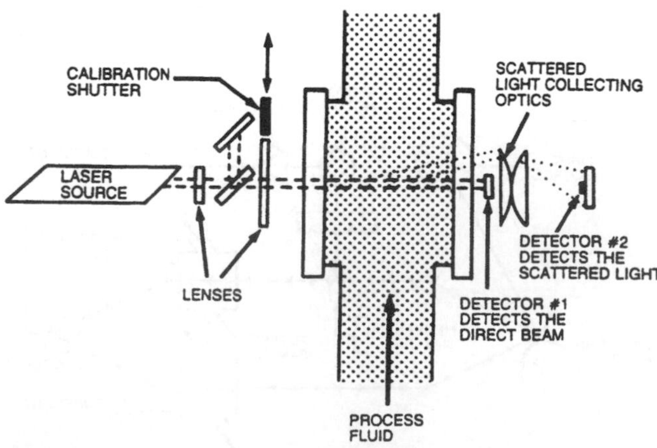

FIG. 2.4.6 Laser-type in-line turbidity meter detecting the total attenuation and the amount of scattering separately. (Reprinted, with permission, from ACSI)

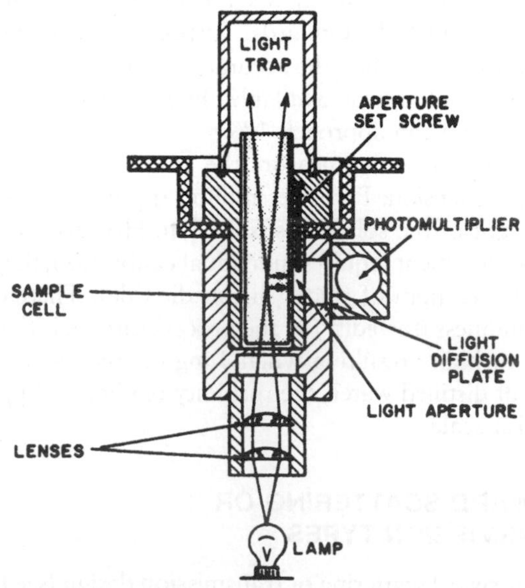

FIG. 2.4.7 Light-scattering turbidity meter.

Probe Design

For wastewater and biological sludge applications, probe-type turbidity transmitters are preferred. One design (see Figure 2.4.8) uses an IR light source and measures the resulting 90° scattered light intensity. These microprocessor-based units are provided with built-in compensators for ambient light variations, and wipers for automatic cleaning of the dip or insertion probe. Cleaning frequency is adjustable between 1 and 6 hr. During the wiper action of the cleaner, the transmitter output signal is held at its last value.

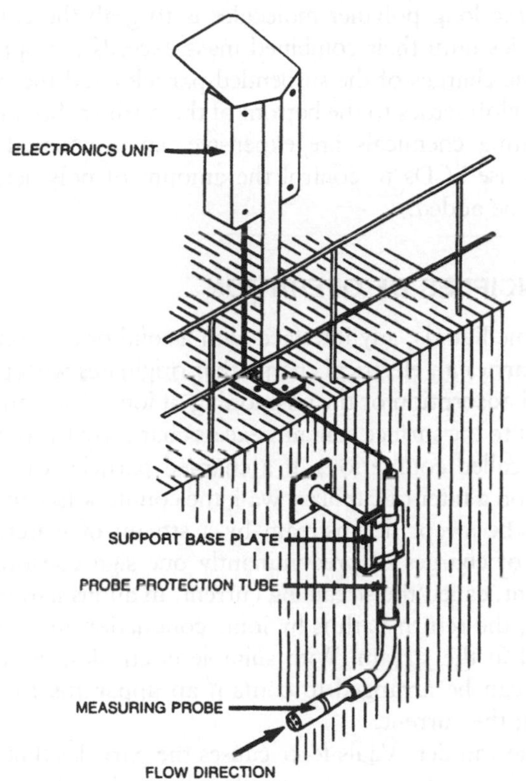

FIG. 2.4.8 Automatically cleaned, 90° scatterer light-detecting turbidity transmitter. (Courtesy of BTG Inc.)

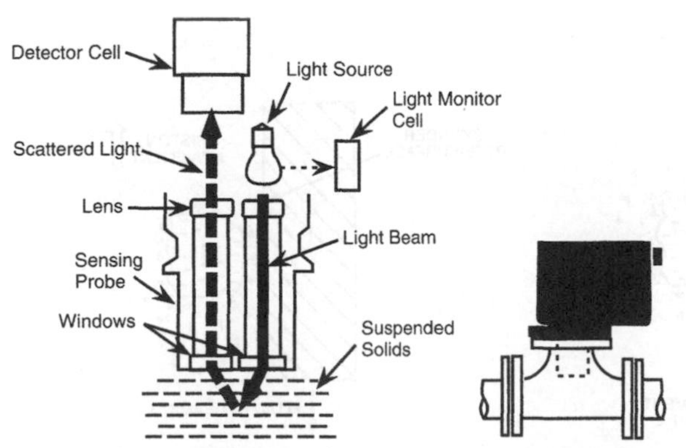

FIG. 2.4.9 In-line, backscatter-type turbidity analyzer. (Courtesy of Gam-Rad Inc.)

BACKSCATTER TURBIDITY ANALYZERS

Figure 2.4.9 shows the in-line version of the backscatter-type turbidity analyzer, which can be installed in either pipes or vessels. Here, the 180° backscatter effect is measured. The units are suited for high-temperature applications (up to 450°F or 232°C) and for high concentrations of solids. Ranges from 10 to 5000 ppm to 5 to 15% on the silica scale are available. A backscattering design using fiber-optic light cables is also available (see Figure 2.4.3).

Summary

Turbidity measurement is fairly simple in theory; the most serious practical problems are posed by light source intensity changes, deposits on optical windows, and the presence of dissolved colors in the sample. Units are available that automatically correct for these effects and variations in the ambient light intensity as well as for gas bubbles. Self-cleaning probe design units are also available. Selection should be made on the basis of information needed (transmission or 90° or 180° scatter), and on the nature and concentration of the solids to be detected and the material of construction for each type.

Installing the Sludge Monitor

Before designing the sludge monitoring installation, environmental engineers must answer two questions:

1. Is the information in the liquid or solid phase of the sludge?
2. Should the measurement be made online (in-place) or should a sample be delivered to the analyzer?

The answers to these questions will direct the environmental engineer to the correct installation.

ONLINE MONITORING

The main advantage of online monitoring is eliminating the sampling system. Without a sampling system, transportation lag time is eliminated, allowing good closed-loop control. Maintenance is also reduced, since many delicate sampling system components are eliminated. Finally, online monitoring detects the unaltered real process; with sampling, sample integrity can be impeded by filtering, condensation, and leakage.

Online monitoring requires an analyzer (usually a probe) that is clean and in good working order. Therefore, probe cleaners (see Tables 2.2.1 and 2.2.2) and placing the probe inside a sight-glass for convenient visual inspection (see Figure 2.2.2) are important. The capability of removing the probe from the pipe or tank without a process shutdown is also beneficial. This is accomplished using retractable probes (see Figure 2.2.8), which are periodically (and automatically) withdrawn from the process for unattended cleaning and recalibration.

If the probe measures only the composition of the liquid phase, a periodically backflushed, porous filter cup (see Figure 2.2.6) can protect it from being coated with solids.

SAMPLING-BASED SLUDGE MONITORING

Before a sludge sample is transported to the analyzer, the environmental engineer must determine if the information of interest is in the liquid or solids phase of the sludge. If only the liquid phase must be monitored, solids can be removed from the sample by self-cleaning filters (see Figures 2.3.2 and 2.3.3). When a sample is collected over a long time period, intermittent sample collectors should be used for monitoring (see Figure 2.3.5).

If the information is in the solids phase and a sampling system is used, the size of the solid particles must be reduced through homogenization (see Figure 2.3.4). Once the slurry is liquified, it can flow through the sampling system without plugging it.

Colloidal Suspension Monitors

APPLICATIONS

Batch operations, titrations, or continuous monitoring; can control the clarification of beverages, dewatering, thickening of suspensions, addition of coagulant chemicals, or treatment demand by measuring the surface charge on particles

MATERIALS OF CONSTRUCTION

Stainless steel, silver, and Teflon

SAMPLE SIZE REQUIRED

About 10 cc

APPROXIMATE COST

$12,000

PARTIAL LIST OF SUPPLIERS

Komline-Sanderson Engineering Corp.; Leeds and Northrup; Mütek GmbH; Panametrics, Inc.

In wastewater treatment, clarification is a major step. Certain materials such as clay do not settle out because the static electric charges of the individual particles keep the clay particles uniformly dispersed. In such *colloidal suspensions,* gravitational forces alone cannot cause settling because the opposing forces caused by the like electrical charges of the particles are stronger than the gravitational forces acting on them.

Clarification of colloidal suspensions involves measuring particle surface charge and then adding coagulating chemicals in proportion to that charge. The surface charge is detected by streaming current detectors (SCDs), while the coagulating chemicals are usually polymers. The role of these long polymer molecules is to grab the colloidal particles until their combined mass exceeds the opposing electric charges of the suspended particles and the coagulated glob settles to the bottom of the clarifier. Because coagulating chemicals are expensive, environmental engineers use SCDs to control the amount of polymers that must be added.

PRINCIPLES OF OPERATION

In ionic liquids, any interface with a solid or a second liquid carries an electrical charge that originates with preferential adsorption or the positioning of ions. The liquid adjacent to the surface contains excess charges of the opposite sign, called counterions. If a charged particle is immobilized on a filter or capillary wall, the counterions can physically be swept downstream by a stream of water. This flow of charges of predominantly one sign constitutes a current, called the streaming current. In an insulating capillary, the return path is by ionic conduction through the liquid in the stream. With suitable electrodes, the return path can be arranged to contain an apparatus for measuring the current.

The van der Waals force causes the particles that carry high charges to be preferentially adsorbed to the cylinder and piston surfaces as shown in Figure 2.4.10. When such a particle moves upward by piston movement, its counter-

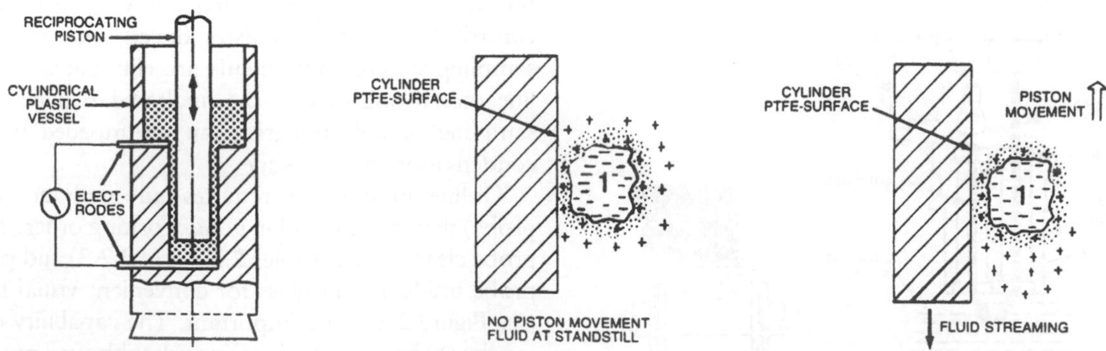

FIG. 2.4.10 A cloud of counterions sheared off by the streaming fluid in the boundary of the diffuse layer at the cylinder surface. (Courtesy of Mütek GmbH)

ions are sheared off as the fluid moves in the opposite direction relative to the piston. The totality of these ions is the streaming potential detected by this analyzer.

APPLICATIONS

Most applications involve either titration of the sample or prior treatment in the plant since a single reading on untreated material provides little information. SCD readings are almost independent of the concentration of SS. Titrations can be made with as little sample as will submerge the active part of the instrument.

To estimate the unit treatment demand of a liquid, the environmental engineer must first titrate a volumetric sample with the contemplated treating chemical at a known concentration until he obtains a zero signal. To compare alternative treating chemicals, the environmental engineer titrates identical samples of material with various chemicals. When the effect of a change in pH on treating requirements is studied, the environmental engineer titrates identical samples at various pH levels. This effect can be significant, with chemicals differing considerably in their tolerance of low or high pH.

In the usual treatment plant, the SCD can continuously control the feed of cationic chemicals. The need for changing the rate of adding these chemicals arises from variations in the stream flow rate, changes in the SS loading, the unit demand of the solids, or any combination of these factors. The main advantage of SCD control is its early response to changes. Charge neutralization occurs almost as soon as the treating chemical is dispersed in the stream; therefore, samples can be taken 1 or 2 min after addition of the chemical.

Batch samples taken to the SCD should be adequate to permit rinsing the apparatus several times. Skimming or decanting removes sand, larger solid particles, or oil globules. Since charge is a surface phenomenon and the fines have most of the surface, removing larger particles has little effect. For a continuous sample, a self-cleaning bypass filter should be used (Figures 2.3.2 and 2.3.3). Periodic backflushing or cleaning can also be required.

For continuous control, the SCD measurement signal is fed to a two-mode controller that modulates the chemical feed pump or control valve (see Figure 2.4.11). The maximum and minimum valve opening is limited as a defense against sample loss, which can cause an open loop. Pressure regulators serve as adjustable settings to limit the controller output to a range between the minimum and maximum expected demand.

The SCD controller set point is based on the downstream turbidity measurement. If an existing flow proportioning controller throttles the chemical feed, the SCD controller can influence its ratio set point in a cascade arrangement. Often, more than one chemical is involved, and the environmental engineer must consider a sequence of additions and attendant interactions.

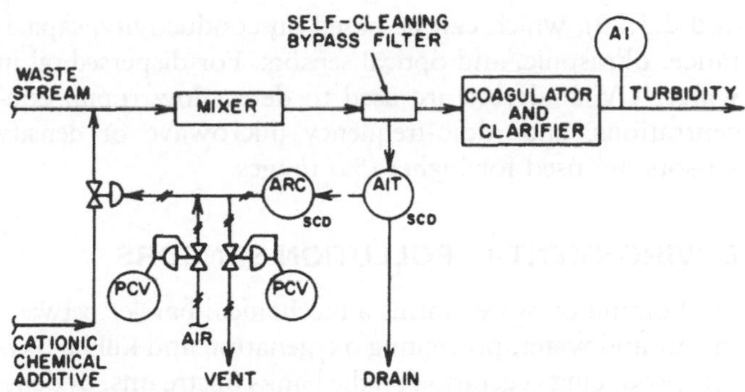

FIG. 2.4.11 Chemical addition control using a streaming current detector.

Oil Monitors

TYPES OF DESIGNS

A. Reflected light oil slick detector (on–off)
B. Capacitance—available in probe form for interface detection, in a flow-through design, or in a floating plate configuration for measuring oil thickness
C. UV
D. Microwave (radio frequency)—available as an interface probe, as a tape-operated tank profiler, or as an oil in the water content detector
E. Conductivity probes for interface detection

RANGE

A. Generally from 0–50 ppm to 0–100%
B. The flow-through, dual-concentric detector from 5 to 15% water in oil
C. 0–10 ppm to 0–150 ppm of oil in water
D. The oil content is detectable from 0 to 100%

INACCURACY

Generally from 1 to 5% of full scale
B. The flow-through, dual-concentric detector has a sensitivity of about 0.05 to 0.1% water.
C. 0.1 ppm for a 0- to 10-ppm range
D. Interface detected to 5%, tank profile to 1%, and water concentration to 0.1%

COSTS

C. $12,000 to $20,000 for dual-wavelength unit with auto-zero and 0–10 ppm to 0–150 ppm range

PARTIAL LIST OF SUPPLIERS

(For capacitance, conductivity, and ultrasonic probe suppliers see *The Instrument Engineers' Handbook: Process Measurement,* Third Ed.; Agar Corp. (D); Amprodux Inc.; Bailey Controls Div.; Bernhard & Scholtissek, Veba Oel AG; Delta C Technologies (B); Du Pont Instrument Systems (C); Endress + Hauser Instruments (B); Foxboro Co. (B); Invalco (C); Spatial Dynamics Applications Inc.; Teledyne Analytical Instruments (C)

Oil and grease must be removed from wastewater plant effluents before they are discharged into the environment. Oil can either float on top of the aqueous effluent or be dispersed in it. With an oil layer, the monitoring task is an interface level measurement (see Figures 2.1.18, 2.1.19, and 2.1.20), which can be based on conductivity, capacitance, ultrasonic, and optical sensors. For dispersed oil in water, UV analyzers are used to detect low (ppm) concentrations, and radio-frequency microwave or density sensors are used for higher (%) ranges.

ENVIRONMENTAL POLLUTION SENSORS

Oil floating on water forms a mechanical barrier between the air and water, preventing oxygenation and killing oxygen-producing vegetation on the banks of streams. By coating the gills of fish, these materials prevent breathing and cause fish to suffocate. Therefore, ships and municipal and industrial waste treatment plants must monitor outfalls and control oil removal to prevent oil-bearing wastes from entering natural waves. Continuous monitors are available to detect any hydrocarbon floating on the surface of water.

Oil in the water is equally undesirable. It contributes to the BOD and can also be toxic to aquatic biota, fish food in water, and fish themselves. Optical detection methods for both types of contamination require regular, conscientious maintenance for continuous, reliable performance. The capacitance approach for monitoring oil film thickness on water appears to require less maintenance but is limited to detecting floating oil. Environmental engineers must evaluate each application separately considering the limited capabilities of available instrumentation.

On–Off Oil-on-Water Detector

This device (an application of nephelometry) detects a visible oil (hydrocarbon) slick on fresh or salt water. It consists of two parts: a sensing head and a controller. The sensing head, in an explosion-proof housing supported on pontoons, floats on the body of water. An S-shaped baffle directs flowing water past the sensing head. A beam of light is focused through a lens onto the water's surface. Reflected light is refocused by a second lens onto a photocell. In the absence of oil on the water, minimum light reflection occurs. In the presence of floating oil, the reflected light intensity increases.

Measurement is based on the difference between the reflected light photocell output and a reference photocell measuring light source output. Alarm functions and an output signal proportional to reflected light intensity are available from the controller.

Oil-Thickness-on-Water Detector

The device just described measures the presence or absence of oil floating on water. The oil-thickness-on-water detector measures the oil layer thickness. It consists of a floating sensing head connected by shielded cable to a remote controller. The sensor measures the thickness of an oil layer on water by capacitance measurement (see Figure 2.4.12).

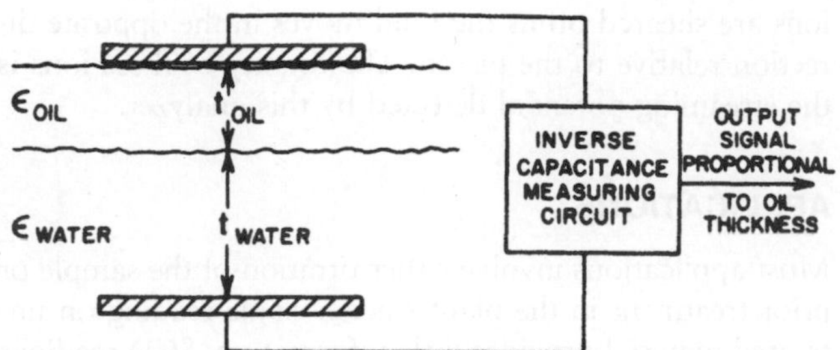

FIG. 2.4.12 Parallel-plate capacitor detecting the thickness of an oil layer on water.

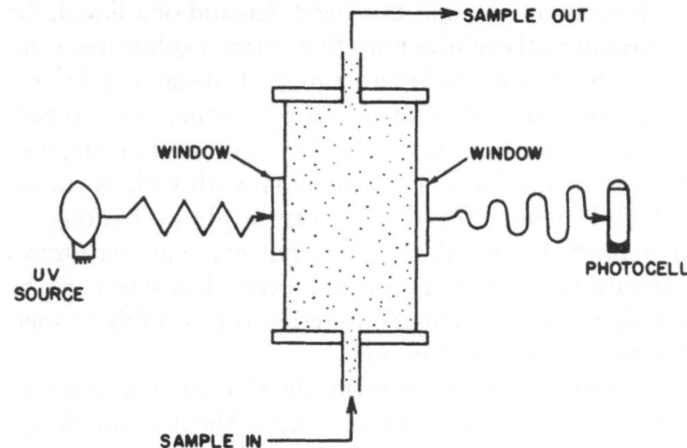

FIG. 2.4.13 Oil-in-water detector.

The inverse capacitance is proportionate to the oil thickness. The circuit generates a dc voltage in proportion to the inverse capacitance, which is in direct proportion to the oil thickness and is available for remote transmission. The sensor depends on the large differential in dielectric constants between oil and water for its operation. Manufacturers claim that the sensor is not confused by emulsified sludge, which has a large dielectric constant, or by oily froth, which cannot pass under the float.

Oil-in-Water Detector

When a contaminated water sample stream is irradiated with UV waves at a peak intensity of 365 nm, the oil contaminant emits visible radiation. This radiation can be measured by a photocell. Visible radiation increases with increasing concentrations of the fluorescent substance. The relationship between the concentration and the visible radiation emitted is substantially linear in low concentrations (below 15×10^{-6}). In higher concentrations, some nonlinearity occurs as a result of a saturation effect.

The most common measurement method is to pass a sample through the sensing head in an upflow direction (see Figure 2.4.13). The head is equipped with two windows set at right angles that minimize the intensity of direct radiation from the source striking the photocell and also reduce the multiple scattering of visible radiation effect. Optical filters at the incident and emergent windows (not shown) reduce this effect to a negligible level.

To detect the oil concentration in water, a falling-stream-type detector is also available. With this device, the sample stream is shaped into a rectangle and falls through the viewing field of the UV beam and the photocell. Efficient optical filtration is important to overcome the unavoidable effects of the direct reflection of incident radiation from the surface of the shaped stream.

UV Oil-in-Water Analyzer

Figure 2.4.14 shows the sampling system of a continuous oil-in-water analyzer used to monitor steam condensate, recycled cooling water, and refinery or offshore drilling effluents. This system uses a single-beam, dual-wavelength UV analyzer, superior to the single-wavelength designs because it compensates for variations in sample's sediment content, turbidity, algae concentration, or window coatings. The cell operates according to Beer's law, which relates oil concentrations to UV energy absorption by the fixed-length cell. The UV measuring band is centered at 254 nm, and the readings are sensitive to 0.1 ppm with a range of 0 to 10 ppm and provide a 90% response in 1 sec.

The automatic-zero feature of the instrument is provided by sending sample water to both the measurement and zeroing sides of the conditioning system. When the sample is in the measurement mode, it is sent through a high-speed, high-shear homogenizer, which disperses all suspended oil droplets and oil adsorbed onto foreign matter so that the sample sent to the analyzer becomes a uniform and true solution.

Once an hour, the analyzer is automatically rezeroed. In this mode, the sample water is sent through a filter that removes all oil and after sparging, the sample water is sent to the analyzer. This oil-free, zero-reference sample still contains the other compounds found in the measurement sample and therefore can be used for zeroing out this background.

Radio-Frequency (Microwave) Sensors

When a cup of water and oil is placed in a microwave oven, the water heats up, while the oil does not. This occurs because shortwave radio-frequency energy is absorbed more efficiently by water than oil. In the radio-wave detector, the transmitter produces fixed-frequency and con-

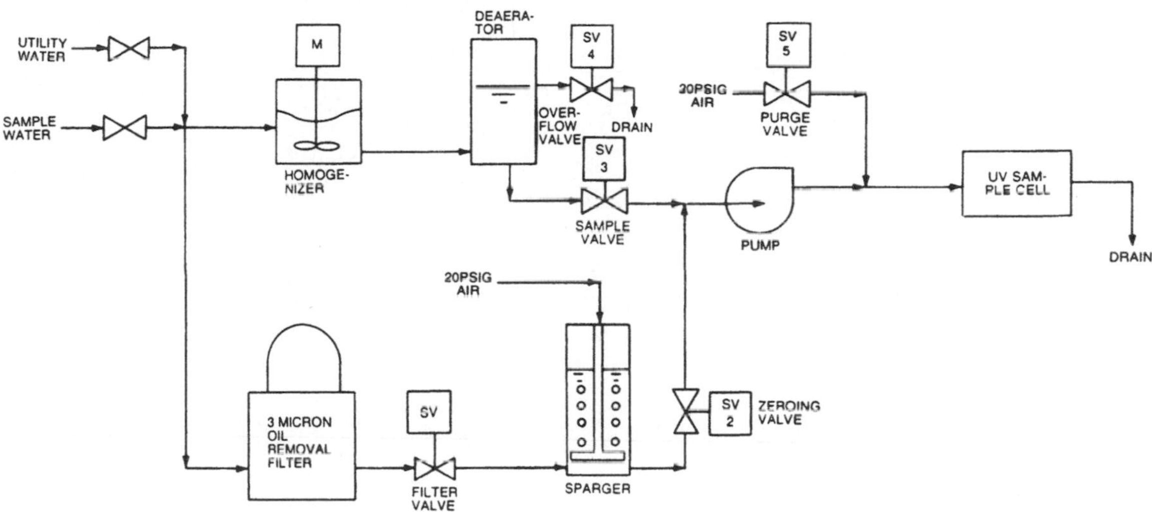

FIG. 2.4.14 UV oil-in-water analyzer with automatic-zero feature. (Courtesy of Teledyne Analytical Instruments).

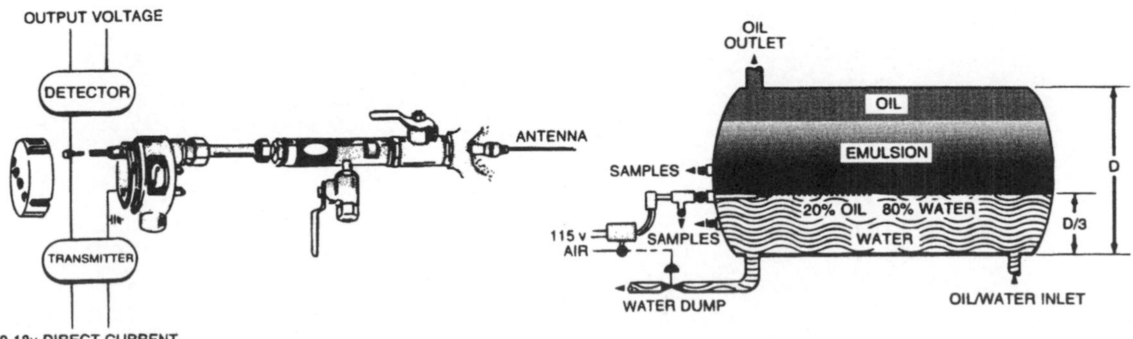

FIG. 2.4.15 Radio-wave oil–water interface detector probe. (Courtesy of Agar Corp.)

stant-energy waves. The more energy is absorbed by the process fluid (the more water in the mixture), the lower the voltage at the detector. The advantages of this design, compared to capacitance systems, include a wider range (0 to 100%), lower sensitivity to buildup, insensitivity to temperature and salinity variations, and suitability for higher temperature operations (up to 450°F or 232°C).

Radio-wave, oil-in-water sensors are available as probe-type sensors for water–oil interface control. A typical application is the free-water knockout (see Figure 2.4.15), where the probe is installed horizontally at one-third of the diameter from the bottom and is set to open the water dump valve when the emulsion concentration drops below 20% oil (80% water). In this way, the emulsion (rug layer) builds up above the probe, while only clean water is dumped. The probe can also provide a 4- to 20-mA transmitted output signal that signals the water concentration within an error of 5%.

An available portable tank profiler also uses the same principle of operation. Here, the radio-wave element supported by a tape, is lowered into the tank, which can be 100 ft (30 m). As the sensor is lowered, it measures both the interface location (within an error of 0.12 in or 3 mm) and the emulsion concentration throughout the tank from 0 to 100% within an error of 1%.

A water-in-oil monitoring probe is also available, which can detect water concentration over a 0 to 100% range within an error of 0.1% in tanks or pipelines. All these devices are available in explosion-proof construction and with digital displays.

—Béla G. Lipták

3

Sewers and Pumping Stations

Robert D. Buchanan | Mark K. Lee | David H.F. Liu

3.1 INDUSTRIAL SEWER DESIGN

Industrial sewers, such as refinery sewer systems do not tie into municipal systems for two reasons:

Refinery waste products are not compatible with prevalent sanitary sewage treatment.
Spent cooling water from a refinery can equal or exceed the flows expected in a municipal sewer system.

To prevent waste products from entering rivers and lakes, almost all large industrial plants have facilities to separate and collect waste products. Refinery and chemical plant sewers flow by gravity and are usually partially filled.

Basic Sewer Systems

Figure 3.1.1 shows a typical process plant sewer system. The waste streams in most large plants can be classified under the following four basic sewer systems:

- The oily water sewer
- The acid (chemical) sewer
- The storm water sewer
- The sanitary sewer

THE OILY WATER SEWER

This system collects all non-corrosive process waste periodically drained from tanks, towers, exchangers, pumps, and other process equipment using open-end drain hubs located adjacent to the equipment. During maintenance shutdowns and at turnarounds, these drain hubs drain water from equipment for hydrostatic testing or washing out towers or tanks.

Pumps and compressors should also have open-end drain hubs located at the ends of foundation blocks. These open-end drain hubs collect drainage from pump bedplates and gland and seal piping at pump and compressor bearings.

Paved and unpaved surface drainage areas adjacent to tanks, towers, exchangers, pumps, and compressors, where process waste spillage can be considerable, should divert drainage to the oily water sewer. This drainage includes heavily contaminated wash water from turnaround or maintenance operations. Rain water runoff can constitute the largest flow quantity in a drainage area and can be the governing factor in sizing sewer pipes.

Fire water from hoses is included in the estimated maximum flow quantities within unit areas containing haz-

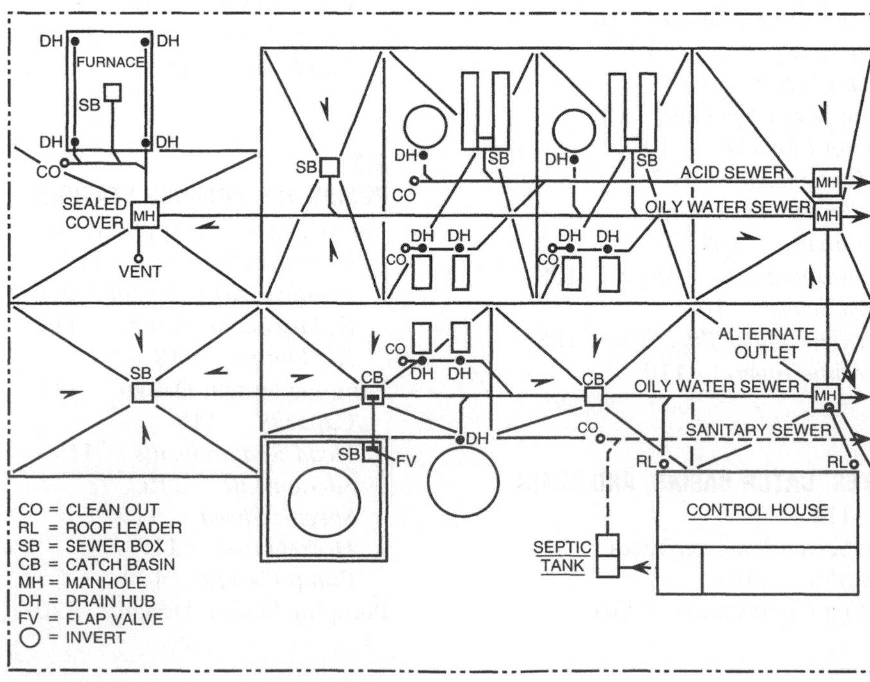

FIG. 3.1.1 Typical process plant sewer system.

ardous hydrocarbon or chemical equipment. Where fire water is included, it can vary from 500 to 1000 gpm. The amount depends on the size and number of equipment pieces as well as the number of sewer boxes or drains within the area.

The oily water sewer main should be run to the battery limit as a separate system. There it should be connected to the oily water trunk sewer that runs to an oil–water separator.

ACID (CHEMICAL) SEWER

This sewer collects heavily contaminated, corrosive, process chemical waste that occurs as spillage, leakage, and valved drains at process equipment and pumps.

Open-end drain hubs located at all tanks, towers, exchangers, and associated equipment facilitate draining. Large drains at towers or tanks can be handled more conveniently by an acid-proof concrete or acid-brick-line sewer box rather than an open-end drain hub.

Pump blocks should have an open-end drain hub to collect pump casing drains, drainage from gland and seal piping at pumps, and drains in pump suction and discharge piping.

Acid areas that collect corrosive process waste usually have acid-resistant curbed paving to confine and collect any acid drainage or spillage within these areas. Curbed and paved areas should be provided in locations where pump groups, storage, and handling areas are subject to spillage and wash-down water.

Wash water collected in these surface drainage areas should be collected in the acid sewer. However, where possible, storm water surface drainage should not be run into the acid sewers.

Acid waste should be run in a separate sewer from alkaline waste. Acid and alkaline waste should be run as two separate sewer systems to the battery limit and the acid treating facility or a neutralizing sump.

STORM WATER SEWER

The storm water sewer collects maximum surface drainage including rainfall, wash water that is not contaminated, and cooling water that is not returned in return headers to cooling water facilities.

Storm water runoff is calculated on the basis of 100% runoff for all paved areas and 50% runoff for unpaved areas. The remaining 50% in unpaved areas is assumed to be absorbed into the ground.

Rainfall data for various geographic locations are readily available from the government, state, and city weather bureau records, and other published data. Storm water accumulation for each in of rainfall/hr/sq ft is equal to 0.0104 gpm.

Fire water from hoses should be included in the estimates of storm water runoff if flooding would cause damage to installations and present a hazard during fire fighting operations. The storm water main should be run to the battery limit and connected to the trunk sewer.

SANITARY SEWER

The sanitary sewer constitutes a separate sewer system into which only wastes of sanitary facilities are permitted. The sanitary sewer discharges into a septic tank. The effluent from the septic tank can be discharged into the oily water sewer if a sanitary sewer main is not provided at the battery limit.

Designing Sewer Systems

The oily water sewer flows to an oil–water separator to remove oil and sediments, which are also removed in a sludge disposal chamber. Conventional chemical treatment is also required. The acid sewer flows to some form of neutralizing sump or acid-treating facility. Acid and alkaline sewer wastes are collected separately at sumps for neutralization or treatment. The storm water sewer has facilities for oil skimmers and a trash screen before final discharge at the point of disposal.

The steps involved in designing a sewer system follow.

DEVELOP PLOT PLAN

The plot plan is a major aid in the layout of sewer systems. The plot plan indicates the locations of all pumps, exchangers, tanks, and towers. It also indicates the extent of paved areas, roadways, and underground utilities (water and electric) and the locations and inverts of sewer tie-ins to the sewer mains at the battery limits.

LAY OUT SYSTEM

An environmental engineer can begin the layout of a sewer system by indicating all major equipment foundations, taking their locations from the plot plan. The layout should indicate all pipe rack columns, lighting poles, and minor footings that can interfere with the sewers. The environmental engineer should integrate underground cooling water systems into the sewer system layout as an additional system, as well as any underground electrical utilities to avoid any interferences.

The task now is to design sewer systems into an intricate layout that has as many as four separate sewer systems underground. The layout must provide gravity flow in each of the systems, maintain the given inverts of the sewer systems at the battery limits, and be free of interferences at the cross-overs of sewer systems and underground water and electrical systems. Existing underground piping, electrical trenches, and other encumbrances further complicate the sewer layout.

CLASSIFY AND TYPE SURFACE DRAINAGE

The plot plan, having furnished the locations of all pumps, exchangers, compressors, tanks, and towers, also shows the extent of paved and curbed areas. The paved and curbed areas adjacent to process equipment should be segregated into proper sewer system classifications, namely, oily water sewer, acid (chemical) sewer, storm water sewer, and sanitary sewer depending on the type of process waste drainage or spillage. These areas must be divided into surface drainage areas that collect and channel drainage to the proper sewer system classification.

Paved or unpaved areas in outlying locations, not adjacent to process equipment or buildings, should also be divided into surface drainage areas. These areas are usually free from contamination and can be channeled into the storm water sewer.

Diked or curbed areas at tank farm and storage locations also require provisions for surface drainage and should be divided into suitable drainage areas. These drains should be run to the oily water sewer main.

ESTABLISH AND CHECK SLOPE

The environmental engineer must design and size surface drainage areas while considering the limits of the permissible paving slope or grade. The slope of all drainage areas should be governed by the following limits so that a hazardous or tripping condition is avoided:

The paving inside a building should have a minimum slope of 1 in over 10 ft and a maximum elevation drop of 3 in for each drainage area.

Paved and unpaved areas outside buildings should have a sewer box or catch basin for each surface drainage area. The maximum difference in elevation between the high point of grade and the grade at the catch basin should not be more than 6 in with a slope of 1 in over 10 ft.

The total number of surface area divisions depends on the drop in elevation of surface drainage areas.

DESIGN SEWER

After the surface drainage areas are divided into areas based on slope and drop in elevation, the environmental engineer must segregate and run them to the proper sewer classification.

The divided areas are provided with a sewer box or catch basin. The separate sewer systems must now be connected to the proper classified sewer. The outlet connections at the sewer box or catch basin can be at the bottom or side depending on the limits of sewer inverts (see Figure 3.1.2). The *invert* is the elevation of the bottom inside surface of the sewer pipe.

The sewer design should provide for ample future expansion of the plant and unit areas. Sewer mains, in particular, should be sized to include the estimated flow of any expansion.

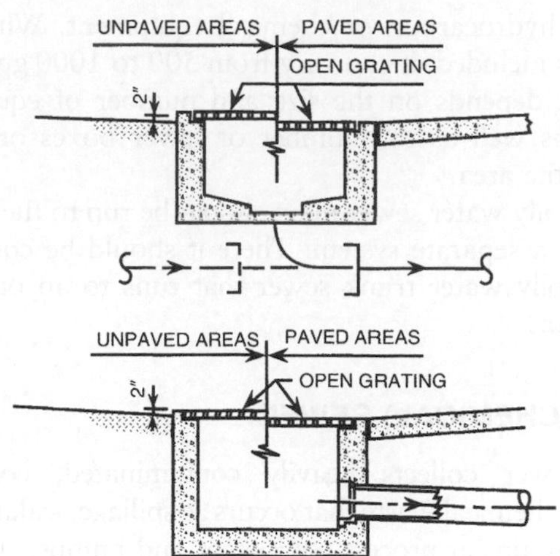

FIG. 3.1.2 Sewer box outlet detail.

Storm water runoff is calculated in rainfall in based on 100% runoff for all paved areas and 50% runoff for unpaved areas. Obtain design rainfall in amount of in per hr for the specific area where the plant is to be located; otherwise, the general chart shown in Figure 3.1.3 can be used.

Storm water for each in of rainfall/hr/sq ft equals 0.0104 gpm.

DETERMINE SEWER PIPE SIZE

The minimum size of underground sewer pipe, branch or drain hub, should be 4 in. The minimum size of the sewer main (collecting two or more 4-in sewer branches or drain hubs) should be 6 in. The minimum size of the sewer pipe in any curbed or diked area should be 8 in.

Sewer size depends on plot size, amount of rainfall, and quantity of process waste, fire water, and any other liquids requiring disposal. As previously stated, the sewer mains should be sized to include the estimated flow of future expansion.

Velocities used in sewer system design should have a minimum of 3 ft per sec and a maximum of 7 ft per sec. Flow capacities, velocities, and slopes (for sewers running $\frac{3}{4}$ full) can be coordinated so that the curves shown in Figure 3.1.4 give the sewer line size required for the slope and velocity for the flow capacity in gallons per minute.

The flow capacities, velocities, and slopes of sewer lines are contingent on plant site grades available. The flat grades necessary at most plant sites are a determining factor in the slope and velocity of the sewer lines. Where possible, and if gradients permit, the maximum velocity should be used.

The following design example determines the sublateral size for a sewer system:

Given: Rainfall = 3 in/hour

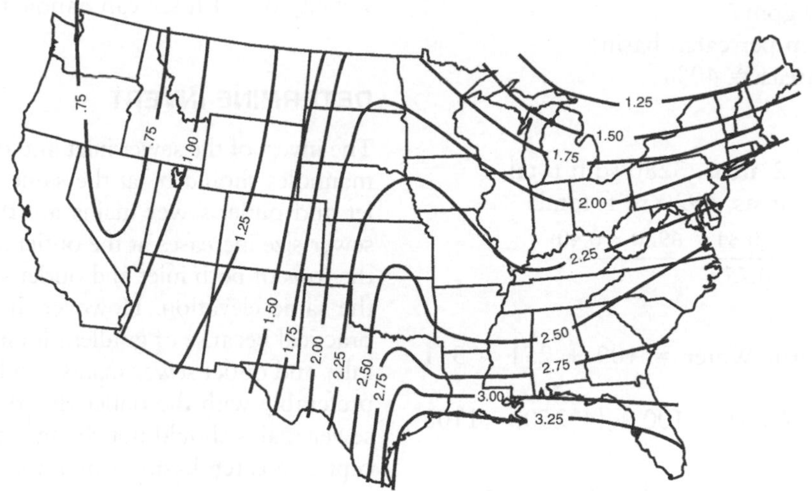

FIG. 3.1.3 General rainfall chart. This chart should be used only when the design rainfall of the job area is not specified.

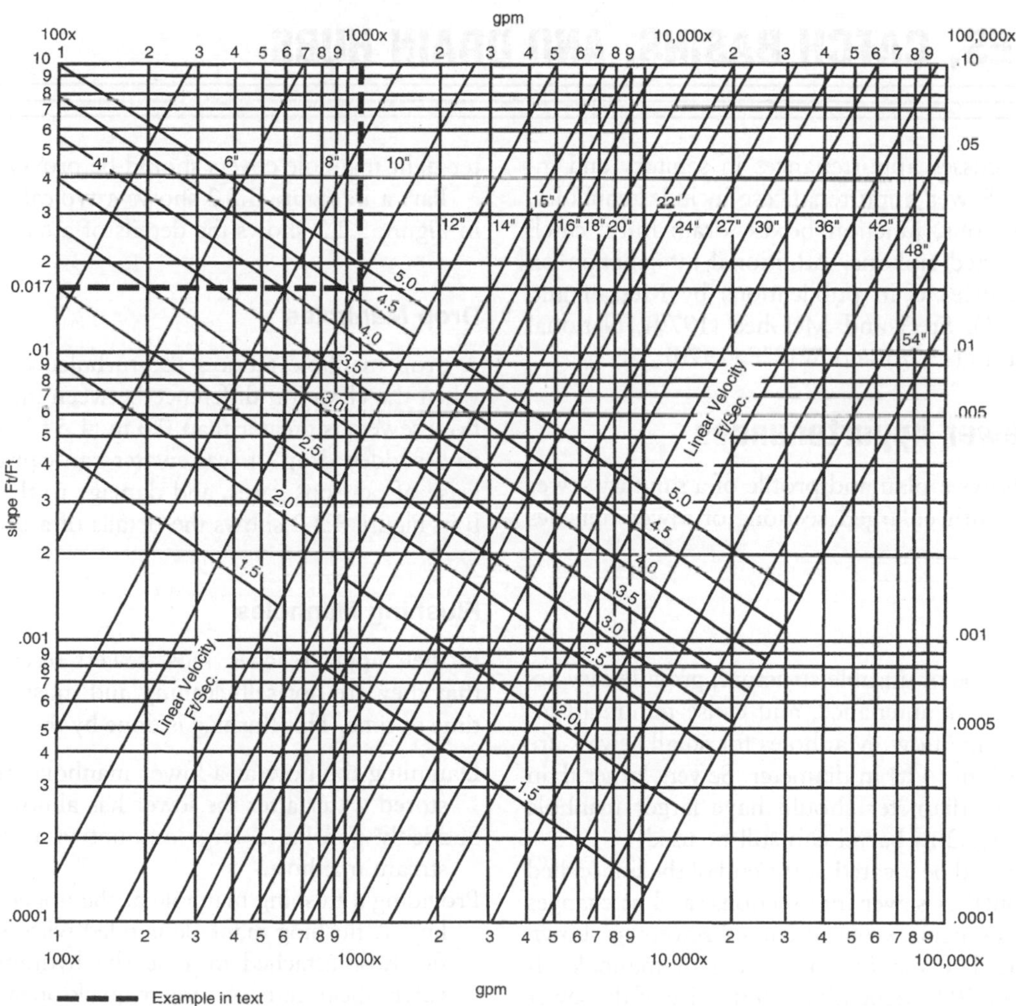

FIG. 3.1.4 Sewer sizing chart. The chart shows the linear velocity and slope for tile, concrete, and cast iron sewer pipe during gravity flow. The chart is based on Manning's formula for circular pipes flowing full for values of N = .015 for 10-in pipe and smaller and N = .013 for 12-in pipe and larger.

Process waste = 100 gpm
Fire water = 250 gpm per catch basin
Runoff in unpaved areas = 40%
Computations:
Storm water
Area involved: Paved, 2 areas, 3280 sq ft total.
Unpaved, 3 areas, 6920 sq ft total.

$$\text{gpm} = \frac{(0.31 * 3280) + (0.31 * 6920 * 0.40)}{0.75}$$

gpm = 251
Process Waste + Storm Water = 100 + 251 = 351 gpm
Process Waste + Fire Water = 100 + (4*250) = 1100 gpm

Using 1100 gpm and the sewer sizing chart (see Figure 3.1.4) shows that a 10-in line sloped .017 ft/ft with a flow velocity of 4.4 ft/sec can handle this quantity.

DETERMINE INVERT

The invert of the sewer inlet and outlet at catch basins and manholes should be at the same elevation when both inlet and outlet sewer mains are the same size. Where the sewer size increases at the outlet and the gradient permits, the tops of both inlet and outlet sewer mains should be at the same elevation. However, if this arrangement is not practical because of gradient limitations, the inverts of the inlet and outlet sewer mains can be at the same elevation, preferably with the outlet slightly lower. Straight runs of sewer mains should not change in pipe diameter (size) except at a catch basin or manhole.

—*Mark K. Lee*

3.2 MANHOLES, CATCH BASINS, AND DRAIN HUBS

This section discusses appurtenances in sanitary and industrial sewers. Sewer appurtenancese include manholes, building connections, junction boxes, drain hubs, catch basins, and inverted siphons. Additional information on sewer appurtenances is in publications by Metcalf and Eddy, Inc. (1991), Steel and McGhee (1979), National Clay Pipe Institute (1978) and WPCF (1970).

Sanitary Sewer Appurtenances

Figure 3..2.1 shows a plan and profile of a sanitary sewer and its laterals with enlarged sections of sewer trenches and manholes.

MANHOLES

Manholes should be of durable structure, provide easy access to sewers for maintenance, and cause minimum interference to sewage flow. Manholes for small sewers are usually about 1.2 m (4 ft) in diameter. Sewers larger than 60 mm (24 in) in diameter should have larger manhole bases although a 1.2-m barrel can still be used.

Manholes should be located at the end of the line (called terminal cleanout), at sewer intersections, and at changes in grade and alignment except in curved sewers as shown in Figure 3.2.1. The maximum spacing of manholes is 90–180 m (300–600 ft) depending on the size of the sewer and the sewer cleaning equipment. Manholes, however, should not be located in low places where surface water can enter. If such locations are unavoidable, special water-tight manhole covers should be provided.

Part a in Figure 3.2.2 shows a typical manhole; part a in Figure 3.2.3 shows the details of a terminal cleanout.

Drop Manholes

A drop manhole reduces the turbulence in the manhole when the elevation difference between incoming and outflow sewers is greater than 0.5 m (1.5 ft). Turbulence due to a sudden drop of wastewater can cause splashing, release of odorous gases, and damage to the manhole. Part b in Figure 3.2.2 shows the details of a drop manhole.

Flushing Manholes

In their upper reaches, most sewers receive so little flow that they are not self-cleaning and must be flushed from time to time. This flushing is done by the following means:

Damming the flow at a lower manhole and releasing the stored water after the sewer has almost filled.

Suddenly pouring a large amount of water into an upstream manhole.

Providing a flushing manhole at the uppermost end of the line. A flushing manhole is filled with water through a fire hose attached to a nearby hydrant before a flap valve, shear gate, or similar quick-opening device leading to the sewer is opened.

Installing an automatic flush tank that fills slowly and discharges suddenly. Apart from the cost and maintenance

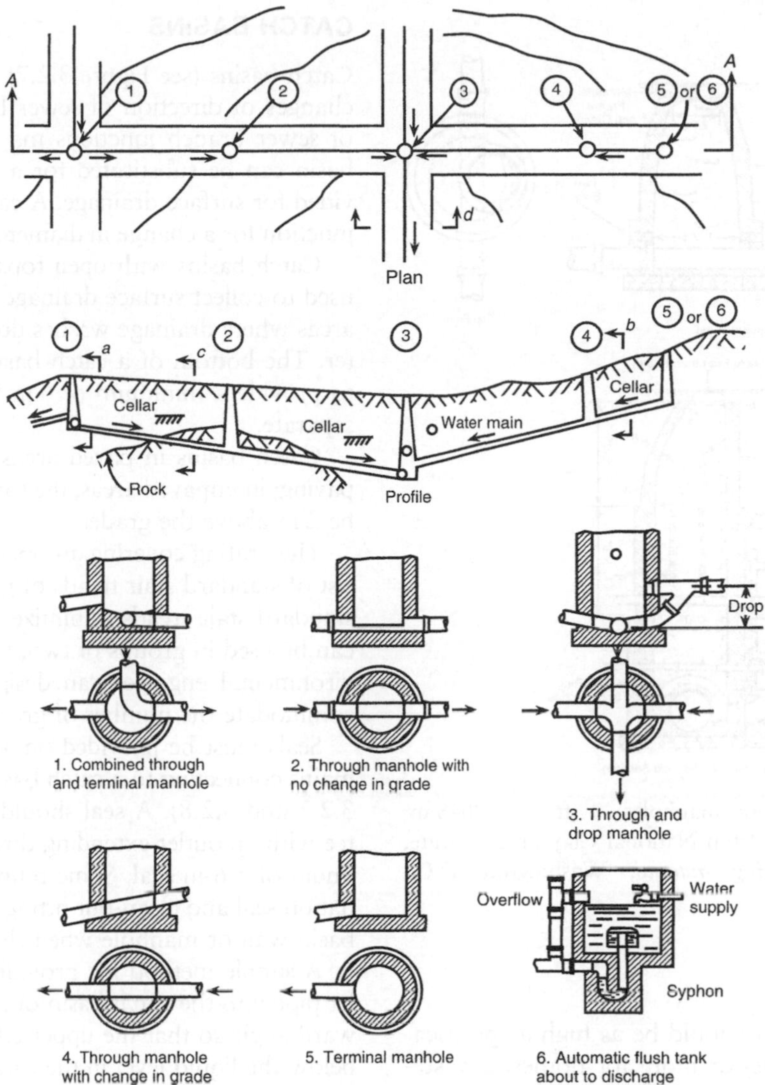

FIG. 3.2.1 Plan, profile, and construction details of sanitary sewers.

difficulties, the danger of backflow from the sewer into the water supply is a negative feature of automatic flush tanks.

BUILDING CONNECTIONS

Building sewers are generally 10–15 cm (4–6 in) in diameter and constructed on a slope of 0.2 m/m. Building connections are also called house connections, service connections, or service laterals. Service connections are generally provided in municipal sewers during construction. While the sewer line is under construction, connections are conveniently located in the form of wyes or tees and plugged tightly until service connections are made. In deep sewers, a vertical pipe encased in concrete (called a chimney) is provided for house connections.

Part b in Figures 3.2.3 and 3.2.4 show the details of house connections.

DEPRESSED SEWERS (INVERTED SIPHONS)

Any dip or sag in a sewer that passes under structures, such as conduits or subways, or under a stream or across a valley, is often called an inverted siphon. It is a misnomer because it is not a siphon. The term *depressed sewer* is more appropriate. Because the pipe constituting the depressed sewer is below the hydraulic grade line, it is always full of water under pressure although little flow may occur in the sewer. Figure 3.2.5 shows a depressed sewer and its associated inlet and outlet chambers.

Due to practical considerations, such as the increased danger of small pipe blockage, the minimum diameters for depressed sewers are usually the same as for ordinary sewers: 150 or 200 mm (6 or 8 in) in sanitary sewers and 300 mm (12 in) in storm water sewers. Since obstructions are more difficult to remove from a depressed sewer, the ve-

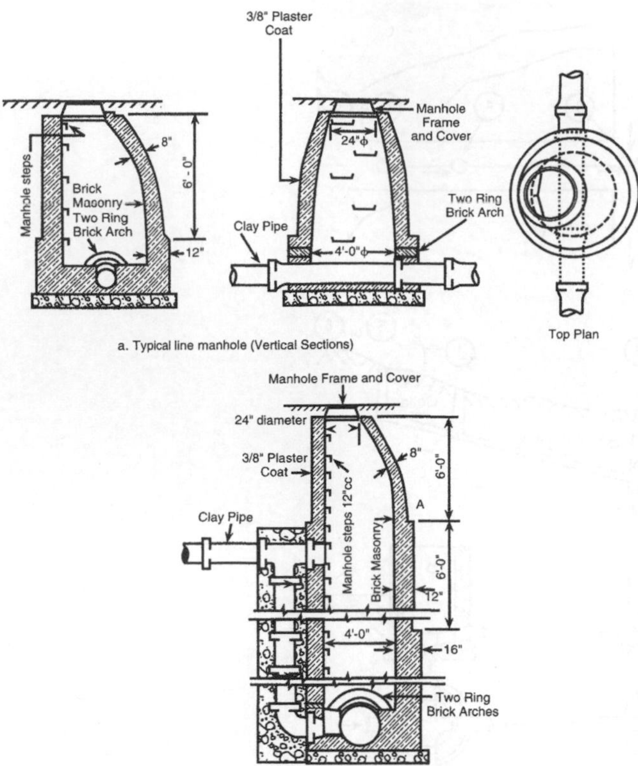

FIG. 3.2.2 Typical designs of manholes. 1 ft = 0.3048 m. (Reprinted, with permission, from National Clay Pipe Institute, 1978, *Clay pipe engineering manual,* Washington, D.C.: National Clay Pipe Institute.)

locity in a depressed sewer should be as high as practicable, about 0.9 m/sec (3 fps) or more for domestic wastewater and 1.25 to 1.5 m/sec (4 to 5 fps) for stormwater. Using several pipes instead of one pipe for a depressed sewer is also advantageous. This arrangement maintains reasonable velocities at all times because additional pipes are brought into service progressively as wastewater flow increases as shown in Figure 3.2.5.

Industrial Sewer Appurtenances

Appurtenances in industrial sewers are shown in Figure 3.1.1 in the preceding section.

DRAIN HUBS

Drain hubs (see Figure 3.2.6) collect drainage from equipment above the grade or paving and run it to the proper classified sewer. Drain hubs need not consist of an actual hub attached to the pipe. A piece of pipe projecting 2 in above grade or paving is sufficient. Drain hubs extending 2 in above the grade or paving prevent surface drainage from entering the sewer systems.

CATCH BASINS

Catch basins (see Figure 3.2.7) are used as a junction for changes of direction of sewer branch lines. The location of sewer branch junctions may coincide so that a catch basin can be substituted for a sewer box, which is provided for surface drainage. A catch basin is also used as a junction for a change in diameter (size) of all sewer mains.

Catch basins with open tops covered with grating are used to collect surface drainage and process waste in unit areas where drainage washes down debris or foreign matter. The bottom of a catch basin should be deep enough to provide a minimum of 6 in for sediment to settle and separate.

Catch basins in paved areas should be flush with the paving; in unpaved areas, the top of the catch basin should be 2 in above the grade.

The grating covering an open-top catch basin can consist of standard stair treads of grating construction. These standard stair treads minimize the cost of gratings. They can be used in groups of two, three, or more; and the environmental engineer can design the catch basin to accommodate the number of gratings used.

Seals must be provided on all sewer branch inlets and mains connecting to a catch basin or manhole (see Figures 3.2.7 and 3.2.8). A seal should consist of an elbow or a tee with an outlet extending downward to provide a minimum of a 6-in seal. Some refineries use a special combination seal and clean-out fitting inserted into the the catch basin wall or manhole when the concrete is poured.

A simple method for providing a seal is to run the inlet pipe into the catch basin or manhole at a sharp downward angle so that the upper edge of the inlet pipe is 6 in below the liquid level in the catch basin or manhole. This method should be used only when the pipeline is short. Fittings should be removable from inside the catch basin or manhole for cleaning and rodding.

A catch basin and manhole can be constructed to have a wire-type wall seal where an application warrants this type of seal.

Areas considered hazardous, where flammable gases accumulate in the sewer mains, must have catch basin and manhole covers sealed or gasketed so that these gases do not escape or leak from the sewer lines. Catch basins and manholes have a 4-in minimum size vent. Areas with furnaces or other fired equipment or ignition sources must run underground vents at least 100 ft from the source of ignition and 10 ft above grade in a safe location.

FLOOR DRAINS

Floor drains inside buildings can be used in floors that only handle water. A sewer box or catch basin must be used where process waste spillage can occur inside a building.

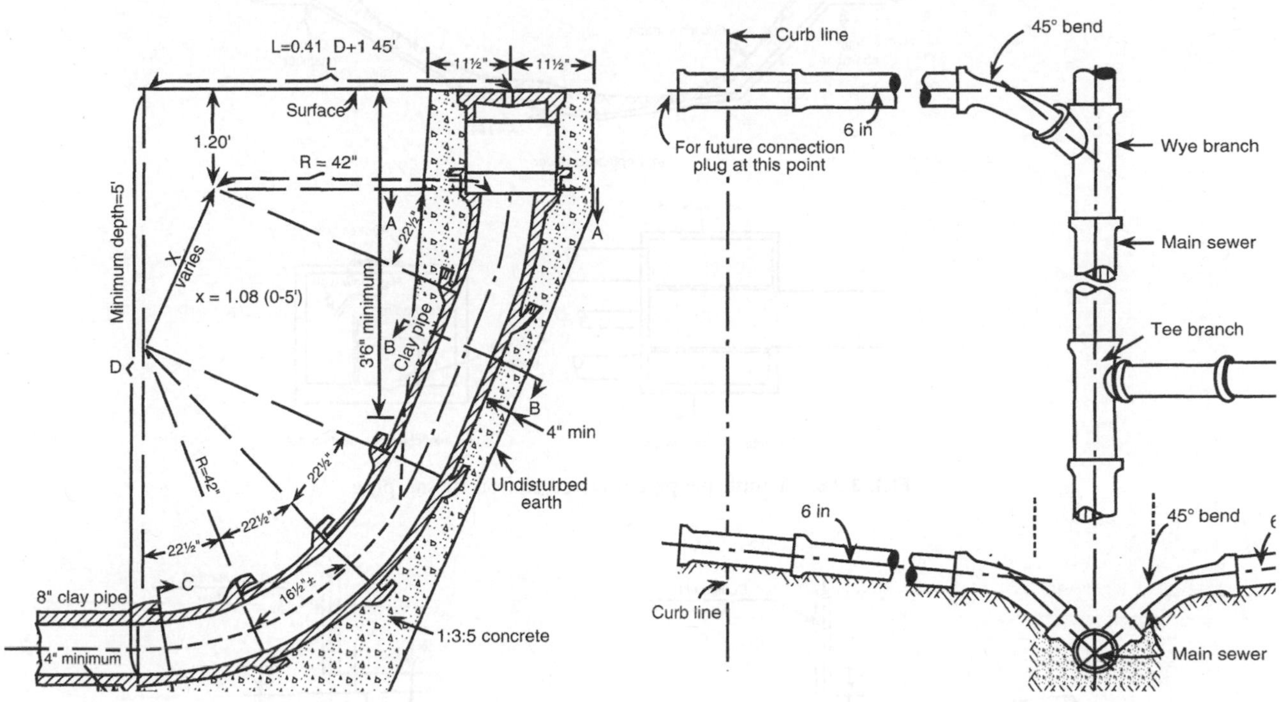

FIG. 3.2.3 Typical terminal cleanouts and house connections. 1 ft = 0.3048 m. (Reprinted, with permission, from National Clay Pipe Institute, 1978; Water Pollution Control Federation and American Society of Civil Engineers).

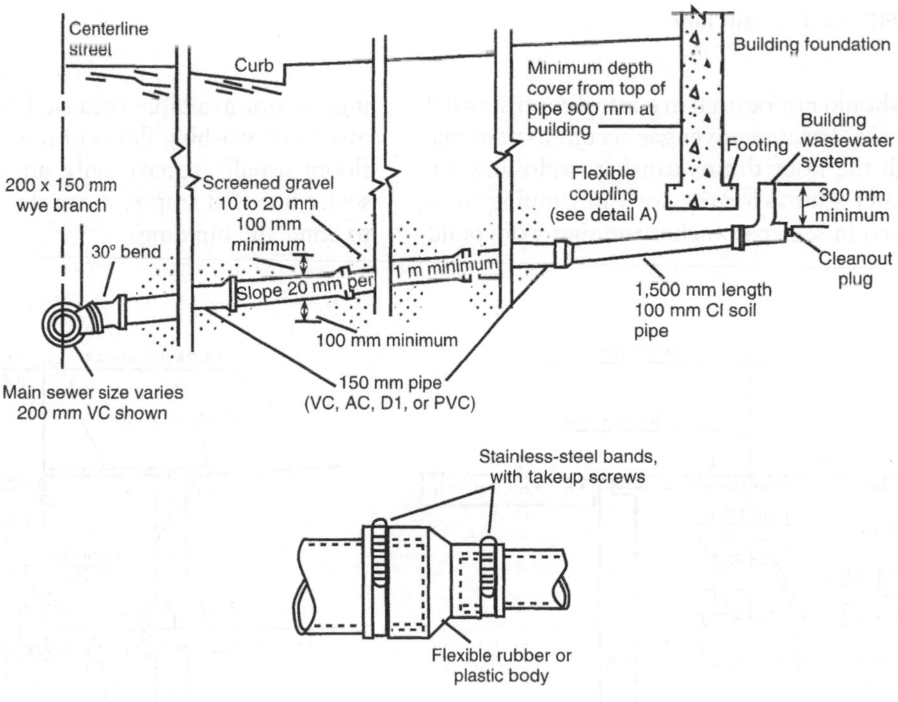

FIG. 3.2.4 Typical house connection. Note: mm × 0.03937 = in.

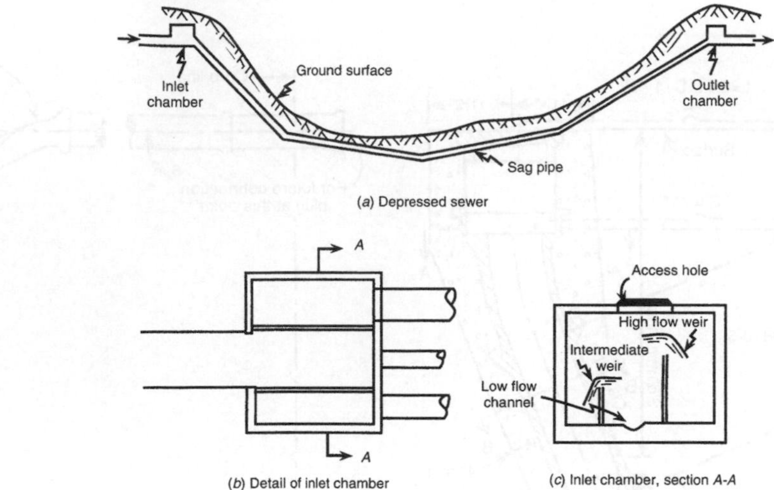

FIG. 3.2.5 A multiple-pipe inverted siphon or sag pipe.

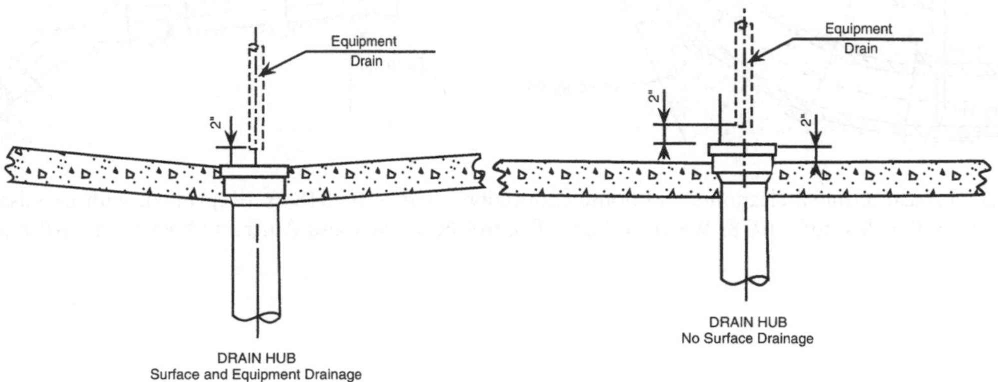

FIG. 3.2.6 Drain hub.

Floor drains should not be used in control rooms, switch rooms, or lavatories, because sewer gas accumulations may back up through the floor drains causing explosions and flash fires. A water source for the seals in running traps, normally provided in sewer branches running from buildings, is not available because little, if any, water is available from washing floors or any other source. Because the floors usually receive only an occasional mopping, the sewer line and trap seal can dry up allowing sewer gases to enter the buildings.

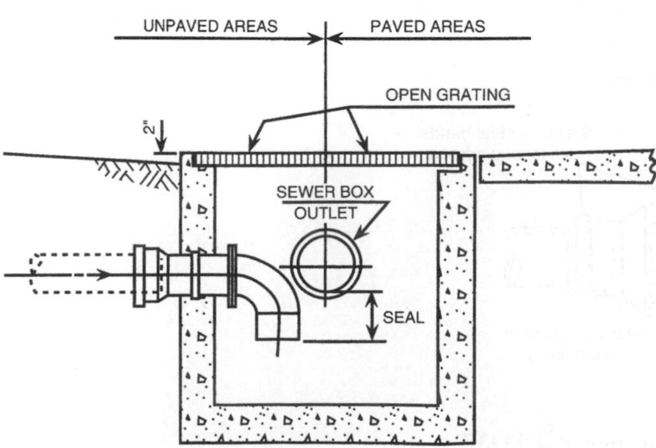

FIG. 3.2.7 Catch basin.

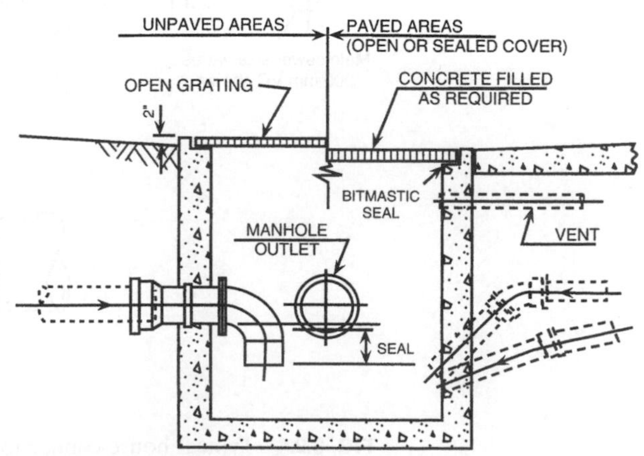

FIG. 3.2.8 Manhole.

Sewer branch mains from a pump house, compressor house, or any enclosed building should have connections to sewer systems at manholes or catch basins with an inlet provided with a seal. Where these connections are not feasible, building drains and separately defined areas should have running clean-out traps (P trap) located outside building walls with accessible clean-out plugs to avoid disturbing the paving.

Running a sewer branch or subbranch in one plane from a drain hub to a catch basin and from a catch basin to a manhole, prevents the sewer line from clogging and facilitates rodding and cleaning. Whenever possible, all sewer system branch connections should be at a catch basin or manhole.

Curbed or diked storage areas should have sufficient catch basins to accommodate surface drainage. The sewer pipe runs from the catch basin through the curb or dike wall to the nearest manhole. The sewer outlet connection inside the catch basin should use a flap valve (see Figure 3.2.9).

A flap valve operates, by chain, at the curb or dike wall. It should be closed at all times except to drain surface water from the enclosure. A gate valve should be installed in the sewer line outside the enclosure and have an extension stem for operation at grade level.

When a leak or break occurs in the storage tank or piping, the flap valve prevents loss of tank contents and allows for the recovery of tank contents retained within the diked area. It also prevents large amounts of hazardous and flammable liquids from entering the sewer system, which can create a hazardous fire condition if tank contents are volatile.

Drain lines that collect hot, noncorrosive waste drainage above 210°F, such as boiler blow-off, steam trap discharges, and hot waste drainage, without receiving cooling quench from other streams should be constructed of steel pipe and fittings. The steel sewer pipe should be run to the nearest catch basin or manhole. Hot acid waste drainage should be run in acid-resistant, alloy material sewer pipe and fittings.

A valve box for underground pipelines should have side walls with drainage only through a gravel bed at the open bottom of the valve box.

Sewer pipes running under roadways and trucking areas must be protected from damage by heavy concrete slabs or protective pipe sleeves. Sewer pipes embedded in concrete foundations should be constructed of steel pipe and fittings.

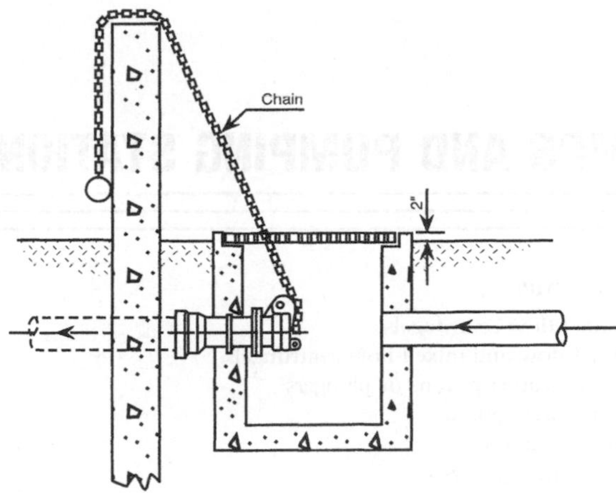

FIG. 3.2.9 Flap valve.

MANHOLES

Manholes should be installed in sewer mains at intervals of 300 ft maximum for sewer sizes to 24 in and at 500 ft maximum intervals for sizes above 24 in (see Figure 3.2.8). Manholes installed at dead ends of sewer mains can act as junctions to connect sewer branches. Manholes are also installed in sewer main runs as junctions, where there are changes in sewer main diameter. As previously explained, open-top manholes covered with grating or catch basins collect surface drainage. For sealed manholes requiring vent connections, the grating cover should be filled with concrete and sealed or bolted down to prevent the escape of sewer gases.

—*Mark K. Lee*

References

Metcalf and Eddy, Inc. 1991. Wastewater engineering: Treatment, disposal collection, and reuse. 3d ed. New York: McGraw-Hill, Inc.

National Clay Pipe Institute. 1978. Clay pipe engineering manual. Washington, D.C.: National Clay Pipe Institute.

Steel, E.W. and T.J. McGhee. 1979. Water supply and sewerage. New York: McGraw-Hill Book Co.

Water Pollution Control Federation (WPCF). 1970.

3.3 PUMPS AND PUMPING STATIONS

TYPES OF PUMPS

a. Radial-flow centrifugals
b. Axial-flow and mixed-flow centrifugals
c. Reciprocating pistons or plungers
d. Diaphragm pumps
e. Rotary screws
f. Pneumatic ejectors
g. Air-lifts

Other pumps and pumping devices are available, but their use in environmental engineering is infrequent.

SELECTION OF PUMPS

See Table 3.3.1.

EFFICIENCIES

Efficiencies range from 85% for large capacity centrifugals (types a and b) to below 50% for many smaller units. For type c, efficiency ranges from 30% up depending on horsepower and number of cylinders. For type d, efficiency is almost 30%, and for types e, f, and g, it is below 25%.

MATERIALS OF CONSTRUCTION

For water using type a or b pumps, normally bronze impellers, bronze or steel bearings, stainless or carbon steel shafts, and cast iron housing; for domestic waste using type a, b, or c pumps, similar except that they are often cast iron impellers; for industrial waste and chemical feeders using type a or c pumps, a variety of materials depending on corrosiveness; type d similar except that the diaphragm is usually rubber; types e, f, and g normally steel components

PUMPING STATIONS

Designed with pump in liquid chamber (wet well design) or pump in dry pit with wet well before it (dry well design). Most are designed for a specific application, but for lower capacity pumping facilities (up to 60 hp), prefabricated stations are common.

PARTIAL LIST OF SUPPLIERS

Allweiler Pump Inc. (e); Aurora Pump Unit, General Signal Corp. (a,b); Barnes Pumps Inc. (a,d); Crane Co. (a,d); Dresser Pump Div. (a,c); Duriron Co. (a,c,d); Fairbank Morse Pump Corp. (a,b); Flygt Corp. (a); Gorman-Rupp Industries Div. (a,b,d); Goulds Pumps (a,c); Ingersoll-Rand Co. (a,c,d); Komline-Sanderson Engineering Corp. (c,d); Lakeside Equipment Corp. (e); Marlow Div., ITT (a,c,d); Moyno Pump, Robbins & Myers Inc. (c,e); Smith & Loveless (a,d,f); Vanton Pump & Equipment Corp. (a,c,d,e); Wallace & Tiernan (c,d); Weil Pump Co. (a,f); Wemco Div., Envirotech (a); Zimpro-Passavant Inc. (c,d,e)

TABLE 3.3.1 PUMP SELECTION

Feature Summary Type Designation	Type of Pump	*For Liquid Pumped*					*For Capacity*	*For Ft of Head*
		Clear Liquids—Low Viscosity	*Clear Liquids—High Viscosity*	*Thin Slurries or Suspensions*	*Raw or Partially Treated Sewage and Heavy Suspensions*	*Viscous or Thick Slurries and Sludges*	*GPM* 0.1 1.0 10 10^2 10^3 10^4 10^5	*PSIG* 0.1 1.0 10 10^2 10^3 / 0.1 1.0 10 10^2 10^3 10^4 *FEET OF HEAD (H_2O)*
a	Radial-flow centrifugals	✔		✔	✔①	✔①		
b	Axial-flow and mixed-flow centrifugal	✔		✔	✔			
c	Reciprocating pistons and plungers	✔②	✔	✔	✔②	✔		
d	Diaphragm pumps	✔②	✔②	✔	✔	✔		
e	Rotary screws		✔②		✔②	✔		
f	Pneumatic ejectors				✔			
g	Airlift pumps			✔	✔			

Notes: ✔Suitable for normal use. ① See text for limitations. ② Not used for this purpose in environmental engineering (with some exceptions). If not checked, either not suitable or not normally used for this purpose.

Pumping applications at wastewater treatment facilities include pumping (1) raw or treated wastewater, (2) grit, (3) grease and floating solids, (4) dilute or well-thickened raw sludge or digested sludge, (5) sludge or supernatant return, and (6) chemical solutions. Pumps and lift stations are also used extensively in the collection system. Each pumping application requires specific design and pump selection considerations.

Pumping stations are often required for pumping (1) untreated domestic wastewater, (2) storm water runoff, (3) industrial wastewater, (4) combined domestic wastewater and storm water runoff, (5) sludge at a wastewater treatment plant, (6) treated domestic wastewater, and (7) circulating water systems at treatment plants.

This section briefly discusses various pumps and their applications. Because centrifugal pumps are commonly used for raw wastewater pumping, centrifugal pumps and pump selection, the design procedure for a raw wastewater pumping station, specifications of pumping, and control equipment are described.

Pump Types and Applications

Figure 3.3.1 shows the basic configurations of some pump installations.

CENTRIFUGAL PUMPS

This classification is the most common type of pump, including radial-flow centrifugals and axial- and mixed-flow centrifugals. In the form of tall, slender, deep-well submersibles, they pump clear water from depths greater than 2000 ft. Horizontal centrifugals with volutes almost the size of a man can pump 9000 gpm of raw sewage through municipal treatment plants. Few applications are beyond their range, including flow rates of 1 to 100,000 gpm and process fluids from clear water to the densest sludge.

Radial-Flow Centrifugals

Radial-flow pumps throw the liquid entering the center of the impeller or diffuser out into a spiral volute or bowl. The impellers can be closed, semiopen, or open depending on the application (see Figure 3.3.2). Closed impellers have higher efficiencies and are more popular than the other two types. They can be readily designed with nonclogging features. In addition using more than one impeller can increase the lift characteristics. These pumps can have a horizontal or vertical design.

Axial- and Mixed-flow Centrifugals

Axial-flow propeller pumps, although classed as centrifugals, do not truly belong in this category since the propeller thrusts rather than throws the liquid upward. Impeller vanes for mixed-flow centrifugals are shaped to provide partial throw and partial push of the liquid outward and upward. Axial- and mixed-flow designs can handle large capacities but only with reduced discharge heads. They are constructed vertically.

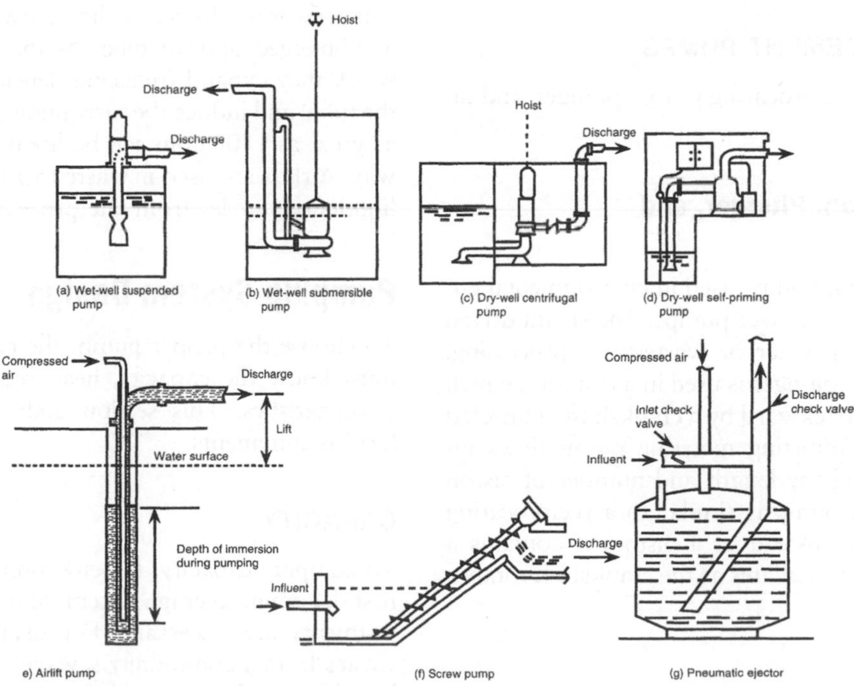

FIG. 3.3.1 Types of pumps and pumping stations.

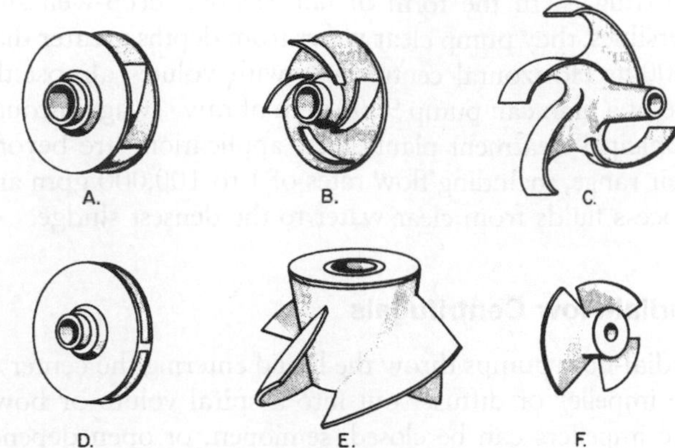

FIG. 3.3.2 Types of centrifugal pump impellers. **A.** Closed impeller; **B.** Semiopen impeller; **C.** Open impeller; **D.** Diffuser; **E.** Mixed flow impeller; **F.** Axial flow impeller.

Applications

Most water and waste can be pumped with centrifugal pumps. Therefore, listing the applications for which they are not suited is easier than listing the ones for which they are. They should not be used for the following: (1) Pumping viscous industrial liquids or sludges. The efficiencies of centrifugal pumps drop to zero, and therefore positive displacement pumps are used. (2) Low flows against high heads. Except for deep-well applications, the large number of impellers needed is a disadvantage for the centrifugal design. (3) Low to moderate liquid flows with high-solids contents. Except for the recessed-impeller type, rags and large particles clog smaller centrifugals.

POSITIVE DISPLACEMENT PUMPS

These pumps include reciprocating piston, plunger, and diaphragm pumps.

Reciprocating Piston, Plunger, and Diaphragm Pumps

Almost all reciprocating pumps used in environmental engineering are metering or power pumps. The steam-driven pump is rarely used in water or wastewater processing. Frequently, a piston or plunger is used in a cylinder, which is driven forward and backward by a crankshaft connected to an outside drive. Adjusting metering pump flows involves merely changing the length and number of piston strokes. A diaphragm pump is similar to a reciprocating piston or plunger, but instead of a piston, it contains a flexible diaphragm that oscillates as the crankshaft rotates.

Applications

Plunger and diaphragm pumps feed metered amounts of chemicals (acids or caustics for pH adjustment) to a water or waste stream. They also pump sludge and slurries in waste treatment plants.

ROTARY SCREW PUMPS

In this type, a motor rotates a vaned screw or rubber stator on a shaft to lift or feed sludge or solid waste material to a higher level or the inlet of another pump.

AIR PUMPS

These pumps include pneumatic ejectors and airlifts.

Pneumatic Ejectors

In this pumping method waste flows into a receiver pot, and an air pressure system then blows the liquid to a treatment process at a higher elevation. A controller is usually included, which keeps the tank vented while it is being filled. When the tank is full, the level controller energizes a three-way solenoid valve to close the vent port and open the air supply to pressurize the tank.

The air system can use plant air (or steam), a pneumatic pressure tank, or an air compressor. With large compressors, a capacity of 600 gpm with lifts of 50 ft can be obtained. This system has no moving parts in contact the waste; thus, no impellers become clogged. Ejectors are normally more maintenance free and operate longer than pumps.

Airlifts

Airlifts consist of an updraft tube, an air line, and an air compressor or blower. Airlifts blow air into the bottom of a submerged updraft tube. As the air bubbles travel upward, they expand (reducing density and pressure within the tube) and induce the surrounding liquid to enter. Flows as great as 1500 gpm can be lifted short distances in this way. Airlifts are used in waste treatment to transfer mixed liquors or slurries from one process to another.

Pumping System Design

To choose the proper pump, the environmental engineer must know the capacity, head requirements, and liquid characteristics. This section addresses the capacity and head requirements.

CAPACITY

To compute capacity, the environmental engineer should first determine average system flow rate, then decide if adjustments are necessary. For example, when pumping wastes from a community sewage system, the pump must handle peak flows roughly two to five times the average flow, depending on community size. Summer and winter flows and future needs also dictate capacity, and population trends and past flow rates should be considered in this evaluation.

HEAD REQUIREMENTS

Head describes pressure in terms of feet of lift. It is calculated by the expression:

$$\text{Head in feet} = \frac{\text{Pressure (psi)} \times 2.31}{\text{Specific gravity}} \qquad 3.3(1)$$

The discharge head on a pump is a sum of the following contributing factors:

STATIC HEAD (h_d)—The vertical distance through which the liquid must be lifted (see Figure 3.3.3).

FRICTION HEAD (h_f)—The resistance to flow caused by friction in the pipes. Entrance and transition losses can also be included. Because the nature of the fluid (density, viscosity, and temperature) and the nature of the pipe (roughness or straightness) affect friction losses, a careful analysis is needed for most pumping systems although tables can be used for smaller systems.

VELOCITY HEAD (h_v)—The head required to impart energy into a fluid to induce velocity. Normally this head is quite small and can be ignored unless the total head is low.

SUCTION HEAD (h_s)—Reduces the pressure differential that the pump must develop when a positive head is on the suction side (a submerged impeller). If the water level is below the pump, the suction lift plus friction in the suction pipe must be added to the total pressure differential required.

TOTAL HEAD (H)—Expressed by the following equation:

$$H = h_d + h_f + h_v \pm h_s \qquad 3.3(2)$$

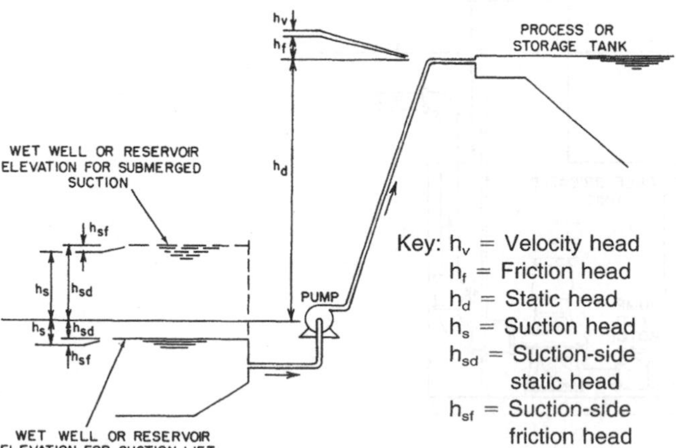

FIG. 3.3.3 Determination of pump discharge head requirements.

SUCTION LIFT

The amount of suction lift that can be handled must be carefully computed. As shown in Figure 3.3.4, it is limited by the barometric pressure (which depends on elevation and temperature), the vapor pressure (which also depends on temperature), friction and entrance losses on the suction side, and the net positive suction head (NPSH)—a factor that depends on the shape of the impeller and is obtained from the pump manufacturer.

SPECIFIC SPEED

The impeller's rotational speed affects the capacity, efficiency, and extent of cavitation. Even if the suction lift is within permissible limits, cavitation can be a problem and should be checked. The specific speed of the pump is determined with the following equation:

$$\text{Specific speed, } N_s = \frac{\text{RPM} \times \sqrt{\text{Capacity (gpm)}}}{H^{3/4}} \qquad 3.3(3)$$

Charts are available showing the upper limits of specific speed for various suction lifts.

HORSEPOWER

The horsepower required to drive the pump is called *brake horsepower* (bhp). The following equation determines the brake horsepower:

$$\text{bhp} = \frac{\text{Capacity (gpm)} \times \text{H (ft)} \times \text{Sp. Gr.}}{3960 \times \text{Pump efficiency}} \qquad 3.3(4)$$

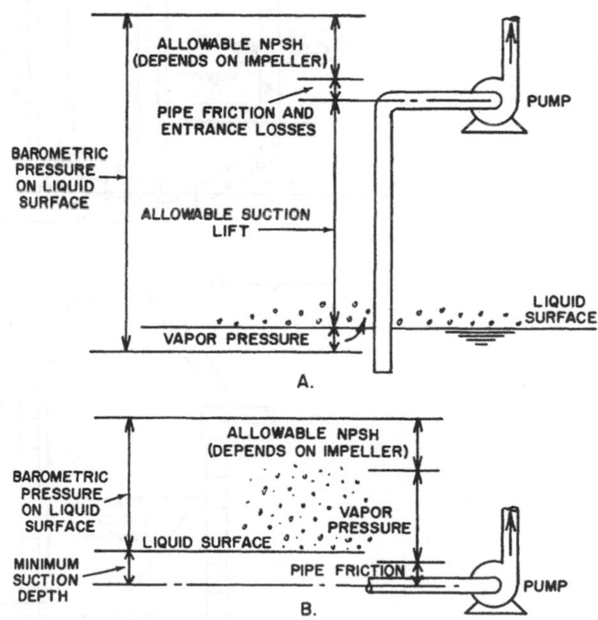

FIG. 3.3.4 Role played by NPSH in determining allowable suction lift. **A.** Pump with suction lift. **B.** Pump with submerged suction but high vapor pressure (possibly hot water).

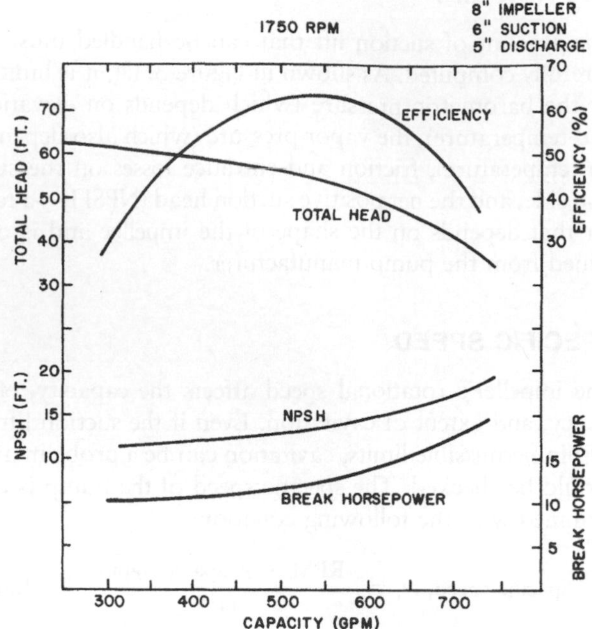

FIG. 3.3.5 Typical pump curve for a single impeller.

PUMP CURVES

Essential pump features are described by performance curves. Charts or tables that summarize pump curve data are also available. Figure 3.3.5 shows a typical centrifugal pump curve.

Pumping Station Design

Figure 3.3.6 shows typical pump station designs. Figure 3.3.7 illustrates a pneumatic ejector package. In selecting the best design for an application, environmental engineers should consider the following factors:

Many gases are formed by domestic waste, including some that are flammable. When pumps or other equipment are located in rooms below grade, the possibility of explosion or gas buildup exists, and ventilation is extremely important.

When wastewater is pumped at high velocities or through long lines, the hammering caused by water can be a problem. Valves and piping should be designed to withstand these pressure waves. Even pumps that discharge to the atmosphere should use check valves to cushion the surge.

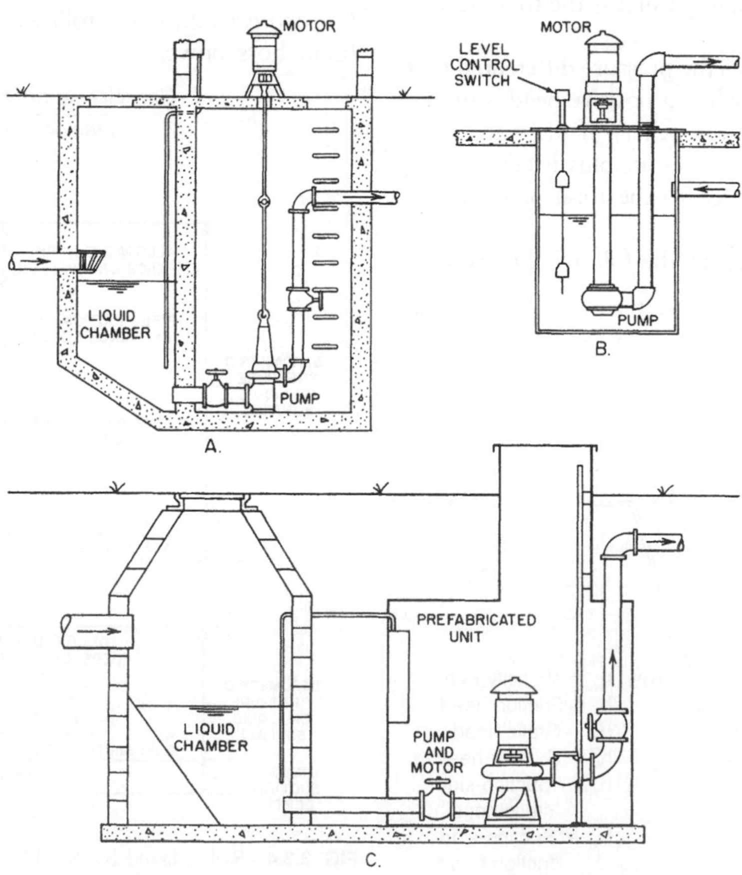

FIG. 3.3.6 Pumping stations. A. Dry-well design; B. Wet-well design; C. Prefabricated pumping station.

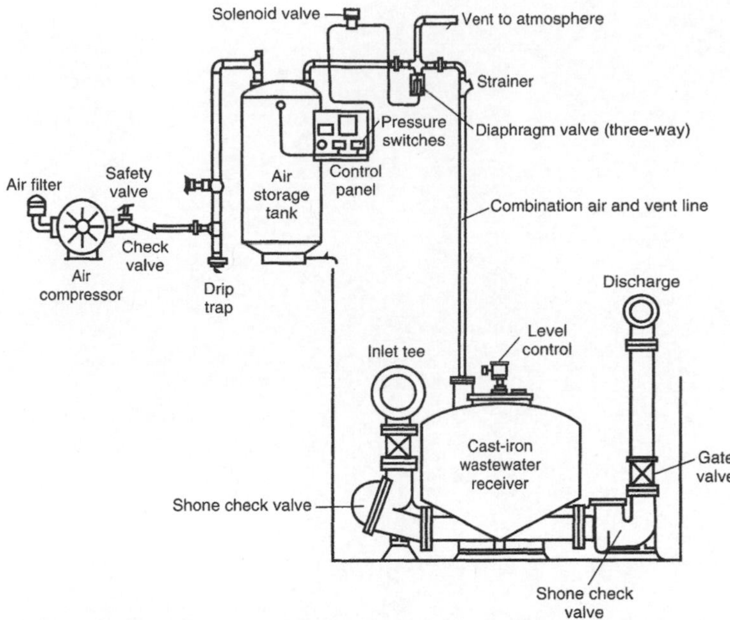

FIG. 3.3.7 Pneumatic ejector and associated piping.

Bar screens and comminutors are not recommended, but for small centrifugal pump stations, they can be necessary.

Pump level controls are not fully reliable because rags can short electrodes and hang on floats. Purged-air systems (air bubblers) require less maintenance but need an air compressor that operates continuously. Therefore, maintenance-free instrumentation must be provided.

Charts and formulas are available for sizing wet wells, but infiltration and runoff must also be considered.

Sump pumps, humidity control, a second pump with an alternator, and a pump hoisting mechanism are recommended.

Most states prefer dry-well designs.

—R.D. Buchanan
David H.F. Liu

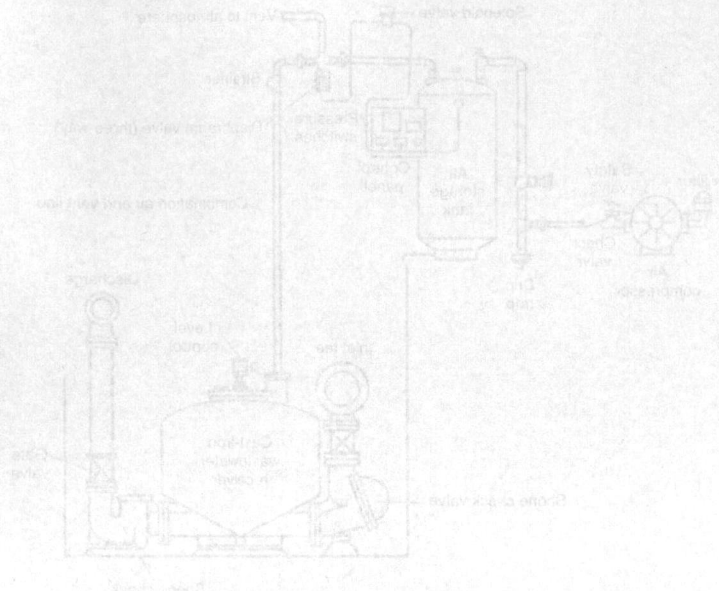

4

Equalization and Primary Treatment

Ronald G. Gantz | János Lipták | David H.F. Liu |

4.1 EQUALIZATION BASINS

Purpose of Flow Equalization

Flow equalization is not a treatment process but a technique that improves the effectiveness of secondary and advanced wastewater treatment processes. Flow equalization levels out operation parameters such as flow, pollutant levels, and temperature over a time frame (normally 24 hr), minimizing the downstream effects of these parameters. Environmental engineers determine the need for flow equalization primarily based on the potential effects of the waste stream on the receiving waters or treatment facility.

This effect is determined by the following key components:

- The variability of operating parameters to be equalized (including toxicity)
- The volume of the flow being discharged

In defining the need for flow equalization, environmental engineers need sufficient background information on these factors as well as information on the relative cost of constructing and implementing effective flow equalization facilities and the cost savings by reducing the effects on downstream equipment.

This section provides information on flow equalization processes used to pretreat industrial waste streams, considers the effects of each process, and provides basic design criteria for each.

The following locations are suitable for flow equalization:

Near the head end of treatment work. Flow equalization usually involves constructing large basins to collect and store wastewater flow, from which wastewater is pumped to the treatment plant at a constant rate. These basins are normally located near the head end of the treatment work, preferably downstream of pretreatment facilities such as bar screens, comminutors, and grit chambers.

Prior to discharge. Wastewater flows have a diurnal variation from less than $\frac{1}{2}$ to more than 200% of the average flowrate. In addition, daily volumes increase from inflows and infiltration into the sewer collection system during wet weather. Municipal waste strength also has a diurnal variation resulting from nonuniform discharges of domestic and industrial waste. Industrial waste entering a municipal system can cause excessive flows and peak organic loads. Therefore, facilities should be installed at industrial sites for flow smoothing prior to discharge.

Prior to advanced waste treatment operations. Many advanced operations, such as filtration and chemical clarification, are adversely affected by flow variation and sudden changes in solid loading. Maintaining a uniform influent improves chemical feed control and process reliability. The costs saved by installing smaller units for chemical precipitation and filtration, together with reduced operating expenses, can compensate for the added costs of flow equalization facilities.

Offline in a collection system. Figure 4.1.1 shows the treatment scheme using side-line flow equalization. This facility uses biological–chemical processing followed by multimedia filters. The flow equalization basin is a circular concrete tank with a volume of 315,000 gal, which is equivalent to 15% of the 2.1 million-gallons-per-day (mgd) design flow. The process pumps transfer a constant preset flow from the wet well for treatment, and variable-speed pumps deliver excess flow to the equalization basin. During periods of low influent flow, wastewater is released from the basin to the wet well to maintain the established flow through the plant.

As in-line units. Equalization chambers can also be in-line units that pass all wastewater through the basins.

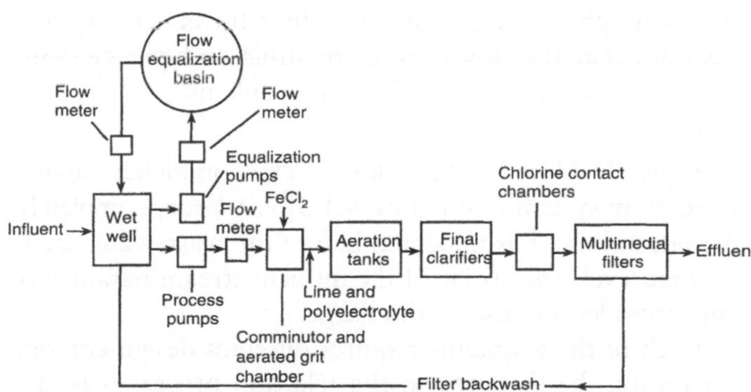

FIG. 4.1.1 Process diagram for biological–chemical treatment followed by filtration using side-line flow equalization, Walled Lake–Novi Waste Water Treatment Plant. (Reprinted from U.S. Environmental Protection Agency (EPA), 1974, *Flow equalization,* Technology Transfer, 19, U.S. EPA [May].)

Although the normal placement is between grit removal and primary settling, holding tanks can be placed at other points in the treatment. For example, a basin serving as a pump suction pit can be located just ahead of the filters to dampen hydraulic surges without providing complete flow equalization.

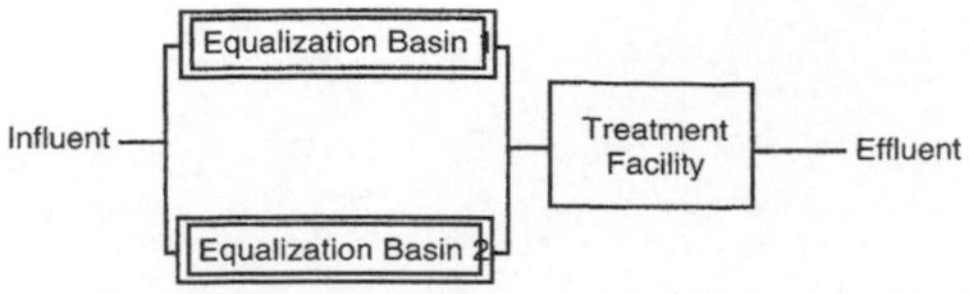

FIG. 4.1.2 Alternating-flow diversion equalization system.

Flow Equalization Processes

Four basic flow equalization processes are as follows:

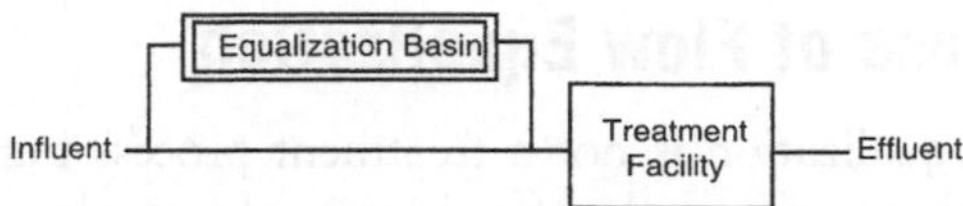

FIG. 4.1.3 Intermittent-flow diversion system.

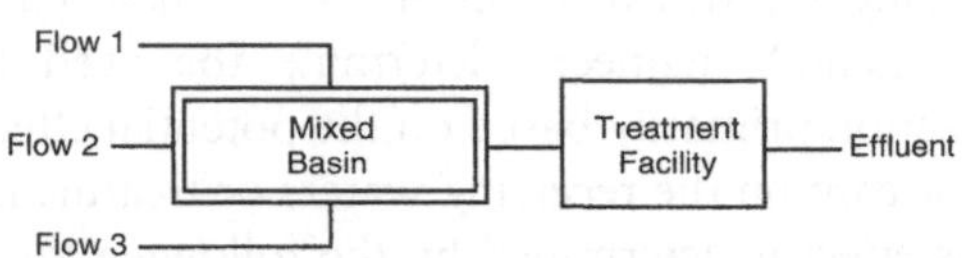

FIG. 4.1.4 Completely mixed, combined-flow system.

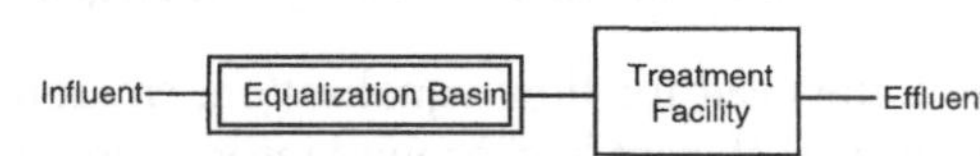

FIG. 4.1.5 Completely mixed, fixed-flow system.

Alternating Flow Diversion. The alternating-flow diversion system (see Figure 4.1.2) collects the total flow of an effluent for a time period (normally 24 hrs) while a second basin is discharging. The basins alternate between filling and discharging for successive time periods. Thorough mixing is maintained so that the discharge maintains constant pollutant levels with a constant flow. This system provides a high degree of equalization for a basin size by leveling all discharge parameters. A disadvantage of this system is the high construction cost associated with storing the waste stream volume for the time period used.

Intermittent Flow Diversion. The intermittent-flow diversion system (see Figure 4.1.3) diverts significant variance in stream parameters to an equalization basin for short durations. The diverted flow is then bled into the stream at a controlled rate. The rate at which the diverted flow is fed back to the main stream depends on the volume and variance of the diverted water, reducing downstream effects.

Completely Mixed, Combined Flow. The completely mixed, combined-flow system (see Figure 4.1.4) completely mixes multiple flows combined at the front end of the facility. This system reduces the variance in each stream by thoroughly mixing with the other flows. This system assumes that the flows are compatible and can be combined without creating additional problems.

Completely Mixed, Fixed Flow. The completely mixed, fixed-flow system (see Figure 4.1.5) is a large, completely mixed, holding basin located before the wastewater facility that levels variations of the influent stream parameters and provides a constant discharge.

Each of these systems requires different design criteria. Therefore, the first step in the selection process is to define the type of variability the system must equalize. Then, the facility can be designed with appropriate criteria.

Design of Facilities

The design of equalization facilities begins with a detailed study to characterize the nature of the wastewater and its variability. This study should also include gathering data on flow and pollutants of consequence.

A primary consideration is the effect of the effluent on downstream facilities. The most significant quantity is the mass flow rate; therefore, data on both flow and concentration (in terms of BOD_5, TSS, or other variables) must be measured on a time-series basis. Previous studies indicate that this type of data is normally distributed; therefore, the average mass flow from the sampled values is an estimate of the true average mass flow.

Because the collected data is time-series, obtaining random samples is difficult. Time-series data are by nature not random. Therefore, the study must contain sufficient samples for proper characterization of the statistical parameters as follows (McKeown and Gellman 1976):

For cyclical data, a minimum of two cycles must be collected. The spacing of data should be small enough to have a reasonable probability of measuring peak or minimum values.

Where seasonable considerations are important, at least one sampling program should be conducted during each season.

For flow equalization design, a minimum recommendation for industry waste sampling is two weeks of data for variables of primary concern (chemical constituents, COD/BOD, or TSS). Environmental engineers should collect samples every hour for the first day using an auto discrete sampler and collect composite 24-hr samples for the remaining thirteen days.

Environmental engineers can also use strip flow chart flow recordings and real-time TOC analyzers to determine variability. If possible, they should gather hourly flow data for the entire two-week sampling period.

While statistical analysis and extrapolation for confidence levels are important in determining the effects of variance (on reducing potential effects), they are not the focus of this section. This section assumes that analysis has occurred and deals with the individual concepts.

ALTERNATING FLOW DIVERSION

Because the alternating flow diversion system is intended to hold the total flow for a fixed time period (normally 24 hr), its design is based strictly on flow. Therefore, the design criteria depend on flow variability, standard deviation, and maximum flow for the time frame.

For example, an industrial facility has a total daily flow and pollution profile as shown in Table 4.1.1. If a thirty-day period is assumed to represent one month, the equalization basin can be designed using Table 4.1.1 and a management design criterion of 110% of maximum flow. Each of the equalization basins is designed to hold 686.98 m³. Therefore, the management design criterion (that is, 110% of maximum flow) becomes the dominant variable in this equation.

This variable is given to the design engineers by plant management or assumed to be based on prior experience. The following equation applies:

$$V_t = D_c F_c T k \qquad 4.1(1)$$

where:

- V_t = Volume of the equalization basin, m³
- T = Time period of equalization
- D_c = Management design criteria, %
- F_c = Management flow criteria (f_a, average flow for a time period, or f_m, maximum flow for a time period), m³/hr
- k = Units conversion constant

INTERMITTENT FLOW DIVERSIONS

Intermittent flow diversion systems are more complex as design criteria include the variance of pollutants being diverted, the average length of the variance, and the rate of discharge back to the system. Environmental engineers must evaluate each of these factors with respect to the effect on downstream processes, especially if biological systems follow. This equalization system is best used when variances are easily detectable and infrequent and can have a dramatic effect on downstream processes. An example of this variance is the phenol levels in an effluent stream.

TABLE 4.1.1 INDUSTRIAL FACILITY DAILY FLOW PROFILE

Day of Month	*Total Flow, m³/d*	*Phenol Levels, µg/l (ppb)*
1	377.44	45
2	471.46	393
3	411.96	421
4	254.35	433
5	350.64	683
6	464.19	822
7	339.29	123
8	624.52	467
9	569.57	682
10	47.15	732
11	420.13	398
12	238.46	541
13	553.42	868
14	487.81	558
15	241.18	656
16	562.75	329
17	272.52	822
18	229.83	771
19	237.55	613
20	348.47	400
21	134.44	821
22	0.00	0
23	143.98	160
24	610.90	214
25	97.65	670
26	398.78	362
27	560.94	303
28	253.44	245
29	574.11	120
30	525.06	251
Average daily	360.59 (0.25 m³/min)	463
Minimum	0.00	0
Maximum	624.52	868

The following steps should be applied to the design of an intermittent flow diversion system.

Step 1: Determine the frequency and duration of the variance to be diverted (to design of the equalization basin)

Step 2: Calculate the controlled release rate of the diverted flow to maintain normal operation.

Step 3: Use the diverted volume to calculate the surge basin volume to maintain continuous flow to the treatment facility.

Step 4: Verify that the equalized flow meets discharge limits.

As stated earlier in this section, data collection and system profiling are keys to an effective design for this type of equalization system. An effective system is automated based on electronic monitoring of the stream, with diversion occurring as necessary. Three examples of this technology are pH sensors to monitor pH for excursions, online gas chromatographs to monitor phenol excursions, and conductivity sensors to monitor TDS. Variations of these parameters can cause substantial damage to biological systems or receiving waters (especially when only primary treatment is used).

For example, in Table 4.1.1, the phenol levels vary substantially from day to day because of variance in plant operation. With the variability of plant operation, diverting the flow at this facility to prevent violation, and bleeding the diverted flow back as concentrations allow is necessary.

The phenol level in Table 4.1.1 shows 24 hr composite samples with a discharge limit of 500 parts per billion (ppb). Further analysis of individual samples indicates that the problem was generated during two periods over the course of the day, lasting about 3 hr each. Also, during this time frame, flow rate increased to 0.473 m/min.

Therefore, the total volume to be diverted is calculated as follows:

$$V_D = F_D T_D f_D k \quad \textbf{4.1(2)}$$

where:

V_D = Volume of flow to be diverted per time period, m
F_D = Flow rate diverted, m/min
T_D = Time of diversion, hr
f_D = Frequency of diversion, number/day
k = Conversion constant for unit, min/hr

Therefore, V_D = (0.473 m^3/min)(3 hr)(2/day) (60 min/hr) = 170.28 m^3/day.

The control discharge rate can be established as follows:

$$f_C = V_D/Tk \quad \textbf{4.1(3)}$$

where:

f_C = Controlled discharge rate, m^3/min
V_D = Volume diverted, m^3
T = Time period for return, hr
k = Conversion factor for unit

Therefore, f_C = (170.28 m^3/24 hr)(1 hr/60 min) = 0.118 m^3/min.

The volume of the surge basin can now be calculated. As calculated in Equation 4.1(2), 170.28 m^3 of the total flow will be diverted and fed back to the stream at a constant rate. Therefore, the average flow for the remainder of the time is (360.09 − 170.28) = 189.81 m^3 for the 18-hr period. This amount equates to 0.1318 m^3/min on a 24-hr basis. Correspondingly, maintaining this flow for the 6-hr diversion period requires a surge basin equal to the volume for the diversion time frame (6 hr in this case) at an average flow rate for the remaining period. This volume can be calculated as follows:

$$V_S = F_A T_D k \quad \textbf{4.1(4)}$$

where:

V_S = Volume of the surge basin, m^3
F_A = Average flow rate without diversion flow, m^3/min
T_D = Diversion time period, hr
k = Conversion factor

Therefore, V_S = (0.175 m^3/min)(6 hr)(60 min/hr) = 63.22 m^3.

Any excesses in design capacity determined by management as part of the design criteria are not represented in calculation.

Combining the return of the diverted flow with the mainstream can be accomplished with in-line mixing or flash mixing just before downstream processes. The total combined flow (f_T) is calculated as follows:

$$f_T = f_A + f_C \quad \textbf{4.1(5)}$$

where:

f_A = Average flow rate without diversion, m^3/min
f_C = Controlled discharge rate, m^3/min

Therefore, f_T = (0.118 + 0.176) m^3/min = 0.294 m^3/min.

COMPLETELY MIXED COMBINED FLOW

The completely mixed, combined flow, equalization system addresses the variability resulting from multiple flows from different sections of a plant. This variability often generates impulse or step input changes to the wastewater treatment facility. The primary purpose of this system is to trim impulse variance or provide a more gradual change in operating parameters.

Again, the volume of the equalization basin is determined based on the effects the change in operating parameters has on downstream systems. Because this situation is more complex, this discussion approaches the design details from the simplest perspective: time and combined flows.

Therefore, the volume of the equalization basins V_e is calculated as follows:

$$V_e = (\Sigma f_i) T_e k \quad \textbf{4.1(6)}$$

where:

f_i = Individual flow rates, m^3/min
T_e = Time for equalization, hr
k = Conversion factor for units

For example, if three flows come into the equalization basin with flow rates of 1.98, 0.567, and 0.189 m^3/min, respectively, and the required equalization time is 1 hr, the following equation applies:

$$\begin{aligned} V_e &= (f_1 + f_2 + f_3)T_e k \\ &= (1.98 + 0.567 + 0.189)(1\ \text{hr})(60\ \text{min/hr}) \\ &= 164.16\ \text{m}^3 \end{aligned} \quad \textbf{4.1(7)}$$

From here, the environmental engineer can calculate the relative change in each operating parameter using the following formulas and converting the variability of the individual stream to the variability in the total flow:

$$Var_T = (Var_{pi}) \frac{f_i}{f_t} \qquad 4.1(8)$$

where:

Var_T = Variance in the concentration of the total stream, ppm or ppb (mg/l or μg/l)
Var_{pi} = Variance in the concentration of the individual stream, ppm or ppb (mg/l or μg/l)
f_i = Flow of the individual stream, m^3/min
f_t = Flow of the total stream, m^3/min

For example, if the concentration of a pollutant in an individual stream changes by 50 mg/l, the total stream changes as follows:

$$Var_T = 50\ (150/700) = 10.7 \text{ mg/l} \qquad 4.1(9)$$

This variance can be used in the calculation as a change in the concentration of the combined stream and a potential effect on the downstream system.

A typical industrial waste problem involves constant flow with wastewater concentration as the only variable. Environmental engineers can use the following method in designing facilities to reduce this kind of concentration variability. The method assumes the data are normally distributed.

For example, if a completely mixed, constant flow tank has a variable concentration input and discrete samples are collected at a uniform interval time interval Δt, the influent variance (s^2) can be estimated as follows:

$$s^2 = [(C_i - C)^2]/(n - 1) \qquad 4.1(10)$$

where:

C_i = Influent concentration at the i time interval
C = Mean concentration
n = Number of samples

The influent coefficient of variation (v_o) is as follows:

$$v_o = s/C \qquad 4.1(11)$$

An estimate of required equalization time based on the variation of concentration and sampling interval is calculated as follows:

$$\Theta = \Delta t/2[(v_o/v_t)/v_e]^2 \qquad 4.1(12)$$

where:

Θ = Required equalization time, hr
Δt = Sampling interval, hr
v_o = Influent coefficient of variation of concentration, mg/l
v_t = Average influent concentration, mg/l
v_e = Effluent variability coefficient, mg/l

Both the influent and effluent coefficients of variability are based on discrete samples collected at uniform time intervals Δt.

Normally, raw wastewater characteristics, providing v_o and Δt, are the only information available. The effluent variability v_e must be related to the downstream requirements and therefore is the primary design variable. The environmental engineer must select it based on subsequent treatment units and effluent standards. Where specific limits on acceptable variability do not exist, engineering judgment must be exercised.

The effluent variability V_e can be estimated as follows:

$$V_e = \{[(C_e \max/C) - 1]/C\}/N \qquad 4.1(13)$$

where:

(C) max = The equalization tank effluent concentration not to be exceeded
C = Mean value of concentration
N = Cumulative standard normal for the required confidence level (confidence level is the probability that a specified concentration will not be exceeded.)

Cumulative standard N can be selected from the abbreviated Table 4.1.2. Application of this method is illustrated in the example based on data in Table 4.1.3: v_t = 698 mg/l and v_o = 158.6 mg/l.

If downstream conditions (for example, the next treatment unit in line) restrict effluent variability to 10% (v_e = 0.1), using Equation 4.1(12) gives the required equalization time as follows:

$$\Theta = 1/2[(158.6/698)/0.1]^2 \qquad 4.1(14)$$

Specific restrictions on variability are uncommon; a more realistic problem is to design an equalization tank so that the effluent does not exceed a specified value.

These analyses can ultimately produce the type of curve shown in Figure 4.1.6. This graph allows the subjective analysis of a particular tank size to determine how well it suits the requirements.

If a detention time of 3 hr has tentatively been selected based on the foregoing analysis and physical considerations at a plant, the effluent from this size tank is not expected to exceed the value of 800 mg/l approximately 5% of the time, or about eight samples per week. Reducing this expectation to fewer than two samples per week (that is, a confidence level of 99%) exceeding 800 mg/l means increasing the detention time to approximately 7 hr.

TABLE 4.1.2 SELECTION OF CUMULATIVE STANDARD NORMAL FOR A DESIRED CONFIDENCE LEVEL

Confidence Level	*Cumulative Standard Normal N*
90.0	1.282
95.0	1.645
99.0	2.327
99.9	3.091
99.99	3.719

TABLE 4.1.3 HOURLY INFLUENT COD (MG/L) DATA FOR FOUR-DAY PERIOD

Hour of Day	*First Day*	*Second Day*	*Third Day*	*Fourth Day*
7	413	565	485	723
8	468	612	409	765
9	510	536	466	864
10	568	637	482	844
11	487	536	507	669
Noon	600	684	631	711
1	674	644	695	879
2	638	615	545	847
3	638	662	660	876
4	648	468	545	890
5	584	752	736	890
6	697	738	666	1030
7	629	752	704	1090
8	606	655	625	920
9	626	695	730	823
10	684	800	679	1030
11	742	738	853	1050
Midnight	729	380	612	1010
1	884	708	504	736
2	638	678	606	882
3	677	648	599	812
4	1210	608	5651	832
5	995	738	590	867
6	780	662	631	775

Source: A.T. Wallace and D.M. Zellman, 1971, Characterization of time varying organic loads, *J. Sanit. Eng. Div., Proc. ASCE* 97:257.

Notes: Average = 698 mg/l
Maximum = 1210 mg/l
Standard deviation = 158.6 mg/l

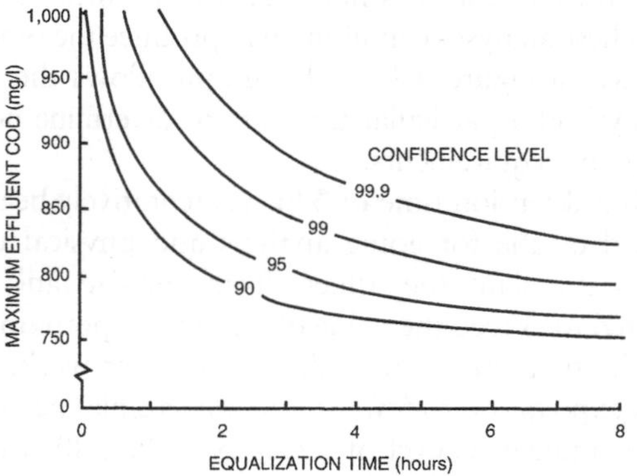

FIG. 4.1.6 Maximum effluent values as a function of equalization time and confidence level.

Similarly, if the confidence level of 90% (one sample in ten or roughly two samples per day exceeding 800 mg/l) is acceptable, the size of the tank can be reduced to yield a detention time of about 2 hr.

CUMULATIVE FLOW CURVE

Equalization basins for individual facilities can be sized based on a cumulative flow or mass diagram. This well-known method has long been used to determine the storage required for water reservoirs. The graphic technique consists of plotting cumulative flow versus time for one complete cycle (24 hr for municipal facilities). Two parallel lines, with slopes representing the rate of pumping or flow of the equalization tank, are drawn tangent to the high and low points of the cumulative flow curve. The required tank size is the vertical distance between the two tangent lines. The method is shown in Table 4.1.4 and Figure 4.1.7.

The preceding procedure provides the tank size for the flow-time trace of one day. The variability and thus the

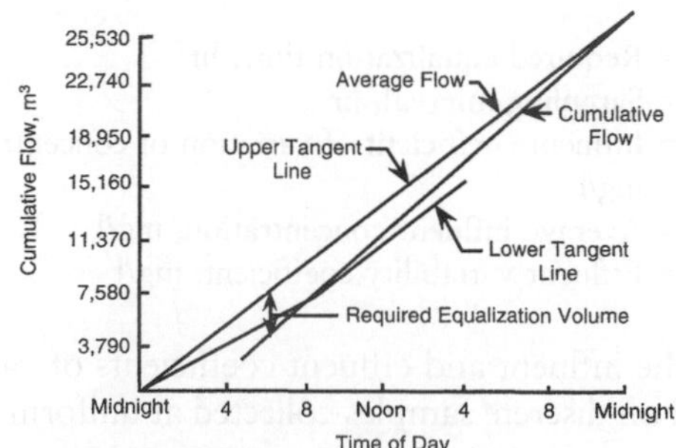

FIG. 4.1.7 Cumulative flow curve.

TABLE 4.1.4 FLOW DATA FOR EXAMPLE PROBLEM

Time	*Flow Rate,* m³/hr*	*Cumulative Flow, m³*	*Time*	*Flow Rate,* m³/hr*	*Cumulative Flow, m³*
Midnight	946	0	1	1110	12,830
1	901	901	2	1400	14,290
2	799	1700	3	1310	15,600
3	753	2453	4	1490	17,090
4	738	3191	5	1350	18,440
5	719	3910	6	1100	19,540
6	749	4659	7	1370	20,910
7	780	5439	8	1420	22,330
8	1000	6439	9	1370	23,700
9	1370	7809	10	1100	24,800
10	1280	9089	11	1270	26,070
11	1230	10,320	Midnight	1230	27,300
Noon	1400	11,720			

Note: Flow from Ewing Township, New Jersey, WWTP.

amount of equalization required changes from day to day. Therefore, environmental engineers must select a day or flow rate that represents the flow conditions to be equalized.

Operational Considerations

This section discusses the operational considerations of equalization systems including mixing and draining and cleaning requirements.

MIXING REQUIREMENTS

The contents of an equalizing vessel must be mixed. Typically, continuous mechanical mixing is best although inlet arrangements sometimes provide the necessary homogeneity of soluble waste constituents. If settleable and floatable solids are present, the wastewater must be mixed to maintain a constant effluent concentration and prevent accumulations. For biodegradable waste, the equalization tank will develop odor problems unless aeration is provided. Aeration and mixing systems can be combined (for example, floating surface aerators).

Although mixing power levels vary with basin geometry, 0.3 l/m^3 sec (18 cfm/1000 cu ft) of basin volume is the minimum to keep light solids in suspension (approximately 0.02 kW/m^3 [0.1 hp/1000 gal]). Heavy solids like grits, swarf from machining, fly ash, and carbon slurries require more mixing energy. The most common approaches to mixing are baffling, mechanical agitation, aeration, or a combination of all three.

Baffling

Although not a true form of mixing and less efficient than other methods, baffling prevents short-circuiting and is the most economical. Over-and-under or around-the-end baffles can be used. Over-and-under baffles are preferable in wide equalization tanks because they provide more efficient horizontal and vertical distribution.

The influent should be introduced at the tank bottom so that the entrance velocity prevents SS in the wastewater from sinking and remaining on the bottom. Additionally, a drainage valve should be located on the influent side of the tank to allow drainage of the tank when necessary. Normally, baffling is not recommended for wastewater that has a high concentration of settleable solids.

Mechanical Mixing

Because of its high efficiency, mechanical mixing is typically recommended for smaller equalization tanks, wastewater with higher suspended concentrations, and waste streams with rapid waste strength fluctuations. Environmental engineers select mechanical mixers on the basis of pilot plant tests or data provided by manufacturers. When pilot plant results are used, geometrical similarity should be preserved, and the power input per unit volume should be maintained. Because power is wasted when water levels change by vortex formation, designers should avoid creating a vortex by mounting the mixer off center or at a vertical angle or by extending baffles out from the wall.

Because both mechanical and diffused-aeration systems must have a minimum depth to maintain mixing, extra volume should be provided below the low water level.

Aeration

Mixing by aeration is the most energy-intensive of the equalization methods. In addition to mixing, aeration provides chemical oxidation of reducing compounds as well as physical stripping of volatile chemical compounds. Some states require an air discharge permit for discharging volatile organic emissions to the atmosphere or classifying an equalization tank as a process tank.

Waste gases can be used for mixing if no harmful substance is added to the wastewater. Flue gas containing large quantities of carbon dioxide can be used to mix and neutralize high-pH wastewater.

DRAINING AND CLEANING

Equalization systems should be sloped to drain, and a water supply should be provided for flushing without hoses; otherwise, the remains after draining can cause odor and health problems.

—*David H.F. Liu*

Reference

McKeown, J.J. and I. Gellman. 1976. Characterizing effluent variability from paper industry wastewater treatment processes employing biological oxidation. *Prog. Water Technol.* 8:147.

4.2 SCREENS AND COMMINUTORS

PARTIAL LIST OF SUPPLIERS

Screens

American Well Works; BIF Sanitrol, a unit of General Signal; Chain Belt Co.; Envirex Inc., a Rexnord Co.; FMC Corp. Materials Handling Div.; Hycor Corp.; Keene Corp., Water Pollution Control; Lakeside Equipment Corp.; Link Belt Co.; LYCO; Walker Process Corp.; Welker Equipment Co.; Wemco

Comminutors

Chicago Pump Co.; Clow Corporation; Infilco; Worthington Pump Corp.; Yeomans

Screens are usually installed at the entrance of the wastewater treatment plant to protect mechanical equipment, avoid interference with plant operations, and prevent objectionable floating materials such as rags or rubber from entering the primary settling tanks. Screening devices intercept floating or suspended larger material. The retained material is then removed and disposed of by burial or incineration or is returned into the waste flow after grinding.

Trash racks or coarse racks are screening devices constructed of parallel rectangular or round steel bars with clear openings, usually 2 to 6 in (5.1 to 15.2 cm). They protect combined sewer systems from large objects and are usually followed by regular bar screens or comminutors.

Bar screens are racks of inclined or vertical flat bars installed in a channel and are basically protective devices. They are used ahead of mechanical equipment such as raw sewage pumps, grit chambers, and primary sedimentation tanks.

Where mechanical screening, comminuting devices, or both are used, manually cleaned, auxiliary bypass bar screens should also be provided. These bypass screens provide automatic diversion of the entire sewage flow should the mechanical equipment fail (see Figure 4.2.1). Velocity

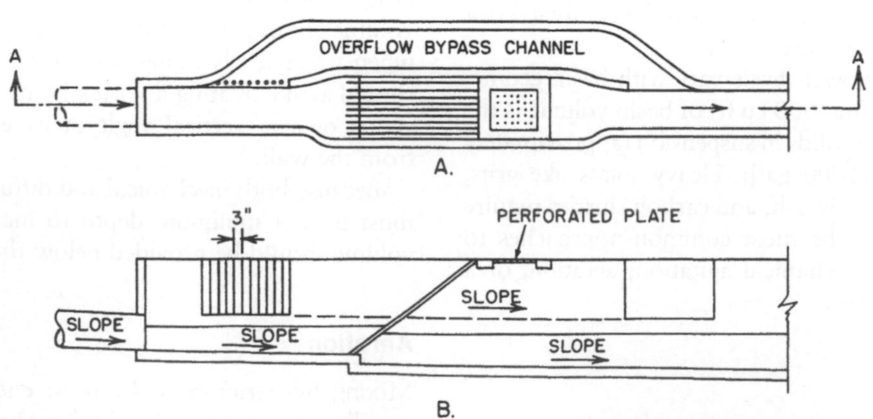

FIG. 4.2.1 Hand-cleaned bar screen with overflow bypass. **A.** Plan view; **B.** A–A section.

distribution in approach and discharge channels is important to screen operation, and a straight arrangement for approach channels is recommended.

The wastewater treatment plant should have gates installed to divert the flow from mechanical screens and comminutors or distribute the flow. Provisions should also be made for dewatering each unit. To compensate for head loss through the racks and prevent jetting action behind the screen, the rack chamber floor should be 3 to 6 in below the approach channel invert.

Environmental engineers often determine the width of the bar screen on the basis that the net submerged area of openings below the crown of the incoming sewer line be not less than 150 to 250% of the cross-sectional area of the influent sewer.

Screen Openings and Hydraulics

Screen openings should be narrow enough to retain sticks, rags, and other trash but wide enough to allow excreta and toilet paper to pass. Table 4.2.1 lists common screen openings. The lower the velocity through the screen, the greater the amount of material removed from the waste. Deposition of solids in the channel, however, prohibits reducing the velocity beyond certain limits.

The *Ten states' standards* (Great Lakes-Upper Mississippi River Board of State Sanitary Engineers 1968) require that the average rate of flow velocity through manually raked bar screens should be approximately 1 fps and the maximum velocity during wet weather periods through mechanically cleaned bar screens should not exceed 2.5 fps. The velocity should be calculated from a vertical projection of the screen openings on the cross-sectional area between the channel invert and the flowline.

Head loss for screens varies with the quantity and nature of the screenings that accumulate between cleanings. Environmental engineers can calculate the head loss created by a clean screen by considering the flow and the effective area of the screen openings as follows:

$$H = \frac{V^2 - v^2}{2g} \times \frac{1}{0.7} \text{ or } H = 0.0222(V^2 v^2) \qquad 4.2(1)$$

where:

H = head loss, ft
V = velocity through screen, fps
v = velocity of incoming waste, fps
g = acceleration due to gravity

The minimum allowance for loss through a hand-cleaned screen is 6 in, assuming frequent attention to the screens by operating personnel. The maximum head loss through clogged racks should be kept below 2.5 ft.

Material collected on screens impedes the flow. Excessive back ups in the incoming line can cause pounding and deposition of putrefying solids. When screens are cleaned, high flow surges occur that can cause hydraulic and treatment problems at the plant. The use of steeper grades in the influent pipe preceding the screen can reduce these problems.

Mechanically cleaned screens are usually protected by enclosures. Efficient ventilation is important and prolongs the life of both the equipment and the enclosure. The enclosure should have separate outside entrances. Convenient access and ample working space are important. Convenient unloading and handling of rackings can be provided by screw conveyors, belt conveyors, containers, or buckets.

Screen Types

HAND-CLEANED BAR SCREENS

Many bar screens have no mechanical cleaning devices and are cleaned periodically by hand rakes (see Figure 4.2.2). Manually cleaned screens, except those used for emergency, should be placed on a slope of 30° to 45° with the horizontal. This positioning increases screening surface by 40 to 100%, facilitates cleaning, and prevents excessive head loss by clogging. The clear opening between bars should be 1 to 1¾ in, and operators should remove screening as often as necessary (from two to five times a day) to secure free flow in the sewer.

When excessive head loss is expected, an overflow bypass channel equipped with a trash rack (vertical bars set 3 to 4 in apart) can be used.

TABLE 4.2.1 TYPICAL SCREEN OPENINGS

Type of Screen	*Opening*	*Remarks*
a) Trash racks	2–6 in (5.1–15.2 cm)	Most commonly 3 in (7.6 cm)
b1) Manually cleaned bar screens	1–1¾ in (2.5–4.4 cm)	
b2) Mechanical screens	⅝–1 in (1.4–2.5 cm)	Most commonly ¾ in (1.9 cm)
e) Fine screen	$\frac{3}{32}$–$\frac{3}{16}$ in (0.24–0.48 cm)	A few have openings smaller than $\frac{3}{32}$ in (0.24 cm).
Comminuting devices	¼–¾ in (1.0–1.9 cm)	The opening is a function of the hydraulic capacity.

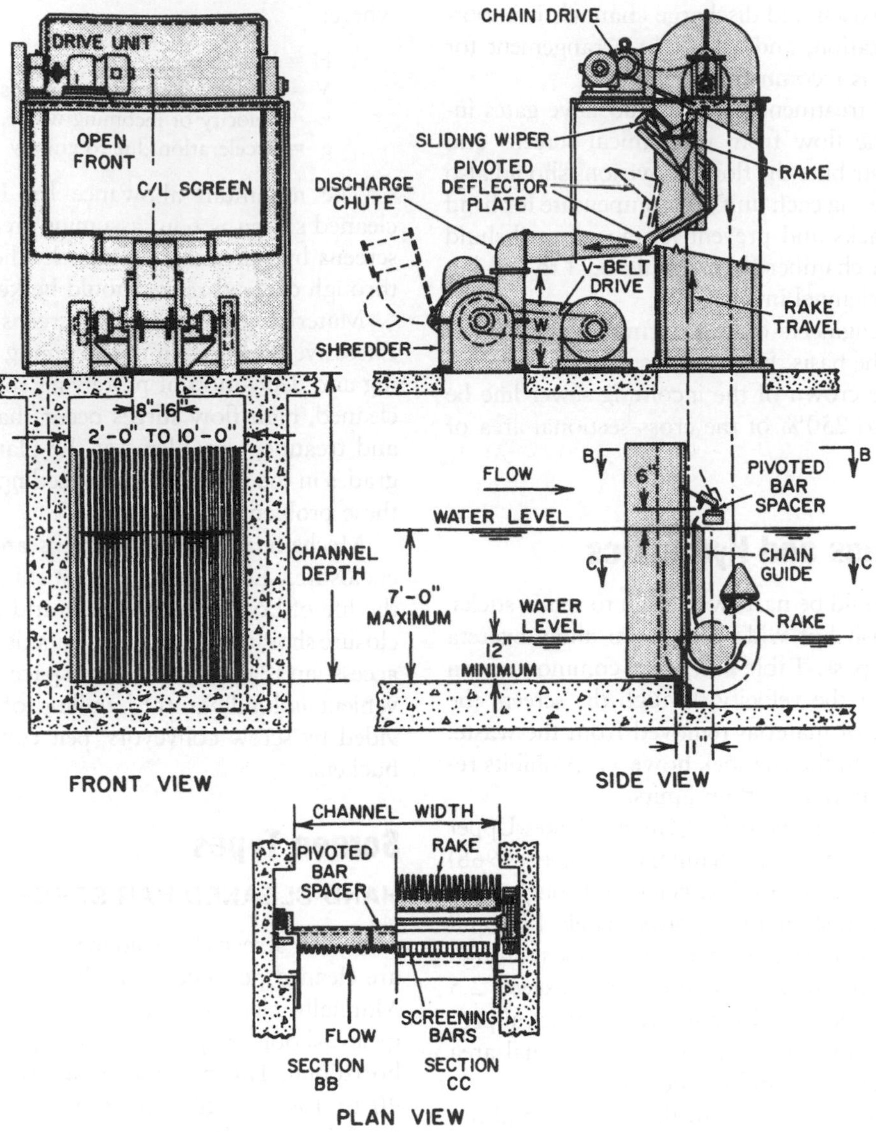

FIG. 4.2.2 Backcleaned flat-bar screen.

MECHANICALLY CLEANED BAR SCREENS

These screens are also called mechanical screens or mechanical rakes. Almost all large- and medium-size treatment plants use mechanically cleaned bar screens. The clear opening for mechanical screens is between 1 and $\frac{5}{8}$ in. This size is why mechanical screens are sometimes called fine racks.

The controls that operate mechanically cleaned screens include a manual start and stop switch, a clock-operated automatic start and stop switch, a high-water-level switch with or without an audible alarm; a head-loss- (screen pressure drop) actuated start and stop switch, and an overload switch with or without an audible alarm.

All motors and controls should be explosion proof. The racks are cleaned with long-tined rakes that fit into the openings. Cross bars or bolts should be located so that they do not interfere with raking.

Mechanical flat-bar screens can be back cleaned or front cleaned. The rakes travel at a rate of 7 to 20 fpm and can be adjusted to rest at the top for a period from 3 sec to 60 min. Backcleaning mechanisms are not subject to jamming at the bottom by trash deposits because they are also protected by the screen which is cleaned.

Discharge of screenings can be at the front or back. The front discharge of screenings is preferable because any raking dropped upstream of the screen can be recovered.

In *backcleaned flat-bar screens,* the raking mechanism consists of a rake or series of rakes with the ends attached to a pair of endless chains. These chains continuously move the rake slowly upward over the back face, or effluent side of the screen, carrying the rakings to the top of the screen. There they are dropped into a conveyor or bucket or onto a screening platform (see Figure 4.2.2).

In *frontcleaned flat-bar screens,* the cleaning mechanism

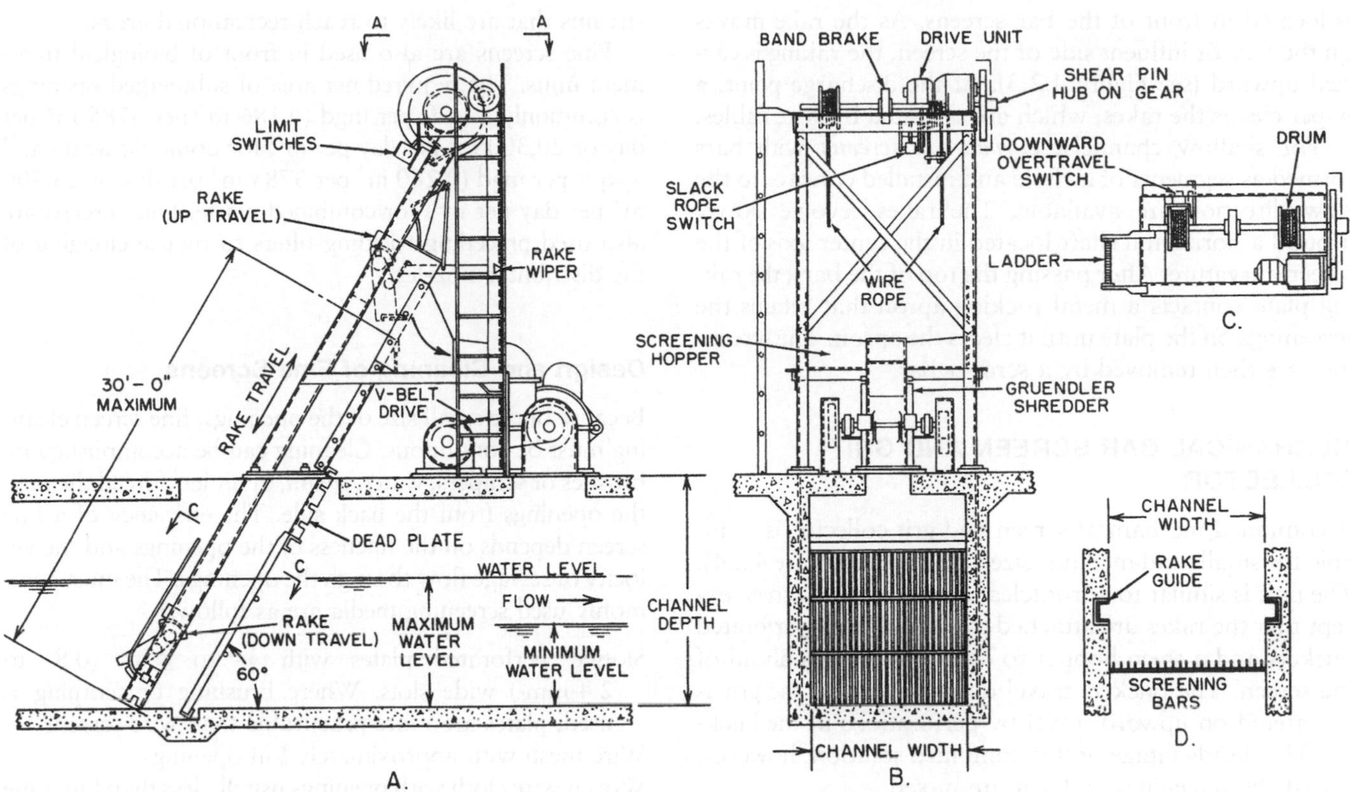

FIG. 4.2.3 Frontcleaned flat-bar screen. A. Side view; B. Rear view; C. Plan at AA; D. Section CC

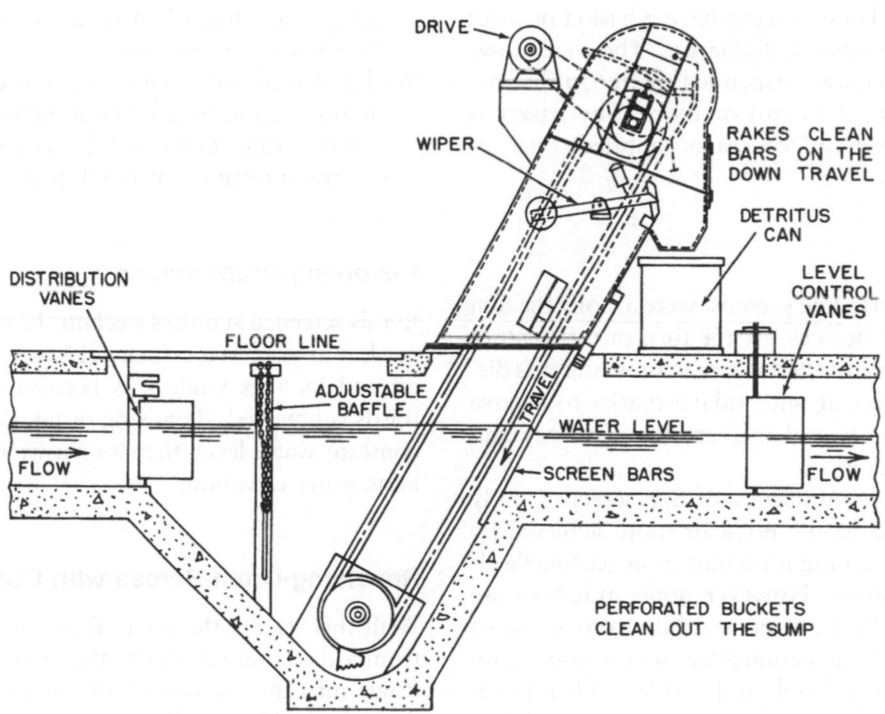

FIG. 4.2.4 Mechanical bar screen and grit collector.

is located in front of the bar screens. As the rake moves on the face or influent side of the screen, the raking is carried upward (see Figure 4.2.3). At the discharge point, a wiper cleans the rakes, which are operated by wire cables.

For shallow channels, *curved-bar screens* with bars formed as segments of a circle and installed concave to the flow direction are available. The rakes revolve slowly around a horizontal shaft located in the center axis of the screen curvature. After passing the top of the bars, the raking plate contacts a metal rocking apron that retains the screenings on the plate until it clears the apron. The screenings are then removed by a scraper bar.

MECHANICAL BAR SCREEN AND GRIT COLLECTOR

A combined mechanical screen and grit collector is available for small- and medium-sized plants (see Figure 4.2.4). The unit is similar to a frontcleaned mechanical screen except that the rakes are attached to one or more perforated buckets and a steep hopper to collect the grit is ahead of the screen. The buckets travel downward, and the grit is dewatered on upward travel by perforations in the buckets. The disadvantage of this combined solution, however, is that the screenings and grit are mixed.

COARSE-MESH SCREENS

Instead of bar screens or comminutors, a few plants use coarse-mesh screens. These screens have a basket of wires or rods into which the waste is discharged. The water flows through, leaving the coarse suspended matter in the basket. Mesh size is 1 in (2.54 cm) or more. The basket is raised at intervals by hand or crane, emptied, then replaced.

FINE SCREENS

Screens, and particularly fine screens, were among the first wastewater treatment devices. At the turn of this century fine screens were often installed where wastewater was discharged into large rivers or wide tidal estuaries to remove unsightly floating matter and increase the efficiency and economy of chlorination.

Since the SS removal efficiency of fine screens is only 10 to 20% compared to the 60% or more achieved by sedimentation, fine screening is no longer an acceptable alternative to sedimentation. However, some industries use fine screens successfully to remove solids from waste of processes for meat packing, canning (causes excessive scum or foaming in digesters), wool, and textiles. When possible, fine screens should be installed at the source of the industrial waste.

Instead of sand filters, some wastewater treatment plants use fine screens to remove fine suspended matter from treatment plant effluent when it is discharged into streams that are likely to reach recreational areas.

Fine screens are also used in front of biological treatment units. The required net area of submerged openings is commonly 2 sq ft per mgd (0.186 m^2 per 3785 m^3 per day or 20,300 m^3 per day per m^2) for domestic waste and 3 sq ft per mgd (0.280 m^2 per 3785 m^3 per day or 13,500 m^3 per day per m^2) for combined waste. Fine screens are also used preceding trickling filters to reduce clogging of the distributor nozzles.

Design and Cleaning of Fine Screens

Because of the small size of the openings, fine screen cleaning must be continuous. Cleaning can be accomplished by brushes or scrapers, water, steam, or air jets forced through the openings from the back side. The efficiency of a fine screen depends on the fineness of the openings and the velocity of sewage flow through the openings. The most commonly used screening media are as follows:

Slotted perforated plates with $\frac{1}{32}$- to $\frac{3}{32}$-in- (0.8- to 2.4-mm-) wide slots. Where brushing or scraping is used, plates are more practicable than wire screens.

Wire mesh with approximately $\frac{1}{8}$-in openings

Woven wire cloth with openings usually less than $\frac{1}{8}$ in. One type is made of bars that are wedge shaped in cross section, with the large end of the bar in the face of the screen. The wedge-shaped bars are set in slots of a series of U-bars to maintain spacing and hold the bars in place. The standard openings range from $\frac{1}{100}$ to $\frac{1}{4}$ in, and the slots are continuous.

Wedge-shaped wire. The wire is pressed into a wedge-shaped section built into flat panels that have openings varying from 0.005 to $\frac{3}{16}$ in. The screens are made of corrosion-resistant material, preferably stainless steel.

Revolving-Drum Screen

In this screen, a stainless-steel no. 12 to no. 20 woven-wire mesh is applied to a cylindrical frame. The cylinder rotates around its axes while it is between one-third and two-thirds submerged. Revolving drum screens require a fairly constant water level; therefore, a weir plate usually maintains water elevation.

Revolving-Drum Screen with Outward Flow

With this screen, the waste flow can approach the drum from a direction parallel to the revolving axis. The liquid flows into the interior of the drum at one end, passes through the filter media, and flows out at a right angle to the axis (see Figure 4.2.5). The solids, which are retained on the inside surface of the screen, are raised above the liquid level as the drum slowly rotates and are usually removed by a water spray.

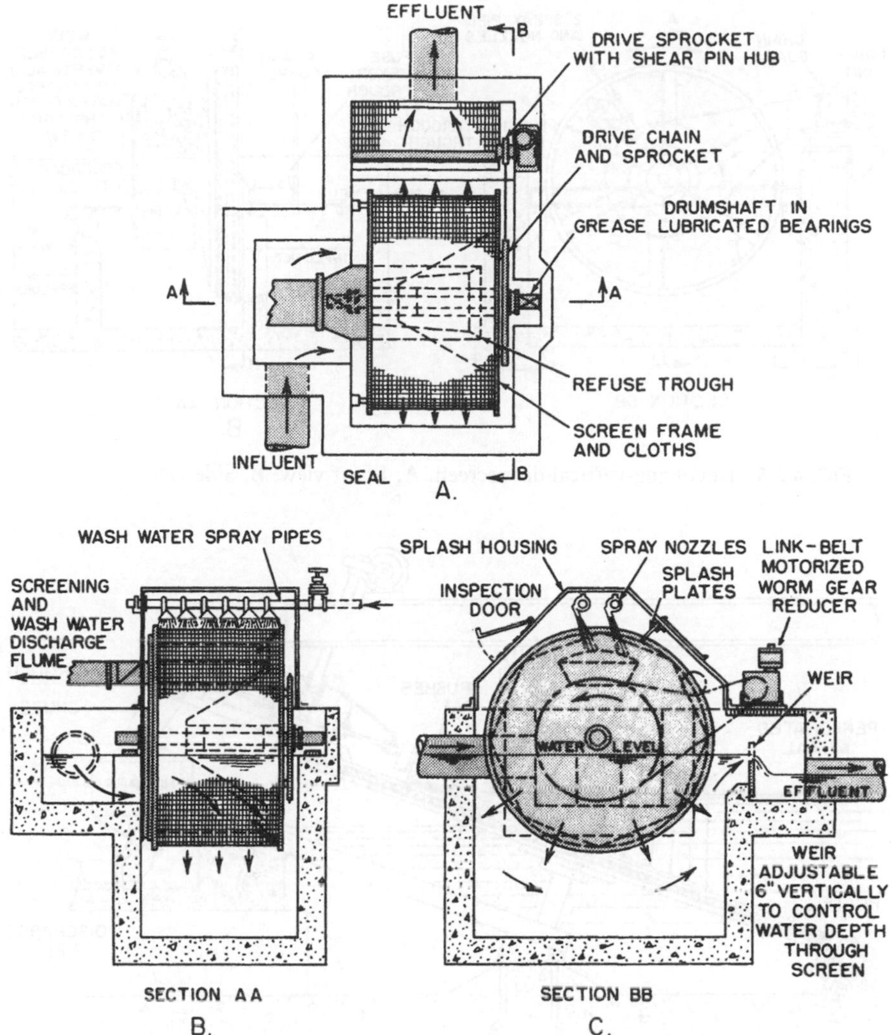

FIG. 4.2.5 Revolving-drum screen with outward flow. **A.** Plan view; **B.** Side view; **C.** End view.

Revolving-Drum Screen with Inward Flow

With this screen, the waste flow approaches the drum perpendicular to its axis. The liquid passes through the screen and flows out at one end. The solids, which are retained on the outside surface of the drum, are raised above the liquid level as the drum rotates and are removed by brushes; scrapers; or backwashing with water, air, or steam.

A disadvantage of removing screenings by water spray is that the removed solids are mixed with large amounts of spray water.

Revolving-Vertical-Disk Screens

In operating principle, this screen is similar to the revolving drum screen except that instead of a drum, a slowly revolving disk screen is placed in the approach channel completely blocking the flow so that it must pass through the screen (see Figure 4.2.6). As the liquid passes through the screen, solids are retained, elevated above the water level, and flushed by a water spray to a trough.

This screen is not suitable to remove larger objects or excessive amounts of suspended matter; neither is it suitable to handle greasy, gummy, or sticky solids. It requires a constant water level, which is usually secured by using a weir. The screening medium can be 2- to 60-mesh stainless steel wire cloth. Screenings are mixed with large amounts of spray water.

Inclined Revolving-Disk Screens

This screen consists of a round, flat plate revolving on an axle inclined 10° to 25° from the vertical (see Figure 4.2.7). The disk consists of several bronze plates containing slots, generally $\frac{1}{16}$ to $\frac{1}{32}$ in (1.6 to 0.8 mm) wide. The waste flows through the lower two-thirds of the plates. As the plate rotates, retained solids are brought above the liquid level where brushes remove them for disposal.

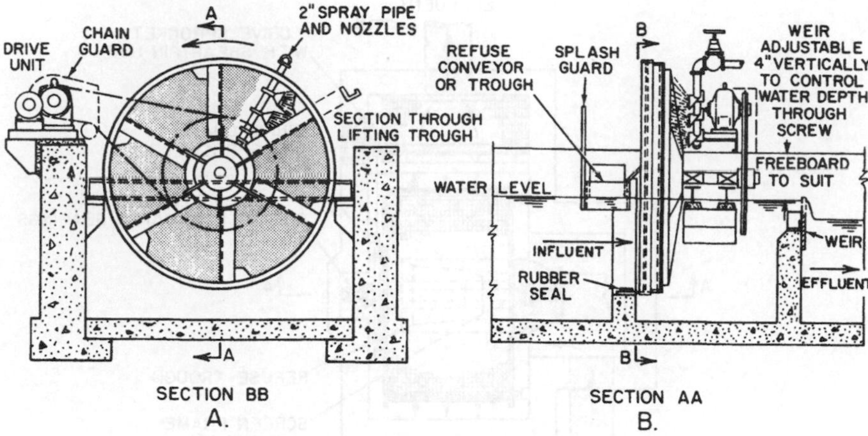

FIG. 4.2.6 Revolving-vertical-disk screen. **A.** Front view; **B.** Side view.

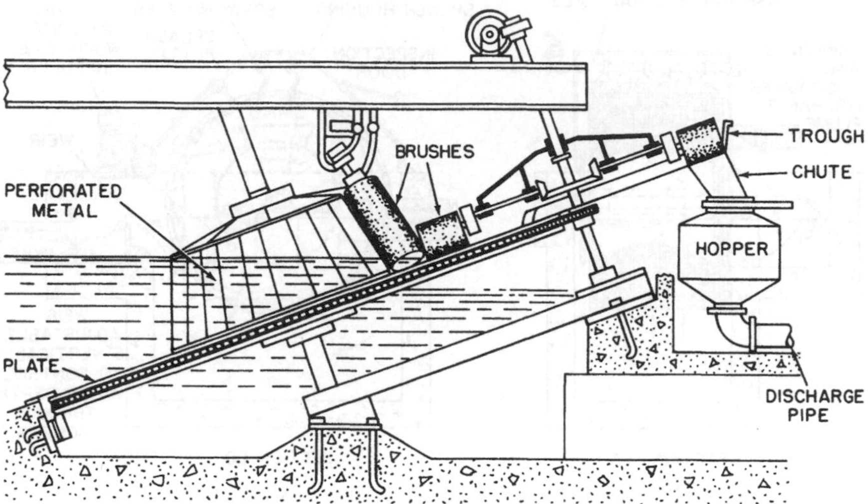

FIG. 4.2.7 Inclined revolving-disk screen.

Traveling-Water Screen

This screen has had limited use in sewage treatment to remove solids from plant effluent. This screen uses a series of tilted or inclined overlapping screen trays mounted on two strands of steel chain. The head wheel is motor-driven, moving screen trays out of the sewage for solids removal by jets of water, then returning the trays into the wastewater flow.

Endless Band Screen

This screen operates in basically the same way as the revolving-vertical-disk screen. Instead of a vertical revolving disk, a screen in the form of an endless band is installed in the approach channel and the incoming flow is forced to pass through it. This endless screen moves slowly around top and bottom drums while half or more of it is submerged.

Vibrating Screens

Vibrating screens are used in the food packing industry to remove grease and meat particles, eliminate manure, recover animal hair, remove feathers from poultry processing, and remove vegetable and fruit particles from canning waste.

Vibrating screens are flat and covered by fine stainless steel cloth of 20- to 100- or even 200-mesh supported by rubber-covered bars or stronger stainless steel, coarse-wire mesh. Vibration reduces the blinding and clogging of the fine screens. A manual or automatic spray washer with steam or detergents can reduce blinding and clogging and facilitate handling of greasy or sticky material.

MICROSCREENS

Microscreens with apertures as small as 20 μ are used to remove fine suspended matter from plant effluent as *tertiary treatment units.* See Section 7.3.

HYDRASIEVES

This unit was developed for industrial purposes (see Figure 4.2.8). It requires no power to operate except for that required to lift the waste or process water to the headbox of the screen. It is reasonably self-cleaning and does not require much attention and maintenance for continuous, trouble-free operation.

Hydrasieves are used in municipal wastewater treatment plants where the manufacturer claims an efficiency of 20 to 35% SS and BOD removal. The most frequently used screen sizes are 0.040 and 0.060 in (1 and 1.5 mm).

Wastewater is fed by gravity or pumping into the headbox of the screen. The screen is constructed with three different slopes, 25°, 35°, and 45° from the vertical. The overflowing waste first hits the 25° screen, where most free water is removed. On the second slope, more water is removed, and the solids start to agglomerate as they roll down. On the last slope, the solids are pushed down by the continuously accumulating new screenings while draining progresses.

VOLUME OF SCREENINGS

Estimating screening volume is difficult because the volume depends on the size of the screen and on infiltration, storm water quantity in combined sewers, the habits of the contributing population (e.g., garbage grinders), and

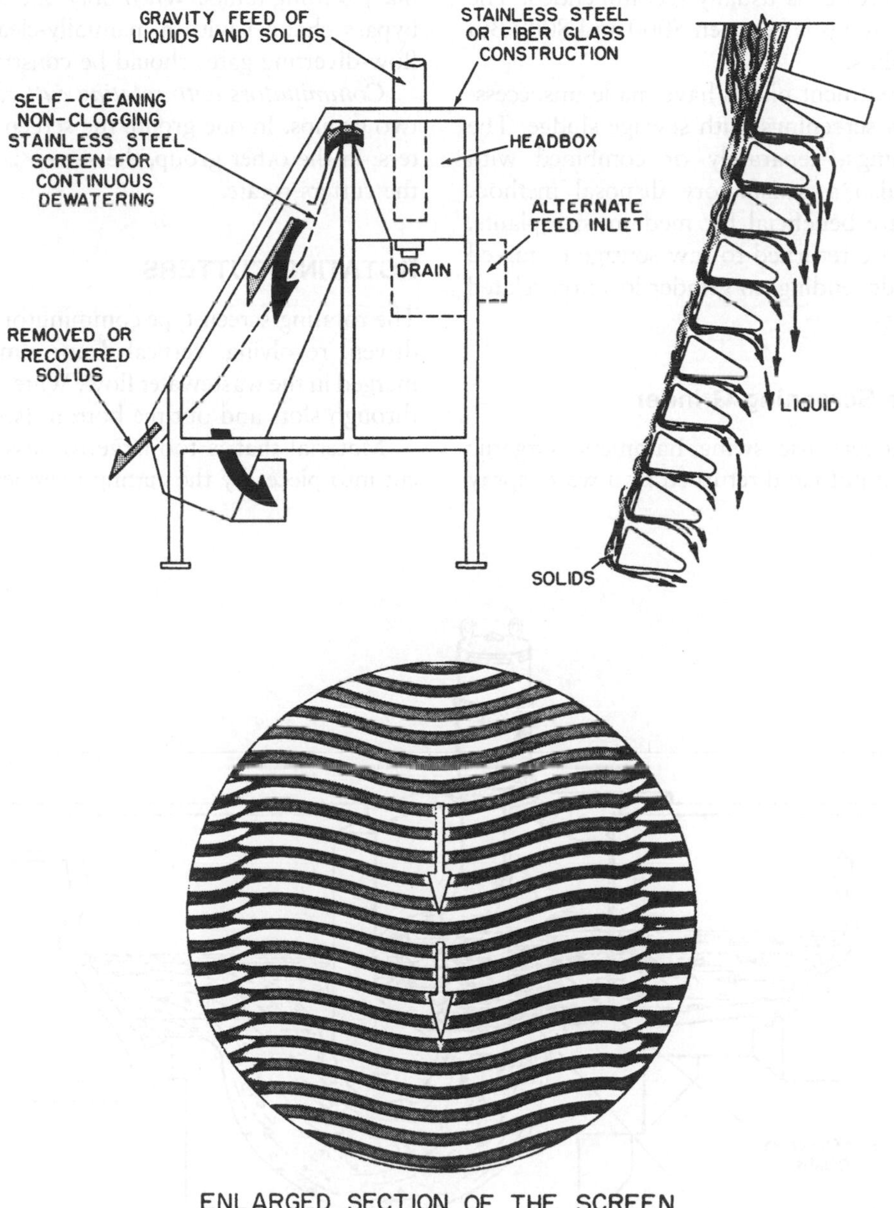

FIG. 4.2.8 Hydrasieve.

the character of industrial waste received.

The solids removed from municipal waste by fine screens ranges from 10 to 35 ft^3 per mgd (0.28 to 0.99 m^3 per 3785 m^3 per day) or from 5 to 20% of the SS in raw waste depending on screen size and the nature of the waste.

DISPOSAL OF SCREENINGS

Disposal methods include burial, incineration, digestion, and grinding. Open-area disposal is prohibited. Most smaller plants dispose their screenings by burial. Each day they add a cover of approximately 6 in over buried screenings to prevent fly and odor problems.

Large plants often use incinerators to dispose screenings alone or mixed with dewatered sludge. Dewatering the screenings with presses is usually recommended. The heating value of screenings is between 5000 and 8000 Btu per pound of dry solids.

Several smaller treatment plants have made unsuccessful attempts to burn screenings with sewage sludge. The digestion of screenings, separately or combined with sewage sludge, is also a satisfactory disposal method. Screening grinders are beneficial for medium-size plants. Reduced-size solids are returned to raw sewage or mixed with sewage sludge depending on grinder location related to the treatment units.

Hammer-Mill-Type Screening Grinder

These screening grinders use swing hammers. Organic solids are ground to a pulp and returned by a water spray into the waste stream. Power requirements are higher than with other designs.

Comminutors

To eliminate the problems associated with collection, removal, storage, and handling of screenings, wastewater treatment plants install devices that continuously intercept, shred, and grind large floating material in the waste flow into small pieces. These cutting and shredding devices are called *comminutors. Comminution* reduces solid particles to smaller sizes.

The use of comminutors reduces odors, flies, and unsightliness. They normally operate continuously and are generally located between the grit chamber and the primary settling tanks. When only one comminutor exists, a bypass channel with a manually-cleaned bar screen and flow-diverting gates should be constructed.

Comminutors with rotating cutters can be divided into two groups. In one group, the screen rotates and has cutters; in the other group, the screen is stationary and only the cutters rotate.

ROTATING CUTTERS

The rotating-screen-type comminutor consists of a motor-driven, revolving, vertical drum almost completely submerged in the wastewater flow. Water passes into the drum through slots and out the bottom (see Figure 4.2.9).

Material that is too large to pass through the slots is cut into pieces by the cutting members acting like shears.

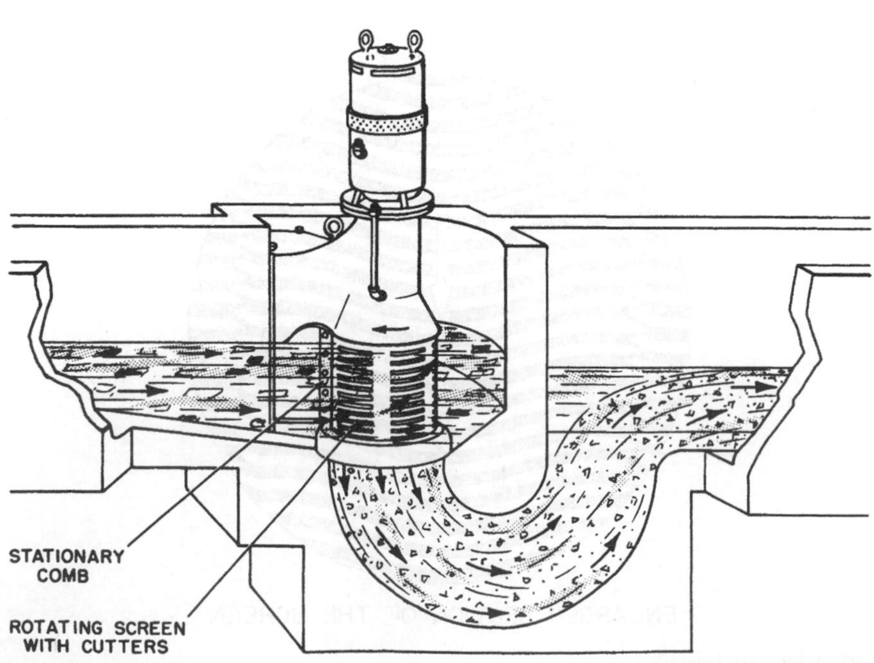

FIG. 4.2.9 Comminutor with rotating-screen cutter.

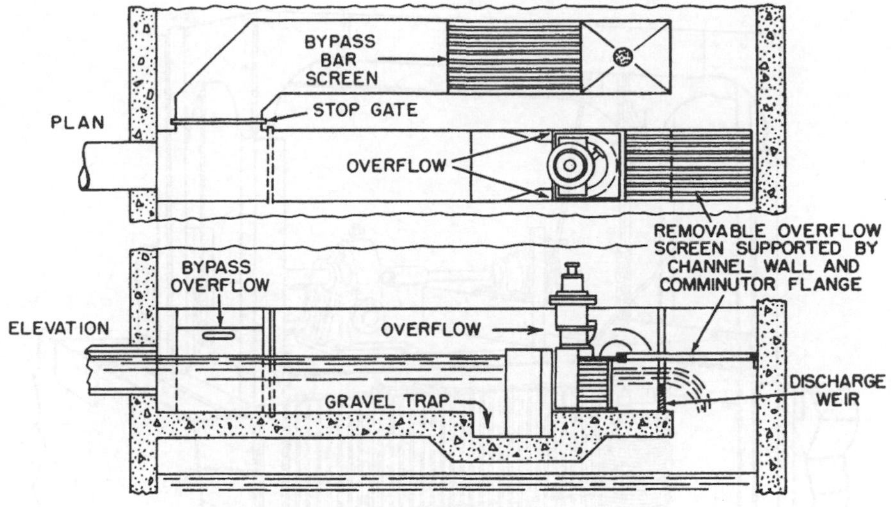

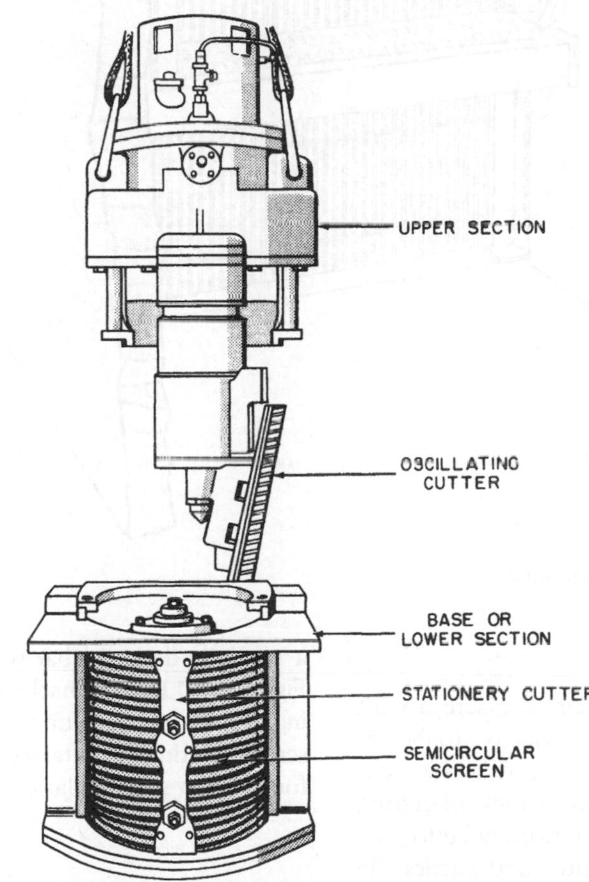

FIG. 4.2.10 Comminutor with stationary screen and oscillating cutter.

The angles between the cutting members are designed to eject iron or other hard materials. This comminuting device requires a special volute-shaped basin to give the proper hydraulic conditions for satisfactory operation. The basin shape makes installation more expensive than that of other devices. Smaller sizes, however, are provided with special cast-iron outlets, which eliminate onsite construction work.

The stationary-screen-type comminutor consists of a stationary semicircular screen and a rotating circular cutting disk. The grid intercepts larger solid particles, whereas smaller solids pass through the space between the grid and cutting disks. Larger units can be installed in a rectangular channel; smaller units are self-contained.

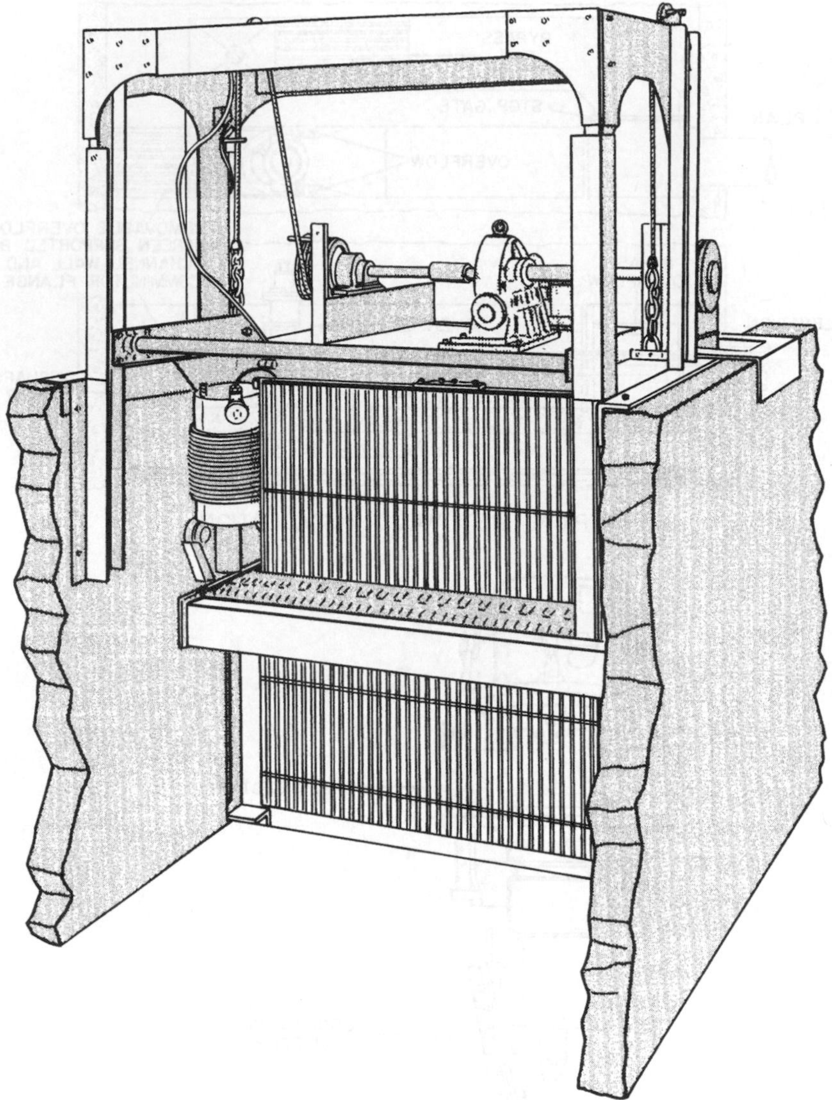

FIG. 4.2.11 Barminutor.

OSCILLATING CUTTERS

This comminutor consists of a semicircular screen with horizontal slots set in a sewage channel concave to the approaching flow and a stationary vertical cutter in its center. A motor-driven, vertical arm with a rack of cutting teeth (which mesh with those of the stationary cutter) oscillates 180° around the screen's center and carries the screenings to the stationary cutter bar where they are shredded (see Figure 4.2.10).

BARMINUTORS

This comminuting unit is used for flows exceeding 10 mgd. This combined screening and cutting machine consists of a specially designed bar screen with small openings (see Figure 4.2.11). The machine has rotating cutters comprising a comminuting unit that travels up and down the screen, cutting the retained solids. The screen is designed for use in a rectangular channel.

—*János Lipták*
David H.F. Liu

Reference

Great Lakes–Upper Mississippi River Board of State Sanitary Engineers. 1968. *Recommended standards for sewage works* (*Ten states' standards*). Health Education Service.

4.3 GRIT REMOVAL

PARTIAL LIST OF SUPPLIERS

Aerators, Inc.; Dorr-Oliver Inc.; Edex Inc.; Eimco Process Equipment; Envirex, Inc., a Rexnord Company; Eutek Systems; Fairfield Engineering Co.; Franklin Miller Inc.; Infilco Degremont Inc.; Krebs Engineers; Laval, Claude, Corp.; Smith & Loveless, Inc.; Techniflo Systems; TLB Corp.; Waste Tech/Centrisys; Wemco Operation of Eimco Process Equipment; Westech Engineering Inc.

Removing grit from waste treatment plant influent prevents wear and abrasion of pumps and other mechanical machinery in the system. High-speed equipment, such as centrifuges, requires the elimination of practically all grit to prevent rapid wear and reduce maintenance. Removing grit from the incoming waste flow also reduces the potential for pipe plugging. Heavy grit loads can also lead to deposition in settling tanks, aeration units, and digesters, which can require frequent tank draining for cleaning.

Usually the removal of the particles greater than 0.2 mm or 65 mesh is accomplished. Clean grit containing no organic material is ideally disposed on land without odor or nuisance problems. When organic or decomposing material is removed with the grit, it must be washed to free the organic material and return it to the treatment process.

When centrifugation is used to thicken and dewater waste sludges, higher degrees of grit removal are required. Most centrifuges used in waste treatment applications are hard-finished and subject to heavy wear. Fine grit can drastically shorten the operating life (period between resurfacings) of these instruments. For such applications, grit of at least 150-mesh (0.10 mm) size should be removed from the waste stream, and occasionally detritus as fine as 325 mesh (0.04 mm) must also be removed to protect centrifugal machines.

Characteristics of Grit

Grit is the heavy mineral material in raw sewage, and may contain sand, gravel, silt, cinders, broken glass, seeds, small fragments of metal, and other small inorganic solids. It is generally nonputrescible. Grit settles more rapidly than organic or putrescible material in sewage, allowing a reasonably clean separation from the waste stream under normal conditions.

Grit is an inert material. Once drained of most of its water, it can be spread on the ground and used on roadways and sand drying beds. Ideally, clean grit contains less than 3% putrescible material. If it is not washed and cleaned, it presents a nuisance problem by causing foul odors, which attract rodents. Grit with high-putrescible levels must be buried after being removed from the grit collecting device. In a few cases, grit is burned before final disposal.

The quantity of grit varies from plant to plant. Storm drainage contains runoff from streets and open land areas and carries large concentrations of soil, sand, and cinders from land construction sites and street sanding. In a system serving only sanitary sewers, the grit load is smaller. This source contains egg shells, coffee grinds, broken glass, and similar materials. Sewer infiltration can also bring in fine silt and sand. The widespread use of home garbage grinders significantly adds to the grit load. Industrial waste discharged to the system can also carry a variety of gritty materials.

For design purposes environmental engineers should conduct studies to determine the quantity of grit carried by the system. In lieu of studies, 2 to 12 ft^3 per mgd of sewage are often used although considerably larger grit loading can be experienced. Environmental engineers should consider the types of sewers used and the amount and types of industrial waste when designing the grit removal facility.

Grit Removal Devices

Grit is selectively removed from other organics in a velocity-controlled grit channel or an aerated chamber. Both unit operations are commonly used. A newer, more efficient approach to grit removal is the use of hydrocylones.

GRAVITY SETTLING

The grit in wastewater has a specific gravity in the range of 1.5–2.7. The organic matter in wastewater has a specific gravity around 1.02. Therefore, differential sedimentation is a successful mechanism for separating grit from organic matter. Also, grit exhibits discrete settling, whereas organic matters settle as flocculant solids (see Section 5.2).

The velocity-controlled grit channel is a long, narrow, sedimentation basin with better flow control through velocity. Some wastewater treatment plants control the velocity by using multiple channels. A more economical arrangement and better velocity control is achieved by the use of control sections on the downstream of the channel. These control sections maintain constant velocity in the channel for a range of flows by using proportional weirs, Parshall flumes, and parabolic flumes.

To design effective grit removal facilities, environmental engineers must know the volume of the sewage flow and quantity of grit. The quantity of grit can be variable; therefore, a safety factor must be allowed. Multiple channels are usually provided when manual grit cleaning is used.

The typical values of detention time, horizontal velocity, and settling velocity for a 65-mesh (0.21-mm diameter) material are 60 sec 0.3 m/sec, and 1.15 m/min, respectively. The theoretical length is 18 m (60 ft). The depth of flow is governed by the volume of sewage flow. The width is not critical but is normally small so that the channels are long and narrow.

The head loss through a velocity-controlled channel is 30–40% of the maximum water depth in the channel. The effect of scouring the settled grit surface at this velocity washes away much of the putrescible material that settles out with the grit. The grit removed by this design usually requires burial to prevent odor.

Grit chambers can be cleaned manually or mechanically. Manual cleaning is usually used only in smaller and older plants where manual methods are used to rake, shovel, or bucket the grit from the chamber. During this operation, flow to the chamber is shut off or diverted to another channel, and the grit tank is drained for cleaning. Designing the grit chamber with the necessary depth provides grit storage.

Mechanical grit removal equipment consists of moving bucket scrapers, horizontal and circular moving rake scrapers, or screw conveyors. Once the grit is removed from the grit collection tank the facility dewaters it using screw or rake classifiers, screens, or similar devices. Hydraulic ejectors, jets, and air lift pumps are also used. Mechanically cleaned tanks require a smaller grit storage volume. Aeration is occasionally used in the grit chamber to wash out organic material from the grit. Figure 4.3.1 shows a mechanical grit removal design.

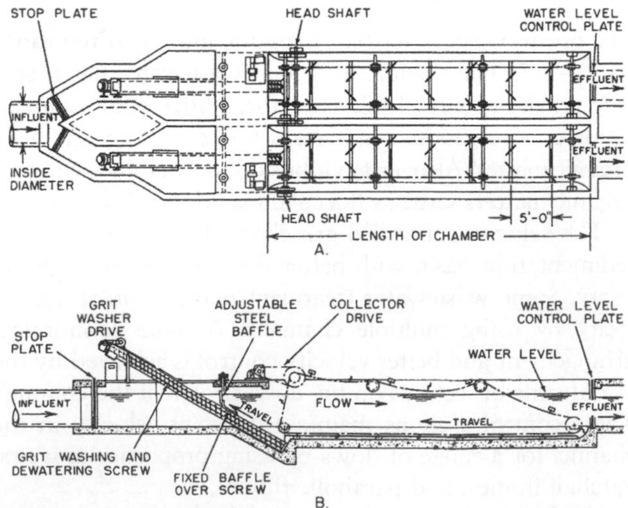

FIG. 4.3.1 Mechanical grit removal facility. **A.** Plan view; **B.** Longitudinal section.

AERATED GRIT CHAMBER

Aerated grit chambers are widely used for selective removal of grit. The spiral roll of the aerated grit chamber liquid drives the grit into a hopper located under the air diffuser assembly (see Figure 4.3.2). The shearing force of the air bubbles strips the inert grit of much of the organic material that adheres to its surface.

Aerated grit chamber performance is a function of roll velocity and detention time. Adjusting the air feed rate controls the roll velocity. Nominal air flow values are 0.15 to 0.45 m^3/min of air per meter of tank length (m^3/min · m). The liquid detention time is usually about 3 min at the maximum flow. Length-to-width ratios range from 2:5 to 5:1 with depths of 2 to 5 m.

The grit that accumulates in the chamber varies depending on the type of sewer system (combined type or separate type) and the efficiency of the chamber. For combined systems, 90 m^3 of grit per million cubic meters of sewage (30 $m^3/10^6$ m^3) is not uncommon; in separate systems, the amount is something less than 30 $m^3/10^6$ m^3. Deposited grit is normally recovered by air lift or screw conveyor. The grit is buried in a sanitary landfill.

Aerated grit chambers are extensively used at medium- and large-size treatment plants. They offer the following advantages over velocity-controlled grit channels:

An aerated chamber can also be used for chemical addition, mixing, and flocculation before primary treatment.
Wastewater is freshened by air, reducing odors and removing additional BOD_5.
Minimal head loss occurs through the chamber.
Grease removal can be achieved if skimming is provided.
By controlling the air supply, the chambers can remove grit of low-putrescible organic content.
By controlling the air supply, the chambers can remove grit of any specified size. However, due to the variable specific gravity and the size and shape of the particles, some limitations on removal may exist.

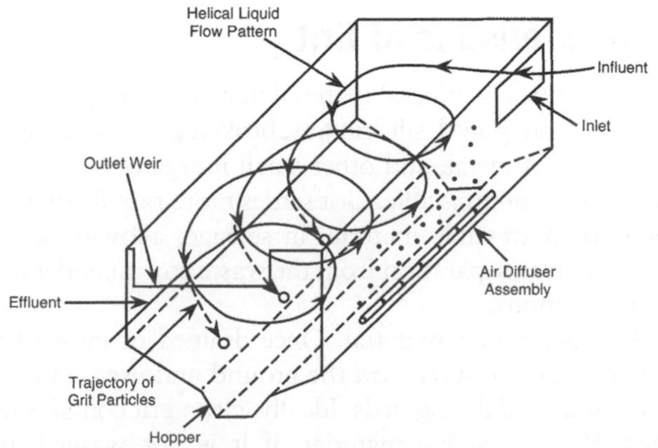

FIG. 4.3.2 Aerated grit chamber. (Reprinted, with permission, from Metcalf & Eddy, Inc. and G. Tchobanoglous. 1979, *Wastewater engineering: Treatment, disposal, reuse* (New York: McGraw-Hill).

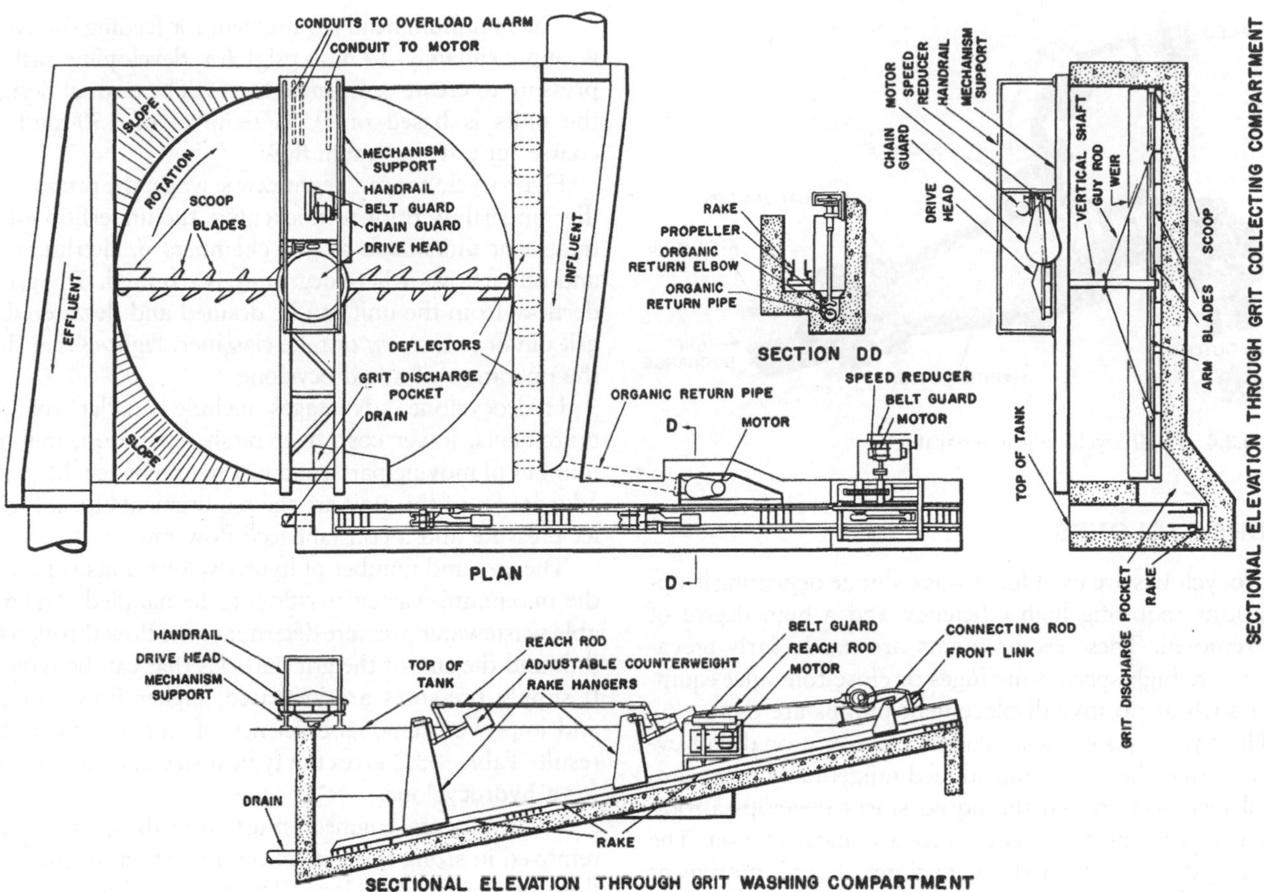

FIG. 4.3.3 Detritus tank and grit washer.

DETRITUS TANKS

Detritus tanks remove a mixture of grit and organic matter. The width and shape of the grit chamber are not critical with this design, but the surface area (which relates to settling velocity) is. The area requirements are proportional to the settling velocity of the grit particle and to the wastewater flow rate.

The settled grit is conveyed to a common collection sump. A raking or conveying mechanism washes and removes the grit from the chamber in a clean and drained condition, and the turbulence created by the raking mechanism washes the organic or putrescible material out of the grit. This rejected material is discharged back into the collection tank. Figure 4.3.3 shows a degritting unit of this design.

For detritus tanks, grit removal capacity is proportional to surface area. Multiplying the unit area given in Table 4.3.1 by the maximum wastewater flow expected gives the total area of the tank.

The normal design of detritus tanks is based on the removal of at least 65-mesh-size particles with a unit area requirement of 38.6 ft^2 per mgd. At this design rate, 95% of all grit coarser than 65-mesh is removed. Minimum velocities are not critical. Settled organic matter is washed from the grit in the classifier and returned to the process.

TABLE 4.3.1 AREA REQUIREMENTS FOR GRIT REMOVAL BY DETRITUS TANKS

Particle Size		*Settling Rate (ft per min)*	*Unit Area Required per mgd (ft)*
Mesh	*mm*		
28	0.595	10.9	8.5
35	0.417	7.8	11.9
48	0.295	5.5	16.8
65	0.208	3.7	25.0
100	0.147	2.4	38.6

Note: Data are based on a grit particle specific gravity of 2.65 in water flowing at a velocity of 1.2 fps or less.

TABLE 4.3.2 HYDROCYCLONE SIZING DATA

Hydrocyclone (size)	12 in	18 in	24 in
Capacity, gpm at 6 psig	205	580	800
Diameter of vortex finder (inches)°	5	8	10
Feed pipe size (inches)	4	6	6
Overflow pipe size (inches)	6	10	10
Mesh of separation°°	270	200	170

Notes: ° Other sizes and capacity ranges are available.
°° Smallest particle size removed; design is based on 95% removal of 150-mesh or larger grit.

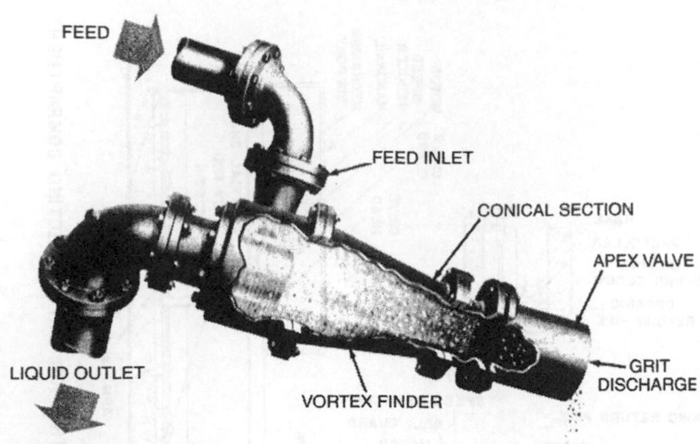

FIG. 4.3.4 Hydrocyclone grit separator.

HYDROCYCLONES

Hydrocyclones are used for sewage sludge degritting in applications requiring high efficiency and a high degree of grit removal. These requirements are particularly prevalent where high-speed centrifuges or close-tolerance equipment such as positive displacement pumps are used.

The hydrocyclone is similar to a conventional dust cyclone in that the feed is introduced tangentially to a cylindrical feed section and the liquid slurry develops a rotational movement and passes into a conical section. The centrifugal force created by cyclonic liquid movement forces heavier solid particles to the outer wall. Solids move along this wall and out the apex of the cone.

A vented overflow opening in the top of the cylindrical section insures that atmospheric pressure exists at the axis of the cyclone. The liquid and lighter solid materials pass up the center of the vortex and out the overflow. Shearing forces are high due to the change in tangential velocity across the diameter of the cyclone, and scouring of the lighter organic material from the grit occurs.

The minimum head requirement for feeding the cyclone is approximately 14 ft (6 psig) for developing sufficient pressure to create centrifugal forces. The normal design of the units is based on 95% removal of 150-mesh and coarser grit at maximum flow.

Hydrocyclones can degrit raw sewage, the primary clarifier underflow prior to thickening, the underflow of pretreatment units such as grit chambers or detritus tanks, and other flows where degritting is required. The grit underflow from the unit can be drained and dewatered in a grit bin or in a screw or rake classifier. Figure 4.3.4 shows the operation of a hydrocyclone.

Hydrocyclone advantages include smaller space requirements, lower cost, finer mesh separation, minimum number of moving parts, and low maintenance. Major disadvantages of this unit are the requirements for a high inlet pressure and a constant feed flow rate.

The size and number of hydrocyclone units is based on the maximum wastewater flow to be handled. The available wastewater pressure determines the flow through each unit and the size of the grit particles that can be removed. If supply pressures are increased, higher flow capacities and improved removal efficiency of smaller grit particles result. Table 4.3.2 gives the typical size and capacity data for a hydrocyclone.

Environmental engineers must know the quantity of grit removed in sizing the grit collection system for the underflow of the hydrocyclone. The hydrocyclone requires a constant feed flow rate for efficient operation. The overflow of the hydrocyclone must be vented for proper unit operation. For variable flow conditions, multiple units (operating as many units as the incoming flow requires) or an on–off timer control should be used.

—János Lipták
David H.F. Liu

4.4 GREASE REMOVAL AND SKIMMING

TYPES OF DESIGNS

a) Grease Interceptors, b) Flotation Units, b1) Aeration Type Units, b2) Pressure Type Units, b3) Vacuum Type Units, b4) Combined Treatment Units, c) Slotted Pipe Skimmers for Square Tanks, d) Rotating Arm Skimmer for Circular Tanks.

The terms grease and oil as used in wastewater treatment denote a variety of materials, including fats, waxes, free fatty acids, calcium and magnesium soaps, mineral oils and other nonvolatile materials that are soluble in and can be extracted by hexane from an acidified sample. In average domestic waste, grease constitutes 10 percent of the total organic matter, and the per capita contribution is estimated to be 0.033 lb. (15 gm) per day. Meat packing, dairy, laundry, garage-machine shop and oil refinery wastes also have high grease-oil content.

It is usually required to reduce the grease content of the industrial wastes below 100 mg per liter before it is discharged to a municipal system. Under quiescent conditions, some portion of the grease settles with the sludge and some floats to the surface, where it may be removed by skimming.

The term scum, as used in wastewater treatment, denotes *all floating* material collected or collectable by skimming, including floating grease, septic sludge raised to the surface, wood pieces, rubber and plastic bottles. Usually, however, floating grease constitutes the bulk and is the most putrescible part of the scum. The terms grease, grease and oil, or scum are often used interchangeably.

GREASE TRAPS AND GREASE INTERCEPTORS

Grease traps or grease interceptors collect grease and floating material from households, garages, restaurants, small hotels, hospitals and the like. Frequent cleaning is essential if operation is to be efficient. Owing to the danger of explosion or fire, motor oil, gasoline and similar light mineral oils should not be allowed to enter sewer systems.

Where the wastewater contains large amounts of greasy kitchen waste, grease traps should be used. Their minimum capacity is 3.0 gal (11.5 l) per capita and should be no less than 30 gal (0.115 m^3) per unit (see Figure 4.4.1). The influent line should terminate at least 6 in (15 cm) below the water line, and the effluent pipe should take off near the bottom of the tank.

Commercial grease interceptors are also available for restaurants and smaller industrial establishments, which are connected to public sewer systems.

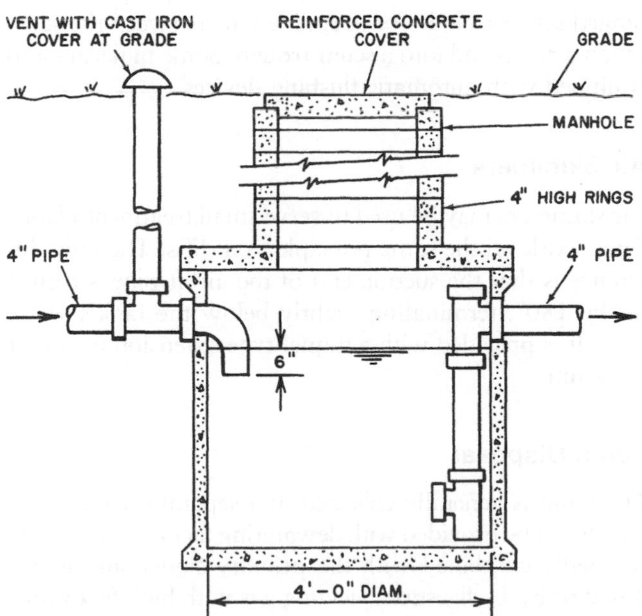

FIG. 4.4.1 Concrete grease trap, 300-gallon capacity.

GREASE REMOVAL BY AIR-AIDED FLOTATION

Grease and finely divided suspended solids may be converted to floating matter by air or gas-aided flotation with or without flotation aids.

Aerated Skimming Tanks, Preaeration

A few decades ago separate aerated skimming tanks with 3 to 5 minutes detention time and 0.05 to 0.08 ft^3 per gallon (0.375 to 0.60 m^3 per 1000 liters) air supply were used to treat large volumes of waste with moderate grease content. Compressed air containing 1 to 1.5 mg per liter chlorine gas was often used to increase efficiency.

The current version of this treatment (preaeration, that is, aeration before primary treatment) accomplishes more than grease removal. It also facilitates sedimentation and helps to refresh septic waste, which combined with increased floating and suspended solid removal improves the BOD reduction.

For grease removal 5 to 15 minutes aeration, using 0.01 to 0.1 ft^3 per gallon (0.075 to 0.75 m^3 air per 100 liters) of air is usually sufficient.

Manufactured Skimming Tanks

Although aeration facilitates grease removal, alone it is not very effective. Therefore, for manufactured skimming units, more efficient, mechanized flotation processes are used.

Pressure and vacuum flotation techniques generate air bubbles by reducing the pressure of a supersaturated air-waste mixture or by applying vacuum to the mixture, which is saturated under atmospheric pressure. The liberated minute bubbles tend to form around and attach themself to suspended particles in the waste. With this type of equipment, capital and operating costs are both relatively high. Therefore this type of flotation unit is most popular for pretreatment of industrial wastes at the source, where the flowrate is less and the concentration is high.

PRESSURE TYPE UNITS

This process consists of pressurizing the wastewater with air, usually in a separate air-saturation tank, to 1 to 3 atmospheres and venting the tank to the atmosphere (see Figure 4.4.2). When the pressure on the liquid is reduced, the dissolved air, in excess of saturation at atmospheric pressure, is released in extremely fine bubbles. The rising air bubbles attach themselves to the solid particles in the waste and carry them to the surface. Overflow rates range from 2000 to 6000 gpd per ft^2. This type of flotation involves high operating power costs.

VACUUM TYPE UNITS

This process consists of saturating the wastewater with air by aerating in a tank, or by permitting air to enter on the suction side of the waste transfer feed pump. Under vac-

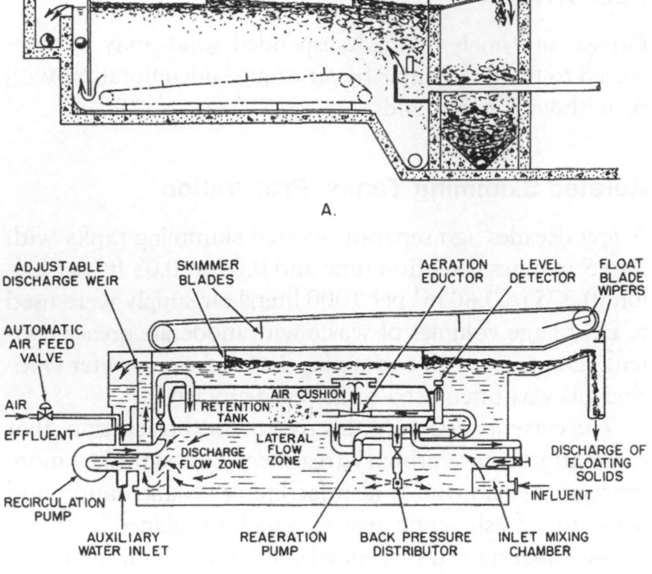

FIG. 4.4.2 Flotation unit with skimmers. A. Flotation unit; B. Pressure flotation unit.

uum the solubility of air in the waste is decreased and air is released in minute bubbles. The rising air bubbles attach themselves to the solid particles in the waste, carrying them to the surface.

Because of the fairly high vacuum levels involved (9 in Hg vacuum), this type of flotation unit requires an expensive, airtight construction. Grease removal of up to 50%; suspended solid removal of 35 to 55%; BOD removal of 17 to 35% may be achieved at surface loadings of 4000 to 6000 gpd per ft^2. The air requirement is 2.5 to 5 ft^3 per 100 gallons.

SCUM COLLECTION

Scum collecting and removal facilities, including baffling are desirable ahead of the outlet weirs on all settling tanks. Horizontal spraying with water under pressure may be employed to collect the scum for hand removal, if no mechanical skimmers are installed.

Square Settling Tanks

The straight scum baffle is installed ahead of the effluent weir-troughs and is submerged to a minimum depth of 18 in. Scum collectors usually move the scum toward the effluent side of the tank. Treatment plants are often equipped with hand-operated revolving slotted pipe skimmers installed horizontally across the tank ahead of the scum baffle (see Figure 4.4.2). When scum is to be removed, the

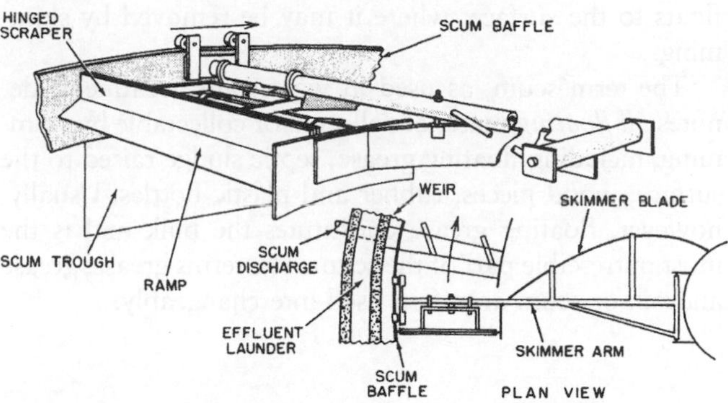

FIG. 4.4.3 Manual scum removal with rotating radial arm skimmer.

skimmer pipe is rotated until one edge of the slot is submerged slightly below the waste surface. The scum, mixed with waste, flows into it and is discharged through the pipe to a scum pit located outside the tank.

If the scum is to be removed *mechanically,* cross collectors consisting of endless chains above the surface are used. These carry flights which move the scum into the scum-trough, or in case of the helicoid spiral skimmer, the collector slowly turns and sweeps (with its rubber blades) the scum over the curving back edge of the tank into the scum-trough.

Circular Settling Tanks

The circular scum baffle is installed ahead of the circular effluent weir-trough and is submerged to a depth of at least 18 in.

Scum removal in circular tanks is usually performed by a *radial arm skimmer,* which is attached to and rotates with the sludge-removal equipment (see Figure 4.4.3). A skimmer blade moves the scum to the periphery, and hinged scraper blades or neoprene wipers sweep the scum up on a ramp and into a scum-trough. Some tanks are also equipped with automatic flushing devices.

Air Skimmers

Air skimmers may be used to serve small treatment plants. They work on the same principle as airlifts. The only difference is that the suction end of the airlift pipe is turned up by 180°, terminating slightly below the tank's water level. It is provided with a funnel-type extension to collect the scum.

Scum Disposal

The scum is generally collected in a separate scum sump, which can be provided with dewatering facilities. The scum is usually combined with the primary sludge and is disposed of by 1) digestion, yielding gas with high fuel value; 2) by vacuum filtration or incineration; and 3) by burial.

The volume of scum in normal domestic waste may range from 0.1 to 6 ft^3 per mg (0.75 to 45 liters per 100 m^3), or 200 ft^3 (5.75 m^3) per 1000 capita per year.

Skimming Devices

TYPE OF SKIMMING DEVICES

a) Rotatable slotted-pipe, b) Revolving roll, c) Belt, d) Flight scrapers, e) Floating pump.

SKIMMING RATE

Types a and d are essentially unlimited. Types b and c handle up to 1 gpm per foot of roll length or belt width; therefore, they typically handle up to 10 and 3 gpm, respectively. Type e handles up to 500 gpm.

WATER CONTENT OF RECOVERED OIL

Oils recovered with types a and e typically contain 80 to 90% water. Oils from b and c contain only 5 to 10% water.

MATERIALS OF CONSTRUCTION

Type a is carbon steel. Types b and e are usually stainless steel, aluminum, fiberglass, or plastic. Type c has a synthetic rubber or nickel belt. Type d has wooden flights.

In hydrocarbon processing industries, oily wastewater usually passes through a gravity oil–water separator (see Figure 4.4.4). In the separator, most free oil rises to the surface of the water, and is continuously or intermittently removed by a skimmer.

ROTATABLE SLOTTED-PIPE SKIMMERS

Slotted-pipe skimmers are used extensively for applications where considerable oil quantities must be removed and the water level does not vary significantly. These units can be purchased, or a plant can fabricate them by cutting slots in a carbon steel pipe (see Figure 4.4.5). Each skimmer is usually long enough to span the width of the separator. The pipe diameter must allow ample capacity for gravity drainage of the skimmed oil to a sump pump. In addition, the diameter must be large enough to allow each edge of the open slots to be rotated from well above and below the liquid level. This diameter allows adjustment or termination of the skimming rate. Due to their simplicity,

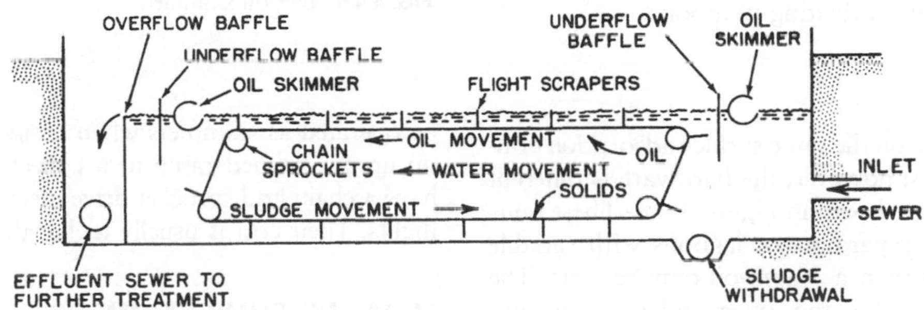

FIG. 4.4.4 Oil–water separator.

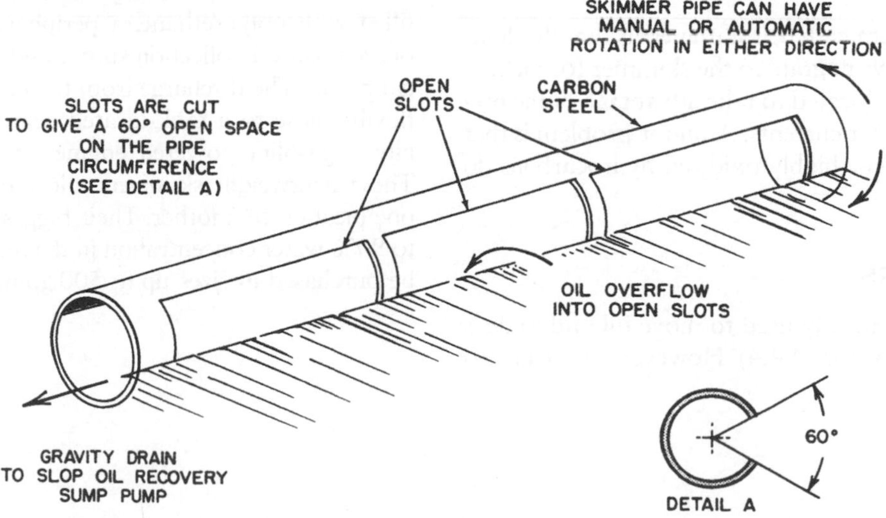

FIG. 4.4.5 Typical rotatable, slotted-pipe oil skimmer.

slotted pipes are inexpensive and essentially maintenance-free.

The major disadvantage of this skimmer is the high percentage of water collected with the skimmed oil. When a thick layer of oil accumulates prior to skimming, the initial oil recovered contains only small quantities of water. However, unless the skimmer is constantly adjusted, the water content averages 80 to 90%. The best solution to this problem is to pump the recovered mixture to a large tank for phase separation. The water phase can then be drained back to the separator inlet, and the oil can be reclaimed for further processing.

REVOLVING ROLL SKIMMERS

Revolving roll skimmers are used for applications where the quantity of oil for recovery is less than 10 gpm and the water collected does not exceed 5 to 10%. The rolls generally have a smooth surface and a shape similar to slotted pipe skimmers except that the diameter is slightly larger. Each cylinder is positioned horizontal to the liquid surface and is immersed at least 0.5 in. As the skimmer revolves, a thin oil film is adsorbed onto the roll surface and is rotated to a scraper blade. The capacity of a roll skimmer depends on oil viscosity, rotation speed, and cylinder length. If the liquid level in the separator varies, roll skimmers can be mounted on floating pontoons.

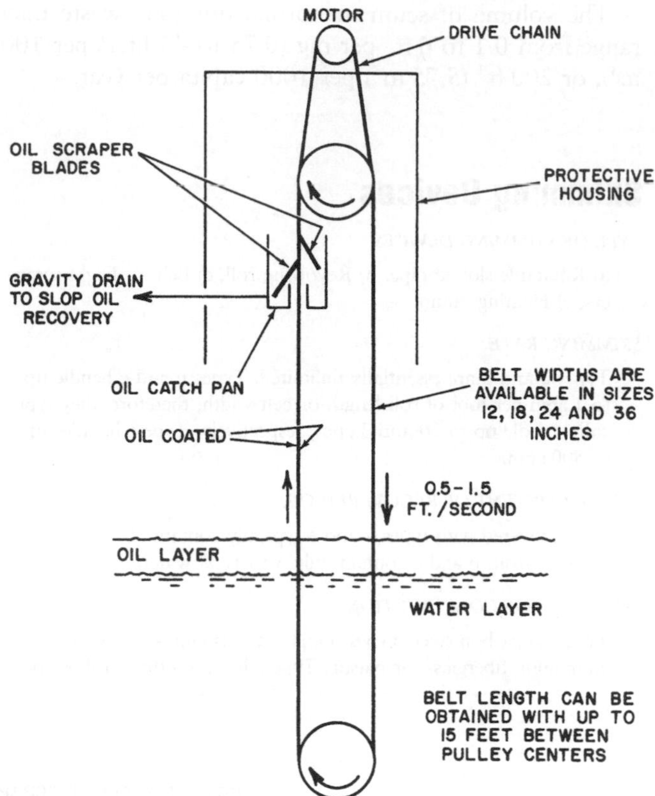

FIG. 4.4.6 Belt oil skimmer.

BELT SKIMMERS

Belt skimmers operate on the same surface adsorption principle as roll skimmers; however, the hardware is different in shape and design as shown in Figure 4.4.6. These units are ideally suited for separators or lagoons with variable liquid levels and less than a 3-gpm-oil-removal rate. The belt length is sized so that the lower end is always immersed. Widths are available in sizes of 12, 18, 24, and 36 in, and the amount of water picked up is only 5 to 10%.

The biggest problem with belt skimmers is that the floating oil does not always migrate to the skimmer for pickup. Skimmers are usually located to take advantage of the prevailing winds and water currents. A minor problem is that heavy greases and some highly oxidized hydrocarbons do not cling to the belt.

FLIGHT SCRAPERS

Flight scrapers are generally used to move oil and sludges to pickup points (see Figure 4.4.4). However, they can also be operated as skimmers when designed to push floating oil up an inclined ramp to a collecting pan. Most units have a chain and sprocket drive mechanism with wooden flights. Their cost is usually higher than other skimmers.

FLOATING-PUMP SKIMMERS

Several types of floating-pump oil skimmers are on the market. Each model usually has a corrosion-resistant float filled with polyurethane, a peripheral overflow weir for oil, a small oil collection sump, and a low-head centrifugal pump. The discharge from the pump passes through a flexible hose to a tank located nearby. The drive is usually a gasoline-powered engine or an electrical motor. These lightweight skimmers allow easy movement from one location to another. Their biggest drawback is the 80 to 90% water concentration in the recovered oil. They can be purchased in sizes up to 500 gpm.

—R.G. Gantz
János Lipták

4.5 SEDIMENTATION

PARTIAL LIST OF SUPPLIERS

Degremont-Cottrell, Inc.; Dorr-Oliver, Inc.; FMC Corp.; Dravo Corp.; Sybron Corp.; Edens Equipment Co.; Eimco Process Equipment; Envirex; General Filter Co.; Great Lakes Environmental; Lakeside Equipment Corp.; Komline-Sandersen Engineering Corp.; Neptune MicroFloc, Inc.; Parkson Corp.; U.S. Filter, Permutit Co.; Walker Process; Zimpro Environmental Inc.; Zurn Industries, Inc.

Sedimentation, sometimes called clarification, is generally used in combination with coagulation and flocculation to remove floc particles and improve subsequent filtration efficiency. Omitting sedimentation prior to filtration results in shorter filter runs, poorer filtrate quality, and dirtier filters that are more difficult to backwash. Sedimentation is particularly necessary for high-turbidity and highly colored water that generates substantial solids during the coagulation and flocculation processes. Sedimentation is sometimes unnecessary prior to filtration (direct filtration) when the production of flocculation solids is low and filtration can effectively handle solids loading.

Sedimentation is sometimes used at the head of a water treatment plant in a presedimentation basin, which allows gravity settling of denser solids that do not require coagulation and flocculation to promote solid separation. The application of a presedimentation basin is most common where surface water has a high silt or turbidity content. Some wastewater treatment plants use coagulation before presedimentation basins.

Types of Clarifiers

The design of most clarifiers falls into one of the following categories: horizontal flow, solids contact, or inclined surface.

HORIZONTAL-FLOW CLARIFIERS

In horizontal-flow clarifiers, sedimentation occurs in specially designed basins. These basins are known as settling tanks, settling basins, sedimentation tanks, sedimentation basins, or clarifiers. They can be rectangular, square, or circular. The most common basins are rectangular tanks and circular basins with a center feed.

In rectangular basins (see part A in Figure 4.5.1), the flow is in one direction and is parallel to the basin's length. This is called *rectilinear flow*. In center-feed circular basins (see part B in Figure 4.5.1), the water flows radially from the center to the outside edges. This is called *radial flow*. Both basins are designed to keep the velocity and flow distribution as uniform as possible so that currents and eddies do not form and keep the suspended material from settling. Other flow patterns are shown in parts C, D, and E in Figure 4.5.1.

Basins are usually made of steel or reinforced concrete. The bottom slopes slightly to make sludge removal easier. In rectangular tanks, the bottom slopes toward the inlet end, whereas in circular or square tanks, the bottoms are

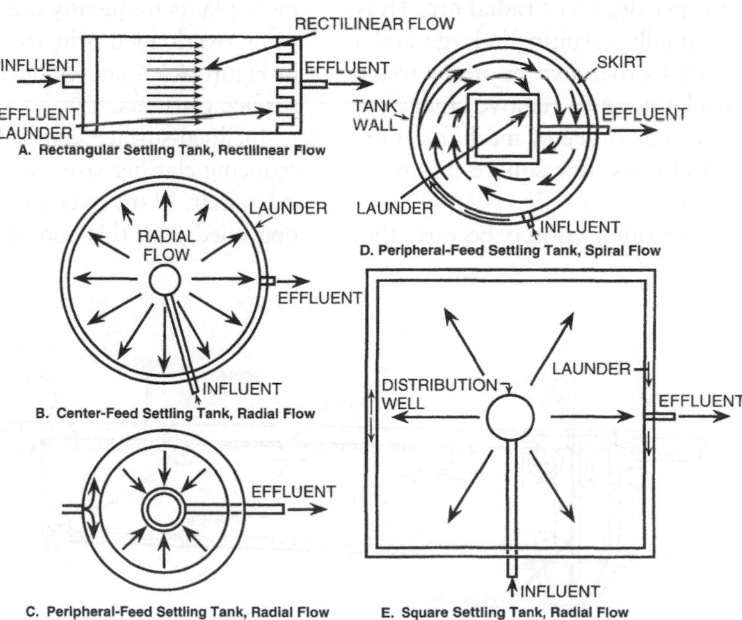

FIG. 4.5.1 Flow patterns in sedimentation basins.

conical and slope toward the center of the basin.

The selection of any shape depends on the following factors:

- Size of installation
- Regulation preference of regulatory authorities
- Local site conditions
- Preference, experience, and engineering judgement of the designer and plant personnel

The advantages and disadvantages of rectangular clarifiers over circular clarifiers follow.

ADVANTAGES:

- Less area occupied when multiple units are used
- Economic use of common walls with multiple units
- Easy covering of units for odor control
- Less short circuiting
- Lower inlet–outlet losses
- Less power consumption for sludge collection and removal mechanisms

DISADVANTAGES:

- Possible dead spaces
- Sensitivity to flow surges
- Collection equipment restricted in width
- Multiple weirs required to maintain low-weir loading rates
- High upkeep and maintenance costs of sprockets, chains, and fliers used for sludge removal

Square clarifiers combine the common-wall construction of rectangular basins with the simplicity of circular sludge collectors. These clarifiers have generally not been successful (Montgomery 1985). Because effluent launderers are constructed along the perimeter of basins, the corners have more weir length per degree of radial arc. Thus, the flow is not distributed equally, resulting in large sludge depositions in basin corners. Corner sweeps added to circular sludge collection mechanisms to remove sludge settling in the corners have been a source of mechanical difficulty. Because of these problems, few square basins are constructed for water treatment.

Circular settling tanks are often chosen because they use a trouble-free, circular sludge removal mechanism and, for small plants, can be constructed at a lower capital cost per unit surface area.

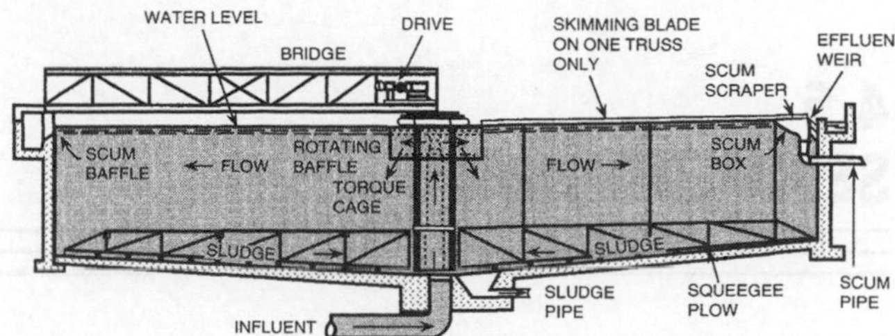

FIG. 4.5.3 Parts of a circular basin. (Courtesy of the FMC Corp., Material Handling Systems Division)

Figures 4.5.2 and 4.5.3 show the details of rectangular and circular horizontal flow clarifiers.

SOLID-CONTACT CLARIFIERS

Part A in Figure 4.5.4 shows the operational principles of solid-contact clarifiers. Incoming solids are brought in contact with a suspended sludge layer near the bottom. This layer acts as a blanket, and the incoming solids agglomerate and remain enmeshed within this blanket. The liquid rises upward while a distinct interface retains the solids below. These clarifiers have hydraulic performance and a reduced retention time for equivalent solids removal in horizontal flow clarifiers.

INCLINED-SURFACE CLARIFIERS

Inclined-surface basins, also known as a high-rate settler, use inclined trays to divide the depth into shallower sections. Thus, the depth of all particles (and therefore the settling time) is significantly reduced. Wastewater treatment plants frequently use this concept to upgrade the existing overloaded primary and secondary clarifiers. Part B in Figure 4.5.4 shows the operating principles of inclined surface clarifiers.

Inclined-surface clarifiers provide a large surface area, reducing clarifier size. No wind effect exists, and the flow is laminar. Many overloaded, horizontal-flow clarifiers are upgraded with this concept. The major drawbacks of the

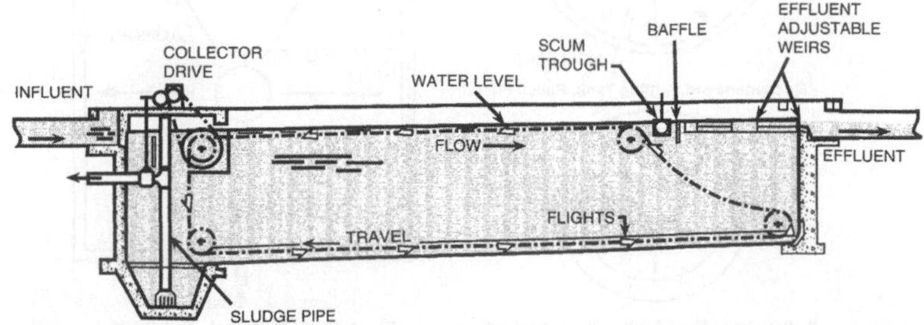

FIG. 4.5.2 Parts of a rectangular basin. (Courtesy of the FMC Corp., Material Handling Systems Division)

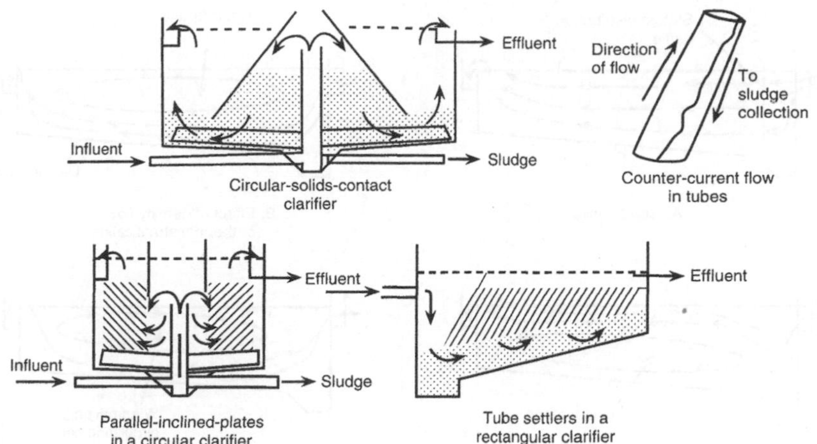

FIG. 4.5.4 Types of clarifiers. **A.** Circular-solids-contact clarifier. **B.** Parallel inclined plates in a circular clarifier. **C.** Tube settlers in a rectangular clarifier. **D.** Counter-current flow in tubes.

inclined-surface clarifiers include:

- Long periods of sludge deposits on the inner walls can cause septic conditions.
- The effluent quality can deteriorate when sludge deposits slough off.
- Clogging of the inner tubes and channels can occur.
- Serious short-circuiting can occur when the influent is warmer than the basin temperature.

Two design variations to the inclined-surface clarifiers are tube settlers and parallel-plate separators.

Tube Settlers

In these clarifiers, the inclined trays are constructed with thin-wall tubes. These tubes are circular, square, hexagonal, or any other geometric shape and are installed in an inclined position within the basin. The tubes are about 2 ft long and are produced in modules of about 750 tubes. The incoming flow enters these tubes and flows upward. Solids settle on the inside of the tube and slide down into a hopper.

The most popular commercially available tube settler is the steeply inclined tube settler. The angle of inclination is steep enough so that the sludge flows in a countercurrent direction from the suspension flow passing upward through the tube. Thus, solids drop to the bottom of the clarifier and are removed by conventional sludge removal mechanisms.

Test results for alum-coagulated sludge indicate that solids remain deposited in the tubes until the angle of inclination increases to 60° or more from the horizontal.

Parallel Plate Separators

Parallel-plate separators have parallel trays covering the entire tank. The operational principles for these separators are the same as those for the tube settlers.

Other Inclined-Surface Separators

Another design of shallow-depth sedimentation uses *lamella plates* (see Figure 4.5.5), which are installed parallel at a 45° angle. In this design, water and sludge flow in the same direction. The clarified water is returned to the top of the unit by small tubes.

Design Factors

The design objective of primary sedimentation is to produce settled water with the lowest possible turbidity. For effective filtration, the turbidity of settled water should not exceed 10 NTU. Since effective sedimentation is closely

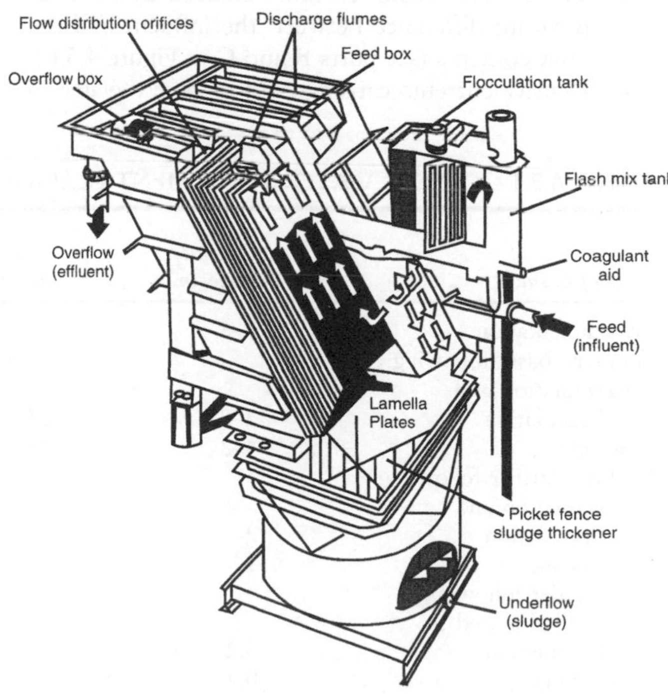

FIG. 4.5.5 Lamella plates. (Courtesy of Parkson Corp.)

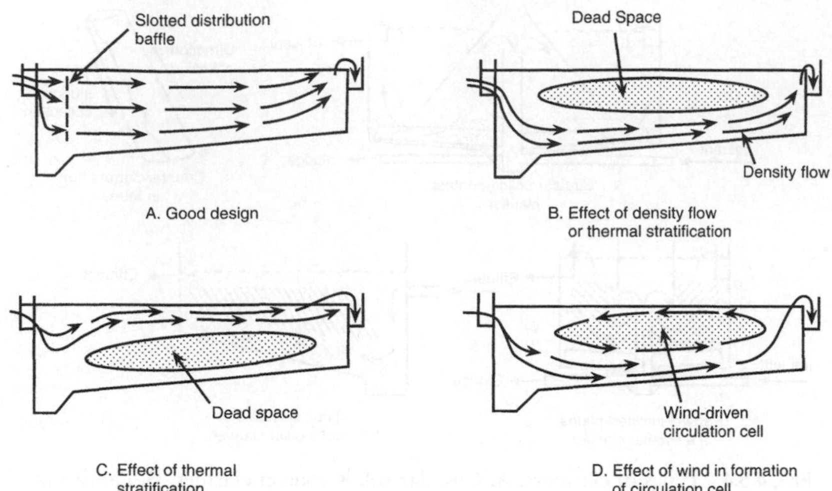

FIG. 4.5.6 Flow patterns in rectangular sedimentation tanks.

linked with coagulation and flocculation, the wastewater treatment plant must ensure that the best possible floc is formed.

The flow should be distributed uniformly across the inlet of the basin (see Figure 4.5.6). The solids removal efficiency of a clarifier is reduced by the following conditions:

- Eddy currents induced by the inertia of the incoming fluid
- Surface current produced by wind action (see part D in Figure 4.5.6). The resulting circulating current can short-circuit the influent to the effluent weir and scour settled particles from the bottom.
- Vertical currents induced by the outlet structure
- Vertical convection currents induced by the temperature difference between the influent and the tank contents (see parts B and C in Figure 4.5.6).
- Density currents causing cold or heavy water to underrun a basin, and warm or light water to flow across its surface (see part B in Figure 4.5.6).
- Currents induced by the sludge scraper and sludge removal system

Therefore, factors such as the overflow rate, detention period, weir-loading rate, shape and dimensions of the basin, inlet and outlet structures, and sludge removal system affect the design of a sedimentation basin (Table 4.5.1).

DETENTION TIME

The detention time depends on the purpose of the basin. In a mechanically cleaned presedimentation basin, the detention time can be sufficient to remove only coarse sand and silt. In a plain sedimentation basin, which depends on removing fine SS, the detention time must be long since small particles settle very slowly.

TABLE 4.5.1 TYPICAL WATER TREATMENT CLARIFIER DESIGN DETAILS

Type of Basin	*Detention Time, hr*	*Weir Overflow Rate*		*Surface Overflow Rate*	
		m³/(m · day)	*gal/(ft · day)*	*m/day*	*gal/(ft² · day)*
Presedimentation	3–8				
Standard basin following:					
Coagulation and flocculation	2–8	250	20,000	20–33	500–800
Softening	4–8	250	20,000	20–40	500–1000
Upflow clarifier following:					
Coagulation and flocculation	2	175	14,000	55	1400
Softening	1	350	28,000	100	2500
Tube settler following:					
Coagulation and flocculation	0.2				
Softening	0.2				

SURFACE OVERFLOW RATE

The surface overflow rate is an important parameter for basins clarifying flocculent solids. It is expressed in cubic meters per day per square meter of the surface area of the tank or gal/ft^2-day. The optimum surface overflow rate depends on the settling velocity of the floc particles. If the floc is heavy (as with lime softening), the overflow rate can be higher than with lighter, alum floc. A typical overflow for alum floc is 500 gpd/sq ft (AWWA 1990).

The surface overflow rate can be determined by jar test studies in which the best coagulant, optimum dosage, and best flocculation are used. However, the environmental engineer must usually rely on past empirical experience and estimate a safe basin overflow rate based on representative water analyses and estimated coagulant use. Changing seasons and changing water quality pose additional problems.

WEIR-OVERFLOW RATE

Weir-loading rates have some effect on the removal efficiency of sedimentation basins. These rates are expressed in cu m/m or gal/ft length of the weir. The higher the weir overflow rate, the more influence the outlet zone can have on the settling zone. To minimize this impact, environmental engineers should not use a rate exceeding 20,000 gpd/ft. For light alum floc, the rate may have to be decreased to 14,000 gpd/ft or 10 gpm/ft. Typical weirs consist of 90°V notches approximately 50 mm (2 in) deep placed from 100 to 300 mm (4 to 12 in) on the center. The length calculated from the weir overflow rate is the total length, not the length over which flow occurs.

Table 4.5.2 is a compilation of typical surface overflow rates, weir overflow rates, and detention times used in water treatment. These values are provided for comparison purposes, not as recommended standards. Individual states normally establish recommended design criteria that environmental engineers can alter by demonstrating that they do not apply to the water being treated or the process being used. The design of water treatment systems should be based on a laboratory evaluation of the proposed system.

TABLE 4.5.2 DIMENSIONS OF RECTANGULAR AND CIRCULAR BASINS

Clarifier	*Range*	*Typical*
Rectangular		
Length, m	10–100	25–60
Length-to-width ratio	1.0–7.5	4
Length-to-depth ratio	4.2–25.0	7–18
Sidewater depth, m	2.5–5.0	3.5
Width, m[a]	3–24	6–10
Bottom slope, %	1	1
Circular		
Diameter, m[b]	3–60	10–40
Side depth, m	3–6	4
Bottom slope, %	8	8

[a]Most manufacturers build equipment in width increments of 61 cm (2 ft). If the width is greater than 6 m (20 ft), multiple bays may be necessary.

[b]Most manufacturers build equipment in 1.5-m (5-ft) increments of diameter.

DIMENSIONS

The dimensions of a sedimentation basin must accommodate standard equipment supplied by the manufacturer. Also, environmental engineers must consider the size of the installation, local site conditions, regulations of local water pollution control agencies, the experience and judgment of the designer, and the economics of the system. Table 4.5.2 summarizes the basic dimensions of rectangular and circular clarifiers.

For any wastewater supply that requires coagulation and filtration to produce safe water, a minimum of two basins should be provided.

INLET STRUCTURE

Water that by-passes the normal flow path through the basin and reaches the outlet in less than normal detention time occurs to some extent in every basin. It is a serious problem, causing floc to be carried out of the basin due to the shortened sedimentation time.

The major cause of short-circuiting is poor inlet baffling. If the influent enters the basin and hits a solid baffle, a strong current and short-circuit result. The ideal inlet reduces entrance velocity to prevent development of currents toward the outlet, distribute water uniformly across the basin, and mixes it with water already in the tank to prevent density current. A near-perfect inlet consists of several small openings (100–200-mm diameter, circular [4–8-in or equivalent]) distributed through the width and depth of the basin. In these openings, the head loss is large compared to the variation in head between the different openings. Figure 4.5.7 shows some typical designs that compromise between simplicity and function.

Based on inlet structures, circular clarifiers are classified as center- and peripheral-feed. In center-feed, circular clarifiers, the inlet is at the center, and the outlet is along the periphery. A concentric baffle distributes the flow

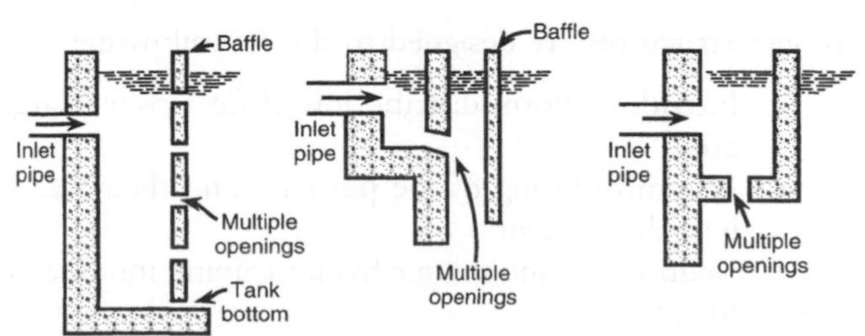

FIG. 4.5.7 Typical sedimentation tank inlets.

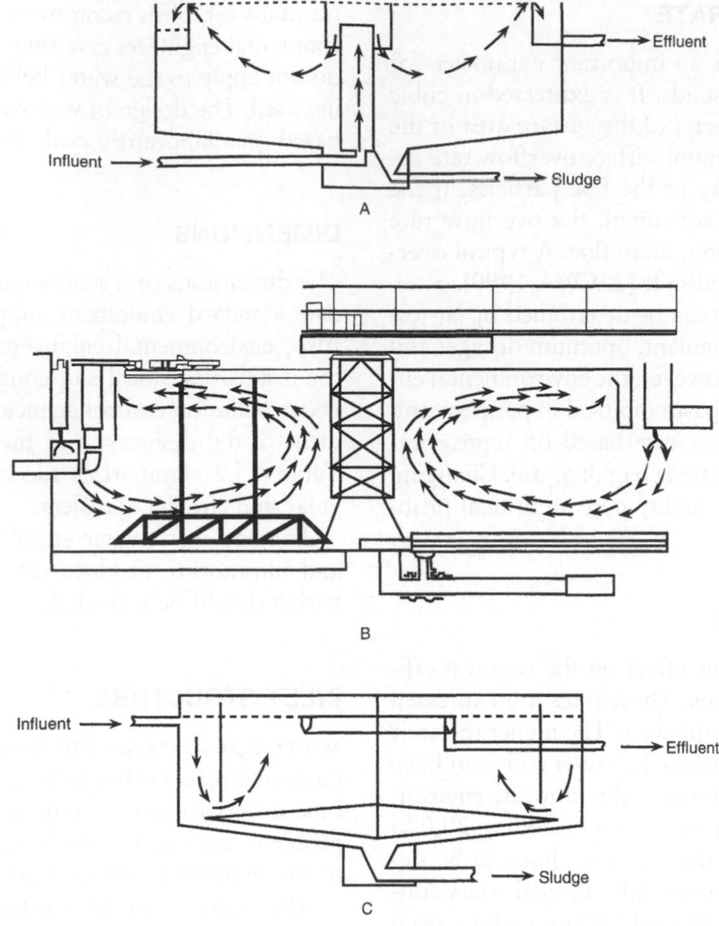

FIG. 4.5.8 Influent and effluent structures for circular clarifiers. (Reprinted, with permission, from Envirex Inc., a Rexnord Company)

equally in radial directions. The advantages of center-feed clarifiers are low upkeep cost and ease of design and construction. The disadvantages include short-circuiting, low detention efficiency, lack of scum control, and loss of sludge into the effluent. Part A in Figure 4.5.8 shows the flow scheme of a center-feed, circular clarifier.

In peripheral-feed clarifiers, the flow enters along the periphery. These clarifiers are considerably more efficient and have less short-circuiting than center-feed clarifiers. Peripheral-feed clarifiers have two major variations. These variations are shown in parts B and C in Figure 4.5.8.

OUTLET STRUCTURES

Effluent structures are designed to do the following:

- Provide uniform distribution of flow over a large area
- Minimize lifting of the particles and their escape into the effluent
- Reduce floating matter from escaping into the effluent

The most common effluent structures for rectangular and circular tanks are weirs that are adjustable for leveling. These weir plates are long enough to avoid the high heads that can result in updraft currents and particle lifting.

Both straight-edge and V notches on either one or both sides of the trough have been used in rectangular and circular tanks. V notches provide uniform distribution at low flows. A baffle in front of the weir stops floating matter from escaping into the effluent. Normally, weirs in rectangular tanks are on the opposite end of the inlet structure. Environmental engineers can use different weir configurations in rectangular basins to obtain a beneficial weir length. Figure 4.5.9 shows typical tank outlets.

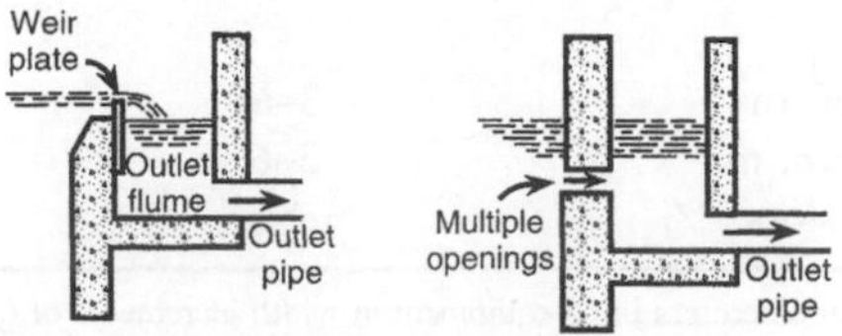

FIG. 4.5.9 Outlet details of sedimentation tanks.

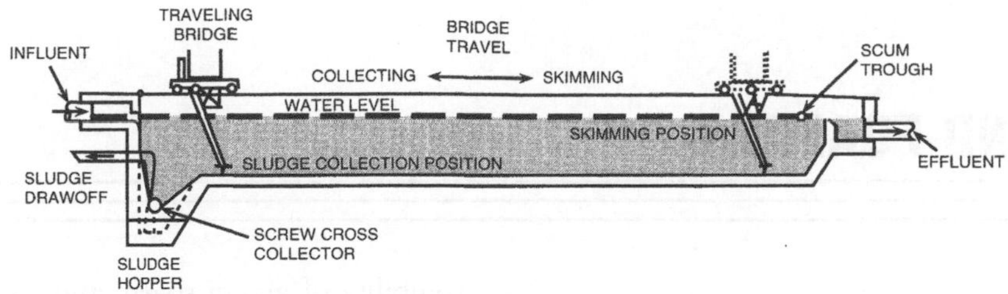

FIG. 4.5.10 Traveling-bridge collector.

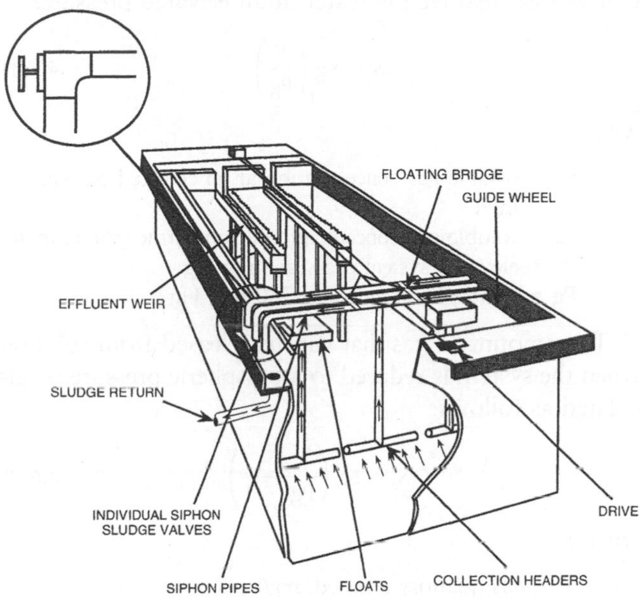

FIG. 4.5.11 Floating-bridge siphon collector. (Courtesy of Leopold Co., Division of Sybron Corp.)

In circular clarifiers, the outlet weir can be near the center of the clarifier or along the periphery as shown in Figure 4.5.8. The center weir generally provides a high-velocity gradient that can result in solids carryover.

SLUDGE REMOVAL

As solids settle to the bottom of a basin, a sludge layer develops. This layer must be removed because the solids can become resuspended or tastes or odors can develop. Wastewater treatment plants can manually remove the sludge by periodically draining basins and flushing the sludge to a hopper and drawoff pipe. This practice is recommended only for small installations or installations where not much sludge is formed. Mechanical removal is usually warranted.

For rectangular basins, sludge removal equipment is usually one of the following mechanisms (AWWA 1990):

A *chain and flight collector* (see Figure 4.5.1) consisting of a steel or plastic chain and redwood- or fiberglass-reinforced plastic flights (scrapers).

A *traveling-bridge collector* (see Figure 4.5.10) consisting of a moving bridge, which spans one or more basins. The mechanism has wheels that travel along rails mounted on the basin's edge. In one direction, the scraper blade moves the sludge to a hopper. In the other direction, the scraper retracts, and the mechanism skims any scum from the water's surface.

A *floating-bridge siphon collector* (see Figure 4.5.11) using suction pipes to withdraw the sludge from the basin. The pipes are supported by foam plastic floats, and the entire unit is drawn up and down the basin by a motor-driven cable system. For suction sludge removal, the velocity can be 1 m/min (3 fpm) because the main concern is not the resuspension of settled sludge but the disruption of the settling process.

To keep solids from returning to the cleaned liquid, scrapers should operate at velocities below 1 fpm. The power requirements are about 1 hp per 10,000 sq ft of tank area, but straight-line collectors must have motors about ten times that strong to master the starting load (Fair, Geyer, and Okun 1968).

Circular basins are usually equipped with scrapers or plows, as shown in Figure 4.5.3. These slant toward the center of the basin and sweep sludge toward the center of the basin, then to the effluent hopper or pipe. The bridge can be fixed as illustrated, or it can move with the truss.

Regardless of the collection method, the sludge is washed or scraped into a hopper. It is then pumped to sludge discharge treatment facilities.

—*David H.F. Liu*

References

American Water Works Association (AWWA). 1990. *Water quality and treatment: A handbook of community water supplies.* American Water Works Association, McGraw Hill, Inc.

Fair, G.M., J.C. Geyer, and D.A. Okun. 1968. *Water and water engineering.* New York: John Wiley & Sons.

Montgomery, J. McKee. 1985. *Water treatment principles and design.* John Wiley & Sons, Inc.

4.6 FLOTATION AND FOAMING

PROCESS PERFORMANCE

Foaming partially removes oil, SS, and 80 to 90% of surface active compounds.
Flotation is 65 to 95% effective in removing SS; 65 to 98% effective for fats, oils, and grease; and 25 to 98% effective in reducing BOD_5.

FLOTATION SYSTEM DATA

Vessel area: 1 to 5000 ft^2
Capacity: 1 to 3600 gpm
Operating pressure: 25 to 90 psig
Most flotation system components are made of surface-coated mild steel or concrete. Stainless steel and other corrosive-resistant metals are available.

Flotation is a unit operation which removes solid or liquid particles from a liquid (such as oil droplets removed from water). Adding a gas (usually air) to the system facilitates separation. Rising gas bubbles either adhere to or are trapped in the particle structure of the SS, thereby decreasing its specific gravity relative to the liquid phase and affecting separation of the suspended particles.

Methods of floation include dispersed- and the dissolved-gas flotation. Dispersed-gas flotation, commonly referred to as froth flotation, is not widely used in wastewater treatment. Although many design criteria and removal mechanisms apply to both dispersed- and dissolved-gas flotation, this section emphasizes dissolved-gas flotation.

Gas–Particle Contact

Before flotation of the SS can be accomplished, the particle must be in contact with the gas. Figure 4.6.1 shows gas–particle contact possibilities. The first type of contact is by precipitation of the gas bubble on the suspended particle or by collision of the rising gas bubble with the suspended particle. The angle of contact between the gas bubble and the suspended particle determines whether the gas bubble attaches or remains attached to the suspended particle. The second mechanism of attachment is trapping the rising gas bubble in a floc structure. The third mechanism is entrapment of a gas bubble within a floc structure as it forms.

Gas Solubility and Release

According to Henry's law, gas solubility in water is directly proportional to partial gas pressure and inversely proportional to water temperature. Table 4.6.1 lists the solubility of air, nitrogen, and oxygen in water and their respective densities with respect to temperature.

The following equation determines the quantity of gas that can be dissolved in water at an elevated pressure:

$$S = Sg\left(\frac{P}{Pg}\right) \qquad 4.6(1)$$

where:

S = soluble gas concentration at an elevated pressure, mg/l
Sg = soluble gas concentration at one atmosphere, mg/l
P = elevated system pressure, mm Hg
Pg = atmospheric pressure, 760 mm Hg

The amount of gas that can be released from solution when the system is reduced to atmospheric pressure is calculated as follows:

$$S_r = Sg\left(\frac{P}{Pg - 1}\right) \qquad 4.6(2)$$

where:

S_r = gas quantity released, mg/l

Modifying Equation 4.6(2) to account for the gas solution and release efficiency of the pressurization process yields the following:

$$S_r = Sg\left(\frac{Pf}{Pg} - 1\right) \qquad 4.6(3)$$

where:

f = pressurization system gas dissolving and release efficiency, fraction

Conventional pressurization systems usually attain 50% gas dissolving efficiency, whereas packed- or mixed-retention tank designs produce up to 90% efficiency.

An additional correction in gas solubility may be required when gases are dissolved in wastewater with a high-dissolved-solids content. Gas solubility reductions of as much as 20% from that listed in Table 4.6.1 have been observed.

Pressurization Systems

Flotation system performance depends not only on supplying sufficient gas for flotation but also on the manner in which gas is delivered to the flotation vessel. The pressurization system generally consists of a pressurization pump, a retention tank, and a gas supply. The pump in-

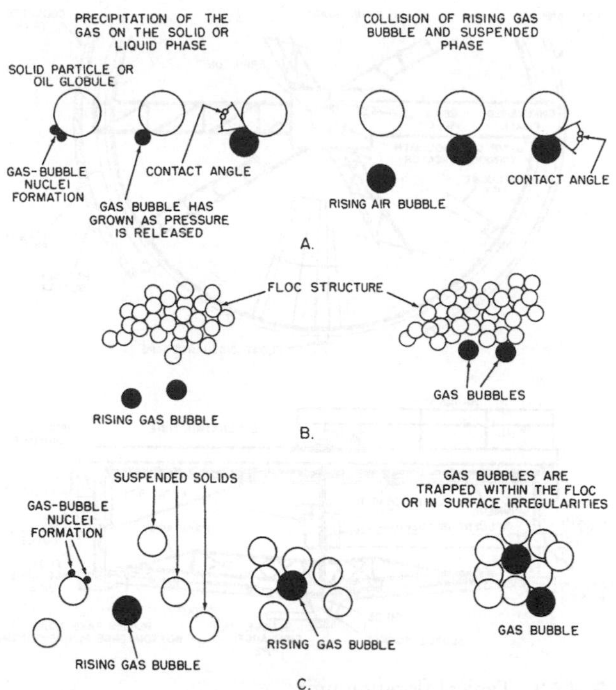

FIG. 4.6.1 Three methods of dissolved-air flotation. **A.** Adhesion of a gas bubble to a suspended liquid or solid phase; **B.** Trapping of gas bubbles in a floc structure as the gas bubbles rise; **C.** Absorption and adsorption of gas bubbles in a floc structure as the floc structure is formed.

creases wastewater pressure while the retention tank provides adequate time for gas to transfer into liquid and also for excess gas applied to the system to be released.

Three general types of pressurization systems can dissolve gases. Figure 4.6.2 shows the position of the pressurization system in each flotation method. Full-flow pressurization transfers gas to the total-feed flow. The pressure of this system is generally 30 to 40 psig. This technique is applied when enough gas can be dissolved for flotation above pressure and when the wastewater flow passing through a centrifugal pump does not impair separation efficiency of the flotation process.

Partial-flow pressurization is used when a portion of the wastewater flow passing through the pressurization system does not impair the separation efficiency of the flotation system and when enough gas can be dissolved to affect flotation with the bypass stream pressurized to 60 to 75 psig. Using a partial rather than a full-flow pressurization system can frequently yield savings in the cost of the pressurization system.

Recycle-flow pressurization is favored when a natural or chemically formed floc is separated from the wastewater. In this system, a portion of the clarified flotation effluent is recycled to the pressurization system. This recycled flow then becomes the carrier of the dissolved gas later released for flotation. Recycle-flow pressurization is being applied increasingly in flotation applications. Recycle-flow pressurization systems are favored when dissolved air flotation is used for thickening biological sludges.

Process Design Variables

In addition to the process variables that govern the design of gravity clarification and thickening equipment, the performance of pressurization flotation equipment is affected by the gas-to-solids weight ratio and the point of chemical addition. Figure 4.6.3 shows a flotation unit. The circular tank shown includes a feed entrance baffling area, bottom sludge scraping, and an effluent discharge overflow weir. The float skimming equipment and float re-

TABLE 4.6.1 GAS SOLUBILITIES AND DENSITIES

Temperature	*Solubility* mg/l*			*Density† g/l*		
C°	*Air*	N_2	O_2	*Air*	N_2	O_2
0	37.2	29.4	69.6	1.293	1.251	1.429
10	29.3	23.3	54.3	1.249	1.209	1.375
20	24.3	19.3	44.3	1.206	1.168	1.330
30	20.9	16.8	37.4	1.166	1.129	1.287
40	18.5	14.9	33.2	1.130	1.091	1.247
50	17.0	13.8	30.2	1.093	1.060	1.208
60	15.9	12.9	28.3	1.061	1.027	1.170
70	15.3	12.4	26.8	1.030	0.997	1.137
80	15.0	12.3	25.8	1.000	0.970	1.105
90	14.9	12.3	25.4	0.974	0.944	1.073
100	15.0	12.3	25.3	0.949	0.918	1.047

*Weight solubilities in water at one atm (760 mm Hg) in the absence of water vapor.
†Gas densities at one atm (760 mm Hg) in the absence of water vapor.

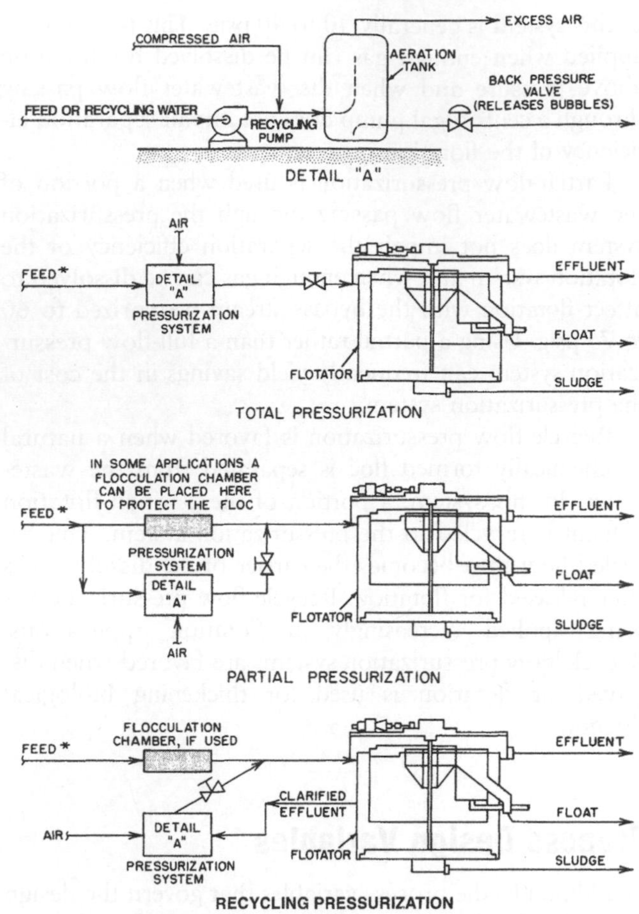

FIG. 4.6.2 Pressurization methods applied in flotation.

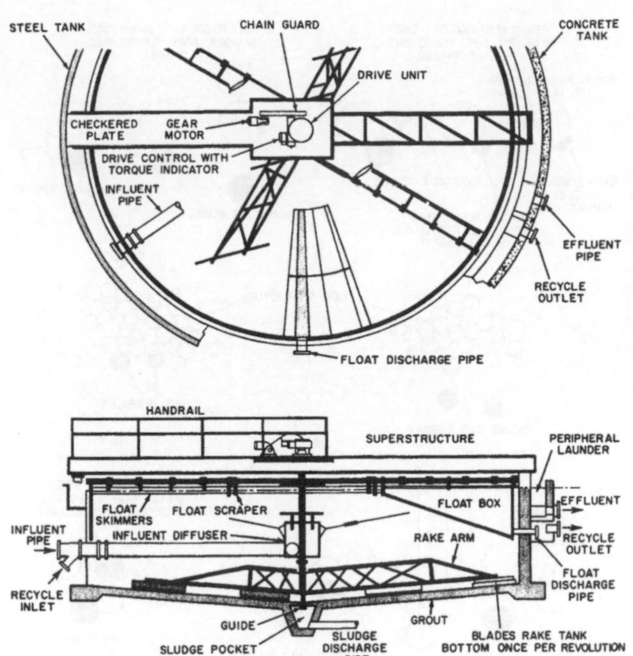

FIG. 4.6.3 Typical flotation unit.

tention baffle are the only additions that are different from a conventional clarification unit.

The gas-to-solids weight ratio is a major design consideration. This ratio determines the operating pressure of the pressurization system. This weight ratio varies from 0.01 to 0.06, but 0.02 is sufficient in most applications. Environmental engineers should use bench-scale tests or, if possible, pilot-plant tests to determine the gas-to-solids weight ratio for each application. An increase in the gas-to-solids weight ratio increases (to a limiting value) the separation rate of the gas–solids mass. This relationship cannot be adequately predicted and must be established by testing.

Full- and partial-flow pressurization systems do not affect the size of the flotation vessel since the flow to the flotation vessel does not increase. However, recycle-flow pressurization systems increase the feed flow rate to the flotation vessel by the recycled amount. Thus, flotation vessel size must be larger for recycle-flow pressurization systems than for full- or partial-flow pressurization systems.

Hydraulic and Solids Loadings

Hydraulic and solids loadings determine flotation vessel surface area as they do in gravity settling. In most cases, hydraulic loading is the limiting design criterion. Table 4.6.2 lists the removal efficiencies and the approximate hydraulic and solids loadings involved in treating industrial and municipal wastewater. The feed flow rate to a flotation unit using recycle-flow pressurization includes the recycled flow in addition to the raw feed flow. Therefore, flotation vessel surface area should be sized for total flow when hydraulic loading is the limiting design criterion.

Solids loading is directly related to raw feed flow rate, solids concentration, and flotation vessel area. Only in cases of sludge thickening by dissolved-gas flotation does solids loading determine flotation vessel surface area. In these cases, only recycle-flow pressurization systems are capable of supplying the gas quantities required at conventional operating pressures to achieve the required gas-to-solids ratio for thickening. A recycle flow-rate of twice the feed flow rate is common.

Chemical Additions

Wastewater treatment facilities are using chemicals in an increasing number of gravity and flotation clarification applications to reduce SS content of the effluent. They add

TABLE 4.6.2 FLOTATION SYSTEM PERFORMANCE DATA

Type of Wastewater	*Hydraulic Loading** gpm/ft²*	*Solids Loading lb/hr · ft²*	*Type System**	*Typical Influent Wastewater Characteristics mg/l*	*Contaminant Removal*	*Chemical Addition Used*
Oil refining	2.0–2.5	—	P or T	200–1000 oil	All free oil, no emulsified oil, 70–80% SS	None
	1.0–1.5	—	R	200–1000 oil	90% of all oil, no soluble oil, 90% SS	Alum Polyelectrolyte
Meat packing	2.5	—	P or T	500–5000 SS 1000–2000 grease	All floating grease 40–60% SS	None
	1.5–1.8	2.0	R	500–5000 SS 1000–2000 grease	90% grease, 90% SS	Alum Lime Polyelectrolyte
Paper mill	1.0–1.5	2.0	R	200–3000 SS	90% fiber	Alum Lime Polyelectrolyte
Poultry processing	1.5–2.0	—	P, T, or R	200–2500 SS 30–1000 grease	40–60% SS, 90% grease	None
	1.0–1.5	—	R	200–2500 SS 30–1000 grease	80–90% SS 90% grease	Alum Lime Polyelectrolyte
Fruit cannery	0.5–1.5	—	R	200–2500 SS	80–90% SS	Polyelectrolyte
Waste-activated sludge thickening municipal	1.25–1.5	2.0	R	2000–10,000 SS	80–90% SS 90–95% SS	None Polyelectrolyte

Notes: *Pressurization System Type
P = Partial flow
T = Total flow
R = Recycle flow
**In nonrecycling flotation units, it is the maximum influent rate per unit of tank surface area.

alum, ferric chloride, lime, and various polymeric compounds to form stable floc particles or break oil emulsions. These chemicals are usually added to pretreatment flash mixers and flocculators ahead of the flotation unit. The chemically treated wastewater then flows by gravity to the flotation system where gas is added by recycle-flow pressurization. Full- and partial-flow pressurization systems are infrequently used because the floc structure in the chemically treated wastewater degrades in the pressurization system and does not reform in the flotation unit.

Foaming Process

Foam fractionation separates a solution containing a surface-active solute into two fractions: the foam fraction containing a high concentration of surface-active solute and a drain fraction depleted of the same solute. Foam formation also collects SS and oils and separates them from the wastewater being treated.

Figure 4.6.4 shows the basic concept of foam fractionation. Gas, usually air, is diffused into the bottom section of the fractionation unit while the feed flow enters above the gas inlet but below the liquid surface. Foam is generated and lifted upward by the gas in the unit. This process discharges the foam from the fractionation unit and collapses it by heating it or spraying it with previously collapsed foam. Heating to facilitate foam collapsing is preferred when SS are separated by this process. The effluent is discharged near the bottom of the fractionation unit.

Process Variables

Variables affecting foam fractionation process efficiency include gas-to-liquid volume ratio, gas-to-solute concentration weight ratio, gas bubble size, type of solute, and foam characteristics. Increasing the gas-to-liquid flow rate ratios results in increased solute removal due to an increased gas surface area but yields a wetter foam.

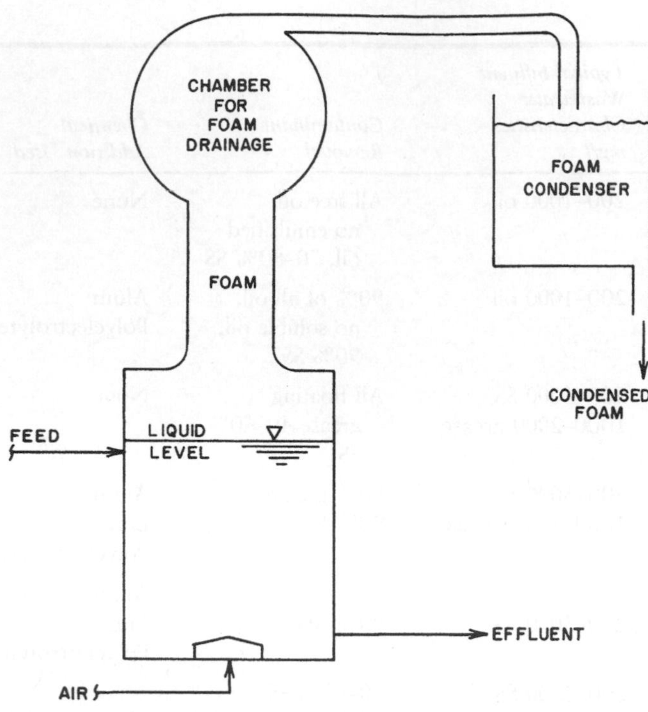

FIG. 4.6.4 Foam fractionation unit.

Decreasing the gas-to-liquid flow rate ratio results in a lower foam volume. Reducing the gas bubble size results in a greater bubble surface area for greater solute removal at lower gas flow rates. Effective removal of surface-active agents with high volume reduction requires a stable foam from which the excess liquid can be drained rapidly. In some cases, the wastewater treatment facility must add a surfactant to the process feed to attain this stable foam. Empirical equations relating process variables and removal efficiencies have been developed.

Equipment geometry affects fractionation unit efficiency. The method of gas dispersion, feed point, foam-discharge point, liquid depth, and vessel shape all affect process efficiency.

—*David H.F. Liu*

4.7 SLUDGE PUMPING AND TRANSPORTATION

Sludges from wastewater treatment vary in composition and ease of dewatering. Because of the rheological characteristics of sludge, transport through closed conduits is a difficult engineering task, and the design is usually based on empirical knowledge.

Solids are removed from clarifier sludge wells using gravity or pumps. In most cases, raw sludge and digested primary sludge are treated by positive displacement pumps, whereas activated sludge is easily moved by less-expensive centrifugal units.

Types of Sludge

Sludge describes a number of slurries in wastewater and water treatment.

RAW SLUDGE

Raw sludge is obtained from the primary clarifier. It is usually a vile, putrescible material, containing from 1 to 12% solids. It cannot be disposed of without further treatment, which can consist of aerobic or anaerobic digestion, incineration, or wet combustion. It dewaters reasonably well in a centrifuge because it consists mostly of large solid particles (smaller solids do not settle in the primary clarifier). Therefore, centrifuges are frequently used ahead of incinerators. Vacuum filtration of raw sludge is not recommended because of odor.

ACTIVATED SLUDGE

Waste-activated sludge results from the overproduction of microbial organisms in the activated sludge process. Wastewater treatment facilities periodically discharge this material to maintain the recommended SS concentration of mixed liquor in the aeration tank.

Waste-activated sludge creates a handling problem in many treatment plants. It is a light, fluffy material, composed of bacteria, rotifers, protozoa, and enough filamentous organisms to make concentration difficult. The underflow from a final clarifier may only contain 1% solids. Pumping this much water to digesters is inefficient and can lead to digester failure. Accordingly, wastewater treatment facilities use thickening devices to concentrate the sludge to 5 to 7% solids. In addition to gravitational thickeners used in the past, flotation and centrifugal thickeners are also used.

FILTER HUMUS

Trickling filter humus is obtained from the final clarifiers following trickling filters. The quantity of filter humus for a facility is significantly lower than the quantity of waste-activated sludge. Usually, wastewater treatment facilities do not thicken filter humus prior to pumping to the digester.

CHEMICAL SLUDGE

Chemical sludge is a new problem for wastewater treatment plant operators. Chemicals like alum or lime are now used for nutrient removal as well as for increasing plant efficiency without adding hardware. Centrifugation effectively removes calcium carbonate formed by the addition of lime. Metal hydroxides (e.g., magnesium hydroxide) are light precipitates that resist removal and compaction. Waste alum sludge from water treatment is difficult to thicken or dewater; only recently has it been recognized as a serious problem. Future water treatment plants must provide the means to treat and dispose of waste alum sludge. Currently, no effective method exists.

Physical Characteristics of Sludge

Sludges are almost always characterized by their solids concentration. Often, environmental engineers use the sludge volume index (SVI) to describe sludge settling characteristics. Unfortunately, the SVI is an unsatisfactory parameter and should only be used for plant control and other applications in which comparing two or more sludges is not necessary. A better means of describing sludge physical characteristics is sludge rheology.

Almost all slurries, especially wastewater treatment sludges, are thixotropic and pseudoplastic fluids (see Figure 4.7.1). As shown in Figure 4.7.2, they exhibit an apparent yield strength, a parameter that environmental engineers can use to describe sludge behavior. Rheograms (see Figure 4.7.2) can be constructed for any sludge provided that the proper tools are used. Unfortunately, rheological data are difficult to obtain because viscometers (Liptak 1995) must be modified before the sludge can be analyzed. Nevertheless, environmental engineers should measure the rheological properties and relate them to other physical or biological characteristics under study.

Transport of Sludge in Closed Conduits

Because of the nonNewtonian nature of sludge (Dick and Ewing 1967), environmental engineers cannot usually ap-

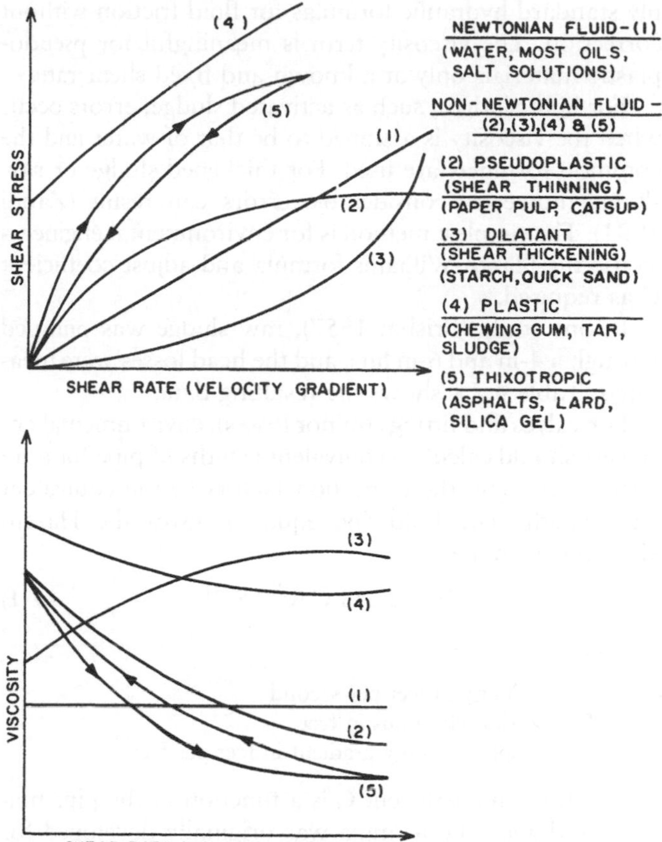

FIG. 4.7.1 Shear diagram of Newtonian and nonNewtonian fluids.

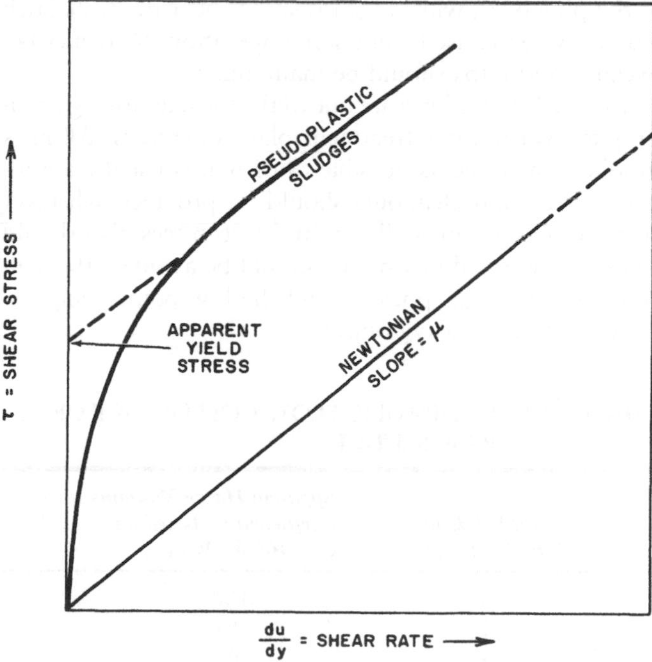

FIG. 4.7.2 Typical rheogram for a sludge. Projecting the straight line portion of the pseudoplastic curve to the zero shear rate gives the apparent yield stress.

ply standard hydraulic formulas for fluid friction without correction. The viscosity term is meaningful for pseudoplastic materials only at a known and fixed shear rate.

For light sludges, such as activated sludge, errors occur when the viscosity is assumed to be that of water and the common formulas are used. For thickened sludge or raw sludge, however, considerable errors can result (Zandi 1971). The simplest method is for environmental engineers to use the Hazen-Williams formula and adjust coefficient C as required.

In one study (Brisbin 1957), raw sludge was pumped through a 4-in and 6-in line, and the head losses were measured. Table 4.7.1 shows the resulting data.

For valves and fittings (minor losses), environmental engineers should calculate equivalent lengths of pipe for pure water and apply the correction factors to the equivalent pipe length. The following equation gives the Hazen-Williams formula:

$$V = 1.318\ C\ R^{0.63}\ S^{0.54} \qquad 4.7(1)$$

where:

V = velocity in feet per second
R = hydraulic radius in feet
S = slope of energy gradient in feet per feet

The friction coefficient C is a function of the pipe material and age. The formula was originally developed for water, and its application for sludge, as shown on Table 4.7.1, requires modifications of C.

Velocity in pipes is also important because at low velocities laminar flow can be attained. For pseudoplastic materials, this velocity forces the flow into the central core of the pipe only, with stagnant sludge near the wall. Such plug flow can cause troublesome operation. Velocities between 5 and 8 fps should be maintained.

Clogged sludge lines are one of the most annoying problems for wastewater treatment plant operators. All lines should be as large as possible (8 or 6 in is usually a minimum size), and cleanouts should be provided wherever possible. Long radius elbows and sweep tees also should be used. High and low points should be avoided; the high points result in gas pockets, and the low points clog easily with large or heavy objects.

TABLE 4.7.1 HYDRAULIC FLOW COEFFICIENT OF RAW SLUDGE

Total Solids in Raw Sludge (%)	*Apparent Hazen-Williams Coefficient C, Based on C = 100 for Water*
0	100
2	81
4	61
6	45
8.5	32
10	25

Sludge Pumps

Sludge pumping is an important part of wastewater treatment plant operations. Raw sludge must be moved to digesters, activated sludge must be returned and periodically discharged, scum must be pumped from scum pits, and pumping sludge is often used for mixing digesters. The type and consistence of sludges vary, and environmental engineers must individually analyze each pumping requirement.

PISTON PUMP

Plunger (positive displacement) pumps are used mostly for pumping raw sludge (see Figure 4.7.3). They can pump heavy solids without clogging and are easy to clean if clogging does occur. Pumping at low rates is possible, and suction lifts can be accommodated. The disadvantages of plunger pumps are that they can be messy and noisy, but their trouble-free service more than compensates for these disadvantages.

DIAPHRAGM PUMP

Diaphragm positive displacement pumps are often used for small installations (see Figure 4.7.4). Changing the length of the strokes varies the capacities. Counting the strokes and knowing the pump volume determines the sludge pumping rate (Zandi 1971). Both diaphragm and plunger pumps can be used as scum pumps and pumps for digested sludge.

SCREW PUMP

Screw-feed and rotary pumps are positive displacement units that produce a steady flow, as opposed to the pulsating flow from diaphragm and plunger pumps (see Figure 4.7.5). Screw and rotary units are always used to pump sludge to the dewatering equipment. The centrifuge requires a steady, uninterrupted flow. A pulsating or im-

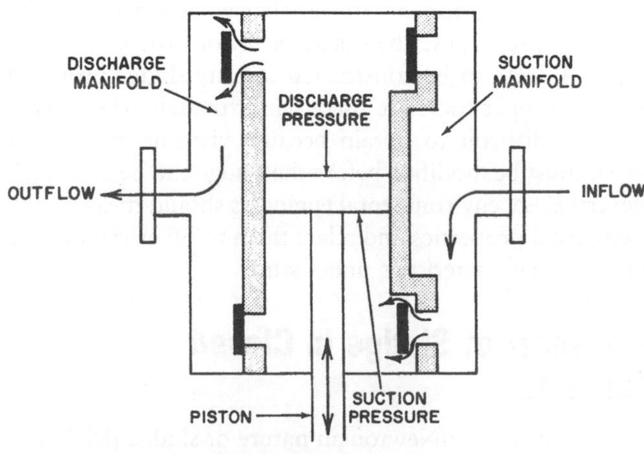

FIG. 4.7.3 Plunger-type sludge pump.

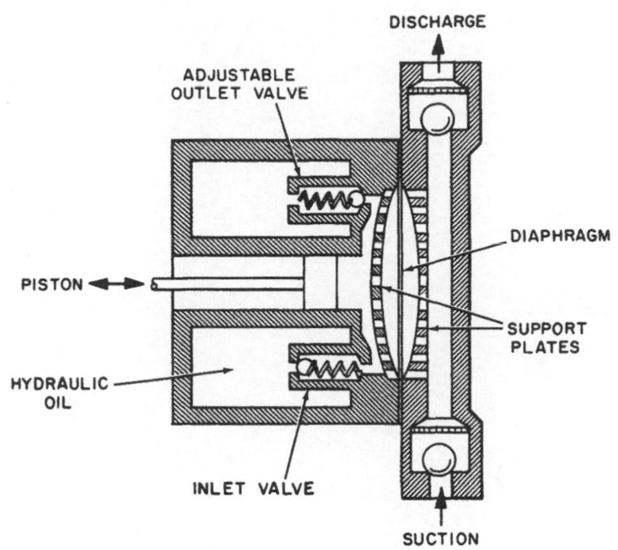

FIG. 4.7.4 Diaphragm-type sludge pump.

peller-type centrifical pump should never be used in front of a centrifuge.

AIR-LIFT PUMP

The air-lift pump is an excellent means of returning activated sludge in small, activated-sludge plants. It is almost trouble-free and requires no extra power.

CENTRIFUGAL PUMP

Centrifugal pumps, although less expensive than positive displacement pumps, suffer from clogging problems, both in the pump and in the line. Centrifugal pumps are used most often with activated sludge since the possibility of fouling is minimized on that service. The use of open, bladeless impellers prevents clogging problems.

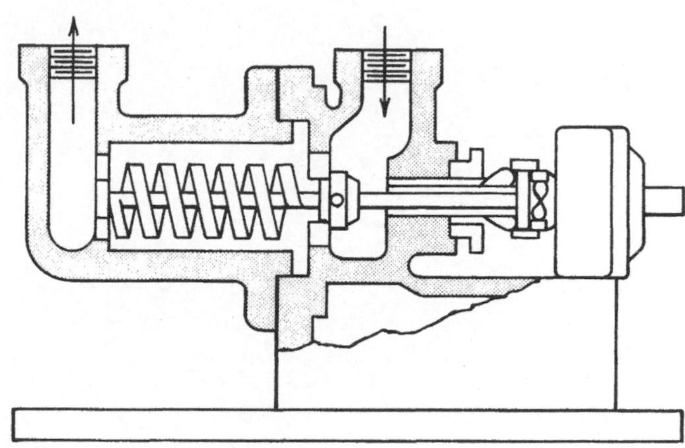

FIG. 4.7.5 Screw-type sludge pump.

References

Brisbin, S.G. 1957. Flow of concentrated raw sewage sludges in pipes. *Journal, Sanitary Engineering Division, ASCE* 83, SA3 (June).

Lipták, B.G. (ed.). 1995. *Instrument engineers' handbook. Process Measurement and Analysis.* Radnor, Pa.: Chilton.

Dick, R.I. and B.B. Ewing. 1967. The rheology of activated sludge. *Journal Water Pollution Control Federation* 39:10.4 (April).

Zandi, I. (ed.). 1971. *Advances in solid-liquid flow in pipes and its application.* New York: Pergamon.

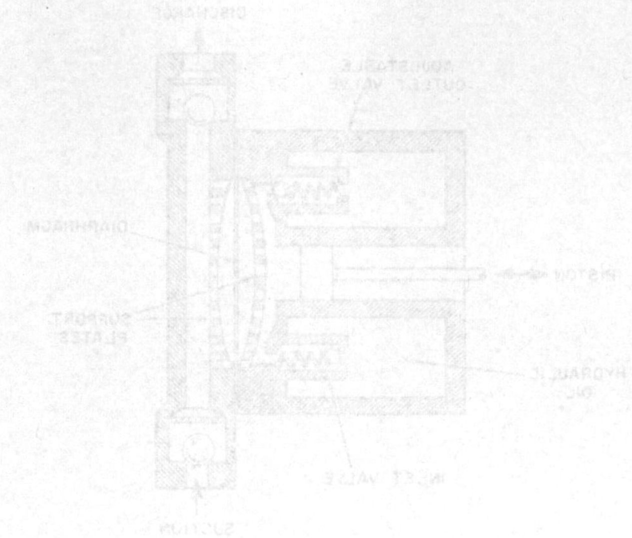

5

Conventional Biological Treatment

Robert D. Buchanan | LeRoy H. Reuter

5.1 SEPTIC AND IMHOFF TANKS

PSEPTIC TANK DESCRIPTION

A settling tank provides treatment for raw sewage. Flow-through settling and sludge digestion take place in the same chamber. Treatment is anaerobic.

IMHOFF TANK DESCRIPTION

A settling tank provides treatment for raw sewage. The flow-through settling chamber is separate from the quiescent sludge digestion compartment. Sludge digestion is anaerobic.

CAPACITIES (GALLONS PER DAY [gpd])

Septic tanks: 500 to 10,000 gpd
Imhoff tanks: 3000 to 300,000 gpd
(These tanks are used for larger installations.)

LOADING RATES TO DISTRIBUTION ON SAND FILTERS

0.1 to 0.15 mgd per acre of Imhoff effluent

EFFICIENCIES

Septic tanks: 15 to 25% removal of BOD and 40 to 60% SS removal.
Imhoff tanks: 25 to 35% BOD removal and 40 to 60% SS removal.

MATERIALS OF CONSTRUCTION

Septic tanks: usually concrete (often prefabricated), occasionally prefabricated steel
Imhoff tanks: concrete

ADDITIONAL TREATMENT REQUIRED

Septic tanks: usually subsurface drainage fields, occasionally intermittent sand filters or lagoons
Imhoff tanks: usually trickling filters, occasionally intermittent sand filters or lagoons

DISPOSAL OR TREATMENT OF SLUDGE

Septic tanks: normally to municipal treatment plant or burial by private contractor
Imhoff tanks: usually sludge drying beds

PRIMARY USE

Septic tanks: dwellings, camps, and small institutions
Imhoff tanks: communities and small cities

Treatment Characteristics

Raw sewage can be treated in either septic tanks or Imhoff tanks.

SEPTIC TANKS

Septic tanks receive raw sewage, allow it to settle, and pass the relatively clear liquid to the adsorption field, which is the next stage of treatment. The remaining solids digest slowly in the bottom of the tank. Septic tanks are inexpensive, but because of their incomplete treatment, they are suitable only for small flows.

Anaerobic decomposition, which takes place in the absence of free oxygen in a septic tank, is a slow process. To maintain practical detention times, the reactions cannot be carried far. Therefore, the effluent is often malodorous, containing a multitude of microorganisms and organic materials that require further decomposition.

IMHOFF TANKS

The process that takes place in an Imhoff tank is similar except that the tank is designed so that the flow-through upper chamber is separate from the lower digestion chamber, resulting in a two-story tank. The upper compartment acts only as a settling zone where little or no decomposition takes place. This chamber often remains aerobic, and its effluent has a lower BOD than the effluent from a septic tank. Anaerobic digestion takes place in the lower chamber. Because the effluent is a higher quality, the process is suitable for communities and small cities. Additional treatment for further decomposition of organic matter in the effluent is required.

Septic Tank Design

In septic tank design, environmental engineers must consider the treatment following the septic tank as a part of the septic tank system (U.S. Department of Health, Education, and Welfare 1967). A two-compartment design arranged in a series is preferred (see Figure 5.1.1). The first chamber should contain two-thirds and the second chamber should contain one-third of the total volume. The liquid depth should be between 4 and 7 ft.

The minimum effective tank capacity should be as follows: for flows up to 1500 gpd, $1\frac{1}{2}$ times the daily sewage flow; for flows in excess of 1500 gpd, the volume V in gallons can be calculated from the following equation:

$$V = 1125 + 0.75\ Q \qquad 5.1(1)$$

where:

Q = The daily sewage flow

Environmental engineers must check soil porosity and base the design of the soil absorption field on the rate of percolation. This rate is the number of minutes required for the effluent to recede 1 in in a test hole that has been

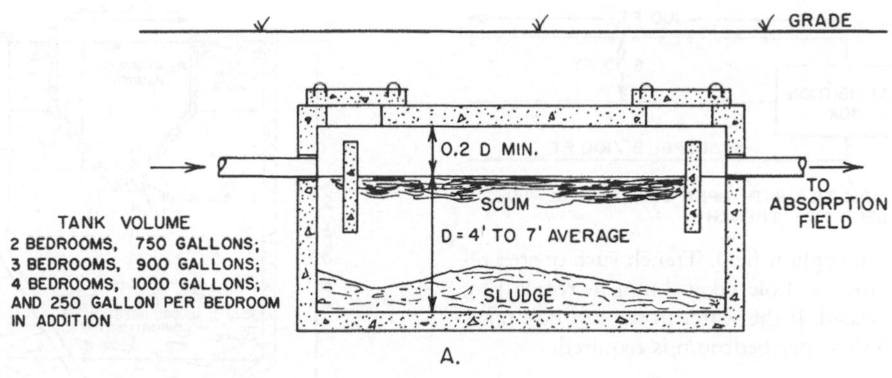

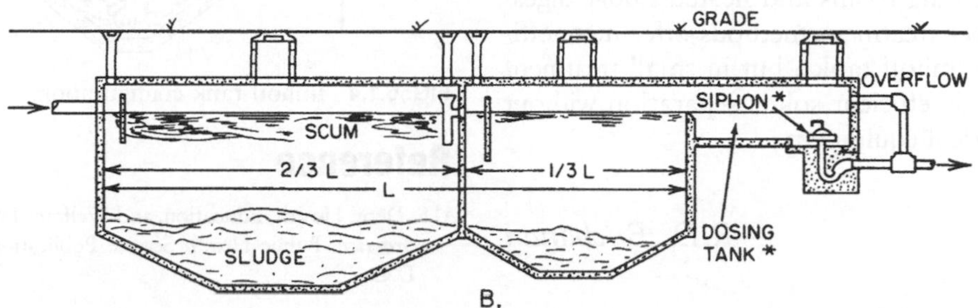

FIG. 5.1.1 Septic tank configurations. **A.** Typical household septic tank; **B.** Typical large institutional septic tank with dosing siphon. For large fields, uniform distribution is obtained by periodic flooding of the field followed by periodic drying. Dosing tanks are used to flood these fields; they collect the sewage, and automatic bell siphons or pumps transport the waste to the field.

bored, filled with water, and allowed to swell the day previous to the test (see Figure 5.1.2). Figure 5.1.3 provides information on absorption area requirements.

If the septic tank must handle wastewater from garbage grinders and automatic washing machines, the soil absorption area (see Figure 5.1.3) should be increased by 60%, and the tank volume should be increased (see Part A in Figure 5.1.1) by 25%. If the soil absorption trench required exceeds 500 ft^2 or the septic tank is larger than 1500 gal, a dosing tank is needed.

IMHOFF TANK DESIGN

Surface loading of the settling zone should be 600 gpd per ft^2 with detention times of $1\frac{1}{2}$ to 2 hr and velocities below 0.75 in per sec. The effective settling zone depth should be about 7 ft and its length can be from 25 to 50 ft. The gas-vent and scum area should be 20% of the total surface area. Total depths average around 30 ft (see Figure 5.1.4).

Septic tanks are suitable only for isolated facilities with low waste flows where the soil can be used as an absorption field. Their use should be avoided except when an alternative is not available and the site conditions are favorable.

The operation of Imhoff tanks is not complex. They are

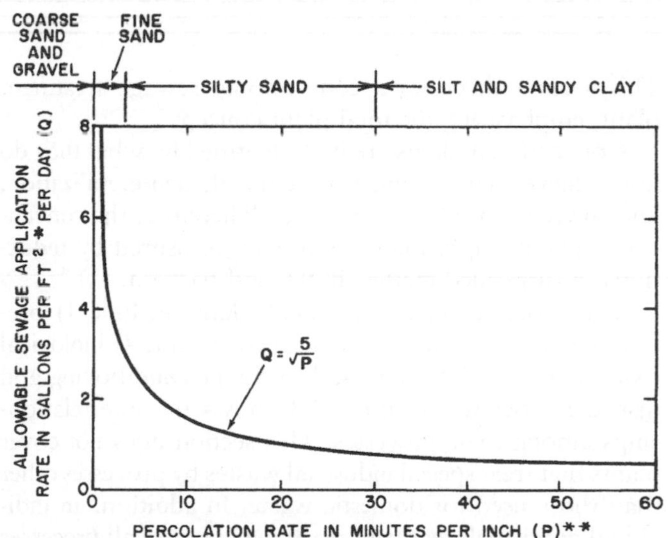

FIG. 5.1.2 Relationship between allowable sewage application rate and soil percolation rates for soil absorption trenches or seepage pits. (Reprinted from U.S. Department of Health, Education, and Welfare, 1967, *Manual of septic tank practice,* Public Health Service Publication no. 526, Washington, D.C.)

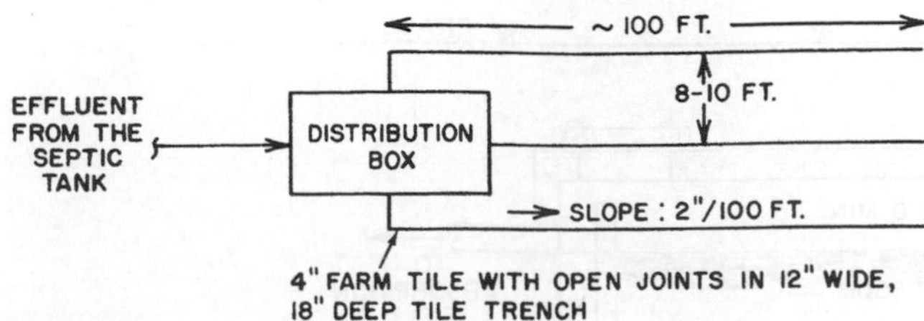

FIG. 5.1.3 Septic tank absorption field. Trench surface area required: If the water in the test hole takes 1 min to recede 1 in, 70 ft^2 per bedroom is needed. If the water in the test hole takes 30 or 60 min, 250 or 330 ft^2 per bedroom is required.

less efficient than settling basins and heated-sludge digestion tanks. The newer treatment methods offer more efficient alternatives to Imhoff tanks, but in small treatment units, they do provide efficient solids separation without mechanical or electrical equipment.

—*R.D. Buchanan*

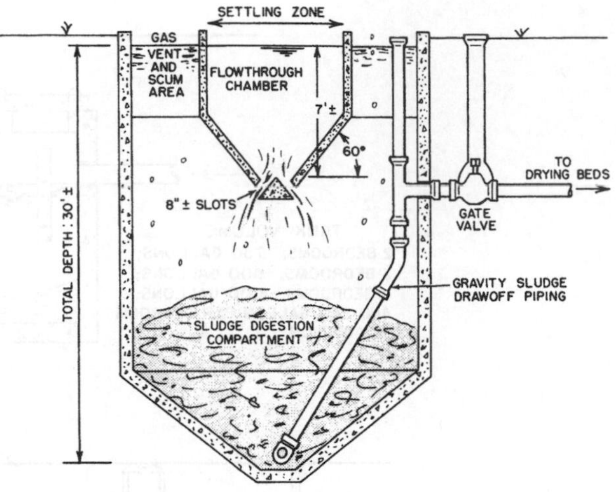

FIG. 5.1.4 Imhoff tank configuration.

Reference

U.S. Dept. Health, Education, and Welfare. 1967. *Manual of septic tank practice.* Public Health Service Publication. No. 526. Washington, D.C.

5.2 CONVENTIONAL SEWAGE TREATMENT PLANTS

This section describes a conventional sewage treatment plant, emphasizing the total plant concept.

Conventional plants are best identified by what they do not achieve, namely nutrient removal, demineralization, and the removal of trace organics. Therefore, the conventional plant's performance is usually measured by reductions in suspended matter, BOD, and bacteria.

The processes in conventional plants include 1) pretreatment, 2) settling, 3) chemical treatment, 4) biological oxidation, 5) disinfection, and 6) sludge conditioning and disposal processes. Figure 5.2.1 shows the interrelationships among these processes. This section does not cover plants that treat special industrial wastes by processes other than those used for domestic waste. In addition, an individual municipal treatment plant may not use all processes discussed in this section.

Selection of Specific Processes

In the selection of specific processes and plant design, environmental engineers must consider the existing and potential regulatory standards; plant operators' requirements and availability; existing and projected sewage flow; flow pattern and waste characteristics; climate, topography, and availability of land; plant location within the community; and all aspects of cost. They must also assess the life-cycle cost of the plant. Lower construction costs can frequently be offset by high maintenance and increased operation costs.

The plant design cannot be optimized for all of these factors, consequently the environmental engineer must make difficult decisions in the face of uncertainty. The problem is complicated because the design life of the facility is normally twenty-five years or more. A typical approach is for the design to minimize cost while achieving a treatment level for a set of constraints represented by the legal and physical aspects of the project.

Although many treatment plants, particularly large facilities, to be constructed include advanced forms of wastewater treatment, most plants in operation use conventional processes. For several more decades, these plants will serve a large percentage of the population. Upgrading existing facilities offers a promising alternative for improving treatment compared to completely new construction and the total loss of an existing plant's capabilities.

A principal consideration in overall plant design is to

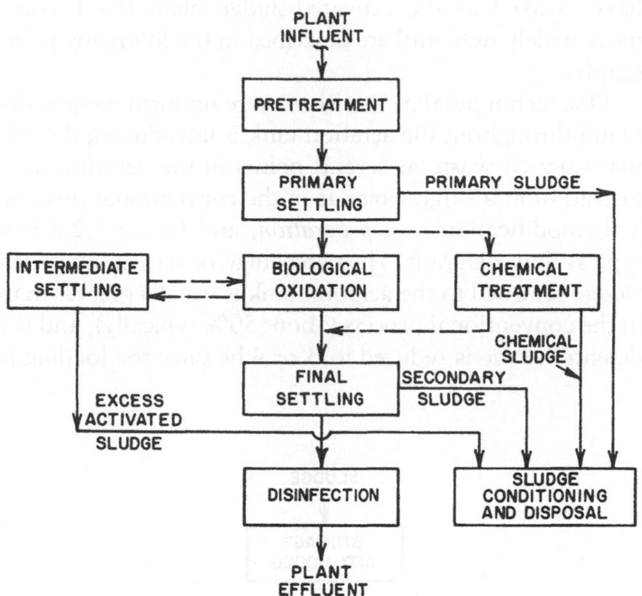

FIG. 5.2.1 Conventional domestic wastewater treatment process flow sheet.

provide flexibility for expansion and upgrading of the initial plant. Flexibility for efficient plant operation allows the operator to overcome problems and provide maintenance and repair with minimum effect on plant performance. The design should also include the capability for implementing new ideas that may improve plant performance. Even minor modifications are sometimes difficult and more costly to make than the same capability built into the initial design and construction.

Anticipating plant expansion can also save money and avoid disruption. Land availability and wise plant layout are two basic considerations. Space allowances in buildings for future needs—pumps, instrumentation, and pipelines in the initial construction—are usually only a minor cost factor. For a small additional cost, environmental engineers can select chemical feeders and chlorinators for the initial project with sufficient capacity to handle anticipated increased rates. Often, the tanks designed for one purpose can be effectively used for a different process and accommodate plant expansion or upgrading.

PRETREATMENT PROCESSES

To protect pumping equipment, control and monitor instruments and prevent clogging filters, environmental engineers routinely include chemical feeders, valves and overflow devices, and physical pretreatment processes in all plants. Treatment units serving individual households are about the only exception to this practice. Pretreatment equipment includes screens, grinders, skimmers, and grit chambers. Flow equalization is also a pretreatment process. Equalization assists in controlling hydraulic overloads that can occur during the day and also balances the incoming waste strength.

SETTLING OR CLARIFICATION

Settling processes remove settleable solids by gravity settling either prior to or after biological or chemical treatment and between multiple-stage biological or chemical treatment steps. In larger tanks, mechanical scrapers accumulate the solids at an underflow withdrawal point, whereas in smaller and some older systems, a hopper bottom is used for solids collection. Solids move down the sloped tank bottom by gravity in hopper-bottom tanks. Both circular and rectangular tank shapes are used. A rectangular or square tank uses the land area more efficiently and environmental engineers can save construction costs by nesting units and using common walls. With circular tanks, this cannot be done.

Settling tanks are commonly designed based on the overflow rate, the unit volume of flow per unit of time divided by the unit of tank area (gallons per day per square foot). Typical overflow rates are 600 gpd per ft^2 for primary settling, 1000 gpd per ft^2 for intermediate settling, 800 to 1000 gpd per ft^2 for final clarifiers after activated-sludge units, and 700 to 1000 gpd per ft^2 for final clarifiers after trickling filters. The detention times for settling range from 1 to 2.5 hr for average flows depending on the processes before or after the settling step.

CHEMICAL TREATMENT

Traditional chemical precipitation uses either iron or aluminum salts to form a floc, which is then settled. Lime also clarifies. This process step can reduce the SS up to 85%. The accumulated chemical sludge is removed by gravity flow or pumping to conditioning or disposal or both. The chemicals and sewage are flash-mixed in a mixing tank that has only a few minutes detention time followed by 30 to 90 min detention in a flocculation tank that is slowly agitated to aid floc growth. As shown in Figure 5.2.1, settling follows the flocculation tank.

BIOLOGICAL OXIDATION

Two basic techniques, fixed-bed and fluid-bed, are used in conventional biological treatment. The trickling filter has a fixed-bed of stone or plastic packing material that provides a growth surface for zoogleal bacteria and other organisms. The intermittent-sand filter and the spray irrigation system are other examples of the fixed-bed technique.

The activated-sludge processes and sewage lagoons are fluid-bed systems. The activated-sludge process uses mechanical aeration and returns a percentage of the active sludge to the process influent. Lagoons or stabilization ponds and oxidation ditches do not routinely waste sludge, but multipond systems can have recirculation. Septic tanks and Imhoff tanks combine the settling and biological oxidation processes in a single tank.

Activated Sludge

Activated-sludge processes use continuous agitation and artificially supplied aeration of settled sewage together with recirculation of a portion of the active sludge that settles in a separate clarifier back to the aeration tanks. These processes vary in detention time, the method of mixing and aeration, and the technique of introducing the waste and recirculated sludge into the aeration tank.

Figure 5.2.2 is a conventional activated-sludge plant flow diagram. A return of activated sludge at a rate equal to about 25% of the incoming wastewater flow is normal; however, plants operate with recirculation rates from 15 to 100%. The mixture of primary clarifier overflow and activated sludge is called *mixed liquor*. The detention time is normally 6 to 8 hr in the aeration tank.

In a conventional plant, the oxygen demand is greatest near the influent end of the tank and decreases along the flow path. Plants built before the process was well understood provided uniform aeration throughout the tank. A conventional plant cannot accommodate variations in hydraulic and organic loadings effectively, and the final clarifier must be sized to handle a heavy solids load. Usually aeration units are in parallel so that a shutdown of one unit does not totally disrupt plant operation. Modifications have evolved as the activated-sludge plant has become more widely used and are described in the following paragraphs.

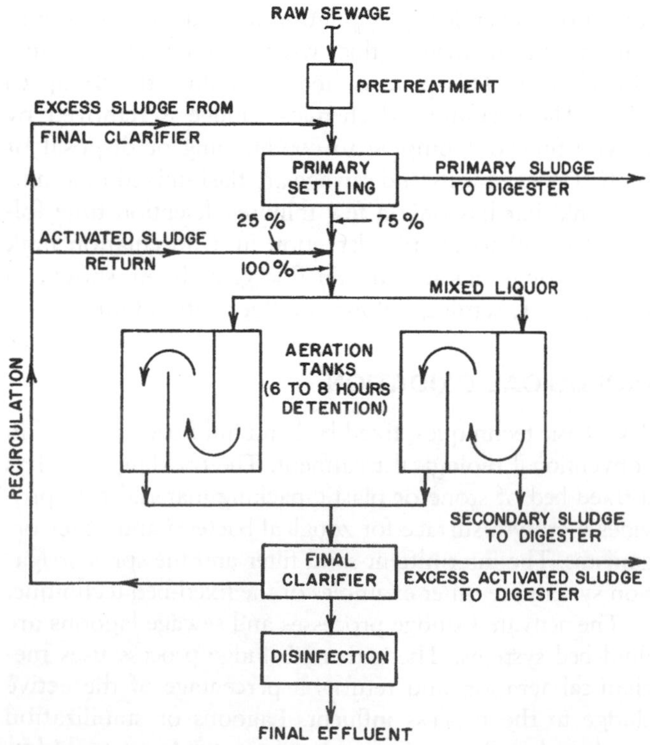

FIG. 5.2.2 Conventional activated-sludge plant flow sheet.

One technique that furnishes more uniform oxygen demand throughout the aeration tank is introducing the primary settled waste at several points in the aeration tank instead of at a single point as in the conventional process. This modification is *step aeration*, and Figure 5.2.3 is a typical flow diagram. The percentage of settled, activated sludge returned to the aeration tank is usually greater than in the conventional process (about 50% typically), and the detention time is reduced to 3 or 4 hr since the loading is more evenly distributed in the tank. Additional piping and pumps are required to distribute the waste to several locations; however, the improved performance is considered to be worth the expense.

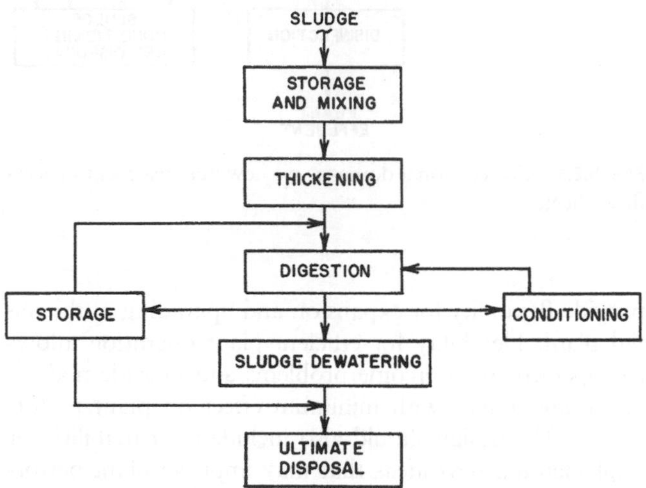

FIG. 5.2.3 Step-aeration type activated-sludge plant flow sheet.

A less popular alternative to distributing the load to the aeration tank is to provide different quantities of oxygen along the tank length, related to the oxygen demand that gradually decreases along the tank length. The flow sheet for *tapered aeration* is the same as that in Figure 5.2.2. The disadvantage of tapered aeration is that although it is more economical due to reduced air quantities, it can only be designed for one loading.

Figure 5.2.4 shows the *extended aeration* treatment process. In extended aeration, as the name implies, the activated-sludge detention time is increased by a factor of 4 or 5 compared to conventional activated sludge. The primary clarifier is eliminated. At a surface settling rate of 350 to 700 gpd per ft^2, 4 hr final settling is typical. The extended-aeration period reduces or eliminates the requirement for disposing excess sludge and is therefore a popular system for small plants.

Figure 5.2.5 is a flow diagram for the *completely mixed,*

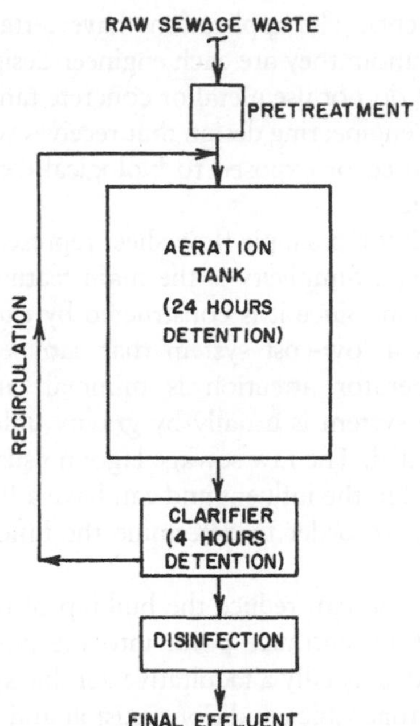

FIG. 5.2.4 Extended-aeration plant flow sheet.

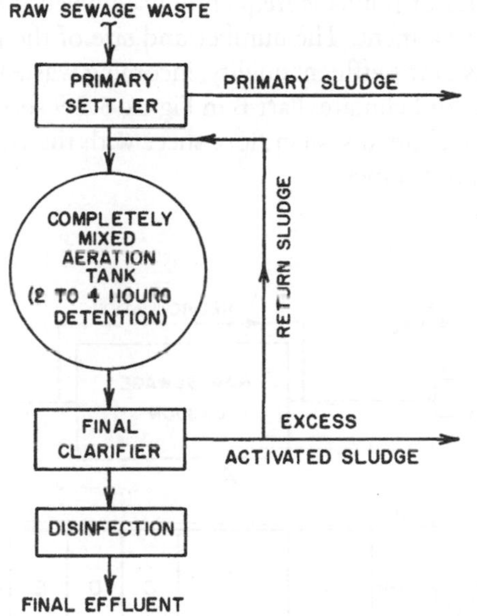

FIG. 5.2.5 Completely mixed, activated-sludge treatment flow sheet.

activated-sludge system. This process is an extension of step aeration and provides a uniform oxygen demand throughout the aeration tank. Mechanical aerators also provide mixing for this unit. The SS concentration in the mixed liquor is two to three times the concentration in most conventional plants. Aeration detention times are reduced to 2 to 4 hr. The sludge recycling ratio is generally high because the greater flow improves mixing.

Figure 5.2.6 is a typical *contact stabilization* process flow diagram. The small contact tank, with detention times

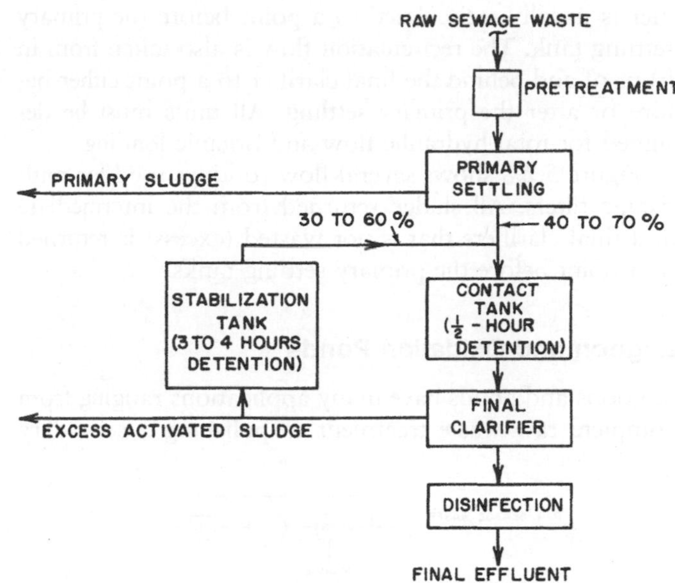

FIG. 5.2.6 Contact stabilization plant flow sheet.

of about ½ hr binds the insoluble organic matter in the activated sludge. Clarification separates the contact-settled sludge from the supernatant. The smaller sludge volume is then aerated for an additional 3 or 4 hr. Since the total sewage flow is aerated for a shorter period (with only the returned activated sludge being aerated for longer periods), the aeration tank capacity is smaller than in conventional plants. With the activated sludge in two tanks, the plant is not out of operation when the contact tank is disabled. Sludge recycling percentages of 30 to 60% are normal with contact stabilization.

Trickling Filters

Trickling filters are the most common treatment units used by municipalities to provide aerobic biological treatment. Trickling filters are classified according to the hydraulic and organic loading applied. Filters are categorized as follows: the low rate is 2 to 4 million gal per acre per day (mgad), the intermediate rate is 4 to 10 mgad, the high rate is 10 to 30 mgad, and super-rate units are greater than 30 mgad.

Only a few fixed-nozzle trickling filters are in operation because the design requires extensive piping and nozzle heads that permit dosing the total filter bed. A pump-discharge, head-driven, rotary distributor and a dosing chamber with siphon or a constant head-box design is more common. Single-stage and multistage filter arrangements are both used because many recirculation schemes and options of intermediate settling between multistage filters are available. The recirculation method is much less a factor in plant performance than the recirculation ratio.

Figure 5.2.7 shows the more common single-stage filter recirculation flow diagrams. Sludge from the final clarifier is usually recirculated to a point before the primary settling tank. The recirculation flow is also taken from in front of and behind the final clarifier to a point either before or after the primary settling. All units must be designed for total hydraulic flow and organic loading.

Figure 5.2.8 shows several flow routings used for multistage filters. All sludge returned from the intermediate and final clarifiers that is not wasted (excess) is returned to a point before the primary settling tanks.

Lagoon and Oxidation Ponds

Lagoons and ponds have many applications ranging from complete raw waste treatment to polishing a secondary plant's effluents. The applications have certain characteristics in common: they are each engineer-designed and uncovered and do not use metal or concrete tanks. A *lagoon* is a pond of engineering design that receives waste that has not been settled or exposed to biological oxidation prior to entering it.

Figure 5.2.9 is a simple flow-sheet representing the raw sewage lagoon. Simplicity is the main feature of the raw sewage lagoon. Since it is constructed by excavation and diking, it is a low-cost system that can be constructed rapidly. Operator attention is minimal, and the flow through the system is usually by gravity unless recirculation is provided. The raw sewage lagoon usually has a bar screen placed in the influent and can have a Parshall flume with a drum recorder to determine the inflow to the lagoon.

Recirculation can reduce the buildup of bottom solids near the inflow entrance point into the pond. The raw sewage pond is usually a facultative aerobic system, which means that anaerobic conditions exist at and near the bottom and aerobic conditions prevail in the upper layers of the pond most of the time. Facultative organisms can function under either aerobic or anaerobic conditions.

A series of ponds is frequently used when it comprises the sole treatment. The number and size of the ponds are functions of the effluent quality, incoming waste load, temperature, and climate. Part B in Figure 5.2.9 shows a multipond facultative system flow sheet with the corresponding detention times.

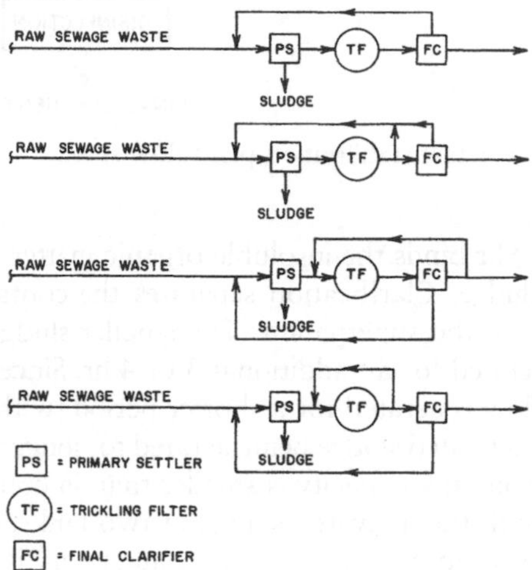

FIG. 5.2.7 Typical single-stage, trickling filter, recirculation flow sheets.

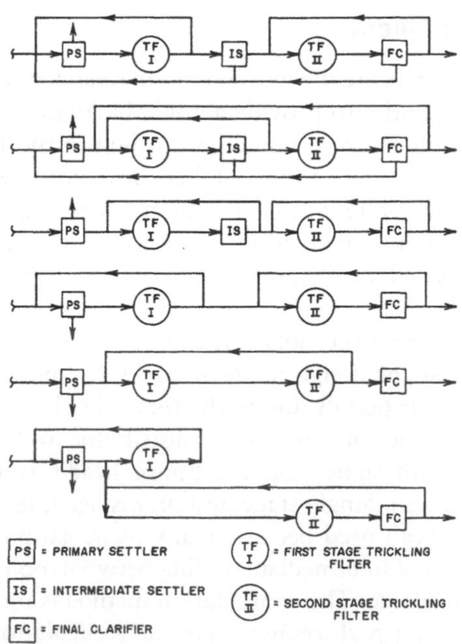

FIG. 5.2.8 Typical multistage, trickling-filter, recirculation flow sheets.

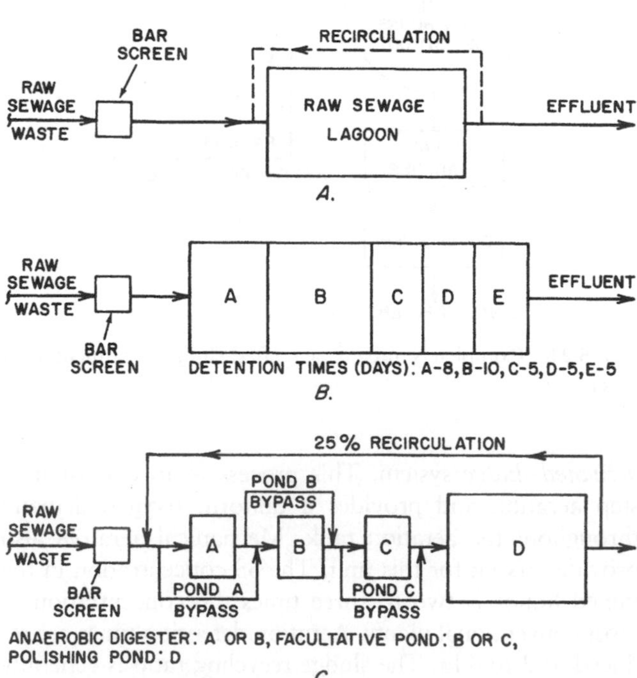

FIG. 5.2.9 Raw sewage lagoon flow sheets. **A**, Single pond system; **B**, Multipond facultative aerobic lagoon system; **C**, Anaerobic-aerobic pond system.

The primary pond designed as an anaerobic pond is becoming more popular. Part C in Figure 5.2.9 is a typical flow sheet for an anaerobic–aerobic pond system. Ponds A, B, and C are each anaerobic ponds, and the flow arrangement provides flow through any two of the anaerobic ponds in the series (AB, AC, and BC). This arrangement permits one pond to serve as an anaerobic digester.

The second anaerobic pond produces a higher quality effluent than does a single pond, thus reducing the load and size of the facultative pond. An anaerobic pond is normally used for six months to a year as the anaerobic digester. A pumped recirculation of about 25% of the total flow is common. The raw sewage lagoon can have additional ponds (D in Figure 5.2.9) in the series after the facultative pond for additional polishing treatment.

The oxidation pond, as opposed to the raw sewage lagoon, receives influent that has undergone primary treatment. A maturation pond provides a final, polishing treatment step that follows some form of secondary treatment. Therefore, the maturation pond is a form of tertiary treatment. Mechanically aerating the oxidation pond improves treatment and reduces the pond size. When mechanical aeration is provided, floating surface aerators are almost universally used. The series flow in ponds buffers against shock loadings.

Oxidation Ditch

The oxidation ditch is a variation of the aerated raw sewage lagoon in that the process combines settling and aerobic biological oxidation in a single unit. Oxidation ditches are effective in treating the waste of small communities. Similar to lagoons, construction and operating costs are low and they can be constructed rapidly. The energy requirement for treatment is small, and operator attention is minimal.

Oxidation ditches operate on higher loadings than aerated ponds. A circulation rate of about 1 ft per sec maintains the solids in suspension. Oxygenation is supplied by an aeration rotor system, which is a power unit of either angle-iron or cage design.

The single-ditch unit in Part A in Figure 5.2.10 operates in the following sequence:

First, the aeration rotor is turned off when the overflow level of the ditch is reached.

After sludge settling occurs in the ditch, additional raw waste is pumped in displacing a like volume of supernatant, and this cycle repeats.

When the detention time of raw waste sewage is at least 24 hr and sufficient oxygen is present, the quantity of excess sludge is small.

Part B in Figure 5.2.10 shows the multiple-ditch configuration. Ditches B and C are alternately used for settling, while ditch A operates continuously. When ditch B is used for settling, the gates connecting pond A with B

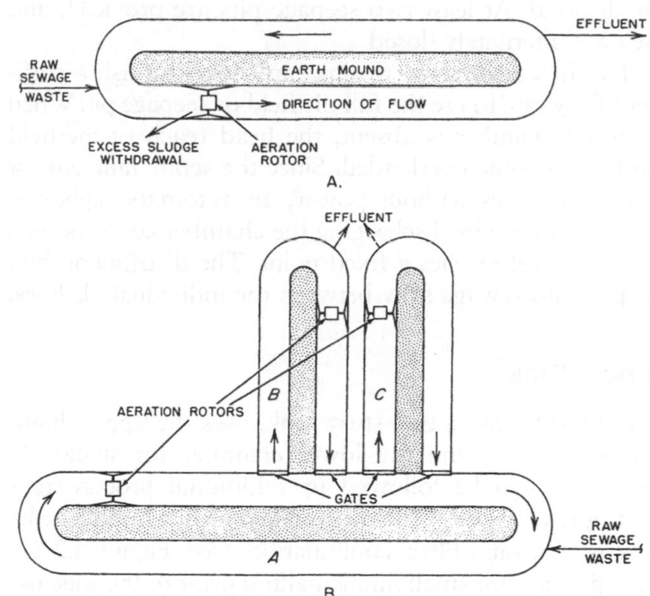

FIG. 5.2.10 Oxidation ditch flow sheets. **A,** Single ditch unit, **B,** Multiple ditch unit.

are closed, and the aeration rotor in ditch B is shut off. When the ditch is not used for settling, the aeration rotor is turned on, and the ditch functions in an auxiliary treatment capacity. After settling occurs in either ditch B or C, the gates to the ditch are opened, and the supernatant is discharged as in the single-ditch unit. After the supernatant is discharged, the settled sludge in the ditch is resuspended and distributed by the aeration rotor in that ditch.

Septic Tank

The septic tank continues to serve as the wastewater disposal system for millions of households and numerous small industries, trailer parks, and recreation areas. Figure 5.2.11 is a typical flow sheet for a septic tank. The system combines settling and anaerobic surface disposal, usually by an open-jointed tile underdrain network. Seepage pits—covered pits lined with open-jointed masonry surrounded by gravel—are occasionally used instead of the tile field

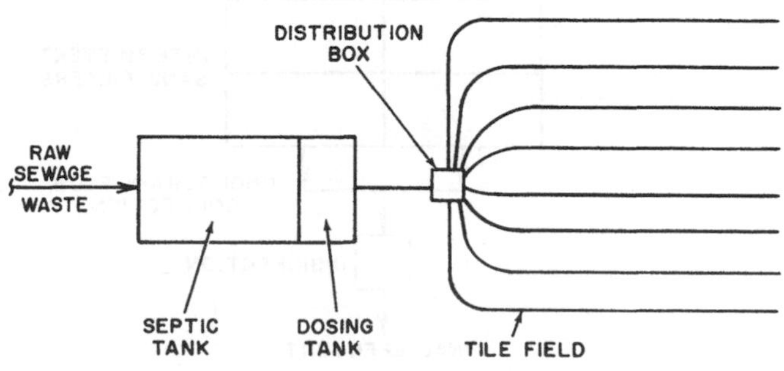

FIG. 5.2.11 Septic tank and disposal field flow sheet.

for disposal. At least two seepage pits are provided, and they are alternately dosed.

The function of the dosing tank is to furnish a sufficient flow rate to use the full tile field or seepage pit. When a dosing chamber is absent, the head reach of the field tends to become overloaded. Since the septic tank almost always operates without power, an automatic siphon is used for dosing by discharging the chamber contents each time the level reaches a fixed point. The distribution box proportions sewage flow between the individual tile lines.

Imhoff Tank

The Imhoff tank, a two-story tank, uses the upper chamber for settling and the lower chamber for sludge digestion. It can be followed by additional process steps to improve plant effluent quality. The Imhoff-tank–intermittent-sand-filter combination (see Figure 5.2.12) were popular for small municipalities prior to the wide use of the trickling filter and activated-sludge processes. Some units serve homes and recreation areas.

The tank has the advantage of being a nonmechanical device; however, it does require deep excavation unless it is built above ground, which requires more expensive construction and pumping of the raw sewage. Since heating of the digester is uneconomical due to the heat losses in the settling section of the tank, the digester is sized for unheated operation. For uniform sludge distribution to the digestion section of the tank, the tank periodically reverses the flow pattern. Both the tank bottom and the division between the settling and digestion sections have steep slopes so that the sludge can slide down the walls.

Sludge is normally withdrawn through a pipeline to open sludge drying beds by hydrostatic pressure. The gas produced by digestion is inhibited from entering the settling chamber by an overlap on the chamber dividing walls. It escapes through ventilation compartments that make up at least 20% of the tank surface area. Gases are usually not collected because the digester is not heated and no fuel is needed. A rectangular tank is the most common, although circular tanks are also used.

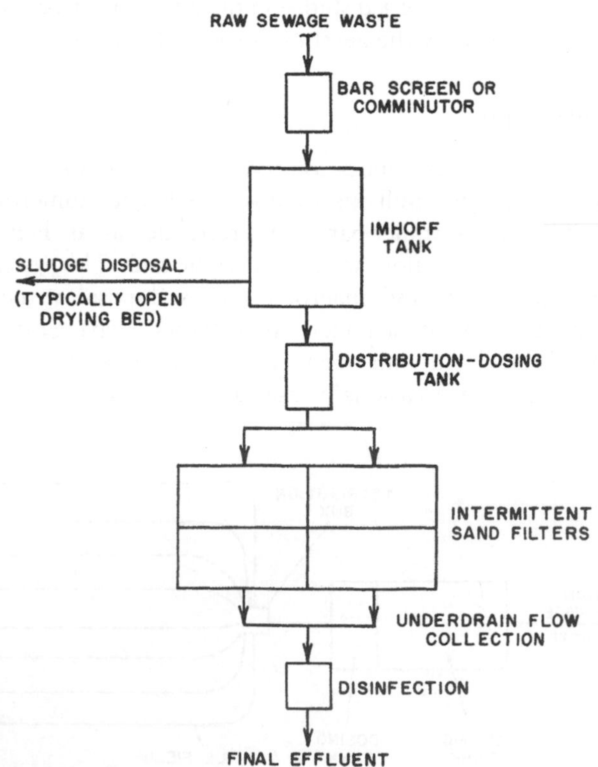

FIG. 5.2.12 Imhoff-tank–sand-filter flow sheet.

Intermittent Sand Filters

Although the intermittent sand filter was popular in small plants early in this century, it is being phased out. Figure 5.2.12 shows a flow diagram of an intermittent sand filter that furnishes additional treatment to an Imhoff tank effluent. Because both sand filters and Imhoff tanks were popular during the same period, their combined use was also common. Requirements for better quality plant effluents and the availability of other processes that do not require as much operator attention or produce the objectional odors associated with sand filters have influenced their being phased out.

Even small plants must use several sand filter units to permit time for drying, cleaning, and replacing the media. They frequently use a dosing tank with an automatic siphon to dose the filter in use, usually two to four times a day. A loading rate of about 100,000 to 150,000 gal per acre per day is practical for Imhoff tank effluent. Where more effective treatment precedes the sand filters, the application rate can be increased. Selecting the sand filter to be dosed is usually a manual operation controlled by valves.

A sand filter usually consists of 3 to 4 in of sand laid over 6 to 12 in of gravel. Tile underdrains collect the effluent, which is discharged to the receiving waters. A uniform flow distribution to sand filters is important and is usually achieved by trough distribution or a rotary-arm distribution device.

DISINFECTION

Conventional treatment plants use chlorination as the final treatment process to reduce bacteria concentration. Prechlorination, performed on the plant influent, is used if the incoming sewage is septic or the flows are low and the holdup time in the plant is long enough that the waste can become septic.

Prechlorination is usually at a fixed dosage, and a residual chlorine level is not maintained. Postchlorination provides 15 to 30 min detention time in a baffled, closed tank to prevent short-circuiting and dissipation of the chlorine. Wastewater treatment facilities do not frequently use chlorination to reduce the BOD in the effluent. Ordinarily, a combined chlorine residual between 0.2 and 1 mg/l is the target for the final effluent.

SLUDGE THICKENING, CONDITIONING, AND DISPOSAL

Settled solid material from various treatment processes is called *sludge*. For excess quantities of sludge to be disposed of economically and with minimal objections, the raw sludge is digested and dewatered. Controlled digestion reduces the quantity of complex organic material, increases the number of ultimate sludge disposal alternatives, and decreases the undesirability of the sludge.

Septic tanks and lagoons are not supplied with separate digesters or sludge conditioning equipment. Imhoff tanks can have a separate digester depending on unit size, waste and effluent character, and plant location. Disposing excess sludge from these units involves either periodic manual cleaning (septic tanks and lagoons) or, more commonly, pumping or the gravity-flow transfer of the sludge to open drying beds for Imhoff tanks. Clarifiers, activated-sludge-type units, and trickling filters all require separate sludge digesters and/or disposal systems.

Sludge Storage

Wastewater treatment facilities route excess quantities of activated sludge and sediments from primary, intermediate, and final clarifiers through some or all of the steps outlined in Figure 5.2.13. Plants can have facilities for disposal only; dewatering and disposal; digestion, dewatering, and disposal; or thickening, digestion, dewatering, and disposal. Wet sludge quantities, depending on the waste and treatment processes, can constitute as much as 1.5% of influent flow.

Some form of short-term storage before the digestion step is essential to prevent overloading, regulate sludge flow to the digesters, and allow collection and mixing of the sludge. Some plants mix some final effluent with the sludge (in the mixing tank) to improve its thickening characteristics. Storage tanks should be open so that gas does not buildup in the tank.

Sludge Thickening

Gravity thickening in a deep, circular, open tank is frequently used before anaerobic digestion. Some wastewater treatment facilities also use polyelectrolytes to improve gravity thickening, but their use for this purpose is not widespread.

Air flotation thickening of activated sludge is an alternative. This process uses dissolved or diffused air and sometimes a coagulant to float the sludge to the surface

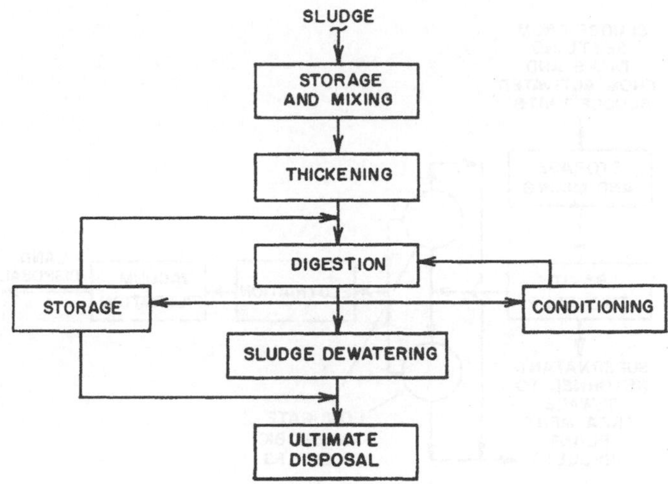

FIG. 5.2.13 Process flow sheet for sludge digestion and disposal.

where it is removed by a mechanically driven skimmer. The effluent drawn near the bottom of the flotation tank is combined with incoming plant influent and passes through the treatment cycle. Most flotation units recycle a portion of the effluent through the flocculation-air chamber because the recycled liquid enhances flocculation.

Although flotation is effective on activated sludge because of its density, it is not used on clarifier sludges because they are more difficult to float.

The primary purpose of mixing tanks is to provide a satisfactory mixture of clarifier sludge and activated sludge when a single thickening process is used. A mixing–storage tank also permits the thickener to operate continuously with a uniform inflow rate.

Sludge Digesters

Anaerobic digestion is more common than aerobic digestion because the anaerobic process does not require an air supply and generally produces enough gas to provide the digester with fuel for heating. Anaerobic digesters can be classified as low-rate, high-rate, and secondary units. The low-rate system merely holds the sludge for long periods (30 days or more). The high-rate system uses some form of mixing and heating to accelerate digestion and consequently has shorter detention periods (10 to 20 days) and can accommodate increased solids loadings.

The high-rate unit is typically followed by a larger secondary digester that furnishes stratification of the supernatant and sludge. At least two digester tanks are provided in addition to a secondary digester tank when such a high-rate unit is used. Multiple tanks furnish flexibility and safety if one of the units has an upset.

Aerobic sludge digestion requires an extensive air supply to maintain a DO surplus in the digester. Aerobic digestion tanks are mixed either by air or through agitators

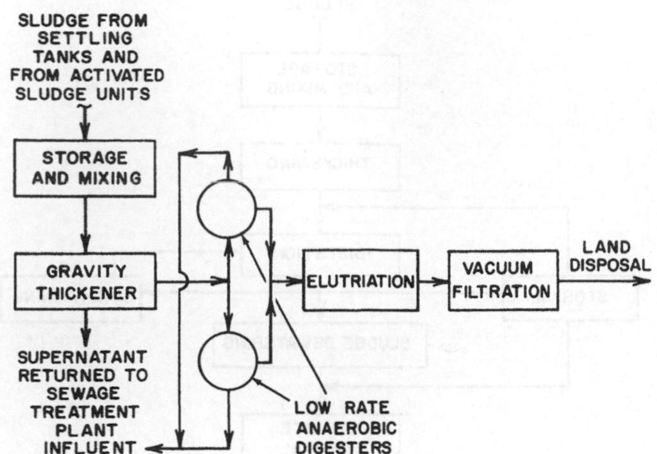

FIG. 5.2.14 Conventional sludge disposal with gravity thickener, elutriation, and vacuum filtration.

to insure aerobic conditions throughout the tank. The supernatant from aerobic digestion, similar to an anaerobic system, is mixed with the raw plant influent. The advantages of aerobic tanks is that they are not subject to upsets and produce a more treatable supernatant. The cost of air makes the operating cost of aerobic digestion higher than that of anaerobic digestion.

Sludge Conditioning

Elutriation and chemical addition are the two sludge conditioning alternatives. Conditioning is an intermediate process between primary and secondary anaerobic digesters. It also improves dewatering of digested sludge. Iron, aluminum salts, and lime are the most common chemicals for conditioning.

Elutriation is the washing of suspended sludge held in suspension by air or stirring. It reduces alkalinity and makes the sludge more filterable. Single and multitank (countercurrent) elutriation are both used. The single-tank system uses repeated sludge washing, whereas in the countercurrent system, fresh wash water is added to the last sludge tank. The wash water overflow from the last-stage tank furnishes the wash for the previous tank. Although the countercurrent system requires additional tanks and piping, it uses less makeup wash water.

Sludge Dewatering

Rotating-drum vacuum filters are the conventional sludge dewatering equipment. The filtrate is returned to the plant influent, and the sludge cake is disposed. Sludge drying beds are common in smaller plants for disposal. Weather conditions are an important factor when open drying beds are used. Some wastewater treatment facilities use glass-covered drying beds to exclude precipitation. Drying beds have underdrain networks of tile fields laid in sand and gravel. The discharge from this network should be disinfected when it is not returned to the treatment plant. Sludge treatment and disposal are usually performed in separate treatment plants (or plant sections). Figure 5.2.14 shows a typical sludge treatment plant.

—L.H. Reuter

6

Secondary Treatment

Van T. Nguyen | Wen K. Shieh

6.1 WASTEWATER MICROBIOLOGY

The objectives of biological wastewater treatment are to coagulate and remove nonsettlable colloidal solids and to stabilize organic matter. For example, in municipal wastewater treatment, the objective is usually to reduce organic content and, if necessary, to remove nutrients such as nitrogen and phosphorus. In some cases, trace concentrations of toxic organic compounds also require removal.

In industrial wastewater treatment, reduction or removal of organic and inorganic compound concentrations is essential. Microorganisms (see Figure 6.1.1) play a major role in decomposing waste organic matter, removing carbonaceous BOD, coagulating nonsettlable colloidal solids, and stabilizing organic matter. These microorganisms convert colloidal and dissolved carbonaceous organic matter into various gases and cell tissue. The cell tissue, having a specific gravity greater than water, can then be removed from treated water through gravity settling. Thus, wastewater treatment facilities use these microorganisms in biological wastewater treatment processes to dispose of wastes in a nontoxic and sanitary manner.

This section discusses the fundamentals of wastewater microbiology by examining microorganisms' nutritional requirements, enzymic reactions involved in their activities, environmental parameters affecting their growth and activities, and microbial groups associated with various biological wastewater treatment processes.

Nutritional Requirements

For microorganisms, nutrients (1) serve as an energy source for cell growth and biosynthetic reactions, (2) provide the material required for synthesis of cytoplasmic materials, and (3) serve as acceptors for the electrons released in energy-yielding reactions. To sustain reproduction and proper function, microorganisms require an energy source, a carbon source for synthesis of new cellular material, and inorganic nutrients such as nitrogen, phosphorus, sulfur, potassium, calcium, and magnesium. In addition, organic nutrients (growth factors) may also be required for cell synthesis. Table 6.1.1 lists the primary nutritional requirements.

The nutritional requirements of microorganisms provide a basis for classification. Microorganisms are classified on the basis of the form of carbon they require:

Autotrophic: These microorganisms use carbon dioxide or bicarbonate as their sole source of carbon, from which they construct all their carbon-containing biomolecules.
Heterotrophic: These microorganisms require carbon in the form of complex, reduced organic compounds, such as glucose.

Microorganisms are also classified on the basis of their required energy source:

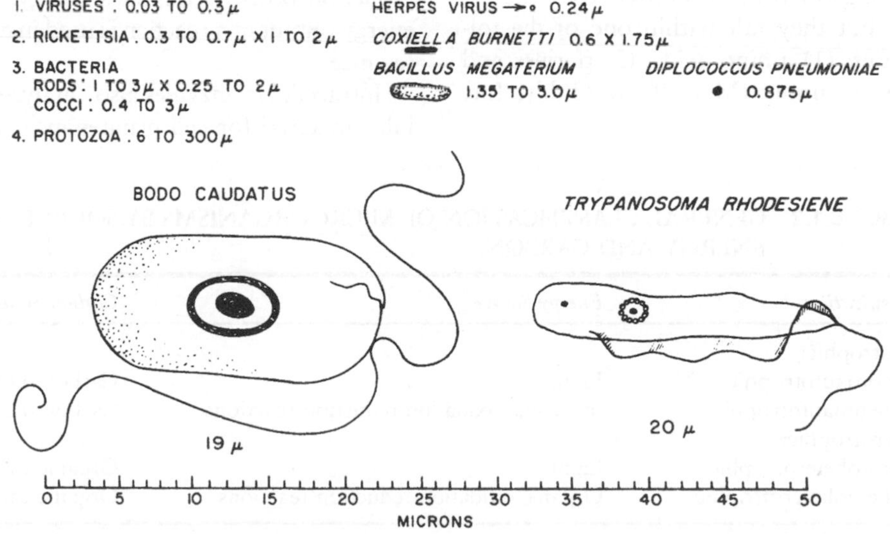

FIG. 6.1.1 Species from classes of organisms.

TABLE 6.1.1 CLASSIFICATION OF NUTRIENT REQUIREMENTS

Function	*Sources*
Energy Source	Organic compounds, inorganic compounds, and sunlight
Carbon Source	Carbon dioxide, bicarbonate, and organic compounds
Electron Acceptor	Oxygen, organic compounds, and combined inorganic oxygen (nitrate, nitrite, sulfate)

Source: Adapted from L.D. Benefield and C.W. Randall, 1980, *Biological process design for wastewater treatment* (Englewood Cliffs, N.J.: Prentice-Hall).

Phototrophs: These microorganisms use light as their energy source.

Chemotrophs: These microorganisms use oxidation-reduction reactions to provide their energy.

Chemotrophic microorganisms can be further classified on the basis of the type of chemical compounds that they oxidize, i.e., on the basis of their electron donor. For example, chemoorganotrophs use complex organic molecules as their electron donors, while chemoautotrophs use simple inorganic molecules such as hydrogen sulfide (H_2S) or ammonia (NH_3^+). Table 6.1.2 summarizes microorganism classification by sources of energy and cell carbon.

In addition to energy and carbon sources, microorganisms require principal inorganic nutrients such as nitrogen, sulfur, phosphorus, potassium, magnesium, calcium, iron, sodium, and chloride. Minor nutrients of importance include zinc, manganese, molybdenum, selenium, cobalt, copper, nickel, and tungsten (Metcalf and Eddy, Inc. 1991). Microorganisms also require organic nutrients, known as growth factors, as precursors or constituents of organic cell material that cannot be synthesized from other carbon sources. These growth factors differ from one organism to the next, but they fall within one of the following three categories: (1) amino acids, (2) purines and pyrimidines, and (3) vitamins (Metcalf and Eddy, Inc. 1991).

Microbial Enzymes

All microbial cell activities depend upon food utilization, and all chemical reactions involved are controlled by enzymes. Enzymes are proteins produced by a living cell that act as a catalyst to accelerate specific reactions in accordance with rate equations. Enzymes are specific in that they catalyze only certain kinds of reactions, and they act on only one kind of substance. Few hundredths of a second elapse while enzymes combine with chemicals undergoing change; chemical reactions occur, and new compounds are formed. Enzymes have little affinity to new compounds; thus, they are free to combine with other molecules of the substance for which they have specificity.

Microbial enzymes catalyze three types of reactions: hydrolytic, oxidative, and synthetic. Hydrolytic reactions involve enzymes hydrolyzing insoluble substrates into simple soluble components that pass through cell membranes into a cell by diffusion. These enzymes are extracellular, that is, they are released into the medium, while intracellular enzymes are released after cell disintegration.

Reactions that yield energy for growth and cell maintenance are catalyzed by intracellular enzymes. These reactions involve oxidation and reductions, that is, the addition or removal of oxygen or hydrogen. Most microorganisms oxidize by the enzymatic removal of hydrogen from molecules. Hydrogen is removed from compounds one atom at a time by dehydrogenases. Then, it is passed from one enzyme system to another until it is used to reduce the final hydrogen acceptor, otherwise known as the electron acceptor.

The electron acceptor is determined by the nature of the surrounding environment and the character of the relevant cells. Thus, in aerobic reactions, oxygen is the electron acceptor, while an oxidized compound is the electron acceptor in an anaerobic reaction resulting in a reduced compound. The oxidation process releases energy, and the reduction process consumes energy. Thus, a positive net energy output in a reaction is used in growth and cell maintenance.

Intracellular enzymes also catalyze the synthesis of cellular material for cell maintenance and new cell produc-

TABLE 6.1.2 GENERAL CLASSIFICATION OF MICROORGANISMS BY SOURCES OF ENERGY AND CARBON

Classification	*Energy Source*	*Carbon Source*
Autotrophic:		
Photoautotrophic	Light	Carbon dioxide
Chemoautotrophic	Inorganic oxidation-reduction reactions	Carbon dioxide
Heterotrophic:		
Photoheterotrophic	Light	Organic carbon
Chemoheterotrophic	Organic oxidation-reduction reactions	Organic carbon

Source: Adapted from Metcalf and Eddy, Inc., 1991, *Wastewater engineering: Treatment, disposal, and reuse,* 3d ed., (New York: McGraw-Hill).

tion. These enzymes are synthetic enzymes, and are required to produce the types of complex compounds found in a microbial cell. Synthetic reactions obtain the required large amount of energy from oxidation reactions occurring during the microorganism's energy metabolism.

Environmental Factors Affecting Microbial Growth

Enzyme activity is affected by environmental conditions, which also affects the activity of the corresponding microorganisms. Environmental parameters influencing the growth and performance of microorganisms include temperature, pH, and oxygen concentration.

TEMPERATURE EFFECTS

Since microbial growth is controlled mostly by chemical reactions, and the nature and rate of chemical reactions are affected by temperature, the rate of microbial growth and total biomass growth are affected by temperature. The microbial growth rate increases with temperature to a certain maximum where the corresponding temperature is the optimum temperature (see Figure 6.1.2). Then, growth does not occur after a small increase in temperature above the optimum value, followed by a decline in the growth rate with an increase in temperature beyond the optimum.

For example, bacteria can be divided into three different classes on the basis of their temperature tolerance: psychrophilic, mesophilic, and thermophilic. Psychrophilic bacteria tolerate temperatures in the range of −10 to 30°C, with the temperature for optimum growth in the range of 12 to 18°C. The mesophilic group tolerates temperatures in the range of 20 to 50°C, with an optimum temperature between 25 and 40°C, while thermophilic bacteria survive in a temperature range of 35 to 75°C and have optimum growth at temperatures in the range of 55 to 65°C (Metcalf and Eddy, Inc. 1991). In their respective classes, facultative thermophiles and facultative psychrophiles are bacteria that have optimum temperatures that extend into the mesophilic range. Optimum temperatures for obligate thermophiles and obligate psychrophiles are outside the mesophilic range.

The van't Hoff rule provides a generalization of the effect of temperature on enzyme reaction rates stating that the reaction rate doubles for a 10°C temperature increase. Also, according to Arrhenius, the following equation describes the relationship between reaction-rate constants and temperature:

$$d(\ln K)/dt = (E_a/R)(1/T^2) \qquad 6.1(1)$$

where:

K = the reaction-rate constant
E_a = the activation energy, cal/mole
R = the ideal gas constant (1.98 cal/mole-°K)
T = the reaction temperature, °K

Integrating Equation 6.1(1) yields the following equation:

$$\ln K = -(K_a/R)(1/T) + \ln B \qquad 6.1(2)$$

where B is an integration constant.

Equation 6.1(2) can be integrated between two temperature boundaries T_2 and T_1 to yield the following relationship that estimates the effect of temperature over a limited range:

$$\ln (K_2/K_1) = (E_a/R)[(T_2 - T_1)/(T_2T_1)] \qquad 6.1(3)$$

In biological treatment, the activation energy E_a can range from 2000 to 20,000 cal/mole. For most biological treatment cases, the term $(E_a/R)/(T_2T_1)$ is constant; therefore, the following equation applies:

$$K_2/K_1 = \Theta^{(T2-T1)} \qquad 6.1(4)$$

where Θ is the temperature coefficient.

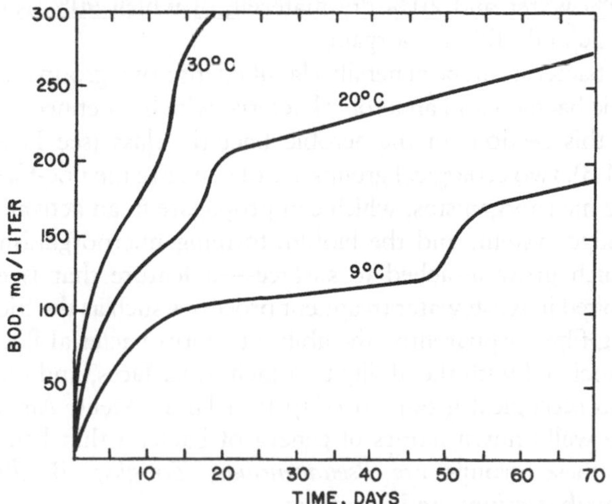

FIG. 6.1.2 Progress of BOD at 9°, 20°, and 30°C. The break in each curve corresponds to the onset of nitrification.

pH EFFECTS

Since enzymes are responsible for microorganism activity, pH effects on enzymes translate to effects on the corresponding microorganism. Some enzymes like acidic environments, while some like a medium environment, and others prefer an alkaline environment. When the pH increases or decreases beyond the optimum, enzyme activity decreases until it disappears. For most bacteria, the extremes of the pH range for growth are 4 and 9, while the optimum pH for growth is within the range of 6.5 to 7.5. Bacteria, in general, prefer a slightly alkaline environment; in contrast, algae and fungi prefer a slightly acidic environment. Biological treatment processes, however, rarely operate at optimum growth. Full-scale, extended-aeration, activated sludge and aerated lagoons can successfully operate at pH levels between 9 and 10.5; however, both systems are vulnerable to a pH less than 6 (Benefield and Randall 1980).

OXYGEN REQUIREMENTS AND MICROBIAL METABOLISM

Microorganisms can also be classified on the basis of whether they use an electron acceptor in the generation of energy. Organisms that generate energy by the enzyme-mediated electron transport from an electron donor to an external electron acceptor carry out respiratory metabolism. Fermentative metabolism, on the other hand, does not involve an external electron acceptor. Fermentation is less efficient in yielding energy than respiration. Hence, heterotrophic microorganisms that are strictly fermentative are characterized by smaller growth rates and cell yields than respiratory heterotrophs.

Microorganisms using molecular oxygen as electron acceptors are called *aerobes,* while those using molecules other than oxygen for electron acceptors are called *anaerobes.* Facultative microorganisms can use oxygen or another chemical compound as electron acceptors. Facultative microorganisms can be divided into two subgroups based on metabolic abilities. True facultative anaerobes can switch from fermentative to aerobic respiratory metabolism depending on the presence of molecular oxygen. Aerotolerant anaerobes, however, have a strictly fermentative metabolism but are insensitive to the presence of molecular oxygen. Obligate aerobes cannot grow in the absence of molecular oxygen, and obligate anaerobes are poisoned by an oxygen presence.

Oxidized inorganic compounds such as nitrate and nitrite can function as electron acceptors for some respiratory organisms in the absence of molecular oxygen. The biological treatment processes that exploit these microorganisms are often referred to as *anoxic.* In addition, those microorganisms that grow best at low molecular oxygen concentrations are termed *microaerophiles.*

The principal significance of the electron acceptors used by microorganisms involves the completeness of the resulting reaction and therefore the amount of energy available for cell growth and maintenance. Aerobes and facultative microorganisms completely oxidize the electron donors, while anaerobes, sometimes referred to as fermenters, do not. Table 6.1.3 gives some typical electron acceptors.

TABLE 6.1.3 TYPICAL ELECTRON ACCEPTORS IN BIOLOGICAL WASTEWATER TREATMENT BACTERIAL REACTIONS

Environment	*Electron Acceptor*	*Process*
Aerobic	Oxygen	Aerobic metabolism
Anaerobic	Nitrate	Denitrification*
	Sulfate	Sulfate reduction
	Carbon dioxide	Methanogenesis

Source: Adapted from Metcalf and Eddy, Inc., 1991, *Wastewater engineering: Treatment, disposal, and reuse,* 3d ed., (New York: McGraw-Hill).
Note: *Also known as anoxic denitrification.

Microbial Populations

Microorganisms are commonly classified on the basis of cell structure and function as eukaryotes, eubacteria, and archaebacteria. Eubacteria and archaebacteria are prokaryotes—cells whose genomes are not contained within a nucleus. Eukaryotes have a membrane-bound nucleus that stores the genome of the cell as chromosomes composed of deoxyribonucleic acid (DNA). Prokaryotes are generally referred to as bacteria. Eukaryotic organisms involved in biological treatment include fungi, protozoa and rotifers, algae, and invertebrates.

BACTERIA

Bacteria are members of a diverse and ubiquitous group of prokaryotic, single-celled organisms. They are the only living organisms that use all possible metabolic pathways. Bacteria can be classified based on their shapes: spherical, cylindrical (rods), and helical (spiral). Most bacteria reproduce by binary fission although some reproduce sexually or by budding. Bacteria range in size from 0.5 to 15 μ depending on their shape: 0.5–1.0 μ for spherical-shaped species, 1.5–3.0 μ for rod-shaped species, and 6–15 μ for spiral-shaped species. The interior of a typical bacteria cell—known as the cytoplasm—contains a colloidal suspension of proteins, carbohydrates, and other complex organic compounds. The cytoplasm also houses ribonucleic acid—responsible for protein synthesis—and the nuclear area that contains the DNA—carrying the information necessary for cell reproduction. Bacteria are approximately 80% water and 20% dry material, of which 90% is organic and 10% is inorganic.

Bacteria can be generally classified into two groups, aerobic bacteria and anaerobic bacteria, which is defined later in this section. In the aerobic bacteria class (see Figure 6.1.3), two ecological groups are of concern: the floc-forming microorganisms, which can propagate in an activated-sludge system, and the biofilm-forming microorganisms, which grow attached to surfaces—a feature that is exploited in wastewater treatment processes such as the trickling filter. Apparently, the ability to form bacterial floc is associated with the ability to attach to surfaces, and these two ecological groups overlap to a large extent. Among the well-known names of genera of bacteria that belong to these groups are *Pseudomonas, Zooglea, Bacillus, Flavobacterium,* and *Nocardia.*

The anaerobic group (see Figure 6.1.4) includes the fermentative bacteria such as *Clostridium, Propionibacter-*

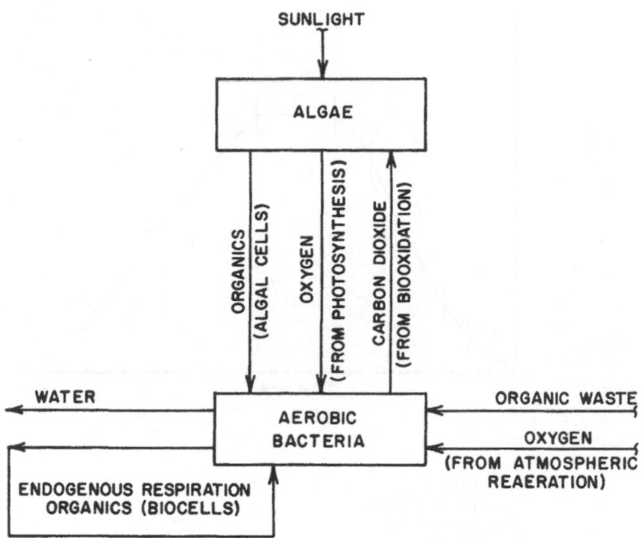

FIG. 6.1.3 Algal–bacterial interplay in an aerobic lagoon.

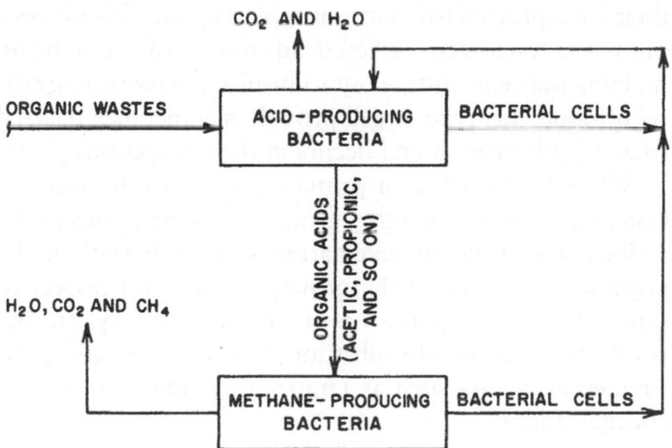

FIG. 6.1.4 Anaerobic degradation process.

ium, Streptobacterium, Streptococcus, Lactobacillus, and *Enterobacter.* Other common genera in the anaerobic group include the sulfate-reducing bacterium, *Desulfovibrio,* and methanogens such as *Methanosarcinia* and *Methanothrix.* Anaerobic degradation of organic matter usually requires a complex, interactive community with many different species.

FUNGI

Another group of decay organisms is fungi. Fungi are diverse, widespread, unicellular (e.g., yeasts) and multicellular (e.g., molds possessing a filamentous mass termed mycelium) eukaryotic organisms, lacking in chlorophyll and usually bearing spores and often filaments. Hence, they are nonphotosynthetic, heterotrophic protists. They are classified on the basis of their mode of reproduction: sexually or asexually, fission, budding, or spore formation. Fungi are strict aerobes that are tolerant of low pH levels and nitrogen-limiting conditions. Because of their ability to degrade cellulose, fungi are important in the biological treatment of some industrial wastes and in composting of organic solid waste.

Compared to the research on waste degradation by bacteria, much less exists concerning the active role of fungi in waste degradation. Fungi are present in suspended-growth systems, but their role is not well known. In attached-growth systems, they are a major component of the biota and may be responsible for forming the base film to which other microorganisms attach. Most of the fungi that have been recovered from wastewater treatment systems are the imperfect stages of Ascomycetes. Microorganisms that can grow as either single cells; yeast; or as filaments; *Candida, Rhodotorula, Oedidendron, Geotrichum,* and *Tricosporon;* are common in waste systems along with many common molds.

PROTOZOA, ALGAE, AND INVERTEBRATES

Three other groups present in wastewater treatment systems include protozoa, algae, and invertebrates. Protozoa are a group of diverse eukaryotic, typically unicellular, nonphotosynthetic microorganisms generally lacking a rigid cell wall. Most protozoa are aerobic heterotroph although some are anaerobic. In general, protozoa are larger than bacteria. They are secondary consumers in the systems, feeding on the bacteria and fungi that degrade organic matter in wastewater or on large particles of organic matter that the bacteria and fungi cannot consume. Thus, they polish the effluents from biological treatment processes.

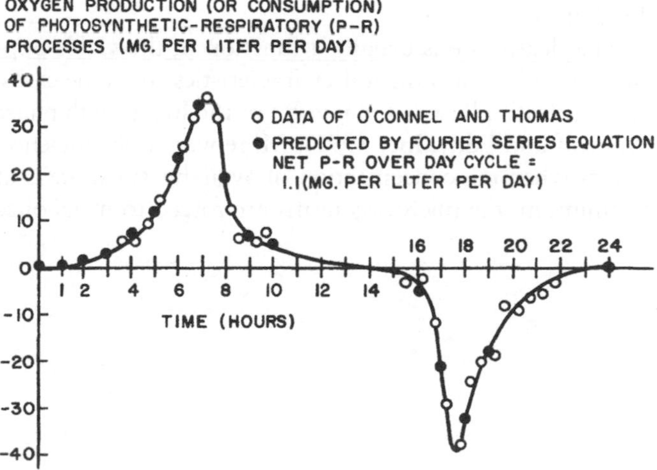

FIG. 6.1.5 Daily cycle of algal activity related to net oxygen production. (Data from R.L. O'Connell and N.A. Thomas, 1965, Effect of benthic algae on stream dissolved oxygen. *Journal of the American Society of Civil Engineers* 91, no. SA3:1).

Algae is a heterogeneous group of eukaryotic, photosynthetic, unicellular, and multicellular organisms lacking true tissue differentiation. In ponds, algae provide oxygen by photosynthesis, benefiting the ecology of the water environment. For example, in waste stabilization ponds, *Chlorella* and *Scenedesmus,* small green algae, produce the oxygen (see Figure 6.1.5) that is required by aerobic, heterotrophic bacteria. However, algae can be a problem in blooms where excessive algal growth in the receiving water can deplete the oxygen supply to the animal population below the water's surface.

Invertebrates are secondary or tertiary consumers. Invertebrates in wastewater treatment systems include rotifers, crustacea, insect larvae, nematodes, and worms. Rotifers are aerobic, heterotrophic, and multicellular animals. A rotifer possesses two sets of rotating cilia on its head, providing mobility and the ability to feed. It is effective in consuming dispersed and flocculated bacteria and small particles of organic matter. The presence of rotifers in an effluent indicates a highly efficient biological purification process.

GENERAL GROWTH PATTERN OF BACTERIA

Figure 6.1.6 shows the general bacterial growth pattern. Bacterial growth is comprised of four phases: lag phase, log-growth phase, stationary phase, and log-death phase. During the lag-phase, microorganisms acclimate to their new environment and begin to reproduce. In the log-growth phase, bacterial cells multiply at a rate determined by their generation time and ability to process the substrate. When the microorganisms enter the stationary phase, they have exhausted the substrate necessary for growth, and their population is at a standstill. If no new substrate is added, the microorganisms begin to die; hence, in the log-death phase, the death rate exceeds the production of new cells.

The death rate is usually a function of the viable population and environmental characteristics. In some cases, the log-death phase is the inverse of the log-growth phase (Metcalf and Eddy, Inc. 1991). Moreover, a phenomenon occurs when the concentration of available substrate is at a minimum: the microorganisms are forced to metabolize their own protoplasm without replacement. This process, known as lysis, occurs when dead cells rupture and the remaining nutrients diffuse out to furnish the remaining cells with food. This type of cell growth is sometimes referred to as cryptic growth and occurs in the endogenous phase.

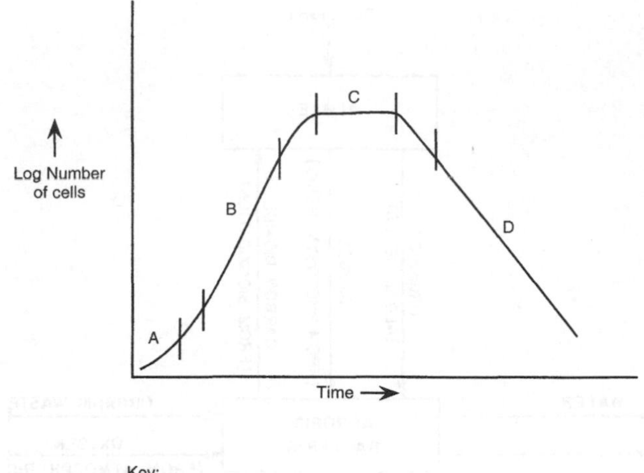

Key:
A: Lag Phase
B: Log-Growth Phase
C: Stationary Phase
D: Log-Death Phase

FIG. 6.1.6 Batch bacterial growth curve.

While bacteria play a primary role in waste degradation and stabilization, other groups of microorganisms described previously also take part in waste stabilization. The position and shape of the growth curve, with respect to time, of a microorganism in a mixed-culture system depend on the available substrate and nutrients and environmental factors such as temperature, pH, and oxygen concentrations.

—*Wen K. Shieh*
Van T. Nguyen

References

Benefield, L.D., and C.W. Randall. 1980. *Biological process design for wastewater treatment.* Englewood Cliffs, N.J.: Prentice-Hall.

Metcalf and Eddy, Inc. 1991. *Wastewater engineering: Treatment, disposal, and reuse.* 3d ed. New York: McGraw-Hill.

6.2 TRICKLING FILTERS

FILTER TYPE

a. Low rate
b. Intermediate rate
c. High rate
d. Super high rate
e. Roughing
f. Two stage

FILTER MEDIUM

a and b: Rock and slag; c: rock; d: plastic; e: plastic and redwood; and f: rock and plastic

HYDRAULIC LOADING (gal/ft²-min)

a: 0.02–0.06; b: 0.06–0.16; c: 0.16–0.64; d: 0.2–1.2; e: 0.8–3.2; and f: 0.16–0.64

BOD_5 LOADING (lb/ft³-day)

a: 0.005–0.025; b: 0.015–0.03; c: 0.03–0.06; d: 0.03–0.1; e: 0.1–0.5; and f: 0.06–0.12

BOD_5 REMOVAL (%)

a: 80–90; b: 50–70; c: 65–85; d: 65–80; e: 40–65; and f: 85–95

DEPTH (ft)

a: 6–8; b: 6–8; c: 3–6; d: 10–40; e: 15–40; and f: 6–8

RECIRCULATION RATIO

a: 0; b: 0–1; c: 1–2; d: 1–2; e: 1–4; and f: 0.5–2

FILM SLOUGHING

a and b: intermittent and c–f: continuous

NITRIFICATION

a and f: well; b: partial; c and d: little; and e: none

Process Description

Trickling filters have been used for wastewater treatment for nearly 100 years. A trickling filter (see Figure 6.2.1) is an attached-growth, biological process that uses an inert medium to attract microorganisms, which form a film on the medium surface. Table 6.2.1 lists the physical properties of trickling filter media.

A rotatory or stationary distribution mechanism distributes wastewater from the top of the filter percolating it through the interstices of the film-covered medium. As the wastewater moves through the filter, the organic matter is adsorbed onto the film and degraded by a mixed population of aerobic microorganisms (see Figure 6.2.2). The oxygen required for organic degradation is supplied by air circulating through the filter induced by natural draft or ventilation.

A light-weight, highly-permeable medium with a large specific surface area (e.g., plastic modules) is conducive to microorganism buildup and ensures unhindered movement of wastewater and air. A porous underdrain system at the bottom of the filter collects treated effluent and circulates air. The filter recirculates and mixes a portion of the effluent with the incoming wastewater to reduce its strength and provide uniform hydraulic loading (Metcalf and Eddy, Inc. 1991).

As the film thickness increases, the region of the film near the medium surface can be deprived of organic matter, reducing the adhesive ability of the microorganisms. Therefore, a thick film is more susceptible to the sloughing effects caused by wastewater flow. Furthermore, the inner portion of a thick film can become anaerobic because oxygen may be unavailable. As a result, the release of gases can weaken the film and increase the sloughing effects. Once the thick film is removed, a new film starts to grow on the medium surface, signaling the beginning of a new growth cycle (Characklis and Marshall 1990).

Process Microbiology

The microorganism population in a trickling filter consists of aerobic, anaerobic, and facultative bacteria, fungi, algae, and protozoans. Also present are higher forms such as worms, insect larvae, and snails. The predominating microorganisms in the trickling filter are the facultative bacteria. *Achromobacter, Flavobacterium, Pseudomonas,* and *Alcaligenes* are among the bacterial species commonly associated with the trickling filter. Filamentous forms such as *Sphaerolitus natans* and *Beggiatoa* are found in the slime layer, while *Nitrosomonas* and *Nitrobacter* are present in the lower reaches of the filter.

Fungi in the filter are responsible for waste stabilization. Their presence becomes important in industrial

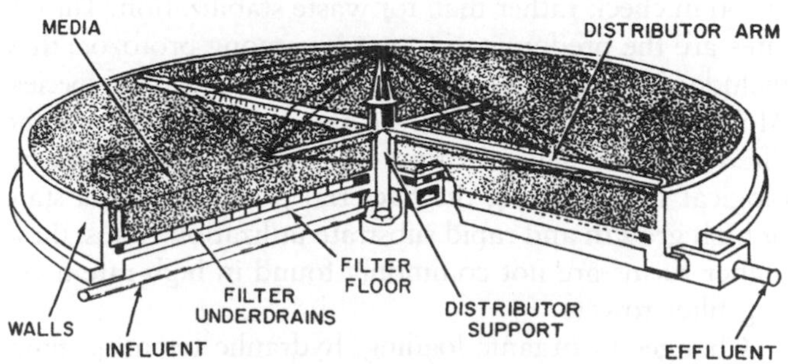

FIG. 6.2.1 Cross section of a stone media trickling filter.

TABLE 6.2.1 COMPARATIVE PHYSICAL PROPERTIES OF TRICKLING FILTER MEDIA

Types of Media	*Nominal Size*	*Units per ft³*	*Unit Weight lb/ft³*	*Specific Surface Area ft²/ft³*	*Void Space %*
Granite	1–3 in	—	90	19	46
	4 in	—	—	13	60
Blast Furnace Slag	2–3 in	51	68	20	49
Aeroblock (vitrified tile)	6 in × 11 in × 12 in	2	70	20–22	53
Raschig Rings (ceramic)	$1\frac{1}{2}$ in × $1\frac{1}{2}$ in	340	40.8	35	68.2
Dowpac 10	21 in × $37\frac{1}{2}$ in	2	3.6–3.8	25	94
Dowpac 20	$21\frac{1}{2}$ in × $38\frac{1}{2}$ in	2	6	25	94

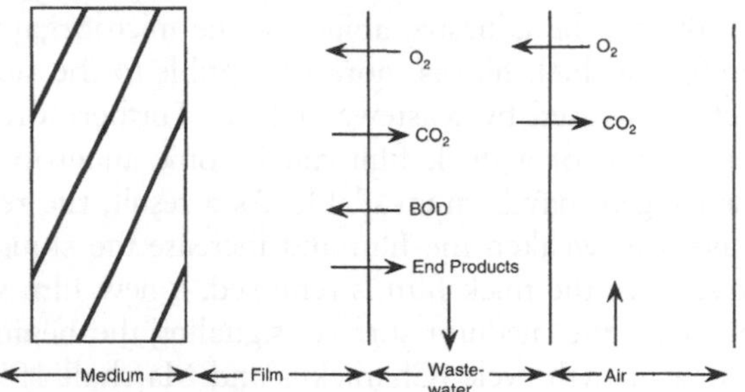

FIG. 6.2.2 A schematic representation of the biological film in a trickling filter.

wastewater treatment where pH levels are low. Various fungal species identified include *Fusazium, Muco, Pencillium, Geotrichum, Sporatichum,* and various yeasts. Fungi, however, are often responsible for clogging filters and preventing ventilation due to their rapid growth.

Algae are also found in trickling filters, albeit only in the upper reaches of the filter where sunlight is available. Their main role is not in degrading the organic matter but in providing oxygen during the daytime to the percolating wastewater. Some of the algae species commonly found in trickling filters include *Phormidium, Chlorella,* and *Ulothrix*. Algae can also clog the filter surface, resulting in undesirable odors.

Protozoans in trickling filters, as in activated-sludge processes, are responsible for keeping the bacterial population in check rather than for waste stabilization. The ciliates are the predominating species among protozoa; they include the *Vorticella, Opercularia,* and *Epistylis* species. Along with the protozoans, the higher animal forms in the filter—such as snails, worms, and insects—feed on the biological film, keeping the bacterial population in a state of high growth and rapid substrate utilization. Thus, these higher forms are not commonly found in high-rate trickling filter towers.

Changes in organic loading, hydraulic retention time, pH, temperature, air availability, influent wastewater composition, and other factors vary the populations of each type of microbial community throughout the filter depth.

Process Flow Diagrams

Figures 5.2.7 and 5.2.8 are examples of common process flow diagrams for single- and multistage trickling filters. Wastewater treatment facilities often recirculate the treated effluent from the clarifier to (Atkinson and Ali 1976; Metcalf and Eddy, Inc. 1991):

- Reduce the possibility of organic shock loadings by diluting the incoming wastewater
- Maintain uniform hydraulic loadings especially under low and intermittent flow conditions
- Achieve an extensive film coverage and a relatively uniform film thickness through the filter
- Reduce the nuisances of odor and flies

However, such benefits are achieved at the expense of higher hydraulic loadings.

Trickling filters are classified by their hydraulic loadings. Typical hydraulic loadings for low-rate (without recirculation) and high-rate (with recirculation) trickling filters are 1.17–3.52 and 9.39–37.55 m^3/m^2-day, respectively. The corresponding loadings for super high-rate trickling filters are as high as 70.41 m^3/m^2-day. The effluent from a low-rate trickling filter is usually low in BOD and well nitrified. Wastewater treatment facilities commonly use two-stage trickling filters for treating high-strength wastewater and achieving nitrification at hydraulic loadings comparable to those for high-rate trickling filters (Tebbutt 1992).

Process Design

Despite recent advances in attached-growth biological wastewater treatment processes, the design and analysis of trickling filters are still largely based empirical models.

Some of these empirical models are presented next (McGhee 1991; Metcalf and Eddy, Inc. 1991).

VELZ EQUATIONS

The following equation is used for a single-stage system and the first stage of a two-stage system:

$$S_{e1} = [(S_i + r_1S_{e1})/(1 + r_1)] \exp [(-kDA^n/Q^n)(1.035^{T-20})] \quad 6.2(1)$$

The following equation is used for the second stage of a two-stage system:

$$S_{e2} = [(S_e + r_2S_{e2})/(1 + r_2)] \exp [-kDA^nS_{e1}/Q^nS_i)(1.035^{T-20})] \quad 6.2(2)$$

where:

S_e = the effluent BOD from the filter, mg/l
S_i = the influent BOD, mg/l
r = the ratio of recirculated flow to wastewater flow
D = the filter depth, m
A = the filter plan area, m^2
Q = the wastewater flow, m^3/min
T = the wastewater temperature, °C
k, n = empirical coefficients (for municipal wastewaters, k = 0.02 and n = 0.5)
subscript i (i = 1,2) = the stage number

NRC EQUATIONS

The following equations apply to a single-stage system and the first stage of a two-stage system:

$$1 - (S_{e1}/S_i) = 1/[1 + 0.532(QS_i/V_1F_1)^{0.5}] \quad 6.2(3)$$

$$F_1 = [(1 + r_1)/(1 + 0.1r_1)^2] \quad 6.2(4)$$

The following equations apply to the second stage of a two-stage system:

$$1 - (S_{e2}/S_{e1}) = 1/[1 + 0.532(Q/S_{e1}V_2F_2)^{0.5}] \quad 6.2(5)$$

$$F_2 = [(1 + r_2)/(1 + 0.1r_2)^2] \quad 6.2(6)$$

where:

V = the filter volume, m^3
F = the recirculation factor

ECKENFELDER EQUATION (PLASTIC MEDIA)

The Eckenfelder equation used for plastic media is as follows:

$$S_e/S_i = \exp [-KS_a^mD(Q/A)^{-n}] \quad 6.2(7)$$

where:

K = the observed rate constant for a given filter depth, ft/day
S_a = the specific surface area of filter, ft^2/ft^3
D = the filter depth, ft
Q = the wastewater flow, ft^3/day
A = the filter plan area, ft^2
m and n = empirical coefficients

GERMAIN/SCHULTZ EQUATIONS (PLASTIC MEDIA)

The Germain/Schultz equations used for plastic media are as follows:

$$S_e/S_i = \exp [-k_{20,i}D_i(Q/A)^{-n}] \quad 6.2(8)$$

$$k_{20,2} = k_{20,1}(D_1/D_2)^x \quad 6.2(9)$$

where:

$k_{20,i}$ = the treatability constant corresponding to a specific filter depth D_i at 20°C, $(gal/min)^n$ft
Q = the wastewater flow, gal/min
n and x = empirical constants (n is usually 0.5; x is 0.5 for rock and 0.3 for cross-flow plastic media)

UNDERDRAINS

The underdrains used in trickling filters support the filter medium, collect the treated effluent and the sloughed biological solids, and circulate the air through the filter. Precast blocks of vitrified clay or fiberglass grating arranged on a reinforced concrete floor can be used as the underdrain system for a rock-media trickling filter. Precast concrete beams supported by columns or posts can be used as the underdrain and support system for a plastic-media trickling filter (McGhee 1991). The floor should be sloped towards either central or peripheral collection channels at a 1 to 5% grade for improved liquid flow. The minimum flow velocity in the collection channel should be 0.6 m/sec (2 ft/sec) at the average daily flowrate (Metcalf and Eddy, Inc. 1991). The liquid flow in underdrains and collection channels should not be more than half full for adequate air flows.

FILTER MEDIA

The ideal medium used in a trickling filter should have the following properties: high specific surface area, high void space, light weight, biological inertness, chemical resistance, mechanical durability, and low cost. Table 6.2.2 summarizes the properties of some commercially available media suitable for trickling filter applications. Plastic media are reported to be highly effective for BOD and SS removal over a range of loadings (Harrison and Daigger 1987). Furthermore, lighter and taller filter structures can

TABLE 6.2.2 PROPERTIES OF FILTER MEDIA

Packing Type	*Nominal Size (in)*	*Density (lb/ft³)*	*Specific Surface Area (ft²/ft³)*	*Porosity (%)*
Redwood	48 × 48 × 20	9–11	12–15	70–80
Blast Furnace	2.0–3.2 (small)	56–75	17–21	40–50
Slag	3.0–5.0 (large)	50–62	14–18	50–60
River Rock	1.0–2.6 (small)	78–90	17–21	40–50
	4.0–5.0 (large)	50–62	12–15	50–60
Plastic				
Surpac*	—	3.6	25	94
Koroseal*	—	2.7–3.5	40	94
Flocor*	—	4.1	29	95
PVC Tubes				
Cloisonyle*	—	—	69	—
Raschig Rings				
Ceramic	0.25–4.00	36–60	14–217	62–80
Carbon	0.25–3.00	23–46	122–212	85–95
Steel	0.50–3.00	25–75	20–122	85–95
Pall Rings				
Ceramic	2.00–3.00	38–40	20–29	74
Steel	1.00–2.00	24–30	31–63	94–96
Polypropylene	1.00–3.50	4.25–5.50	26–63	90–92
Ceramic Berl Saddles	0.25–2.00	39–56	32–274	60–72
Ceramic Intalox Saddles	0.25–3.00	37–54	28–300	75–80

Source: Adapted from A.Y. Li, 1984, *Anaerobic processes for industrial wastewater treatment,* Short Course Series, no. 5 (Taichung, Taiwan: Department of Environmental Engineering, National Chung Hsing University).

Note: *Trade name.

be constructed to house plastic media, reducing land requirements.

CLARIFIERS

Clarifiers used in trickling filters remove large and heavily sloughed biological solids or humus without providing thickening functions. Therefore, the design of these clarifiers is similar to the design of primary settling tanks. The overflow rates are 400–600 gal/ft^2-day at average flow and 1000–1200 gal/ft^2-day at peak flow, respectively. The overflow rate is based on the influent flow plus the recirculation flow.

Clarifier depth ranges from 10 to 15 ft. Details on the design of secondary clarifiers are presented in Section 6.8.

DESIGN PROCEDURES

Trickling filters are used extensively in the treatment of municipal wastewater and to a lesser extent in industrial wastewater. Using synthetic media has increased the capability of the trickling filter, and using multistage, synthetic-media filters has achieved a high degree of treatment in industrial wastewater.

Table 6.2.3 lists the treatability factor obtained on settled sewage by trickling filters with various media. The treatability factor (K) is a characteristic of the wastewater, and the n value is a characteristic of the trickling filter media. However, Figure 6.2.3 indicates that the treatability factor is influenced by the n value. If filter media with higher n values are used, the treatability constant is reduced for settled sewage.

The treatability factor varies with wastewater type. Selected organic compounds such as phenol and compounds containing the cyanide complex show high treatability factors, but typical industrial wastewater with organics in solution has a treatability factor considerably below that of domestic sewage. This variability indicates the need for confirming the K factor with a pilot-plant evaluation prior to the final design of the trickling filter. Suitable pilot facilities are usually available from manufacturers of synthetic media.

TREATABILITY FACTOR DETERMINATION

The method of obtaining treatability factors and n values (media specific) involves the use of complex equations. For practical design purposes, a graph is more convenient. Figure 6.2.4 is a design chart for a trickling filter that uses synthetic media with an n value of 0.5 for various treatability factors.

TABLE 6.2.3 BOD TREATABILITY FACTORS OF SETTLED SEWAGE IN TRICKLING FILTERS WITH VARIOUS MEDIA

Media Used	*Depth (ft)*	*Range of Influent BOD Concentration (mg/l)*	*Applied Hydraulic Loading (gal/min/ft²)*	*n Factor*	*Treatability Factor (K)* at 20°C (min⁻¹)*
$1\frac{1}{2}$ in Flexirings	8	65–90	0.196–0.42	0.39	0.09
$2\frac{1}{2}$ in Clinker	6	220–320	0.015–0.019	0.84	0.021
$1\frac{1}{2}$–$2\frac{1}{2}$ in Slag	6	112–196	0.08–0.19	1.00	0.014
$2\frac{1}{2}$ in Slag	6	220–320	0.015–0.019	0.75	0.029
$2\frac{1}{2}$–4 in Rock	12	200	0.48–1.47	0.49	0.036
1–3 in Granite	6	186–226	0.031–0.248	0.4	0.059
$\frac{3}{4}$ in Raschig Rings	6	186–226	0.031–0.248	0.7	0.031
1 in Raschig Rings	6	186–226	0.031–0.248	0.63	0.031
$1\frac{1}{2}$ in Raschig Rings	6	186–226	0.031–0.248	0.306	0.078
$2\frac{1}{4}$ in Raschig Rings	6	186–226	0.031–0.248	0.274	0.08
Straight Block	6	186–226	0.031–0.248	0.345	0.048
Surfpac	21.6	200	0.49–3.9	0.5	0.05
Surfpac	12.0	200	0.97–3.9	0.45	0.05
Surfpac	21.5	—	—	0.50	0.045†
Surfpac	21.5	—	—	0.50	0.088‡

Notes: *The treatability factor is calculated with formula $\frac{L_e}{L} = e - \frac{KD}{Q_i^n}$. This formula gives the K and n values. The treatability factor relates to the degree of ease of treating wastewater. The treatability factor K has the unit min⁻¹ when the flow rate is expressed as gal per min per ft². The n factor is related to the type of media and is function of the specific surface and configuration of media. The n factor is a dimensionless constant.
†Dissolved BOD only
‡Total BOD

Environmental engineers must obtain the treatability factors by field testing for specific wastewater. Figure 6.2.4 relates the raw hydraulic loading (Q, without recirculation) for a depth (D) to the design performance required. For example, obtaining a BOD removal of 80% with a specific raw hydraulic dosage rate (Q) and depth (D) a wastewater with a K factor of 0.020 requires four times as much filter area as a wastewater with a treatability factor of 0.08.

Trickling filters can provide roughing or complete treatment of wastewater. Roughing treatment does not reduce the treatability of the filter effluent in a subsequent, activated-sludge, process step. A synthetic-media trickling filter is suited for roughing treatment when high temperature is involved because the cooling effect of the filter makes the effluent more amenable to activated-sludge treatment.

Synthetic-media trickling filters are not economically suitable to obtain high treatment levels for soluble organic matter with low treatability factors (below 0.05) because of the large filter volume required.

LIST OF ABBREVIATIONS

The following abbreviations apply to the design of trickling filters:

E_1 Percent BOD removal efficiency through first-stage filter and clarifier

E_2 Percent BOD removal efficiency through second-stage filter and clarifier

W BOD loading, in lb per day, to first-stage filter, not including recycling

W_1 BOD loading, in lb per day, to second-stage filter, not including recycling

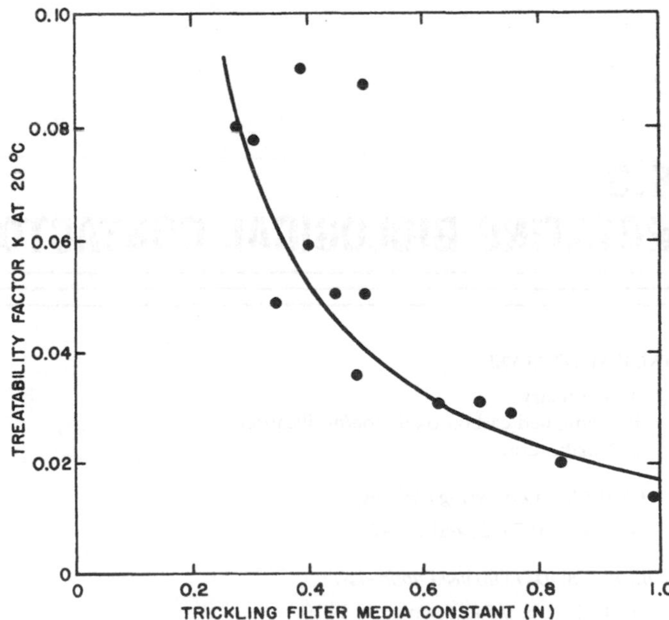

FIG. 6.2.3 Effect of the n value on the treatability factor of settled sewage.

L_D Removable BOD at depth D, in mg/l
L Total removable BOD, in mg/l
D Depth of filter, in ft
K_1 Constant
L_e Unsettled filter effluent BOD, in mg/l
L_i Filter influent BOD, in mg/l
i Influent flow, in mgd
r Recirculation flow, in mgd
a Filter radius, in ft
T Wastewater temperature, in °C
Q_1 Hydraulic load, in gal per min per sq ft (not including recirculation)
K_3 or K Treatability constant
Q Flow, in mgd
L_o Influent BOD (including recirculation), in mg/l
A Area of filter, in acres
m,n Constants for media
K_{20} Treatability constant at 20°C
V Volume of filter, in acre-feet
F Recirculation factor

$$F = \frac{1 + R}{(1 + 0.1R)^2} \qquad 6.2(10)$$

where R is the recirculation ratio

—*Wen K. Shieh*
Van T. Nguyen

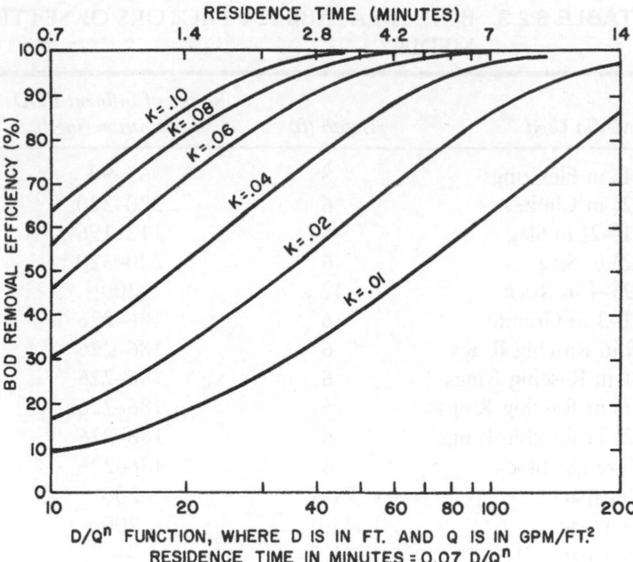

FIG. 6.2.4 Effect of residence time on BOD removal efficiency. The n is a dimensionless exponent function of the trickling filter media. Its value is 0.67 for conventional rock media, 0.50 for most synthetic media, and intermediate values for other types of trickling filter media.

References

Atkinson, B., and M.E. Ali. 1976. Wetted area, slime thickness and liquid phase mass transfer in packed bed biological film reactors (trickling filters). *Trans. Instr. Chem. Engrs.* 54, no. 239.

Characklis, W.G., and K.C. Marshall, eds. 1990. *Biofilms.* New York: John Wiley & Sons.

Harrison, J.R., and G.T. Daigger. 1987. A comparison of trickling filter media. *J. Water Poll. Control Fed.* 59, no. 679.

McGhee, T.J. 1991. *Water supply and sewerage.* 6th ed., New York: McGraw-Hill.

Metcalf and Eddy, Inc. 1991. *Wastewater engineering: Treatment, disposal, and reuse.* 3d ed. New York: McGraw-Hill.

6.3 ROTATING BIOLOGICAL CONTACTORS

TREATMENT LEVEL

a. Secondary
b. Combined carbon oxidation/nitrification
c. Nitrification

HYDRAULIC LOADING (gal/ft²-day)

a: 2–4; b: 0.75–2; and c: 1–2.5

SOLUBLE BOD$_5$ LOADING (lb/ft²-day)

a: 0.75–2; b: 0.5–1.5; and c: 0.1–0.3

LIQUID RETENTION TIME (LRT) (hrs)

a: 0.7–1.5; b: 1.5–4; and c: 1.2–2.9

BOD$_5$ REMOVAL (%)

a–c: 85–95

EFFLUENT NH_3 (mg/l)

b: <2 and c: 1–2

Process Description

A rotating biological contactor (RBC) is an attached-growth, biological process that consists of a basin(s) in which large, closely spaced, circular disks mounted on horizontal shafts rotate slowly through wastewater (see Figure 6.3.1). The disks are made of high-density polystyrene or PVC for durability and resistance. Corrugation patterns increase surface area and structural integrity (Metcalf and Eddy, Inc. 1991).

Bacterial growth on the surface of the disks leads to the formation of a film layer that eventually covers the entire wetted surface of the disks. The rotating disks are partially submerged in the wastewater. In this way, the film layer is alternatively exposed to the wastewater from which the organic matter is adsorbed and the air from which the oxygen is absorbed.

The mechanisms of organic degradation in an RBC film layer are similar to those shown in Figure 6.2.2. Rotation also provides a means for removing excess bacterial growth on the disks' surfaces and maintaining suspension of

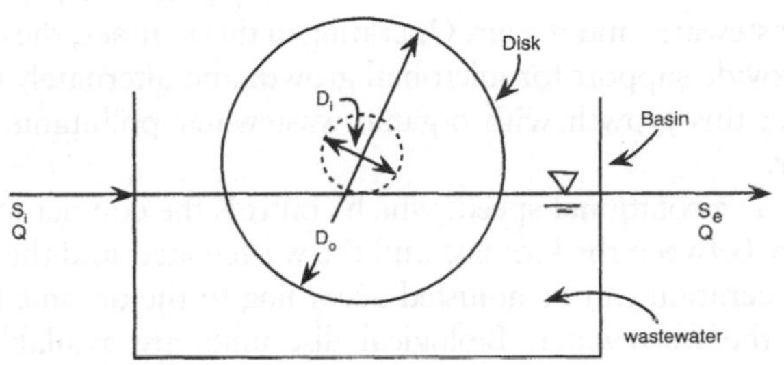

FIG. 6.3.1 A schematic of an RBC system.

sloughed biological solids in wastewater. A final clarifier removes sloughed solids.

Partially submerged RBCs are used for carbonaceous BOD removal, combined carbon oxidation and nitrification, and nitrification of secondary effluent (Grady and Lim 1980; Metcalf and Eddy, Inc. 1991). Completely submerged RBCs are used for denitrification (Grady and Lim 1980).

Process Flow Diagrams

Figure 6.3.2 shows typical arrangements of RBCs. In general, an RBC system is divided into a series of independent stages or compartments by baffles in a single basin (see Part A in Figure 6.3.2) or separate basins arranged in series (see Part B in Figure 6.3.2).

Compartmentalization creates a flow pattern with little longitudinal mixing in the flow direction (i.e., a plug-flow pattern), increasing overall removal efficiency of an RBC system (Tchobanoglous and Schroeder 1985). It can also promote separation of bacterial species at different stages, achieving optimal performance. For example, autotrophic bacteria responsible for nitrification can concentrate at later stages in an RBC system designed for combined carbon removal and nitrification where the mixed liquor BOD is low. Consequently, nitrification performance is more reliable and stable.

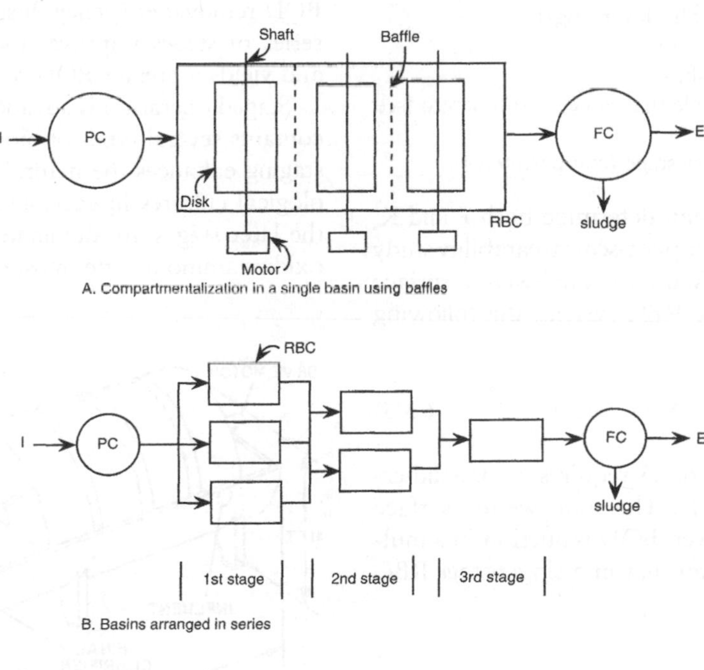

FIG. 6.3.2 Typical arrangements of RBCs. **A,** Compartmentalization in a single basin using baffles; **B,** Basins arranged in series.

Process Design

RBC design is often based on empirical design curves supplied by RBC manufacturers. Once the environmental engineer estimates the surface loading L (gal/ft^2-day) required to achieve a BOD removal efficiency, the required disk surface area A (ft^2) for a total flow Q (gal/day) is calculated as follows:

$$A = Q/L \qquad 6.3(1)$$

A more rational design approach, which considers the mass balances for both the substrate and biomass at a specific stage, has been proposed by Ramalho (1983). The resulting design equations for a single-stage RBC are as follows:

$$Q(S_i - S_e) = PAS_e/(K_s + S_e) \qquad 6.3(2)$$

$$P = kX_A\delta \qquad 6.3(3)$$

$$A = \pi(D_o^2 - D_i^2)/2N \qquad 6.3(4)$$

where:

- Q = wastewater flowrate, m^3/day
- S_i = influent BOD, mg/l
- S_e = effluent BOD, mg/l
- A = required wetted surface area to achieve the required BOD reduction from S_i to S_e, m^2
- K_s = half-velocity constant, mg/l (see Section 6.4)
- k = maximum BOD removal rate, l/day (see Table 6.4.4)
- X_A = dry density of the film layer, mg/l
- δ = film layer thickness, m
- D_o = diameter of the disk, m
- D_i = diameter of the circle that is never submerged (see Figure 6.3.1), m
- N = number of disks per stage (compartment)

Environmental engineers can determine both P and K_s by performing a laboratory- or pilot-scale treatability study using the same wastewater. With an equal wetted surface area per stage for an n-stage RBC system, the following equation applies:

$$S_{n-1} - S_n = PAS_n/(K_s + S_n) \qquad 6.3(5)$$

The solution of Equation 6.3.5 requires a trial-and-error approach (McGhee 1991). The total wetted surface area required to achieve a given BOD reduction in a multistage RBC system is less than that in a single-stage RBC system.

OPERATING PROBLEMS

Many RBC operating problems are caused by shaft failures, disk breakage, bearing failures, and organic overloadings. By adopting proper design, operation, and maintenance practices, wastewater treatment facilities can mitigate many of these problems. For example, many RBC systems are enclosed to eliminate disk exposure to UV light, reduce temperature effects, and protect the equipment. These facilities can control odor problems by reducing organic loading or increasing the oxygen supply using supplemental air diffusers in the basin.

BIOLOGICAL DISCS

A biological disc unit consists of a series of closely spaced, large-diameter, expanded polystyrene discs mounted on a horizontal shaft. The discs are partially immersed in wastewater and rotated.

From the microorganisms present in the wastewater, a biological growth develops on the surface of the discs. As the discs rotate, the bacteria alternately passes through the wastewater and the air. Operating in this manner, the discs provide support for microbial growth and alternately contact this growth with organic wastewater pollutants and air.

The rotational speed, which controls the contact intensity between the biomass and the wastewater, and the rate of aeration can be adjusted according to the organic load in the wastewater. Biological disc units are available in sizes up to 12 ft in diameter.

The residence time of the wastewater in the disc sections and the rotational speed of the discs determine unit BOD removal efficiency. Installing a number of discs in a series of stages improves the residence time distribution and yields a greater BOD removal efficiency.

Staged operation is advantageous when the wastewater contains several types of biodegradable materials because staging enhances the natural development of different biological cultures in each stage. For example, the discs in the later stages are dominated by nitrifying bacteria that oxidize ammonia after most of the carbonaceous BOD has

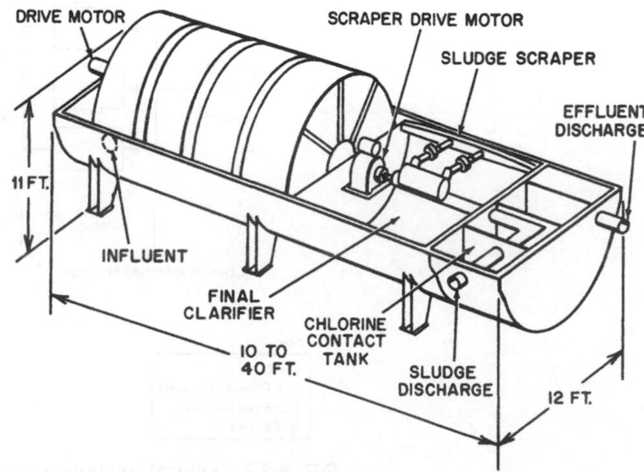

FIG. 6.3.3 Packaged biological disc unit.

been removed. Staged operation also permits the use of intermediate solids separation units at strategic points. The process is claimed to be stable under hydraulic surges and intermittent flows.

The approximate cost of a bio-disc unit is 25¢ per gal per day of domestic sewage, excluding site work. Because of the high buoyancy of the disc materials and the low rotational speeds, the power consumption is low. For domestic wastewater, power consumption can be 10 hp per mgd. This rate is approximately equivalent to removing 4.2 lb of BOD per hp-hr invested.

For sewage flow capacities of 5000 to 120,000 gpd, a package unit is available (see Figure 6.3.3). It includes a feed mechanism, a section of bio-disk surfaces, an integral clarifier tank with sludge removal mechanism, and a chlorine contact section. Depending on the nature of the influent sewage and on the total flow capacity, the length of the package unit varies from 10 to 40 ft.

—*Wen K. Shieh*
Van T. Nguyen

References

Grady, C.P.L., Jr., and H.C. Lim. 1980. *Biological wastewater treatment: Theory and applications.* New York: Marcel Dekker.

McGhee, T.J. 1991. *Water supply and sewerage.* 6th ed. New York: McGraw-Hill.

Metcalf and Eddy, Inc. 1991. *Wastewater engineering: Treatment, disposal, and reuse.* 3d ed. New York: McGraw-Hill.

Tchobanoglous, G., and E.D. Schroeder. 1985. *Water quality.* Reading, Mass.: Addison-Wesley.

6.4 ACTIVATED-SLUDGE PROCESSES

PROCESS TYPE

a. Conventional
b. Completely mixed
c. Step feed
d. Contact stabilization
e. High-purity oxygen
f. Oxidation ditch
g. Sequencing batch reactor
h. Deep shaft

LRT (hrs)

a: 4–8; b: 3–5; c: 3–5; d: 0.5–1 (contact tank), 3–6 (stabilization tank); e: 1–3; f: 8–36; g: 12–50; and h: 0.5–5

SOLIDS RETENTION TIME (SRT) (days)

a–d: 5–15; e: 3–10; f: 10–30; and g: not applicable

F/M (lb BOD_5/lb MLVSS-day)

a & c: 0.2–0.4; b & d: 0.2–0.6; e: 0.25–1; f & g: 0.05–0.3; and h: 0.5–5

VOLUMETRIC LOADING (lb BOD_5/ft^3-day)

a: 0.02–0.04; b: 0.05–0.12; c: 0.04–0.06; d: 0.06–0.075; e: 0.1–0.2; f: 0.005–0.03; and g: 0.005–0.015

BOD_5 REMOVAL (%)

85–95

MLSS (mg/l)

a: 1500–3000; b: 2500–4000; c: 2000–3500; d: 1000–3000 (contact tank), 4000–10,000 (stabilization tank); e: 2000–5000; f: 3000–6000; and g: 1500–5000

RECYCLING RATIO

a & c: 0.25–0.75; b: 0.25–1; d: 0.5–1.5; e: 0.25–0.5; f: 0.75–1.5; and g: not applicable

AERATION TYPE

Diffused aeration and mechanical aeration.

Process Description

The activated-sludge process, first developed in England in 1914, has been used widely in municipal and industrial wastewater treatment. Although many process variations have been developed for specific applications, biodegradation of organic matter in the activated-sludge process can be illustrated using a typical flow diagram as shown in Figure 6.2.2.

Clarified wastewater discharged from the primary clarifier is delivered into the aeration basin where it is mixed with an active mass of microorganisms (referred to as activated sludge) capable of aerobically degrading organic matter into carbon dioxide, water, new cells, and other end products (see Figure 6.4.1). Diffused or mechanical aeration maintains the aerobic environment in the basin and keeps reactor contents (referred to as mixed liquor) completely mixed.

After a specific treatment time, the mixed liquor passes into the secondary clarifier, where the sludge settles under quiescent conditions and a clarified effluent is produced for discharge. The process recycles a portion of settled sludge back to the aeration basin to maintain the required activated-sludge concentration (expressed in terms of mixed-liquor, volatile SS [MLVSS] concentration). The process also intentionally wastes a portion of the settled

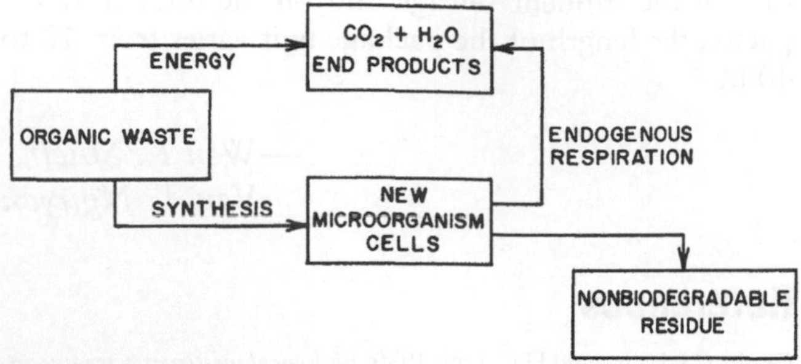

FIG. 6.4.1 Aerobic biological oxidation of organic wastes.

sludge to maintain the required SRT for effective organic (BOD) removal.

Process Microbiology

The activated-sludge process is an aerobic, continuous-flow, secondary treatment system that uses sludge-containing, active, complex populations of aerobic microorganisms to break down organic matter in wastewater. Activated sludge is a flocculated mass of microbes comprised mainly of bacteria and protozoa.

In the activated-sludge process, bacteria are the most important microorganisms in decomposing the organic material in the influent. During treatment, aerobic and facultative bacteria use a portion of the organic matter to obtain energy to synthesize the remaining organic material into new cells. Only a portion of the original waste is actually oxidized to low-energy compounds such as nitrate, sulfate, and carbon dioxide; the remainder of the waste is synthesized into cellular material. In addition, many intermediate products are formed before the end products.

The group of bacteria involved in activated-sludge systems belongs primarily to the Gram negative species, including carbon oxidizers and nitrogen oxidizers, floc-formers and nonfloc-formers, and aerobes and facultative anaerobes. In general, the bacteria in the activated-sludge process include those in the genera *Pseudomonas, Zoogloea, Achromobacter, Flavobacterium, Nocardia, Bdellovibrio, Mycobacterium, Nitrosomonas,* and *Nitrobacter.* An adequate population of the nitrifying bacteria, *Nitrosomonas* and *Nitrobacter,* must be maintained. These are slow-growing species; therefore, maintaining the sludge wasting rate ensures that they do not wash out.

Although floc-formers are mainly selected by the settling and recycling process, activated sludge can become dominated by filamentous bacteria. This situation is frequently associated with poor settlement characteristics. Researchers have shown that increasing the mean residence time of the cells enhances settling characteristics of biological floc (Forster 1985).

Another bacteria group found in activated sludge is the actinomycetes group, in particular *Nocardia* and *Rhodococcus.* These species are blamed for the formation of stable foams on activated-sludge tanks. The reason for the proliferation of these species is not known, and control methods have yet to be established (Forster 1985).

The protozoan population in activated sludge includes flagellates, amoebae, and ciliates. Over 200 different species of protozoa have been found in activated sludges (Forster 1985). Ciliates are the most prevalent type in activated sludge, with species such as *Vorticella* and *Opercularia* comprising up to one-third of the ciliate population. These species attach themselves to the sludge flocs. Another significant type of ciliates includes *Aspidisca* and *Trachelophylum*—species that creep over the sludge surface.

The balance of bacteria and protozoans in activated sludge depends on the nature of the wastewater and the plant operation. Protozoans are more susceptible to toxins and heavy metals than bacteria, and disruption of the protozoan population has been attributed to poor plant operation (Forster 1985). The role of the protozoan is not to stabilize the waste but to control the bacterial population, feeding on free-swimming bacteria that would otherwise produce a turbid effluent. However, carnivorous ciliates maintain a check on the bacteria-feeding population. Hence, protozoans are important in determining effluent quality.

Other microorganisms in activated sludge include fungi, nematodes, and rotifers. Fungi appear to have two roles: consumers of organic matter and predators for nematodes and rotifers. The role of fungi as a consumer of organic matter is the more common, especially in systems with low pH where bacterial growth is inhibited.

A proliferation of fungi usually imparts poor settleability to the sludge. Nematodes, like protozoans, also consume bacteria, while rotifers ingest sludge flocs, removing small particles that would otherwise cause turbidity. Rotifers also break up large flocs, providing available adsorption sites. Nevertheless, the effluent from an activated-sludge system can be high in biological solids as a result of poor design of the secondary settling unit, poor operation of the aeration units, or the presence of filamentous microorganisms such as *Sphaerotilus, E. coli,* and fungi (Metcalf and Eddy, Inc. 1991).

Process Flow Diagrams

Figure 6.4.2 shows a conventional activated-sludge process flow diagram. This process is primarily used in the treatment of municipal wastewater. The process uses long, rectangular aeration tanks with minimal longitudinal mixing that creates plug-flow patterns (see Figure 6.4.3). The wastewater is mixed with the recycled sludge at the head end of the aeration tank and then flows through the tank where organic matter is progressively removed. As a result, a BOD concentration profile is established through

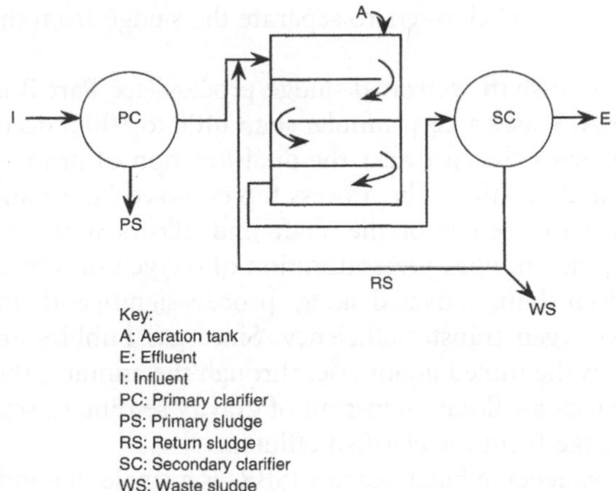

FIG. 6.4.2 Conventional activated-sludge process.

the tank that can diminish when the recycled sludge flow is significant. Air application is generally uniform through the tank.

The conventional activated-sludge process is susceptible to shock and toxic loading conditions since longitudinal mixing is absent in aeration tanks. The tapered-aeration, activated sludge process (see Figure 6.4.4) and the step-feed-aeration, activated-sludge process (see Figure 5.2.3) are two process variations of the conventional activated-sludge process. The aeration rate decreases along the tank length in the tapered-aeration, activated-sludge process and matches the BOD concentration profile to improve process economy. The aeration equipment is spaced unevenly through the tank.

Settled wastewater enters at several points in the aeration tank in the step-feed-aeration, activated-sludge process, equalizing loading and oxygen demand. This operation mode increases the flexibility of the process to handle shock and toxic loading conditions.

On the other hand, mixing intensity in the aeration tank of the completely-mixed, activated-sludge process (see Figure 5.2.5) is sufficiently high to yield a uniform mixed liquor that can smooth out and dilute load variations. As a result, the completely-mixed, activated-sludge process is resistant to shock and toxic loadings and is used widely for treating industrial wastewater. The aeration equipment is equally spaced for good mixing.

The contact-stabilization, activated-sludge process (see Figure 5.2.6) uses two separate tanks or compartments (contact and reaeration) to treat wastewater. This process first delivers the wastewater (usually without primary settling) into the aerated contact tank where it mixes with the stabilized sludge that rapidly removes suspended, colloidal, and a portion of the dissolved BOD (entrapment of suspended BOD in sludge flocs and adsorption of colloidal and dissolved BOD by sludge flocs). These reactions yield approximately 90% removal of BOD within 15 min of contact time (Eckenfelder 1980).

The mixed liquor then passes into the secondary clarifier where sludge is separated from clarified effluent. The settled sludge is recycled back to the reaeration tank where organic matter stabilization occurs. The resulting total aeration basin volume is typically 50% less than that of the conventional activated-sludge process (Metcalf and Eddy, Inc. 1991).

The oxygen-activated-sludge process uses high-purity oxygen instead of air (see Figure 6.4.5). The aeration tanks are usually covered, and the oxygen is recirculated, re-

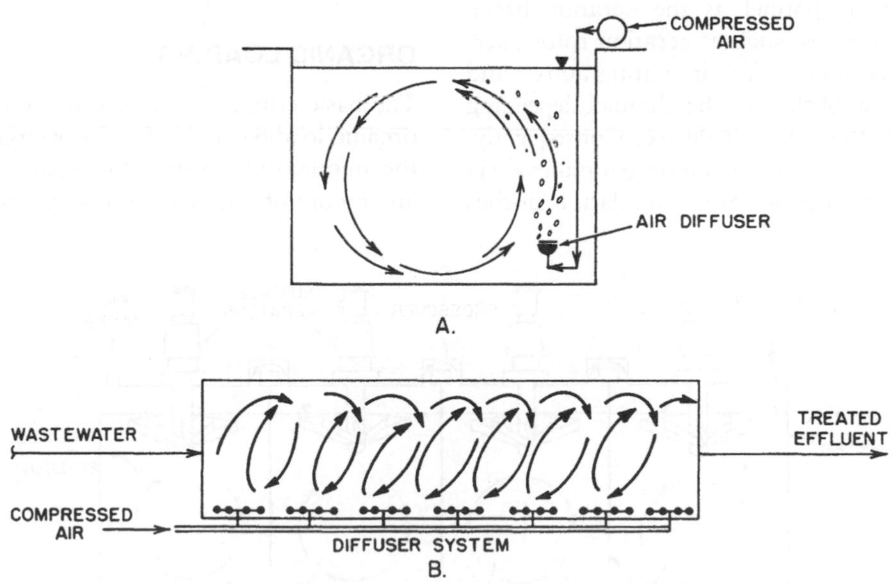

FIG. 6.4.3 Conventional activated-sludge process showing plug-flow and spiral-flow diffused aeration. **A,** End view; **B,** Top view.

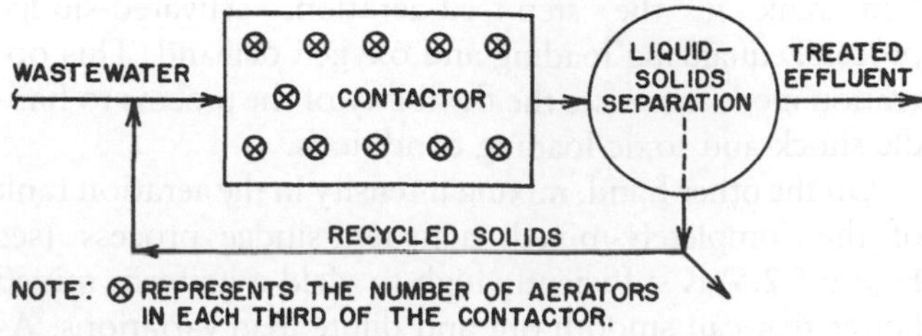

FIG. 6.4.4 Tapered-aeration, activated-sludge process.

ducing the oxygenation requirements. This process must vent a portion of the gas accumulated inside the aeration tank to remove carbon dioxide. However, adjusting the pH of the mixed liquor may still be needed.

Since the amount of oxygen added in the oxygen-activated-sludge process is approximately four times greater than that available in the conventional activated-sludge process, the BOD loading applied is higher, yielding a small aeration basin volume. Experimental evidence also indicates that oxygen-activated sludge settles better than air-activated sludge. A facility for generating and supplying high-purity oxygen is needed at a treatment site.

The extended-aeration process (see Figure 5.2.4) is similar to the conventional activated-sludge process except it operates in the endogenous respiration phase to reduce excess process sludge. As a result, the aeration basin is generally much larger. Only preliminary wastewater treatment to remove coarse materials is needed to protect treatment equipment. The extended-aeration process is designed for the treatment of wastewater generated from small installations and communities. Section 6.3 presents a detailed discussion on the extended-aeration process.

The oxidation ditch (see Part A in Figure 6.4.6) is a process variation of the extended-aeration process that uses a ring- or oval-shaped channel as the aeration basin. Mechanical aeration devices, such as aeration rotors, aerate and mix the mixed liquor. An alternating anoxic and oxic environment is established in the channel depending on the distance from the aeration device. Consequently, the oxidation ditch can achieve good nitrogen removal via nitrification and denitrification. Some oxidation ditches use intrachannel clarifiers to separate the sludge from the mixed liquor.

The deep-shaft, activated-sludge process (see Part B in Figure 6.4.6) uses a deep annular shaft (400 to 500 ft deep) as the reactor that provides the dual function of primary settling and aeration. The process forces mixed liquor and air down the center of the shaft and allows it to rise through the annulus. Oversaturation of oxygen occurring in the deep-shaft, activated-sludge process significantly increases oxygen transfer efficiency. Since gas bubbles are formed as the mixed liquor rises through the annulus, this process uses air flotation instead of gravity settling to separate sludge from the clarified effluent.

The sequencing batch reactor (SBR) is a single, fill-and-draw, completely-mixed reactor that operates under batch conditions. Recently, SBRs have emerged as an innovative wastewater treatment technology (Irvine and Ketchum 1989; U.S. EPA 1986). SBRs can accomplish the tasks of primary clarification, biooxidation, and secondary clarification within the confines of a single reactor. A typical treatment cycle consists of the following five steps: fill, react, settle, draw, and idle (U.S. EPA 1986). Depending on the mode of operation, SBRs can achieve good BOD and nitrogen removal. SBRs are uniquely suited for wastewater treatment applications characterized by low or intermittent flow conditions.

Design Concepts

In designing activated-sludge processes, environmental engineers must consider the organic loading, microorganism concentration, contactor retention time, artificial aeration, liquids–solids separation, effluent quality, and process costs.

ORGANIC LOADING

The basic criterion of design is the organic loading. The organic loading or food to microorganism (F/M) ratio is the amount of biodegradable organic material available to an amount of microorganisms per unit of time. This ratio

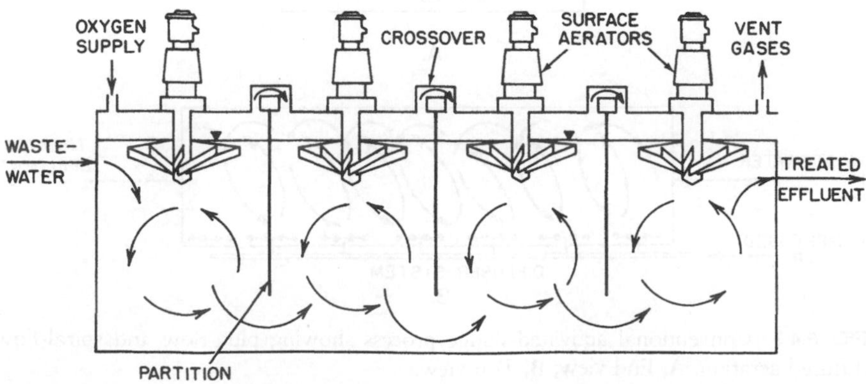

FIG. 6.4.5 Activated-sludge contactor using pure oxygen.

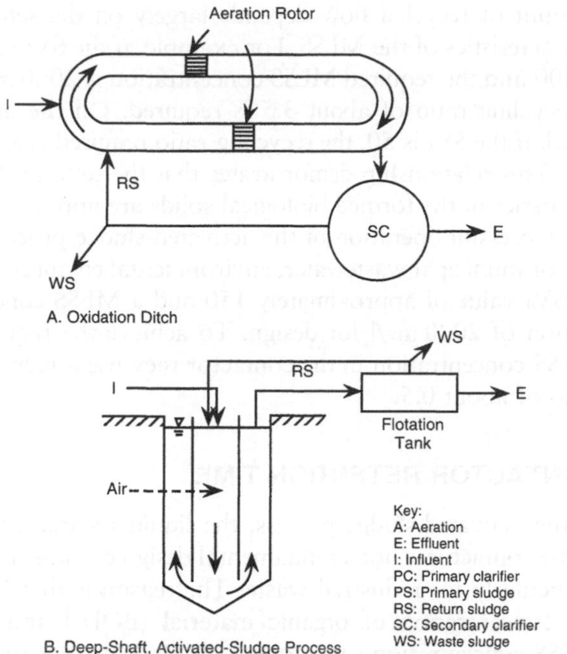

FIG. 6.4.6 Activated-sludge process flow diagrams. A, Oxidation ditch; B, Deep-shaft activated sludge process.

can be expressed more concisely as follows:

$$F/M = \frac{\text{(Organic concentration)(Wastewater flow)}}{\text{(Microorganism concentration)(Contactor volume)}} \quad 6.4(1)$$

or

$$F/M = \frac{(BOD_5)Q}{(MLSS)V} \quad 6.4(2)$$

where:

F/M = Organic loading, lb BOD_5 per lb mixed-liquor SS (MLSS) day
BOD_5 = Biological oxygen demand, mg/l
MLSS = Mixed-liquor SS, mg/l
V = Contactor volume, million gal
Q = Wastewater flow, mgd

To use this analytical expression of organic loading, environmental engineers must collect or assume data on the wastewater to be treated. The concentration of biodegradable organic material is expressed as BOD_5. For municipal wastewater, the BOD_5 ranges from 100 to 300 mg/l. The volume of wastewater to be treated is based on historical flow measurements plus an estimation of any increase or decrease anticipated during the life of the treatment plant.

The viable microorganisms in the activated-sludge process are expressed in terms of MLSS. MLSS is not the concentration of viable microorganisms but an indication of the microorganisms present in the system. Environmental engineers use the MLSS concentration because measuring the actual number of viable organisms in the system is difficult. The organic loading equation represents the ratio of the weight of organic material fed to the total weight of microorganisms available for oxidation.

Environmental engineers choose the organic loading on the basis of the desired effluent quality. If the organic loading is maintained at a high level, the effluent quality is poor, and solids (excess microorganisms) production is high. As the organic loading is reduced, however, the quality of the effluent improves, and the sludge production decreases. Table 6.4.1 shows the effect of organic loading.

MICROORGANISM CONCENTRATION

Since the concentration of microorganisms (MLSS) maintained in the contactor has a direct effect on the oxidation of organic pollutants, the liquid–solids separation characteristics of these solids are important. The SVI value indicates the ability of microorganisms separate from the wastewater after contact.

The SVI is defined analytically as the volume in milliliters occupied by 1 g of MLSS after a 1-l sample has settled in a graduated cylinder for 30 min. The SVI value for an activated-sludge system varies with the concentration of microorganisms maintained in the contactor. Table 6.4.2 reflects this point by listing identical settling characteristics as indicated by SVI values for various MLSS concentrations.

TABLE 6.4.1 EFFECT OF ORGANIC LOADING ON ORGANICS REMOVAL EFFICIENCY AND EXCESS SLUDGE PRODUCTION

Design Parameter	*Organic Loading (lb BOD_5/lb MLSS-day)*				
	0.1[(a)]	*0.3*[(b)]	*0.5*[(b)]	*1.0*[(c)]	*1.5*[(c)]
BOD_5 removal efficiency	95	90	90	75	70
Excess sludge produced (lbs/lb BOD_5 removed)	0.2	0.4	0.5	0.6	0.7

Notes: [(a)]Extended-aeration, activated sludge
[(b)]Conventional activated sludge
[(c)]High-rate, activated sludge

TABLE 6.4.2 EFFECT OF MLSS CONCENTRATION ON SETTLED VOLUME FOR A CONSTANT SVI VALUE

MLSS mg/l	*SVI (ml/g of settled MLSS)*	*Volume* (ml)*
500	100	50
1000	100	100
2000	100	200
4000	100	400
8000	100	800

Note: *MLSS volume after 30 min settling.

The table shows that the same SVI value of 100 can be observed for MLSS concentration from 500 to 8000 mg/l, yet the volume occupied by the MLSS after 30 min of settling is in the same proportion as the MLSS concentration. Therefore, the SVI value is meaningful only in indicating separation characteristics of solids at a particular concentration. If the same 30-min MLSS volume were required for a concentration of 8000 mg/l compared to 500 mg/l, the SVI value would have to be 6 compared to 100 at 500 mg/l MLSS.

The SVI value is of operational importance since it reflects changes in the treatment system. Any increase of SVI with no increase of MLSS concentration indicates that the solids settling characteristics are changing and a plant upset can occur.

Figure 6.4.7 shows the relationship between the MLSS concentration, SVI, and the recycling ratio (R/Q). The amount of recycled flow depends largely on the settling characteristics of the MLSS. For example, if the SVI value is 400 and the required MLSS concentration is 2000 mg/l, a recycling ratio of about 3.5 is required. On the other hand, if the SVI is 50, the recycling ratio required is about 0.2. This relationship demonstrates that the settling characteristics of the formed biological solids are important to the successful operation of the activated-sludge process.

For municipal wastewater, environmental engineers use an SVI value of approximately 150 and a MLSS concentration of 2000 mg/l for design. To achieve the required MLSS concentration in the contactor they use a recycling ratio of about 0.5.

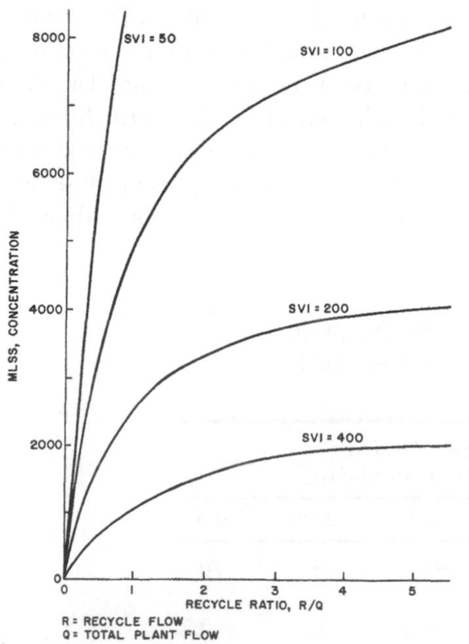

FIG. 6.4.7 Relationship between SVI, recycling ratio, and MLSS concentration.

CONTACTOR RETENTION TIME

In the activated-sludge process, the liquid retention time in the contactor is not a fundamental design consideration, particularly for industrial waste. The reason is that both the concentration of organic material (BOD_5) and the MLSS concentration can vary greatly for a wastewater or activated-sludge system, and both have a more pronounced effect on the process results than does the liquid retention time. The basic design parameters are the organic strength (BOD_5) of the wastewater and the MLSS concentration. Of these two parameters, only the MLSS can be varied by operation. Environmental engineers base their designs on the organic loading or F/M ratio, incorporating both the BOD_5 and MLSS concentration.

ARTIFICIAL AERATION

The activated-sludge process design must provide oxygenation and mixing to achieve efficient results. Current methods of accomplishing both oxygenation and mixing include 1) compressed-air diffusion, 2) sparge-turbine aeration, 3) low-speed surface aerators, and 4) motor-speed surface aerators.

Air diffusers were the earliest aeration devices used (see Figure 6.4.8). These devices compress air to the hydrostatic pressure on the diffuser (3 to 10 psig) and release it as small air bubbles. The larger the number and the smaller the size of the air bubbles produced, the better the oxygen transfer. Releasing air bubbles beneath the surface also results in airlift mixing of the contactor contents.

Combining compressed-air and turbine mixing eliminates the problems of clogging experienced with diffusers and adds versatility to the mixing and oxygen transfer. With the sparge-turbine aerator, the mixing and oxygenation can be varied independently within an operating range.

The additional development of aeration devices resulted in the elimination of compressors. The low-speed surface aerator uses atmospheric oxygen by causing extreme liquid turbulence at the surface. It is nearly twice as efficient in oxygen transfer as diffusers or sparge turbines.

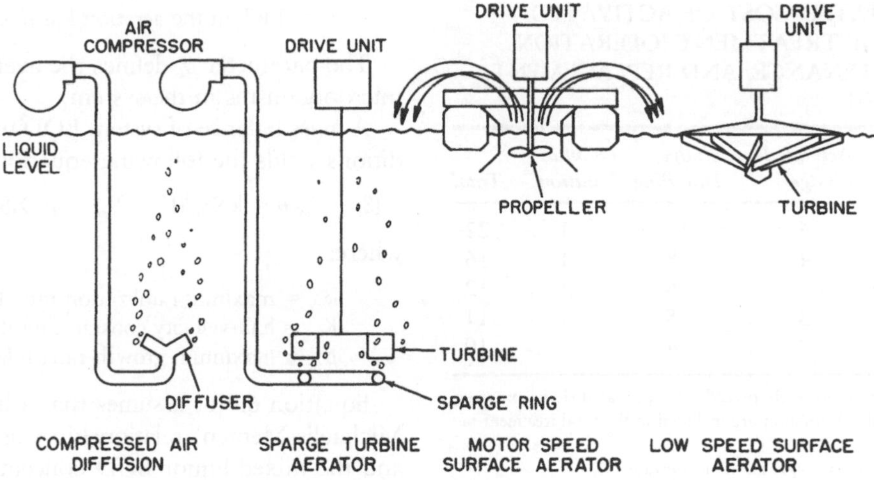

FIG. 6.4.8 Artificial oxygenation and mixing devices.

The motor-speed surface aerator is the latest aeration device. This device operates at the liquid surface but does not have a gear reducer between the motor and impeller. Because no gear reducer is used, the cost is significantly less than the low-speed surface aerator. Unfortunately, the oxygen transfer efficiency and liquid pumpage rate are also significantly reduced. The device has been used extensively to supplement oxygen requirements for oxidation ponds.

LIQUID–SOLIDS SEPARATION

Since the key to the activated-sludge process is maintaining a high concentration of microorganisms in the contactor, an efficient liquid–solids separation device must be employed. A major operational problem associated with the activated-sludge process is sludge bulking, a condition in which the settling characteristics of the solids make the liquid–solids separation inordinately difficult.

Among the environmental conditions causing sludge bulking are a high concentration of carbohydrates in the wastewater or a nutrient or oxygen deficiency in the system. These conditions can be rectified if they are quickly identified. Unfortunately, sludge bulking and contributing factors are not always easily identifiable, and difficult liquid–solids separation develops periodically. During these critical periods, the liquid–solids separation device must effectively separate bulking material from the wastewater and allow the solids to recycle back to the contactor. For gravity settling, a hydraulic separation rate of 250 to 500 gal per day per sq ft should be used. At a low hydraulic overflow rate, critical periods of sludge bulking can usually be handled without loss of gross solids into the effluent.

EFFLUENT QUALITY

Activated-sludge process design is based on the desired effluent quality. Successful operation of the activated-sludge process with an F/M ratio of 0.35 lb BOD_5 per lb MLSS per day and efficient liquid–solids separation should yield effluent containing an average of 20 mg/l of SS and 20 mg/l of BOD_5. For municipal wastewater, activated-sludge treatment removes the following major pollutants in the percentages listed:

90+% BOD_5 (biological oxygen demand)
70+% COD (chemical oxygen demand)
90+% SS (suspended solids)
30+% P (phosphorus)
35% N (nitrogen)

If a more efficient liquid–solids separation device, such as a granular-media filter, removes the remaining effluent solids (~20 mg/l), an effluent quality of 5 mg/l or less of BOD_5 and SS can be achieved.

PROCESS COSTS

The costliest item in the activated-sludge process is the artificial aeration device in the contactor. This equipment represents a large initial capital outlay, with high operation and maintenance costs. Treatment associated with most activated-sludge treatment plants includes primary clarification, sludge handling, and chlorination. Table 6.4.3 shows an estimated cost breakdown for the total activated-sludge treatment facility for plants varying in size from 1 to 100 mgd. The table shows the approximate cost distribution in such a facility together with the approximate cost for various plant sizes and each phase of the total treatment facility.

Process Kinetic Models

Environmental engineers have applied the principle of reactor engineering in the analysis and design of the activated-sludge process. In general, three models are widely used: continuous-flow, stirred-tank (CFSTR), plug-flow

TABLE 6.4.3 ESTIMATED COST OF ACTIVATED-SLUDGE TREATMENT (OPERATION, MAINTENANCE, AND REPLACEMENT COSTS)†

Flowrate mgd	*Primary Treatment**	*Activated-Sludge*	*Sludge Handling*	*Chlorination*	*Total*
1	4	6	11	1	22
3	3	4	8	1	16
10	2	3	6	1	12
30	2	3	5	1	11
100	2	3	4	1	10

Notes: *Primary clarification normally precedes the activated-sludge process; sludge handling, disposal, and chlorination are included in the total treatment facility.

†Treatment costs are in cents per 400 gal of wastewater.

(PF), and CFSTR-in-series. However, some modifications may be required depending on the specific process diagram selected.

CFSTR MODEL

The CFSTR model is derived based on the flow diagram shown in Figure 5.2.2. The model assumes that waste stabilization occurs only in the aeration basin, although the secondary clarifier is an integral part of the activated-sludge process. Furthermore, the amount of sludge in the secondary clarifier is assumed to be small compared to that in the aeration basin. A mass balance on the microorganisms in the system yields the following equation:

$$(dX/dt)V = -Q_wX_r - Q_eX_e + VR_g \qquad 6.4(3)$$

where:

X = MLVSS concentration in the aeration basin, mg/l
V = aeration basin volume, l
Q_w = sludge waste rate, l/day
Q_e or $(Q - Q_w)$ = effluent rate, l/day
X_r = MLVSS concentration in the recycled sludge flow, mg/l
R_g = net growth rate of microorganisms, mg MLVSS/l-day

Under steady-state conditions, Equation 6.4(3) can be simplified to the following equations:

$$(Q_wX_r + Q_eX_e)/VX = 1/\theta_c = YU - k_d \qquad 6.4(4)$$

$$U = Q(S_i - S_e)/XV = (S_i - S_e)/X\theta \qquad 6.4(5)$$

where:

Y = bacterial yield
U = specific utilization rate, l/day
k_d = bacterial decay rate, l/day
S_i = influent BOD (after primary settling), mg/l
S_e = effluent BOD (soluble), mg/l
θ_c = SRT, days
θ = LRT in the aeration basin, days

The parameter θ_c defines the average residence time of microorganisms in the system.

A mass balance of system BOD under steady-state conditions yields the following equation:

$$(S_i - S_e)/\theta = kXS_e/(K_s + S_e) = \mu_m XS_e/Y(K_s + S_e) \qquad 6.4(6)$$

where:

k = maximum utilization rate, l/day
K_s = half-velocity constant, mg/l
μ_m = maximum growth rate, l/day

Equation 6.4(6) assumes that a hyperbolic (Monod or Michaelis-Menten) relationship exists between k (or μ_m) and the mixed-liquor BOD concentration (S) (see Figure 6.4.9). From Equations 6.4(4) to 6.4(6) the following equations can be written:

$$X = \theta_c Y(S_i - S_e)/\theta(1 + k_d\theta_c) \qquad 6.4(7)$$

$$S = K_s(1 + k_d\theta_c)/[\theta_c(Yk - k_d) - 1] \qquad 6.4(8)$$

$$Y_o = Y/(1 + k_d\theta_c) \qquad 6.4(9)$$

where Y_o = observed bacterial yield. The F/M ratio is a widely used term in the analysis and design of the activated-sludge process and is calculated as follows:

$$F/M = S_i/X\theta \qquad 6.4(10)$$

$$U = (F/M)E/100 \qquad 6.4(11)$$

where E = BOD removal efficiency, %.

PF MODEL

The following equations describe the PF model proposed by Lawrence and McCarty (1970):

$$1/\theta_c = Yk(S_i - S_e)/[S_i - S_e + (1 + r)K_s \ln(S^*/S_e)] - k_d \qquad 6.4(12)$$

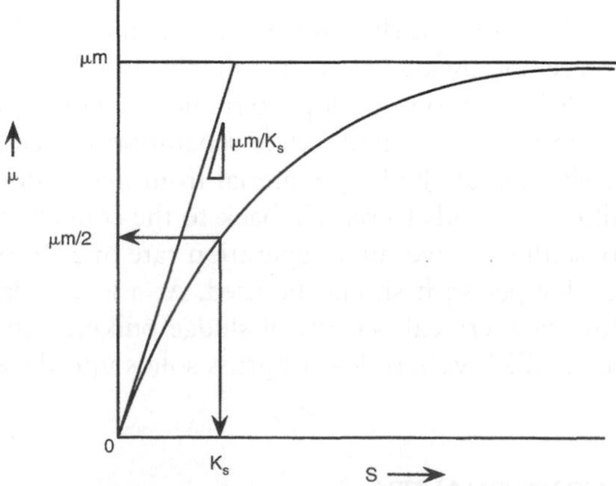

FIG. 6.4.9 Plot showing the effects of the substrate (BOD) concentration (S) on growth rate (μ). Subscript m indicates the maximum value.

$$S^* = (S_i + rS_e)/(1 + r) \qquad 6.4(13)$$

where r = recycling ratio (Q_r/Q). Equation 6.4(12) is valid for $\theta_c/\theta > 5$.

CFSTR-IN-SERIES MODEL

The CFSTR-in-series model is used as an approximation to the PF model. A PF reactor is divided into n compartments arranged in series, with each compartment modelled as a CFSTR. Since S_e is usually small compared to K_s, the hyperbolic relationship depicted in Figure 6.4.9 can be simplified to a linear relationship with $K = k/K_s$. Therefore, Equation 6.4(6) becomes the following equation:

$$S_e/S_i = 1/(1 + KX\theta) \qquad 6.4(14)$$

Applying Equation 6.4(14) to each compartment of a PF reactor yields the following equations:

$$S_{e1}/S_i = 1/(1 + K_1X_1\theta_1)$$

$$S_{e2}/S_{e1} = 1/(1 + K_2X_2\theta_2) \qquad 6.4(15)$$

or

$$S_{e2}/S_i = 1/(1 + K_1X_1\theta_1)(1 + K_2X_2\theta_2) \qquad 6.4(16)$$

$$S_{en}/S_i = S_e/S_i = 1/(1 + K_1X_1\theta_1)(1 + K_2X_2\theta_2)..(1 + K_nX_n\theta_n) \qquad 6.4(17)$$

If all n compartments are sized for equal volume and mean K and X values are adopted for all compartments, then Equation 6.4(17) can be simplified to the following equation:

$$S_e/S_i = 1/[1 + KX(\theta/n)]^n \qquad 6.4(18)$$

Table 6.4.4 summarizes typical kinetic coefficients for municipal wastewater applications. Environmental engineers must correct all kinetic coefficients used in design equations to account for temperature effects (see Equation 6.1[4]).

The kinetic models previously derived can be applied directly in designing a variety of activated-sludge processes.

TABLE 6.4.4 TYPICAL ACTIVATED-SLUDGE KINETIC COEFFICIENTS (MUNICIPAL WASTEWATER)

Coefficient	*Range**	*Typical**
k (l/day)	2–10	5
K_s (mg BOD_5/l)	25–100	60
(mg COD/l)	15–70	40
Y (mg MLVSS/mg BOD_5)	0.4–0.8	0.6
k_d (l/day)	0.025–0.075	0.06

Source: Adapted from Metcalf and Eddy, Inc., 1991, *Wastewater engineering: Treatment, disposal, and reuse,* 3d ed. New York: McGraw-Hill.
*20°C values. Equation 6.1(4) accounts for temperature effects. A typical θ value for municipal wastewater is 1.04.

However, some modifications may be necessary since the flow diagram involved can be different. The next subsections use the design of contact-stabilization and step-aeration activated-sludge processes as examples to show the modifications involved (Ramalho 1983).

CONTACT-STABILIZATION ACTIVATED-SLUDGE PROCESS

A steady-state mass balance on the microorganisms around the stabilization tank (see Figure 5.2.6) yields the following equations:

$$rQX_r + YQ(1 - \alpha)S_{iT} + YrQS_e - k_dV_sX_s - rQX_s = 0$$

or

$$V_s = [rQ(X_r - X_s + YS_e) + YQ(1 - \alpha)S_{iT}]/k_dX_s \qquad 6.4(19)$$

$$r = X_c/(X_s - X_c) \qquad 6.4(20)$$

where:

S_{iT} = total influent BOD, mg/l
V_s = stabilization tank volume, l
X_s = MLVSS concentration in the stabilization tank, mg/l
X_c = MLVSS concentration in the contact tank, mg/l
α = soluble fraction of influent BOD

All other terms are defined previously. The example assumes that all insoluble BOD (suspended and colloidal) entrapped and adsorbed by microorganisms is stabilized in the stabilization tank.

STEP-AERATION ACTIVATED-SLUDGE PROCESS

A two-stage, step-aeration activated-sludge process (see Figure 5.2.3) shows the design procedures involved (Ramalho 1983). This example assumes that the feed stream is equally divided between the two stages and the MLVSS concentration is constant throughout the tank. The steady-state mass balances on the microorganisms yield the following equations:

$$rQX_r - Q(r + 0.5)X = YQ(S_i - S_{e1}) - k_dXV_1 \qquad 6.4(21)$$

$$Q(r + 1)X - Q(r + 0.5)X = 0.5QX = YQ(S_{e1} - S_{e2}) - k_dXV_2 \qquad 6.4(22)$$

A steady-state mass balance on the microorganisms around the entire system yields the following equations:

$$(Q - Q_w)X_e + Q_wX_r = YQ[(S_i - S_{e1}) + (S_{e1} - S_{e2})] - 2k_dXV \qquad 6.4(23)$$

or

$$(Q - Q_w)X_e + Q_wX_r = YQ(S_i - S_e) - 2k_dXV \qquad 6.4(24)$$

where $V_1 = V_2 = V$ (equal volumes for each stage) and

$S_{e2} = S_e$.

The SRT and recycling ratio (r) of the process are as follows:

$$\theta_c = 2XV/[Q_wX_r + (Q - Q_w)X_e] \quad 6.4(25)$$

$$r = [Y_o(S_i - S_e) - X]/(X - X_r) \quad 6.4(26)$$

Equation 6.4(25) can be generalized for an n-stage, step-aeration, activated-sludge process as follows:

$$1/\theta_c = [YQ(S_i - S_e)/nXV] - k_d \quad 6.4(27)$$

whereas Equation 6.4(26) is valid for any numbers of stages.

Process Design

The design of a completely mixed, activated-sludge process illustrates the general procedures involved (Metcalf and Eddy, Inc. 1991). When a BOD_5 (5-day BOD) is used and the effluent produced is to have ≤20 mg/l BOD_5, the following equation applies:

$$S_e = 20 - \alpha X_e \quad 6.4(28)$$

where αX_e = the biodegradable portion of effluent biological solids. An $\alpha = 0.63$ is generally applicable (Metcalf and Eddy, Inc. 1991). Furthermore, the values of the following parameters in the design exercise that follows are S_i (influent BOD_5), Q (influent flow rate), X (MLVSS), θ_c (SRT), and X_r (recycled sludge concentration). A wastewater temperature of 20°C is assumed. The following exercise summarizes the design of a completely mixed activated-sludge process:

1. Compute the aeration basin volume (V) using Equation 6.4(7). In this case, the values of X and θ_c are assumed, and the values of kinetic coefficients are taken from Table 6.4.4. Note that θ (LRT) is defined as V/Q, where Q is the influent flow rate.
2. Compute the amount of sludge to be wasted per day. The amount of MLVSS produced due to the removal of BOD (P_x) is as follows:

   $$P_x = Y_oQ(S_i - S_e)10^{-6} \text{ (kg/day)} \quad 6.4(29)$$

   Y_o is defined in Equation 6.4(9). The amount of sludge lost in the effluent is QX_e10^{-6} (kg/day). Therefore, the amount of sludge to be wasted per day is $P_x - QX_e10^{-6}$ (kg/day).
3. Compute the sludge waste rate (Q_w) using Equation 6.4(4).
4. Compute the LRT (θ) of the aeration basin.
5. Check the F/M ratio and BOD removal efficiency (E) using Equations 6.4(10) and 6.4(11), respectively.
6. Compute the oxygen requirements. The amount of BOD_L (ultimate BOD) removal is $Q(S_i - S_e)10^{-6}/f$, where f is the conversion factor for converting BOD_5 to BOD_L. (Note that f = 0.65–0.68 for municipal wastewater.) Therefore, the amount of oxygen required (kg/day) is $(Q/f)(S_i - S_e)10^{-6} - 1.42P_x$.
7. Compute actual air requirements (diffused aeration). The design air requirements (m^3/day) are as follows:

   $$[(SF)(Q/f)(S_i - S_e)10^{-6} - 1.42P_x]/(\gamma_a)(0.232)(\beta_d) \quad 6.4(30)$$

   where:

   SF = safety factor (usually 2)
   γ_a = specific weight of the air, kg/m^3
   β_d = oxygen transfer efficiency of the diffused aeration equipment
   γ_a = p/RT
   p = sum of atmospheric pressure and air-diffuser discharge pressure
   R = universal gas constant
   T = absolute temperature

   For mechanical aeration equipment, the power requirement (kW) is calculated as follows:

   $$[(SF)(Q/f)(S_i - S_e)10^{-6} - 1.42P_x]/24N \quad 6.4(31)$$

   where N = field oxygen transfer capacity of the mechanical aeration equipment (kg O_2/kW-hr). Table 6.4.5 summarizes the typical ranges of field oxygen transfer capacities of various mechanical aerators.
8. Calculate the required recycling ratio (r) as $r = X/(X_r - X)$.

The secondary clarifier design is an integral part of the overall activated sludge process design. Details on the secondary clarifier design are presented in Section 7.2.

Operational Problems

Major operational problems of the activated-sludge process are caused by bulking sludge, rising sludge, and Nocardia foam (Metcalf and Eddy, Inc. 1991). A bulking sludge has poor settleability and compactability and is usually caused by excessive growth of filamentous microorganisms. Factors such as waste characteristics and composition, nutrient contents, pH, temperature, and oxygen

TABLE 6.4.5 OXYGEN TRANSFER EFFICIENCIES OF MECHANICAL AERATORS UNDER FIELD CONDITIONS

Aerator	*Oxygen Transfer Rate (kg O_2/kW-hr)*
Surface low-speed	0.73–1.46
Surface low-speed with draft tube	0.73–1.28
Surface high-speed	0.73–1.22
Surface downdraft turbine	0.61–1.22
Submerged turbine with sparger	0.73–1.09
Submerged impeller	0.73–1.09
Surface brush and blade (aeration rotor)	0.49–1.09

Source: Adapted from Metcalf and Eddy, Inc., 1991, *Wastewater engineering: Treatment, disposal, and reuse,* 3d ed., (New York: McGraw-Hill).

Note: Field conditions are wastewater temperature, 15°C; altitude, 152 m (500 ft); oxygen-transfer correction factor, 0.85; salinity-surface-tension correction factor, 0.9; and operating DO concentration, 2 mg/l.

availability can cause sludge bulking. The absence of certain components in the wastewater such as nitrogen, phosphorus, and trace elements can lead to the development of a bulking sludge. This absence is critical when industrial wastes are mixed with municipal wastewater for combined treatment.

Wide fluctuations in pH and DO are also known to cause sludge bulking. At least 2 mg/l of DO should be maintained in the aeration basin under normal operating conditions. Wastewater treatment facilities should check the F/M ratio to insure that it is within the recommended range. They should also check the additional organic loads received from internal sources such as sludge digesters and sludge dewatering operations to avoid internal overloading conditions, especially under peak flow conditions.

Chlorination of the return sludge effectively controls filamentous sludge bulking. Chlorine doses in the range of 2–3 mg/l of Cl_2 per 1000 mg/l of MLVSS are suggested. However, high doses can be necessary under severe conditions (8 to 10 mg/l of Cl_2 per 1000 m/l of MLVSS).

Rising sludge is usually caused by the release of gas bubbles entrapped within sludge flocs in the secondary clarifier. Nitrogen gas bubbles formed by denitrification of nitrite and nitrate under anoxic secondary clarifier conditions are known to cause sludge rising. Oversaturation of gases in the aeration tank can also cause sludge rising in the secondary clarifier, especially when aeration tank depth is significantly deeper than that of the secondary clarifier (Li 1993). Reducing the sludge retention time in the secondary clarifier is effective in controlling rising sludge. Close monitoring and control of aeration in the aeration tank can also reduce rising sludge in the secondary clarifier.

Nocardia foam is associated with a slow-growing, filamentous microorganisms of the *Nocardia* genus. Some factors causing Nocardia foaming problems are low F/M ratio, long SRT, and operating in the sludge reaeration mode (Metcalf and Eddy, Inc. 1991). Reducing SRT is the most common means of controlling Nocardia foaming problems.

—*Wen K. Shieh*
Van T. Nguyen

References

Eckenfelder, W.W., Jr. 1980. *Principles of water quality management.* Boston: CBI Publishing Co. Inc.

Forster, C.F. 1985. *Biotechnology and wastewater treatment.* Cambridge: Cambridge University Press.

Irvine, R.L., and L.H. Ketchum, Jr. 1989. Sequencing batch reactor for biological wastewater treatment. *CRC Crit. Rev. Environ. Control* 18, no. 255.

Lawrence, A.W., and P.L. McCarty. 1970. Unified basis for biological treatment design and operation. *J. San. Eng. Div., ASCE* 96, no. 757.

Li, A.L. 1993. Dynamic modeling of the deep tank aeration process. Ph.D. diss., Department of Systems, University of Pennsylvania, Philadelphia.

Metcalf and Eddy, Inc. 1991. *Wastewater engineering: Treatment, disposal, and reuse.* 3d ed. New York: McGraw-Hill.

Ramalho, R.S. 1983. *Introduction to wastewater treatment processes.* 2d ed. New York: Academic Press, Inc.

U.S. Environmental Protection Agency (EPA). 1986. *Sequencing batch reactors.* EPA/625/8-86/011. Cincinnati: U.S. EPA, Center for Environmental Research Information.

6.5 EXTENDED AERATION

LRT (hr)

18–36

SRT (days)

20–30

F/M (lb BOD_5/lb MLVSS-day)

0.05–0.15

VOLUMETRIC LOADING (lb BOD_5/ft^3-day)

0.01–0.25

BOD_5 REMOVAL (%)

85–95

MLSS (mg/l)

3000–6000

RECYCLING RATIO

0.5–1.5

AERATION TYPE

Diffused aeration and mechanical aeration

Process Description

The extended-aeration process (see Figure 5.2.4) is a modification of the conventional activated-sludge process. It is commonly used to treat the wastewater generated from small installations (e.g., schools, resorts, and trailer parks) as well as small and rural communities.

In extended aeration activated-sludge detention time is increased by a factor of four or five compared to conventional activated sludge. A final settling of 4 hr is typical at a surface settling rate of 350 to 700 gpd per sq ft. The main advantage of the extended-aeration process is that the amount of excess biological solids (sludge) produced is eliminated or minimized. Wastewater treatment facilities minimize this amount by operating the process in the endogenous respiration phase with the SRT maintained in the range of 20–60 days. As a result, the cost incurred with sludge disposal is reduced.

The extended-aeration process is further simplified since only preliminary influent wastewater treatment is required to remove coarse materials; the primary clarifier is eliminated. However, the size of the aeration basin is much larger than that of the conventional activated-sludge process. This larger basin accommodates a longer LRT in the aeration basin (i.e., 16–36 hr). The effluent produced is generally low in BOD and well nitrified (Ramalho 1983). An operational problem related to nitrification is a drop in pH which treatment facilities can correct by adding lime slurry to the aeration basin.

Although the amount of excess sludge in the extended-aeration process is significantly reduced, secondary clarification is needed to remove the accumulated non-biodegradable portion of sludge and the influent solids that are not degraded or removed. The design of secondary clarifiers is discussed in Section 6.8.

Process Flow Diagrams

In addition to the conventional extended-aeration process shown in Figure 5.1.4, a variation known as the oxidation ditch (see Figure 5.2.10) is also widely used. The aeration rotor provides the dual function of aeration and flow velocity. An alternating anoxic or oxic environment occurs in the oxidation ditch depending on the distance from the aeration rotor. As a result, an oxidation ditch achieves good nitrogen removal via nitrification and denitrification. The process separates sludge from the treated effluent by using either an external secondary clarifier or an intrachannel clarifier. Primary clarification is usually not provided.

Several manufactured extended-aeration units are available (Viessman and Hammer 1993).

Process Design

The following equations show the design of an extended-aeration process without nitrification and with zero-sludge yield (Ramalho 1983):

$$0.77YQ(S_i - S_e) = k_dXV \quad 6.5(1)$$

$$\theta = 0.77Y(S_i - S_e)/k_dX \quad 6.5(2)$$

where:

Y = sludge yield
Q = influent flow rate, l/day
S_i = influent BOD, mg/l
S_e = effluent BOD, mg/l
k_d = sludge decay (endogenous respiration) rate, l/day
X = MLVSS concentration in the aeration basin, mg/l
V = basin volume, l
θ = LRT, day

These equations incorporate 0.77 since approximately 77% of the sludge produced is biodegradable.

The recycling ratio (r) is calculated as follows:

$$r = [X - 0.23Y(S_i - S_e)]/(X_r - X) \quad 6.5(3)$$

where X_r = MLVSS concentration in the recycled sludge flow, mg/l.

—Wen K. Shieh
Van T. Nguyen

References

Ramalho, R.S. 1983. *Introduction to wastewater treatment processes.* 2d ed. New York: Academic Press, Inc.

Viessman, W., Jr., and M.J. Hammer. 1993. *Water supply and pollution control.* 5th ed. New York: Harper Collins.

6.6 PONDS AND LAGOONS

TYPE

a. Aerobic (low-rate)
b. Aerobic (high-rate)
c. Aerobic (maturation)
d. Facultative
e. Anaerobic
f. Aerated lagoon

FLOW REGIME

a–c: intermittent mixing; d: mixed (surface layer); e: no mixing; and f: completely mixed

SURFACE AREA (acres)

a: <10; b: 0.5–2; c: 2–10; d: 2–10; e: 0.5–2; and f: 2–10

DEPTH (ft)

a: 3–4; b: 1–1.5; c: 3–5; d: 4–8; e: 8–16; and f: 6–20

LRT (days)

a: 10–40; b: 4–6; c: 5–20; d: 5–30; e: 20–50; and f: 3–10

BOD_5 LOADING (lb/acre-day)

a: 60–120; b: 80–160; c: ≤15; d: 50–180; and e: 200–500

BOD_5 REMOVAL (%)

a, b, d, & f: 80–95; c: 60–80; and e: 50–85

ALGAL CONCENTRATION (mg/l)

a: 40–100; b: 100–260; c: 5–10; d: 5–20; and e: 0–5

EFFLUENT SS (mg/l)

a: 80–140; b: 150–300; c: 10–30; d: 40–60; e: 80–160; and f: 80–250

Process Microbiology

This section briefly discusses two low-cost, suspended-growth wastewater treatment systems—stabilization ponds and aerated lagoons.

Stabilization ponds possess a similar biological community as activated-sludge with the addition of an algal population. Oxygen is supplied in an aerobic photosynthetic pond by natural reaeration from the atmosphere and algal photosynthesis. The oxygen released by photosynthetic algae is used by bacteria to degrade organic matter (see Figure 6.1.3). Degradation by bacteria releases carbon dioxide and nutrients used by algae. Higher life forms such as rotifers and protozoa are also present in the pond and function primarily as polishers of the effluent. Temperature has a significant effect on aerobic pond operation. Organic loading, pH, nutrients, sunlight, and degree of mixing also affect each microbial group's population throughout the pond.

In facultative ponds, two different biological communities exist. The microbial community of the pond's upper layer is similar to that of an aerobic pond, whereas microorganisms in the lower and bottom layers are facultative and anaerobic (see Figure 6.6.1). Respiration occurs in the presence of sunlight; however, the net effect is oxygen production, i.e., photosynthesis. During photosynthesis, algae uses carbon dioxide, resulting in high pH levels in low alkalinity wastewater. In facultative ponds, algae can use bicarbonate as a carbon source for cell growth; when this occurs, there is a diurnal fluctuation in pH. However, at high-pH levels, carbonate and hydroxide species predominate. In wastewater containing a high concentration of calcium, calcium carbonate precipitates preventing the pH from rising any higher.

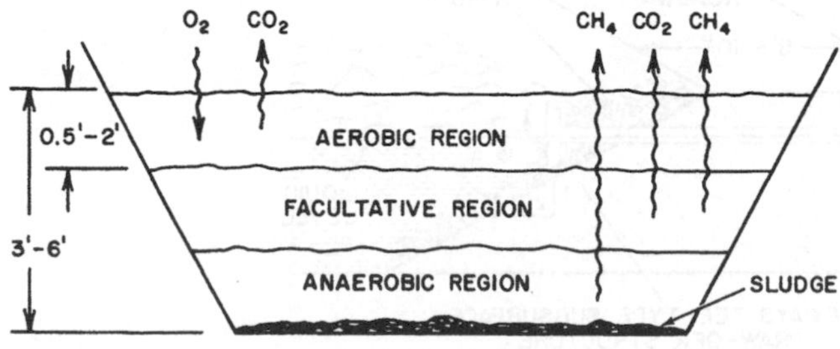

FIG. 6.6.1 Elevation diagram of facultative lagoon strata and operation.

The microbiology involved in an aerated-lagoon process is similar to that of an activated-sludge process. However, differences arise because the large surface area of lagoons can cause more temperature effects than normally encountered in conventional activated-sludge processes. Aerobic digestion—a process that treats organic sludges produced from various treatment operations—is similar to the activated-sludge process.

When the available substrate supply is depleted, microorganisms consume their own protoplasm to obtain energy for cell maintenance and enter the endogenous phase. Cell tissue is aerobically oxidized to carbon dioxide, water, and ammonia. The cell tissue actually oxidized is 70 to 80%, with inert components and nonbiodegradable organic matter remaining. The ammonia produced in cell tissue oxidation is eventually oxidized to nitrate digestion proceeds.

Stabilization Ponds

A stabilization pond is a low-cost treatment process widely used in small communities and industrial facilities. It is a shallow body of wastewater contained in an earthen basin, using a completely mixed biological process without solids return. Mixing is usually provided by natural processes such as wind, heat, or fermentation; however, mixing can be induced by mechanical or diffused aeration.

One of three types of environmental conditions—fostering three corresponding types of biological activity—can prevail in a stabilization pond process: aerobic, aerobic–anaerobic, and anaerobic. Aerobic ponds are used primarily for treating soluble organic wastes and effluents from wastewater treatment plants. Facultative ponds (see Figure 6.6.1), in which aerobic–anaerobic conditions exist, are the most common type and are used to treat domestic waste and a variety of industrial waste. Anaerobic ponds are applied where rapid stabilization of strong organic waste is required (see Figure 6.1.4). Wastewater treatment facilities commonly use anaerobic ponds in series with facultative ponds to provide complete treatment.

Aerobic and facultative ponds are biologically complex. Figure 6.6.2 shows the general reactions that occur. Part of the organic matter in the influent is oxidized by bacteria, producing ammonia, carbon dioxide, sulfate, water, and other end products of aerobic metabolism. These products are subsequently used by algae during daylight to produce oxygen. Bacteria use this supplemental oxygen and the oxygen provided by wind action to decompose the other part of the organic matter.

In states where stabilization-pond-treatment processes are commonly used, regulations govern pond design, installation, and operation. A minimum retention time of 60 days is often required for flow-through facultative ponds receiving untreated wastewater (Metcalf and Eddy, Inc. 1991). Frequently, retention times as high as 120 days are

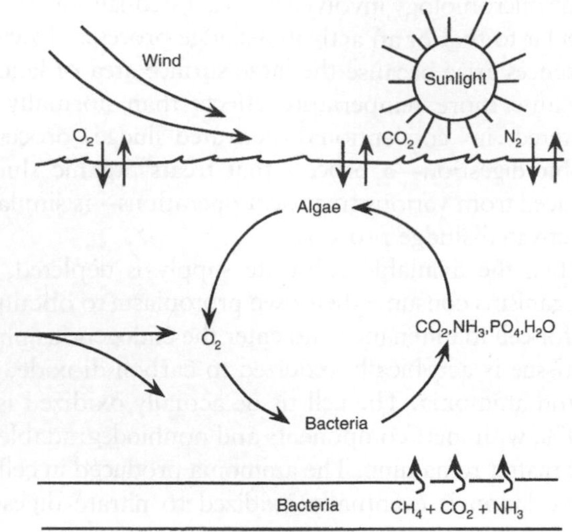

FIG. 6.6.2 Schematic of a stabilization pond.

specified. However, even with a low retention time of 30 days, a high degree of coliform removal is ensured (Metcalf and Eddy, Inc. 1991). Other typical standards (see Figure 6.6.3) include embankment slopes (1:3 to 1:4), organic loading rate (2.2 to 5.5 g BOD/m^2-day, depending on climate), and permissible seepage through the bottom (0 to 6 mm/day). In some climates, treatment facilities can operate ponds without discharge to surface waters (McGhee 1991).

AEROBIC PONDS

Of all the biological treatment process designs, the stabilization pond design is the least defined. Aerobic stabilization ponds contain bacteria and algae in suspension under aerobic conditions. Aerobic ponds are of two basic types. In one type, the objective is to maximize algae production. These aerobic ponds generally operate at depths of 0.15 m to 0.45 m. In the other type of aerobic ponds, the amount of oxygen produced is maximized, and depths range to 1.5 m. Shallower depths encourage rooted aquatic plant growth, interfering with the treatment process. However, greater depths can interfere with mixing and oxygen transport from the surface. To achieve the best results with aerobic ponds, wastewater treatment facilities should provide mixing with pumps or surface aerators.

Environmental engineers usually base the aerobic pond process design on the organic loading rates and hydraulic retention times derived from pilot-plant studies and observations of operating systems. They adjust the pond loading rate to reflect the oxygen available from photosynthesis and atmospheric reaeration. Frequently, environmental engineers design large aerobic pond systems as completely mixed reactors, with two or three reactors in series.

Another design approach involves the use of a first-order, removal-rate equation developed by Wehner and Wilhelm (Metcalf and Eddy, Inc. 1991). This equation describes the substrate removal for an arbitrary flow-through pattern that lies somewhere between completely mixed and plug-flow as follows:

$$S/S_o = 4ae^{-(1/2d)}/[(1 + a)^2e^{-(a/2d)} - (1 - a)^2e^{-(a/2d)}] \quad 6.6(1)$$

where:

S = effluent substrate concentration, mg/l
S_o = influent substrate concentration, mg/l
a = $(1 + 4ktd)^{1/2}$
d = dispersion factor (D/uL)
D = axial dispersion coefficient, (m^2/h)
u = fluid velocity (m/h)
L = characteristic length (m)
k = first-order reaction constant (h^{-1})
t = retention time (h)

The term kt in Equation 6.6(1) can be plotted as a function of S/S_o for various dispersion factors—varying from zero for PF reactors to infinity for completely mixed reactors—to yield a graph that facilitates the use of the equation in designing ponds (Metcalf and Eddy, Inc. 1991). The dispersion factor ranges from 0.1 to 2.0 for most stabilization ponds. For aerobic ponds, the dispersion factor is approximately 1.0 since completely mixed conditions usually prevail in these ponds for high performance. Depending on the operational and hydraulic characteristics of the pond, typical values for the overall first-order BOD_5 removal-rate constant k range from 0.05 to 1.0 per day (Metcalf and Eddy, Inc. 1991).

Although aerobic pond efficiency is high—up to 95%—

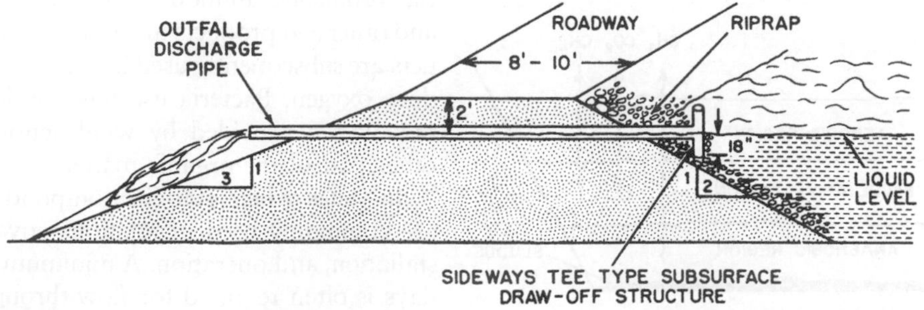

FIG. 6.6.3 Levee and outfall structure.

and most soluble BOD_5 is removed from influent wastewater, bacteria and algae in the effluent can exert a BOD_5 higher than that of the original waste. Hence, wastewater treatment facilities must apply methods of removing biomass from the effluent.

FACULTATIVE PONDS

In facultative ponds, waste conversion is performed by a combination of aerobic, anaerobic, and facultative bacteria. As shown in Figures 6.6.1 and 6.6.2, the facultative pond is comprised of three zones: (1) a surface zone where algae and bacteria thrive symbiotically, (2) an anaerobic zone at the bottom sludge layer where accumulated organics are decomposed by anaerobic bacteria, and (3) an aerobic–anaerobic zone in the middle where facultative bacteria are responsible for waste conversion. Using the oxygen produced by algae growing near the surface, aerobic and facultative bacteria oxidize soluble and colloidal organics, producing carbon dioxide. This carbon dioxide is used by the algae as a carbon source. Anaerobic waste conversion in the bottom zone produces dissolved organics and gases such as CH_4, CO_2, and H_2S that are either oxidized by aerobic bacteria or released to the atmosphere.

Facultative stabilization pond designs (see Figure 5.2.9) are similar to those of aerobic ponds; i.e., they are usually based on loading factors developed from field experience. Unlike aerobic ponds, facultative ponds promote settling of organics to the anaerobic zone. Therefore, quiescent conditions are required, and dispersion factors in facultative ponds vary from 0.3 to 1.0 (Metcalf and Eddy, Inc. 1991).

The sludge accumulation in facultative ponds calls for another deviation from aerobic pond design. In cold climates, a portion of BOD_5 is stored in the accumulated sludge during the winter months. In the spring and summer as the temperature rises, accumulated BOD_5 is anaerobically converted. The end products of conversion—gases and acids—exert an oxygen demand on the wastewater. This demand can exceed the oxygen supply provided by algae and surface reaeration in the upper layer of the pond. In this case, wastewater treatment facilities should use surface aerators capable of satisfying 175 to 225% of the incoming BOD_5. The accumulation of sludge in the facultative pond can also lead to a higher SS concentration in the effluent, reducing overall pond performance.

ANAEROBIC PONDS

Anaerobic ponds (see Figure 5.2.9) treat high-strength wastewater with a high solids concentration. These are deep earthen ponds with depths to 9 m to conserve heat energy and maintain anaerobic conditions. Influent waste settles to the bottom, and partially clarified effluent is discharged to another treatment process for further treatment. Anaerobic conditions are maintained throughout the depth of the pond except for the shallow surface zone. Waste conversion is performed by a combination of precipitation and anaerobic metabolism of organic wastes to carbon dioxide, methane and other gases, acids, and cells. On the average, anaerobic ponds achieve BOD_5 conversion efficiencies to 70%, and under optimum conditions, 85% efficiencies are possible (Metcalf and Eddy, Inc. 1991).

Aerated Lagoons

Aerated lagoons are basins where wastewater can be treated in a flow-through only manner or without solids recycling. Lagoon depths vary from 1 to 4 m (Ramalho 1983). Oxygenation of the wastewater in lagoons is usually accomplished by surface, turbine, or diffused aeration. The turbulence created by aeration keeps lagoon contents suspended. Depending on the retention time, the aerated lagoon effluent contains approximately one-third to one-half the value of the incoming BOD in the form of cellular mass. Wastewater treatment facilities can use a settling basin or tank for solids removal, by settling, from the effluent prior to discharge.

In designing an aerated lagoon, environmental engineers must incorporate the following parameters: (1) BOD removal, (2) effluent characteristics, (3) temperature effects, and (4) oxygen requirements. The design basis for a lagoon can be the mean cell residence time since the aerated lagoon is a completely mixed reactor without recycling. Selected mean cell residence time should ensure that the suspended biomass has good settlement properties, and be high enough to prevent cell wash-out. Typical design mean cell residence time for lagoons treating domestic waste varies from 3 to 6 days.

From the mean cell residence time, environmental engineers can estimate soluble substrate concentration in the effluent and determine the removal efficiency from substrate utilization equations used in activated-sludge process design. Alternatively, they can assume a first-order removal function for the observed BOD_5 removal in a single aerated lagoon (Metcalf and Eddy, Inc. 1991) as follows:

$$S/S_o = 1/[1 + k(V/Q)] \qquad 6.6(2)$$

where:

S = effluent BOD_5 concentration, mg/l
S_o = influent BOD_5 concentration, mg/l
k = overall, first-order, BOD_5, removal-rate constant, day^{-1}
V = volume, l
Q = flow rate, l/day

The values for the removal-rate constant k vary from 0.25 to 1.0. Effluent characteristics of significance are the BOD_5 and the SS concentration. Environmental engineers can estimate both of these characteristics using the equations presented in Section 6.4 for calculating similar parameters in an activated-sludge effluent.

The effect of temperature on biological activity is described in Section 6.1. When influent wastewater temperature, ambient temperature, lagoon surface area, and wastewater flow rate are known, environmental engineers can estimate the resulting temperature in the lagoon using the following equation (Metcalf and Eddy, Inc. 1991):

$$(T_i - T_w) = [(T_w - T_a)fA]/Q \qquad 6.6(3)$$

where:

T_i = influent wastewater temperature
T_w = lagoon wastewater temperature
T_a = ambient temperature
f = a proportionality factor that incorporates heat transfer coefficients and the effects of surface area increase due to aeration, wind, and humidity (typical value for the eastern United States is 0.5 in SI units)
A = lagoon surface area
Q = wastewater flow rate

Oxygen requirements are computed as outlined in the design calculations for aeration in the activated-sludge process (see Section 6.4).

Anaerobic Lagoons

Anaerobic lagoons produce noxious odors that result when the acid-producing bacteria reduce sulfate compounds to hydrogen sulfide (H_2S). As the acid producers deplete the nonsulfate sources of combined oxygen, they start to reduce sulfate for oxygen with the liberation of H_2S, which has the odor of rotten eggs. At low concentrations, H_2S is merely obnoxious and therefore a nuisance, but at higher concentrations, it attacks painted surfaces and is also deleterious if inhaled for an extended period. To operate an anaerobic lagoon, wastewater treatment facilities must minimize the liberation of H_2S to eliminate these effects. They can accomplish this task by controlling the concentration of sulfate compounds in the waste contents.

If a sulfate concentration of less than 100 mg/l is maintained in the influent to the lagoon, no significant odor problems occur. If odor is a problem, the facility can add nitrate to the lagoon to alleviate it temporarily. When nitrate is applied, the acid producers switch to nitrate for oxygen. Sulfate reduction is thus stopped. This measure is only temporary; the only real long-term solution is to limit sulfate concentration in the influent. Due to the odor problem, anaerobic lagoons must be located in remote areas if sulfate starvation is not practiced.

DIMENSIONS

An anaerobic lagoon is similar in construction to an aerobic lagoon in levee dimensions and construction materials. The anaerobic lagoon, however, usually requires less surface area than the aerobic facility. Since oxygen transfer from the atmosphere is not important, the anaerobic lagoon can be as deep as is practical. A depth of at least 15 ft is recommended whenever groundwater considerations and area geology permit. The relative depth of an anaerobic lagoon provides improved heat retention. The lagoon should be as long as practical (an efficient length to width ratio is 2:1).

The BOD loading rate in anaerobic lagoon design is 500 to 1000 lb BOD per acre per day with an expected BOD removal efficiency of 50 to 80%. The required detention time is between 30 and 50 days. The ideal pH range for the anaerobic process is 6.6 to 7.6, but lagoon efficiency is not significantly hampered if pH is gradually increases to 9.0. Above pH 9.0, the efficiency drops off rapidly. Sudden bursts of high and low pH also hinder lagoon performance.

Because of the buffering effect provided by liberating carbon dioxide in the anaerobic process, the lagoon can also act as an effective neutralization system. It is capable of neutralizing approximately 0.5 lb of caustic per lb of BOD removed while the lagoon is buffered at a pH of roughly 8.0.

The anaerobic process functions optimally over two temperature ranges: the mesophylic range of 85° to 100°F and the thermophilic range of 120° to 135°F. Only the mesophylic range, however, applies to an unheated lagoon. The lagoon is optimally effective when the temperature range for mesophylic operation is not violated. However, as temperatures decrease below 85°F, the lagoon efficiency decreases only slightly until a temperature of about 60°F is reached, at which point the efficiency drops off rapidly. This temperature requirement is why the lagoon should be as deep as possible, i.e., to maximize heat retention.

APPLICATIONS

The aerobic lagoon is applicable to lower strength wastes (usually with a BOD of less than 200 mg/l), which are not toxic to an algal system. The anaerobic lagoon is applicable to high-strength wastes (usually greater than 500 mg/l of BOD) and applications in which a highly purified effluent is not required. In anaerobic lagoons, either the sulfate concentration must be low (less than 100 mg/l), or the lagoon must be in a remote location. The facultative lagoon is applicable to wastes of approximately 200 to 500 mg/l BOD concentration. The waste cannot be toxic to algae or contain a large sulfate concentration.

In all three lagoons, a large amount of land must be available since each lagoon requires many acres for construction. Lagoons are much less susceptible to upsets from accidental discharges or large loading variations than other methods of biological treatment. Therefore, they are applicable in these situations.

Frequently, more than one type of lagoon is used. For example, additional effluent treatment from an anaerobic lagoon can be provided in a facultative or aerobic lagoon. The initial treatment in an anaerobic lagoon often renders the waste more amenable to aerobic treatment. The use of two lagoons in series further purifies the effluent and requires less land area.

COSTS

The primary investment associated with constructing a lagoon is the cost of the land and the excavation and earthmoving costs in constructing the basin. If the soil where the lagoon is constructed is permeable, an additional cost for lining is incurred.

In the midwest region of the United States, except for major population centers, the price of land is about $1500 per acre. Excavation costs vary and depend on whether dirt must be introduced or hauled away. If the dirt removed from the lagoon floor can be used for levee construction, excavation costs are roughly $2.00 per cu yd of dirt excavated. Levees are frequently compacted sufficiently by earthmoving equipment, but compacting equipment, if required, costs from $3 to $5 per cu yd. Synthetic lining material is expensive, and its use should be avoided wherever possible. The price for most plastic liners is about $1 per sq yd.

Operating costs are almost zero. In most cases, neither pumps nor any other electrically operated device is required. Therefore, power costs are usually nonexistent. Although some analytical work is required to assure proper operation, the extent of such a program is minimal compared to other methods of biological and chemical treatment. An extensive sampling system is usually not required to obtain samples for analysis. Due to the equalization effect of a large facility, a daily grab sample usually produces the necessary operational information.

Operational personnel are not required except for sampling, analysis, and general upkeep; the system is virtually maintenance-free.

—*Wen K. Shieh*
Van T. Nguyen

References

McGhee, T.J. 1991. *Water supply and sewerage.* 6th ed. New York: McGraw-Hill.

Metcalf and Eddy, Inc. 1991. *Wastewater engineering: Treatment, disposal, and reuse.* 3d ed. New York: McGraw-Hill.

Ramalho, R.S. 1983. *Introduction to wastewater treatment processes.* 2d ed. New York: Academic Press, Inc.

6.7 ANAEROBIC TREATMENT

REACTOR TYPE

a. Anaerobic contact process
b. Anaerobic upflow sludge blanket (USB) reactor
c. Anaerobic filter (downflow or upflow)
d. Anaerobic fluidized-bed reactor (AFBR)

LRT (hrs)

a: 2–10; b: 4–12; c: 24–48; and d: 5–10

ORGANIC LOADING (lb COD/ft³-day)

a: 0.03–0.15; b: 0.25–0.75; c: 0.06–0.3; and d: 0.3–0.6

COD REMOVAL FOR INDUSTRIAL WASTEWATER (%)

a: 75–90; b: 75–85; c: 75–85; and d: 80–85

OPTIMAL TEMPERATURE (°C)

30–35 (mesophilic) and 49–57 (thermophilic)

OPTIMAL pH

6.8–7.4

OPTIMAL TOTAL ALKALINITY (mg/l as $CaCO_3$)

2000–3000

OPTIMAL VOLATILE ACIDS (mg/l as acetic acid)

50–500

Process Description

Anaerobic treatment applies to both wastewater treatment and sludge digestion. This section discusses only anaerobic wastewater treatment. Anaerobic wastewater treatment is an effective biological method for treating many organic wastes. The microbiology involved in the process includes facultative and anaerobic microorganisms, which, in the absence of oxygen, convert organic materials into gaseous end products such as carbon dioxide and methane.

Anaerobic wastewater treatment was discovered in the middle of the last century; however, environmental engineers have only seriously considered it in the last twenty years (Forster 1985). Despite intense research in this field in the past few decades, much research is still needed in several areas. These areas include (Forster 1985):

Microbiology—Further research on the biochemistry and genetics related to the anaerobic microbial species is required.

Start up procedures—Optimal procedures to minimize the lag time between the commissioning of a reactor and its placement into full operation must be investigated.

Optimization of process engineering—Further optimization of the anaerobic treatment process is required, especially involving ancillary equipment, small-scale reactors, and support media (where applicable).

The major advantages of anaerobic treatment over aerobic treatment are as follows:

The biomass yield for anaerobic processes is much lower than that for aerobic systems; thus, less biomass is produced per unit of organic material used. This reduced biomass means savings in excess sludge handling and disposal and lower nitrogen and phosphorus requirements.

Since aeration is not required, capital costs and power consumption are lower.

Methane gas produced in anaerobic processes provides an economically valuable end product.

The savings from lower sludge production, electricity conservation, and methane production range from \$0.20 to \$0.50 per 1000 gal of domestic sewage treatment (Jewell 1987). The reduction of sludge and aeration energy consumption each result in savings that are greater than the cost of the energy required by the anaerobic process (Jewell 1987). In addition, a substantial part of the energy requirements for anaerobic processes can be obtained from exhaust gas.

Higher influent organic loading is possible for anaerobic systems than for aerobic systems because the anaerobic process is not limited by the oxygen transfer capability at high-oxygen utilization rates in aerobic processes.

However, some disadvantages are associated with the anaerobic process as follows:

Energy is required by elevated reactor temperatures to maintain microbial activity at a practical rate. (Generally, the optimum temperature for anaerobic processes is 35°C.) This disadvantage is not serious if the methane gas produced by the process can supply the heat energy.

Higher detention times are required for anaerobic processes than aerobic treatment. Thus, an economical treatment time can result in incomplete organic stabilization.

Undesirable odors are produced in anaerobic processes due to the production of H_2S gas and mercaptans. This limitation can be a problem in urban areas.

Anaerobic biomass settling in the secondary clarifier is more difficult to treat than biomass sedimentation in the activated-sludge process. Therefore, the capital costs associated with clarification are higher.

Operating anaerobic reactors is not as easy as aerobic units. Moreover, the anaerobic process is more sensitive to shock loads (Benefield and Randall 1980).

Process Microbiology

The end products of anaerobic degradation are gases, mostly methane (CH_4), carbon dioxide (CO_2), and small quantities of hydrogen sulfide (H_2S) and hydrogen (H_2). The process involves two distinct stages: acid fermentation and methane fermentation.

In acid fermentation, the extracellular enzymes of a group of heterogenous and anaerobic bacteria hydrolyze complex organic waste components (proteins, lipids, and carbohydrates) to yield small soluble products. These simple, soluble compounds (e.g., triglycerides, fatty acids, amino acids, and sugars) are further subjected, by the bacteria, to fermentation, β-oxidations, and other metabolic processes that lead to the formation of simple organic compounds, mainly short-chain (volatile) acids (e.g., acetic [CH_3COOH], propionic [CH_3CH_2COOH], butyric [CH_3-CH_2-CH_2-$COOH$]) and alcohols. In the acid fermentation stage, no COD or BOD reduction is realized since this stage merely converts complex organic molecules to short-chain fatty acids, alcohols, and new bacterial cells, which exert an oxygen demand.

In the second stage, short-chain fatty acids (other than acetate) are converted to acetate, hydrogen gas, and carbon dioxide—a process referred to as *acetogenesis.* Subsequently, several species of strictly anaerobic bacteria bring about *methanogenesis*—a process in which hydrogen produces methane from acetate and carbon dioxide reduction. In this stage, the stabilization of the organic material truly occurs. Figure 6.7.1 shows the two stages of anaerobic treatment as sequential processes; however, both stages occur simultaneously and synchronously in an active, well-buffered system.

The main concern of a wastewater treatment facility in operating an anaerobic system is that the various bacterial species function in a balanced and sequential way (Forster 1985). Hence, although other types of microorganisms may be present in the reactors, attention is focussed mostly on the bacteria.

The major groupings of bacteria, as numbered in Figure 6.7.1, and the reactions they mediate are as follows (Pavlostathis and Giraldo-Gomez 1991): (1) fermentative bacteria, (2) hydrogen-producing acetogenic bacteria, (3) hydrogen-consuming acetogenic bacteria, (4) carbon-dioxide-reducing methanogens, and (5) aceticlastic methanogens. Two common genera of aceticlastic methanogens are *Methanothrix* and *Methanosarcina*; and species from the *Methanobacterium* group are commonly known to produce methane by hydrogen reduction of carbon dioxide.

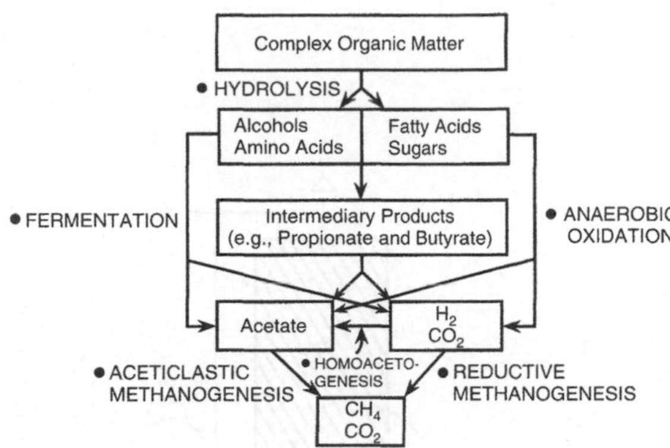

FIG. 6.7.1 Reaction pathways of anaerobic treatment of complex organic matter.

Environmental Factors

Facultative and anaerobic bacteria associated with the acid fermentation process are tolerant to changes in pH and temperature and have a higher growth rate than the methanogenic bacteria from the second stage. Hence, the methane fermentation stage is the rate-limiting step in anaerobic processes. Since methane fermentation controls the process rate, maintaining optimal operating conditions in this stage is important.

Within the pH range of 6.0–8.5, the rate of methane fermentation is somewhat constant; outside this range, the rate drops dramatically (Benefield and Randall 1980). Other research has shown that the optimum pH range is 6.8 to 7.4 (Ramalho 1983). The alkalinity produced from the degradation of organic compounds in the anaerobic process helps control the pH by buffering the anaerobic system. The alkalinity, at typical fermentation pH levels of approximately 7, is primarily in the form of bicarbonates. Carbon dioxide comprises 30–40% by volume of the off gas from anaerobic treatment. Thus, within the operating pH range of 6.6–7.4, the alkalinity concentration can vary from 1000 to 5000 mg/l as calcium carbonate.

Another parameter requiring control is the reactor retention time for methanogens; it must be adequate to prevent cell wash-out. Research shows that the required retention time varies from 2 to 20 days (Ramalho 1983).

In the methanogenesis stage, approximately 70% of methane produced is formed from the methyl group of the acetate by acetophilic methanogens, while the remainder of the methane is formed from the oxidation of hydrogen by hydrogenophilic methanogens. The partial pressure of hydrogen is thought to regulate both intermediate fatty acid catabolism and methane formation (Forster 1985). Thus, the methanogens must maintain a low hydrogen concentration.

Hydrogen is also an inhibitory substance in methane production when a high concentration of sulfate ions is present. Sulfate-reducing bacteria, such as *Desulfovibrio,* compete for acetate and hydrogen and use them more effectively than methanogens to convert sulfate to sulfide. Therefore, the methane production is diminished. A secondary inhibition of methanogenesis occurs if the soluble sulfide ion concentration becomes greater than 200 mg/l (Forster 1985).

Cation concentration has been shown to affect the rate of methane formation (Benefield and Randall 1980). At low concentrations, cations stimulate the fermentation rate. However, the rate decreases when the optimum concentration is exceeded. The intensity of rate reduction depends on the extent that the optimum concentration is exceeded. For example, concentrations of calcium within the range of 100–200 mg/l have a stimulatory effect, while concentrations between 2500–4500 mg/l are moderately inhibitory, and concentrations of 8000 mg/l or higher are strongly inhibitory to methane fermentation (Benefield and Randall 1980).

Ammonia concentrations have a similar effect on the rate of methane fermentation as cation concentrations. However, one distinction is that the process pH determines the distribution between free ammonia and the ammonium ion. High pH levels favor free ammonia—the toxic form of ammonia. Table 6.7.1 shows some optimum environmental factors for methane fermentation.

Treatment Processes

The anaerobic wastewater treatment processes discussed in this section include the anaerobic contact process, the USB reactor, the anaerobic filter, and the AFBR.

ANAEROBIC CONTACT PROCESS

The anaerobic contact process is a suspended-growth process, similar in design to the activated-sludge process except that anaerobic conditions prevail in the former process. Part A in Figure 6.7.2 shows the process schematic.

The anaerobic contact process is comprised of two parts. The contact part involves thorough mixing of the wastewater influent with a well-developed anaerobic sludge culture. The separation part involves the settling out

TABLE 6.7.1 ENVIRONMENTAL FACTORS FOR METHANE FERMENTATION

Parameter	*Optimum*	*Extreme*
Temperature (°C)	30–35	25–40
pH	6.8–7.4	6.2–7.8
Volatile acids (mg/l as acetic acid)	50–500	2000
Alkalinity (mg/l as $CaCO_3$)	2000–3000	1000–5000

Source: Adapted from L.D. Benefield and C.W. Randall, 1980, *Biological process design for wastewater treatment* (Englewood Cliffs, N.J.: Prentice-Hall).

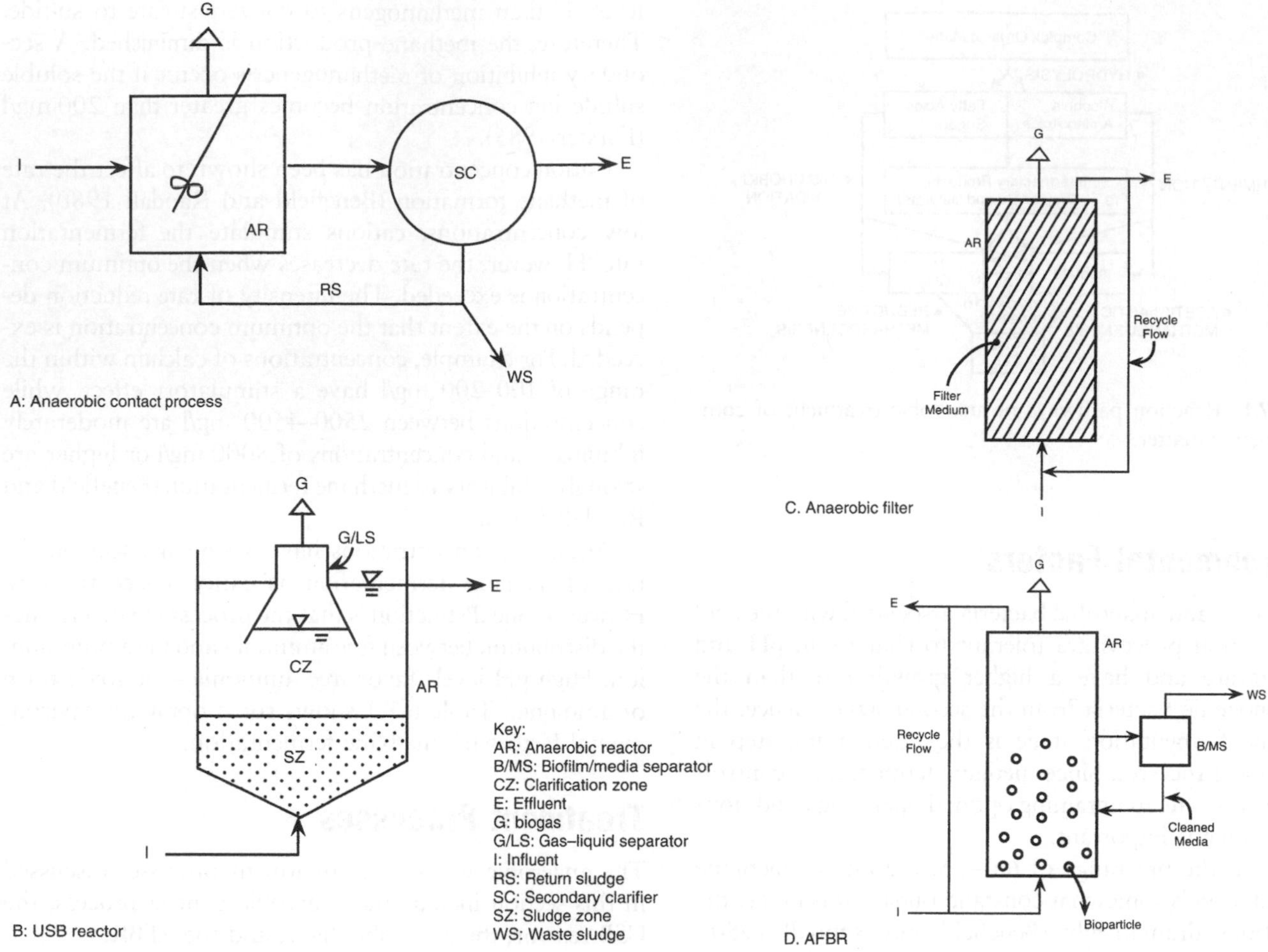

FIG. 6.7.2 Process schemes of anaerobic treatment processes. **A,** Anaerobic contact process; **B,** Upflow sludge blanket reactor; **C,** Anaerobic filter; **D,** Anaerobic fluidized bed reactor.

of anaerobic sludge from the treated wastewater and recycling back to the contact reactor. The process usually has a vacuum degasifier placed following the aerobic reactor to eliminate gas bubbles that cause SS in the clarifier to float.

BIOENERGY and *ANAMET* are two commercially available, proprietary anaerobic contact processes. *BIOENERGY* is a conventional anaerobic contact process that uses a thermal shock procedure to facilitate sludge separation. As the mixed liquor, at 35°C, flows from the contact reactor to the settling unit, a series of heat exchangers rapidly decreases its temperature to 25°C. This temporarily interrupts gasification allowing effective sludge–solids separation by gravity. The temperature of the recycled sludge is increased before it is returned to the contact unit.

In the *ANAMET* process, an aerobic biological treatment polishing step follows the anaerobic contact process to provide near-complete organics removal. The process recycles the sludge produced in the aerobic treatment process back to the anaerobic reactor to reduce excess sludge production across the entire system and increase biogas yield. Also, recirculation of the nutrient-containing sludge from the aerobic reactor reduces external nutrient requirements in the anaerobic reactor (Shieh and Li 1987).

Design considerations for the anaerobic contact unit are similar to those described for the activated-sludge process in Section 6.4. In addition, separation unit design can follow the prescriptions for secondary clarifiers described in Section 6.8.

USB REACTOR

USB reactor is essentially a suspended-growth reactor, but it is also a fixed-biomass process. Part B in Figure 6.7.2 shows the process schematic. This USB system is based on the development of a sludge blanket. In this sludge blanket, the component particles are aggregated to withstand the hydraulic shear of the upwardly flowing wastewater without being carried upwards and out of the reactor. The sludge flocs must be structurally stable so that hydraulic shear forces do not break them into smaller portions that can be washed out, and they should also have good settlement properties.

The wastewater is fed at the bottom of the reactor, and active anaerobic sludge solids convert the organics into methane and carbon dioxide. The anaerobic biomass is distributed over the sludge blanket and a granular sludge bed. The sludge solids concentration in the sludge bed is high—100,000 mg/l SS—and does not vary over a range of process conditions (Shieh and Li 1987). The sludge solids concentration in the sludge blanket is lower and depends on process conditions (Li 1984). The reactor can include an internal baffle system, usually referred to as a gas–liquid separator, above the sludge blanket to separate the biogas, sludge, and liquid. A patented USB reactor called the *BIOTHANE* process was developed by the Biothane Corporation in the United States.

In general, the USB process can achieve high COD removal efficiency at volumetric COD loadings up to 2.0 $lb/ft^3/day$ and hydraulic retention times of 4 to 24 hr for a variety of wastewater (Shieh and Li 1987). The formation of granular sludge particles is essential to adequate reactor performance (Shieh and Li 1987). Granular sludges have a good settlement rate and can form a compact sludge bed with a solids concentration of 40–150 g/l (Forster 1985). Researchers report that the presence of calcium ions, adequate mixing in the sludge zone, and a low concentration of poorly flocculating suspended matter in the wastewater contribute to the formation of granular sludge particles with favorable qualities (Shieh and Li 1987).

The design of an USB reactor must provide an adequate sludge zone since most of the biomass is retained there. The sludge zone is completely mixed because the wastewater is fed into the reactor through a number of regularly spaced inlet ports (Shieh and Li 1987). Hence, the volume of the sludge zone can be determined with Equation 6.4(6).

The sludge blanket zone is another completely mixed contact unit in sequence with the sludge zone. If no waste conversion occurs in the gas–liquid separator and because the sludge blanket zone receives the effluent COD concentration from the USB reactor, the volume of the sludge blanket zone can also be determined with Equation 6.4(6). Thus, the total volume of the USB reactor, not including the gas–liquid separator, is the sum of the sludge zone volume and the sludge blanket zone volume.

The use of a gas–solids separation system in the upper portion of a USB reactor is claimed to be an essential feature, regardless of the settlement characteristics of the sludge (Forster 1985). An investigation of the effects of hydraulic and organic loading rates on the solids and design values of a particular plant is necessary to prove this claim. An increase in organic loading results in increased gas production, reduced floc density, and a greater tendency for floc flotation. The net result is a greater probability of solids wash-out (Forster 1985). This condition is exaggerated at high hydraulic loading rates. Hence, an evaluation of whether solids wash-out significantly depletes sludge solids and whether the solids concentration is tolerated in the final effluent is necessary to determine if a gas–liquid separator is required.

Researchers have reported that USB reactor performance is limited by the ability of the gas–liquid separator to retain sludge solids in the system (Hamoda and Van der Berg 1984). They maintain that the amount of sludge solids retained in the reactor increases with an increased gas–liquid volume in the separator. Lettinga et al. (1980) detail design information for a gas–liquid separator and the start up guidelines for the USB reactor.

ANAEROBIC FILTER

In an anaerobic filter reactor, the growth-supporting media is submerged in the wastewater. Anaerobic microorganisms grow on the media surface as well as inside the void spaces among the media particles. The media entraps the SS present in the influent wastewater that can be fed into the reactor from the bottom (upflow filter) or the top (downflow filter) as shown in the process schematics in Part C of Figure 6.7.2. Thus, the flow patterns in the filter can be either PF or completely mixed depending on recirculation magnitude. Periodically backwashing the filter solves bed-clogging and high-head-loss problems caused by the accumulation of biological and inert solids. *BACARDI* and *CELROBIC* are two proprietary anaerobic filter processes currently available.

The anaerobic filter process is effective in treating a variety of industrial wastewater (Shieh and Li 1987). An advantage of using the filter process for industrial wastewater treatment is that the filter reactor can retain the active biomass within the system for an extended time period. The long sludge-retention time maintained by the reactor allows ample time for aerobic microorganisms to remove organics in the wastewater, and there is no appreciable loss of the active biomass from the system until the filter is saturated (Shieh and Li 1987). In addition, the anaerobic filter minimizes operational concerns of sludge wasting and disposal because the synthesis rate of excess biomass under anaerobic conditions is low (Young and McCarty 1968).

Because it can retain a high concentration of active biomass within the system for an extended time period, the anaerobic filter can easily adapt to varied operating conditions (e.g., without significant changes in effluent quality and gas production due to fluctuations in parameters such as pH, temperature, loading rate, and influent composition). Also, intermittent shutdowns and complications in industrial treatment will not damage the filter since it can be fully recovered when it is restarted at a full load (Shieh and Li 1987).

A problem associated with the filter's ability to retain the biomass for a long time period is the close control of biomass holdup. Although periodic backwashing of the filter is a feasible method for maintaining the biomass holdup at the required level, more efficient techniques are needed.

Environmental engineers can determine the size of an anaerobic filter from the volumetric loading approach or from the biofilm kinetic theory approach described next. From the design approach commonly used in heterogeneous catalytic processes, the following expressions describe the overall substrate utilization rate for a completely mixed anaerobic filter:

$$R_o = (kSX_s)/(K_s + S) + (\eta k'S)/(K_s + S) \quad 6.7(1)$$

$$k' = \rho k A \delta \quad 6.7(2)$$

where:

- R_o = the overall substrate utilization rate, mass/volume-time
- X_s = suspended biomass concentration, mass/volume
- η = the effectiveness factor that defines the degree of diffusional limitations of the biofilm
- k = the maximum substrate utilization rate in the biofilm, mass/volume-time
- ρ = the biofilm dry density, mass/volume
- A = total biofilm surface area per unit filter volume, l/length
- δ = biofilm thickness, length

Expressions for the effectiveness factor have been developed by researchers Atkinson and How (1974) as follows:

$$\eta = 1 - [\tanh(B)/B]\{[\Phi/\tanh(\Phi)] - 1\} \quad \text{for } \Phi \le 1 \quad 6.7(3)$$

$$\eta = 1/\Phi - [\tanh(B)/B]\{[1/\tanh(\Phi)] - 1\} \quad \text{for } \Phi \ge 1 \quad 6.7(4)$$

$$\Phi = [0.709(S/K_s)]/[1 + (S/K_s)]\,\{(S/K_s) - \ln[1 + (S/K_s)]^{-0.5}\} \quad 6.7(5)$$

$$B = [(\delta^2 k')/(DK_s)]^{0.5} \quad 6.7(6)$$

where D = the effective diffusivity of substrate in the biofilm, area/time.

Hence, the filter volume can be calculated from the following equation:

$$V = (QS_o)/R_o \quad 6.7(7)$$

For a PF anaerobic filter, environmental engineers can calculate the overall substrate utilization rate using the CFSTR-in-series model (see Section 6.4) as follows:

$$V_i = (QS_{i-1})/(R_{oi}) \quad \text{for } i = 1, 2, 3, \ldots, n \quad 6.7(8)$$

$$V = \Sigma V_i \quad 6.7(9)$$

Applying Equations 6.7(1) to 6.7(6) to each reactor i calculates R_{oi}.

AFBR

The AFBR is an expanded-bed reactor that retains media in suspension from drag forces exerted by upflowing wastewater. Part D in Figure 6.7.2 shows the process schematic. Fluidization of the media particles provides a large surface area where biofilm formation and growth can occur.

The media particles have a high density resulting in a settling velocity that is high enough so that high-liquid-velocity conditions can be maintained in the reactor. However, the media particles' overall density decreases as biomass growth accumulates on the surface area. The decrease in density can cause the bioparticles to rise and be washed out of the reactor. To prevent this situation, the reactor controls fluidized-bed height at a required level by wasting a corresponding amount of overgrown bioparticles. The wasted bioparticles can then be received by a mechanical device that separates the biomass from the wasted media particles. The cleaned particles can then be returned to the reactor, while the separated biomass is wasted as sludge.

The AFBR combines a suspended-growth system and an attached-growth system since biomass growth attaches to the media particles which are suspended in the wastewater. The reactor recycles a portion of the effluent flow ensuring uniform bed fluidization and sufficient substrate loading.

An advantage of the AFBR is that it employs small fluidized media that provide a high biomass holdup in the reactor, reducing hydraulic retention time. The AFBR also prevents bed-clogging and high-pressure drops—complications associated with anaerobic filters. Due to the flexibility provided by bed-height control in an AFBR, a constant biomass concentration can be maintained in the reactor independent of substrate loadings (Shieh and Keenan 1986). Another advantage of the AFBR is that it is insensitive to variations in influent pH, temperature, and waste loading because it maintains a high biomass holdup and completely mixed conditions inside the reactor.

Some commercially available AFBR processes include the *ANITRON* system developed by Dorr-Oliver, Inc.; the *BIOJET* process, which employs an AFBR with an enlarged top section; and the *ENSO-FENOX* process, which combines an AFBR with a trickling filter. The AFBR has been applied to a variety of industrial treatment processes with substrates such as molasses, synthetic sucrose, sweet whey, whey permeate, glucose, and acid whey.

Since high circulation rates are used in AFBR operation (especially for treating high-strength industrial wastewater) the reactor can be designed as a completely mixed, heterogenous process. Because most of the active biomass is retained in the fluidized-bed, the contribution of suspended biomass growth to overall reactor performance is insignificant. Thus, the first term in Equation 6.7(1) can be removed, and the remaining expression calculates the overall substrate utilization rate in an AFBR as follows:

$$R_o = (\eta k'S)/(K_s + S) = (\eta kXS)/(K_s + S) \quad 6.7(10)$$

$$V = (QS_o)/R_o \quad 6.7(11)$$

Environmental engineers can use Equations 6.7(3) through 6.7(6) to determine the effectiveness factor associated with an AFBR in a similar manner as for the anaerobic filter. To determine the reactor biomass concentration (X), they can use solid–liquid fluidization correlations. These correlations link the particle concentration in the fluidized state to the measurable physical characteristics of a fluidized-bed process. The following Richardson–Zaki correlation is widely used:

$$U/U_t = \varepsilon^n \quad 6.7(12)$$

where:

- U = superficial upflow velocity of the wastewater through an AFBR, distance/time
- U_t = bioparticle terminal settling velocity, distance/time
- = bed porosity
- n = the expansion index

Shieh and Chen (1984) propose two empirical correlations relating U_t and n to the Galileo number (N_{Ga}) that defines the physical characteristics of an AFBR as follows:

$$U_t = 5753.71(N_{Ga})^{-0.8222} \quad 6.7(13)$$

$$n = 47.36(N_{Ga})^{-0.2576} \quad 6.7(14)$$

$$N_{Ga} = [8(r_p)^3(\rho_p - \rho)\rho g]/\mu^2 \quad 6.7(15)$$

$$\rho_p = \rho_m(r_m/r_p)^3 + [\rho/(1 + P)][1 - (r_m/r_p)^3] \quad 6.7(16)$$

where:

- ρ_p = bioparticle density, mass/volume
- ρ = wastewater density, mass/volume
- g = gravitational acceleration, distance/time^2
- μ = wastewater dynamic viscosity, mass/time–distance
- ρ_m = media density, mass/volume
- P = biofilm moisture content
- r_m = support media radius, length
- r_p = bioparticle radius, length

The following equation calculates the AFBR biomass concentration:

$$X = \rho(1 - \varepsilon)[1 - (r_m/r_p)^3] \quad 6.7(17)$$

The choice for media types should be based on the following media characteristics:

- A large surface area for microbial growth
- A large void space to accommodate the accumulation of biological and inert solids and minimize short-circuiting
- Inertness to biological and chemical reactions
- Resistance to abrasion and erosion
- A light weight

Small media should be used since they provide large surface-to-volume ratios, and thus, a greater surface area for biofilm growth without increasing reactor volume. Small media are also easier to fluidize, reducing the circulation requirements which decreases the shearing effects and allows a more quiescent environment for optimal biofilm growth. Silica sand, anthracite coal, activated carbon, stainless-steel wire spheres, and reticulated polyester foams are some of the media that can be considered for AFBR applications. Yee (1990) reports on the effects of various media types on AFBR performance and kinetics.

ANAEROBIC LAGOONS

Anaerobic microorganisms do not require DO in the water to function. They obtain their oxygen requirement from the oxygen chemically contained in organic materials.

Anaerobic decomposition involves two separate but interrelated steps. First, the acid-producing bacteria decompose the dissolved organic waste to organic acids, such as acetic, propionic, and butyric acid (see Figure 6.1.4). The organic acids are then further decomposed by methane-producing bacteria to the end products of methane, carbon dioxide, and water. Effective operation requires a balance between acid production and breakdown because methane producers are sensitive to the concentration of volatile acids.

As a general rule, inhibition occurs at volatile acid concentrations in excess of 2000 mg/l. This tolerance level also depends on the concentration of ammonia and other cations. The maximum alkalinity concentration is approximately 2000 mg/l as $CaCO_3$. As an operational guide, the alkalinity concentration should be greater than 1.67 times the volatile acids concentration.

The decomposition of organic acids to methane and carbon dioxide can be generalized as follows:

$$C_nH_aO_b + \left(n - \frac{a}{4} - \frac{b}{2}\right) H_2O \longrightarrow \left(\frac{n}{2} - \frac{a}{8} + \frac{b}{4}\right) CO_2 + \left(\frac{n}{2} + \frac{a}{8} - \frac{b}{4}\right) CH_4 \quad 6.7(18)$$

At a standard temperature and pressure, 1 lb of BOD removed yields 5.62 ft^3 methane. Anaerobic decomposition processes are summarized in Figure 6.1.4.

A portion of the waste material is used by the anaerobic biosystem as a source of energy and in the synthesis of new bacterial cells. Cell synthesis is affected by the type of waste being treated, but generally, for every pound of BOD destroyed by the anaerobic process, approximately 0.1 lb of new cells is produced compared to 0.5 lb in the aerobic process. Therefore, the sludge or solids buildup is less in the anaerobic system. When anaerobic lagoon contents are black in color, this indicates that the lagoon is functioning properly.

—Wen K. Shieh
Van T. Nguyen

References

Atkinson, B., and S.Y. How. 1974. The overall rate of substrate uptake reaction by microbial films. II. Effect of concentration and thickness with mixed microbial films. *Trans. Instr. Chem. Engrs.* 52, no. 260.

Benefield, L.D., and C.W. Randall. 1980. *Biological process design for wastewater treatment.* Englewood Cliffs, N.J.: Prentice-Hall.

Forster, C.F. 1985. *Biotechnology and wastewater treatment.* Cambridge: Cambridge University Press.

Jewell, W.J. 1987. Anaerobic sewage treatment. *Environ. Sci. Technol.* 21, no. 14.

Lettinga, G., A.F.M. Van Velsen, S.W. Hobma, W. de Zeeuw, and A. Klapwijk. 1980. Use of the upflow blanket (USB) reactor concept for biological wastewater treatment, especially for anaerobic treatment. *Biotechnol. Bioeng.* 22, no. 669.

Li, A.Y. 1984. *Anaerobic process for industrial wastewater treatment.* Short Course Series, no. 5. Taichung, Taiwan: Department of Environmental Engineering, National Chung Hsing University.

Pavlostathis, S.G., and E. Giraldo-Gomez. 1991. Kinetics of anaerobic treatment: A critical review. *CRC Crit. Rev. Environ. Control* 21, no. 411.

Ramalho, R.S. 1983. *Introduction to wastewater treatment processes.* 2d ed. New York: Academic Press.

Shieh, W.K., and J.D. Keenan. 1986. Fluidized bed biofilm reactor for wastewater treatment. In *Advances in biochemical engineering/ biotechnology 33,* edited by A. Flechter. Berlin: Springer-Verlag.

Shieh, W.K., and A.Y. Li. 1987. High-rate anaerobic treatment of industrial wastewaters. In *Global bioconversions 3,* edited by D.L. Wise, 41–79. Boca Raton, Fla.: CRC Press, Inc.

Yee, C.J. 1990. Effects of microcarriers on performance and kinetics of the anaerobic fluidized bed biofilm reactor. Ph.D. diss., Department of Systems, University of Pennsylvania, Philadelphia.

Young, J.C., and P.L. McCarty. 1968. The anaerobic filter for wastewater treatment. *J. Water Poll. Control Fed.* 41, no. R160.

6.8 SECONDARY CLARIFICATION

REATMENT TYPE

a. Following air-activated-sludge
b. Following oxygen-activated-sludge
c. Following extended-aeration
d. Following trickling filter
e. Following secondary RBC
f. Following nitrification RBC

TANK GEOMETRY

Rectangular, circular, or square

OVERFLOW RATE (gal/ft²-day)

AVERAGE

a, b, & e: 400–800; c: 200–400; d & f: 400–600

PEAK

a, b, d, & e: 1000–1200; c: 600–800; f: 800–1000

SOLIDS LOADING (lb/ft²-day)

AVERAGE

a & e: 0.8–1.2; b: 1–1.4; c: 0.2–1; d & f: 0.6–1

PEAK

a, b, & e: 2; c: 1.4; d & f: 1.6

DEPTH (ft)

a–c: 12–20; d–f: 10–15

An important aspect of biological wastewater treatment is secondary clarification. First, biological solids (sludge) produced during biological wastewater treatment must be separated from treated effluent prior to final discharge. Therefore, a secondary clarifier must have an adequate clarification capacity to insure that SS discharge requirements are met. Second, since maintaining proper SRTs is important in the operation of activated-sludge processes, secondary clarifiers in the activated-sludge process must have an adequate thickening capacity to produce the required underflow density for sludge recirculation.

Process Description

Biological solids removal is essentially accomplished by gravity settling (see Figure 4.4.5). However, biological solids can settle differently depending on their origins and characteristics. The sloughed solids produced from trickling filters and RBCs are generally large and heavy. Therefore, their settling motion is discrete (i.e., not influenced by the motion of neighboring particles, as shown in Figure 4.4.6) and can be described by Stokes' Law (Metcalf and Eddy, Inc. 1991).

On the other hand, biological flocs produced in activated-sludge processes undergo some flocculation with neighboring particles during the settling process. As flocculation occurs, the mass of particles increases and settles faster (see Figure 4.4.7). As a result, the settling process is classified as flocculant settling (Metcalf and Eddy, Inc. 1991). Since many complex mechanisms are involved during flocculation and their interactions are difficult (if not impossible) to define, the analysis of flocculant settling requires experimental data obtained from settling tests (see Figure 4.4.8).

Design

TRICKLING FILTERS AND RBCs

Secondary clarifiers used with trickling filters and RBCs provide effluent clarification by removing large, sloughed solids. Therefore, the design criteria are based on particle size and density.

Environmental engineers can formulate the gravitational force (F_G) and the frictional drag force (F_D) acting on a spherical particle settling through a liquid using the classic laws of Newton and Stokes, respectively, as follows:

$$F_G = (\rho_s - \rho)gV_s \quad 6.8(1)$$

$$F_D = C_D A_s \rho v^2/2 \quad 6.8(2)$$

where:

ρ_s = particle density
ρ = liquid density
g = gravitational acceleration
V_s = particle volume
C_D = drag coefficient
A_s = projected area of the particle perpendicular to the direction of settling
v = particle settling velocity

The particle settling velocity v becomes the terminal settling velocity v_c when $F_G = F_D$. Therefore, the following equation applies

$$v_c = [4gd(\rho_s - \rho)/3\rho C_D]^{0.5} \quad 6.8(3)$$

where d = particle diameter ($1.5V_s/A_s$).

Under laminar flow conditions, Equation 6.8(3) can be modified as follows:

$$v_c = gd^2(\rho_s - \rho)/18\mu \quad 6.8(4)$$

where μ = liquid viscosity.

For the design of continuous-flow secondary clarifiers, v_c is designated as the surface overflow rate as follows:

$$v_c = (Q + Q_r)/A = D/\theta_{sc} \quad 6.8(5)$$

where:

Q = influent wastewater flow rate
Q_r = recirculation flow rate
A = surface area of the secondary clarifier
D = depth of the secondary clarifier
θ_{sc} = LRT in the secondary clarifier

The total fraction of biological solids removed F is calculated as follows:

$$F = (1 - \chi_s) + \int_0^{\chi_s} v_s dx \quad 6.8(6)$$

where $(1 - \chi_s)$ = fraction of particles with settling velocities $\geq v_c$, and the second term at the right side of Equation 6.8(6) indicates the fraction of particles removed with settling velocities $v_s < v_c$.

ACTIVATED-SLUDGE PROCESSES

The minimum surface area required for clarification in the secondary clarifier A_c can be calculated as follows:

$$A_c = (Q + Q_r)/v_z \quad 6.8(7)$$

where v_z = the zone settling velocity, which can be determined from the following procedure. The mixed liquor with a MLVSS concentration of X is placed in a settling column. Activated-sludge begins to settle under quiescent conditions, and an interface forms between the surface of the blanket of settling sludge and the clarified liquid above. Plotting the height of this interface as a function of settling time generates a settling curve corresponding to the MLVSS concentration X. The slope of a tangent drawn to the initial portion of the settling curve yields v_z.

Applying the technique of solid flux analysis (Dick and Ewing 1967; Dick and Young 1972; Dick 1976) determines the minimum surface area required for thickening in the secondary clarifier A_T. Figure 6.8.1 is a schematic representation of a secondary clarifier operated under steady-state conditions. The mass flux of biological solids due to gravity settling N_G is calculated as follows:

$$N_G = X_i v_i \quad 6.8(8)$$

where:

X_i = biological solids concentration at location i

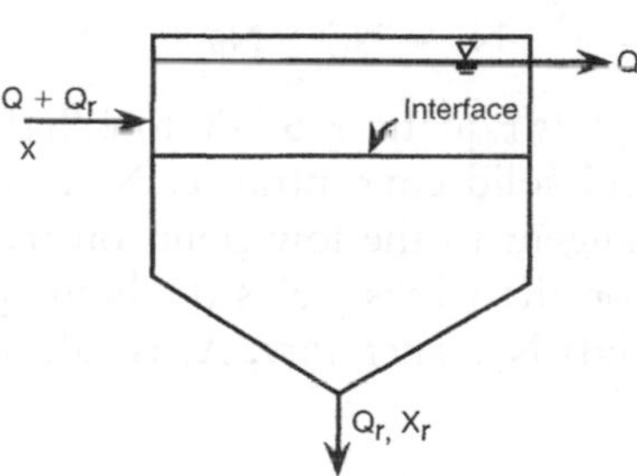

FIG. 6.8.1 Schematic representation of a continuous-flow, secondary clarifier under steady-state conditions.

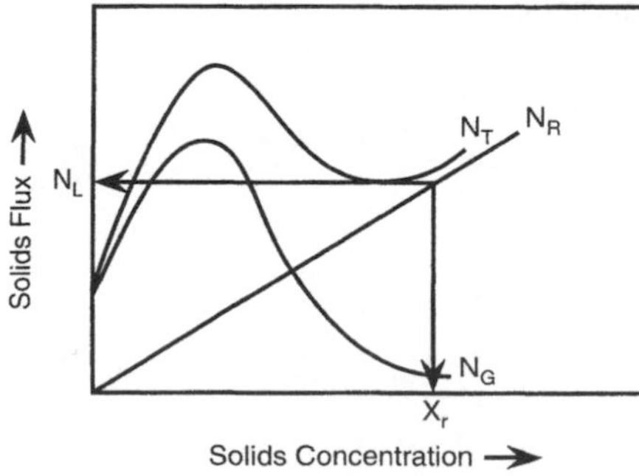

FIG. 6.8.2 Definition sketch of the solids flux analysis technique.

TABLE 6.8.1 DESIGN CRITERIA FOR SECONDARY CLARIFIERS

Biological Treatment	*Overflow Rate (gal/ft²-day)*	*Solids Loading (lb/ft²-day)*	*Depth (ft)*
Air-Activated Sludge	400–800 (avg) 1000–1200 (peak)	19.2–28.8 (avg) 48.0 (peak)	12–20
Oxygen-Activated Sludge	400–800 (avg) 1000–1200 (peak)	24.0–33.6 (avg) 48.0 (peak)	12–20
Extended Aeration	200–400 (avg) 600–800 (peak)	12.0–24.0 (avg) 33.6 (peak)	12–20
Trickling Filters	400–600 (avg) 1000–1200 (peak)	14.4–24.0 (avg) 38.4 (peak)	10–15
RBCs			
Secondary Effluent	400–800 (avg) 1000–1200 (peak)	19.2–28.8 (avg) 48.0 (peak)	10–15
Nitrified Effluent	400–600 (avg) 800–1000 (peak)	14.4–24.0 (avg) 38.4 (peak)	10–15

Source: Adapted from Metcalf and Eddy, Inc., 1991, *Wastewater engineering: Treatment, disposal, and reuse*, 3d ed. (New York: McGraw-Hill).

v_i = settling velocity of biological solids at concentration X_i

The mass flux of biological solids due to sludge recirculation N_R is calculated as follows:

$$N_R = X_i Q_r / A \qquad 6.8(9)$$

Therefore, the following equation calculates the total mass flux of biological solids N_T:

$$N_T = N_G + N_R \qquad 6.8(10)$$

Figure 6.8.2 plots Equations 6.8(8) to 6.8(10) as a function of biological solid concentration X. If a horizontal line is drawn tangent to the low point on the N_T curve, its intersection on the y-axis yields the limiting mass flux of biological solids N_L. Therefore, A_T is calculated as follows:

$$A_T = (Q + Q_r)X/N_L \qquad 6.8(11)$$

The larger value between A_C and A_T is the design value. The required sludge concentration in the recycled flow is X_r. Table 6.8.1 summarizes typical design information for secondary clarifiers.

GEOMETRIC CONFIGURATIONS

Secondary clarifiers have various geometric configurations; the most common ones are either circular (see Figure 4.4.9) or rectangular (see Figure 4.4.10). Other types include tray clarifiers, tube settlers (see Figure 4.4.11), lamella (parallel-plate) settlers, and intrachannel clarifiers (in oxidation ditches). Wastewater treatment facilities can increase existing clarifier capacity by installing inclined tubes or parallel plates.

—*Wen K. Shieh*
Van T. Nguyen

References

Dick, R.I. 1976. Folklore in the design of final settling tanks. *J. Water Poll. Control Fed.* 48, no. 633.

Dick, R.I., and B.B. Ewing. 1967. Evaluation of activated sludge thickening theories. *J. San. Eng. Div., ASCE 93,* no. 9.

Dick, R.I., and K.W. Young. 1972. Analysis of thickening performance of final settling tanks. Presented at the 27th Annual Purdue Industrial Conference, West Lafayette, IN.

Metcalf and Eddy, Inc. 1991. *Wastewater engineering: Treatment, disposal, and reuse.* 3d ed. New York: McGraw-Hill.

6.9 DISINFECTION

DISINFECTION MEANS

a. Chemical (chlorination, ozonation, and acid and alkaline treatments)
b. Physical (heating, UV irradiation, filtration, and settling)
c. Radiation (electromagentic and acoustic)

CHLORINE DOSAGE (mg/l)

8–15 mg/l (secondary effluent)

CONTACT TIME (min)

>30 minutes (peak hourly flow)

CHLORINATION TANK FLOW CONFIGURATION

PF

MAXIMUM CHLORINE RESIDUALS

0.1–0.5 mg/l (undiluted effluent)

Disinfection is the selective destruction of disease-causing organisms. Disinfection of effluents prior to discharge insures that bacteria, viruses, and amoebic cysts are reduced to acceptable levels. Many means can accomplish the disinfection of effluents: chemical agents, physical agents, mechanical means, and radiation.

The most common chemical agents used in disinfection are chlorine and its compounds. Ozone is highly effective but it does not leave residual. Acids and alkalies are sometimes used since pH >11 or pH <3 are toxic to most bacteria. Bromine, iodine, phenols, alcohols, and hydrogen peroxide are other common chemical disinfection agents. Heat and light (especially UV light) are effective physical disinfection agents. However, using heat and UV light to disinfect large quantities of effluents is cost prohibitive. The presence of suspended matter in effluents can also reduce the efficacy of UV radiation.

Preliminary and primary treatment processes used in wastewater treatment (e.g., coarse and fine screens, grit chambers, and primary clarifiers) are capable of removing or destroying a large number of bacteria (Metcalf and Eddy, Inc. 1991). Up to 75% of the bacteria in incoming wastewater can be removed or destroyed by the settling mechanism alone. However, removal and destruction of bacteria are the by-products instead of the primary functions of these treatment processes. Wastewater treatment facilities can use electromagnetic, acoustic, or particle radiation to disinfect water, wastewater, and sludge. However, these applications are limited due to the high costs involved.

Bacteria and viruses are removed or killed by disinfection and sterilization. Numerous disinfection and sterilization techniques are available. This section discusses disinfection with chlorine since it is the most common disinfectant used.

Chlorine Chemistry

Chlorine is an active element that reacts with many chemical compounds in water and wastewater to form new and often less offensive components. Hydrolysis and ionization occur when chlorine gas is added to water which forms HOCl and OCl^-, the free available chlorine. Hypochlorite salts such as $Ca(OCl)_2$ and Na(OCl) can be added to water to form free chlorine. HOCl is predominant at a pH <7.0 which is beneficial since its disinfection power is approximately 40–80 times of that of OCl^-.

HOCl reacts with ammonia in wastewater to form various chloramines (NH_2Cl, $NHCl_2$, and NCl_3), the combined available chlorine (Sawyer and McCarty 1978). Since chloramines are less effective disinfectants, additional chlorine is needed to insure the presence of a free chlorine residual. Figure 6.9.1 is a schematic illustration of the stepwise reaction phenomena that result when chlorine is added to wastewater containing ammonia (breakpoint chlorination curve).

An even more important characteristic of chlorine is that it is toxic to most pathogenic microorganisms (see

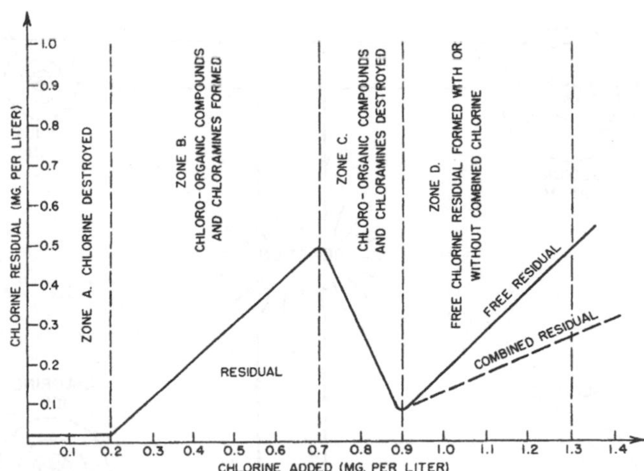

FIG. 6.9.1 Breakpoint chlorination.

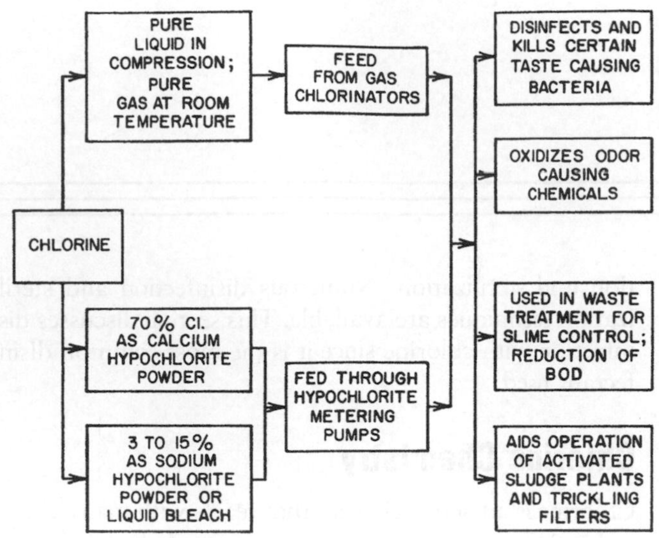

FIG. 6.9.2 Utilization of chlorine.

Figure 6.9.2). Chlorine acts as an oxidizing agent to change the character of an offending chemical. Chlorine is a strong oxidizing agent because its atoms are constructed with three shells of electrons and the outer shell (with seven electrons) has a strong tendency to acquire an eighth electron for stability. Oxidation is a process in which an atom loses electrons (see Figure 6.9.3).

PRECHLORINATION/POSTCHLORINATION

Conventional treatment plants use chlorination as the final treatment process to reduce bacterial concentrations. Prechlorination is performed on plant influent if the incoming sewage is septic or the flows are low and plant holdup time allows waste to become septic. Prechlorination is usually at a fixed dosage, and a residual chlorine level is not maintained. Postchlorination provides 15 to 30 min detention time in a baffled, closed tank to prevent short-circuiting and dissipation of the chlorine. Chlorination is not frequently used to reduce the BOD in the plant effluent. Ordinarily, a combined chlorine residual between 0.2 and 1 mg/l is the target for the final effluent.

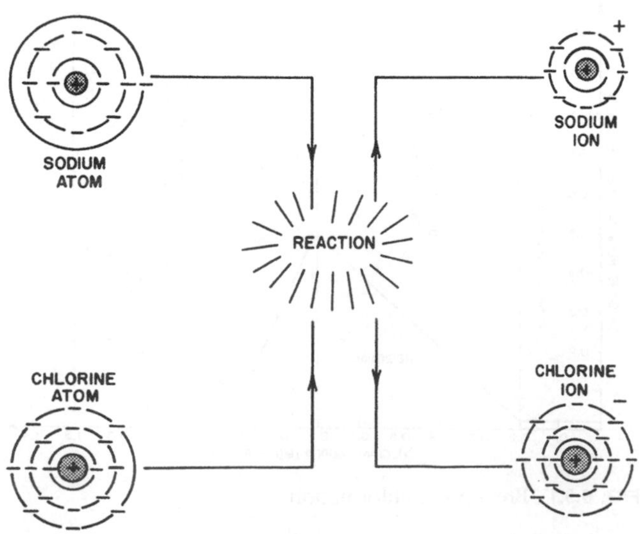

FIG. 6.9.3 A typical oxidation reaction.

Disinfection should kill or inactivate all disease-producing (pathogenic) organisms, bacteria, and viruses of intestinal origin (enteric). These microscopic entities can survive in water for weeks. The amount of chlorine required depends on the chlorine demand of the water being disinfected, the amount of disinfectant required as residual, and the detention time during which the disinfectant acts on the organisms. A free-chlorine residual of at least 2 ppm (mg/l) should be attained. Chlorinator capacities should be based on at least 30 min contact time, and water flow rates should coincide with anticipated maximum chlorine demands.

Chlorinators should be accurate over the entire feed range, and standby machines should be available. The wastewater treatment facility should determine the chlorine demand of the raw water frequently enough to adjust the chlorine feed. This procedure is best accomplished by automatic equipment that continuously determines the residual chlorine content of treated water and adjusts the chlorine dosage accordingly.

A free-chlorine residual of not less than 0.3 ppm should be maintained in the active parts of the water distribution system. Where large, open (uncovered) reservoirs of finished water exist, auxiliary disinfection must be provided. The disinfection of newly laid or extensively repaired water mains is also required, and procedures for this operation are prescribed in standards.

Disinfection by chlorine, in addition to eliminating pathogenic organisms, also controls taste and odor through breakpoint chlorination. This process is the addition of sufficient chlorine to destroy or oxidize all substances that create a chlorine demand with excess chlorine remaining in the free-residual state. Figure 6.9.1 shows this process and its effects. A free-chlorine residual is that part of the total residual remaining in the water (after a specified contact period) that reacts chemically and biologically as hypochlorous acid or a hypochlorite ion.

The effectiveness of chlorine disinfection is a function of contact time, pH, and temperature. Figure 6.9.4 shows these relationships. Providing adequate contact time and dosage is especially important with respect to viruses. Virus particles have been isolated that have escaped treatment processes, but clarification and disinfection of water afford a high measure of protection against viruses.

FACTORS AFFECTING CHLORINE DISINFECTION

The factor Ct determines chlorine disinfection efficacy, where C is the disinfectant concentration in mg/l and t is

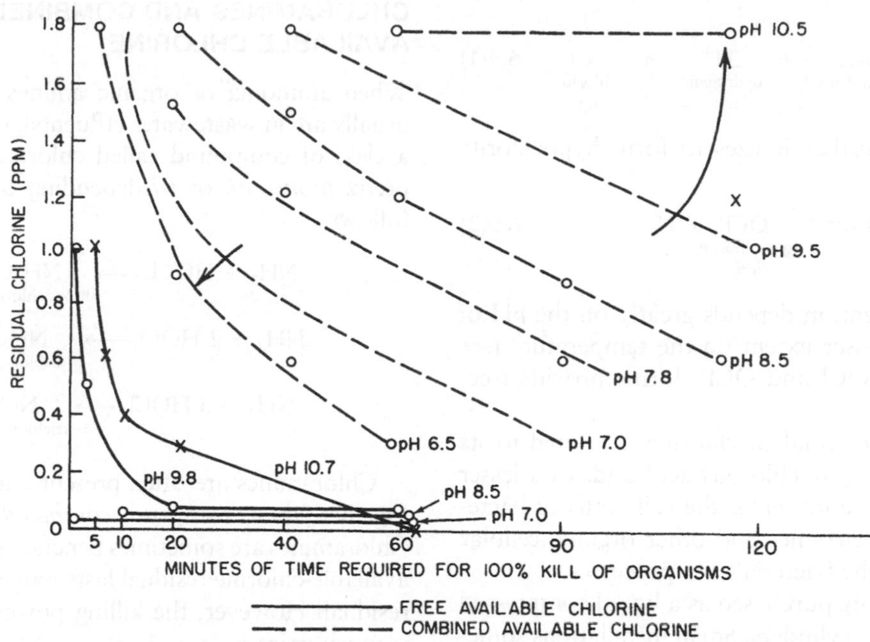

FIG. 6.9.4 Relationship between time, amount of chlorine, and bactericidal action. (Reprinted from *USPHS report.*)

the contact time in minutes. The degree of disinfection remains unchanged at a constant Ct value. This relationship is true when either of the two variables is increased while the other one is simultaneously decreased. Consequently, increasing the contact time increases the efficiency of a less effective disinfectant. Table 6.9.1 summarizes the ranges of Ct values of disinfectants for 99% inactivation of various microorganisms at 5°C.

Effective mixing of chlorine with wastewater is essential for effective bacterial kill. Diffusers inject chlorine directly into the path of the wastewater flow for effective mixing. Mechanical mixing devices insure rapid and complete chlorine mixing with effluent.

Because of the importance of contact time, wastewater treatment facilities usually use a PF chlorination tank with a back-and-forth-flow pattern similar to that shown in Figure 6.4.2 to insure that at least 80 to 90% of the effluent is retained in the tank for the specified contact time. A minimum contact time of 30 min at the peak hourly flow is recommended. A chlorine dosage of 8–15 mg/l is adequate for a well-designed tank for chlorinating secondary effluents.

Wastewater treatment facilities should keep the maximum chlorine residuals in undiluted effluents at 0.1–0.5 mg/l to protect receiving surface water systems. Consequently, dechlorination can be required to reduce chlorine residual toxicity. Sulfur dioxide added at the end of the chlorination tank oxidizes both free chlorine and chloramines to chloride. Activated-carbon adsorption of free- and combined-chlorine residuals is effective but expensive.

FREE-AVAILABLE CHLORINE

When chlorine is injected into water, it dissolves quickly, hydrolyzing to form hypochlorous acid and chloride ions

TABLE 6.9.1 RANGES OF CT VALUES OF DISINFECTANTS FOR 99% INACTIVATION OF MICROORGANISMS AT 5°C

Microorganism	*Free Chlorine (pH 6–7)*	*Chloramines (pH 8–9)*	*Chlorine Dioxide (pH 6–7)*	*Ozone (pH 6–7)*
E. coli	0.034–0.05	95–180	0.4–0.75	0.02
Polio 1	1.1–2.5	770–3700	0.2–6.7	0.1–0.2
Rotavirus	0.01–0.05	3800–6500	0.2–2.1	0.006–0.06
G. lamblia cysts	47–>150	—	—	0.5–0.6
G. muris cysts	30–630	—	7.2–19	1.8–2.0

Source: Adapted from W. Viessman, Jr. and M.J. Hammer, 1993, *Water supply and pollution control,* 5th ed. (New York: Harper Collins).

as follows:

$$Cl_2 + H_2O \rightleftharpoons \underset{\text{hypochlorous acid}}{HOCl} + \underset{\text{hydrogen ion}}{H^+} + \underset{\text{chloride ion}}{Cl^-} \quad 6.9(1)$$

(chlorine gas)

Hypochlorous acid further ionizes to form hypochlorite ions as follows:

$$HOCl \rightleftharpoons \underset{\text{hypochlorite ion}}{OCl^-} + H^+ \quad 6.9(2)$$

The extent of ionization depends greatly on the pH of the water and to a lesser extent on the temperature (see Figure 6.9.5). The HOCl and OCl^- forms provide free-available chlorine.

The disinfection potential of chlorine is related to its oxidation properties. Hypochlorous acid and, to a lesser extent, the hypochlorite ion enter the cell walls of bacteria, oxidizing certain enzymes and other organic cellular material essential to the bacteria's life processes.

Chlorine is ordinarily purchased as a liquid compressed in pressurized tanks or cylinders. Small installations sometimes use solutions of sodium or calcium hypochlorite. These chemicals also ionize to produce the hypochlorite ion. For calcium hypochlorite, the following reaction occurs:

$$\underset{\text{calcium hypochlorite}}{Ca\,(OCl)_2} \longrightarrow Ca^{2+} + 2\,OCl^- \quad 6.9(3)$$

CHLORAMINES AND COMBINED AVAILABLE CHLORINE

When ammonia or organic amines are present (as they usually are in wastewater effluents), chlorination produces a class of compound called chloramines. They have the prefix *mono, di,* or *tri* depending on their final form as follows:

$$NH_3 + HOCl \longrightarrow \underset{\text{monochloramine}}{NH_2Cl} + H_2O \quad 6.9(4)$$

$$NH_3 + 2\,HOCl \longrightarrow \underset{\text{dichloramine}}{NHCl_2} + 2\,H_2O \quad 6.9(5)$$

$$NH_3 + 3\,HOCl \longrightarrow \underset{\text{trichloramine}}{NCl_3} + 3\,H_2O \quad 6.9(6)$$

Chloramines are often present with hypochlorites, and together they comprise *combined available chlorine.* Chloramines are sometimes beneficial because a combined-available-chlorine residual lasts longer than a free-chlorine residual. However, the killing power of the free chlorine is much greater.

RATE OF KILL

Figure 6.9.6 shows the rate of kill (microorganisms) for combined- and free-chlorine residuals. This curve is based on the rate of kill expressed in the following equation (unique for chlorine):

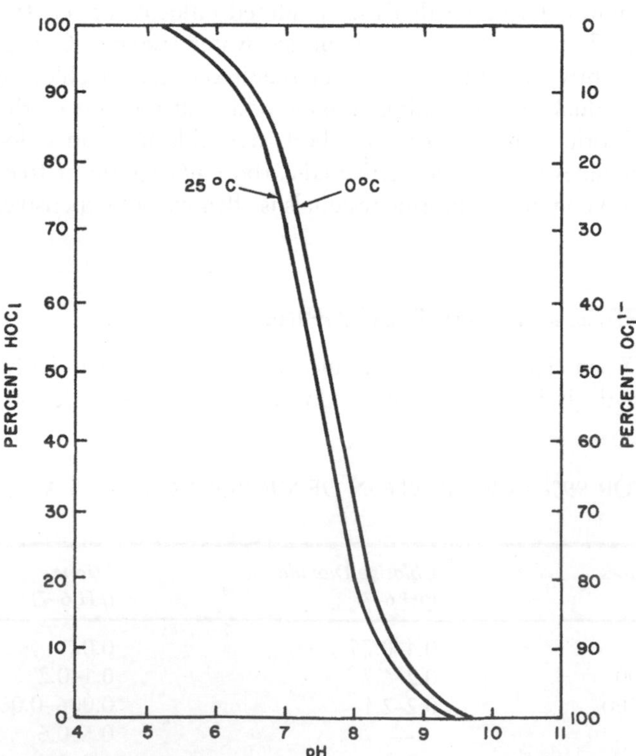

FIG. 6.9.5 Extent of hypochlorous acid (HOCl) ionization into ClO^{1-} as a function of pH and temperature.

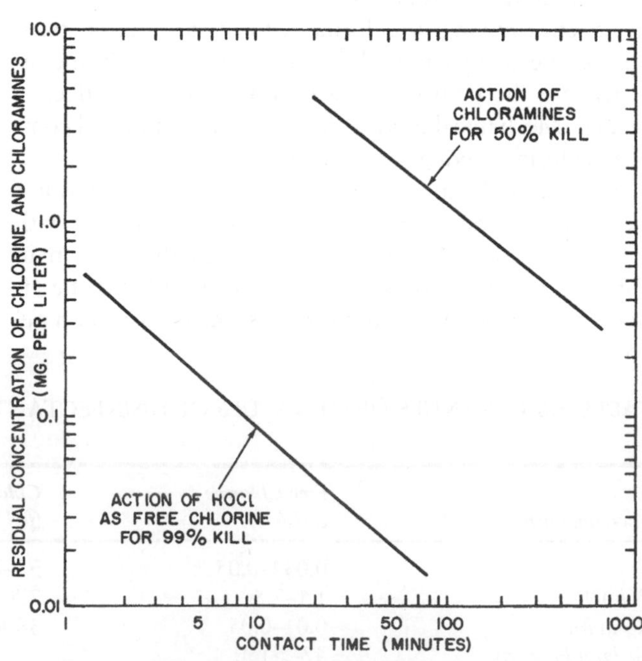

FIG. 6.9.6 Toxicity to microorganisms as a function of hypochlorous acid and chloramine contact time and concentration. Actual curves depend on pH and temperature.

$$\frac{dN}{dt} = -kNt \qquad 6.9(7)$$

where:

N = population of living microorganisms
t = time
k = rate constant

Integrating and converting to base 10 yields the following equation:

$$t_1^2 = \frac{4.6}{k} \log \frac{N_0}{N_1} \qquad 6.9(8)$$

where:

N_0 = initial population
N_1 = population at t_1

Kill rates are usually measured in terms of coliform bacteria present. Figure 6.9.7 shows the relationship between coliform bacteria and chlorine residuals for sewage; Figure 6.9.8 shows the relationship between chlorine dosages and detention times.

CHLORINE DOSAGE

In determining the amount of chlorine required for disinfection, environmental engineers must consider the pH. The state of New York recommends that for drinking water at a pH of 7.0, the concentration of free-available-chlorine residual after a 10-min detention time should be 0.2 mg/l. For a pH of 8.0 with the same detention time, the concentration of free-chlorine residual should be 0.4 mg/l. At pH values of 7 and 8, the recommended combined-available-chlorine residuals are 1.5 and 1.8 mg/l, respectively, after a contact period of 60 min.

For waste treatment plant effluents, a combined-chlorine residual of 0.5 mg/l after 15 min is generally acceptable (although some states require as much as 2 mg/l). Providing a residual of 0.5 mg/l requires a much larger dosage, depending on the efficiency of prior treatment, because the organic matter in the waste exerts an immediate chlorine demand. Table 6.9.2 lists the dosages required to produce a residual of 0.5 to 1.0 mg/l.

Chlorination System Design

In designing a chlorination system, environmental engineers must select the chlorine form, the method of feeding chlorine, the size of the chlorinator, the control system, and the auxiliary components.

Chlorine is commercially available in several forms including almost pure chlorine liquid, compressed in cylinders (or tanks for larger plants). When the liquid is released at room temperature, it vaporizes, and the gas is fed into the water stream where it readily goes into solution. This form of chlorine is the least expensive.

Another form is a powder, usually calcium hypochlorite, which is mixed with water in a plastic drum. The third form is sodium hypochlorite. Commercial liquid bleaches can also be used but are expensive since they contain only 15% or less of available chlorine.

METHODS OF FEEDING

Chlorine can be fed into the wastewater by several methods, including direct feed, solution-feed equipment, hypochlorite feeders, and vacuum-feed chlorinators.

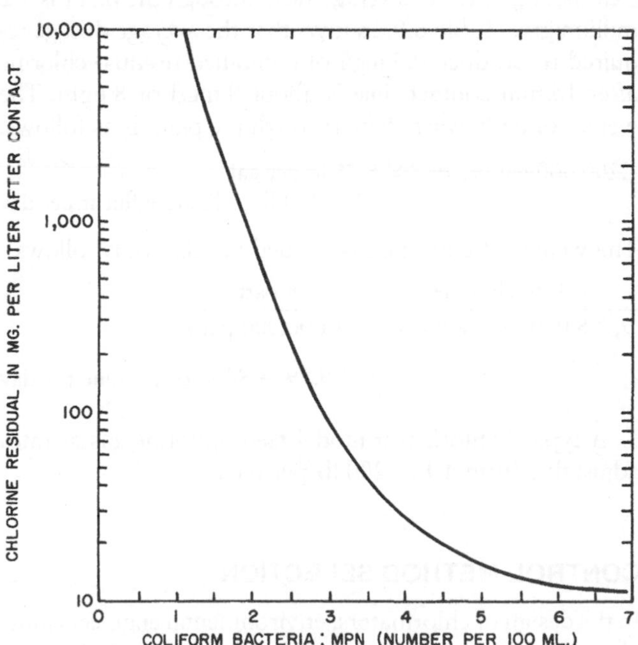

FIG. 6.9.7 Coliform kill versus chlorine residual concentration for biological waste treatment plant effluent.

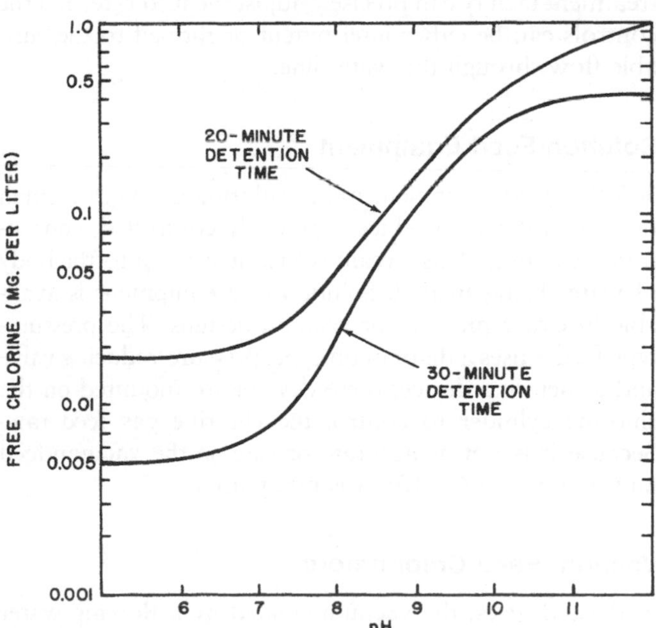

FIG. 6.9.8 Chlorine required to kill 99% of coliform bacteria at two detention times and various pH levels.

TABLE 6.9.2 CHLORINE DOSAGES FOR WASTE TREATMENT PLANT EFFLUENTS

Type of Treatment	*Approximate Dosage* (mg/l)*
Primary plant effluent	20 to 25
Trickling filter plant effluent	15
Chemical precipitation effluent	15
Activated-sludge plant effluent	8
Sand filter effluent	6

Note: *Dosages vary from plant to plant and are those required to produce 0.5 to 1.0 mg/l combined residual chlorine after 15 min.

Direct Feed

In *direct feed,* chlorine gas is fed directly to the stream or body of water being treated, and the need for electric or hydraulic power is eliminated. However, because of the risk of leakage, it is a dangerous technique and is used only in remote locations or under special circumstances.

Hypochlorite Feeders

Hypochlorite feeders are positive displacement diaphragm or metering pumps. Because of the cost of the chlorine compounds (five to eight times the equivalent cost of liquid chlorine), their use is limited to isolated areas where chlorine cylinders are difficult to obtain and the water flows are low, usually less than 100 gpm at peaks. The hypochlorite powder is mixed in plastic drums to the necessary solution strength (see Figure 6.9.9). The wastewater treatment facility can precisely adjust the feed rate, and the controls can be either intermittent or ratioed to the variable flow through the water line.

Solution-Feed Equipment

Solution-feed equipment mixes chlorine gas with a small flow of water to produce a precisely controlled, concentrated solution. This stream is then injected into the body of water being treated. Solution-feed equipment is available in either pressure or vacuum designs. The pressure-type feeder uses a diaphragm-type, pressure-reducing valve and a metering device; these devices are mounted on the chlorine cylinder to control the chlorine gas feed rate. Because it is not as accurate or safe as the vacuum-feed unit (see Figure 6.9.10), it is rarely used.

Vacuum-Feed Chlorinators

In these devices, the vacuum created by a flowing water stream through an ejector motivates the chlorine flow from the pressure cylinder. The chlorine regulator and chlorine flowmeter are between the ejector and the chlorine cylinder (see Figure 6.9.10). The partial vacuum developed by the ejector pulls the chlorine gas through the flowmeter. The regulator prevents the flowmeter inlet pressure from reaching atmospheric, eliminating the hazard of chlorine leakage. The rotameter and the rate setting valve are normally located on the face of the chlorine control cabinet, allowing for ready adjustment.

The chlorine enters the throat of the ejector venturi, where it dissolves in the flowing water. This water solution is fed to the stream or tank being treated. Most states allow only this type of gas chlorinator to be used because the regulator and meter in this design have few moving parts and are trouble free for long periods. Also, because the chlorine (except at the cylinder) is under vacuum, the chance of it escaping to the atmosphere is reduced.

This type of gas chlorinator can be obtained with feedrates of less than 1 to 8000 lb per 24 hr. The meters can be manifolded for any combination of capacities.

In addition to manual control with adjustment of the rotameter outlet valve, automatic remote control from a residual chlorine analyzer or other controller is also feasible. The control signal (pneumatic) is introduced over the middle (signal) plate of the regulator and causes a variation of the chlorine vacuum as it enters the rotameter.

CHLORINATOR SIZING

Chlorination system sizing involves determining the dosage required. This amount depends on the chlorine demand of the water or waste and on its flow rate.

The following example shows the chlorine dosage required for the effluent from an activated-sludge waste treatment plant. The average flow through the plant is 1.2 million gpd. Table 6.9.2 shows that the average dosage required to produce 0.5 mg/l of combined residual chlorine after 15 min contact time is about 8 mg/l or 8 ppm. The weight of daily waste flow through the plant is as follows:

$$1{,}200{,}000 \text{ gal per day} \times 8.34 \text{ lb per gal} = 9{,}998{,}000 \text{ lb of liquid effluent per day}$$

The weight of chlorine gas needed per day is as follows:

$$\frac{x \text{ lb of chlorine}}{9{,}998{,}000 \text{ lb of effluent}} = \frac{8 \text{ parts}}{1{,}000{,}000 \text{ parts}}$$

$$x = 8 \times 9.99 \cong 80 \text{ lb of chlorine per day}$$

A typical chlorinator model feeds chlorine gas at rates adjustable from 10 to 200 lb per day.

CONTROL METHOD SELECTION

In the design of chlorinators, environmental engineers must also select the control system. The simplest control system is a manual rate of feed control. Almost as simple is a control system with an intermittent start and stop feature. This

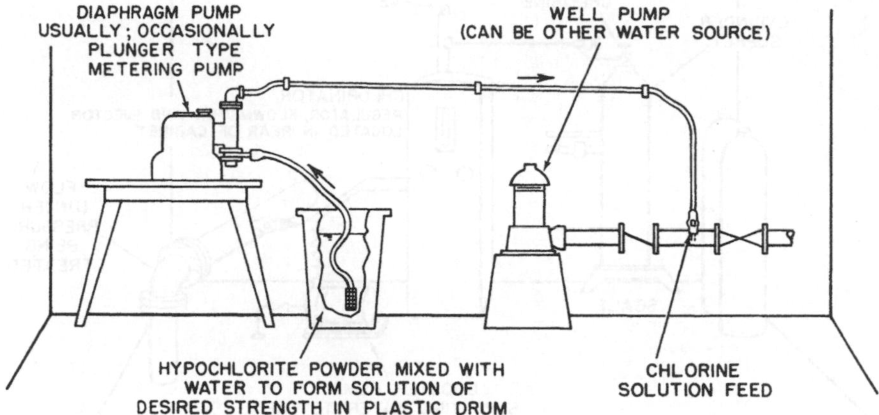

FIG. 6.9.9 Calcium hypochlorite chlorination system.

method is unsatisfactory unless the stream being treated has a constant flow rate when started (such as from a pump).

When the flow varies, as it usually does through a sewage plant, chlorinator feed should be proportionate. A ratio controller can measure the effluent flow rate, to furnish automatic modulation of chlorinator feed rate.

If the average flow is 1,200,000 gpd or 50,000 gph, the hourly peak flow can be twice that, while the minimum flow at four o'clock in the morning can easily be one-tenth that amount. Therefore, a chlorinator with an adjustable feed rate of 8 to 160 lb of chlorine per day throttled by an automatic flow ratio controller is needed.

A more complete control system includes a continuously recording, residual-chlorine analyzer installed downstream from the chlorination unit. If the residual chlorine level drops below the standard, the analyzer controller adjusts the chlorine flow upward. The wastewater treatment facility can vary the rate of feed (see the rate valve opening on Figure 6.9.10) to compensate for changes in the flow through the plant and adjust the dosage (see the control signal port on Figure 6.9.10) to compensate for changes in effluent chlorine demand.

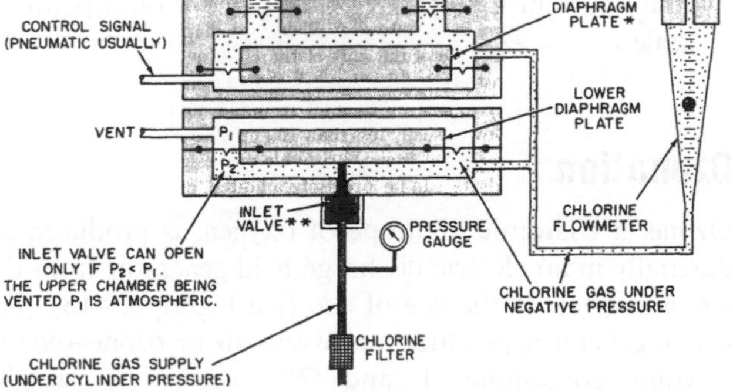

FIG. 6.9.10 Vacuum chlorinator regulator and meter. *Required when remote control signal is used; **this valve closes the chlorine supply if the pressure in the lower chamber rises to atmospheric.

AUXILIARY COMPONENTS

Environmental engineers should also consider the auxiliary components of the system, including the following:

1. A separate room or building should be designed for chlorination. Gas chlorinators should be isolated from other equipment units or areas where personnel work. (Hypochlorite feeders need not be isolated.) The chlo-

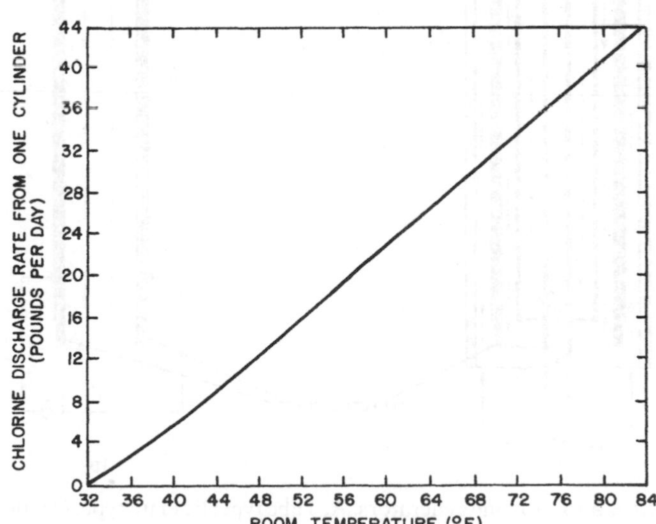

FIG. 6.9.11 Maximum chlorine feed rates at various temperatures for 100- or 150-lb cylinders (can be exceeded for short periods).

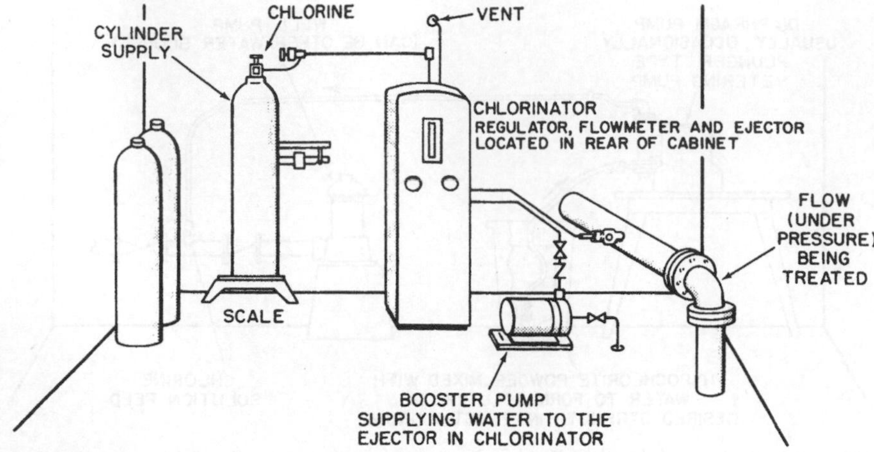

FIG. 6.9.12 Chlorinator with booster pump.

rination room should be heated to 60°F or higher to assure that chlorine remains in vapor form when released from the cylinder. Figure 6.9.11 shows the relationship between the cylinder withdrawal rate and room temperature. In the earlier example, several cylinders connected by a manifold were needed to provide the capacity.

2. The necessity of a booster pump is to inject the chlorine solution into the stream to be treated should be determined. The pressure of the chlorine solution line at the point of injection should be three times the pressure of the effluent being treated. Figure 6.9.12 shows such an installation.
3. Adequate detention should be determined. This provision is important for effective disinfection. Therefore, the environmental engineer must calculate the waste or water storage available after the point of chlorination. The storage can be in the pipeline after the chlorine injection point although it is normally in a chlorine contact tank sized with regard to pH and temperature effects.
4. Adequate mixing should be provided. This insures that the chlorine solution reaches all parts of the flow.
5. Chlorination in two stages should be provided, particularly for wastewater streams. This type provides better results than a single-stage system.
6. Adequate ventilation must be provided. Chlorine gas is extremely poisonous. Adequate mechanical ventilation of the chlorine room and venting of the chlorinator is mandatory. A gas mask in a case by the door and leak detection chemicals are also beneficial.
7. The remainder of the chlorine should be weighed. For smaller plants, a simple way to keep track of the chlorine liquid in the cylinder is to place it on a platform scale.

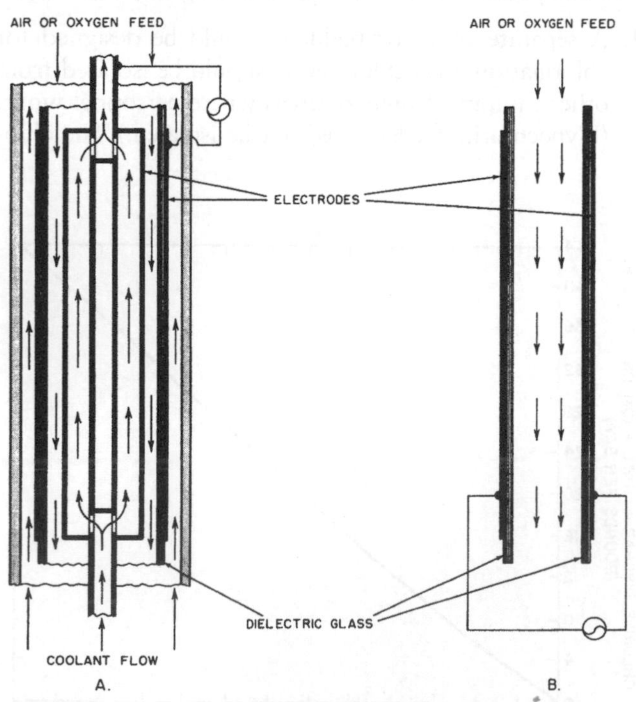

FIG. 6.9.13 Ozone generators. **A,** Tube type; **B,** Plate type. Ozone is generated in an electric discharge field by passing air or oxygen between two electrodes charged with high-voltage alternating current. A dielectric material, usually glass, is placed between the two electrodes to prevent direct discharges.

Ozonation

Ozone, a triatomic allotrope of oxygen, is produced industrially in an electric discharge field generator from dry air or oxygen at the site of use (see Figure 6.9.13). The ozone generator produces an ozone–air or ozone–oxygen mixture containing 1 and 2% ozone by weight. Wastewater treatment facilities introduce this gas mixture into the water by injecting or diffusing it into a well-baffled mixing chamber or scrubber or by spraying the water into an ozone atmosphere.

TABLE 6.9.3 OXIDATION POTENTIALS OF CHEMICAL DISINFECTANTS

	Disinfectant	*Oxidation Potential (volts)*
Ozone	$O_3 + 2H^+ + 2e^- \longrightarrow O_2 + H_2O$	2.07
Permanganate	$MnO_4^- + 4H^+ + 3e^- \longrightarrow MnO_2 + 2H_2O$	1.67
Hypobromous acid	$HOBr + H^+ + e^- \longrightarrow \frac{1}{2}Br_2 + H_2O$	1.59
Chlorine dioxide	$ClO_2 + e^- \longrightarrow ClO_2^-$	1.50
Hypochlorous acid	$HClO + H^+ + 2e^- \longrightarrow Cl^- + H_2O$	1.49
Hypoiodous acid	$HIO + H^+ + e^- \longrightarrow \frac{1}{2}I_2 + H_2O$	1.45
Chlorine gas	$Cl_2 + 2e^- \longrightarrow 2Cl^-$	1.36
Oxygen	$O_2 + 4H^+ + 4e^- \longrightarrow 2H_2O$	1.23
Bromine	$Br_2 + 2e^- \longrightarrow 2Br^-$	1.09
Hypochlorite	$ClO^- + H_2O + 2e^- \longrightarrow Cl^- + 2OH^-$	0.94
Chlorite	$ClO_2^- + 2H_2O + 4e^- \longrightarrow Cl^- + 4OH^-$	0.76
Iodine	$I_2 + 2e^- \longrightarrow 2I^-$	0.54

OZONE CHARACTERISTICS

Ozone is a powerful oxidizing agent (see Table 6.9.3). The mechanism of its bactericidal action is believed to be diffusion through the cell membrane followed by the irreversible oxidation of cell enzymes. The disinfection is unusually rapid and requires low ozone concentrations.

The viricidal action of ozone is even faster than its bactericidal effect. The mechanism by which the virus is destroyed is not yet understood. Ozone is also more effective than chlorine against spores and cysts such as *Entamoeba histolytica.*

Ozonation can accomplish disinfection and color, taste, and odor control in a single treatment step. Ozone reacts rapidly with all oxidizable organic and inorganic materials in water.

The ozone dosage for disinfection depends on the pollutant concentration in raw water. An ozone dose of 0.2 to 0.3 ppm is usually sufficient for bactericidal action only. The ozone dosage for secondary activated-wastewater-treatment-effluent disinfection is 6 or more ppm.

Ozonation leaves no disinfection residue; therefore, ozonation should be followed by chlorination in drinking water supply treatment applications. For optimum drinking water, the raw water should first be ozonated to remove color, odor, and taste and to destroy bacteria, viruses, and other organisms. Then, the water should be lightly chlorinated to prevent recontamination.

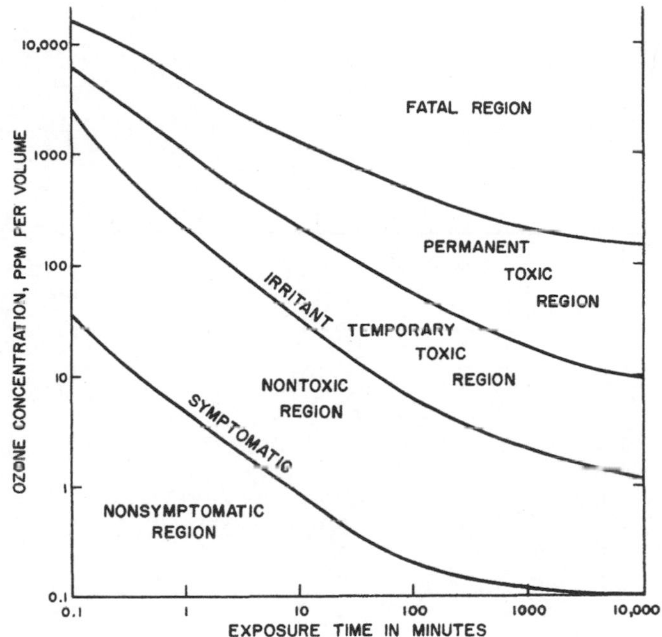

FIG. 6.9.14 Human tolerance for ozone.

DOSAGE AND PERFORMANCE

Wastewater treatment facilities can introduce the ozone-containing air or oxygen mixture produced by the ozone generator (1 or 2% ozone) into the water by injecting or diffusing it into a mixing chamber, spraying the water into an ozone-rich atmosphere, or discharging the ozone into a scrubber. Disinfection is faster and less influenced by pH and temperature variations than with chlorine.

The ozone concentration needed for disinfection depends on the chemicals and contaminants in the water and the concentration of microorganisms. A dosage of 0.2 to 0.3 ppm ozone is sufficient to kill all coliform bacteria in clean water if it is free of all oxidizable chemicals. Municipal water treatment facilities often use a dosage of 1.5 ppm to disinfect as well as remove taste, odor, and color. For the disinfection of tertiary biological sewage treatment plant effluents, a dosage of 6 ppm is sufficient. For secondary effluent, 15 ppm is required. This dosage also reduces BOD and COD.

Ozone kills viruses even more rapidly than bacteria. Electron microscopic examination indicates that viruses appear to have exploded. Ozone is also more effective than chlorine in destroying other hard to kill organisms, such as spores.

Ozone is also toxic to humans. The maximum allowable concentration in air for an 8-hr period is 0.1 ppm by volume (see Figure 6.9.14).

—*Wen K. Shieh*
Van T. Nguyen

References

Metcalf and Eddy, Inc. 1991. *Wastewater engineering: Treatment, disposal, and reuse.* 3d ed. New York: McGraw-Hill.

Sawyer, C.N., and P.L. McCarthy. 1978. *Chemistry for environmental engineering.* 3d ed. New York: McGraw-Hill.

7

Advanced or Tertiary Treatment

Don E. Burns | George J. Crits | Donald Dahlstrom | Stacy L. Daniels | Bernardo Rico-Ortega | Chakra J. Santhanam | E. Stuart Savage | Frank P. Sebastian | Gerry L. Shell | Paul L. Stavenger

7.4 COAGULATION AND EMULSION BREAKING 256

7.1 TREATMENT PLANT ADVANCES

ADVANCED OR SPECIAL PROCESSES

a) Biological—1) Completely mixed, activated sludge, 2) Ultrafiltration membrane in activated sludge, 3) Use of pure oxygen, 4) Multistage activated sludge

b) Physical—1) Electrodialysis, 2) Reverse osmosis, 3) Carbon adsorption, 4) Centrifugation

c) Chemical—1) Hydrolysis, 2) Ozone, 3) Precipitation, 4) Coagulation, 5) Flocculation

d) Heat Treatment—1) Heat treatment alone, 2) Wet-air oxidation

e) Recovery Methods—1) Recalcination, 2) Activated carbon recovery

Advanced treatment removes nutrients, trace organics, and dissolved minerals. Physicochemical treatment techniques used in conjunction with biological processes are an effective means of guaranteeing high-quality, reusable effluents. Because physicochemical methods are not overly effective in reducing soluble organic BOD on such wastewater biological processes are universally applied.

Advanced wastewater treatment plants usually consist of several unit processes operating in series. Each unit process represents an additional treatment step on the way to a treated, reusable effluent. By combining differing unit processes, these plants can attain almost any purity of effluent to meet the use of the treated wastewater.

A convenient method of describing advanced treatment plants is to use engineering flow sheets to graphically present the combination of unit processes and the direction of waste flow through them. The flow sheets in this section show treatment processes that are practical methods of treating wastewater. Some of these processes may only have been used in pilot plants; others are operating in full-scale applications.

Treatment processes can be classified according to the degree of treatment required. Advanced treatment plants have high treatment efficiencies, i.e., they produce water suitable for reuse or nonpolluting disposal.

One of the most widely used indicators of treatment efficiency is BOD. Because organic material usually exerts the largest component of BOD, BOD removal is a direct measure of the organic material removed from wastewater. Another important parameter is phosphate removal. Phosphates act as fertilizers in natural water and promote algae and plant growth. Such growth ultimately leads to anaerobic and polluted conditions.

Common categories of treatment levels are primary, secondary, and tertiary and the physicochemical treatment, which combines primary and tertiary processes. Conventional treatment usually consists of primary (SS removal) and secondary (BOD reduction) steps. Advanced treatment refers to specialized or tertiary processes. The process categories discussed in this section are subdivided into secondary equivalent (80 to 90% BOD reduction), secondary equivalent (80 to 90% BOD reduction) with phosphate removal, and tertiary equivalent (95% BOD reduction with phosphate removal).

Secondary Equivalent (80 to 90% BOD Reduction)

Most of these treatment processes were discussed in Section 5.2 because while the effluent produced is of high quality, it is not to be directly reused. Conventional primary and secondary treatment implies a primary stage of settling followed by biological treatment of a trickling filter or the activated-sludge-process type. The biological stage is usually followed by another settling step, and sludge disposal is by incineration, biological digestion, or landfill.

The San Mateo, California sewage treatment plant is an example of a primary system updated with sludge incineration. Because of odor problems from digesters and increased loads, the city converted a digester to a building to house a multiple-hearth incinerator and installed a 54-in concrete pipeline to carry the effluent into the deep channel of San Francisco Bay. While this treatment plant is not advanced, the city plans to include secondary and tertiary treatment.

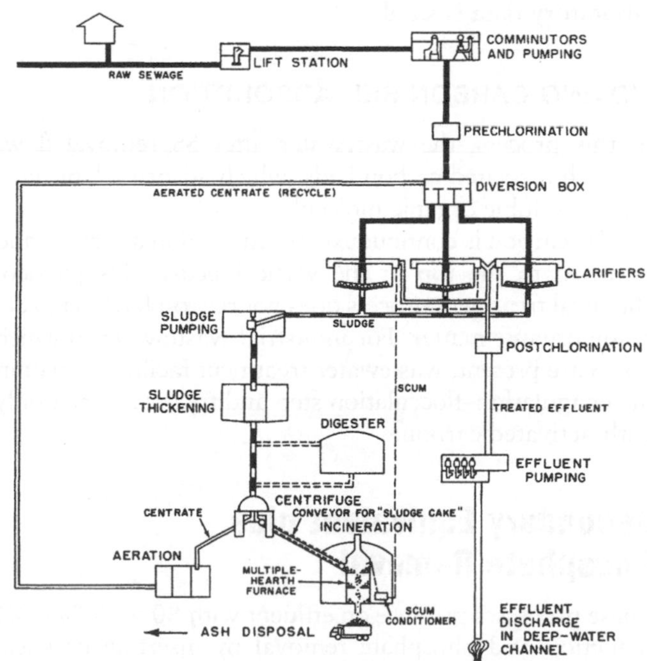

FIG. 7.1.1 Sewage treatment plant at San Mateo, California.

Figure 7.1.1 shows this upgraded, 13-mgd primary plant. The sewage is pumped into clarifiers. The clarifier scrapes floating grease and scum from the surface and feeds it by pumping to a multiple-hearth incinerator. Sludge removed from the bottom of the clarifiers is thickened (250 gpm) to about 6% and dewatered by centrifuges to 30% solids. A combination of three parts raw sludge and one part digested sludge is pumped to the centrifuges. A conveyor belt moves sludge to the top of the multiple-hearth furnace—housed in an obsolete 65-ft diameter tank. Natural gas burners supplement the heating value of the sludge plus grease and skimmings. The sterile, odor-free ash is cooled and trucked to a landfill.

PURE OXYGEN

Conventional secondary treatment usually involves an extensive aeration step, mixing wastewater with a bacterial seed. The use of pure oxygen makes the aeration steps more efficient by providing an oxygen-rich environment for the entire process.

The principle used in this process is that gas solubility in water increases linearly with the partial pressure of that gas (Henry's law). Where normal aeration can only maintain a DO level of 1 to 2 mg/l, pure oxygen provides 8 to 10 mg/l of oxygen accompanied by an increase in the driving force to replenish any oxygen used. Thus, less aeration time is needed without large increases in sludge production as occurs with other high-rate systems. The system is also less susceptible to overloading. In addition, less mixing energy is needed, and a better settling flow is formed due to the reduced turbulence. Equipment manufacturers recommend a normal scaleup factor of 0.5 when equipment for full-scale applications based on laboratory data is sized.

MOVING CARBON BED ADSORPTION

In this process, the wastewater after SS removal flows through activated carbon beds, which adsorb a large portion of soluble organic molecules.

The carbon is continuously reactivated in a furnace and returned to the top of the vertical beds. This physicochemical treatment process does not rely on bacteria to degrade organic matter. For industrial wastewater in which no SS are present, wastewater treatment facilities can omit the coagulation–flocculation step and treat the water only with activated carbon.

Secondary Equivalent plus Phosphate Removal

These processes produce an effluent with 80 to 90% BOD reduction and phosphate removal by inserting chemical and physical treatment steps either in place of or in addition to secondary treatment. The quality of effluent produced is often suitable for reuse in industry. The processes are generally more specialized and must be selected and designed for specific wastewater influents and the required effluents.

LIME PRECIPITATION OF PHOSPHORUS

Figure 7.1.2 shows a phosphate precipitation system, which is a modification of the primary settling step in a primary–secondary treatment plant. Raw wastewater is mixed with lime and flocculated to precipitate the phosphate in the form of an insoluble calcium salt. Other chemicals, such as ferric chloride, can also be used in place of lime.

The process shown in Figure 7.1.2 recirculates lime sludges for economy of lime usage. A wastewater treatment facility can also reclaim lime from the sludge by incinerating the solids. This process provides between 80 and 95% phosphate removal together with substantial BOD removal. Lime tends to soften the water, and the large amounts of lime sludge generated present greater sludge handling problems.

FERROUS PRECIPITATION OF PHOSPHATE

Figure 7.1.3 shows the use of a metal ion, such as in ferrous chloride (pickling waste), to convert soluble phosphate in wastewater into an insoluble form. A polyelectrolyte flocculant is used to help settle out phosphate complexes in the primary settling basin. Either ferric chloride or sodium aluminate can also be used in place of ferrous chloride.

The concentration of metal ion added is between 15 and 20 mg/l, and the polymer addition is about 0.4 mg/l. The process removes from 70 to 80% of the influent phosphate. When combined with biological treatment, this process achieves removals as high as 90% in full-scale trials.

The economy of the process depends largely on the cost of the metal salts. Ferrous chloride, a waste from steel

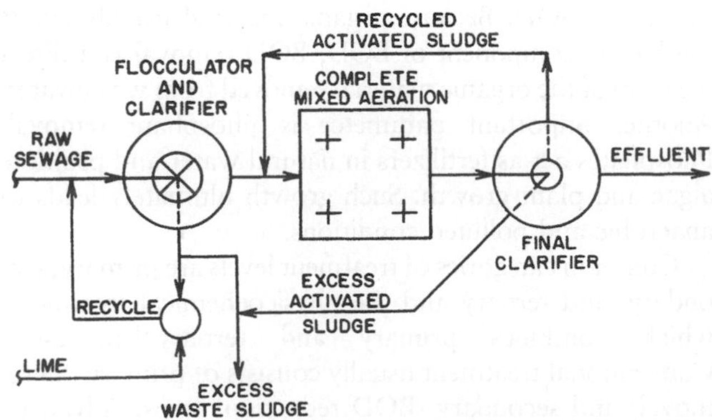

FIG. 7.1.2 Lime precipitation of phosphorus.

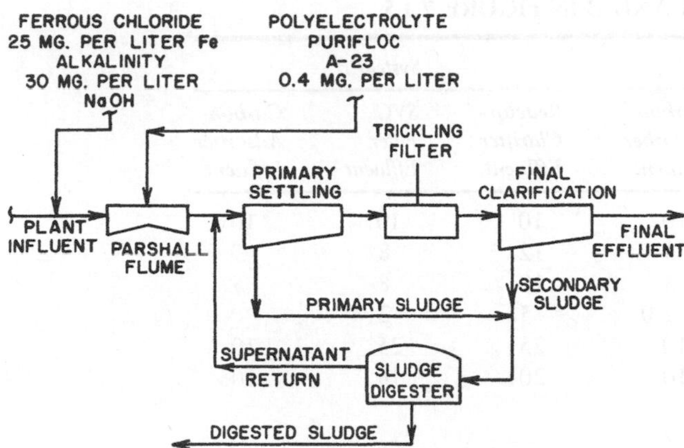

FIG. 7.1.3 Ferrous precipitation of phosphate.

plants, is economical if it does not have to be transported far from its source. This process also reduces the BOD load on secondary treatment units. Corrosion-resistant sludge handling equipment is required due to ferric or ferrous sludges. The chemical sludge produced is not compatible with all methods of dewatering, and the chloride ion content of the effluent increases as a result of this treatment process.

ALUM AND FERRIC PRECIPITATION OF PHOSPHATE

Figure 7.1.4 shows a process in which metal salts (alum or ferric) added to the secondary effluent produce phosphate precipitates that are trapped in the granular media filter. Periodic backwashing of the filter yields a waste phosphorus sludge that is uniformly mixed in an equalization tank. This sludge is then recycled back to the primary settling basin in proportion to the plant influent flow

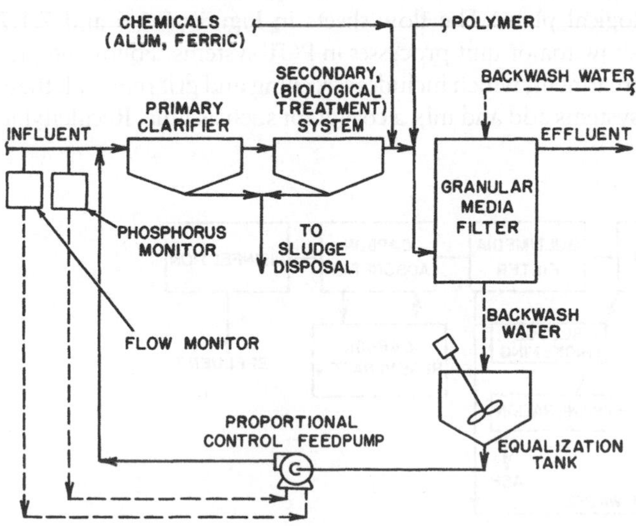

FIG. 7.1.4 Alum or ferric precipitation of phosphate.

rate. The waste sludge adsorbs more phosphorus and thereby reduces the phosphate load on the filter.

Because of the reduced phosphorus concentration in the filter influent, chemical costs can be reduced by as much as 30%. Flow- and phosphorus-monitoring instruments can be used to automate this process. Laboratory and pilot-plant operations have produced effluent phosphorus concentrations of less than 1 mg/l.

Figure 7.1.5 and Table 7.1.1 describe the components and performance of two additional wastewater treatment systems. In system A for the 10-mgd flow shown, two 34-ft diameter by 10-ft high granular media filters are required primarily for SS removal, but they can also remove phosphorus if precipitating chemicals are added. Additional BOD and COD removal is provided by five 20-ft diameter by 20-ft high carbon columns plus one standby column of the same size. Granular carbon can be regenerated in a furnace. Furnaces can also recalcine the lime sludge (burn the $CaCO_3$ into CaO).

In System B, chemical coagulation takes place in two 90-ft diameter by 17-ft sidewall depth columns, which provide high-rate solids-contact treatment and inorganic phosphorus removal.

SLUDGE HEAT TREATMENT PROCESS

Heat treatment of sludges to facilitate dewatering is discussed in Section 10.6 and shown in Figure 10.6.1. After the sludge has been conditioned with heat and pressure, it

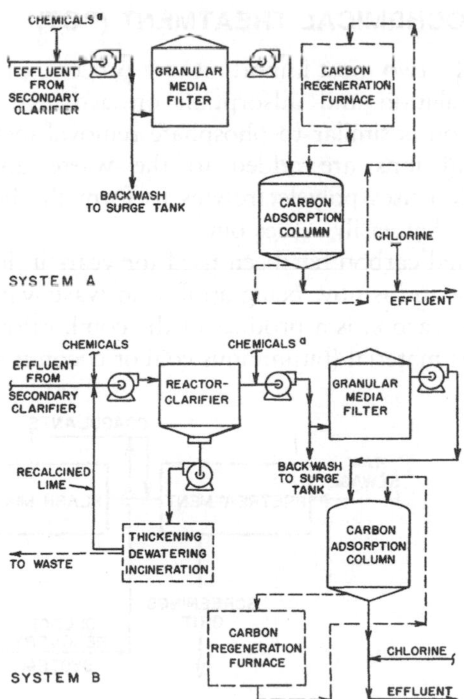

FIG. 7.1.5 Granular filtration, carbon adsorption, and chemical coagulation units. Chemicals can be added to achieve greater BOD, SS, and phosphorus removal.

TABLE 7.1.1 PERFORMANCE OF SYSTEM A AND B IN FIGURE 7.1.5

	System A		*System B*		
Parameters	*Secondary Clarifier Effluent*	*Carbon-Adsorber Effluent*	*Reactor-Clarifier Effluent*	*SVG Filter Effluent*	*Carbon-Adsorber Effluent*
Flow, mgd	10	10	10	10	10
Total BOD, mg/l	25	5	12	8	5
Soluble BOD, mg/l	10	5	10	8	5
SS, mg/l	30	2 0	5	2	
Total COD, mg/l	50	10	25	25	10
Soluble COD, mg/l	20	10	20	16	10
Inorganic Phosphorus, mg/l					
Without chemical	7	7	—	—	—
With chemical	7	0.5	0.5	0.4	0.4

has a lower COD and can be readily dewatered without chemicals in a decanting and vacuum filter unit.

This process grinds the sludge and pumps through a heat exchanger into a reactor where it is held for 30 to 60 min while steam heats it to 380°F and maintains a high pressure of 160 to 250 psig. The system is a continuous process, and the high-temperature sludge leaving the reactor is used to heat the incoming sludge. The process was developed in Europe by Porteous in the 1930s and is now being applied in the United States. No chemical additives are required to make the treated sludge compatible with vacuum filtering.

PHYSICOCHEMICAL TREATMENT (PCT)

In PCT, the two most important unit processes are chemical coagulation and adsorption on activated carbon. Coagulation is similar to phosphate removal methods in that metal ions are added to the wastewater flow. Flocculation uses polyelectrolytes to form the heavy organic floc that easily settles out.

Activated carbon has been used for years in industrial processing and is now being applied in wastewater treatment. The carbon is a product of the combustion of carbonaceous material (bituminous coal or coconut shells) at a high temperature in an oxygen-starved atmosphere. This char has the characteristics of an enormous surface area per granule of carbon—100 acres or more per lb. This large surface area is a result of each particle containing an intricate network of inner channels and accounts for the carbon's great adsorptive capacity.

The treatment uses granular carbon in the form of a bed. Wastewater flows through this bed much like a sand filter. Organic molecules are attracted by the carbon and adsorbed on its surface. When the carbon becomes loaded with the adsorbed molecules and loses its adsorptive ability, it can be removed from the system and regenerated.

Figure 7.1.6 shows a PCT flow sheet. This figure shows all waste flow routes together with the backwash lines and the regeneration equipment shown in Figure 7.1.7. This treatment has great flexibility with interconnected carbon columns because any number of columns can be used in series while others are being refilled with fresh carbon.

In evaluating some processes, the EPA reports that the products of these processes (including carbon treatment) equal or exceed that of a well-operated conventional biological plant. The flow sheets in Figures 7.1.6 and 7.1.7 show major unit processes in PCT systems. Following pretreatment, which includes screening and grit removal, these systems add and mix a coagulant such as lime. Recalcifying

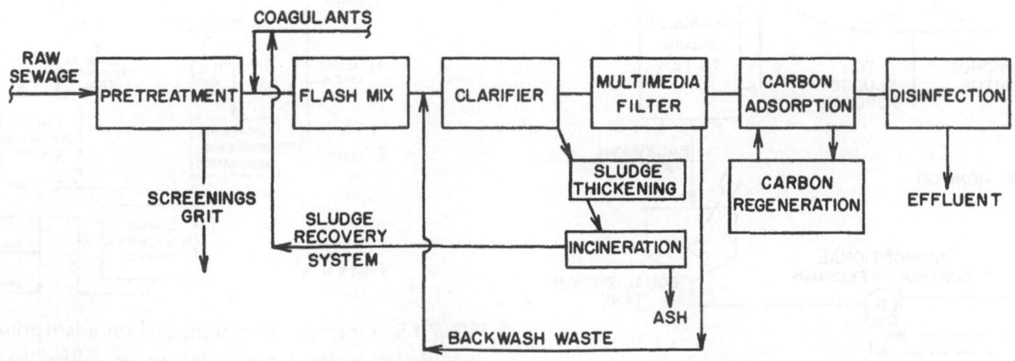

FIG. 7.1.6 PCT of wastewater.

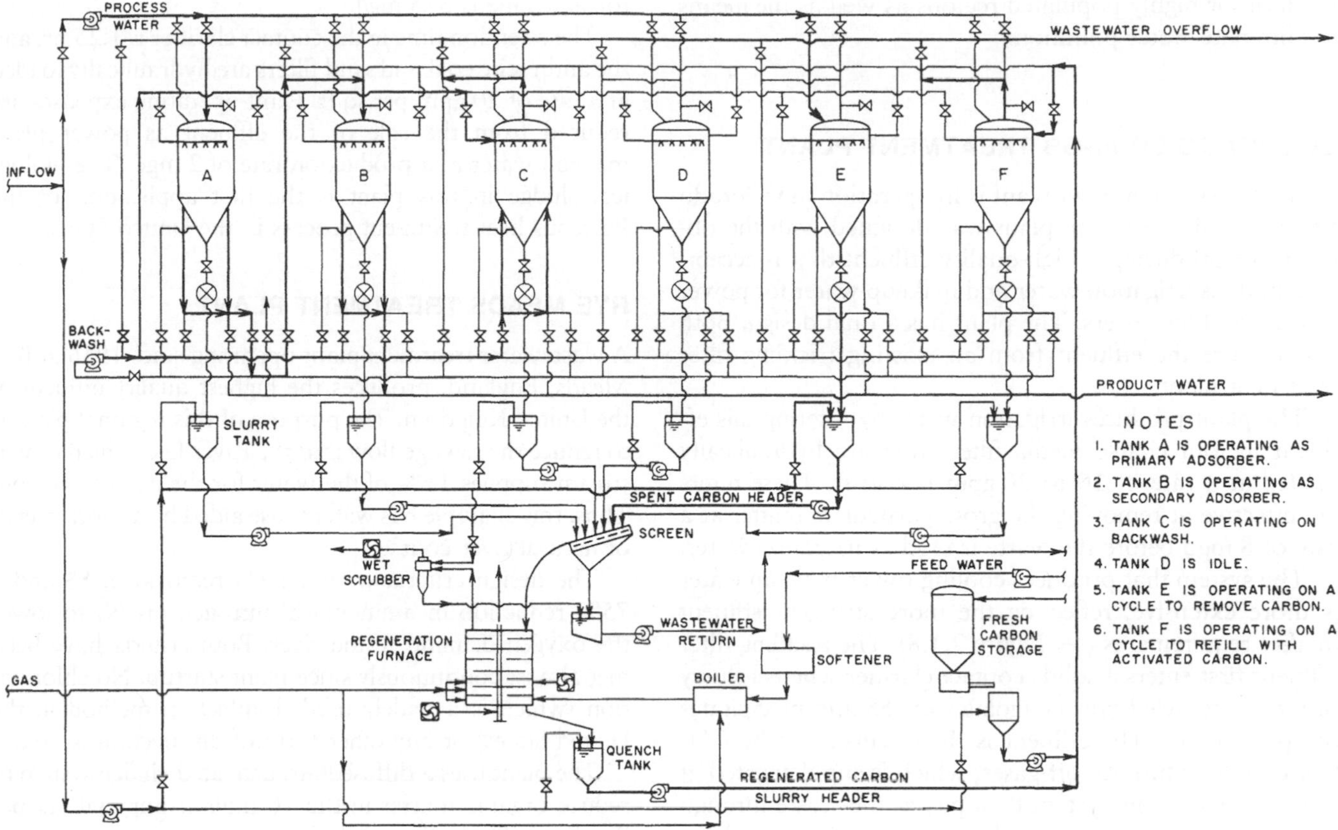

FIG. 7.1.7 Multiple-carbon columns in the PCT of wastewater.

the sludge in a furnace can supply part of the lime coagulant dosage needed. The supernatant from the clarifier passes through a multimedia filter of sand and anthracite coal to remove the remaining SS, which is an inexpensive way to protect the carbon from solids carryover. The activated-carbon beds remove refractory colloidal and some soluble organic molecules.

After disinfection with chlorine, the treated effluent can be discharged. The removal efficiencies of this type of plant are 95 to 99% of organic material and 85 to 99% of phosphorus. The costs are significantly higher if a coagulant other than lime is used since lime can be recovered on site.

HYDROLYSIS–ADSORPTION PROCESS

This process uses lime to hydrolyze large organic molecules into molecules small enough to be adsorbed by activated carbon. The dosage of lime (200 to 600 mg/l) is governed by the molecular nature of the wastewater contaminants and the pH level (11.1 to 12.2 pH), which produces hydrolysis of high-molecular-weight molecules.

Carbon dioxide from furnace stack gases lowers the higher pH of the hydrolyzed waste in two stages, neutralizing hyperalkaline water and precipitating calcium compounds. The hydrolysis of large molecules into smaller ones allows the carbon to adsorb more organic materials, which lowers its detention time to between 7.5 to 15 min. The hydrolysis–adsorption process can operate at 90% BOD and COD removal efficiencies and at 97% phosphorus removal efficiencies.

PCT plants are not subject to upsets and efficiency losses from toxic wastes, and they use many recycling methods to recover some of the chemical treatment agents. Their land requirements are also less, and phosphorus removal is part of their total treatment performance.

Recycling calcium in the sludge can reduce the sludge disposal problems associated with the large quantities of chemicals. Heavy metals are also precipitated into the lime sludge. Wastewater treatment facilities can expand their plant size by adding modular units such as carbon columns or reactor-clarifiers.

Tertiary Equivalent (95% BOD Reduction plus Phosphate Removal)

A tertiary treatment plant removes practically all solid and organic contaminants from wastewater, thereby producing drinkable water direct from sewage. Many tertiary treatment plants include nitrogen removal systems using either bacterial nitrification, air stripping, or ion exchange. In times of scarce water resources, the treatment systems described may represent the future source of water for semiarid or highly populated regions as well as the means to eliminate water pollution.

COLORADO SPRINGS TREATMENT PLANT

A full, tertiary treatment plant is in operation in Colorado Springs, Colorado. The plant was designed with the objective of producing a high-quality effluent that is acceptable both as irrigation water and makeup water for power station cooling towers. The plant has a dual design. Both systems use the effluent from an existing trickling filter treatment plant.

The plant produces irrigation water by filtering this effluent in four coarse media filters that are hydraulically loaded at a rate of 10 to 20 gpm per sq ft. These filters are effective at removing the gross particulate matter at a rate of 8 mgd before the waste is used as irrigation water.

The system that produces cooling tower makeup water is more extensive, reflecting the more stringent effluent quality requirements (see Figure 7.1.8). The trickling filter effluent first enters a solids-contact clarifier where a slurry of mostly recycled lime coagulates the SS and precipitates the phosphates. The effluent is then neutralized by CO_2 from the lime furnace off-gases, which is supplemented at times by acid. This system then passes the waste through dual media filters to remove the solids remaining after the chemical treatment.

The final stage of treatment is by granular, activated-carbon columns. The columns operate in an upflow configuration and remove residual organic contaminants. The actual water quality of the effluent is BOD, 11 mg/l; COD, 17 mg/l; SS, less than 1.5 mg/l; and PO_4^{3-}, less than 3.0 mg/l.

Both the lime and the activated-carbon systems have recycling loops that use multiple-hearth furnaces to regenerate some of the chemicals used. The recycled lime was found to be more effective in raising the pH than fresh lime. The recycled lime dosage is 280 mg/l, and the dosage for new lime is 325 mg/l.

The retention time in the contact clarifier is 1.25 hr, and the anthracite coal and sand filters are hydraulically loaded at a rate of 20 gpm per sq ft. Plant operating expenses are reduced from the sale of the effluent as power plant makeup water at a production rate of 2 mgd. The biological sludge in this plant is the first application of the Porteous heat treatment process in the United States.

RYE MEADS TREATMENT PLANT

A wastewater treatment plant operating since 1956 at Rye Meads, England, produces the highest quality effluent in the United Kingdom. The purpose of this regional plant is to reduce the sewage flow into the River Lee, which downstream supplies 19% of the water for the city of London. Thus, this example has water reuse aided by a short stretch of natural river course.

The design criteria were a 99% removal of SS and a 75% reduction in ammoniacal nitrogen (as N) to lower the oxygen demand in the river. Both criteria have been met almost continuously since plant startup. No chlorination (which is a widely used disinfection method in the United States) or any other form of disinfection is used.

The plant uses a diffused-air, activated-sludge system to reduce organic matter in the 10 million imperial gal per day flow. The activated-sludge system operates in an extended aeration mode to provide biological nitrification of the ammonia as well. The plant then polishes the effluent with rapid sand filters to achieve high-quality water. The sludge is treated by biological digestion and is then trucked to surrounding farms. Part of the wastewater flow is from industrial sources and digester upsets, and nitrifying organisms have occurred during the plant's operation. The plant has corrected such occurrences by carefully monitoring all influent lines to trace the source of the toxic materials and by establishing trade waste treatment agreements. Goldfish in tanks in the maintenance shop are used as monitors for toxic materials.

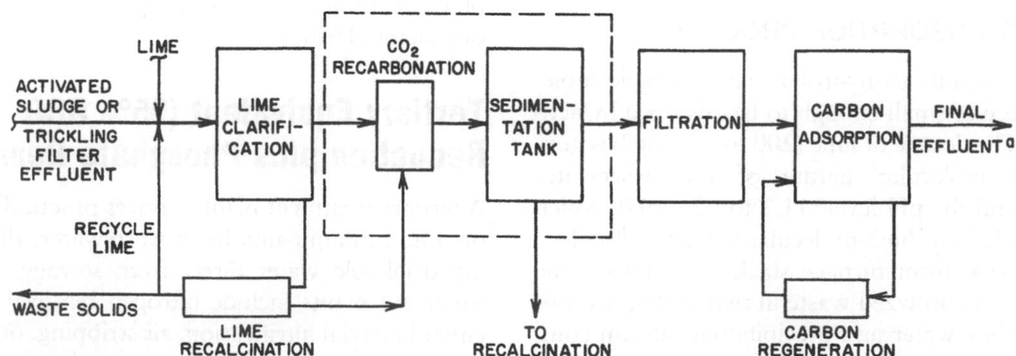

FIG. 7.1.8 Colorado Springs treatment plant section producing an effluent quality that is acceptable as power plant makeup water.

SOUTH LAKE TAHOE RECLAMATION PLANT

The water reclamation plant at South Lake Tahoe, California, is a large and advanced treatment plant. Lake Tahoe in northern California and Nevada is in a natural basin that has been largely undeveloped until recent years. The lake, one of the three clearest lakes in the world, was destined to become another polluted body of water unless the nutrient inflow from sewage disposal was stopped and other sources of pollution were greatly retarded.

To meet this challenge, the South Lake Tahoe Public Utility District, with the cooperation of the U.S. government and industry, built this plant. The treated effluent water, which meets U.S. and World Health Organization (WHO) drinking water standards, is pumped out of the basin and creates the Indian Creek Reservoir, which has been approved by the California Department of Public Health for water sports and has been stocked with rainbow and rainbow hybrid trout by the California Department of Fish and Game.

The flow sheet in Figure 7.1.9 shows the unit processes in the 7.5 mgd South Lake Tahoe plant. The first treatment step is conventional primary settling followed by activated-sludge treatment. The sludges from these steps are centrifuged, dewatered, and incinerated to an ash. The plant next adds lime to the overflow stream for coagulation and to precipitate phosphates and raise the pH. The high pH converts the ammonium nitrogen into ammonia form. In the next process, a cooling tower strips it out. The plant pretreats lime sludges and puts them through a recalcining furnace that burns $CaCO_3$ into CaO. The furnace exhaust gas provides carbon dioxide for the recarbonation (neutralization of the high pH) of the water immediately following the ammonia stripping. A two-stage centrifuging station sends about one-fourth of the recycled lime back to the chemical treatment stage for reuse.

Following nitrogen removal, the wastewater effluent passes through separation beds (mixed-media filters) that remove any remaining SS. Finally, the nearly pure water undergoes adsorption in upflow activated-carbon columns to remove soluble organic contaminants, including pesticides and ABS. Each of the eight columns contains 24 tn of carbon, providing about 4.8 million acres of adsorption surface for the removal of pollutants. Before being discharged to the Indian Creek Reservoir, the sparkling clear effluent is chlorinated to insure against bacterial contamination.

The plant regenerates granular carbon in a complete system using a multiple-hearth furnace with a capacity of 6000 lb or 6,000,000 surface acres per day. The plant scrubs and cools all furnace gases from the organic-sludge and lime-sludge furnaces to eliminate any air pollution.

One of the keys to the economy of the Lake Tahoe plant is the regeneration of the activated granular carbon. Larger plants can also find economy in a lime recovery system.

The median BOD concentration of the effluent produced is 0.98 mg/l, and the median COD is 10.83 mg/l. SS consists mostly of carbon fines in the effluent, amounting to 0.53 mg/l in concentration. The lowest phosphorus concentration (0.09 mg/l) was achieved when waste streams were recycled to the lime basin for 6 months. All systems are efficient, and testing continues to generate more knowledge on further improvements to operation.

The carbon system is loaded at a rate of 6.2 gpm per sq ft and has a short contact time of 17 min.

The greatest difficulties after startup have stemmed from the ammonia air stripper, which initially was sized for 50% of the required capacity. The plant now incorporates a second stripping tower. This custom-built unit incorporating thousands of redwood slats has 50 to 60% efficiency at low winter temperatures. This efficiency is due to the dependence of the transfer mechanism on temperature. To achieve 90% ammonia removal in the winter, the plant must increase the air-to-water ratios from 250 cubic feet per minute (cfm)/gpm to about 800 cfm/gpm. The temperature effect becomes negligible at high pH levels.

Calcium carbonate sludges have precipitated in the tower and caused a heavy, sticky residue to develop. The plant must periodically wash this limestone residue to remove the precipitate. The removable slats in the second unit overcome this problem.

WINDHOEK RECLAMATION PLANT

The Windhoek Water Reclamation system in South-West Africa (see Figure 7.1.10) can function as a water reclamation plant for direct municipal reuse. The plant produces drinking water for a city. The primary and secondary treatment system uses biofilters and digesters. Effluent from the biofilters is held for about 14 days in algae maturation ponds where nitrogen and phosphorus are used by the algae.

Reusing the treated sewage in the ponds was determined to be the most economical source of drinking water. The algae-laden pond water is pumped to a purification plant at a rate of 1.2 mgd. Recarbonation (lowering of the pH) to 7.2 is accomplished by the submerged combustion of propane gas. Then, a unique flotation system removes 90% of the algae by adding alum and mixing it rapidly. Detergents are removed by a foam fractionator.

The plant adds alum, lime, and chlorine to the treated water which is settled and then filtered by sand and adsorption on granular activated carbon to remove trace amounts of organic molecules. Currently, the carbon is not regenerated. Breakpoint chlorination provides additional nitrogen removal. Salt buildup in the drinking water has been kept to a maximum of 180 mg/l. This specialized system for Windhoek is particularly suitable for hot climates because the algae beds do not function well at low temperatures.

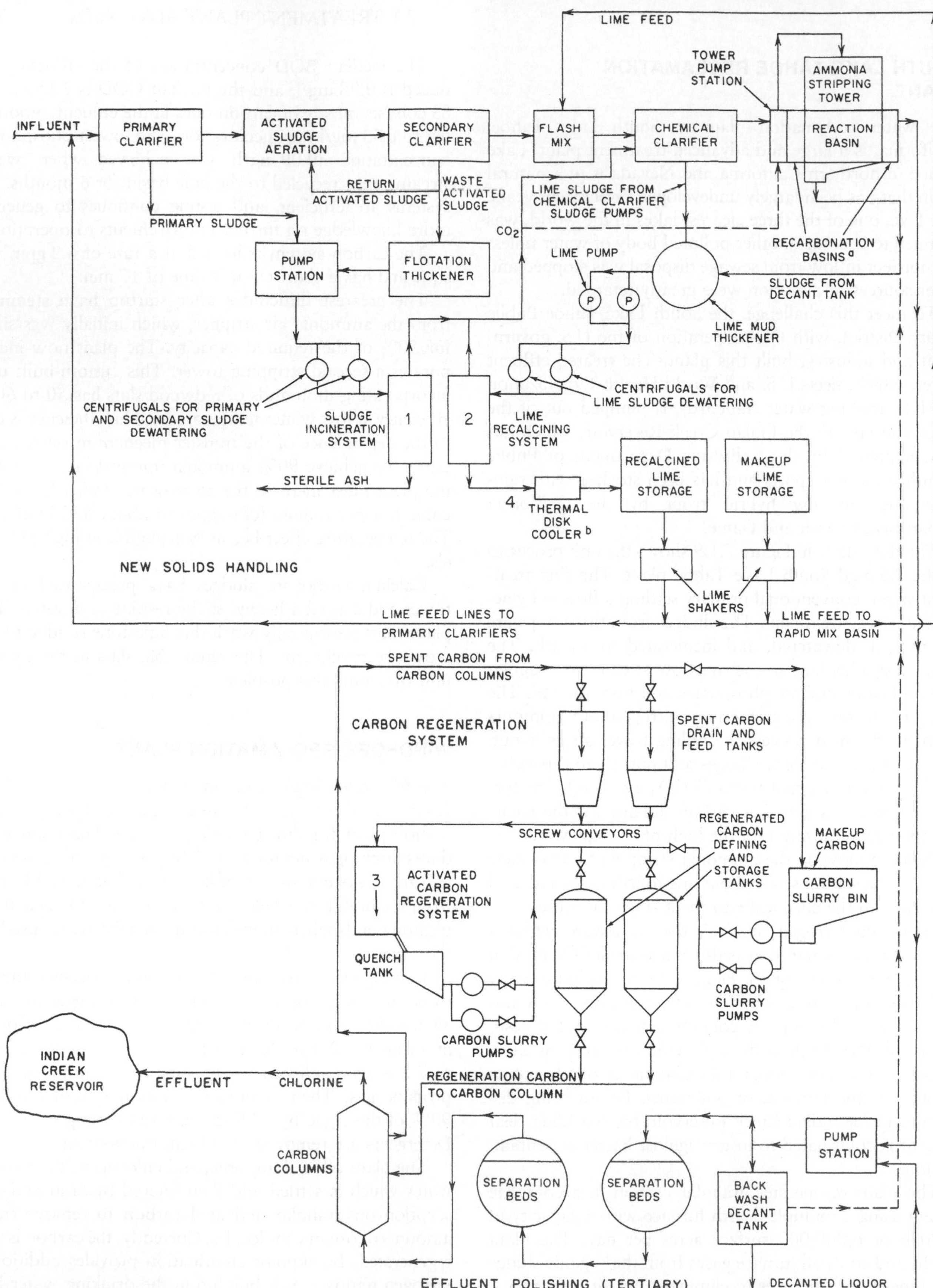

FIG. 7.1.9 Public utility district water reclamation plant at South Lake Tahoe, California. [a]Carbon dioxide is added to water in the reaction basin after it has passed through the ammonia-stripping column. [b]Thermal disk is a processing unit with a series of hollow disks filled with a heat-transfer medium. The solid, reclaimed lime is cooled by the disks as it passes between them.

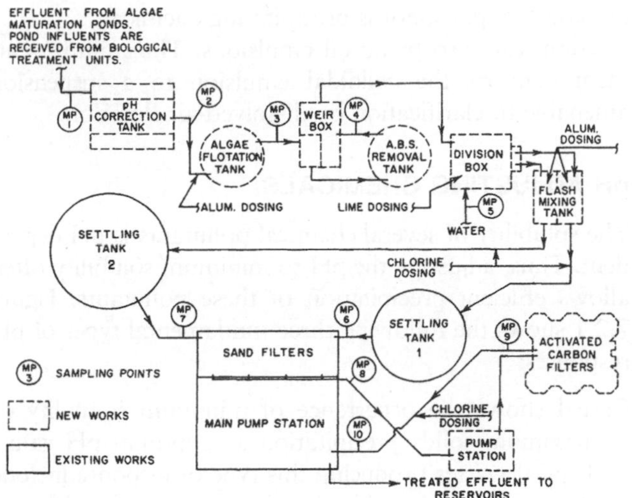

FIG. 7.1.10 Water reclamation plant at Windhoek, South-West Africa.

TABLE 7.1.2 WASTEWATER CHARACTERISTICS FOR A 10-MGD PLANT

Parameters	*Influent Quality*	*Effluent Quality*
Flow, mgd	10	9.9
BOD, mg/l	250	5.0
SS, mg/l	250	5.0
COD, mg/l	500	10.0
Total Inorganic Phosphorus, mg/l	10	0.3

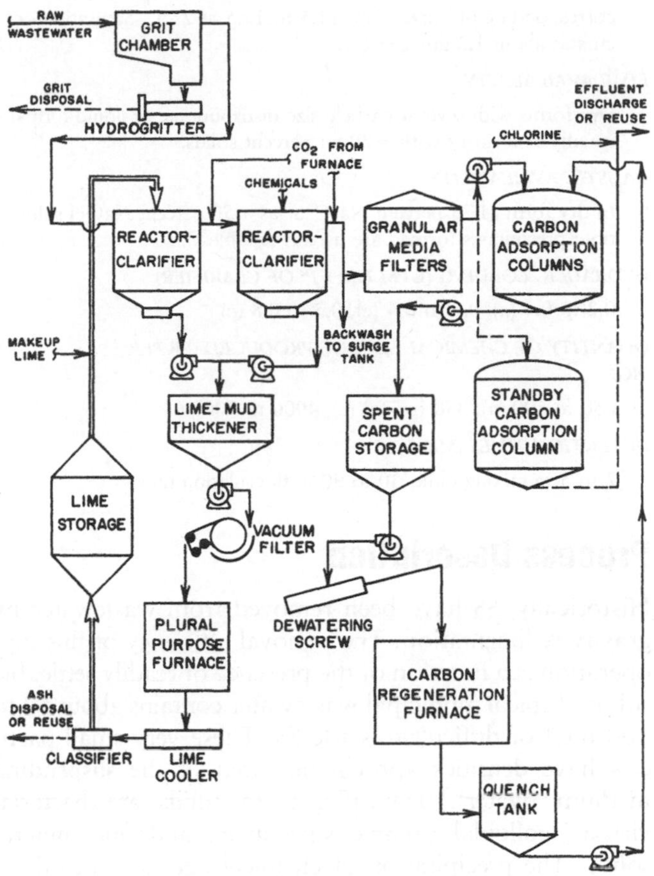

FIG. 7.1.11 Hydrolysis adsorption process for water reclamation.

HYDROLYSIS ADSORPTION PROCESS

The hydrolysis adsorption process produces drinking quality water. Table 7.1.2 and Figure 7.1.11 provide a more detailed version of system B in Figure 7.1.5. Achieving the higher quality water (drinkable) only requires a longer contact time in the carbon adsorbers. Nitrogen removal can be easily added since the high pH established by lime treatment is also a requirement for the ammonia-stripping unit process.

—F.P. Sebastian

7.2 CHEMICAL PRECIPITATION

TYPES OF CHEMICAL COAGULANTS

a) Alum, b) Lime, c) Hydrated lime, d) Sulfuric acid, e) Anhydrous ferric chloride

SS REMOVAL EFFICIENCY

Up to 85%

FLOCCULATION AIDS

f) Polyelectrolytes

CHEMICAL DOSE REQUIRED (mg/l)

80 to 120 (e), 100 to 150 (a), 350 to 500 (c)

NEUTRALIZING REAGENTS

a) Acidic—a1) sulfuric acid, a2) hydrochloric acid, a3) carbon dioxide, a4) sulfur dioxide, a5) nitric acid; *b) Basic*—b1) caustic soda, b2) ammonia, b3) soda ash, b4) hydrated lime, b41) dolomitic hydrated quicklime, b42) high calcium hydrated quicklime, b5) limestone.

SOLUBILITY AND pH OF LIME SOLUTIONS

For a concentration range of 0.1 to 1.0 gm. of CaO per liter the corresponding pH rises from 11.5 to 12.5 at 25°C. Saturation occurs at about 1.2 gm. per liter.

LIME AVAILABILITY

Dry forms with several particle size distributions, or liquid forms usually as a slurry with +20 wt percent solids.

CAUSTIC AVAILABILITY

In dry form as 75 percent Na_2O or as a 50 percent NaOH solution; precipitates formed are usually soluble.

HYDRAULIC LOADING (GPM PER FT² OF CLARIFIER)

0.2 to 0.4 (a), 0.3 to 0.4 (e), 0.5 to 0.8 (c)

QUANTITY OF CHEMICAL SLUDGE PRODUCED (LB PER MG)

250 to 500 (a), 350 to 700 (e), 4000 to 7000 (c)

DETENTION TIMES (MIN)

2 to 5 in mixing tank, 30 to 90 in flocculation tank

Process Description

Historically, SS have been removed from wastewater by gravity sedimentation. The removal efficiency of this unit operation is a function of the presence of readily settleable solids. Typical municipal wastewater contains about 50 to 100 mg/l of difficult-to-settle SS. These very small particles have densities approaching that of the suspending medium (water). Typically, these solids are bacteria, viruses, colloidal organic substances, and fine mineral solids. The precipitation of chemical agents causes these difficult-to-settle solids to flocculate (particle growth) and become settleable.

Chemical treatment can also precipitate certain dissolved pollutants, forming a settleable suspension. For example, phosphate is precipitated by aluminum (Al^{3+}), and heavy metals are precipitated as hydroxides when the pH is raised.

During the middle and late 1800s, chemical treatment was implemented in about 200 sewage treatment plants in England. During the 1930s, the United States became interested in chemical treatment. However, these users recognized that chemical treatment alone does not remove putrescible dissolved organics and is therefore not a complete treatment process. Interest in chemical treatment of municipal wastewater increased in the 1960s due to new and stringent requirements concerning phosphorus removal.

Types of Chemicals Used

Table 7.2.1 shows the properties of several common chemicals used in the removal of SS from wastewater.

Hydrolyzable trivalent metallic ions of aluminum and iron salts are established coagulation chemicals. The hydrolysis species of Al^{3+} or Fe^{3+} destabilize colloidal pollutants and render them amenable to flocculation (particle growth), which enhances settleability. If excessive amounts of these metallic ions are used, the gelatinous floc precipitated can enmesh some small pollutant particles, allowing them to be removed by settling. Both Al^{3+} and Fe^{3+} are effective phosphorus precipitating chemicals.

Alum can also break oil emulsions. Thus, alum treatment converts the colloidal emulsion to a suspension amenable to clarification by dissolved-air flotation.

pH ADJUSTING CHEMICALS

The solubility of several chemical pollutants is pH dependent. Thus, adjusting the pH to minimum solubility often allows efficient precipitation of these pollutants. Figure 7.2.1 shows the following three fundamental types of pH responses:

Case I shows the occurrence of minimum solubility or maximum solids precipitation at optimum pH value. Typical wastes producing this type of response include iron contained in acid mine drainage and fluoride and arsenic-bearing wastes.

Case II shows a neutral or acidic waste containing certain heavy metals. The optimum pH is at the knee of the precipitated solids generation curve. Lime treatment of metal finishing waste is typical of a Case II response.

Case III is the inverse of Case II; the chemical pollutant is precipitated due to pH depression, usually with sulfuric acid. A latex-bearing waste is typical of a Case III response.

Adding lime to municipal wastewater elevates the pH and causes insolubilization of calcium phosphate, calcium carbonate, and magnesium hydroxide. The hydroxide precipitated at an elevated pH acts as a coagulant, destabilizing and enmeshing colloidal pollutants. The chemistry of lime treatment is generally described by water softening reactions. However, the presence of dissolved organic matter and condensed phosphates in municipal wastewater causes some interference with water softening reactions, such as the precipitation of calcium carbonate. Figure 7.2.2 shows the solubility of the total and calcium hardness ions in a typical municipal wastewater.

FLOCCULATION AIDS

Organic polyelectrolyte flocculation aids are effective in promoting SS removal. The addition of polyelectrolytes does not produce a chemical precipitation per se but can drastically promote particle growth. Improvement in the removal of finely divided solids by gravity settling is the result. Polyelectrolytes are effective both for wastewater SS and for precipitates formed by chemical treatment.

Polyelectrolytes in suspensions become attached to two or more particles and provide bridging between them.

Chemical Sludge Production

An inherent burden with the improved SS removal by chemical treatment is the production of chemical sludge. The thickening and dewatering properties of chemical–sewage sludge are worse than those of the sewage

TABLE 7.2.1 PROPERTIES OF CHEMICAL COAGULANTS

Common or Trade Name	*Chemical Name*	*Chemical Symbol*	*Shipping Containers*	*Weight lb/ft³*	*Suitable Handling Materials*	*Commercial Strength*
Alum	Aluminum sulfate	$Al_2(SO_4) \cdot 14\ H_2O$	300–400 lb bbl, carload bulk; carload, barrels	60–63 62–67	Dry: Iron and steel Solution: lead, rubber, silicon, iron, and asphaltum	15–22% Al_2O_3
Quicklime	Calcium oxide	CaO	50 lb bags, 100 lb barrels, bulk carload	40–70 26–48	Rubber, iron, steel, cement, and asphaltum	63–73% CaO
Hydrated lime	Calcium hydroxide	$Ca(OH)_2$	50 lb bags, 100 lb barrels, bulk carload	40–70 26–48	Rubber, iron, steel, cement, and asphaltum	85–99% $Ca(OH)_2$ 63–73% CaO
Oil of vitriol	Sulfuric acid	H_2SO_4	Drums, bulk	—	Concentrated: Steel and iron Dilute: lead, porcelain, glass and rubber	93% H_2SO_4 78% H_2SO_4
Anhydrous ferric chloride	Ferric chloride	$FeCl_3$	300 lb bbl (crystals) 500 lb casks, 300–400 lb kegs	—	Glass, stoneware, and rubber Glass, stoneware, and rubber	59–61% $FeCl_3$ 98% $FeCl_3$

sludge alone because of the presence of hydroxide sludges and the increased amounts of colloidal pollutants.

The addition of alum to a basic wastewater containing alkalinity produces a chemical floc. Generally, 1 lb of SS (chemical floc) is produced for each 0.25 to 0.40 lb of aluminum added. Figure 7.2.2 is a typical example of the sludge production and SS reduction by alum treatment.

The chemical sludge produced by adding lime to municipal wastewater depends on the chemical characteristics of the water, the pH level, and the method of operation. The net chemical sludge produced is a result of an interaction of these three parameters.

Lime treatment of municipal wastewater with low hardness ($\leq$200 mg/l as $CaCO_3$) and low alkalinity ($\leq$150 mg/l as $CaCO_3$) should be accomplished in two stages, with the corresponding pH levels at 11 to 11.5 and 9.5 to 10, respectively. The chemical sludge produced under these conditions is about 4000 to 5500 lb per mg. For municipal wastewater, with high hardness ($\geq$350 mg/l as $CaCO_3$), and high alkalinity ($\geq$250 mg/l as $CaCO_3$), single-stage treatment at a pH of 10.5 to 11.0 is recommended. The chemical sludge produced under these conditions is about 5500 to 6500 lb per mg.

The use of organic polyelectrolytes does not produce significant amounts of chemical sludge.

The nature and quantity of the chemical sludge produced by adjusting the pH of industrial waste depends mainly on the initial concentration of the chemical pollutants and the efficiency of the liquid–solids separation step. Generating the data shown in Figure 7.2.2 provides data on sludge production.

Unit Operations

SS removal through chemical treatment is accomplished by three series unit operations: rapid-mixing, flocculation, and settling.

RAPID MIXING

The chemical reagent must first be completely dispersed throughout the wastewater. This requirement is especially important when an inorganic coagulant such as alum is used because the precipitation reactions occur immediately. In lime treatment, the lime slurry should be dispersed throughout the wastewater in the presence of the previously formed precipitate (recycled sludge). The sludge provides an abundant surface area on which large amounts of chemical precipitates can form. When the sludge is not recycled, gross deposits (scaling) of calcium carbonate develop on the tank walls and other surfaces.

Rapid mixing occurs in 10 to 30 sec in a basin with a turbine mixer. About 0.25 to 1 hp per mgd is used for rapid mixing. A mean temporal velocity gradient in excess of 300 ft per sec per ft is recommended.

FLOCCULATION

After effective coagulation-precipitation reactions (rapid mix) occur, promoting particle size growth through flocculation is the next step. The purpose of flocculation is to bring coagulated particles together by mechanically inducing velocity gradients within the liquid.

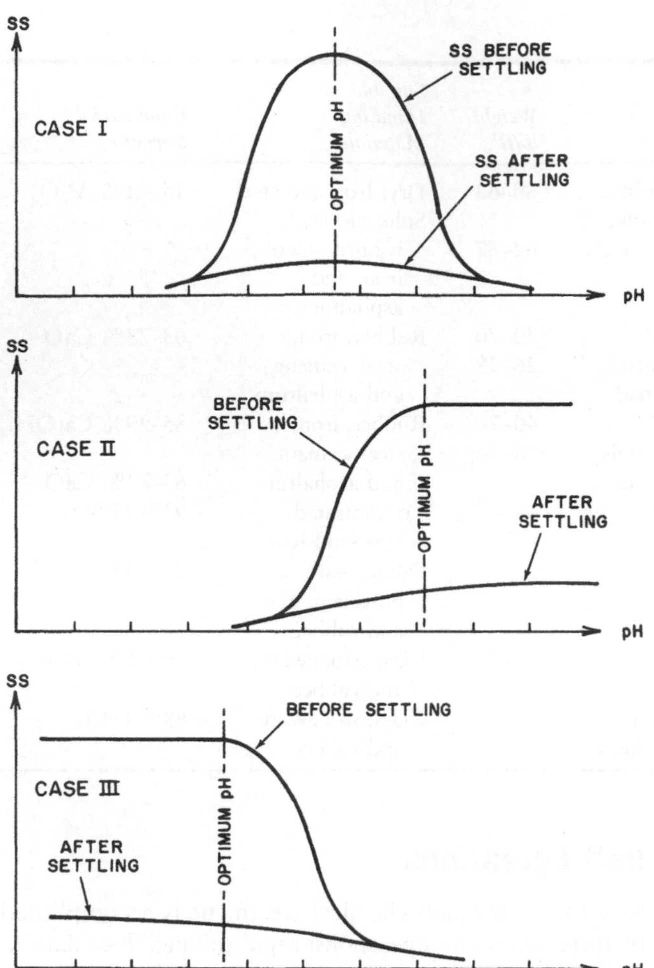

FIG. 7.2.1 Precipitation of chemical pollutants by pH adjustment.

Flocculation takes 15 to 30 min in a basin containing turbine or paddle-type mixers. Mean temporal velocity gradients of 40 to 80 ft per sec per ft are recommended. The lower value is for fragile floc (aluminum or iron floc), and the higher value is for lime-treatment floc.

SOLIDS CONTACT

Solids contacting is especially beneficial for lime treatment because it reduces the deposition problems inherent in once-through, rapid-mix, flocculation systems. Wastewater treatment facilities provide solids-contacting by maintaining or recycling large amounts of previously formed precipitates in contact with the wastewater and adding lime.

Several types of solids-contact treatment units are available. These units were originally developed for lime-softening water treatment and are effective for lime treatment of wastewater.

LIQUID–SOLIDS SEPARATION

After flocculation, the final step is clarification by gravity settling. The conventional clarifier design is suitable for this purpose. However, wastewater treatment facilities should provide positive-sludge withdrawal to prevent problems associated with the formation of septic sewage sludge. Figure 7.2.3 shows a once-through and a solids-contacting system for enhancing SS removal with chemical treatment.

If a highly clarified effluent of less than 10 mg/l SS is needed, an additional liquid–solids separation step is re-

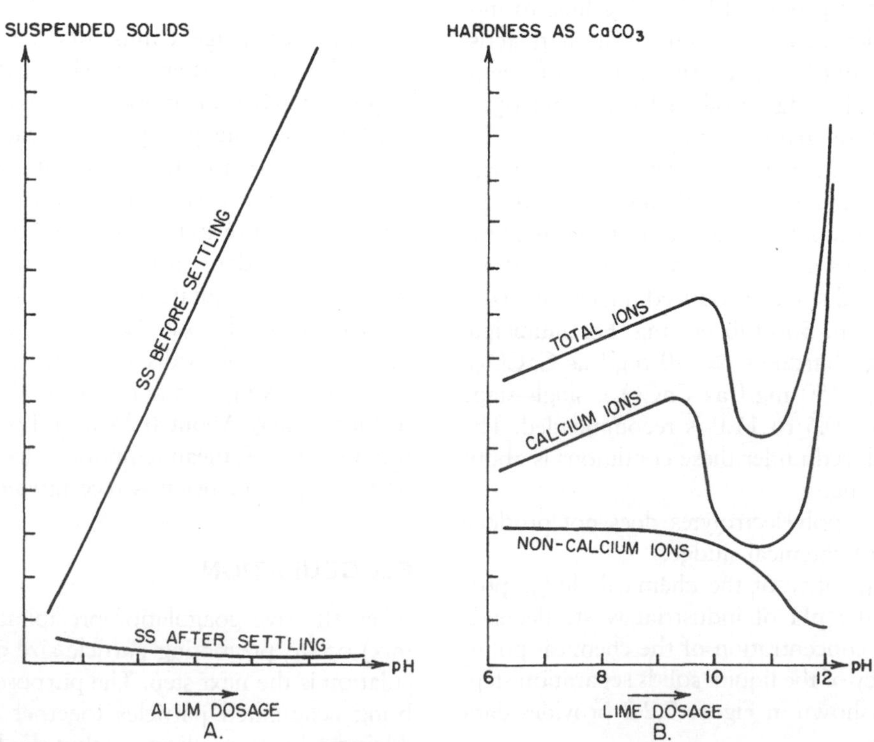

FIG. 7.2.2 Effects of alum and lime additions. **A**, Alum; **B**, Lime.

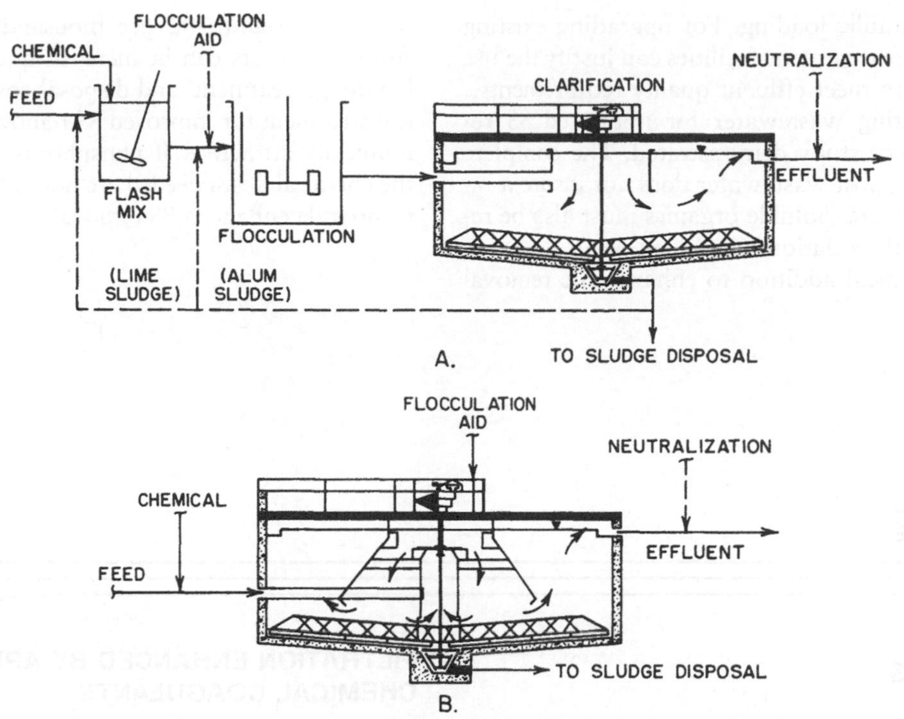

FIG. 7.2.3 Chemical treatment systems. **A,** Once-through system; **B,** Solids-contact system.

quired. The unit operation recommended for effluent polishing is granular media filtration, e.g., dual-media filtration. Here, the clarifier effluent containing 20 to 40 mg/l of finely divided floc is passed through the filter. For peak hydraulic loadings of less than 5 gpm per ft^2, the filtered effluent contains not more than 5 to 10 mg/l of SS. The filter bed in this case consists of 2 ft of 1.0- to 1.5-mm anthracite coal over 1 ft to 0.5- to 0.8-mm sand.

Design Considerations

The principal cost in enhancing SS removal by chemical treatment is the chemicals. Environmental engineers can estimate the chemical dose requirements from laboratory jar tests and/or pilot-plant studies. Daily, weekly, and seasonal variations in wastewater characteristics require that wastewater treatment facilities must adjust the chemical dose during plant operation to minimize additive use while providing the required solids removal. Table 7.2.2 lists the typical chemical doses.

Generally, the hydraulic loading of clarification equipment determines the SS removal efficiency. Because high removal efficiencies are sought, conservative hydraulic loadings should be used. Table 7.2.2 also lists the recommended average hydraulic loadings for 80 to 90% average SS removal from raw municipal wastewater.

Although polyelectrolyte flocculation aids at nominal doses of 0.25 mg/l can at least double the chemical-sewage floc settling rates, providing twice the clarifier capacity is normally less expensive than using the polyelectrolyte. For new plant construction, environmental engineers should make a thorough analysis of polyelectrolyte use versus re-

TABLE 7.2.2 DESIGN CRITERIA FOR 80 TO 90% SS REMOVAL FROM RAW MUNICIPAL WASTEWATER BY CHEMICAL TREATMENT

Criteria	*Ferric Chloride* $FeCl_3$	*Alum*	*Hydrated Lime* $Ca(OH)_2$
Dose, mg/l	80–120	100–150	350–500
Hydraulic loading,° gpm/ft^2 of clarifier	0.3–0.4	0.2–0.4	0.5–0.8
Chemical sludge production, lb/mg	350–700	250–500	4000–7000
Chemical cost, ¢/lb of chemical	4–5	3–4	1
Treatment chemical cost, ¢/1000 gal of wastewater	2½–5	2½–5	3–4

Note: °Without use of polyelectrolytes

duced clarifier hydraulic loading. For upgrading existing facilities, wastewater treatment facilities can justify the use of polyelectrolytes to meet effluent quality requirements.

Chemically treating wastewater for improved SS removal has been successfully demonstrated. The complete removal of SS from most wastewater does not insure a biologically stable effluent. Soluble organics must also be removed by biological oxidation or adsorption.

The cost of chemical addition to enhance the removal of SS is around 10¢ per thousand gal of wastewater. Additional costs can be incurred due to increased sludge handling, treatment, and disposal costs. Therefore, chemical treatment for improved SS removal is not usually economically attractive. If phosphorus removal is required, the chemical doses needed are normally more than enough to provide enhanced SS removal.

—*D.E. Burns*

7.3 FILTRATION

Special Filters

FILTER TYPES

a) Diatomite
b) Microstrainers
c) Gravity
d) Pressure
e) Deep-bed
f) Multilayers
g) Cartridge
h) Continuous Moving Bed
i) Membrane

HYDRAULIC LOADINGS (GPM/FT²)

0.5 to 3 (a,g); 2 to 6 (b,c,cf,d,df,ef); 6 to 10 (b,d,df,e,ef,f); and 10 to 30 (e)

SOLIDS REMOVAL CAPACITY (LB PER SQ FT OF FILTER SURFACE)

0.1 to 0.5 (a,g); 0.3 to 0.5 (c,d); 0.5 to 1 (c,cf,e,ef); 1 to 2 (e,ef,f); and 2 to 5 (e,ef,f)

FILTER BED DEPTH (FT)

1 to 2 (cf,df); 2 to 3 (c,d,df); 2 to 6 (df,ef); and 3 to 12 (e)

FILTER MEDIA USED

Acetate (g); Barium Sulfate (e); Cellulose (a,g); Coal (c,cf,d,df,f); Coke (e); Diatomite Fibers (a); Garnet (b); Gravel (ef); Nylon (g); Polypropylene (g); Sand (c,cf,d,df,f); and Stone (g)

FILTER OPENINGS OR FILTER MEDIA SIZES (MM)

0.5 to 100 μ (g); 0.05 to 0.2 (a); 0.4 to 0.6 (c,cf,d,df,f); 0.6 to 0.8 (c,d); 0.8 to 1.4 (cf,df,f); 0.8 to 4.0 (e); and 6 to 12 (e-coarse layer)

FILTER BACKWASH RATE

15 gpm per ft² with multilayer design having siliceous media

AIR PURGE RATE RECOMMENDED FOR CLEANING

3 to 7.5 scfm per sq ft

This section begins with a discussion of special filters, concentrating on water filters operating on influent SS concentrations of 100 mg/l or less.

FILTRATION ENHANCED BY APPLIED CHEMICAL COAGULANTS

Wastewater treatment facilities can obtain completely clear filter effluents only by feeding chemical coagulants, such as alum, iron, or polyelectrolytes, to the wastewater prior to filtration. Adding chemicals to clarifiers reduces the solids load on the clarifier overflow effluent filters.

When a facility is considering adding chemicals into the filter influent, they should perform coagulation tests or pilot-filtration tests. Chemical coagulation can result in 100% solids removal including colloidal particle sizes as small as 0.05 μ. When stain-free filter effluent is required, chemical feeds are absolutely necessary. Without chemical feeds, the SS removal efficiency of filters generally ranges from 50 to 80%.

However, adding chemical coagulants increases the solids load; therefore, filter runs are shortened. Surface-type and ultra-high-rate filters are drastically affected in this regard. Adding coagulants can be costly not only because of shortened runs, but also because of the cost of chemicals, pumping, and labor.

In multilayered or in-depth filter applications, the influent must contain flocculant particles ranging from $\frac{1}{8}$ to $\frac{1}{4}$ in, whether these particles are natural particles, biological flocs, or freshly coagulated flocs. The larger flocs are removed by the coarse filter layers, and the smaller particles are removed by the fine sand layers below. If only smaller flocs are present, they are deposited primarily in the fine lower layers, resulting in a solids removal capacity no better than that of single-layer filters.

SELECTION AND OPERATION OF FILTERS

Waste flows have cyclic variations; for example, the peak flow of most domestic wastewater sources is about twice the average flow. Equalizing or surge basins are needed as part of the filter system to accommodate these flow variations.

When domestic waste is mixed with industrial waste containing metals or inorganic salts such as iron, copper, or aluminium, the filtration characteristics of the secondary effluents (after biological processes) can be enhanced. In such cases, fewer colloidal particles and larger or stronger flocs can be expected. These effluents can be treated with high-rate filtration or surface-type microscreening with removal efficiencies of over 80%, without the use of additional chemical coagulants. When domestic waste is aerated for more than 10 hr, an easily filtered floc generally forms without the need for chemicals.

FILTRATION ENHANCED BY EQUIPMENT DESIGN

When the SS concentration in filter influents approaches 120 mg/l, single-layer or surface filtration equipment is usually quickly overloaded, and the run lengths between cleanings become short, ranging from 1 to 3 hr with solids removal capacities of about $\frac{1}{4}$ to $\frac{1}{2}$ lb per sq ft.

Multilayer or deep-bed filters were developed for such applications. They enhance the filter capacity by providing more or deeper coarse layers where more settling and storage of solids can take place. Roughly, an extra $\frac{1}{2}$ lb of solids capacity is obtained with each additional layer or foot of depth in the filter.

GENERAL DESIGN PARAMETERS

Horizontal or vertical pressure filters are recommended in wastewater applications because they handle higher solids loads and pressure heads and are more compact and less costly. The freeboard (open space above the filter bed) for anthracite- (carbon and plastic) containing filters should be a minimum of 50%. For sand filters, a 30% minimum is required.

Environmental engineers can best determine the filter backwash rate based on the operating temperature and available bed expansion. For dual or multilayered filters containing siliceous media, a minimum of 15 gpm per sq ft backwash rate should be used. An air scour (purging), applied from the bottom, gives superior cleaning to surface washers or subsurface washers. Air purging also saves 30 to 50% of the wash water requirement. The recommended air rate is 3 to 7.5 scfm per sq ft. Underdrain graded gravel or siliceous layers should be a minimum of 16 in deep, with sizes ranging from $1\frac{1}{2}$ in to 6×10 mesh.

Plastic strainer underdrain nozzles screwed into flat steel decks, cemented into glazed blocks, or screwed into header laterals offer underdrains for either gravity or pressure filters without graded gravel layers. The application of these nozzles for wastewater must be carefully considered because of the possibility of clogging the fine strainer openings.

Filter strainers can also be fitted with long stems or air-metering tubes for uniform air distribution during the scouring (backwash) cycle. The air is introduced under the filter deck and forms a cushion, as shown in Figure 7.3.1. After the air pocket forms it allows air seepage to flow through the stem slots in proportion to the back pressure that develops.

The total bed depth of single or dual layers should be at least 24 in, with the sand or anthracite layers preferably 16 in each (12 in minimum). Deeper beds offer more storage space and thus longer runs. Table 7.3.1 gives the media specification ranges.

Backwash Water Source

A clear backwash water storage tank should be provided for at least a 7-min-backwash period at 15 gpm per sq ft or more. For a battery of three or more filters, the clear backwash water supply can be the filtered effluent from other filters.

Filter runs are terminated when (1) the head pressure loss across the filter reaches a predetermined value (10 ft H_2O), (2) the effluent turbidity exceeds the acceptable

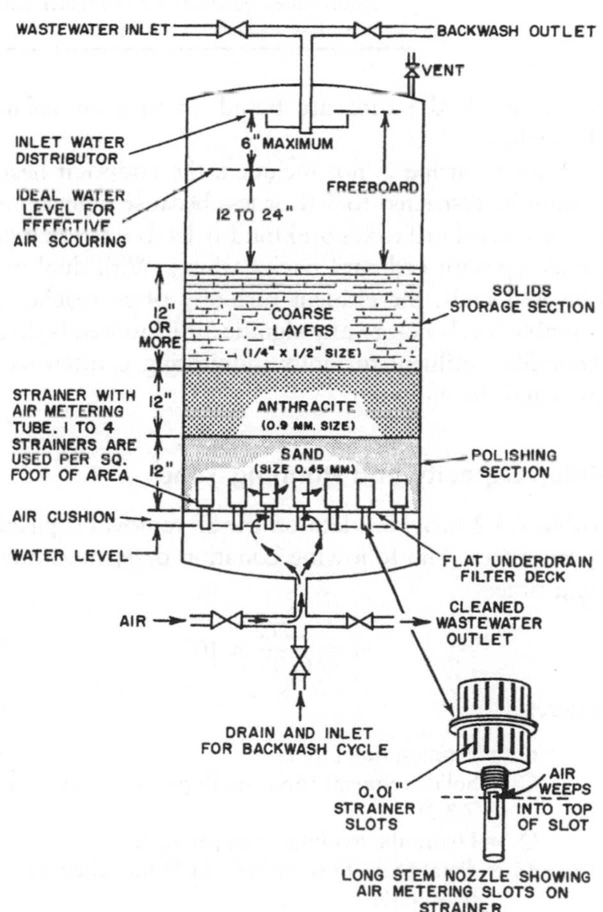

FIG. 7.3.1 Multilayered filter showing a flat underdrain deck with long-stem nozzles for washing with backwash and air.

TABLE 7.3.1 FILTER MEDIA SELECTION

	Filter Media Size (mm unless otherwise noted)				
Type of Filter Design	*Sand*	*Anthracite Coal*	*Garnet*	*Coarse, Activated Carbon*	*Plastic Granules*
Single-Layer Beds	0.4–0.6	0.6–0.8	—	—	—
Multilayer Beds	0.4–0.6	0.8–1.4	—	10 × 4 mesh[a]	$\frac{1}{4} \times \frac{1}{8}$ in[b]
Ultra-high Rates	0.8–4.0	0.8–4.0	—	—	—
Garnet for Tightening Sand Beds			0.15–0.3	—	—

Notes: [a]Bulk weight of 16 lb per cu. ft or less.
[b]Specific gravity of 1.02 to 1.07.

TABLE 7.3.2 CAPACITY OF FILTER TYPES

Filter Type[a]	*SS Removal Capacity (lb per sq ft)*
Surface filtration, septum or leaf	$\frac{1}{8}$ to $\frac{1}{4}$
Single-layer granular, sand or coal	$\frac{1}{4}$ to $\frac{1}{2}$
Mixed or dual-layer, sand and coal	$\frac{1}{2}$ to 1
Multilayered with coarse top layer (and upflow type)	1 to 2
Deep, high-rate, coarse media without chemical flocs	$\frac{1}{2}$ to 2

Note: [a]Values are based on 2 to 3 ft deep beds. For deeper layers or beds, the solids removal capacity increases by 0.5 lb per ft of additional coarse layer.

limit, or (3) the runs are based on time or volumetric throughput.

If air scouring is not included, the endpoint head loss should be restricted to 5 ft or less because higher pressure drops cause hard cakes and mud-balls to form that do not break up with ordinary backwashing. With dual or multilayered beds, the effluent turbidity often reaches unacceptable levels before any significant head loss is detected. Therefore, effluent turbidity monitoring is often required to signal the endpoint.

Filter Capacity and Running Time

Table 7.3.2 provides data on the SS removal capacities of some filters. The following equation determines the running time:

$$t = \frac{1.2C}{Q\ Ts} \times 10^5 \qquad 7.3(1)$$

where:

t = running time, min
C = Solids removal capacity, lb per sq ft (see Table 7.3.2)
Q = Hydraulic loading, gpm per sq ft
Ts = Total SS in filter influent including chemical additives, mg/l

Table 7.3.3 provides data on hydraulic loading based on a 4-hr filtering period.

DIATOMACEOUS EARTH FILTERS

The diatomaceous earth filter is available both as a vacuum and as a pressure filter. Both are designed with either septums or leaves covered with screening or fine slots that have openings of 0.003 to 0.005 in. These septums are precoated with a matt of 0.10 to 0.15 lb per sq ft of diatomaceous earth.

The head loss at the end of the run can be 35 to 100 psi with the pressure units. The solids removal capacity is rather low at about $\frac{1}{8}$ lb per sq ft. For a feed with an SS concentration of 10 mg/l, the running time at 1 gpm per sq ft hydraulic loading is about 24 hr.

Diatomaceous earth filtration produces clear effluents with removal efficiencies of over 90%. However, the costs are also high. Colloidal substances are usually not removed (as with ordinary granular filters) unless coagulants are added. Coagulants are seldom added to diatomaceous earth filter influents because they shorten the runs.

MICROSCREENING

Microscreening has been applied in domestic water treatment, sewage waste water filtration, and filtering industrial effluents. Microscreening uses a special woven metallic or plastic filter fabric mounted on the periphery of a revolving drum provided with continuous backwashing. The drum operates submerged in the flowing wastewater to approximately two-thirds of its depth.

TABLE 7.3.3 MAXIMUM RECOMMENDED HYDRAULIC LOADING OF FILTERS WITH 4 HR OF RUNNING TIME IN GPM-PER-SQ-FT UNITS

Total (including alum) SS Concentration of Filter Influent (mg/l)	Type of Filter Design			
	Standard Single Layer	*Mixed or Dual Layers*	*Multilayers*	*Deep Beds*[a]
10	5	10	10	15
25	5	8	10	15
50	4	6	8	10
100	2.5	4.5	8	8
150	2.0	3	5	6
200	—	2.5	4	5
250	—	2	3	4
300	—	—	3	4

Note: [a]Beds at least 48 in deep with noncoagulated feeds and a SS removal efficiency of 50 to 80%.

TABLE 7.3.4 MICROSTRAINER COST AND PERFORMANCE ON THE OVERFLOW FROM SECONDARY SEWAGE TREATMENT UNITS

Microscreen Size Diameter × Width (ft)	*Range of Total Capacity (mgd)*	Required bhp	
		Drive	*Backwash Pump*
5 × 1	0.1–0.5	0.5	1
5 × 3	0.3–1.5	0.75	3
7.5 × 5	0.8–4.0	2	5
10 × 10	3–10	5	7.5

Screen Size (ft)	*Screen Opening (μ)*	*Hydraulic Loading of Submerged Area (gpm/ft)*	*Total Maximum Capacity (mgd)*	Removal Efficiency (%)	
				SS	*BOD*
7.5 × 5	23	10	3	70–80	60–70
7.5 × 5	35	13	3	50–60	40–50
10 × 10	23	10	50	70–80	60–70
10 × 10	35	13	50	50–60	40–50

Wastewater enters through the open upstream end of the drum and flows radially outward through the microfabric leaving behind the SS. The deposited solids are carried upward on the inside of the fabric beneath a row of wash water jets. From there, they are flushed into a waste hopper mounted on a hollow axle of the drum.

Water for backflushing is drawn from the filtered water effluent and pumped through the jets spanning the full width of the screen fabric. Depending on the rotation speed and the size of the screen openings, only about one-half of the applied wash water actually penetrates the screen.

The drum rotation and backwash are continuous and adjustable. Either manual or automatic control based on the differential pressure can be provided. The pressure head develops due to the intercepted solids, which build up on the inside of the microfabric and create a filtration mat capable of removing particles smaller than the mesh aperture size.

Microscreen openings vary between 23 and 60 μ, which corresponds to 165,000 to 60,000 openings per sq in of surface area. The stainless steel wire cloth used in microstrainers is generally more successful than the plastic type.

The flow capacity of a size of microscreen depends on the rate of fabric clogging, drum speed, area of submergence, and head loss. The rate of screen blockage under standard head and flow conditions is called the *filterability index,* which can be determined experimentally. The amount of backwash water used ranges from 2 to 5% of the total hydraulic loading, which is in the range of 5 to 30 gpm per sq ft. Table 7.3.4 provides performance data for such units.

MOVING-BED FILTERS

Moving-bed filters are also applied to wastewater filtration. The filter involves the intermittent removal of the most heavily clogged portion of the sand filter media from the filtration zone without interrupting the filtration process.

As the influent wastewater passes through the face of the sand bed, the entire filter bed is periodically pushed in the opposite direction. The face or clogged portion of the sand bed is then periodically washed into a sludge hopper by a stream of cutter water. From there, the sludge plus the dirty sand is moved by eductors to a washing section and storage tower. The clean sand is gravity fed into the filter after each stroke.

The largest filter unit available is rated at a maximum of 250,000 gpd and has a 350-sq-ft total area. The wash water requirements represent about 7½% of the influent flow rate. A 9 hp motor is required for this unit.

The moving-bed filter operates at a maximum hydraulic loading of 7 gpm per sq ft and can remove 70 to 90% of SS from secondary, primary, and rural wastewater.

Because the intermittent movement of sand creates periodic upsets in the effluent quality and SS content, these units are classed as rough filtration or straining devices.

MEMBRANE FILTRATION

Surface filtration at high pressures (50 to 1000 psig) and low flow rates through the films or dynamically formed membranes is termed *membrane filtration.*

In membrane filtration, porous membranes with flux rates (hydraulic loadings) over 500 gpd per sq ft at 50 psig are used for polishing effluents from other filters. Membranes with accurately controlled porosities of 0.01, 0.1, 0.22, 0.45 and higher μ openings are available. Environmental engineers use the 0.45-μ membrane in evaluating filter effluents for trace concentrations of colloids, color, metallic oxides, and bacteria.

In *ultrafiltration,* tighter or less porous ultramembranes, with flux rates (hydraulic loadings) initially ranging from 50 to 300 gpd per sq ft at 50 psig, are capable of rejecting high-molecular-weight (2000 and above), soluble, organic substances, but not salt.

In *hyperfiltration* (reverse osmosis), specially prepared membranes or hollow fibers with flux rates at 5 to 50 gpd per sq ft at 400 to 800 psig affect salt, soluble organic matter, colloidal or soluble silica, and phosphate removal at 80 to 95% efficiency.

All membrane processes are considered to be final polishing filters, with common particulate removals in excess of 99%. In so doing, they foul easily, and their flux flow rate declines logarithmically with running time. Therefore, wastewater treatment facilities must protect membrane filters from fouling by pretreating the feeds using coagulation and rough filtration.

Ultrafiltration Membranes in Activated-Sludge Processes

BOD REMOVAL EFFICIENCY

+95%

FILTER EFFLUENT QUALITY

BOD below 1 mg/l, COD between 20 and 30 mg/l, no SS or coliform bacteria, essentially colorless and odorless, and removal of molecules with molecular weights of 8000 to 45,000 together with some viruses

MEMBRANE FILTER PROTECTION AGAINST FOULING

By screening of the influent and maintaining high-flow velocities at the membrane surface

AVAILABLE TREATMENT CAPACITY RANGE

3000 to 30,000 gpd

MEMBRANE SURFACE VELOCITY REQUIRED

3 to 8 fps provided by recirculation.

PORE SIZES

3 to 100 Å

MEMBRANE PRESSURE DROP

2 to 40 psid

HYDRAULIC LOADING (FLUX) RANGE

5 to 30 gpd per sq ft

Continuous biological oxidation processes combined with ultrafiltration membranes are used as a means of effluent separation. The membranes filter out the biological cells while allowing passage of the treated effluent.

These membranes can achieve BOD removals in excess of 99% on a commercial scale with an influent containing 200 to 600 mg/l BOD. The membrane filter produces an essentially colorless and odorless effluent having both zero SS and coliform bacterial counts. These systems consist of an activated-sludge reactor and a membrane filter and are available in treatment capacities ranging from 3000 to 30,000 gpd.

ULTRAFILTRATION MEMBRANES

Ultrafiltration membranes are thin films cast from organic polymer solutions. The film thickness is 5 to 10 mils (0.005 to 0.01 in). The film is anisotropic, i.e., it has thin separation layer on a porous substructure. The thin working or separation layer has a thickness of 0.1 to 10 μ. Figure 7.3.2 shows a film cross section. The separation layer contains pores of closely controlled sizes ranging from 3 to 100 Å.

ULTRAFILTRATION PROCESS

In ultrafiltration devices, the separation layer is adjacent to a pressurized chamber containing the filter influent. When pressure is applied, small molecules pass through the membrane and exit on the other side; larger molecules

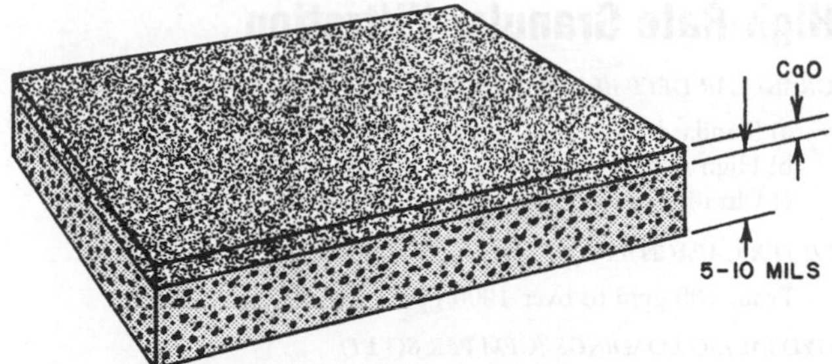

FIG. 7.3.2 Anisotropic, diffusive ultrafilter.

are retained within the pressurized chamber. The pressure drop across the membrane ranges from 2 to 40 psid. If the chamber is continuously fed with new influent, the concentration on the feed side gradually increases. This concentrate is continuously bled from the pressurized side of the membrane.

Typical ultrafiltration membranes used in these systems exhibit useful operating fluxes (hydraulic loadings) from 5 to 30 gpd per sq ft at pressure drops of 2 to 30 psid. The membranes filter out protein molecules with molecular weights from 8000 to 45,000; consequently, most viruses are also retained.

ULTRAFILTRATION DEVICES

The objective in the mechanical design of an ultrafiltration device is to provide the largest working area of membrane surface per unit of filter volume. Provisions must be made for a pressurized channel on the feed side of the membrane, support of the membrane film, draining and collecting the filtered effluent that permeates the membrane, and mechanically supporting the whole structure.

Environmental engineers determine the dimensions of the feed channel based on the size of the particles contained in the feed streams and hydrodynamic considerations to provide sufficient flow past the membrane surface to minimize concentration polarization (a concentrated layer developing at the membrane surface). Flow velocities in the range of 3 to 8 ft per sec are used. Because of the high solids content in the reaction systems and the presence of large particles in the feed, large feed channels are required. The membranes are packaged in either plate configurations having channel dimensions of approximately 0.090 in or tubular configurations having inside diameters of $\frac{1}{4}$ to 1 in. Figure 7.3.3 is a schematic representation of both types.

The plate device is comprised of sheets of porous support material on which the membrane is cast. The sheets are in a parallel array and terminate in a collection header or manifold. Feed material (influent) passes between the sheets, and the effluent permeates the membrane and passes into and up the porous support member to the exit header.

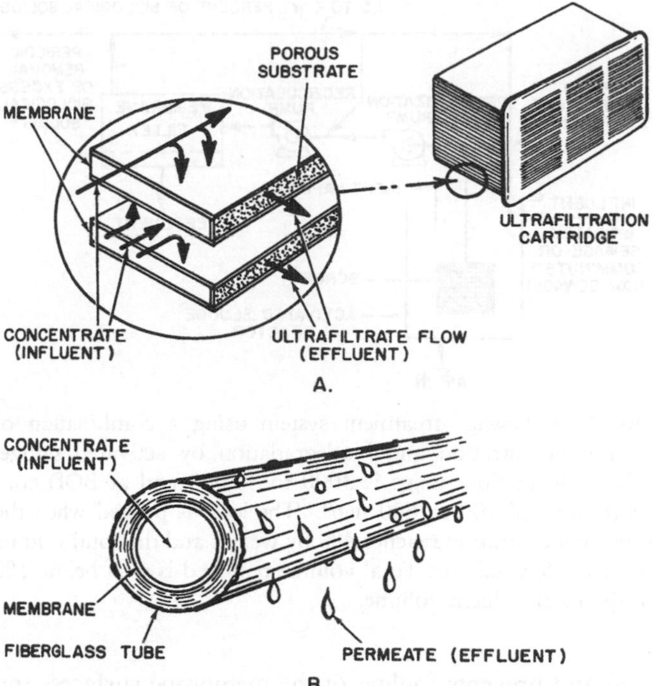

FIG. 7.3.3 Ultrafilter membrane packaging configurations. A. Plate device; B. Tubular device.

With the tubular configuration, a support tube manufactured from sintered, porous, polymeric materials or fabricated as a composite from fiberglass and polyester or epoxy materials forms the pressure vessel. The membrane is cast or placed on the inside of the tube. Feed material (influent) flows through the inside of the tube, and due to operating pressure, the effluent permeates the membrane, passes into the porous supporting substrate, and is collected in a manifold. Series and parallel arrays of tubes are available (as in a shell and tube-heat exchanger) guaranteeing adequate flow past the membrane to minimize concentration polarization. Groupings of membrane modules in series and parallel can also provide the feed recirculation rates required to minimize concentration polarization.

SEWAGE TREATMENT APPLICATIONS

As shown in Figure 7.3.4, the effluent from the activated-sludge reactor is continuously withdrawn through a screening device to the recirculation pump of a membrane loop. The membranes separate the large molecules, and the retained solids are returned to the activated-sludge reactor. Because only small molecules can pass through the membrane separation device, a high concentration of biological solids develops in the activated-sludge reactor. Wastewater treatment facilities characteristically run this type of activated-sludge process with biological solids concentrations from 15,000 to 40,000 mg/l. Operating with such high solids content has two advantages: (1) the large biomass quickly degrades organics that can enter with the

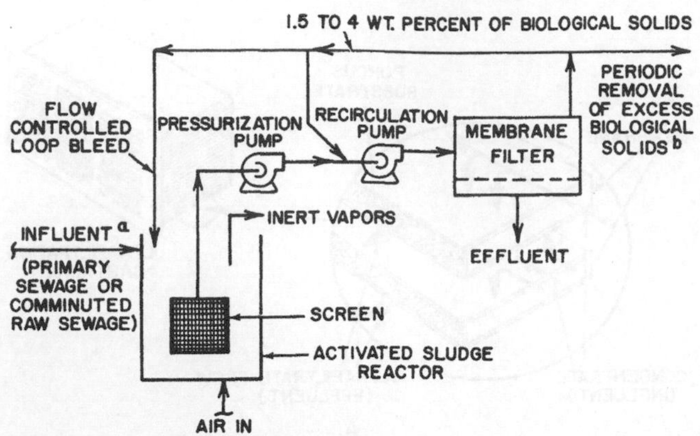

FIG. 7.3.4 Sewage treatment system using a combination of membrane filtration and biodegradation by activated sludge. [a]The influent flow range is 3000 to 30,000 gpd at BOD concentrations of 200 to 600 mg/l. [b]The loop is purged when the solids concentration reaches 4% by weight and the total volume is only a few gallons. Total volume removed is well below 1% of the total influent volume.

feed and prevents fouling of the membrane surfaces, and (2) the activated-sludge reactor size is much reduced.

The membrane filter in the system guarantees a practically infinite detention time for the slow biodegradable components in the sewage because they cannot exist from the system. The biological solids are also totally contained by the membranes. Such a sewage treatment system operates with almost no discharge of excess activated-sludge solids. Practically all feed materials are converted to carbon dioxide, water, and inorganic salts. Some inert materials do accumulate in the reactor, and purging (see Figure 7.3.4) of the contents of the reaction system is periodically necessary.

Performance of Sewage Treatment Applications

Commercial-scale systems with capacities of 3000 to 30,000 gpd have operated for thousands of hours. These systems handle sanitary waste containing 200 to 600 mg/l BOD. Hydraulic loading (flux) of the membranes is sustained at approximately 10 gpd per sq ft.

These systems achieve essentially 95% BOD removal. The BOD that does pass through is in the form of dissolved solids. The BOD content of the effluent is below 1 mg/l, with COD ranging from 20 to 30 mg/l. Coliform bacteria counts in the effluent have been zero, and additional sterilization of the effluent is not practiced.

The processes are stable when occasionally exposed to toxic materials in the feed that would upset other, less-concentrated, biological systems. This stability can be explained partially by the large biomass in the process and partially by the fact that the membrane is a positive barrier preventing the exit of the dead biomass.

High-Rate Granular Filtration

GRANULAR DEEP-BED FILTER TYPES

a) Standard
b) High rate
c) Ultra-high rate

FILTER CAPACITIES

From 100 gpm to over 1000 gpm of water

HYDRAULIC LOADINGS (GPM PER SQ FT)

2 to 5 (a), 5 to 15 (b), and over 15 (c)

BED DEPTHS

4 to 8 ft

SS REMOVAL EFFICIENCY

50 to 75% (c), 80 to 90% (b), and 90 to 98% (a).

SOLIDS LOADINGS (LB PER SQ FT)

1 to 10

MAXIMUM ACCEPTABLE SOLIDS CONCENTRATIONS IN INFLUENT

100 mg/l

MAXIMUM ACCEPTABLE FIBER CONCENTRATION IN INFLUENT

10 to 25 mg/l

MAXIMUM ACCEPTABLE SOLIDS PARTICLE SIZE IN INFLUENT

200 μ

MAXIMUM ACCEPTABLE OIL CONCENTRATION IN INFLUENT

25 to 75 mg/l

High-rate, granular filtration systems are used for SS removal from a variety of water and wastewater streams. Applications include industrial process water; municipal potable water; final polishing of sewage treatment effluents; removal of mill scale and oil from hot rolling-mill cooling water; removal of residual oil from American Petroleum Institute (API) separator and dissolved-air flotation effluents; and pretreatment of water and wastewater for advanced forms of treatment, such as carbon adsorption, reverse osmosis, ion exchange, electrodialysis, and ozonation.

Recent applications include the simultaneous removal of SS and suspended or dissolved phosphorus and nitrogen nutrients using combination biological–physical–chemical processes in granular-filtration-type systems. The basic purpose of high-rate granular filtration processes is to remove low concentrations of SS from large volumes of water (from 100 gpm to usually more than 1000 gpm).

DEFINITIONS

The following terms apply to high-rate granular filtration:

HIGH-RATE FILTRATION—A filtration process designed to operate at unit flow rates (hydraulic loadings) between 5 and 15 gpm per sq ft.

GRANULAR (DEEP-BED) FILTRATION—A filtration process that uses one or more layers of granular filter material coarse enough for the SS to penetrate into the filter bed to a depth of 12 in or more. Total depth of the single or composite beds is defined arbitrarily as a minimum of 4 ft.

MONO-MEDIA—A deep-bed filtration system that uses only one type of granular media for the filtration process, exclusive of gravel support layers.

DUAL-MEDIA—A deep-bed filtration system that uses two separate and discrete layers of dissimilar media, e.g., anthracite and sand, placed on top of each other for filtration.

MIXED-MEDIA—A deep-bed filtration system that uses two or more dissimilar granular materials, e.g., anthracite, sand, and garnet, blended by size and density to produce a composite filter media graded hydraulically after backwash from coarse to fine in the direction of the flow.

SPECIFIC SOLIDS LOADING—The weight of the SS that can be removed by a filter before backwashing is required; usually expressed as lb per sq ft per cycle. Specific loading is a function of filter application, media size, media depth, unit flow rate (hydraulic loading), and influent solids concentration.

OPTIMIZATION—Designing and operating a filtration system to produce the maximum quantity of acceptable filtrate at the minimum capital and operating cost. In an operationally optimized system, SS breakthrough occurs at the maximum available pressure head loss.

REMOVAL MECHANISMS

The predominant mechanisms responsible for removing SS in granular filtration systems have been previously discussed. As shown in Figure 7.3.5, surface screening is not a predominant removal mechanism in this type of system, and solids are retained deep within the voids of the media. This system requires intensive backwashing (see Figure 7.3.6) to dislodge the entrapped solids that accumulate during the filtration process.

Surface wash systems are acceptable only if screening is the predominant removal mechanism. Otherwise, high-energy backwashing is required throughout the depth of deep-bed granular filters to completely clean the media after each filtration cycle.

REMOVAL EFFICIENCY

The removal efficiency of a granular filtration system is a function of the hydraulic loading (unit filtration rate), the grain size and depth of the media used, and the filterability of the solids to be removed. Concentration, particle size, density, and shape of the SS and water temperature also affect filtration efficiency.

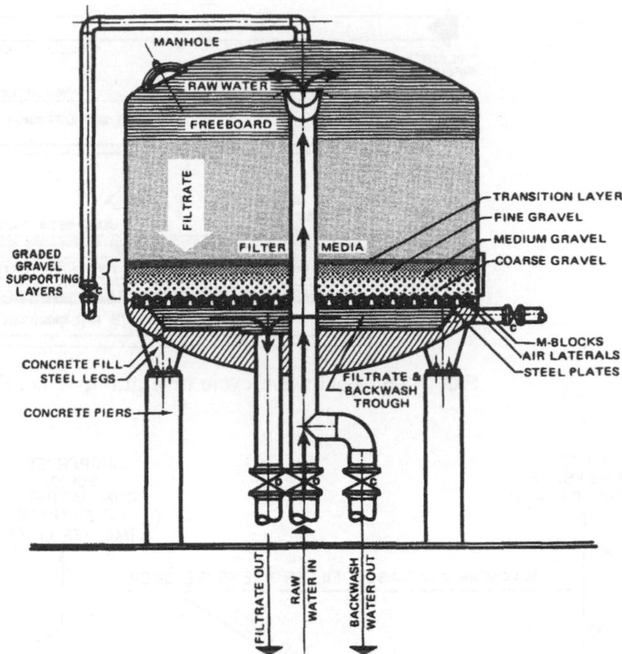

FIG. 7.3.5 Pressure-type, deep-bed, granular filter in forward flow operation. (Forward flow corresponds to the filtering, and back flow corresponds to the wash cycle.)

Ultra-high-rate filters remove only 50 to 75% of applied solids. Filters with less than 15-gpm-per-sq-ft-hydraulic loading provide at least 90% solids removal, and 98% or more is also possible.

Coarse media are used in the diameter range of 0.5 to 5 mm, with most of the media in the 1- to 3-mm range. The specific solids loading ranges are between 1 and 10 lb per sq ft. Coagulants and polyelectrolytes can be fed directly into the filter.

For example, if a specific loading of 5 lb per sq ft can be obtained, a 12.5-ft-diameter filter can remove 600 lb of SS before backwashing is required. This specific loading permits the addition of chemical coagulants into the filter influent to improve the removal efficiency without excessively reducing the filtration cycle period.

SYSTEM COMPONENTS AND DESIGN

The use of air scouring (purging to assist the cleaning action) during backwashing is almost universal practice (see Figure 7.3.7). With mono-media filters, wastewater treatment facilities can use high air and backwash water rates without disrupting the bed layers or carrying the lighter components of the bed out of the filter.

Downflow filters eliminate the tendency of upflow units to fluidize or break through, especially near the end of the filtration cycle when pressure drop across the filter is substantial. Both gravity and pressure-motivated filters are used in waste treatment, but pressure units have deeper beds, higher capacities, and the ability to be completely

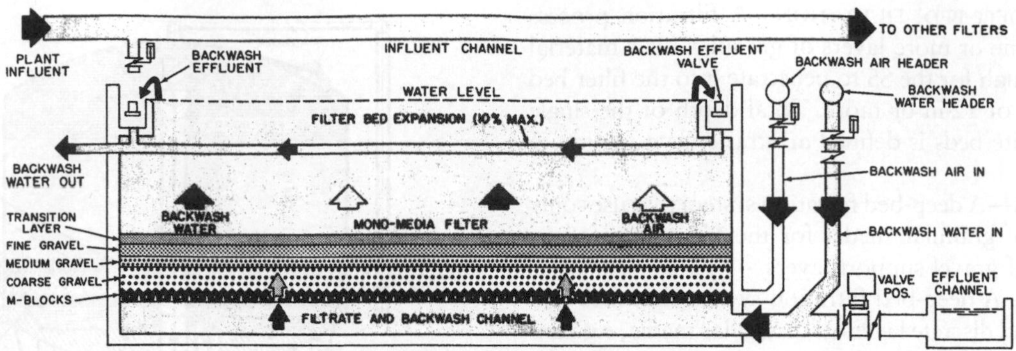

FIG. 7.3.6 Backwash cycle of a gravity-type, deep-bed, granular filter.

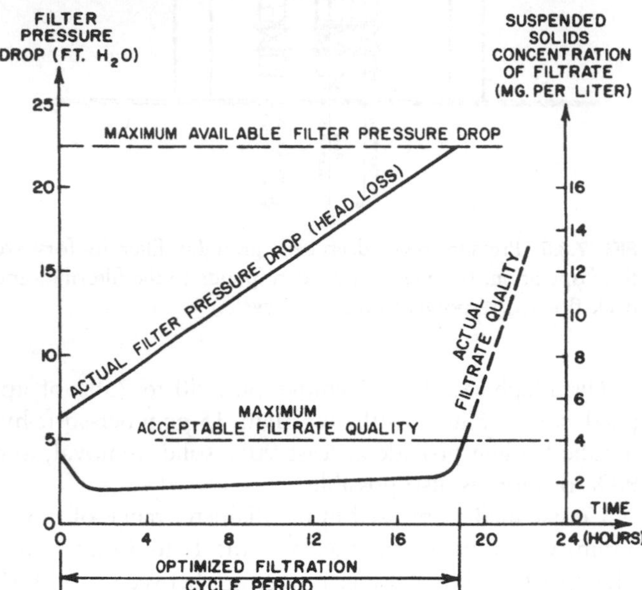

FIG. 7.3.7 Optimization (economic) of filter operation.

automated. Some regulatory agencies, however, prefer gravity systems because the filters can be observed during filtration and backwashing.

OPTIMIZATION

For many years, filters were used as insurance against contamination of potable water. Historically, in wastewater treatment, the effluent quality obtained from clarifiers was sufficient to meet discharge requirements, and effluent filtration was unnecessary. However, with the advent of effluent standards requiring more than 90% removal of the SS, wastewater filtration has become common practice.

Once the physical system is designed and installed, environmental engineers can try to optimize the operational cost. The criterion is that the maximum amount of acceptable quality filtrate is produced per unit of operating cost.

During filtration, the pressure required to maintain flow across the filter increases gradually due to solids accumulating in the filter media. The effluent quality is slightly better at the start of each filtration and remains relatively constant until media voids are full. Then, a sudden breakthrough of solids into the filtrate occurs. A filter is optimized when the maximum available filter pressure drop coincides in time with the breakthrough of the SS (see Figure 7.3.7).

APPLICATIONS

High-rate, granular filtration is used for effluent polishing, pretreatment, phosphate removal, and nitrogen removal.

Effluent Polishing

Deep-bed, granular filters are applicable to polishing effluents from physical, chemical, or biological wastewater treatment systems. These filters use coarse-bed media and permit high hydraulic and solids loadings (see Table 7.3.5). The effluent from a typical municipal or industrial activated-sludge plant contains approximately 20 mg/l of SS. Effluent polishing by granular filtration can remove another 80 to 90% of this contaminant. Similarly, these filters can remove 50 to 90% of the free oil (and its associated BOD) from API or rolling-mill system effluents.

Pretreatment

Many advanced wastewater treatment processes require pretreating the waste to reduce its SS content. This requirement is especially true for the adsorption processes (activated carbon and ion exchange) and of some of the membrane processes (reverse osmosis and electrodialysis).

Prefiltration prior to the carbon or resin columns can eliminate the need for backwashing and reduce the loss by attrition. Occasionally, wastewater treatment facilities can reduce the number of columns because taking a column out of service for backwashing is unnecessary. Reverse osmosis membrane systems reduce membrane fouling by eliminating SS from the feed.

In pretreatment, essentially complete removal of SS is necessary; consequently, deep beds of fine media, e.g., 4 ft

TABLE 7.3.5 FILTRATION APPLICATIONS FOR GRANULAR, MONO-MEDIA, DEEP-BED UNITS

Applications	*Coarse Bed Media Size (mm diameter)*	*Bed Media Depth (ft)*	*Hydraulic Loading or Filter Rate (gpm/ft)*
Activated-sludge effluents, gravity	2–3	4–6	3–5
Activated-sludge effluents, pressure	2–3	6	5–10
API effluents (oil removal)	1–2	6–8	3–5
Chemical treatment effluents	1–2	4–6	5–8
Denitrification with effluent polishing	3–6	6–10	2–5
Hot-strip-mill effluents, carbon steel	2–3	6	8–14
Hot-strip-mill effluents, alloy steels	1–2	6–8	6–10
Impounded supply, process water	2–3	4–6	10–20
Phosphate removal	2–3	4–6	5–8
Pretreatment	1–2	4–8	5–12
Primary sewage effluent, storm water	2–3	4–6	5–15
River water filtration, process water	1–2	6–8	4–8
Side arm filtration, cooling towers	2–3	4–6	10–20
Trickling filter effluents	1–2	4–6	4–10

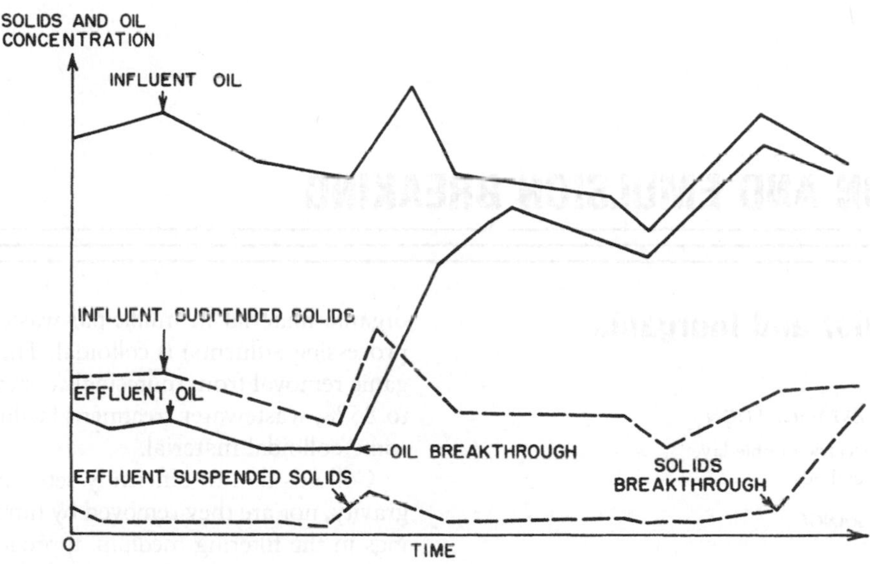

FIG. 7.3.8 The relationship between oil and SS breakthrough in granular filters.

of mixed media or 6 ft of 1.0-mm, mono-media, and low filter rates (2 to 4 gpm per sq ft) are used.

Phosphate Removal

Phosphate removal by chemical precipitation, either as a separate process or in combination with biological processes, does not require filtration. However, when either low residual levels of phosphorus or maximum utilization of chemicals is desired, filtration of the chemically treated effluent is desirable. As much as 2 mg/l phosphorus can be removed by granular filtration of the effluent from the phosphate precipitation processes.

Nitrate Removal

Nitrogen removal with biological denitrification is enhanced by granular effluent filtration. Slow growing denitrifying organisms are easily washed out of a biological system by hydraulic surges or clarifier upsets. Granular filtration prevents the loss of nitrifying or denitrifying organisms and returns them to the system with the backwash water. This form of postfiltration is especially important when wastewater temperatures are below 65°F because biological reaction rates slow down at such temperatures.

Wastewater treatment facilities can combine granular filters for SS removal with columnar denitrification units in a single system. In such combined systems, larger media are used, and special backwashing procedures are required to obtain 80% removal of the nitrate without substantially reducing the SS removal efficiency.

LIMITATIONS

Oil, fiber, and SS can be tolerated in granular filtration systems, within limits. The scale-forming tendencies of the wastewater must also be eliminated or controlled. These systems generate backwash water as a natural consequence of the process and wastewater treatment facilities must provide for its proper handling.

Fiber in concentrations greater than about 10 to 25 mg/l can cause operating problems in granular filters. Long fibers mat and blind off the filter surface. Short fibers accumulate in the underdrains and plug the bottom of the filter unless special designs are used to prevent it.

Oil can plug the underdrains or prevent complete backwashing, especially when fine-bed media (less than 1.0 mm diameter) are used. Water-wash filters can have difficulty operating with 25 mg/l of free oil, while heavy-duty, air-scouring backwash systems can operate satisfactorily with up to 50 to 75 mg/l of oil. Oil reduces the specific solids loading and breaks through the filter sooner than SS. Figure 7.3.8 shows this effect.

An SS concentration in the influent of about 100 mg/l can be tolerated in a properly designed granular filter provided that the size of the solids does not prevent penetration into the media. The maximum feasible particle size is about 200 μ although minor concentrations (5 mg/l) of oversized solids, such as leaf fragments and bits of paper, can be tolerated. High SS concentrations reduce the filter cycle time, which becomes a problem when the operating cycle is so short that the filters cannot be washed in time to keep the system in operation. For example, if six filters are in a continuous system and a total of $\frac{1}{2}$ hr is required to wash a filter, the minimum usable cycle time is 3 hr.

—*G.S. Crits*
P.L. Stavenger
E.S. Savage

7.4 COAGULATION AND EMULSION BREAKING

Colloidal Behavior and Inorganic Coagulants

METHODS OF COLLOID DESTABILIZATION

a) Modification of the electric double layer
b) Polymer bridging of colloids

COLLOID MATERIAL SIZE RANGE

0.01 to 10 μ

COLLOID CONCENTRATION IN SEWAGE

15 to 25% of all organic material

COAGULANTS

Aluminum salt, ferric iron salts, and magnesium

OPTIMUM pH LEVELS

5 to 7 pH for aluminum

CHEMICAL SLUDGE PRODUCTION BY LIME TREATMENT

4500 to 6500 lb per mg of water

COLLOIDAL POLLUTANTS

Colloids are generally defined as particulate matter in the 0.01 to 10 μ size range. Approximately 15 to 25% of the organic material in municipal wastewater (more in food processing effluents) is colloidal. Therefore, to achieve organic removal from municipal wastewater greater than 75 to 85%, wastewater treatment facilities must also remove some colloidal material.

Colloids do not efficiently settle under the influence of gravity, nor are they removed by filtration unless the openings in the filtering medium approach the size of the colloids themselves (see Section 7.3). For effective removal of colloidal material from wastewater by gravity settling or in-depth filtration, wastewater treatment facilities must use coagulation and flocculation. Understanding the significance of these unit processes requires an understanding of the physical, chemical, and electrical properties of colloidal suspensions.

COLLOIDAL PROPERTIES

Two broad classes of colloidal material are inorganic and organic. Inorganic colloids generally consist of inert mineral particles such as silt, clay, and dust. These pollutants do not generally pose a hazard to human health, but they do have an undesirable esthetic quality when they are discharged to receiving waters in high concentration.

TABLE 7.4.1 RELATIVE SIZE OF POLLUTANTS IN WASTEWATER

Particle Diameter (meters)	*Designation*	*Typical Substance*	*Relative Size*
$<10^{-8}$	Ions and molecules	Glucose and chloride	1
$>10^{-8}$ $<10^{-5}$	Colloids	Bacteria, phages, clay, and macro-molecules	1 to 1000
$>10^{-5}$ $<10^{-3}$	Fine particulates	Silt, fine sands, and clays	1000 to 100,000
$>10^{-3}$	Coarse particulates	Coarse sand	Greater than 100,000

Organic colloids generally consist of bacteria, viruses, phages, fragments of cellular material from living organisms, waste food, and a variety of materials from the chemical and food processing industries. These organic colloidal pollutants can be deleterious. In addition, they also consume DO and create a potential for putrefaction during their biological degradation.

The most important property of colloids is their very small size. Table 7.4.1 shows the size range in which colloidal materials are normally classed. Because of their size, colloids have a large surface area per unit volume, or per unit mass of material. The properties of this surface, or more specifically the interfacial region between the solid colloid and the bulk liquid (water) phases, governs the action of a colloidal suspension.

The electrical properties at the solid–liquid interface depend on the origin of the solid surface and the physicochemical properties of the solid and liquid phases. Solid particles encountered in wastewater treatment originate from three general sources including degradation of larger particles, biological agents, and condensation of small particles forming larger ones.

When suspended in water, the surfaces of these solids exhibit a surface charge that can arise from the specific adsorption of potential determining ions, dissociation of ionic species at the surface, internal atomic defects in the solids phase, or other causes. This surface charge is counterbalanced by oppositely charged ions in the liquid adjacent to the solid–liquid interface. The distribution of ions in the region adjacent to the interface is different from that in the bulk of the solution and is described by the electrical double-layer theory.

Figure 7.4.1 is a diagram of an electrical double layer based on Stern's modification of the Gouy–Chapman model. A layer of fixed, nearly immobile ions adjacent to the negatively charged surface reduces the surface potential to the Stern potential. Outside this fixed layer (called the Stern layer) is a diffuse layer of ions, whose concentration is described by a Boltzman distribution. In the diffuse layer, an excess of positive counterions exists so that the total charge in the diffuse layer equals the total surface charge minus the total charge in the Stern layer.

When two charged surfaces are brought together so that their diffuse layers overlap, an electrical force exists between the two surfaces. For surfaces with similar potentials, a repulsive force exists; for surfaces with opposite potentials, an attractive force exists. The magnitude of the repulsive or attractive force is related to the magnitude of their respective potentials and the distance of separation between the surfaces.

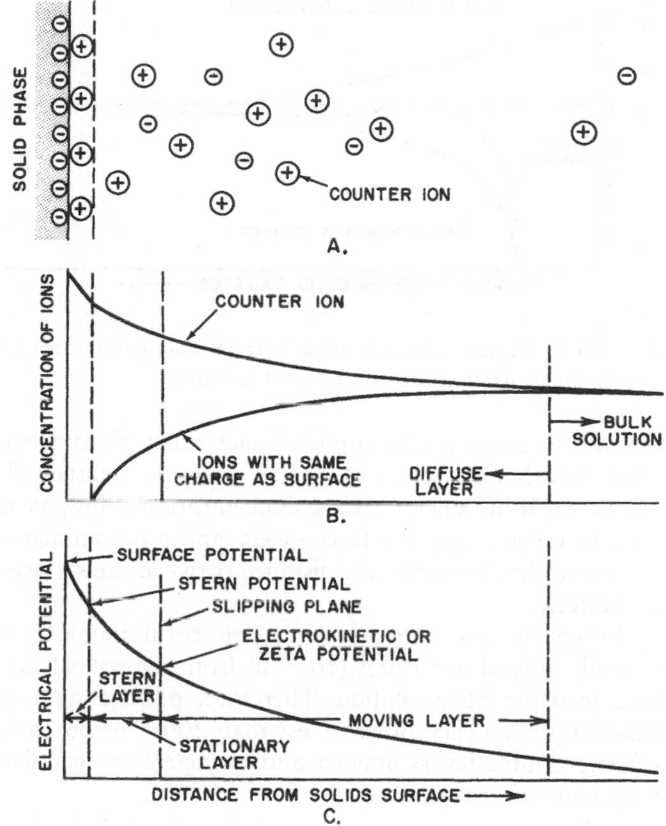

FIG. 7.4.1 Model of electrical double layer for an electronegatively charged surface.

In addition to electrical forces between surfaces, van der Waals universal attractive forces between atoms are also significant. These attractive forces arise from interacting, fixed and induced atomic dipoles and fundamental dispersion forces.

Figure 7.4.2 shows the results of an interaction of charged surfaces according to the electrical double-layer model. Part A shows surfaces with a high charge concen-

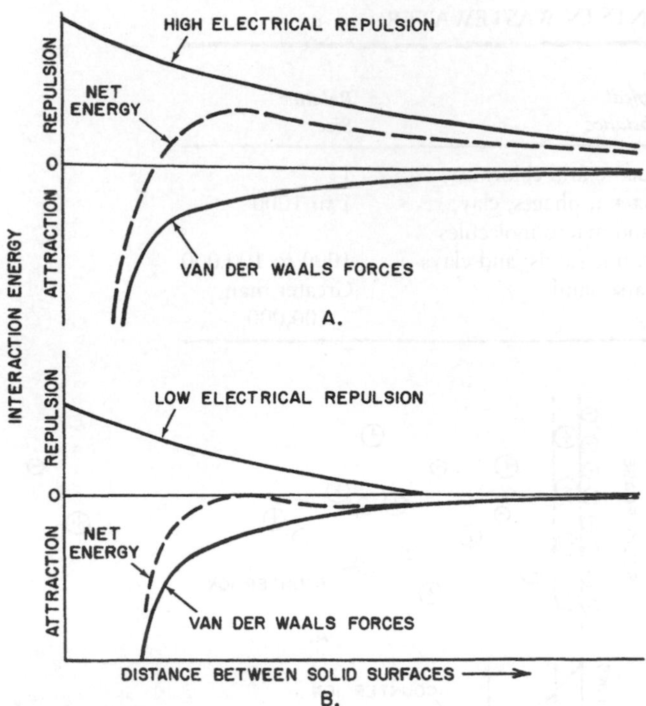

FIG. 7.4.2 Typical electrical double layer interaction. Surface charge concentration is high in A and low in B.

tration. As two particles approach each other, the distance over which net repulsive forces exist is substantial. However, if the surface charge concentration is low, as in Part B, no net repulsive force exists, and a net attractive force can develop when the distance between the surfaces is reduced.

In typical wastewater, the diffuse electrical double layer extends only about 100 Å (10^{-8} m) from the colloid surface into the bulk solution. Therefore, particle surfaces must approach each other to less than 200 Å before their electric double layers interact and attractive or repulsive forces manifest themselves.

DESTABILIZATION OF COLLOIDAL SUSPENSIONS

For definition purposes, a stable colloidal suspension exists when particles subjected to flocculation, i.e., bringing particles together, do not adhere to one another to form agglomerates of primary colloidal particles. The most important property causing colloidal stability is the existence of net repulsive forces when electrical double layers interact. Conversely, a destabilized colloidal suspension exists when particles subjected to flocculation adhere to one another, forming agglomerates of primary colloidal particles.

Modification of the Electric Double Layer

This method of destabilizing colloidal suspensions modifies the electrical double layer. Reducing the surface or Stern potential or compressing the diffuse electrical double layer diminishes the repulsive interaction forces, which can result in destabilization.

In practice, many colloids have hydrogen- and hydroxyl-potential-determining ions. Adjusting the solution pH alone reduces their surface potential, thus diminishing the repulsive electrical forces and resulting in destabilization. This condition is shown by both curves in Figure 7.4.3.

An indifferent electrolyte does not contain surface-charge-potential-determining ions. Adding an indifferent electrolyte to a stable colloidal suspension causes destabilization by compressing the electric double layer, which reduces or eliminates a net repulsive interaction barrier. Figure 7.4.3 shows the effect of indifferent electrolyte concentration changes where moving from point A to B requires the addition of an electrolyte and results in destabilization.

According to the Stern modification of the Gouy-Chapman electric, double-layer model, destabilizing a col-

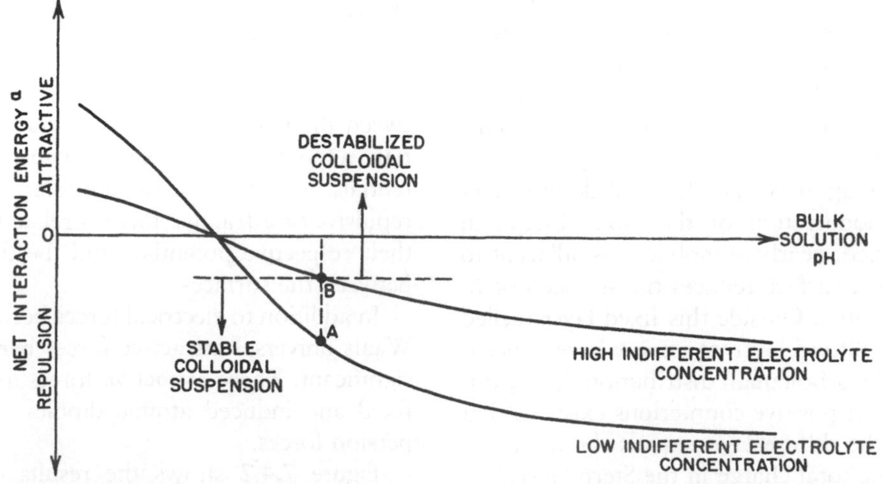

FIG. 7.4.3 Effect of pH and an indifferent electrolyte on colloidal suspension stability (for surfaces with hydrogen- and hydroxide-potential-determining ions). [a]At a given separation distance where electric double layers interact.

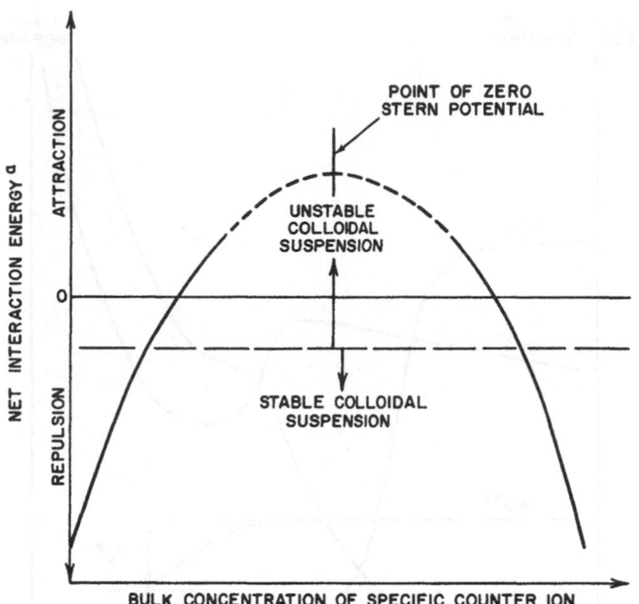

FIG. 7.4.4 Effect of specific counter ion concentration on colloidal suspension stability. [a]At some given separation distance where electric double layers usually interact.

loidal suspension by specific adsorption of specific counter ions at the colloid surface is possible. Here, the surface potential remains unchanged, but the Stern and zeta potential can be reduced or even reversed in charge. In this process, adsorptive rather than electrostatic, forces must be operative. Figure 7.4.4 shows the effect of Stern potential reduction and reversal and gives a range of specific counter ion bulk solution concentrations where destabilization occurs.

Chemical Bridging of Colloids

Since the early 1950s, environmental engineers have used natural and synthetic polymers to agglomerate or flocculate finely divided, suspended material in water and wastewater treatment. From the beginning, they observed the anomalous behavior of the electric double-layer model. Therefore, they developed the chemical bridging model to explain the action of polymers on colloidal suspension stability.

In its simplest form, the chemical bridging model suggests that a polymer can attach itself to the surface of a colloid at one or more sites, with a significant length of the polymer extending into the bulk solution. Reaction A in Figure 7.4.5 shows this condition. That such action causes destabilization is suggested in reaction B, where two colloids with attached and extended polymers agglomerate when flocculated.

The key aspect of the bridging model is that adsorption of polymers on colloid surfaces involves more than coulombic forces. Postulated interactions include hydrogen bonding, coordinate covalent bonding and linkages, van der Waals forces, and polymer–solvent solubility considerations.

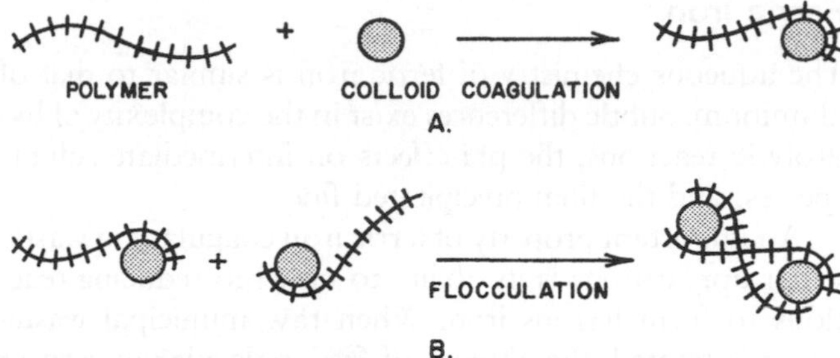

FIG. 7.4.5 Colloidal suspension destabilization with high-molecular-weight organic polymers. **A.** Destabilization; **B.** Agglomeration.

INORGANIC COAGULANTS

Significant coagulants in wastewater treatment are aluminum, ferric iron salts, and magnesium, which is active in lime treatment.

Aluminum

When aluminum is added to wastewater, usually in the form of dissolved aluminum sulfate, it undergoes several reactions with the naturally present alkalinity and various anionic ligands. A series of hydrolytic reactions occurs, proceeding from the formation of a simple hydroxo complex, $Al(OH)^{2+}$, through the formation of soluble polyhydroxy polynuclear inorganic polymers, to the formation of a colloidal precipitate. These reactions are rapid, occurring within a fraction of a second.

The soluble, polymeric, kinetic intermediates are generally believed to be the causative agent, i.e., coagulant species, in destabilizing colloidal suspensions. Because the adsorption of these coagulating species on the colloid surface involves more than coulombic effects, the electrical double-layer model cannot be explicitly invoked.

Since OH^- is the most important ligand in the soluble polymeric aluminum hydroxo complex, the effectiveness of aluminum coagulation is strongly pH dependent. Generally, an optimum pH is in the range of 5 to 7. The presence of significant amounts of the competing ligand PO_4^{3-} reduces optimum coagulation pH.

Since aluminum reacts with wastewater alkalinity (usually HCO_3^- results), wastewater treatment facilities must consider the possible excessive depression of pH in a weakly buffered water. They may have to add alkalinity, usually in the form of hydrated lime ($Ca[OH]_2$), to prevent excessive pH depression.

Using aluminum as a coagulant in wastewater treatment involves dosages in excess of the coagulation requirements, which results in precipitation of excess alum floc. This excess floc results in substantial removal of finely divided SS by entrapment in the floc structure during flocculation and gravity settling.

Ferric Iron

The aqueous chemistry of ferric iron is similar to that of aluminum. Subtle differences exist in the complexity of hydrolytic reactions, the pH effects on intermediate soluble species, and the final precipitated floc.

An important property of ferric iron coagulant in wastewater applications is its ability to undergo reducing reactions to form ferrous iron. When raw municipal wastewater is treated, the absence of free molecular oxygen or any other oxidation agent results in a reducing environment. Research shows that precipitated ferric-iron hydrolysis species release iron as a reduced ferrous compound when they are maintained in an anaerobic environment. This property is significant in applications where removal of phosphorus compounds and the destabilization of colloidal pollutants is required. However, its significance for destabilization only is unknown.

Magnesium

Lime treatment for phosphorus precipitation and SS removal from raw and biologically treated municipal wastewater has become an accepted sanitary engineering unit process. The chemistry of lime treatment related to colloidal suspension destabilization is not a well-developed technology. However, certain observations based on laboratory and full-scale plant operating results can be made.

Adding hydrated lime [$Ca(OH)_2$] to wastewater simply increases the solution pH. At increasing pH levels, calcium phosphate, calcium carbonate, and magnesium hydroxide are insoluble as shown in Figure 7.4.6. The phosphate and carbonate precipitates undergo some agglomerative and adsorptive interactions with soluble and colloidal pollutants.

However, the coagulant species of significance is the magnesium hydroxide precipitate. Thus, wastewater treatment facilities must adjust the pH upward to a value that causes magnesium precipitation. This pH level varies with different wastewaters. For example, the lower the initial magnesium concentrations, the higher the pH required to precipitate the amount of magnesium necessary for destabilization. The incremental precipitation of the magnesium coagulant is shown as Mg^{2+} in Figure 7.4.6.

Two factors are significant in the lime treatment of wastewater. The first is the excessive amount of chemical sludge (precipitates) produced. Depending on the amount of Ca^{2+}, HCO_3^-, and PO_4^{3-} present, chemical sludge production ranges from 4500 to 6500 lb per MG. Second, a variety of precipitates are formed. These precipitates range in character from colloidal hydroxylapatite [$Ca_5OH(PO_4)_3$] to dense $CaCO_3$ to voluminous $Mg(OH)_2$ floc. The proportions of these precipitates and the originally present SS determine the net floc settling rate, concentration of settled sludge, and sludge dewatering properties.

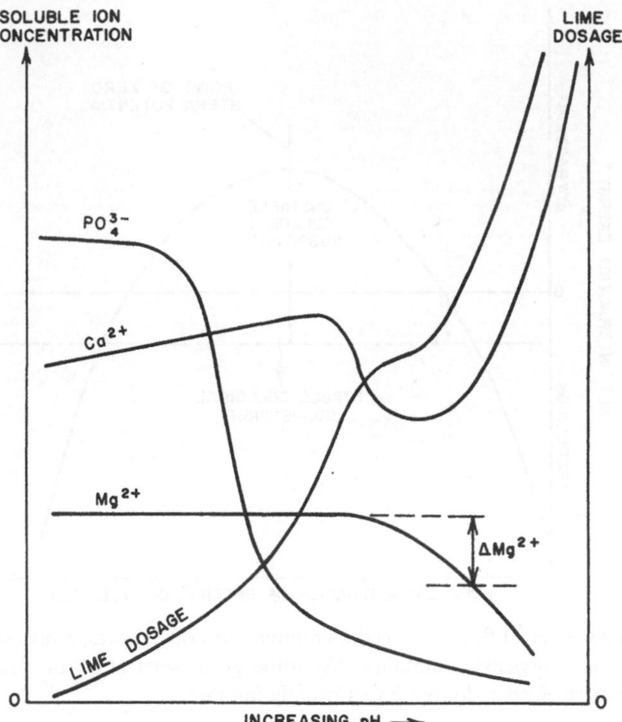

FIG. 7.4.6 Soluble ion distribution at increasing pH values.

PRACTICAL APPLICATIONS

The inorganic coagulant doses required to achieve colloid destabilization depend on the mixing and the liquid–solids separation method.

Coagulant Mixing

The hydrodynamics of inorganic coagulant–wastewater mixing are important. The rapid reaction of aluminum and ferric iron and the fact that soluble kinetic intermediates, the effective coagulating species, are adsorbed on colloid surfaces necessitate rapid, intense dispersion of coagulants into the wastewater. Inadequate mixing causes localized pH and metal ion concentrations, which require increased coagulant dosages to achieve colloid destabilization. Wastewater treatment facilities must add pH adjusting chemicals, e.g., $Ca(OH)_2$, and have them completely reacted before coagulant dispersion to assure the proper pH coagulation level. In addition, they must consider seasonal variations in the wastewater temperature. As the temperature decreases, they must increase the mixing energy to maintain the level of mixing intensity.

Liquid–Solids Separation

Definitive inorganic coagulant dosage is required to effect colloidal suspension destabilization. However, in the practical application of coagulation–flocculation to remove colloidal material from wastewater, the inorganic coagulant dose required depends on the method of flocculation and the liquid–solids separation employed.

Figure 7.4.7 shows the condition that exists when the unit process serves to coagulate and clarify biologically treated wastewater. The inorganic coagulant dose required to coagulate the colloidal material prior to removal by granular media filtration, e.g., coal-sand filters, is one-half to one-sixth that required for removal by mechanical flocculation followed by gravity settling. The reason for this phenomenon is that the deep-bed filter is an efficient flocculation device (bringing destabilized colloids together). On the other hand, gravity settling requires an excess of metal hydroxide precipitate (above that required for destabilization) to produce a settleable floc.

LABORATORY DETERMINATION OF COAGULANT DOSES

No universally accepted laboratory procedure exists for determining the inorganic coagulant doses required for colloidal suspension destabilization in a plant-scale operation. The main problem is that a laboratory model of prototype unit operations of continuous flow through flash mixing, mechanical flocculation, and gravity settling is not available.

Jar Test

A qualitative method used extensively in the water treatment industry is the jar test. In this test, environmental engineers add different coagulant doses to several rapidly mixing samples of wastewater (usually ≤1.5 l) and continue mixing for about 1 min. The sample is then slowly mixed (to simulate flocculation) for 10 to 30 min and allowed to settle quiescently for an additional 10 to 30 min. Environmental engineers then make qualitative observations such as time for visible floc formation, floc size, and floc settling rates.

They also make a direct or indirect measurement of the supernatant SS concentration. They approximate the coagulant dose requirement based on their judgement of these observations. The effective use of jar test information is an art and does not represent a true model of prototype operations.

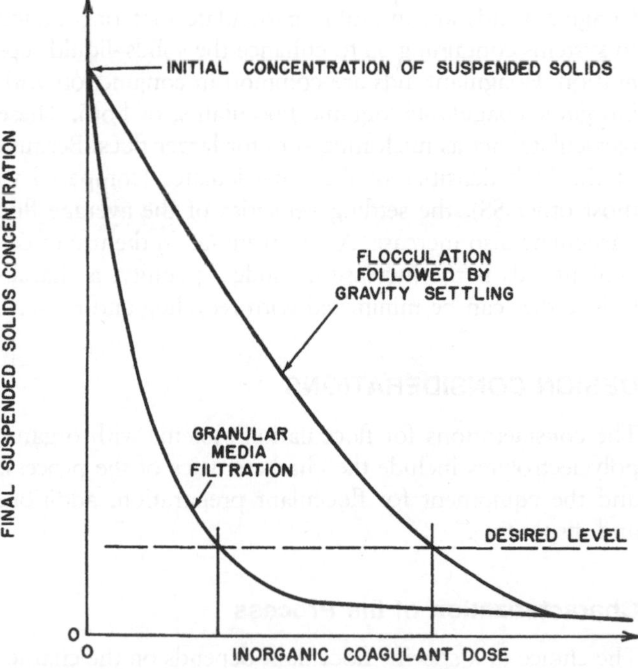

FIG. 7.4.7 The effect of the liquid–solids separation technique on the coagulant dose required for efficient suspended colloidal material removal.

Zeta Potential Test

A more theoretically based method of determining the required inorganic coagulant dose involves the use of a microelectrophoretic device. Laboratory technicians intensely mix varying doses of an inorganic coagulant into samples of wastewater. They then place aliquots (fractions of the sample) in a small glass chamber and apply an appropriate voltage gradient across the solution.

Laboratory technicians measure the rate of particle migration by visual observation using a calibrated eyepiece in a compound microscope. They then compute the rate of migration divided by the voltage gradient, microns per second per volts per centimeter. This resulting value is the electrophoretic mobility and is proportional to the electrokinetic or zeta potential (see Figure 7.4.1).

Based on the electrical double-layer model, a certain coagulant dosage results in a zero electrophoretic mobility corresponding to a zero electrokinetic potential. At zero electrokinetic potential, the electrical repulsive interaction of the electric double layers is minimized. In practice, the optimum destabilization can exist at a slightly positive or negative electrophoretic mobility.

Since hydrogen and hydroxyl ions are potential determining ions for most wastewater colloids as well as the inorganic coagulant species, an optimum coagulation pH exists. Figure 7.4.8 shows the coagulant-dose–pH-interaction on electrophoretic mobility. From these data, environmental engineers can estimate an optimum economic combination of pH adjustment chemical and coagulant.

Flocculation with Organic Polyelectrolytes

POLYELECTROLYTE FLOCCULANT TYPES

a) Anionic
b) Cationic
c) Nonionic
d) Variable charge

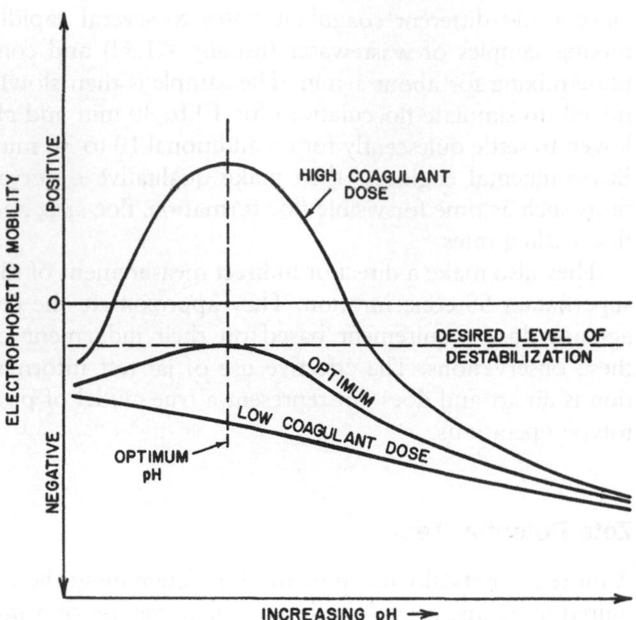

FIG. 7.4.8 Determining the optimum coagulant dose and pH by electrokinetic measurements.

MINIMUM SOLIDS CONCENTRATION FOR EFFECTIVE FLOCCULATION

50 mg/l

FLOCCULANT FEED SOLUTION CONCENTRATION BY WEIGHT

0.25 to 0.5%

TIME REQUIRED FOR FLOCCULATION PRIOR TO SETTLING

Under 5 min

POLYELECTROLYTE ADDITION RATES FOR FLOCCULATION

0.1 to 1 mg/l of raw sewage or 1 to 10 lb per tn of dry solids in sludge treatment applications

FLOCCULANTS

Flocculants are water-soluble, organic polyelectrolytes that are used alone or in conjunction with inorganic coagulants or coagulant aids to agglomerate solids suspended in aqueous systems. The large dense flocs resulting from this process permit more rapid and efficient solids–liquid separation.

Separating SS from raw water and wastewater for purification generally involves gravity settling in large clarifiers operating at low velocity gradients prior to a secondary biological process, a tertiary physicochemical process, or both. This primary physical process is enhanced by coagulation and flocculation of the initially fine colloidal particles into larger and more dense aggregates that settle more rapidly and completely. Coagulation and flocculation are sequential processes distinguished primarily by the types of chemicals used for initiation and the size of the particles developed.

Coagulation is the conversion of finely dispersed colloids into small floc with the addition of electrolytes like inorganic acids, bases, and salts. The salts of iron, aluminum, calcium, and magnesium are inorganic electrolytes. Partial coagulation can also result from naturally occurring processes, such as biological growth, chemical precipitation, and physical mixing. Flocculation is the agglomeration by organic polyelectrolytes (or by mechanical means) of the small, slowly settling floc formed during coagulation into large floc that settles rapidly.

Polyelectrolyte flocculants are linear or branched organic polymers. They have high molecular weights and are water soluble. Compounds similar to polyelectrolyte flocculants include surface-active agents and ion-exchange resins. The former are low-molecular-weight, water-soluble compounds used to disperse solids in aqueous systems. The latter are high-molecular-weight, water-soluble compounds that selectively replace certain ions in water with more favorable or less noxious ones.

Polyelectrolytes can be natural or synthetic in origin. Naturally occurring polyelectrolytes include various starches, polysaccharides, gums, and other plant derivatives. Table 7.4.2 lists various types of synthetic polyelectrolyte flocculants. A variety of products are available among the individual types, which are shipped either as dry granular powders in bags or in bulk or as concentrated viscous liquids in drums or tank cars.

COAGULANT AIDS

Coagulant aids are insoluble particulate materials added to systems containing SS to enhance the solids–liquid separation. Coagulant aids are common in conjunction with inorganic coagulants, organic flocculants, or both. These particulates act as nucleating sites for larger flocs. Because of the high densities of these particulates (compared to most other SS), the settling velocities of the average floc particulate also increase. A disadvantage in the use of coagulant aids is the increase in sludge quantity, a characteristic that can be minimized with recycling and reuse.

DESIGN CONSIDERATIONS

The considerations for flocculation systems with organic polyelectrolytes include the characteristics of the process, and the equipment for flocculant preparation, addition, and dispensing.

Characterization of the Process

The choice of a specific flocculant depends on the characteristics of the process system to be flocculated. The density of the suspending liquid (usually water) and the effective density of the suspended particles must be

TABLE 7.4.2 CLASSIFICATION OF POLYELECTROLYTE FLOCCULANTS

Type	*Ionic Charge*	*Examples*
Anionic	Negative	Hydrolyzed polyacrylamides, polyacrylic acid, polyacrylates, and polystyrene sulfonate
Cationic	Positive	Polyalkylene polyamines, polyethylenimine, polydimethylaminomethyl polyacrylamide, polyvinylbenzyl trimethyl ammonium chloride, and polydimethyl diallyl ammonium chloride
Nonionic	Neutral	Polyacrylamides and polyethylene oxide
Miscellaneous	Variable	Alginic acid, dextran, guar gum, and starch derivatives

sufficiently different to permit separation. Sand and grit particles are heavy and compact. They can have effective densities more than twice that of water. Biological solids and hydrated inorganic precipitates are hydrophilic, i.e., associated with surface-bound and internally contained water. Their densities can be only slightly greater than that of water. The density of water is affected slightly by temperature and more significantly by salt content.

The salt content and pH of suspending water affect the surface charge of the SS. The sign, magnitude, and distribution of this surface charge strongly influence the type and quantity of the flocculant to be used. Negatively charged solids can be flocculated by cationic flocculants; positively charged solids can be flocculated by anionic flocculants. Negatively charged solids can also be coagulated by inorganic cations and then flocculated by an anionic flocculant.

The size, shape, and concentration of solid particles also affect flocculation and settling. Large particles settle faster than small particles. Irregularly shaped particles settle slower than smooth, spherical particles. Flocculation effectiveness is reduced if the solids concentration is low (<50 mg/l) since the probability for contact among particles is reduced. Settling is hindered at high solids concentrations (>2000 mg/l) because of excessive interparticle contact. Most suspensions subjected to settling in water and wastewater treatment settle freely, with the exception of concentrated biological sludges.

Equipment for Flocculant Preparation and Addition

Polyelectrolyte flocculants are soluble in water and compatible with most dissolved materials at the concentrations normally found in tapwater. To prepare flocculant solutions, wastewater treatment facilities should use water that has a low solids content to avoid the formation of insoluble sludges. The water's pH should be nearly neutral.

Concentrated solutions of metallic salts and anionic polyelectrolytes should never be prepared in the same tank, since sufficient concentrations of trivalent cations or certain divalent cations can cause partial precipitation of anionic flocculants. Flocculant solutions are essentially noncorrosive, and standard materials of construction, such as PVC, black iron, and mild steel, can be used for all equipment.

High-molecular-weight polymers require time to completely solubilize since water must hydrate the long entwined molecules before they uncoil in solution. The preparation time decreases when dry particles are evenly distributed throughout the solvent water. Wastewater treatment facilities can distribute the dry product with a mechanical mixer and more easily with an eductor. Figure 7.4.9 shows a typical disperser for preparing solutions of dry flocculant. This disperser operates on an aspiration principle and distributes individual flakes of flocculant throughout the water where they are readily dissolved.

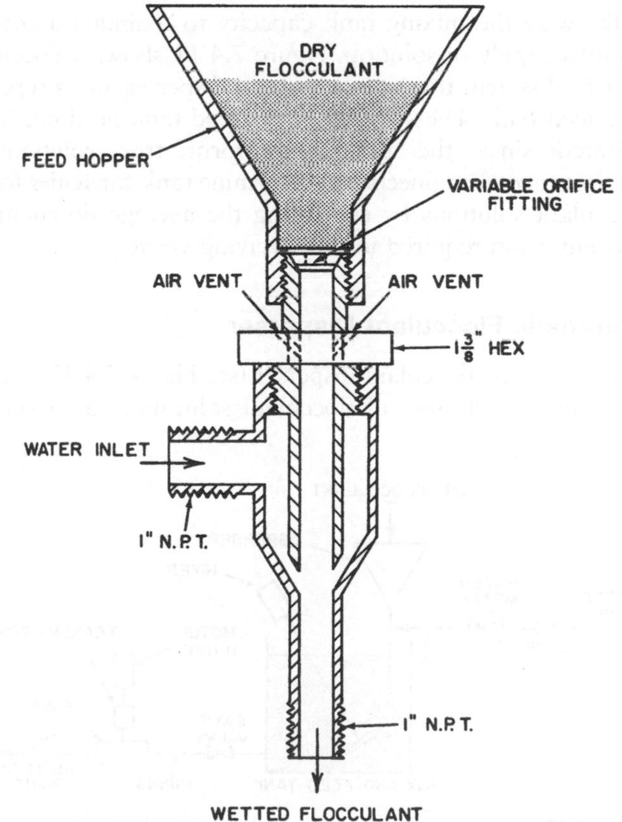

FIG. 7.4.9 Manual flocculant disperser.

After dispersing the solid flocculant, wastewater treatment facilities need only use minimum agitation to assure a uniform solution concentration. A low flow of compressed air or a mechanical mixer can be used. Mechanical mixers or air spargers can be used for agitation during solution preparation. Dissolving dry flocculants does not require violent agitation by high-speed mechanical mixers; the mixing equipment need only generate a moderate rolling action throughout the makeup tank. Figure 7.4.10 shows a typical flocculant feed system using a manual disperser.

Flocculant solutions are highly viscous, and ordinary flow regulating valves and meters are usually not adequate to control the small volumes of these solutions. Wastewater treatment facilities should feed flocculant solutions with accurate chemical metering pumps. Positive displacement pumps such as progressive cavity, rotary gear, or piston pumps are all suitable.

The chemical feed pump should have a variable flow rate control mechanism. For treating less than 700 gpm of water or waste, wastewater treatment facilities can use pumps equipped with dial-controlled, variable-speed drive that can be adjusted while in operation. They should select the pump size so that normal operations use 30 to 50% of the pump capacity. This size provides freedom to decrease or increase the pumping rates as required. Wastewater treatment facilities can use automatic flow ratio control systems for treating larger volumes.

Wastewater treatment facilities often use a feed tank with twice the mixing tank capacity to maintain a continuous supply of solution. Figure 7.4.11 shows a flocculant feed system that uses a manual disperser and a separate feed tank. The solution in the feed tank need not be agitated since the flocculants form true solutions. Environmental engineers can determine tank capacities for flocculant solutions by estimating the average flocculant concentration required in the receiving waste.

Automatic Flocculant Disperser

The automatic flocculant disperser (see Figure 7.4.12) prepares up to 200 gpm of flocculant solution at a concentration of 0.25% by weight. A screw feeder adds dry flocculant at a variable rate (≤10 lb per min) into the vortex formed within a mixing bowl. Water at 40 psig and 25 to 100 gpm first passes through a strainer and a water meter. Then it reaches a solenoid valve activated by a float control that monitors the liquid level in the feedtank receiving the prepared flocculant solution.

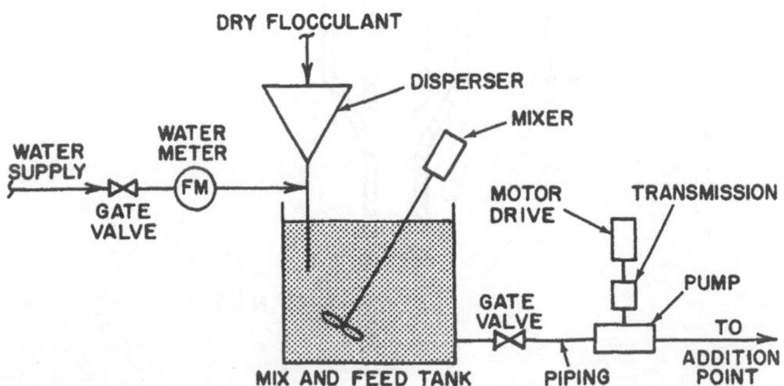

FIG. 7.4.10 Flocculant feed system with manual dispersing equipment.

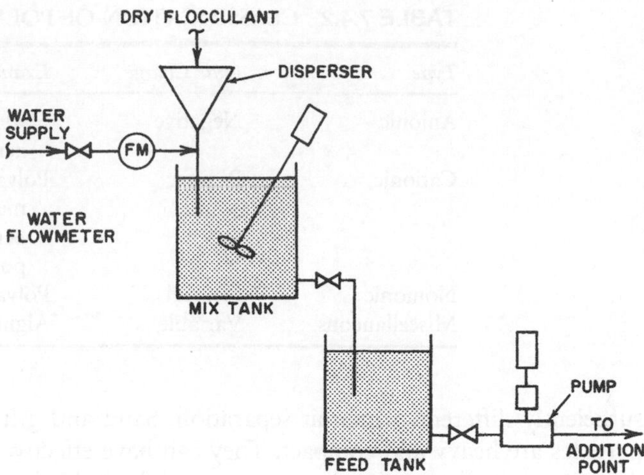

FIG. 7.4.11 Flocculant feed system with manual dispersing equipment and separate feed tank.

The water flow is then split, with most of it passing through a valve and rotameter controlling the flow through the mixing bowl. The remaining water passes through a small internal reservoir equipped with a float valve. This maintains the required water level in the bottom of the mixing bowl for optimum dry flocculant dispersion. The combined flow from the mixing bowl and the reservoir is then pumped through a pump at 25 to 100 gpm directly to the feed tank or combined with another stream of dilution water (≤100 gpm) before delivery to the feed tank.

The flocculant solution is further mixed in the feed tank before it is displaced by the incoming flow and delivered to the point of addition in the treatment plant. The flocculant disperser can be operated either manually or mechanically. Other simplified units are commercially available for preparing flocculant solutions in the intermediate range of 0.5 to 25 gpm. The manual disperser is limited to the preparation of small quantities of flocculant solution.

Wastewater treatment facilities should consider an automatic flocculant disperser (see Figure 7.4.12) when treating larger volumes of water flow. The hopper is filled manually with dry flocculant on a daily to weekly basis, depending on plant flow rate. An automatic dry flocculant addition system should be incorporated when large volumes of water (>100 mgd) are treated. Figure 7.4.13 shows a flocculant feed system with an automatic disperser that incorporates a level control on the feed tank.

Flocculants are ordinarily prepared as dilute solutions that are then dispersed into the water or waste to be

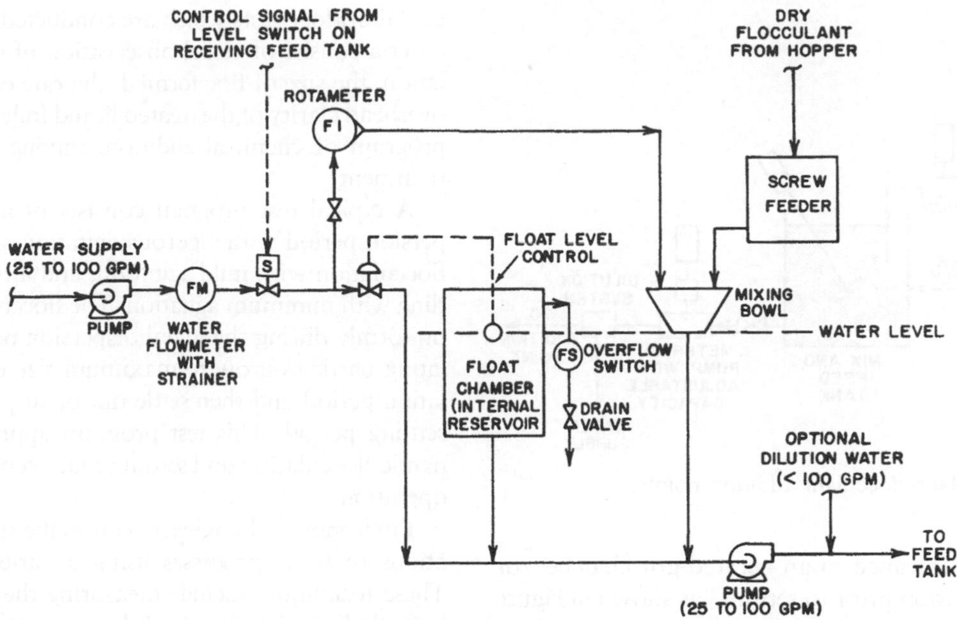

FIG. 7.4.12 Automatic flocculant disperser.

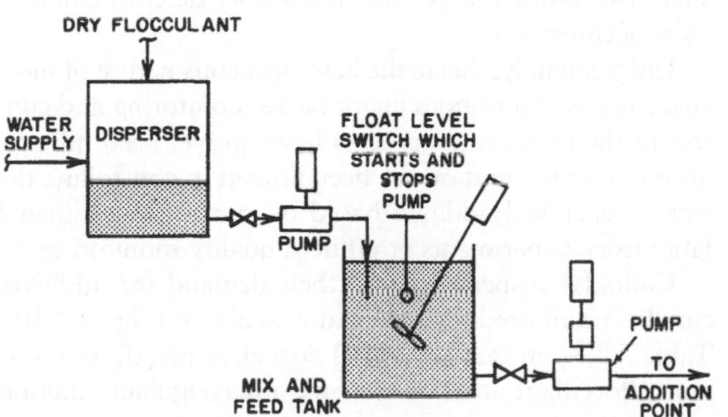

FIG. 7.4.13 Flocculant feed system with automatic dispersing equipment.

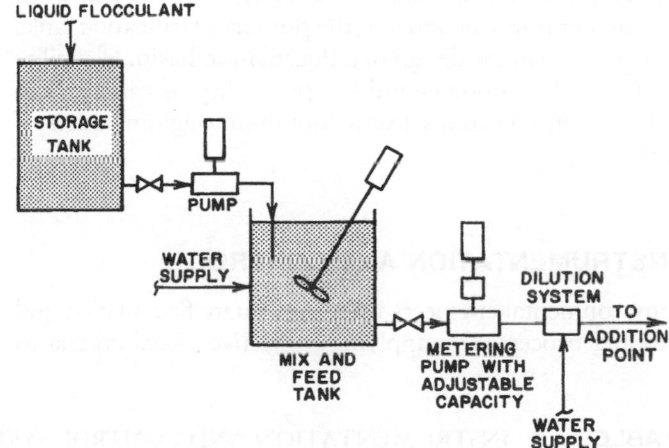

FIG. 7.4.14 Liquid flocculant feed system.

treated. The average feed solution concentration for most flocculants is 0.25 to 0.5% by weight. In applications where liquid flocculants are added at high concentrations, the wastewater treatment facility prepares the feed solution by diluting the concentrated bulk solution as shipped. Figure 7.4.14 shows a liquid flocculant feed system.

Wastewater treatment facilities can apply flocculants by diluting a concentrated solution and then feeding the resulting dilute solution to the stream to be treated. Preparing large volumes of dilute solutions is not necessary. A small volume of the concentrated feed solution can be metered and diluted to a lesser concentration immediately before it is added to the stream to be treated. Environmental engineers should consider the chemical time requirement for completing the dilution step, the storage tank size, and the chemical pump capacities when determining the feed solution concentration.

The flocculation of SS, precipitated inorganic salts, and organic complexes adsorbed or absorbed to particulates is not subject to the same design criteria as conventional water treatment that uses inorganic coagulants alone. The time for flocculation after the flocculant addition and prior to settling can be considerably less (under 5 min) than the time recommended for water treatment plants, which frequently exceeds 30 min.

Experience shows that prolonging gentle mixing after the initial rapid dispersion of the flocculant does not increase the removal and can partially destroy the floc particles. Environmental engineers should also consider the charging period of adding a metal coagulant and adding the flocculant if both agents are used.

The flocculant is usually added at some turbulent point in the plant flow, such as at the throat of a flume, ahead

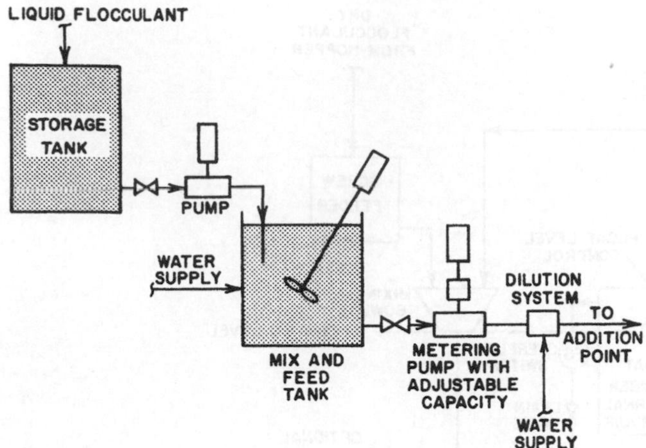

FIG. 7.4.15 Coagulant–flocculant addition points.

of a weir, at the entrance to an aerated-grit chamber, or at some other location prior to settling, as shown in Figure 7.4.15.

Adding the flocculant before low-lift pumps can cause floc destruction. However, the suction side of these pumps is acceptable for the addition of inorganic coagulants. An aerated-entrance channel to the primary clarification tanks is beneficial in the design of a flocculation basin. Flexibility in the addition points and the possibility of varying flocculant concentrations insure optimum agglomeration of solids.

INSTRUMENTATION AND CONTROL

Environmental engineers usually evaluate flocculation and settling processes by applying subjective visual criteria to the laboratory tests which are conducted in parallel. These criteria consist of visual observation of the rate of flocculation, the size of floc formed, the rate of settling, and the overhead clarity of the treated liquid following a prescribed program of chemical addition, mixing, and duration of treatment.

A typical test program consists of a short initial dispersion period with vigorous agitation, a longer period of flocculation with mild agitation, and a final period of settling with minimum agitation. The flocculant is distributed uniformly during the initial dispersion period. The flocculating particles grow to maximum size during the flocculation period and then settle out of suspension during the settling period. This test program approximates the dynamic flocculation and settling that occur during full-scale operation.

Environmental engineers can make quantitative evaluations of these processes using a variety of techniques. These techniques include measuring the SS concentration by turbidimetric means and the surface charges of the particles by streaming current or zeta potential. For control purposes, these techniques are preferred to subjective visual observations or periodic laboratory determination of SS concentrations.

Unfortunately, due to the heterogeneous nature of most suspensions, continuous quantitative monitoring and control of the flocculation process have not yet been fully realized. Instrumentation has been limited to controlling the rate of chemical addition based on previously estimated laboratory experiments or effluent quality monitoring.

Colloidal suspensions and their demand for additives can be monitored by automatic analyses (Fig. 2.4.10). Table 7.4.3 summarizes typical flocculant preparation systems, flocculant addition systems, and coagulant addition

TABLE 7.4.3 INSTRUMENTATION AND CONTROL SYSTEMS FOR THE FLOCCULATION PROCESS

Average Plant Flow (mgd)	*Flocculant Preparation System*	*Flocculant Addition System*	*Coagulant Addition System*
<1	Manual operation; batch preparation in tank	Manual operation; variable-speed device with pump calibration or rotameter, or both	Similar to the flocculant addition system
1–10	Automatic operation; flocculant disperser holding tank	Automatic operation; variable-ratio control of feed to sewage flow	Similar to the flocculant addition system
10–25	Same as above	Same as above with feedback correction and flow totalization	Similar to the flocculant addition system
25–100	Same as above	Same as above plus tank-level indicators and malfunction alarms	Similar to the flocculant addition system plus automatic influent analyzers and transmitters
>100	Same as above; alternate bulk handling	Same as above	Similar to the flocculant addition system plus automatic analyzer and transmitter for both influent and effluent: density transmitter and use of empirical design equations

systems for five ranges of total plant flow. The overall accuracy of instrumentation should increase with increased plant size.

Average Flow: <1 mgd

For this size flow, the wastewater treatment facility manually prepares the flocculant solution in a small storage tank. The addition rates are manually set by variable-speed devices such as positive displacement pumps. For flow rate indication, this system uses the pump calibration curve or a glass tube rotameter. The coagulant solution is fed by manually set, variable-speed pumps similar to the flocculant controls.

Average Flow: 1 to 10 mgd

For this flow, the flocculant solution is automatically prepared and added to the influent flow by a variable-ratio control system that varies the speed of the pump proportional to the influent wastewater flow rate. The coagulant solution addition system is similar. The influent flow should be sensed by a magnetic flowmeter or a Venturi tube.

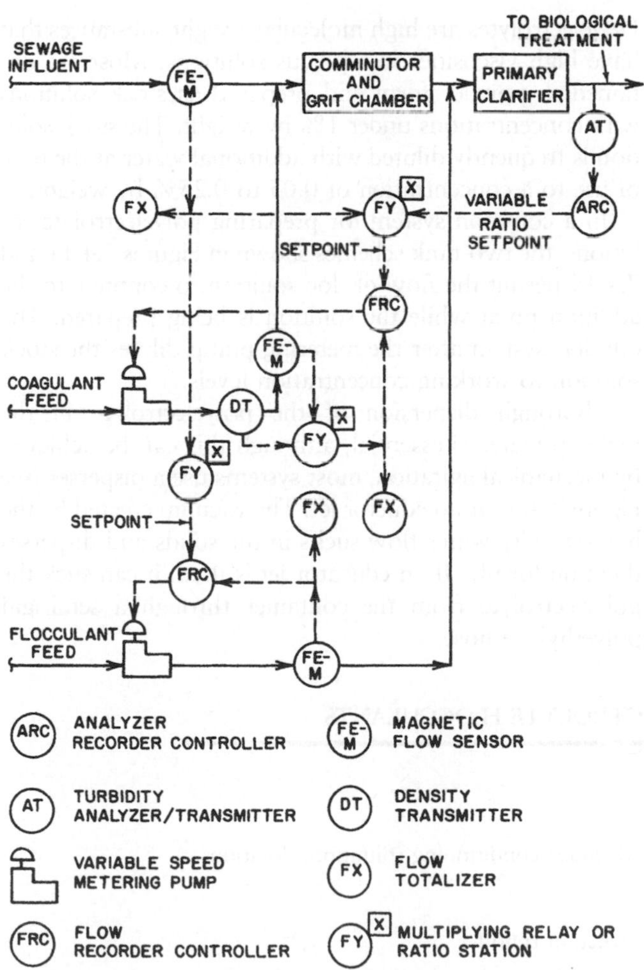

FIG. 7.4.16 Control system for automatic flocculant and coagulant additions.

Average Flow: 10 to 25 mgd

For this flow, the flocculant solution is automatically prepared as in the 1 to 10 mgd system and automatically added to the influent flow by a variable-flow-ratio controller that varies the speed of the pump proportional to the influent flow rate. The actual amount of flocculant added is detected and used as a feedback signal. The controller determines the difference between what should be added and what is being added, and adjusts the pump speed to eliminate the deviation. The coagulant addition system is similar. Flow totalization of each stream is recommended to provide data for material balances.

Average Flow: 25 to 100 mgd

This size plant automatically prepares the flocculant solution as in the 10 to 25 mgd system. It automatically adds the flocculant solution to the influent based on the detected influent flow rate multiplied by the required variable ratio as described for 10 to 25 mgd plants.

This system can automatically control the coagulant addition on the basis of influent composition analyses and influent flow rate. Automatic analyzers and transmitters are usually required. A variable-ratio controller controls the coagulant addition rate. The system can include continuous level transmitters on coagulant storage tanks with high- or low-level switches, or both, to simplify loading and unloading. Flow recorders and totalizers can be included to provide data for material balances.

Average Flow: >100 mgd

Flocculant solutions for systems of this size are automatically prepared using an automatic flocculant disperser and suitable holding tanks. A bulk handling system is an alternative to batch flocculant transfer. This system automatically proportions the flocculant solution to the raw waste flow by multiplying influent flow rate by a variable ratio.

The coagulant feed is automatically proportioned to influent waste flow or it can also be based on empirical design equations for a specific waste stream. A density transmitter on the coagulant feed determines the weight concentration of the coagulant in solution which is multiplied by the coagulant flow rate, resulting in a coagulant mass-flow-rate signal. The performance of the system is detected usually by a turbidity analyzer (ARC in Fig. 7.4.16). The required coagulant addition rate is adjusted by this ARC, which adjusts the set-point of a variable-ratio controller (FRC) that receives the coagulant mass-flow-rate as its measurement and operates a variable-speed pump to deliver the calculated amount of coagulant.

Flow recorders and totalizers provide a material balance (see Figure 7.4.16). Electronic DCS instrumentation is preferable because of the long transmission distances in many waste treatment plants.

APPLICATIONS

Environmental engineers have used organic flocculants to improve solids–liquid separations in the applications listed in Table 7.4.4. Municipal water treatment performance improves when flocculants supplement inorganic coagulants in removing solids from raw water and in improving the solids removal capabilities of vacuum filters and centrifuges. Municipal wastewater treatment uses flocculants to remove SS, flocculate precipitated nutrients such as phosphorus, and condition sludges for dewatering processes such as flotation, elutriation, filtration, centrifugation and sand-bed dewatering. Environmental engineers treat surface waters with flocculants to remove solids that are periodically received from storm runoff, dredging, or construction. Treating industrial waste with flocculants removes a variety of suspended pollutants and enhances the separation and removal of some emulsified and dissolved pollutants.

Coagulation Systems

TYPES OF COAGULATION–FLOCCULATION SYSTEMS

a) Gravity Settling
b) Flotation

COAGULATION APPLICATIONS

Neutralize and agglomerate oil droplets or particles in the 50 μ or smaller size range

MATERIALS OF CONSTRUCTION FOR FLOCCULATION EQUIPMENT

Stainless steel or epoxy-reinforced fiberglass

SYSTEM DESCRIPTION

A coagulation system has the following functions:

1. Preparation of solutions of coagulants, flocculants, and coagulant aids. Usually employed are alum, ferric chloride, or other inorganic coagulants; cationic polyelectrolyte flocculants; or anionic or nonionic polyelectrolyte flocculants.
2. Rapid mixing of coagulants with wastewater.
3. Growth of floc to produce large agglomerates under gentle mixing conditions.
4. Separation of floc from water, usually by settling or flotation.

Figure 7.4.16 shows a complete, fully automated coagulation–flocculation system. The example shown involves coagulation aided by an inorganic coagulant followed by the addition of a polyelectrolyte flocculant. This system disperses the polyelectrolyte in the main wastewater line by injecting the solution into the pipeline. At high velocities in the line, rapid mixing is produced. The polyelectrolyte flocculant is added just before the feed well of the clarifier, which has a slow-moving rake that keeps the settled sludge distributed and induces gentle mixing to enhance floc formation.

The design and nature of coagulation systems are based on the coagulation characteristics of the wastewater and the nature of the coagulant used.

PREPARATION OF COAGULANT SOLUTIONS

Polyelectrolytes are high-molecular weight substances that have high viscosities in aqueous solutions. Most are obtained in powder form and prepared as stock solutions with concentrations under 1% by weight. The stock solution is frequently diluted with additional water at the time of use to a concentration of 0.05 to 0.25% by weight.

In a common system for preparing polyelectrolyte solutions, the two-tank schemes shown in Figures 7.4.11 and 7.4.14 permit the flow of floc solution to continue to the addition point while the solution is being prepared. The dilution system after the metering pump dilutes the stock solution to working concentration levels.

Thorough dispersion of the polyelectrolyte in the makeup water is essential, although this can be achieved by mechanical agitation, most systems use a disperser (see Figure 7.4.9) or an eductor jet. The vacuum created by the high-velocity water flow sucks in the solids and disperses them uniformly. If an educator jet is used, it can suck the polyelectrolyte from the container through a semirigid polyethylene hose.

TABLE 7.4.4 APPLICATIONS FOR POLYELECTROLYTE FLOCCULANTS

Municipal Water Treatment
Primary flocculation and sludge conditioning
Municipal Waste Treatment
Raw waste flocculation, phosphorus removal, and sludge conditioning (filtration, flotation, and centrifugation)
Surface Water Treatment
River clarification, dredging operations, and construction runoff
Industrial Water Treatment
Raw water (influent), process water (internal), and wastewater (effluent)

After the polyelectrolyte is dispersed in water, it must be mechanically agitated to dissolve the polyelectrolyte. A typical agitator for a 400-gal tank requires a 1-hp motor and a marine propeller with a 200- to 400-rpm agitation velocity. The agitation time varies with the concentration of the stock solution and can range from 30 to 120 mins. Stock solution can remain in a storage tank for only several days because stock solutions are not stable for more than 3 to 4 weeks.

Because most polyelectrolytes are acidic, flocculant solutions should be prepared in lined steel, epoxy-reinforced fiberglass, 316SS, or other corrosion-resistant tanks. However, construction materials in the wastewater system downstream of the flocculant solution injection point are not affected by the corrosive effect of the polyelectrolytes since their dilution is in the ppm range or less.

ENGINEERING DESIGN CONSIDERATIONS

The coagulation process starts with the rapid, high-intensity mixing of the coagulant with wastewater. Since complex reactions are involved, environmental engineers must use laboratory tests such as the zeta potential and jar tests to establish proper pH and dosage ranges.

Wastewater treatment facilities can rapidly mix inorganic coagulants with the wastewater either inline or in a separate flash-mixing basin. The detention time requirement ranges from 10 to 60 sec. Polyelectrolyte flocculants are added after the rapid-mix phase or downstream from the coagulant addition point because rapid mixing can break up the floc. Mixing during the flocculation stage should be gentle, with a mean velocity gradient of 50 to 175 determined as follows:

$$\sqrt{\frac{W}{\mu}} = \text{Mean velocity gradient (sec}^{-1}) \qquad 7.4(1)$$

where:

W = power per unit volume of fluid (lb/sec-sq ft)
μ = absolute viscosity of wastewater (lb/sq ft)

SETTLING OR FLOTATION

The separation mechanism is the same for settling and flotation. Stokes' law governs both methods of separation. When the settling velocity (Vs) is positive, the particle settles; when it is negative, the solid particle rises.

When a coagulated waste can be treated by either settling or flotation, wastewater treatment facilities can achieve higher separation rates and higher solid concentrations through dissolved air flotation. This process requires smaller basins and results in smaller sludge volumes with greater water recovery. On the other hand, flotation systems require more operator attention and in some cases extra maintenance.

In addition to their use with settling and flotation, wastewater treatment facilities often use flocculation or coagulation to enhance other separation operations, including filtration and centrifugation.

COAGULATION FOLLOWED BY SETTLING

Coagulation systems used in combination with settling are available in two basic design configurations: conventional systems and sludge-blanket systems. In conventional system, the rapid-mix step is completed before the water enters the large settler, where the flocculation and clarification steps are completed. In a sludge-blanket-type unit, the coagulant mixing, flocculation, and settling steps all take place in a single unit.

Figure 7.4.15 shows a conventional system where rapid mixing of the cationic inorganic coagulant is done in the comminutor (a mix tank can also be used), and the water is then fed to the primary clarifier. The anionic polyelectrolyte flocculant is injected into the wastewater line before the clarifier unit. A moving scraper collects the settled floc and removes it to the sludge outlet. Recycling part of the floc usually improves clarification efficiency.

The flocculation step can also occur in a separate flocculation tank with slow-moving, horizontal paddles located upstream to the conventional settler.

Sludge-blanket units combine coagulation, settling, and upward filtration in one compact unit (see Figure 7.4.17). In this design, influent water is mixed with coagulation chemicals as it is fed to the inner chamber. Floc formation begins in this chamber. The formed floc slurry then moves to the outer chamber where the water rises through a bed of previously formed floc which functions as a filter bed, effectively retaining the floc.

The velocity of the water decreases as it moves up the outer chamber because the cross-sectional area increases. This decreased velocity assists in the separation. The clear effluent water flows out through a weir at the top while draining after blowoff removes the sludge collected at the bottom. This type of unit has a small space requirement.

COAGULATION FOLLOWED BY FLOTATION

Total pressurization is used when the material to be separated is not broken down by shearing forces in the pressurization system or the floc reforms quickly after the pressure is released. This system (see Figure 7.4.18) produces a maximum amount of air bubbles and requires a small flotation compartment. Wastes with heavy solid loads or those that rapidly form strong flocs are compatible with this method.

Chemical additives can be added upstream or in the pressurization system itself. The system operates by pressurizing the wastewater stream to 30 to 60 psig and mix-

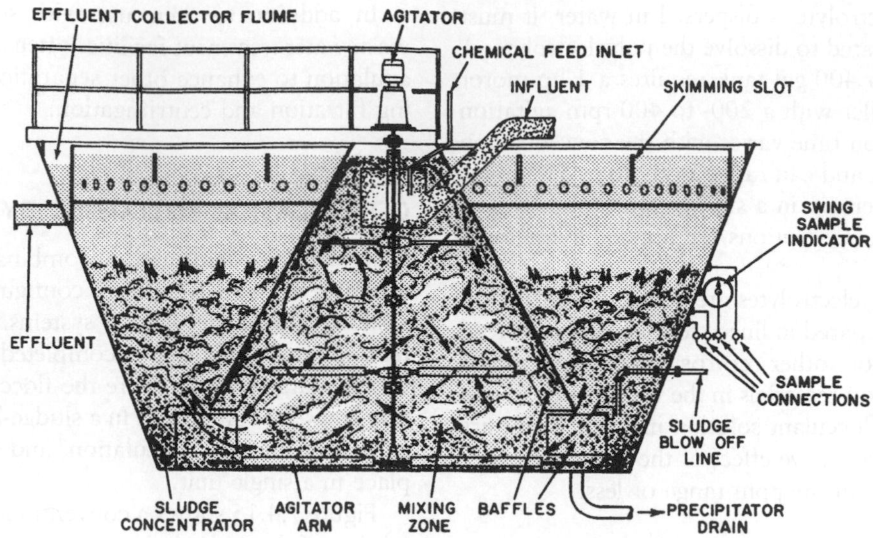

FIG. 7.4.17 Combined coagulation and settling apparatus. (Reprinted, with permission, from Permutit Co.)

ing it with compressed air. A DO level approaching saturation is achieved in the retention tank. The pressure of the air-saturated stream is reduced in a backpressure valve on the tank discharge. This reduced pressure initiates the formation of microscopic air bubbles which assist in the flotation process.

Partial pressurization is used when the solid load is light. As shown in Figure 7.4.18, only a fraction of the waste stream is pressurized by air. This scheme frequently requires a separate flocculation chamber in the bypassed stream. In this system, the pressurized stream solids must be absorbed on the preformed floc when mixed in the inlet compartment. Otherwise, a secondary addition of flocculant solutions is required in the inlet compartment of the flotator.

Recycling pressurization is used when the floc formed cannot be pressurized and large quantities of air must be dissolved. In this scheme (see Figure 7.4.18), the addition of coagulants and flocculants precedes the flotation step. A side stream of the clarified effluent is air-pressurized. When extended floc formation time is required, this method is particularly applicable. The pressurization components are less prone to solids build-up in this system.

COAGULATION IN SLUDGE HANDLING

Coagulation is also used to condition slurries or sludges. Coagulation is a highly effective sludge conditioner and is often used ahead of vacuum or gravity-type filters. It also aids the separation in solid-bowl centrifuges. In these applications, the coagulant solution is added to the pool either within the centrifuge or externally in the feed line.

Coagulant solutions can be used to increase the drying rate on sludge drying beds. The coagulant solution can be added to the sludge as it is applied to sand beds.

Emulsion Breaking

TYPES OF EMULSIONS

a) Oil-in-water
b) Water-in-oil

VOLUME CONCENTRATION OF EMULSIONS

For oil–water systems, between 26 and 74% volume concentrations

TREATMENT TECHNIQUES USED

Emulsion breaking or oily waste treatment is applied at over 1% oil concentration. Below 1%, clarification and coalescing techniques apply

SPECIFIC UNIT OPERATIONS USED

Gravity separation, air flotation, centrifugation, filtration, electrical dehydration, and chemical treatment

Emulsions are stable, heterogeneous systems consisting of at least one immiscible liquid dispersed in another in the form of microscopically visible droplets. Oil–water emulsions are formed in sewage systems as a result of contact between oil, water, and emulsifying agents, or formed directly as industrial by-products.

In refineries, the direct formation of emulsions can result from chemical treatment of lubricating oils, waxes, and burning oils; barometric condensers; tank drawoffs; desalting operations; acid sludge recovery processes; and wax deoiling. Oily wastes from general petrochemical plants are usually treated in oil–water separators. The effluent from separators that still contains small amounts of emulsion is treated before final discharge to the sewer.

Steel manufacturing and finishing operations produce mixtures of waste-soluble oils containing cleaning solutions, which are stable emulsions. The automobile industry and steel mills discharge spent machining and cutting oils, coolants, and drawing compounds.

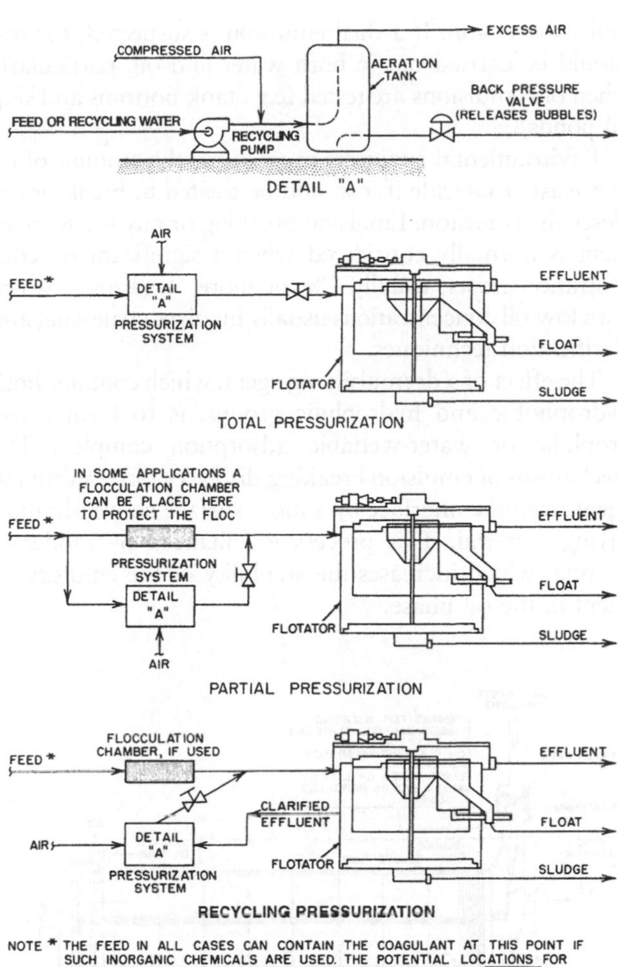

FIG. 7.4.18 Pressurization methods applied in flotation.

TYPES OF EMULSIONS

Emulsions consist of two liquid phases: the disperse and continuous phases. The phase in the form of finely divided droplets is the disperse or internal phase; the phase forming the matrix in which these droplets are suspended is the continuous or external phase. Based on the liquid phases, a typical water–oil emulsion can exist either as an oil-in-water (oil is the disperse phase) or water-in-oil emulsion. These types are abbreviated as o/w and w/o, respectively.

When an assembly of spheres of equal radii are in a position of densest packing, they occupy 74% of the total volume; the remaining 26% is empty space. Theoretically, this relationship means that for a given system, o/w and w/o emulsions are both possible between the phase volume concentrations 0.26 and 0.74. Below 0.26 and above 0.74, only one form can exist. In the intermediate range of volume concentrations, no one form of emulsion is favored. These systems are described as *multiple* or *dual* emulsions with the disperse phase containing globules of the other phase.

STABILITY OF EMULSIONS

The stability of an emulsion is based on its physical and electrical properties.

Physical Properties

The properties of an emulsion depend largely on its *composition* and its mode of *formation*. Physical properties also control the stability of these systems. The formation of an emulsion is a function of the boundary tensions between the two liquid phases determining the type of emulsion as either w/o or o/w. *Interfacial tensions* also exist between the liquid phases and between the liquid and solid phases. The latter is usually a lower magnitude than that of a liquid–liquid interface.

The flow resistance of an emulsion is one of its most important properties. Viscosity measurements provide considerable information about the structure of emulsions and their stability. The viscosity of the continuous or external phase is of prime importance in overall emulsion viscosity. The viscosity of the oil components in a w/o emulsion is usually indicative of emulsion stability but has little significance in an o/w system.

Miscibility determines the emulsion type. An emulsion is readily dilutable by the liquid that constitutes the continuous phase. Therefore, o/w emulsions are readily miscible with water; conversely, w/o systems are readily miscible with oil. An emulsion remains stable as long as the interfacial film and emulsifying agents are not materially affected.

Electrical Properties

Since oil is an insulator, o/w emulsions conduct an electrical current, whereas w/o emulsions ordinarily do not. Researchers have studied the conductivity of emulsions by measuring the current flowing between two fixed platinum electrodes immersed in the emulsion. The dielectric properties of emulsion systems are different from the average of the individual phases. The dielectric constant is important because of its intimate relationship to emulsion stability. The dielectric properties of an emulsion can be measured in a single parameter defined as the zeta potential (see Figure 7.4.1).

Degree of Stabilization

The concentration of droplets of the disperse phase toward the top or bottom of the emulsion results in a difference in density between the liquid phases. This phenomenon is known as *creaming*. Creaming does not necessarily represent a breaking of the emulsion, but with large droplet sizes, it can eventually lead to emulsion breaking.

Another type of emulsion instability results from flocculation or clumping of individual droplets to form larger aggregates. In this case, the emulsion has inverted due to a sudden change from o/w to w/o and vice versa. The progressive coarsening of the dispersion leads to a complete separation or breaking of the emulsion. Emulsion breaking is an irreversible process, often preceded by creaming or inversion.

EMULSION BREAKING METHODS

Several physical methods can separate oils and SS from wastewater, including gravity separation, dissolved air flotation, centrifugation, filtration, and electrical dehydration. Selecting the method depends on the nature of the wastewater and the degree of treatment required. Chemical methods of breaking water–oil emulsions are based on the addition of chemicals that destroy the protective action of hydrophobic or hydrophilic emulsifying agents and allow the water globules and oil to coalesce.

Figure 7.4.19 shows a typical API separator. This two-stage separator can be expanded to include additional sections. Figure 7.4.20 shows a common treatment arrangement that uses a combination of mechanical process and a chemical process. System operation is as follows:

1. Piping design and feeding facilities allow the operator to pump into either or both tanks and add an emulsion-breaking chemical to either tank or to the waste enroute from the API separator.
2. Steam coils are in both tanks since heat improves the speed and efficiency of phase separation after emulsion resolution.
3. This system fills one tank with waste from a bottom inlet until an upper-level limit is reached. The flow then switches to the other tank. The filled tank is allowed to settle for about 60 min and is then inspected.
4. The system pumps any oil that has risen to the top to reclaimed oil storage. It draws water from the bottom and routes it to the separator inlet. After this step, normally about three-fourths of a tank of w/o emulsion are left.
5. The system heats the emulsion with the steam coils. After the proper temperature is reached, it mixes it by rolling the tank with gas. The emulsion-breaking chemical is added during the mixing.
6. After thorough mixing, this system shuts off the gas and steam so that the treated emulsion can cool and settle until phase separation is complete.
7. The water phase, containing some SS but virtually oil-free, is routed to a settling pond. The oil goes to reclaimed-oil storage and is recycled to the refinery crude unit at a steady rate.

TREATMENT OF WASTE EMULSIONS

Environmental engineers can determine the emulsion type by using the phase dilution method. The procedure is to place one drop of the emulsion in about 20 cc. of water with gentle stirring. A w/o emulsion shows no dispersion, and the water remains clear. An o/w emulsion forms a milky dispersion. If a dual emulsion is suspected, the test should be carried out in both water and oil, particularly when old emulsions are tested (e.g., tank bottoms and slop oil ponds).

Environmental engineers must know the amount of oil in a waste to decide if it should be treated to break or coalesce the emulsion. Emulsion breaking or oily waste treatment is normally considered when a significant oil concentration exists, usually 1% or more. Treating a waste with low oil concentrations usually involves coalescing and clarification techniques.

The effect of a deemulsifying agent, which contains both hydrophobic and hydrophilic groups, is to form a hydrophilic or water-wettable adsorption complex. The mechanism of emulsion breaking displaces the emulsifying agent from the interface by a more surface-active, deemulsifying material. This process is enhanced by moderate heating, which increases the solubility of the emulsifying agent in the oil phase.

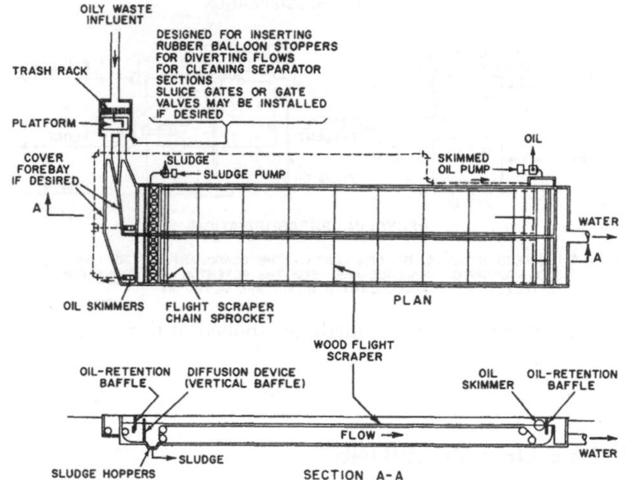

FIG. 7.4.19 API oil-water separator.

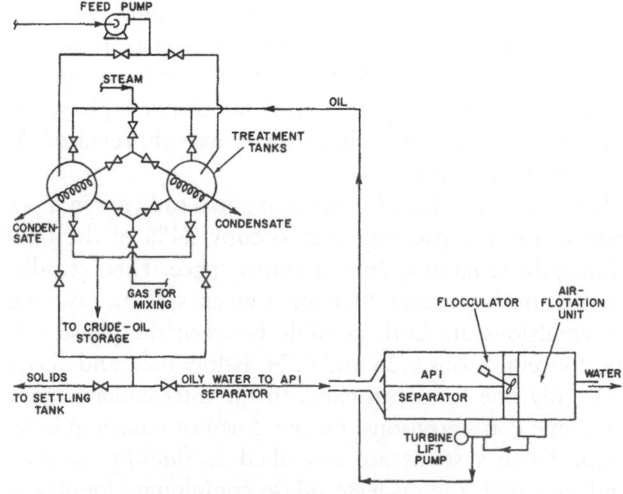

FIG. 7.4.20 Emulsion-breaking installation.

FUTURE OF EMULSION BREAKING

The removal of suspended and floating oil—generally known as primary treatment—can be accomplished by physical means. More advanced treatment is necessary to remove either dispersed or chemically emulsified oil. Chemical treatment can be an economical means of waste emulsion handling. A properly designed, emulsion-breaking process produces recoverable oil and treated water of adequate quality. The small volume of chemical sludge produced during the emulsion-breaking process is amenable to ultimate disposal.

—*D.E. Burns*
Donald Dahlstrom
Gerry L. Shell
S.L. Daniels
C.J. Santhanam
B. Rico-Ortega

8

Organics, Salts, Metals, and Nutrient Removal

R. David Holbrook | Sun-Nan Hong | Derk T.A. Huibers | Francis X. McGarvey | Chakra J. Santhanam

8.1 SOLUBLE ORGANICS REMOVAL

The activated-sludge process, developed in England in the early 1900s, is remarkably successful at removing soluble organics from wastewater (Junkins, Deeny, and Eckhoff 1983; Tchobanoglous and Burton 1991; Vesilind and Pierce 1982). In an air-sparged tank (see Section 6.4), live microorganisms rapidly adsorb, then slowly oxidize these organics to carbon dioxide and water. At the same time, these organisms reproduce. The process removes the microorganism sludge by settling, while digestion of adsorbed organics continues, which activates the sludge for recycling.

Excess activated sludge is disposed after a dewatering attempt. Its disposal is the most difficult and costly aspect of this wastewater treatment. Even the best sludge dewatering equipment produces a sludge cake with not more than 14–18% dry solids.

Improvements are aimed at making the activated-sludge process more efficient. Advances in biotechnology and fluid dynamics allow stricter environmental standards to be attained. Three major advances are as follows:

1. The use of oxygen instead of air facilitates maintaining the required DO levels. It also makes sludge separation easier (see Figure 8.1.1). Therefore, an oxygen-sparged plant can operate at higher sludge levels with a smaller aeration tank, and sludge disposal is reduced.
2. Replacing the air or oxygen-sparged tank with a fluidized-bed reactor allows the biomass to be adsorbed on small particles kept in suspension by the circulating effluent. This biofilm promotes high solids retention. The biomass concentration in the fluidized-bed reactor (15,000 ppm) is ten times greater than in the standard activated-sludge process. This increased concentration allows the fluid bed reactor to operate at a high, contaminant-removal efficiency.
3. Eliminating oxygen and using anaerobic methane fermentation of municipal wastewater in a fluidized-bed reactor is an even more efficient process. It is capable of operating with high biomass levels (30,000–100,000 ppm) at COD removal rates of 5–50 kg/m^3/day.

Using Oxygen Instead of Air

EFFLUENT BOD CONCENTRATION

11 to 23 mg/l

RETENTION TIME

1 to 3 hr

BOD REDUCTION

88 to 92%

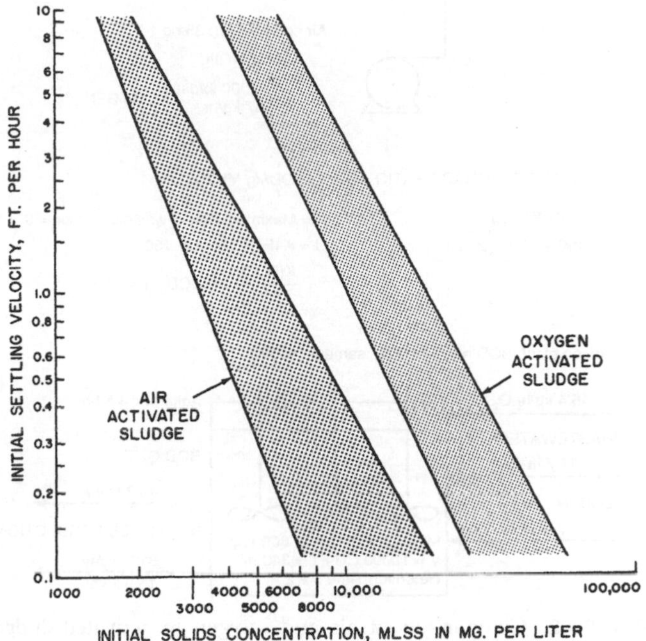

FIG. 8.1.1 Settling velocities of aerated and oxygenated activated-sludge particles.

ORGANIC LOADING PER 10,000 GAL OF AERATOR TANK

150 to 250 lb of BOD per day

COD REDUCTION

71 to 84%

RECYCLED SLUDGE SOLIDS CONCENTRATION

20,000 to 40,000 mg/l

EXCESS ACTIVATED SLUDGE PRODUCTION

0.3 to 0.45 lb of VSS per pound of BOD removed

OXYGEN UTILIZATION

+90%

POWER REQUIREMENTS OF OXYGEN PRODUCTION FROM AIR

3.6 to 4.5 lb of O_2 per hp-hr

POWER REQUIREMENTS FOR MIXING AND RECIRCULATION

3 to 5 hp-hr per mgd of feed or about 2 lb O_2 transferred per hp-hr

TANK SIZES IN OXYGENATION UNITS

The aeration tank is smaller (by up to a factor of three) but needs to be gas tight; the clarifier is the same size.

GAS VOLUME USED IN OXYGENATION PROCESSES

1% of gas volume in aeration systems

OPERATING DO LEVEL IN MIXED LIQUOR

6 to 10 mg/l

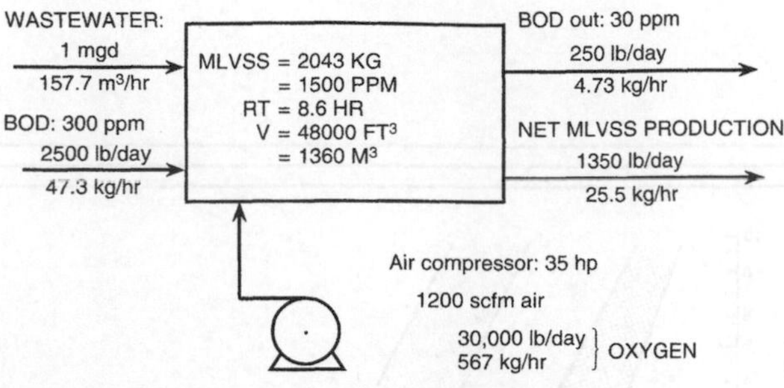

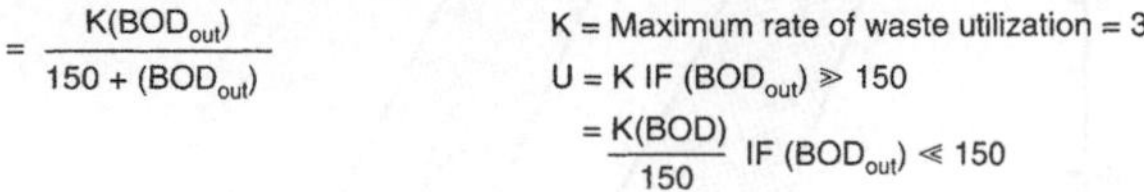

U = FOOD-TO-BIOMASS RATIO = 0.5 kg BOD/kg MLVSS day
= K(BODout) / 150 + (BODout)
K = Maximum rate of waste utilization = 3
U = K IF (BODout) ≫ 150
= K(BOD)/150 IF (BODout) ≪ 150

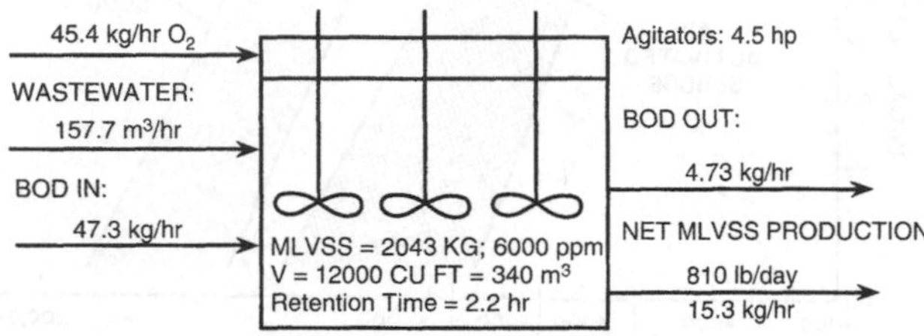

FIG. 8.1.2 Comparison of air and oxygen in activated-sludge wastewater treatment.

Figure 8.1.2 compares the use of air and oxygen in activated-sludge wastewater treatment. An air-sparged plant treating 1 mgd of wastewater with a BOD level 300 ppm and a BOD removal efficiency of 90% removes about 2250 lb/day of BOD. It produces about 1350 lb/day of MLVSS in 10,000 lb/day of sludge cake.

In this system a 35-hp compressor injects 1200-scfm air in the 48,000-cu-ft aeration tank (Nogaj 1972). This air contains 1250 lb/hr of oxygen or 13.33 lb of oxygen per lb of BOD removed. The low oxygen efficiency of 7.5% shows that mixing rather than dissolving oxygen is the determining factor of air sparging. Good mixing is essential for keeping bacteria in suspension, breaking up bacteria flocs, and promoting maximum contact of bacteria with the organics they use as a food source. A well-mixed aeration basin prevents raw wastewater from flowing directly from inlet to outlet. Sludge activation works better with pure commercial oxygen.

Pure oxygen accelerates the oxidative processes by at least a factor of four. Rapid oxygen transfer allows the plant to operate at high sludge concentrations. With oxygen, the sludge is less slimy and settles more readily, as shown in Figure 8.1.1. The SS can be increased threefold to 6000 ppm for the same settling velocity of 7 ft/hr. Figure 8.1.2 shows that at 6000 ppm MLVSS, a 1-mgd plant needs an aeration tank of only 12,000 cu ft. The clarifier can be operated at mass loadings of 45–65 lb/day/sq ft and hydraulic overflow rates of 600–1200 gpd/sq ft. In effect, the plant settler using air and the one using oxygen are the same size in spite of different MLVSS concentrations. For 1-mgd plants, these settlers have a surface of 1100 sq ft, a volume of 16,000 cut ft, and a mean hydraulic residence time of 2.9 hr. Better sludge settling also reduces sedimentation chemical use. Often these savings can pay for the cost of oxygen.

The Linde Division of Union Carbide (now independently operated as Praxair Inc.) demonstrated 95% oxygen utilization in the 1968 study of the Batavia, New York sewage works (Albertson et al., 1971; Gross 1976). (In 1980, the process named UNOX was acquired by Lotepro Corp., the U.S. subsidiary of Linde A.G. of Munich, Germany.) The Batavia plant covered its conventional aeration basins to prevent costly oxygen loss (see Figure 8.1.3). Efficient, upward-pumping, slow-speed, low-shear, agitator impellers kept the activated-sludge microbes in suspension. These agitators reduced energy needs for the

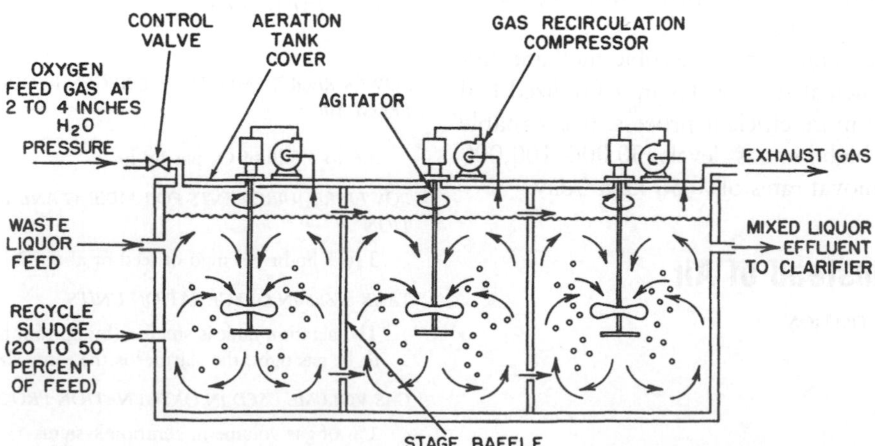

FIG. 8.1.3 Three-stage oxygenation system that uses rotating spargers and recirculation compressors.

1-mgd plant to 4–5 hp, down from the previously estimated 35 hp for air sparging.

Table 8.1.1 shows test results of the UNOX process applied to industrial wastewater streams (Gross 1976). Oxygen is used in more than 250 wastewater treatment plants with an average capacity of 32 mgd.

The four following factors accelerate oxygen use:

1. Tighter legal water and air discharge limits. Tightening the legal discharge limits requires increased bacterial activity, better clarification, and lower excess sludge production. By maintaining optimum metabolic rates, pure oxygen cuts sludge production in half. With pure oxygen, odor problems are greatly reduced or eliminated.
2. New air separation technologies. Noncryogenic air-separation technologies, particularly membrane and pressure swing adsorption (PSA), make lower-cost oxygen available to customers that do not require the highest purity.
3. Better oxygen application techniques. These techniques are discussed later.
4. Wastewater treatment plant capacity shortages. Many plants are looking for effective ways to expand capacity or handle periodic overloads.

Oxygen Application Techniques

TYPES OF SPECIAL AERATORS

a) U-tube
b) Jet
c) Draft tube air lifts
d) Self-priming mechanical aerator
e) Biological discs

APPLICATIONS

Industrial wastewater (b,c,d,e); domestic wastewater (a,b,c,e); lakes and reservoirs (c), and rivers (a)

OPERATING EXPERIENCE IN YEARS

More than 10 (a,b,e); less than 10 (c); and in development (d).

OXYGEN TRANSFER EFFICIENCY

In units of theoretical lb of oxygen per hp-hr of power invested: 0.3 to 3 (a), 2 to 6 (c), 3 to 5 (b), 15 to 17 (d). In units of lb of BOD removed per hp-hr of power invested: 2 to 8 (e), 15 (d).

In 1971, Air Reduction (now part of the British Oxygen Company) introduced a pipeline reactor as an efficient means of contacting wastewater with activated sludge and oxygen (Hover, Huibers, and Serkanic 1971). Turbulent flow causes effective mixing of wastewater and sludge with oxygen, especially in the froth flow regime (Baker 1958).

Praxair's affiliate Societa Italiana Acetilene and Derivati developed the MIXFLO system (see Figure 8.1.4). This 2- to 4-atm, side stream pipeline pumping system dissolves up to 90% of the injected oxygen (Storms 1993); 60% in the pipeline and 30% in the bulk liquid after exiting the dispersion ejector. This system exchanges ease of operation for a high-power input (38 hp at an oxygen-use rate of 100 lb/hr). The system is used in over 150 activated-sludge plants.

Liquid Air Corporation (L'Air Liquide) supplies the VENTOXAL system and the AIROXAL process for dissolving oxygen (Matson and Weinzaepfel 1992) that does not use a pipeline (see Figure 8.1.5). The submersible pump discharges directly via a venturi mixer or primary ejector, where the oxygen is introduced into a distributor sleeve. This configuration halves power input, to 20 hp.

The new Praxair In-Situ Oxygenator system, as shown in Figure 8.1.6, is more complex but uses even less power. The downward pumping impeller in the draft tube uses only 6 hp for adding 100 lb/hr of oxygen. It has a flotation ring and a wide hood to capture off gas. Oxygen use rates are above 90%. With its effective solids mixing capability, it can also be used in ponds.

TABLE 8.1.1 TEST RESULTS OF UNOX PROCESS APPLIED TO INDUSTRIAL WASTEWATER

Type of Production Facility and Character of Wastewater	*Influent BOD (mg/l)*	*Retention Time (hr)*	*MLVSS (mg/l)*	*Food-to-Biomass Ratio (U), lb BOD/day-lb MLVSS*	BOD_R, %	*lb Oxygen Required/lb. BOD*
Pesticides—Organic solvents and alcohols plus other biologically resistant organics	530	5.0	6000	0.41	90	2.45
Petrochemical—Low-molecular-weight organic acids and alcohols	330	3.3	4300	0.58	92	1.02
Petrochemical—Low-molecular-weight organic acids	1750	16	8900	0.31	96	0.98
Petrochemical—Glycols, glycol ethers, alcohols, alkanol amines, and acetone	570	3.5	4800	0.82	84	1.12
Food Processing—citric acid	4824	32.0	6200	0.67	98	1.04
Brewery Distillery	1010	6.1	6100	0.7	98	1.0
Pulp & Paper—Kraft pulp mill	291	2.0	5000	0.69	89	1.23

Source: R.W. Gross Jr., 1976, The UNOX process, *Chem. Eng. Progr.* 72, no. 10.

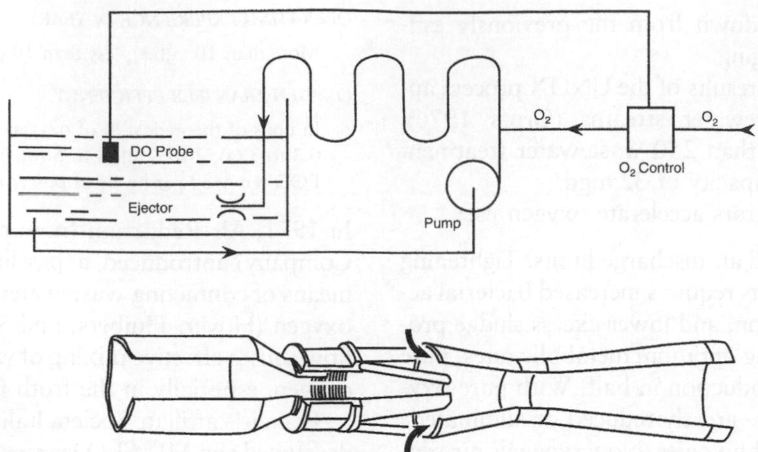

FIG. 8.1.4 Schematic diagram of the MIXFLO oxygenation system and ejector. (Reprinted, with permission, from G.E. Storms, 1993, *Oxygen dissolution technologies for biotreatment applications,* Tarrytown, N.Y.: Praxair Inc.)

Use of Oxygen in Bioremediation

The MIXFLO system was successfully used for the \$54-million bioremediation of the petroleum and petrochemical waste lagoon at the French Limited Superfund Site in Crosby, Texas (Bergman, Greene, and Davis 1992; *French Ltd.* 1992; Sloan 1987). In about five months, the chlorinated organics and mono- and polyaromatics were reduced from 1000+ to 100− ppm. The average rate of this aerobic treatment was 0.36 kg BOD/m^3/day, which is about half the rate of air or aerobic municipal sewage digestion. This rate is remarkable, considering contaminant character. Benzopyrene was reduced from 2000 ppm at day 100 to 20 ppm at day 200, benzene from 800 to 20, vinylchloride from 300 to 25, and total PCBs from 80 to 9. In November 1992 at day 300, all levels were below 8 ppm.

Project costs included \$1.3 million for 18,000 tn of liquid oxygen delivered to the site at \$72/tn; \$0.7 million for pumping power at \$0.05/kWh, and \$0.8 million for other power costs. Onsite incineration would have cost \$125 million.

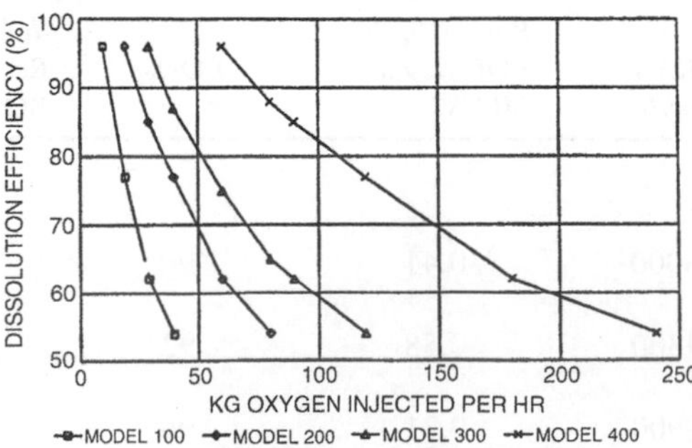

FIG. 8.1.5 Schematic diagram of the VENTOXAL system and its oxygenation efficiency. (Reprinted, with permission, from M.D. Matson and B. Weinzaepfel, 1992, Use of pure oxygen for overloaded wastewater treatment plants: The Airoxal Process, *Water Environment Federation, 65th Annual Conference and Expo, New Orleans, LA, Sept. 25, 1992,* paper AC92-028-007.)

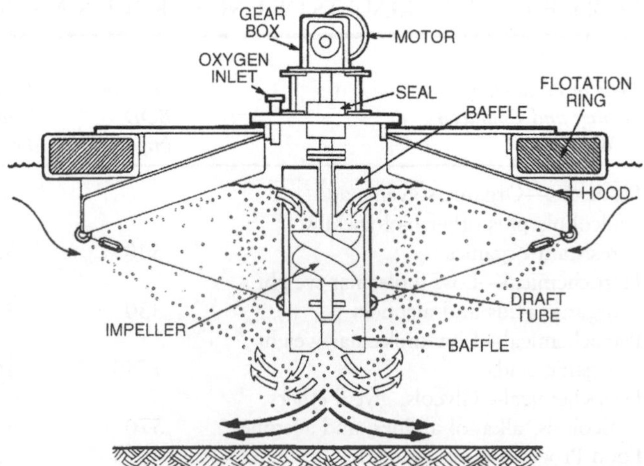

FIG. 8.1.6 The Praxiar Inc. In-Situ Oxygenator (Reprinted, with permission, from G.E. Storms, 1993, *Oxygen dissolution technologies for biotreatment applications,* Tarrytown, N.Y.: Praxair Inc.)

Use of Oxygen in Handling Industrial Wastewater

The VENTOXAL system effectively handles overload conditions, as illustrated by a case study of a paper mill wastewater treatment plant in Fors, Sweden. This paper mill produces newsprint and linerboard from thermo-mechanical pulp (TMP) and chemico-thermo-mechanical paper (CTMP). The periodic production of CTMP caused an effluent overload.

The activated-sludge plant had been designed for 14,000 m³/day of clarified TMP effluent with 1590 kg/day BOD. CTMP production raised this BOD load to 5000 kg/day. The schematic of the treatment plant in Figure 8.1.7 shows that the increased BOD load reduced its removal efficiency from 90 to 70%. More than twice the BOD was removed at the overload conditions than at the design conditions; 3500 versus 1431 kg/day. The mean hydraulic residence time fell from 8.6 to 3.9 hr, and the MLVSS level fell from 1500 to 1063 ppm. The effluent settler was less effective in handling the increased sludge load.

Four 300 m³/hr VENTOXAL systems were capable of adding 180 kg/hr of oxygen. This addition doubled the aeration capacity, raising the BOD removal efficiency to 95%. Figure 8.1.8 shows that the MLVSS concentration in the aeration tank now reached almost 7000 ppm. As

ORIGINAL PLANT: BOD REMOVAL 90%: 1431 kg/day

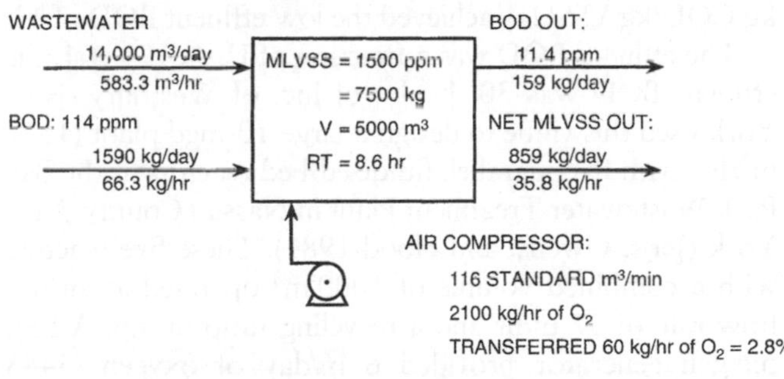

$U = 1431$ kg/day BOD/7500 kg MLVSS = 0.19

$$= \frac{k\,(BOD_{out})}{150 + (BOD_{out})} = 0.071\,k \Big\}\, k = 2.7$$

OVERLOAD: BOD REMOVAL 70%: 3500 kg/day

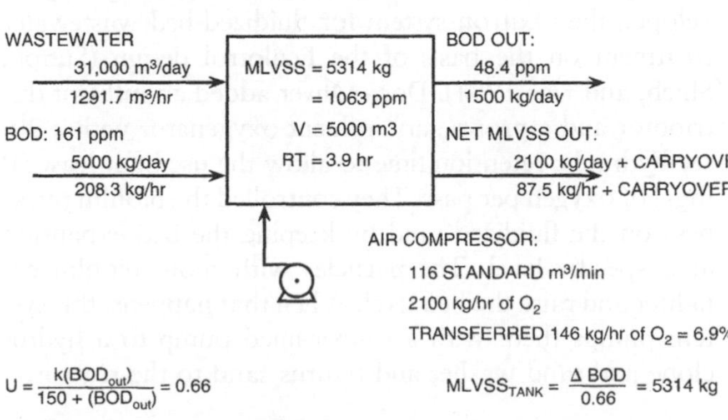

$$U = \frac{k(BOD_{out})}{150 + (BOD_{out})} = 0.66 \qquad MLVSS_{TANK} = \frac{\Delta\, BOD}{0.66} = 5314 \text{ kg}$$

FIG. 8.1.7 Activated-sludge waste wastewater treatment at a paper mill in Fors, Sweden.

previously discussed and shown in Figure 8.1.1, the use of pure oxygen enhances sludge settling velocities. Moreover, at this plant, it had an added benefit—a 30% reduction of sedimentation chemicals from 4540 to 3180 kg/day. These savings covered the oxygen costs, as shown in Table 8.1.2. This table summarizes the capital and operating costs of the plant upgrade.

PLANT UPGRADE:

PLANT EQUIPPED WITH FOUR VENTOXAL 300 m³/hr OXYGEN INJECTORS

BOD REMOVAL 95%: 4750 kg/day

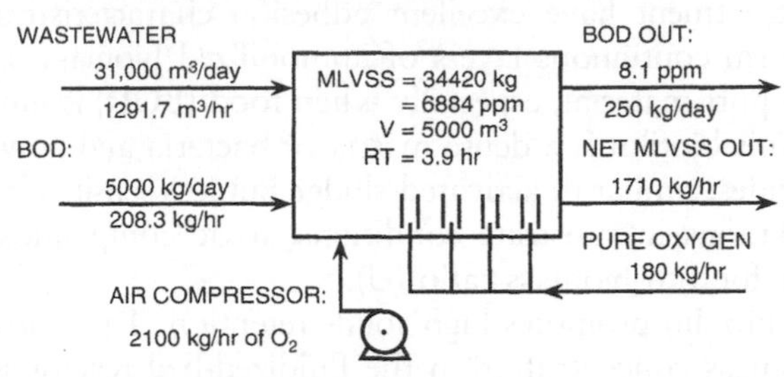

$$U = \frac{K(BOD_{out})}{150 + (BOD_{out})} = \frac{2.7 \times 8.1}{150 + 8.1} = 0.138$$

$$MLVSS_{TANK} = \frac{\Delta\, BOD}{0.138} = 34420 \text{ kg}$$

O_2 TRANSFERRED FROM PURE OXYGEN	151 kg/hr (84%)
O_2 TRANSFERRED FROM AIR	47 kg/hr (2.2%)
BOD REDUCTION	198 kg/hr (95%)

FIG. 8.1.8 Activated-sludge wastewater treatment of a paper mill in Fors, Sweden upgraded with four VENTOXAL oxygen injectors.

TABLE 8.1.2 COST SUMMARY OF THE WASTEWATER TREATMENT PLANT UPGRADE AND ITS ANNUAL OPERATION AT A PAPER MILL IN FORS, SWEDEN

Cost of Plant Upgrade:	
Four VENTOXAL units, installed (including control panels)	$150,000
Second sludge dewatering unit	$100,000
Total	$250,000
Annual Operating Costs:	(350 days/yr)
Amortization (5 yr/12% interest)	$67,000
Oxygen (1512 metric tn @$86/tn)	$130,000
Energy (500,000 kWh @7¢/kWh) (0.33 kWh/kg O_2)	$35,000
Operation and maintenance	$25,000
Chemical credit (sedimentation) (1589 − 1113 = 476 metric ton @ $250/tn)	($119,000)
Total	$138,000
Extra BOD Removal: 3319 kg/day; 1162 metric tn/yr @ $120/metric tn BOD removed	

Source: Matson and Weinzaepfel, 1992.

Biological Fluidized-Bed Wastewater Treatment

The biological, fluidized-bed, wastewater treatment process is a new adaptation of the fixed-film, biological reactor—the trickling filter. In this process, the adsorbent particles are small and kept in suspension by the circulating effluent (see Figure 8.1.9). This improved process is considered the most significant development in wastewater treatment in the last fifty years.

The mixed microbial cultures associated with wastewater treatment have excellent adhesion characteristics. They form continuous layers of immobilized biomass on any support material, especially when food (BOD) is limiting. This biofilm is a dense matrix of bacteria and polysaccharides, similar to activated-sludge but less sensitive to perturbations in substrate conditions, toxic compounds, and the food-to-biomass ratio (U).

The biofilm promotes high solids retention. Therefore, the biomass concentration in the fluidized-bed reactor is typically 15,000 ppm, ten times greater than in the standard activated-sludge process. This concentration allows the fluidized bed reactor to operate at a high contaminant removal efficiency. In the standard activated-sludge process, a low U leads to poor floc formation.

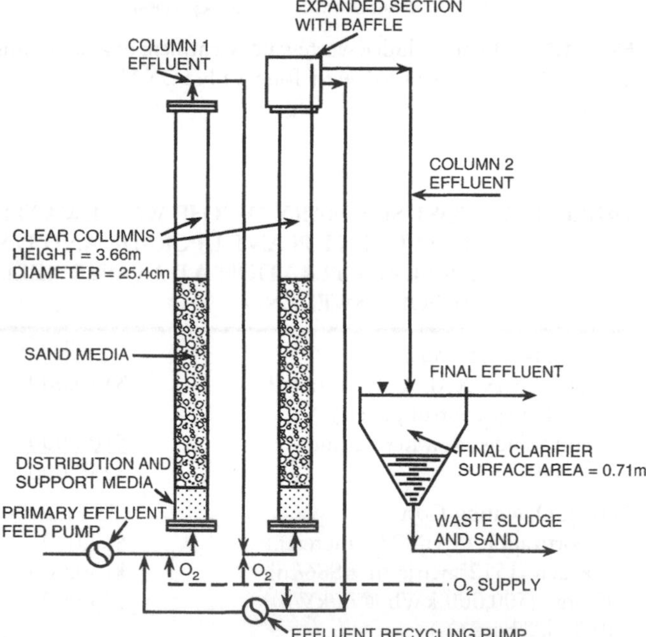

FIG. 8.1.9 Schematic diagram of two fluidized-bed reactors in series for biological wastewater treatment. (Reprinted, with permission, from E.T. Oppelt and J.M. Smith, 1981, U.S. Environmental Protection Agency research and current thinking on fluidised-bed biological treatment, in *Biological fluidised bed treatment of water and waste water,* edited by P.F. Cooper and B. Atkinson, Chichester: Ellis Horwood Ltd., Publishers.)

Aerobic Fluidized-Bed Treatment of Municipal Wastewater

From 1970 to 1973, the EPA used supported microbial cultures in expanded or fluidized-bed reactors to study municipal wastewater treatment at its Lebanon Research Pilot Plant near Cincinnati, Ohio (Oppelt and Smith 1981). Fine 0.5 mm sand particles, like those used in sand filters, supported a 0.25 mm biomass film.

The total COD removal was 26% at 16 min and 65% at 47 min residence time. A COD removal of 65% was the best that the two-column system could do. The minimum upflow rate in the columns was 10 m/hr. The maximum rate was 39 m/hr, which gave a residence time of 7 min per two-column pass. Sand loss was a major problem. It occurred mainly from the transport of oxygen bubbles.

The EPA obtained better results with eight reactors in series. They installed bubble trap devices at the top of each column to collapse the bubbles and allow the sand to drop back. Oxygen was added in fine bubble diffusers at the base of the downleg between the reactor stages.

In a once-though operation with an empty-bed retention time of 44 min, COD removal increased to 75%, and BOD removal increased to 89% (effluent COD 48.8 ppm from 196.3 and effluent BOD 13 from about 118). The high biomass concentration of MLVSS = 15,000 ppm increased the COD treatment efficiency from 3.0 kg/m^3/day to 4.8 kg/m^3/day for a standard oxygen-activated sludge process. The low food-to-biomass mass ratio of U = 0.3 kg COD/kg VSS/day achieved the low effluent BOD of 13.

The effluent BOD was a function of U. At U = 1.1, the effluent BOD was 30. Ecolotrol Inc. of Westbury, New York used this value to design a large 10-mgd plant (1577 m^3/hr) with five, parallel, fluidized-bed reactors at the Bay Park Wastewater Treatment Plant in Nassau County, New York (Jeris, Owens, and Flood 1981). These five reactors with a combined volume of 1000 m^3 operated at an upflow rate of 37 m/hr and a recycling ratio of 4.6. A PSA oxygen generator provided 6 tn/day of oxygen (5443 kg/day). The plant had various operating difficulties; controlling microbial growth was the most problematic. The plant was shut down in 1991.

Dorr–Oliver, a Connecticut engineering company, developed the Oxitron system for fluidized-bed wastewater treatment on the basis of the Ecolotrol design (Sutton, Shieh, and Kos 1981). Dorr–Oliver added an influent distributor and a proprietary influent oxygenator with a 20-sec hydraulic retention time to allow the use of at least 50 mg/l of oxygen per pass. They controlled the biofilm thickness on the fluidized sand by keeping the bed expansion at a specific level. The particles with more biofilm are lighter and raise the bed level. When that happens, this system pumps them with a rubber-lined pump to a hydroclone and sand washer and returns sand to the reactor.

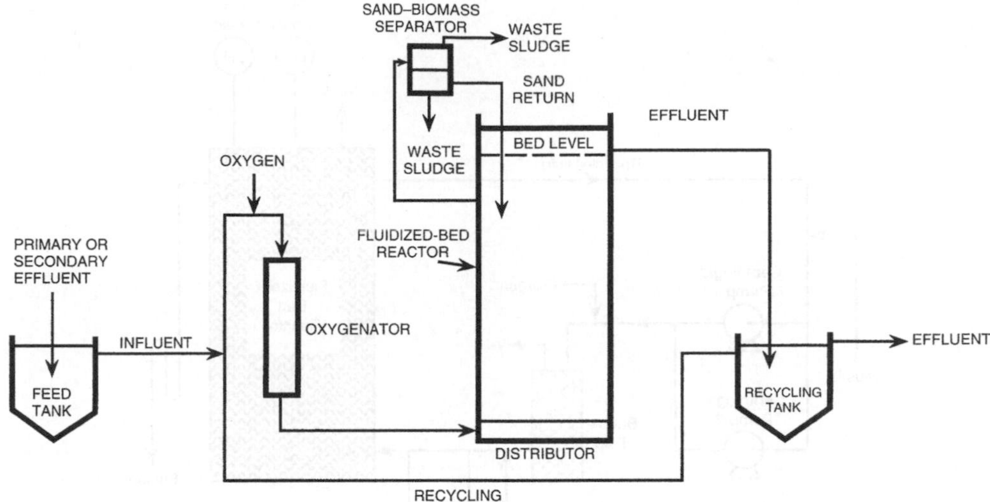

FIG. 8.1.10 Schematic diagram of the Dorr–Oliver Oxitron System pilot plant in Orillia, Ontario. (Reprinted, with permission, from P.M. Sutton, W.K. Shieh, and P. Kos, 1981, Dorr–Oliver's Oxitron system fluidised-bed water and wastewater treatment process, in *Biological fluidised-bed treatment of water and wastewater,* edited by P.F. Cooper and B. Atkinson, Chichester: Ellis Horwood Ltd., Publishers.)

In the pilot plant shown in Figure 8.1.10, tests were conducted with wastewater having a median BOD of 67 ppm and a median COD of 175 ppm. The best results, 78% BOD removal to 15 ppm, were obtained at an upflow rate of 20.3 m/hr, a hydraulic retention time of 37 min, and a food-to-biomass ratio of U = 0.16 kg BOD/kg VSS/day.

Dorr–Oliver designed full-scale plants for municipal wastewater treatment at Haywood, California (70,000 m^3/day) and Sherville, Indiana (35,000 m^3/day). Based on operating costs, the OXITRON system is favored for plants with flows above 20,000 m^3/day. At this size, the operating costs, including the present value of the facility discounted at 7% for 20 yr, is $4.15 million. These costs do not include sludge disposal; they are based on a BOD reduction of 90% from 200 to 20 ppm and a SS reduction from 140 to 30 ppm. The Haywood, California plant experienced operating problems similar to the Nassau County, New York plant. This plant was also shut down.

The OXITRON system design is now available from Envirex in Waukesha, Wisconsin. It has been particularly useful for denitrification in wastewater treatment in Reno, Nevada and in industrial wastewater treatment at a General Motors plant (Jeris and Owens 1975).

An interesting new application of aerobic, fluidized-bed, wastewater treatment is groundwater remediation. With continued recycling, this treatment can reduce effluent concentrations to levels below 5 ppm. The fluidized biomass acts like an organics equalization tank by adsorbing organics and gradually digesting them. No biomass washout occurs at low concentrations. This advantage is useful in the treatment of hazardous wastes, as described next.

Aerobic Fluidized-Bed Treatment of Industrial Wastewater

Recently, Celgene Corporation introduced the fluidized-bed reactor with its *in-process, biotreatment system* (Gruber 1993; Sommerfield and Locheed 1992). Celgene's expertise was in selecting and applying the most suitable microbes for metabolizing specific organics, e.g., trichloroethylene, methylene chloride, other organic chlorides, ketones, and aromatics on EPA's list of seventeen hazardous materials targeted for 50% industrial emission reduction by 1995.

Figure 8.1.11 shows the Celgene fluidized-bed reactor. This reactor suspends carbon particles with the immobilized biomass in an upflow of 29 m/hr. This flow gives a 35–50% bed expansion (measured with a reflective IR-level detector). The time per pass is 8 min. With a total wastewater residence time of 400 min, the recycling ratio is 50.

The reactor injects oxygen from a PSA unit into the reactor recycling loop. A reactor effluent of DO = 35 ppm controls oxygen addition. Particles with excess biomass rise to the top of the reactor. Here, the biomass is knocked off with a slowly rotating peddle and carried with the effluent to the clarifier.

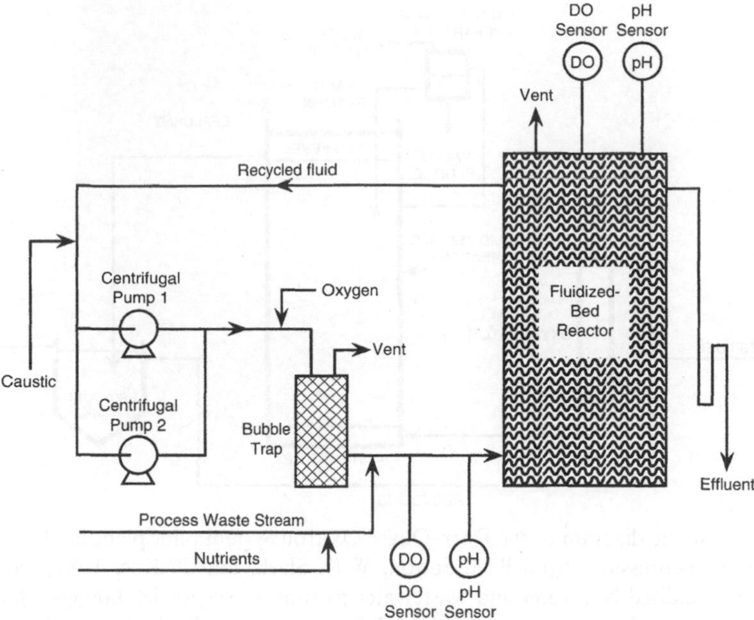

FIG. 8.1.11 Schematic diagram of the Celgene biotreatment process. (Reprinted, with permission, from W. Gruber, 1993, Celgene's biotreatment technology: Destroying organic compounds with an in-line, on-site treatment system, *EI Digest* [November].)

Celgene conducted seven large-scale pilot demonstrations, of which five involved industrial process streams. In the 43,200-gpd pilot plant located at General Electric's Mt. Vernon, Indiana facility, the methylene chloride was reduced from 1260 to <5 ppm. At a Gulf Coast petrochemical plant the effluent COD was reduced from 210 to 40 ppm with 0.3 ppm phenols, <5 ppm aromatics, and <10 ppm SS. Treatment costs are about $10 per 1000 gal. The commercialization of the process has been taken over by Sybron Chemicals.

Manville and Louisiana State University used a fluidized-bed system at Ciba–Geigy's St. Gabriel plant site to lower the sodium chloroacetate level in a 3–4% saline waste stream from 6000 to 10 ppm (Attaway et al. 1988). They used a 0.25-inch, diatomaceous-earth carrier with a pore structure optimized for microbe immobilization in two bioreactors in series with a volume of 141.3 gal each. At a throughput of 0.25 gpm, they observed biological activity in both reactors. The effluent of the first reactor had a sodium chloroacetate level of 2400 ppm.

Anaerobic Wastewater Treatment with Attached Microbial Films

Twenty years ago, no evidence existed to suggest that dilute wastewater could be treated anaerobically at an ambient temperature. Prodded by the 1973 energy crisis, Jewell (1981) and co-workers converted a fluidized-bed reactor for anaerobic methane fermentation and studied municipal sewage treatment (Jewell, Switzenbaum, and Morris 1979).

This study led to the discovery of an efficient process capable of operating at high biomass concentrations of 30,000 ppm. At retention times of <30 min, this anaerobic process removed most biodegradable organics from municipal wastewater, reducing the COD to <40 ppm and the SS to <5 ppm. The volumetric density of the anaerobic film exceeded 100 kg VSS/m^3. The net yield of the biomass produced was Y = 0.1 kg VSS/kg COD. This density gave a solids residence time (SRT) that was three to eight times that of the aerobic process. Ultimately, the biomass concentration of the anaerobic attached-film-expanded bed (AFEB) could approach 100,000 ppm. This concentration raises the COD removal rates to >50 kg/m^3/day and still maintains an SRT >30 days.

An aerobic AFEB cannot operate at these high-rate conditions without washout. However, an aerobic AFEB can ultimately produce a lower effluent COD and nitrify ammonia. A series treatment with an anaerobic AFEB followed by an aerobic AFEB, with each unit having a 15-min retention time, resulted in a superb effluent quality with a COD = 10, an SS = 1, and a turbidity of 2.

Figure 8.1.12 contrasts the aerobic and anaerobic AFEB processes. Up to an organic loading of 2 kg/m^3/day, the two processes are about equal. At higher loading rates, the anaerobic AFEB process produces a better effluent COD.

—*Derk T.A. Huibers*

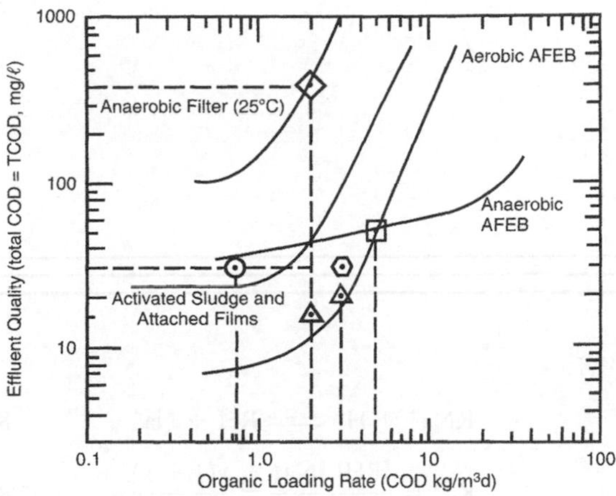

Key:

⊙ = Effluent BOD from conventional activated-sludge treatment in Figure 7.35.1

⊙ = Effluent BOD from conventional activated-sludge process using oxygen instead of air

⊡ = Effluent TCOD of EPA's AFEB pilot plant (Oppelt and Smith 1981)

△ = Effluent BOD of Oxitron AFEB plant (Sutton, Shieh, and Kos 1981)

◇ = Effluent BOD of industrial Biotower (Fouhy 1992)

FIG. 8.1.12 Relationship of COD removal efficiency and organic loading capacity for standard and AFEB aerobic and anaerobic treatment processes.

References

Albertson, J.G., J.R. McWirter, E.K. Robinson, and N.P. Valdieck. 1971. *Investigation of the use of high purity oxygen aeration in the conventional activated sludge process.* FWQA Department of the Interior program no. 17050 DNW, Contract no. 14-12-465 (May) and Contract no. 14 12 867 (September).

Attaway, H., D.L. Eaton, T. Dickenson, and R.J. Portier. 1988. Waste stream detoxified with immobilized microbe system. *Pollution Engineering* (September): 106–108.

Baker, O. 1958. Multiphase flow in pipelines. *The Oil and Gas Journal* (10 November): 156–167.

Bergman Jr., T.J., J.M. Greene, and T.R. Davis. 1992. An in-situ slurry-phase bioremediation case with emphasis on selection and design of a pure oxygen dissolution system. *Air & Waste Management Association and EPA Risk Reduction Laboratory, In-Situ Treatment of Contaminated Soil and Water Symposium, Cincinnati, Ohio Feb. 4–6, 1992.*

Fouhy, K. 1992. Biotowers treat highly contaminated streams. *Chem. Eng.* (December): 101–102.

French Ltd.: A successful approach to bioremediation. 1992. *Biotreatment News* (October, November, and December).

Gross Jr., R.W. 1976. The Unox Process. *Chem. Eng. Progr.* 72, no. 10: 51–56.

Gruber, W. 1993. Celgene's biotreatment technology: Destroying organic compounds with an in-line, on-site treatment system. *EI Digest* (November): 17–20.

Hover, H.K., D.T.A. Huibers, and L.J. Serkanic Jr. 1971. *Treatment of secondary sewage.* U.S. Patent 3,607,735 (21 September).

Jeris, J.S., R.W. Owens, and F. Flood. 1981. Secondary treatment of municipal wastewater with fluidized-bed technology. In *Biological fluidised bed treatment of water and wastewater,* edited by P.F. Cooper and B. Atkinson, 112–120. Chichester: Ellis Horwood Ltd., Publishers.

Jeris, J.S., and R.W. Owens. 1975. Pilot-scale high-rate biological denitrification. *J. Water Pollution Control Fed.* 47:2043.

Jewell, W.J. 1981. Development of the attached film expanded bed (AFEB) process for aerobic and anaerobic waste treatment. In *Biological fluidised bed treatment of water and wastewater,* edited by P.F. Cooper and B. Atkinson, 251–267. Chichester: Ellis Horwood Ltd., Publishers.

Jewell, W.J., M.S. Switzenbaum, and J.W. Morris. 1979. Sewage treatment with the anaerobic attached film expanded bed process. *52nd Water Pollution Control Federation Conference, Houston, Texas, Oct. 1979.*

Junkins, R., K. Deeny, and T. Eckhoff. (Roy F. Weston, Inc.). 1983. *The activated sludge process: Fundamentals of operation.* Ann Arbor, Mich.: Ann Arbor Science Publishers (The Butterworth Group).

Matson, M.D. and B. Weinzaepfel. 1992. Use of pure oxygen for overloaded wastewater treatment plants: The Airoxal process. *Water Environment Federation, 65th Annual Conference & Expo. New Orleans, LA, Sept. 25, 1992.* Paper AC92-028-007.

Nogaj, R.J. 1972. Selecting wastewater aeration equipment. *Chem. Eng.* (17 April): 95–102.

Oppelt, E.T. and J.M. Smith. 1981. U.S. Environmental Protection Agency research and current thinking on fluidised-bed biological treatment. In *Biological fluidised bed treatment of water and waste water,* edited by P.F. Cooper and B. Atkinson, 165–189. Chichester: Ellis Horwood Ltd., Publishers.

Sloan, R. 1987. Bioremediation demonstrated at hazardous waste site. *Oil & Gas Journal* (14 September): 61–66.

Sommerfield, T., and T. Locheed. 1992. Hungry microorganisms devour methylene chloride: Fluidized bed bioreactor reduces emissions to less than 100 ppb, keeps operating costs down. *Chem. Proc.* 55, no. 3: 41–42.

Storms, G.E. 1993. Oxygen dissolution technologies for biotreatment applications. Tarrytown, N.Y.: Praxair Inc.

Sutton, P.M., W.K. Shieh, and P. Kos. 1981. Dorr–Oliver's Oxitron system fluidised-bed water and wastewater treatment process. In *Biological fluidised bed treatment of water and wastewater,* edited by P.F. Cooper and B. Atkinson, 285–300. Chichester: Ellis Horwood Ltd., Publishers.

Tchobanoglous, G., and F.L. Burton (Metcalf & Eddy Inc.). 1991. *Wastewater engineering: Treatment, disposal and reuse.* 3d ed. New York: McGraw-Hill.

Vesilind, P.A., and J.J. Peirce. 1982. *Environmental engineering.* Boston: Butterworth.

8.2 INORGANIC SALT REMOVAL BY ION EXCHANGE

PARTIAL LIST OF SUPPLIERS

Advanced Separation Technologies; Aquatech International Company; Arrowhead Industrial Water; Cochrane Division of Crane Corp.; Culligan International; Degramont Infilco; Dow Chemical; Gregg Water Conditioning Inc.; Craver Water Conditioning; HOH Systems; Hungerford and Terry; Illinois Water Treatment; Ionics Inc.; Ion Pure; Permutit; Purolite; Rohm and Haas; Sybron Chemicals Inc.; Vaponics; U.S. Filter Company.

The ion-exchange reaction is the interchange of ions between a solid phase and a liquid surrounding the solid. Initially, ion exchange was confined to surface reactions, but these reactions were gradually replaced by gel-type structures where exchange sites were available throughout the particle. Figure 8.2.1 shows this process graphically.

Ion-Exchange Reaction

Exchange sites exhibit an affinity for certain ions over others. This phenomenon is helpful in removing objectionable ionic materials from process streams. Environmental engineers have studied the affinity relationships and have identified certain simple rules. First, ions with multiple charges are held more strongly than those of lower charges. Ions with the same charge are held according to their atomic weight with heavier elements held more strongly. The affinity relationships for cation and anion exchangers are as follows:

Cation Exchangers:
- Monovalent—Cs > Rb > K > Na > Li
- Divalent—Ra > Ba > Sr > Ca > Mg >> Na

Anion Exchangers:
- Monovalent—I > Br > NO_3 > Cl > HCO_3 > F > OH
- Divalent—CrO_4 > SO_4 > CO_3 > HPO_4

The affinity relationship can also be expressed with the following equilibrium (selectivity) equations based on the reversibility of ion-exchange reactions and the law of mass action:

$$RNa^+ + H^+ \rightleftharpoons RH + Na^+ \qquad 8.2(1)$$

$$K_H^{Na} = \frac{[RH]\,[Na]}{[RNa]\,[H]} = \frac{y(1-x)}{x(1-y)} \qquad 8.2(2)$$

The following equations are for the divalent–monovalent reactions:

$$2RNa + Ca^{++} \rightleftharpoons RCa + 2Na^+ \qquad 8.2(3)$$

$$K_{Na}^{Ca} = \frac{[RCa]\,[Na^+]^2}{[RNa]^2\,[Ca^{++}]} \qquad 8.2(4)$$

$$\frac{KQ}{C_o} = \frac{y(1-x)^2}{x(1-y)^2} \qquad 8.2(5)$$

In these equations, the brackets represent the ion concentration in the resin and the liquid phase. The y, x notation expresses the reactions as equivalent ratios. For the divalent–monovalent reaction, Q is the resin capacity, and C_o is the total concentration of the electrolyte in solution. Plotting these equations, usually as y versus x plots, shows the exchange processes that occur in the exchange zone or a batch contactor.

Structural Characteristics

Modern ion-exchange materials are prepared from synthetic polymers such as styrene-divinyl-benzene copolymers that have been sulfonated to form strongly acidic cation exchangers or aminated to form strongly basic anion exchangers. Weakly basic anion exchangers are similar to the strong base except for the choice of amines. Weakly acidic cation exchangers are usually prepared from cross-linked acrylic copolymers.

Figure 8.2.2 shows the reactions involved in the preparation of anion exchange resins. Figure 8.2.3 shows the reactions for cation exchange resins. Figure 8.2.4 shows the structure of a typical chelation resin.

Ion-Exchange Characteristics

The resins prepared with the proper synthetic procedures have characteristics related to the percentage cross-linkage in the copolymers. They also have particle size ranges consistent with the 16–50 U.S. mesh size with a uniformity

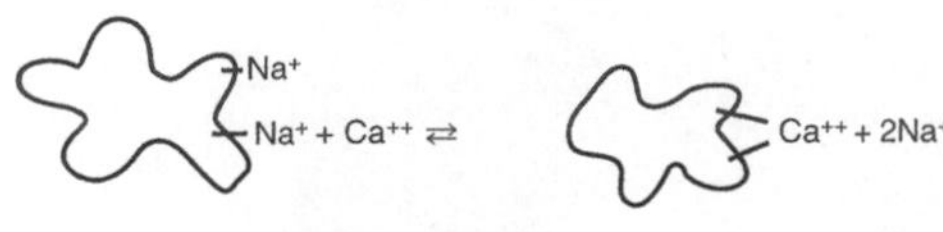

FIG. 8.2.1 The ion-exchange reaction.

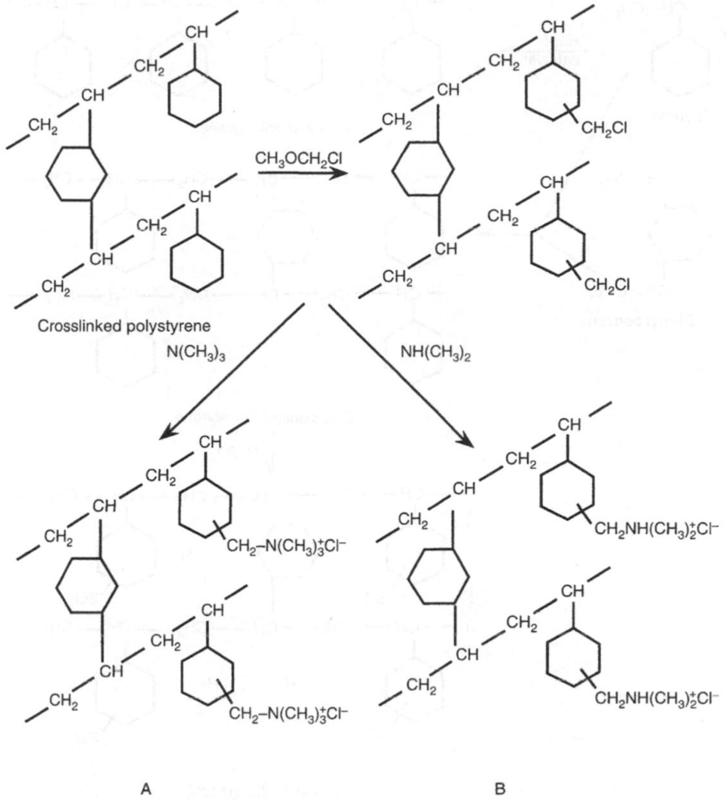

FIG. 8.2.2 Preparation of anion exchange resins. (Reprinted, with permission, from T.V. Arden, 1968, *Water purification by ion exchange,* 22, New York: Plenum Press.

coefficient not exceeding 1.5. (The uniformity coefficient is the 40% size divided by the 90% size.) The effective size (90% size) should be not less than 0.35 mm as measured from a probability plot of size as a function of the accumulative percentage.

The values summarized in Table 8.2.1 are averages. The total number of functional groups per unit weight or volume of resins determines the exchange capacity, whereas the types of functional groups influence ion selectivity and equilibrium. Many specifications deal with these resins, including the whole-bead content, bead strength, attrition values, special sizes, effluent quality, and porosity. The resin supplier must address the specification requirements for various conditions.

Process Applications

Environmental engineers base their selection of the proper process for a water treatment on criteria mentioned previously. Their flow sheet is based on resin properties. Resins are used in the following processes:

- Softening
- Dealkalization
- Desilicizing
- Organic scavenging
- Deionization
- Metal waste treatment

Role of Resin Types

The characteristics of the functional group determine the application in the process.

Generally, the ion-exchange process involves a columnal contact of the liquids being treated. The design of ion-exchange process equipment provides good distribution under design conditions, but environmental engineers must be careful in situations where the flow is decreased or increased beyond design conditions. The flow direction for service and regeneration is an important consideration. Two designs are offered: cocurrent and countercurrent. Generally, cocurrent is practiced when service and regenerant flows are in the same direction, usually downward.

Countercurrent normally services downflow and regeneration upflow. For example, the performance of a two-bed system is closely related to the way the cation exchange resin is regenerated. Alkalinity and sodium control are also closely associated with leakage. The advantages and disadvantages of these process steps can be summarized as

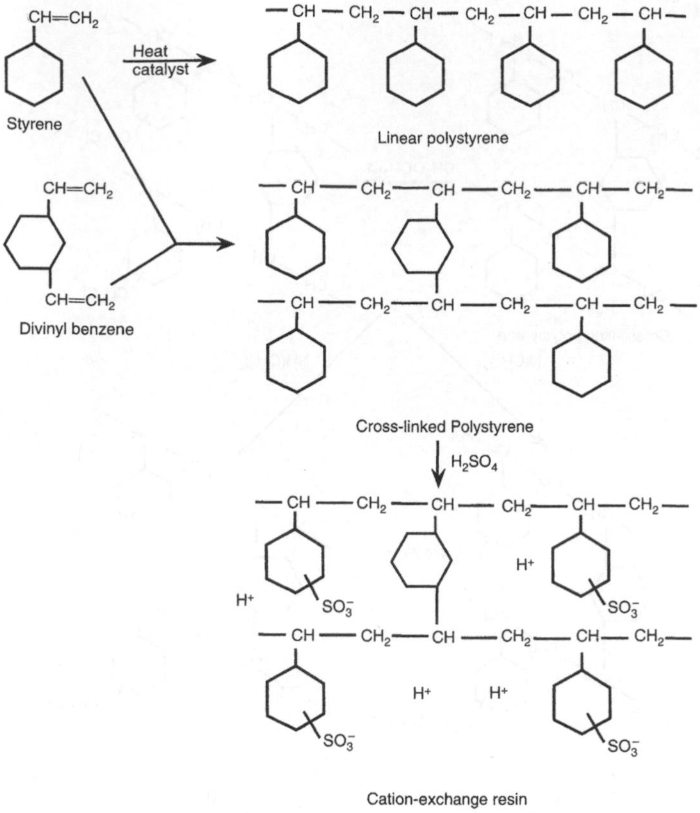

FIG. 8.2.3 Preparation of sulphonic-acid, cation-exchange resin. (Reprinted, with permission, from T.V. Arden, 1968, *Water purification by ion exchange,* New York: Plenum Press.)

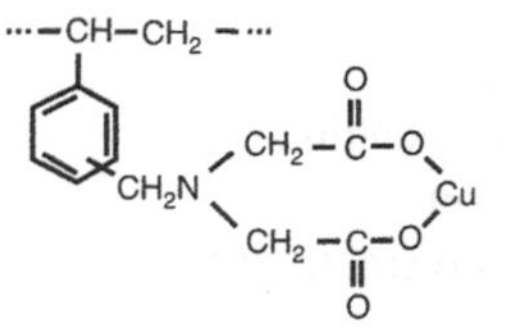

FIG. 8.2.4 Typical structure of a chelating resin—copper form. (Reprinted, with permission from K. Dorfner, *Ion exchange,* 38.)

follows:

STRONG-ACID-CATION-EXCHANGE RESIN—Removes all cations regardless of with which anion they are associated. This resin has moderate capacity and requires a strong acid regenerant such as hydrochloric or sulfuric acid.

WEAK-ACID-CATION-EXCHANGE RESIN—Removes only cations associated with alkalinity, such as bicarbonate, carbonate, or hydroxide. Any cations associated with chloride and sulfate are not removed.

STRONG-BASE-ANION-EXCHANGE RESIN—Removes all dissociated anions such as bicarbonate, sulfate, chloride, and silica. This resin exhibits low to moderate capacity and must be regenerated with a strong alkali such as sodium hydroxide. This resin can remove CO_2 and silica from water.

WEAK-BASE-ANION-EXCHANGE RESIN—Removes only anions associated with the hydrogen ion that are strong acid formers such as sulfate, chloride, or nitrate. Any anions associated with cations other than hydrogen pass through unaffected.

COCURRENT

Figure 8.2.5 shows a typical cocurrent operation for a strong-acid-cation-exchange bed operated in the acid cycle. In this operation, there is a residual band of unregenerated resin at the bottom of the bed. This band is stripped by the acidity, and a small amount of monovalent cations in the bottom band strips as shown in Figure 8.2.6. The ratio in this figure refers to the influent concentration of the ion of least affinity divided by the influent concentration.

Cocurrent operation has a fixed bed, so that particle motion is unlikely. Once the bed is classified, it stays in place, and the facility can backwash during each cycle without disturbing flow patterns.

COUNTERCURRENT

Figure 8.2.7 shows the location of the bands where regenerant passes upward, and influent water passes down-

TABLE 8.2.1 ION-EXCHANGE RESIN CHARACTERISTICS

Property	*Strong Acid*	*Strong Base Type 1*	*Weak Base*	*Weak Acid*
Water Retention, %	44–48	48–53	45–50	46–52
Capacity, meq/g dry	4.25	4.0	3.8	11
meq/ml	2.2	1.5	1.6	3.5–4.0
True Density, gm/cc	1.26–1.30	1.06–1.10	1.15	1.22–1.26
Apparent Density, lb/ft^3	50–52	44–46	44–47	48–50
Particle Range, U.S. Mesh	16–50	16–50	16–40	16–40
Effective Size, mm	0.45–0.60	0.4–0.48	0.36–0.48	0.35–0.55
Swelling, %	5	10	25	50
Shipping Weight, lb	52	45	45	48

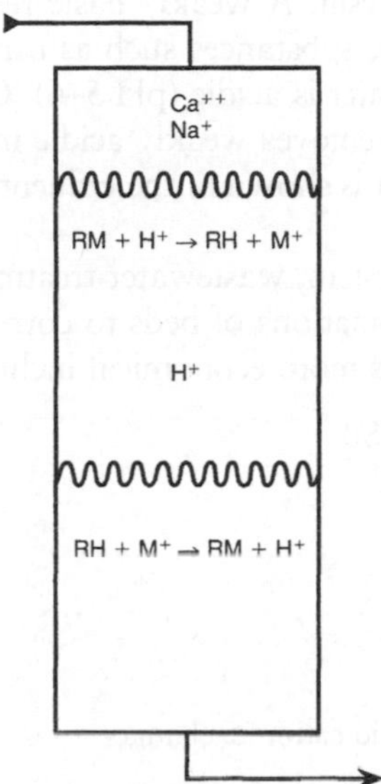

FIG. 8.2.5 Cation exchange with cocurrent regeneration. (Reprinted, with permission, from Frank McGarvey. *Introduction to industrial ion exchange*, Sybron Chemicals Inc.)

ward. In this operation, the residual band is at the top, and stripping of the band does not occur. Leakage depends on the end point selected. However, the bed must be held in place without mixing so that a clean band exists in the bottom of the bed. This clean band is usually achieved either by blocking water or a gas dome above the bed. Blocking water flow is usually 0.5–0.75 times the regenerant flow. This water can have a significant volume in waste control.

Figure 8.2.8 gives the breakthrough curves at various flows.

Application of Ion Exchange

As indicated previously, the ion-exchange process is normally practiced in columns. The actual arrangement is influenced by the following factors:

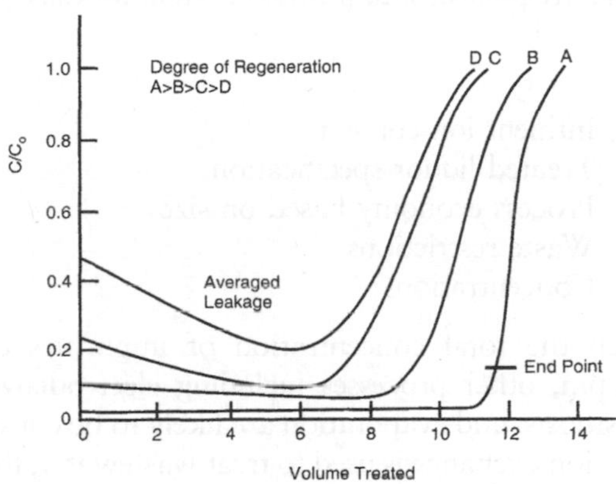

FIG. 8.2.6 Exhaustion curves for cation exchange with cocurrent operation—sodium chloride influent. (Reprinted, with permission, from McGarvey.)

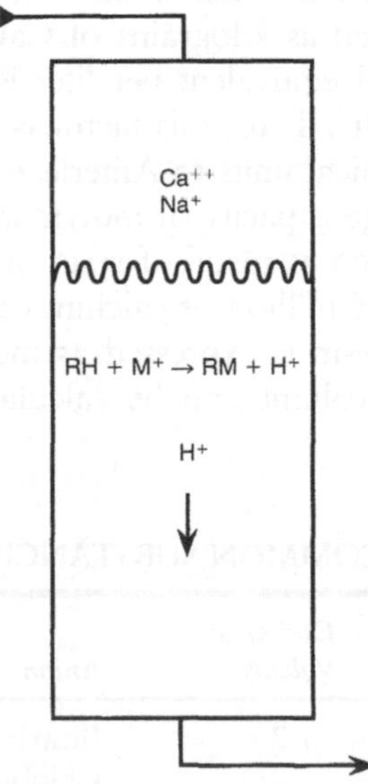

FIG. 8.2.7 Countercurrent operation of a cation-exchange resin. (Reprinted, with permission, from McGarvey.)

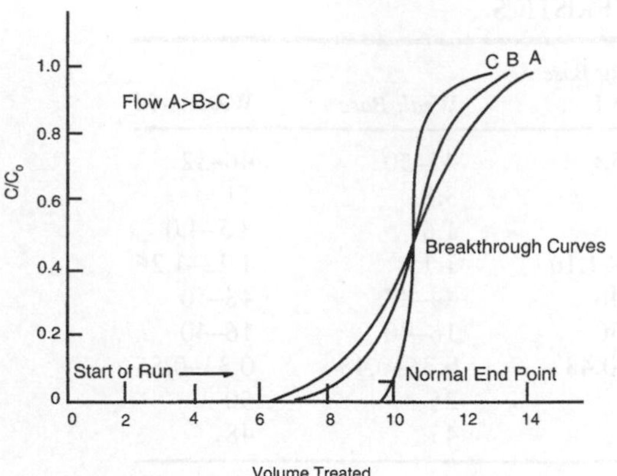

FIG. 8.2.8 Typical breakthrough curves for the countercurrent operation. (Reprinted, with permission, from McGarvey.)

- Influent ion content
- Treated liquor specification
- Process economy based on size
- Waste restrictions
- Concentrations

When the total concentration of impurities exceeds 1000 ppm, other processes including electrodialysis, reverse osmosis, and evaporation are likely to be considered.

Since ion exchange is used to treat wastewater, the technology related to the process is based on the performance of ion exchange with the components in water. Table 8.2.2 shows the substances generally involved. These compounds usually occur in ppm (mg/l) concentration. The common compounds in water can also be a background in a metal waste application.

Understanding the technical units used in the United States is useful. For water treatment, the capacity of the resin is expressed as kilograins of $CaCO_3/ft^3$. A capacity of one chemical equivalent per liter is expressed as 21.8 kgr of $CaCO_3/ft^3$. Thus, this factor is used in the conversion from chemical units to American units.

The exchange capacity of individual ions in feed solutions is expressed as meq/l of resin or in the English system as parts per million of calcium carbonate. When the capacity of a resin is expressed as meq/unit volume, the required resin volume can be calculated from the daily equivalent of ions to be removed and the length of cycle per regeneration. The regenerant consumption is available from the resin manufacturer. The rinse water and spent regenerant volumes are also available from the manufacturer.

TABLE 8.2.2 COMMON SUBSTANCES IN WATER

Cation	*Exchange Valence*	*Anion*	*Exchange Valence*
Calcium	2	Bicarbonate	1
Magnesium	2	Chloride	1
Sodium	1	Sulfate	2
Potassium	1	Nitrate	1
		Silica	1

The selection of resins used in wastewater treatment depends on the ions in solution. A common arrangement is the two-bed system. In this system, strong acid is followed by a weak base or strong base as shown in Figure 8.2.9. Sodium leakage can be controlled by the regeneration level and countercurrent operation. Wastewater treatment facilities can handle an increased concentration of a metal by using a merry-go-round where three or more units operate in series as shown in Figure 8.2.10.

The water quality is also influenced by the selection of the anion-exchange resin. A weakly basic resin does not pick up weakly acidic substances such as carbon dioxide and silica, and the water is acidic (pH 5–6). On the other hand, a strong base removes weakly acidic materials and splits salts. This effect is shown in the effluent by a pH increase.

With a two-bed system, wastewater treatment facilities can use several combinations of beds to correct problems and make the process more economical including:

- WA-SA-WB-SB
- SA-DEG→SB
- SA-SB-MB
- SA-SB-WA

where:

WA = Weak-acid-cation exchanger

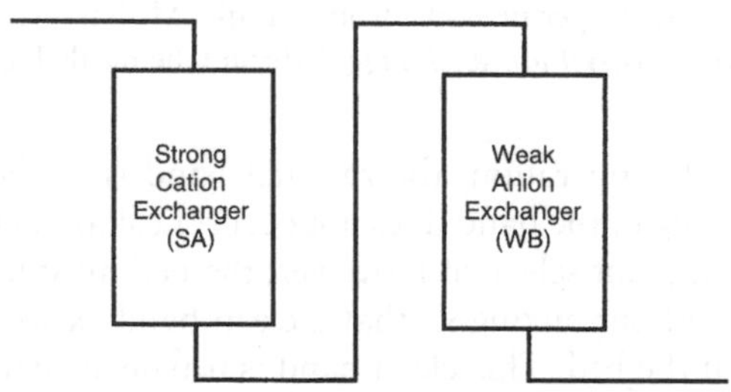

FIG. 8.2.9 Typical two-bed system. (Reprinted, with permission, from McGarvey.)

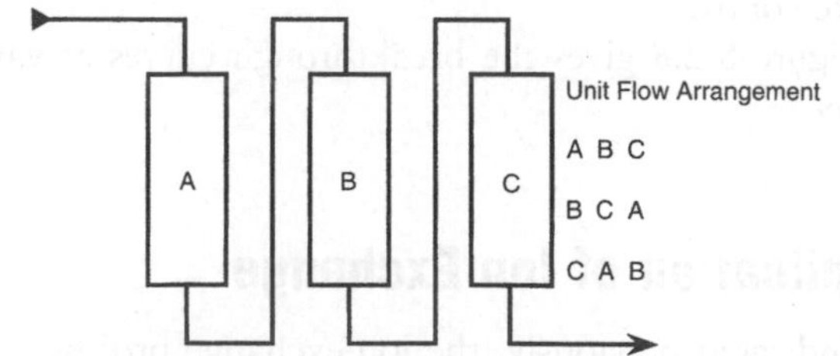

FIG. 8.2.10 Merry-go-round processes. (Reprinted, with permission, from McGarvey.)

TABLE 8.2.3 OPERATIONAL AND DESIGN PARAMETERS (VALUES RELATE TO DEIONIZATION AND SOFTENING DESIGN)

Parameter	*Value*
Hydraulic loading, gal/min/sq ft	6–15
Resin bed depth, ft	2–7
Expansion allowance, % of bed depth	75–100
Resin life expectancy, Cation exchanger	2–5 yr
Anion exchanger	1–3 yr
Softener	5–10 yr
Regeneration frequency, hr	8–12
Resin capacity, eq/l, Cation exchanger	0.70–1.2
Anion exchanger WB	0.8–1.2
Anion exchanger SB	0.5–1.0
Cation exchanger WA	1–2
Regeneration level,	
lb./ft^3 H_2SO_4, Cation exchanger SA	3.6–7.6
lb./ft^3 HCl, Cation exchanger SA	2–6
lb./ft^3 NaOH, Anion exchanger SB	3–10
lb./ft^3 NaOH, Anion exchanger WB	2.5–4
lb./ft^3 HCl, Cation exchanger WA	3.0–6.5
Wastewater (rinse and backwash), gal/ft^3	30–80
Chemical regenerant consumption, % theory	
Cation and anion SA/SB	200%
Cation and anion WA/WB	100%
Resin Cost, \$/ft^3, Strong-acid-cation exchanger	\$75–100
Strong-base-anion exchanger	\$150–200
Special resin	\$300–500

SA = Strong-acid-cation exchanger
WB = Weak-base-anion exchanger
SB = Strong-base-anion exchanger
MB = Mixture SA-SB
DEG = Degasification

Weakly acidic and weakly basic ion exchangers are used since they regenerate efficiently with an acid and base, respectively. Information on ion exchange resin capacity and other performance is available from ion-exchange resin manufacturers. Capacity and leakage values are based on water treatment components. These values are generally available on computer programs. Since waste control is important, most ratings give neutralization estimates so that wastewater treatment facilities can balance acid and alkaline usage to give neutral waste. This consideration is important in plant design.

The engineering parameters listed in Table 8.2.3 show hydraulic and rate values for a typical ion-exchange design.

Waste Metal Processing

The ion-exchange process using ion-exchange resins is well known, and complete process designs have been developed for many processes including softening brackish waters, condensate polishing, and processing ultrapure water used in the semiconductor industry. Processes for metal recovery have also been developed, but the information is still far from complete.

A complete analysis of a waste stream must be available before any process considerations can be made. An understanding of the solution chemistry for a metal is also required before a test program can be started. Copper is a common waste metal and is an example of a divalent metal. Copper is usually found as a divalent cation having good affinity for cation exchange resins. It also forms complexes with amines. It also occurs as an anion when complexed with EDTA or other chelating agents.

A waste liquor with copper, calcium, and sodium shows that the capacity is reduced by the presence of divalent cations other than copper. A selective resin based on EDTA removes copper from high concentrations of sodium chloride and to a lesser extent from calcium chloride. Cyanide complexes form stable complexes with strong base resins, and the metals can be recovered after the complex is destroyed with acid. In process engineering, environmental engineers should avoid such reactions unless they make provision for handling free hydrogen cyanide.

An experimental program to develop process information is assisted by reference-to-affinity orders that are available for most resins. The affinity series helps to establish the likely displacement of metals when they are present as the metals, but does not give the potential for complex formations particularly when chlorides are involved. For example, Fe^{3+} forms a complex with chlorides in high concentration so that iron can be removed from concentrated hydrochloric acid as an anion, $FeCl_4$. By dilution, the complex is destroyed, and the iron comes off the resin bed. Other cations including aluminum can also be removed from a concentrated solution. Amphoteric elements also become anions at elevated pH values.

Since there are several possibilities in a mixture of metals in a waste solution, environmental engineers can establish the presence of metals, cations, or anions by performing a single screening test. In this test, they prepare two small columns of anion- and cation-exchange resins. The anion-exchange resins should be in hydroxide form, and the cation exchange resins should be in acid form. The waste and the effluent sample for the metals should be fed into the columns at about 10 bed volumes per hr. Environmental engineers must be careful not to generate cyanide gas.

Once the valence of the metal is established, environmental engineers should start process estimates. Destroying metal complexes before the ion exchange resin is used may be useful. The amount of competing ions should be reduced if possible. If rinse water is deionized prior to the process, economic saving in the recycling of waste rinse waters and the recovery of valuable metals can be achieved.

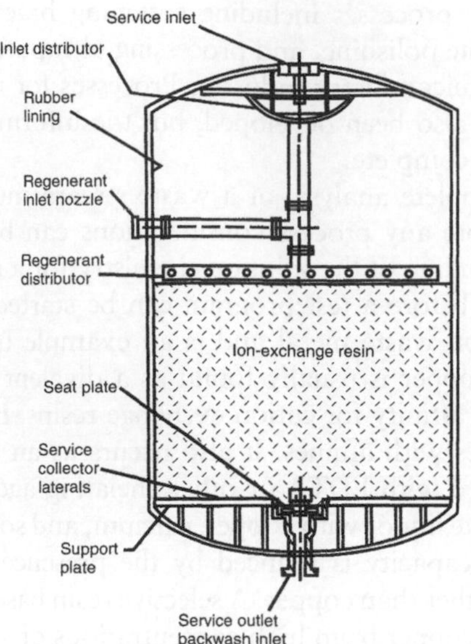

FIG. 8.2.11 Internal assembly of an ion-exchange column. (Reprinted, with permission, from Bolto and Pawkowski, 1982, *Waste water treatment by ion exchange.*)

When precipitation of toxic metals is the preferred process, environmental engineers usually use ion-exchange resins to polish the effluent to meet regulatory requirements. This process is a good place to use chelation resins that can pick up metals with a large salts background. However, precipitation can cause colloidal metal formation which is not easily removed by resins. Adjusting the pH and retention time can prevent this problem.

Metal recovery can be accomplished by electrolysis of the spent regenerant. Since ion exchange only concentrates trace metals, regeneration conditions are important for economical recovery. For example, using sulfuric acid strips the cation exchange of copper so that electrolysis can be achieved without generating chlorine if chlorides are present.

Ion-Exchange Equipment

Ion-exchange resins are installed in tanks that have support plates and distribution piping. This piping distributes fluids across a packed bed of resin that rests on a support plate. A collector is installed in the support plate. All components should be constructed of corrosion-resistant materials. Figure 8.2.11 shows a cross section of a typical unit.

Many detail differences exist among manufactured equipment but they all have common flow streams that require control. Figure 8.2.12 details the operation of an ion-exchange plant. The end point can be determined with resistivity measurement and chemical metals analysis such as calcium, magnesium, and copper. Plants sometimes operate with two beds in series with the second bed operating as a polisher.

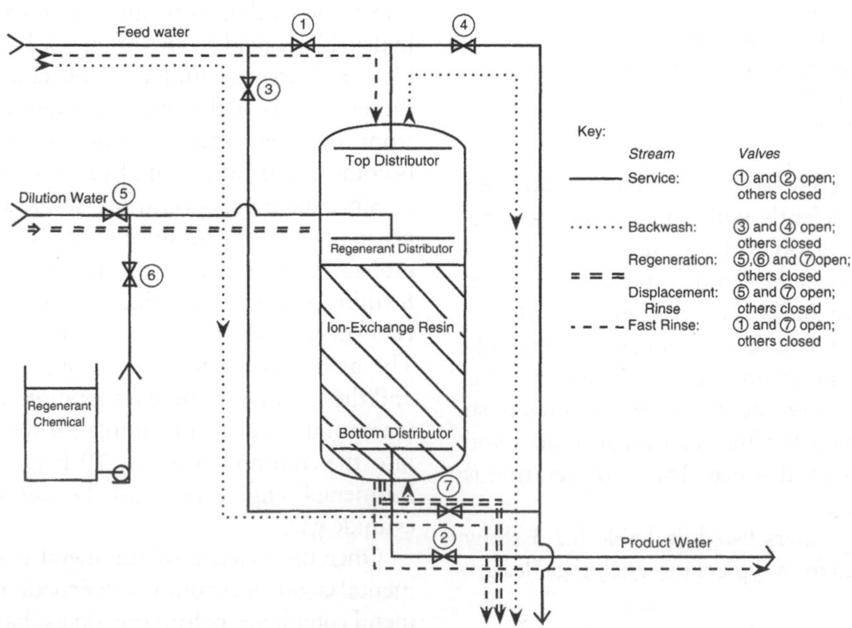

FIG. 8.2.12 Simplified ion-exchange operations cycle. The water used for backwash, dilution water, or displacement rinse can be feed water, softened water, decationized water, or DI water depending on the ion-exchange resin used and the quality of water produced in the service cycle. (Reprinted, with permission, from Dean Owens, 1985, *Principles of ion exchange and ion exchange water treatment,* 89.)

Process Conclusion

Once environment engineers have collected the information outlined previously, they can design a successful waste treatment plant. However, they should also consider the possible contamination of resin beds with oil, special organic chemicals, and SS, e.g., sludge. Generalizing from individual plants is difficult since many plating and anodizing systems operate differently.

Environmental engineers must recognize that ion exchange is a concentrating process. The elution and recovery of metals becomes part of other procedures such as electrodialysis, precipitation, and filtration. They must also consider varying local discharge regulations.

—*Francis X. McGarvey*

8.3 DEMINERALIZATION

Demineralization by Distillation

TYPES OF DISTILLATION SYSTEMS

a) Small units for high-purity water production
al) Single stills
a2) Multistage stills
b) Large desalination systems for potable water production
bl) Multistage flash
b2) Vertical tube evaporator
b3) Vapor compression

DEMINERALIZED WATER QUALITY (ppm TDS)

Below 1 (a2), down to 1 (al), and 500 (b)

APPLICATIONS

Small sizes (al), up to 1 mgd (b3), and over 1 mgd (b1,b2)

MATERIALS OF CONSTRUCTION

Tin (a) and cupronickel (b)

THEORETICAL MINIMUM ENERGY REQUIREMENT

3 kWh per 1000 gal with 3.5% NaCl feedwater

ACTUAL PROCESS OPERATION EFFICIENCY

Less than 10% of thermodynamic optimum

RATIO OF WATER PRODUCED PER POUND OF STEAM INVESTED

Up to 20 (b1)

The Office of Saline Water (OSW), U.S. Department of Interior, funded much development work on sea water desalination. This office is a clearinghouse for technical information on these efforts and maintains a list of companies that have engineered desalting facilities under its program.

Small stills used for producing high-purity water are usually under 100 gph in capacity. They are single-stage stills, usually made of tin. Such units are expensive and unlikely to have significant application in wastewater treatment. Distillation processes used for large-scale desalination offer savings in energy compared to a simple still.

Distillation is an expensive method of demineralization and is not recommended except when one of the following conditions exist: (1) Potable water is required, and the only source is sea water; (2) a high degree of treatment is required; (3) contaminants cannot be removed by any other method; or (4) inexpensive waste heat is available.

Large scale systems have been tested and are being commercially used in brackish and sea water desalination. They have not been applied to wastewater treatment, but demonstration plants have been evaluated in treating mine drainage waters.

DISTILLATION PROCESSES

Distillation operations are as varied as evaporator types and methods of using and transferring heat energy. The following types have been studied or used: (1) Boiling with submerged tube heating surface; (2) boiling with long-tube, vertical evaporator; (3) flash evaporation; (4) forced circulation with vapor compression; (5) solar evaporation; (6) rotating-surface evaporation; (7) wiped-surface evaporation; (8) vapor reheating process; (9) direct heat transfer using an immiscible liquid; and (10) condensing-vapor-heat transfer by vapor other than steam.

Of these types only (2), (3), and (4) are commercially important in desalination. The theoretical minimum energy required for a completely reversible process to obtain pure water from 3.50% NaCl salt water is about 3 kWh per 1000 gal at 25°C. Unfortunately, the thermodynamic minimum energy requirement has little practical relevance due to the many irreversibilities in an actual distillation process. These include pressure and force differences to overcome friction, temperature differences in heat exchangers and between system and the surroundings, and concentration differences for mass transfer. Actual processes operate at less than 10% of the optimum thermodynamic efficiency.

CORROSION AND SCALING

Due to temperature increases, inorganic salts come out of solution and precipitate on the inside walls of pipes and equipment. Scales due to calcium carbonate, calcium sulfate, and magnesium hydroxide are the most important in desalination processes. Controlling the pH minimizes carbonate and hydroxide scales. Most inorganic solutions are corrosive. Cupronickel alloys are most commonly used in sea water desalination. Other metals used are aluminum, titanium, and monel.

A significant development in distillation processes is the use of flash enhancers. These devices permit a closer approach to equilibrium flashing and can substantially improve the overall efficiency by increasing the liquid–vapor contact area through thermosiphon techniques.

All distillation processes reject part of the influent water as waste. Hence, all of these processes have concentrated waste disposal problems. The permissible maximum concentration in the waste depends on the solubility, corrosion, and vapor pressure characteristics of the wastewater. Therefore, the waste concentration is an important process optimization criterion.

FLASH DISTILLATION SYSTEMS

Multistage flash evaporation systems have been used commercially in desalination for many years. Conceptual designs for 1000-mgd plants are based on the flash principle. In the multistage flash process (see Figure 8.3.1), after the influent water has the SS removed and is deaerated, it is pumped through heat transfer units in several stages of the distillation system. Evaporating influent water condenses on the outside of the tubes. The concentrated waste water cascades from one stage to the next as a result of the pressure differential maintained. In each stage, the flashed water condenses on the tubes and is collected in trays (see Figure 8.3.1). When the concentrated wastewater reaches the lowest pressure stage, it is pumped out.

From a thermodynamic point of view, multistage flash is less efficient than ordinary evaporation. On the other hand, it has the advantage of many stages combined into a single unit, resulting in less expensive construction and the elimination of external piping. The largest flash units have a performance ratio (lb water produced per lb steam used) of 20.

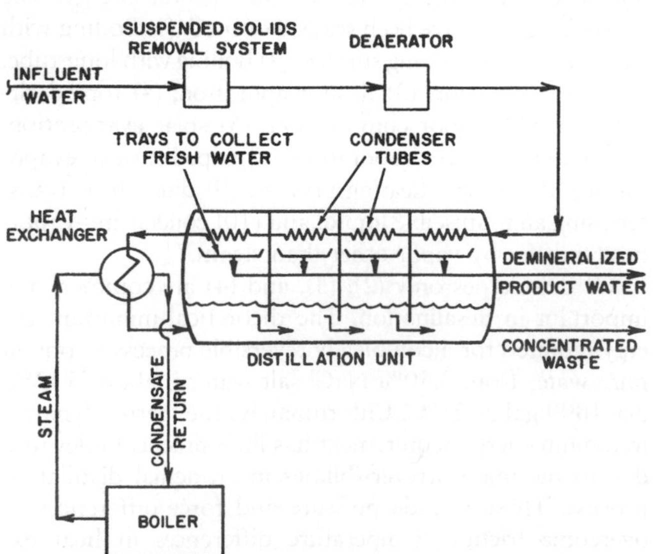

FIG. 8.3.1 Multistage flash process.

VERTICAL TUBE EVAPORATORS

Vertical-tube evaporators give 15 to 20% higher performance ratios and have fewer scale problems. They are used in large desalination plants.

In a three-stage, vertical-tube evaporator (see Figure 8.3.2), after the influent water is pretreated, it enters the heat exchanger in the last stage (No. 2) and progressively warms as it goes through the heat exchangers in the other effects (other stages). As the water moves through the heat exchangers, it condenses the water vapor emanating from the various effects. When the progressively warmed influent water reaches the first stage, it flows down the internal periphery of vertical tubes in a thin film, which is heated by steam. The wastewater feed to the second effect comes

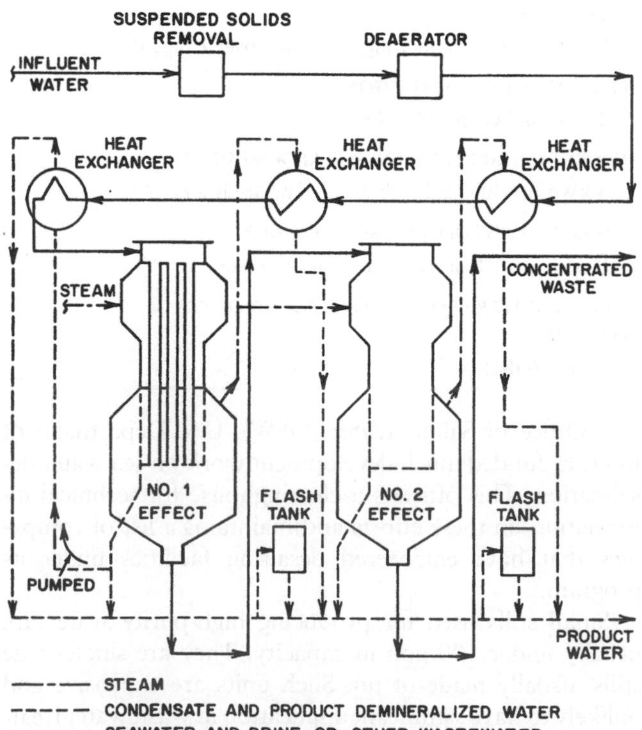

FIG. 8.3.2 Vertical-tube evaporator.

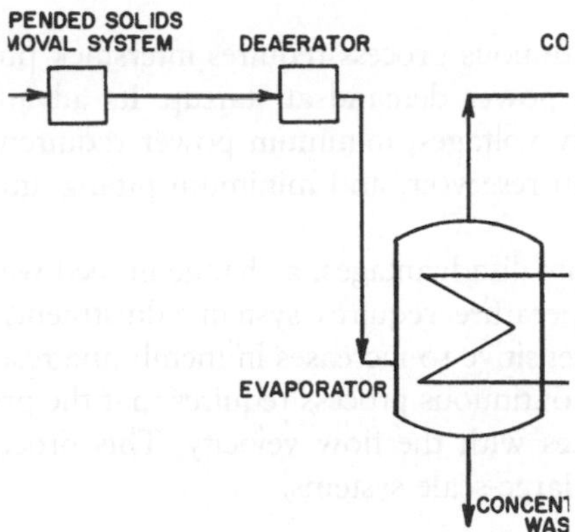

FIG. 8.3.3 Vapor-compression evaporator.

from the bottom of the first effect. Up to 15 effects are used in desalination plants.

VAPOR COMPRESSION EVAPORATORS

In these evaporators, the influent water is heated under nonboiling conditions, and the evaporated vapor is compressed and returned to serve as the heating medium (see Figure 8.3.3).

The vapor compression method has been tested and evaluated on brackish water, and the commercial applicability of such designs is limited by the capital costs of large vapor compressors. For the removal of inorganic salts with distillation, solar energy can also be used in regions where this source is continuously available.

Demineralization by Electrodialysis

TYPES OF SYSTEM DESIGNS

a) Continuous
b) Batch
c) Feed and bleed
d) Internally staged

MATERIALS OF CONSTRUCTION

Cationic membranes made of sulfonic acid derivatives and anionic membranes made of quaternary amine compounds

APPLICATIONS

Demineralization of brackish waters with up to 10,000 ppm salt concentration

This technique separates only ionized materials from water. Electrodialysis involves the use of electromotive forces to transport ionized material through semipermeable membranes that separate two or more solutions. The development of membranes made of ion exchange materials has led to the commercialization of electrodialysis systems, particularly for applications in the desalination of brackish water. Ion-exchange membranes transport ions of one charge only and have low electrical resistances.

PRINCIPLES OF ELECTRODIALYSIS

Electrodialysis units consist of several chambers made up of alternating anionic and cationic membranes arranged between two electrodes (see Figure 8.3.4). The solution containing cations and anions is fed through the chambers, and electromotive forces move the cations toward the cathode and the anions toward the anode.

Alternating anode and cathode membranes permit the passage of only one type of ion. Hence, after passing from one feed chamber to the next, the ions are blocked by an impermeable membrane. With this process, concentrated waste accumulates in every second chamber, and feed streams are purified in the others.

Electrodialysis requires that the membranes have sufficient ion-exchange capacity in addition to small-size (30Å) pores so that they repel electrostatically oppositely charged ions. The system controls the rate of water and current flow for optimum salt removal. Excessive current density results in acidic solutions being collected on the cathode side and basic solutions on the anode side of the membranes.

Properties of Membranes

The membrane properties include electrical conductivity and selectivity for ion transport. Ion-exchange membranes are usually prepared from strong, hydrated electrolytes. Some common resins are polystyrene cross-linked with divinyl benzene and polyethylene- or fluorocarbon-base resins.

The electrical area resistance of membranes is up to 40 ohm-cm^2. Selectivity is characterized by permselectivity, which is defined with reference to transport numbers of ions as follows:

$$Sp = \frac{tm - ts}{1 - ts} \quad 8.3(1)$$

where:

tm = Transport number of ions within the membrane

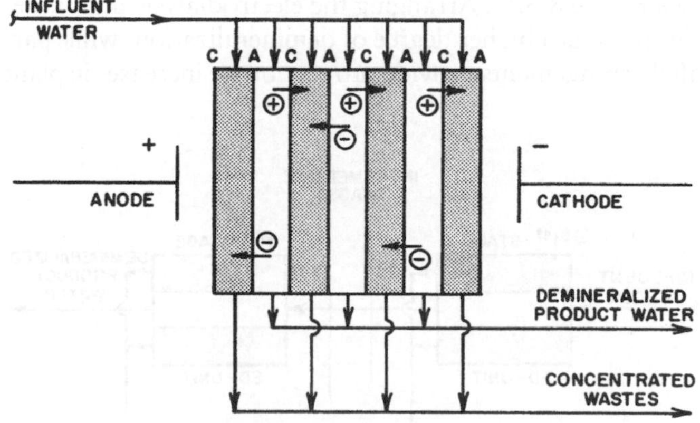

FIG. 8.3.4 Electrodialysis system. A = anion-permeable membrane; C = cation-permeable membrane.

ts = Transport number of ions in free solution
Sp = Permselectivity

Thus, the permselectivity indicates the permeability of the ion to which the membrane is impervious. This property of selectivity is caused by equilibrium between fixed ionic groups in the membranes and the solution wetting the membranes.

System Performance

The electrodialysis system in Figure 8.3.4 is inefficient for several reasons. As the ions migrate outward from the product chamber, the chamber progressively becomes depleted of ions, and its electrical resistance rises rapidly. This resistance determines the lower salinity limit in the product.

Increasing the operating temperature decreases resistance and thus improves the system performance. The build-up of solutes at the membrane–solid interface (concentration polarization) also increases electrical resistance. Ions in the feed water that precipitate due to a change in pH or are irreversibly adsorbed by the membranes can cause fouling. The back diffusion of salts due to an increase in salt concentration in one chamber can also reduce the separation efficiency. In addition, in all electrodialysis systems, disposing the concentrated waste generated must be considered.

ELECTRODIALYSIS PROCESSES

Electrodialysis system performance is influenced by the relationship between the solution concentration and membrane stack resistance and between the current density and solution velocity. Pretreatment including deaeration, filtration, and other operations is often required, depending on the feed characteristics.

Continuous Process

Fully continuous systems give demineralization performances and capacities beyond the range of modular units (see Figure 8.3.5). Arranging the electrodialysis units in series permits a higher degree of demineralization, while parallel arrangements provide proportionate increases in plant capacity.

The continuous process requires interstack pumps and has a peak power demand at startup. Its advantages include steady voltages, minimum power requirements, no recirculation reservoir, and minimum piping and process control.

Among its disadvantages, a change in feed water salinity or temperature requires system adjustment, and the process is sensitive to increases in membrane resistance to flow. The continuous process requires that the production rate balances with the flow velocity. This process is applicable to large-scale systems.

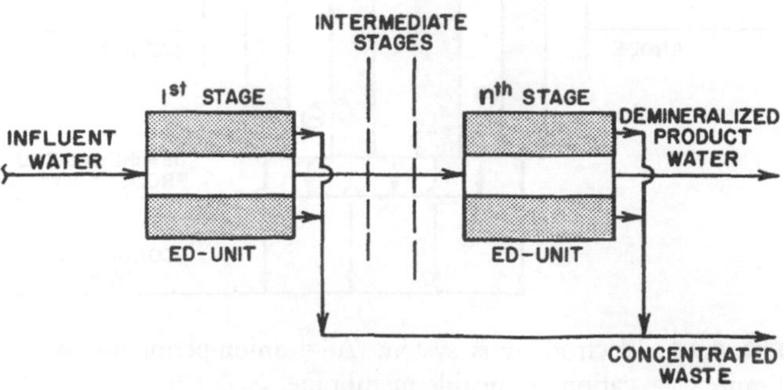

FIG. 8.3.5 Continuous electrodialysis.

Batch-Recirculation Process

This process pumps a fixed volume of feed solution from a reservoir through a membrane stack and back to the reservoir until the required salt removal is achieved (see Figure 8.3.6). The power requirement for such a batch process depends on the degree of recirculation and mixing.

The advantages of this process include the following: (1) changes in feed-water salinity or temperature and changes in membrane properties only modify the production rate but not the effluent quality; and (2) optimum velocity is independent of the production rate.

This process has the following disadvantages: (1)

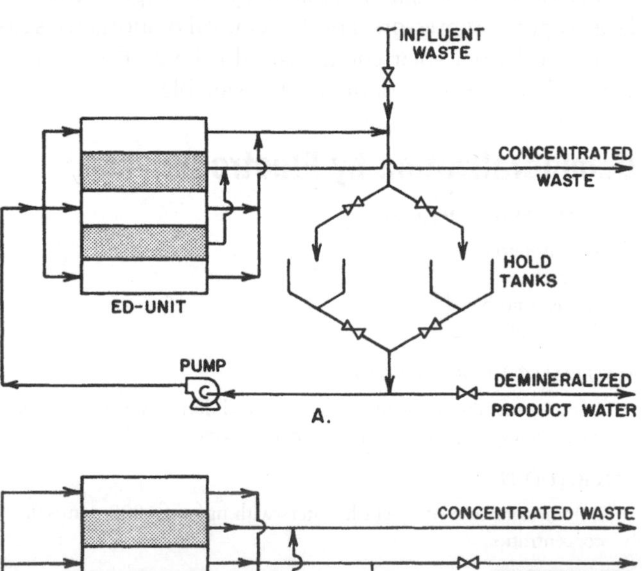

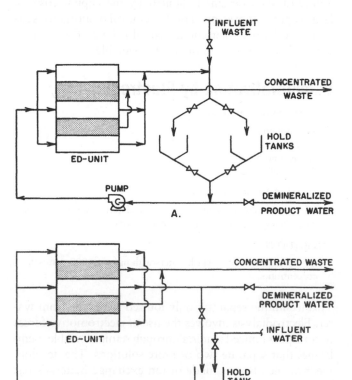

FIG. 8.3.6 Batch-recirculation electrodialysis. A, No mixing; B, Complete mixing.

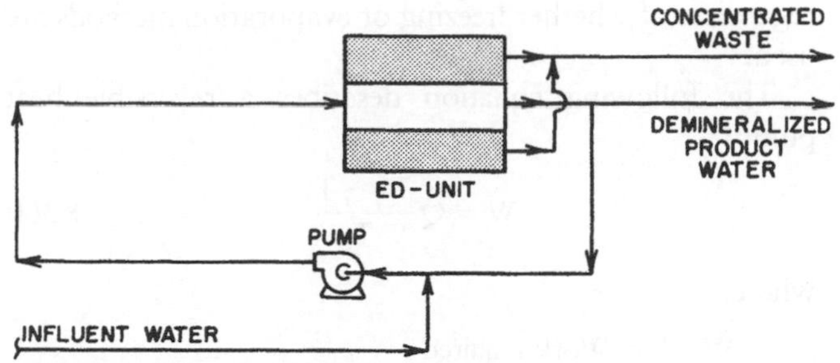

FIG. 8.3.7 Feed-and-bleed electrodialysis.

Higher power is needed; (2) recirculation reservoirs are required; (3) membranes do not operate at one equilibrium point; (4) current density through membrane also varies; and (5) more piping and control hardware are needed.

Feed-and-Bleed Process

The feed-and-bleed process (see Figure 8.3.7) can be used when continuous product flow is required, the degree of demineralization is small, and changes in the feed concentration are substantial.

This process continuously blends and recycles a portion of the product stream with the fresh feed. The power demand of this electrodialysis system is constant under equilibrium conditions. The advantages of the process include the following: (1) it provides a continuous product and can accept feed water with any TDS concentration; (2) the membranes are at equilibrium condition; and (3) a minimum current density is required.

The disadvantages of this system include high power consumption; and the requirement for sophisticated process control. The recirculation rate can be high with this process, and the process can have several feed-and-bleed loops arranged in series to give the effect of multiple-staging with lower recycling rates.

Internally Staged Process

For small units, internal staging within a stack provides all the advantages of a multistage series process without requiring recirculating pumps or interconnecting piping (see Figure 8.3.8). In this design, the product stream makes several passes in series between a single set of electrodes, and therefore the degree of demineralization is higher than with a single pass. The process is continuous with a constant power demand.

Its advantages include a high degree of demineralization, no need for repressurizing pumps, and the need for only a single set of electrodes. This design has the following disadvantages:

A large membrane area is required per unit of product capacity.
The performance is sensitive to changes in flow rate.
The pump must overcome a high pressure drop.

SIZING OF ELECTRODIALYSIS UNITS

The following equations approximate the inorganic salt removal capacity and the power consumption of electrodialysis units:

$$T_\rho = 8.22 \times 10^{-5} \times \left(\frac{I}{A}\right) nA \qquad 8.3(2)$$

$$E = \frac{12.2I(R_p + R_s)}{A} \qquad 8.3(3)$$

where:

E = Energy consumption, kWh lb-equivalent weight
A = Membrane area, ft^2
I = Current density, amps per ft^2
n = Total number of cell pairs
R_p = Membrane resistance of one cell pair, ohm-ft^2
R_s = Solution resistance, ohm-ft^2
= Current efficiency, a fraction
T_ρ = Plant capacity, lb equivalent per hr

MEMBRANES

Ion-exchange resins comprise 60 to 70% or more of the membranes. These membranes are solid, hydrated, strong electrolytes. Many membranes are produced by graft polymerization of ionic monomers on a film base. Homogeneous membranes do not require a separate support and can be made of polyethylenes, aminated copolymers, or modified styrene-divinyl benzene compounds.

Heterogeneous membranes require a separate support. Dynel is the most common support material, while glass and other materials have also been used. These membranes are either fluorinated polymers or vinyl-divinyl benzene copolymers.

Extensive bibliographies and discussions on the methods of preparing membranes are available. Inorganic ion-exchange membranes are also being developed.

APPLICATIONS

The desalination of brackish waters is the main commercial use of electrodialysis systems. Units serving a range of users from individual homes to whole communities have been commercialized. The OSW has conducted extensive

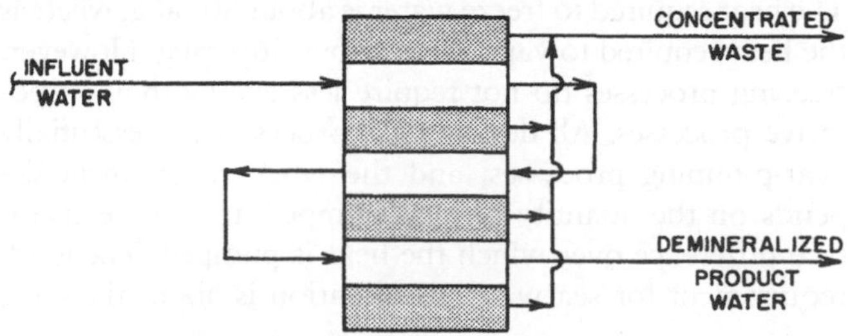

FIG. 8.3.8 Internally staged electrodialysis.

tests on the electrodialysis of brackish waters at salt concentrations of 10,000 ppm or below.

The EPA evaluated the use of electrodialysis in removing inorganic compounds from sewage effluents and found it potentially attractive. The treatment involves the use of diatomaceous earth filtration, granular carbon adsorption, and electrodialysis.

Other areas in which electrodialysis is used or appears to be promising include the following:

- Demineralizing whey by electrodialysis to produce the desalted whey used in infant formulas
- Desalting wastewater from fishmeal plants to recover protein
- Treating sulfonic acid pickle liquors to permit water recycling
- Recovering pulping chemicals and lignin products in the pulp and paper industry

Electrodialysis systems do not become much less expensive as the size becomes larger. Therefore, they are more likely to be used in small- and moderate-size applications. Advances in polymer science are likely to improve membrane selectivity and performance.

Demineralization by Freezing

TYPES OF FREEZING PROCESSES

a) Indirect contact
b) Direct contact
c) Hydrate process

APPLICATIONS

Food concentration, oil dewaxing; salination; and wastewater treatment applications are not yet economically feasible. Freezing by bubbling butane through sludge improves its dewatering characteristics. Natural freezing can be an inexpensive method of alum sludge dewatering in colder climates.

MINIMUM THEORETICAL POWER REQUIREMENT

6.3 kWh to purify 1000 gal

Inorganic salt removal by freezing involves the production of ice crystals by partially freezing saline water. The ice crystals are pure water and can be separated in solid–liquid separation equipment. Freezing processes have been investigated for the desalination of sea water by the OSW U.S. Department of Interior.

ENERGY REQUIREMENTS

The heat required to freeze water is about 80 cal/g, whereas the heat required to vaporize is above 560 cal/g. However, freezing processes do not require less energy than evaporative processes. All desalination processes are essentially heat-pumping processes, and the work requirement depends on the quantity of heat pumped and the temperature difference over which the heat is pumped. The work requirement for sea water desalination is about the same regardless of whether freezing or evaporation methods are used.

The following equation describes a reversible heat pump:

$$W = Q\frac{T_1 - T_2}{T_2} \qquad 8.3(4)$$

where:

W = Work required
Q = Heat to be pumped taken at T_2
T_1, T_2 = Absolute temperature levels

If T_0 is the ambient temperature, the following equation describes the ideal freezing process:

$$W = Q\frac{(T_1 - T_2)}{T_2}\frac{(T_0)}{T_1} \qquad 8.3(5)$$

If 50% of the water from a 3.5% NaCl solution is removed as pure water, the concentrated waste contains 7% NaCl. The minimum (theoretical) energy requirement for such a process is about 6.3 kWh per 1000 gal. Actual values are far higher because of process inefficiencies.

INDIRECT-CONTACT FREEZING PROCESS

In this process, heat transfer takes place through a metal wall. The process uses scraped-surface heat exchangers for both heat transfer and crystallization (see Figure 8.3.9). The feed solution is cooled in the surface crystallizers to below 32°F, and ice crystals are formed. Refrigeration is supplied by either ethylene glycol or an evaporative refrigerant, such as butane, on the shell side of the crystallizer.

The slurry containing ice crystals is sent to a continuous centrifuge where the ice crystals are separated from the mother liquor and sent to the melter tank after being washed. This process can obtain the heat for melting by precooling the incoming feed stream, thereby reducing the load on the refrigeration unit.

This process is used to concentrate coffee and fruit juices, but because the scraped-surface exchangers and other components are expensive, it is not used for wastewater treatment.

DIRECT-CONTACT FREEZING PROCESSES

Direct-contact processes have been investigated for large-scale seawater desalination applications (see Figure 8.3.10). Direct-contact processes use either a wash column or a vacuum-freezing, vapor-compression process.

The direct-contact freeze process uses direct contact with a refrigerant, such as butane, to effect crystallization. Therefore, the expensive scraped-surface units are unnecessary, and the process can produce larger crystals by using lower ΔT for crystallization.

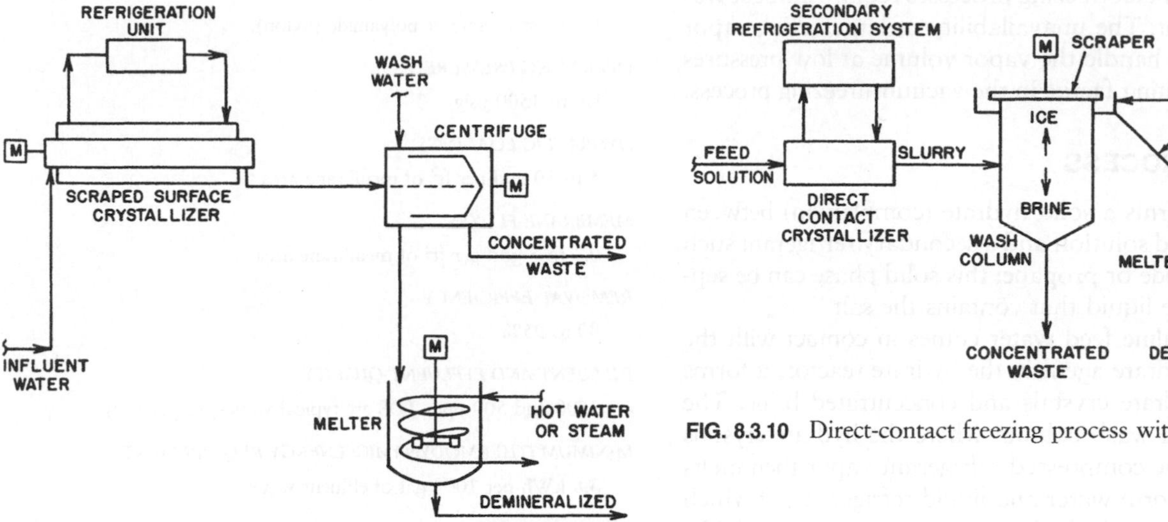

FIG. 8.3.9 Indirect-contact freezing process.

FIG. 8.3.10 Direct-contact freezing process with wash column.

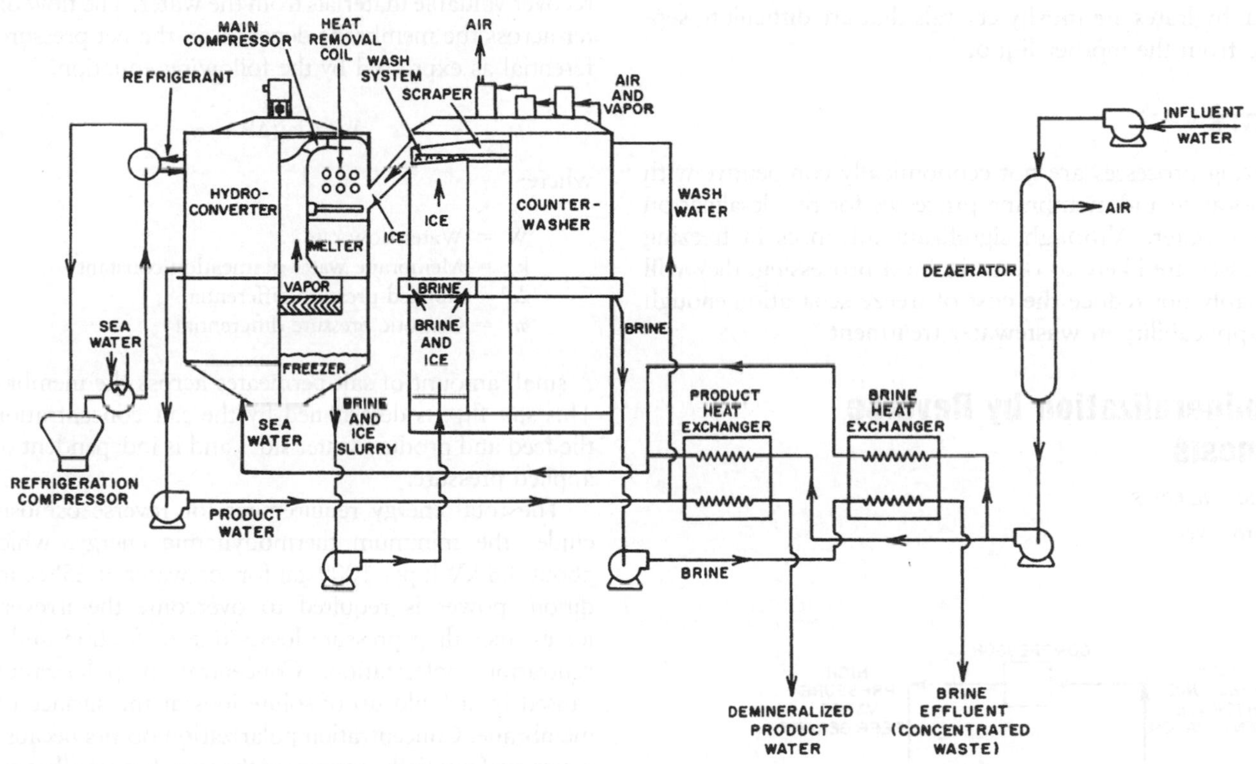

FIG. 8.3.11 Desalination process using vacuum freezing and vapor compression.

The slurry containing the crystals moves to the wash column where the ice crystals rise due to their buoyancy, and countercurrent washing of the ice bed by fresh water takes place. The pure ice crystals are harvested off the top of the column and melted to produce demineralized water.

The vacuum-freezing, vapor-recompression process uses a hydroconverter and a washer (see Figure 8.3.11), and the feed solution enters the hydroconverter freezing chamber at the bottom. The refrigerant compressor causes vaporization in the vacuum chamber (3 to 5 mm Hg absolute), and the removal of the vaporization heat causes the formation of ice crystals in the water.

The slurry from the hydroconverter goes to the washer where the brine drains through the bottom while the crystals are returned to the hydroconverter at the top. They are melted by compressed vapors into pure demineralized water. The compressed vapors condense during this step and are drained with the melted ice as product effluent.

All direct-contact freezing processes require influent water pretreatment. The unavailability of large, axial, vapor compressors to handle the vapor volume at low pressures is the main limiting factor in the vacuum-freezing process.

HYDRATE PROCESS

This process forms a solid hydrate (complexion) between water in the feed solution and a secondary refrigerant such as carbon dioxide or propane; this solid phase can be separated from the liquid that contains the salt.

When the saline feed water comes in contact with the evaporating hydrate agent in the hydrate reactor, it forms a slurry of hydrate crystals and concentrated brine. The slurry goes to a wash column where the hydrate crystals are washed. The compressed refrigerant vapor then melts the crystals to form water and liquid refrigerant, of which the latter returns to the hydrate reactor (see Figure 8.3.12).

Hydrates are formed at temperatures above 32°F and require less energy than freezing processes. On the other hand, hydrates are mushy crystals that are difficult to separate from the mother liquor.

COSTS

Freezing processes are not economically competitive with evaporative and membrane processes for the desalination of sea water. Although significant advances in freezing processes are likely to occur in food processing, they will probably not reduce the cost of freeze separation enough for applicability in wastewater treatment.

Demineralization by Reverse Osmosis

MEMBRANE LIVES

1 to 2 yr

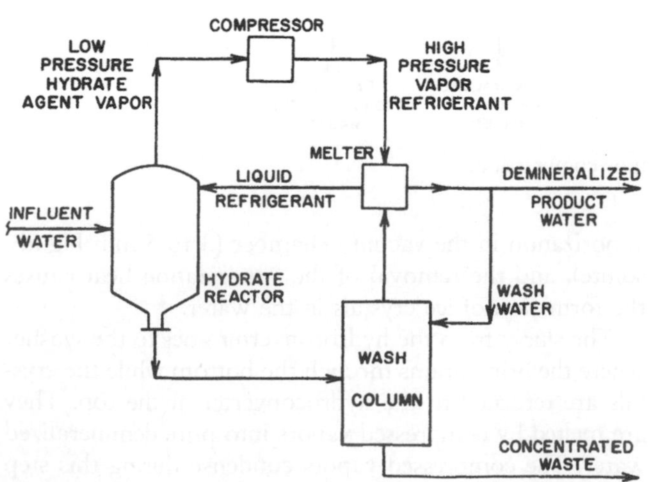

FIG. 8.3.12 Hydrate process.

MEMBRANE MATERIALS

Cellulose acetate or polyamide (nylon)

OPERATING PRESSURE

Up to 1500 psig

HYDRAULIC LOADINGS

5 to 50 gpd per ft^2 of membrane area

MEMBRANE FLUXES

0.3 to 5 gpd per ft^2 of membrane area

REMOVAL EFFICIENCY

80 to 95%

INFLUENT AND EFFLUENT QUALITY

3500 and 500 ppm TDS are typical values, respectively.

MINIMUM (THERMODYNAMIC) ENERGY REQUIREMENT

3.8 kWh per 1000 gal of effluent water

While seldom practiced, wastewater treatment facilities can use reverse osmosis for wastewater dewatering and to recover valuable materials from the water. The flow of water across the membrane depends on the net pressure differential as expressed by the following equation:

$$W = k_w(\Delta P - \pi) \quad 8.3(6)$$

where:

W = Water flow rate
k_w = Membrane water permeation constant
ΔP = Applied pressure differential
π = Osmotic pressure differential

A small amount of salt permeates across the membranes. This salt flux is determined by the salt concentration on the feed and product water sides and is independent of the applied pressure.

The total energy requirement for reverse osmosis includes the minimum thermodynamic energy, which is about 3.8 kWh per 1000 gal for sea water at 25°C. In addition, power is required to overcome the irreversible losses, including pressure losses due to friction and concentration polarization. Concentration polarization is caused by a build up of solute ions at the surface of the membrane. Concentration polarization occurs because the water preferentially permeates the membrane, whereas the solute can only move from the membrane surface by back diffusion. This process depletes some pressure energy and is a major cause of inefficiency.

The following equation gives the salt permeation rate:

$$S = K_s(C_f - C_p) \quad 8.3(7)$$

where:

S = Salt flow rate
K_s = Salt permeation constant
C_f = Concentration of salt in the influent solution
C_p = Concentration of salt in the effluent solution

Equations 8.3(6) and 8.3(7) show that the water effluent produced per unit of membrane surface increases with the pressure differential across the membrane, whereas the salt flux remains constant. Thus, an increase in operating pressure increases the ratio of water to salt and improves the efficiency of demineralization.

PRETREATMENT

Reverse osmosis usually requires pretreatment to remove contaminants including SS in sizes above 3 to 5μ; removing these contaminants prevents fouling the membrane surfaces. Because many saline waters contain salts such as calcium carbonate in concentrations near saturation, lowering the pH of the water is necessary to prevent scaling. Wastewater treatment facilities usually adjust the pH by adding a mineral acid in the range of 5 to 6.

Wastewater treatment facilities often add biocides, such as chlorine, to prevent microbial growth that could foul membrane surfaces. They also add threshold inhibitors to prevent salt precipitation from fouling membrane surfaces.

After pretreatment, the feed solution is pumped by a high-pressure pump to the reverse osmosis unit at about 500 to 800 psig (see Figure 8.3.13). Wastewater treatment facilities can use a number of individual reverse osmosis modules in series and parallel-flow configurations. The back-pressure valves on each assembly control the pressure within the membrane chamber. The pressure energy from the pressurized reject brine can be recovered in a turbine generator.

Wastewater treatment facilities can recover 60 to 75% of the water as pure effluent from the dilute wastes, leading to a three- to four-fold increase in the solute concentration in waste streams.

DESALINATION EXPERIENCE

Most membranes are made of cellulose acetate or polyamides. Cellulose acetate is the most common material because of its high salt rejection rate and high water flux. Except for hollow-fiber types, all design configurations require separate support for membranes. Hollow-fiber-polyamide membranes are compact but require factory fabrication of the modules and a higher level of pretreatment to remove SS.

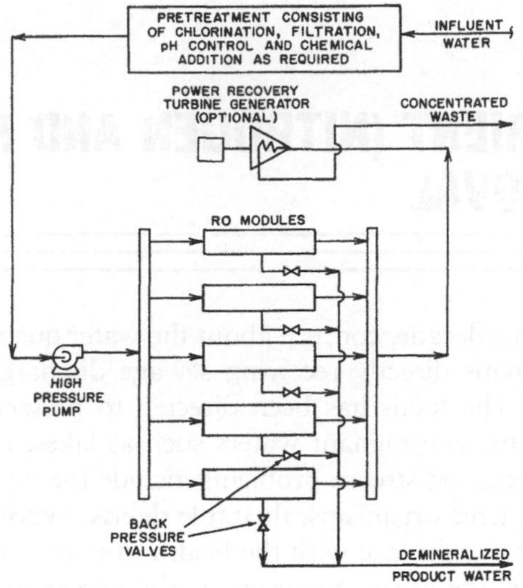

FIG. 8.3.13 Reverse osmosis.

WASTEWATER APPLICATIONS

Reverse osmosis has been considered for the production of fresh water from acid mine wastes containing 2000 to 5000 ppm dissolved acids and salts. In addition to recovering pure water, the reverse osmosis unit produces a concentrated waste that is easier to neutralize. Other applications include the concentration of fruit juices and sulfite pulp liquor. The use of reverse osmosis to remove organic matter from sewage effluents is being tested. Problems yet to be resolved include surface fouling and the internal breakdown of the membranes.

—*C.J. Santhanam*

8.4 NUTRIENT (NITROGEN AND PHOSPHORUS) REMOVAL

In the past decade, concern about the water quality of natural systems directly receiving sewage discharge has increased. The focus has been directed to preserving nonflowing or semistagnant waters such as lakes, inlets, and bays. Receiving stream problems include the introduction of pathogenic organisms, floatable debris, hypoxic conditions, or interference with the health of marine resources. However, the greatest concern is the acceleration of eutrophic conditions in these surface waters.

Eutrophication is the natural aging process of a body of water as biological activity increases (Water Pollution Control Federation 1983). Eutrophic waters are characterized by high concentrations of aquatic weeds and algae. These organisms eventually die, sink to the bottom, and decay. Consequently, this cycle increases the sediment oxygen demand which decreases the DO in the lower water levels. Additionally, eutrophication is enhanced by the large day–night cycling of DO that accompanies increased photosynthesis and respiration (Metcalf and Eddy, Inc. 1991). The acceleration of eutrophication is directly linked to increased nutrient loadings from sewage treatment plant discharges.

Phosphorus and nitrogen are the two major nutrients contributing to eutrophication. In most cases, these nutrients are growth-limiting; algae can no longer grow if these nutrient pools are depleted. Therefore, environmental engineers consider the removal of phosphorus and nitrogen from point sources, such as sewage treatment plants, a cost-effective and appropriate method for controlling the level and extent to which eutrophication occurs.

However, eutrophication is not the only problem caused by these nutrients. Ammonia is toxic in small concentrations to some aquatic life. The oxidation of ammonia to nitrite/nitrate can severely deplete the DO concentration in a body of water. Nitrite, which has a greater affinity for hemoglobin than oxygen and thus replaces it in the bloodstream, has been found to cause methemoglobinemia, or "blue baby" disease, in infants (U.S. EPA 1975; Peavy, Rowe, and Tchobanoglous 1985). Phosphates, in concentrations as low as 0.2 mg/l, interfere with the chemical removal of turbidity in drinking water (Walker 1978).

Due to increased nutrient loadings as well as elevated public awareness and consequent demand for protection of the world's water resources, the research and development of processes that remove phosphorus and nitrogen from wastewater have advanced considerably. Most of the interest has been in the manipulation of ambient conditions to enhance biological mechanisms responsible for nutrient removal. Consequently, both municipal and industrial facilities use many wastewater treatment processes to comply with federal and state regulations.

Nutrient removal processes can be grouped into two main categories: biological and physiochemical systems. Biological processes can be further divided into fixed-film and suspended-growth systems. Recently, many treatment facilities have been required to incorporate some degree of nutrient removal since the majority of plants built in the United States during the 1970s were for organic and SS (BOD and TSS, respectively) removal only. Although applications exist for fixed-film and physiochemical nutrient removal processes, suspended-growth systems have received the most attention. Therefore, this section focuses on suspended-growth activated-sludge processes. It reviews the biological mechanisms responsible for nutrient removal as well as different treatment processes.

Nutrient Removal Mechanisms

Biological nutrient removal in wastewater involves manipulation of the process environment. Bacteria responsible for waste treatment either proliferate or decay as ambient conditions change. The degree of treatment depends on the number of specific genera of bacteria. Therefore, environmental engineers must account for certain influencing parameters during process design and operation. This section reviews the theories of phosphorus, ammonia, and nitrogen removal. Design parameters are reviewed with specific attention to actual operating experience.

PHOSPHORUS REMOVAL

The phenomenon of enhanced biological phosphorus removal (EBPR) by activated sludge in excess of normal metabolic requirements was documented as early as the 1960s and was referred to as luxury uptake (Bargman et al. 1971; Levin and Shapiro 1965; Connell and Vacker 1967; Wells 1969; Borchardt and Azad 1968). Since the early 1970s, numerous studies have been conducted on various aspects of the mechanisms that control this biological process (McLaren et al. 1976; Hong et al. 1982; Spector 1977; Tracy and Flammino 1987; Fuchs and Chen 1975; Davelaar et al. 1978; Nicholls and Osborn 1979; Berber and Winter 1984; Aruw et al. 1988; Claete and

Steyn 1988). Even though environmental engineers have not obtained a confirmed, quantitative knowledge based on fundamental behavioral patterns and kinetics, they have designed activated-sludge systems to achieve the required degree of biological phosphorus removal (Hong 1982; Wentzel et al. 1985; 1988; Hong et al. 1989; Kang et al. 1985; DeFoe et al. 1993). Many full-scale plants, especially in the United States, have been installed in the past ten years.

To induce enhanced phosphorous uptake by activated sludge, the system must subject the process biology to a period of anaerobiosis prior to aeration; a prerequisite for EBPR is the proper selection of organisms. Environmental engineers can accomplish anaerobiosis by subjecting the sludge or mixed liquor to anaerobic/aerobic cycling (Davelaar et al. 1978; Nicholls and Osborn 1979; Berber and Winter 1984). Consequently, microorganisms capable of storing phosphorus as polyphosphate in their cell mass proliferate in the system.

Figure 8.4.1 shows typical concentration profiles for BOD and phosphorus. The release of orthophosphate mirrors the rapid uptake of organic substrate (BOD) in the anaerobic environment. In the aerobic zone, orthophosphate uptake to very low concentrations and continued removal of BOD occur.

Several explanations for these concentration profiles have been offered. Among them, the hypothesis offered by Hong et al. (1982) gives a plausible account of the biological mechanisms. The breakdown of polyphosphate in the anaerobic zone to generate energy for active transport of substrate into the cells of phosphate-accumulating organisms explains the phosphorus and BOD concentration changes in the anaerobic environment. Consequently, this environment serves as a selection zone; phosphate-accumulating bacteria, identified as the genus *Acinetobacter,* have a competitive advantage and proliferate in the system (Fuchs and Chen 1975; Claete and Steyn 1988).

As phosphate-accumulating and other organisms reach the aerobic environment, most readily available BOD is contained inside the high-phosphate bacteria. As the transported BOD is oxidized, the resulting energy resynthesizes polyphosphate and forms new cells. The new cells are responsible for net phosphorous removal in the system. The result of the alternating anaerobic/aerobic operation is wastewater with a low-effluent concentration of BOD and soluble phosphorus, and a waste sludge high in bound-polyphosphate content.

Tracy and Flammino (1987) postulated a more detailed biochemical pathway to further explain the apparent relationship between substrate metabolism and polyphosphate storage. Figure 8.4.2 shows a simplified version of this postulated pathway. The stored organics are reported to be polyhydroxybutyrate (PHB) (Bordacs and Chiesa 1988; Hong et al. 1983).

Environmental engineers have conducted microscopic examinations on mixed liquor samples from EBPR and conventional activated-sludge processes (Hong et al. 1982). They treated the samples with Neisser stain, which detects polyphosphate-containing volutin granules. Figure 8.4.3 shows photomicrographs of the stained samples (U.S. EPA 1987). With the Neisser treatment, darkly stained matter indicates the presence of polyphosphate. As the figure shows, the mixed liquor sample from the EBPR process contains a denser volume of stored polyphosphate.

The rate and extent of phosphate removal is related to the type and quantity of soluble substrate in the influent wastewater. Studies show that low-molecular-weight fatty acids (VFAs), such as acetate, are the preferred carbon source (Wentzel et al. 1985; 1988). Additionally, cationic species such as potassium and magnesium, are required for polyphosphate synthesis (Hong et al. 1983). Typical municipal wastewater contains sufficient soluble substrates and cations for biological phosphorus removal.

With given influent wastewater characteristics, an EBPR system designed for a higher F/M ratio generally achieves a higher degree of phosphorus removal. A system operating with a higher F/M ratio produces a larger amount of cell mass (sludge) to incorporate phosphorus. Since sludge production and wasting are responsible for net phosphorus removal, this parameter is key to successful biological phosphorus removal.

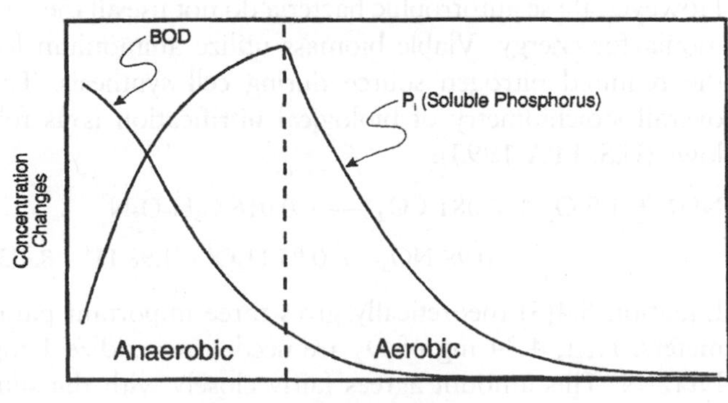

FIG. 8.4.1 Biological explanation for changes in BOD and phosphorus in an EBPR process.

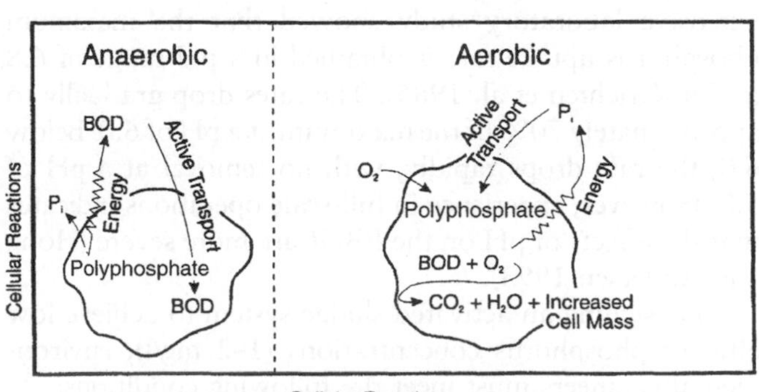

FIG. 8.4.2 A simplified biological pathway of a nutrient removal process.

FIG. 8.4.3 Photomicrographs of stained, mixed liquors. *top,* EBPR sludge showing high concentration of polyphosphate deposits (1000×); *below,* control sludge showing absence of polyphosphate deposits (1000×).

The temperature and pH effects on EBPR processes have seldom been studied. Sell et al. (1981) postulated that the organisms responsible for phosphorus removal are facultative psycrophiles. Laboratory results indicate that phosphorus removal is enhanced as the operating temperature decreases.

Like most biological reactions, the organisms in the EBPR process favor a near-neutral pH. The results of an extensive laboratory study showed that the maximum phosphorus uptake rate is obtained in a pH range of 6.8 to 7.4 (Krichten et al. 1985). The rates drop gradually to approximately 70% of the maximum at a pH of 6.0. Below 6.0, the rate drops rapidly, with no removal at a pH of 5.0. However, experience in full-scale operations indicates that the effects of pH on the EBPR are more severe (Hong and Andersen 1993).

In designing an activated-sludge system to achieve low effluent phosphorus concentrations (1–2 mg/l), environmental engineers must meet the following conditions:

1. An anaerobic zone with a detention time of 1 to 2 hr
2. An influent soluble BOD/soluble P ratio of 12 or greater
3. Operation at the highest F/M and shortest SRT that still permits nitrification, if nitrification is required

NITROGEN REMOVAL

The nitrogen in municipal wastewater is predominately in the organic and ammonium forms. Based on the amount of degradation prior to reaching the treatment plant, the respective fractions vary. Typically, municipal wastewater contains 60% ammonium nitrogen, 40% organic nitrogen, with negligible concentrations of nitrate or nitrite (Joint Task Force of the Water Environment Federation and ASCE 1991). Untreated domestic wastewater typically contains 20–50 mg/l of total nitrogen (Metcalf and Eddy, Inc. 1991).

Ultimately, all activated-sludge treatment processes realize some degree of net nitrogen removal since nitrogen is required for the synthesis of a new, viable cell mass. Both ammonium and nitrate can be used as the nitrogen source. The degree of net nitrogen removal due to cell growth is a function of the organic loading, sludge age (SRT), and endogenous respiration (Marais and Ekema 1976). Typically, assimilation removes 20–30% of the total influent nitrogen (Van Haandel, Ekama, and Marais 1981). However, all biological nitrogen removal systems use two processes to achieve the required effluent quality: nitrification and denitrification. Nitrogen can also be removed from wastewater by ion exchange.

Nitrification

Biological nitrification oxidizes nitrogen from ammonia (NH_4^+) to nitrate (NO_3^-). The overall reaction is a two-step process. In the first step, ammonia is oxidized to nitrite (NO_2^-) and mediated by the genus *Nitrosomonas*. The second step converts nitrite to nitrate, controlled by the genus *Nitrobacter*. The conversional process is as follows:

$$NH_4^+ + 1.5\ O_2 \xrightarrow{\textit{Nitrosomonas}} NO_2^- + 2H^+ + H_2O \qquad 8.4(1)$$

$$NO_2^- + 0.5\ O_2 \xrightarrow{\textit{Nitrobacter}} NO_3^- \qquad 8.4(2)$$

However, these autotrophic bacteria do not use all the ammonia for energy. Viable biomass utilize ammonium for the required nitrogen source during cell synthesis. The overall stoichiometry of biological nitrification is as follows (U.S. EPA 1993):

$$NH_4^+ + 1.9\ O_2 + 0.081\ CO_2 \longrightarrow 0.016\ C_5H_7O_2N + 0.98\ NO_3- + 0.95\ H_2O + 1.98\ H^+ \qquad 8.4(3)$$

Equation 8.4(3) theoretically gives three important parameters. First, 4.34 mg of O_2 are needed to oxidize 1 mg NH_4^+-N. This amount agrees fairly closely with the sum of the two preceding equations (4.57 mg O_2/mg NH_4^+-N). Secondly, 7.07 mg of alkalinity (as $CaCO_3$) are consumed per mg of NH_4^+-N nitrified. Finally, 0.13 mg of viable biomass are produced per mg NH_4^+-N converted. These parameters vary according to process operation and conditions, i.e., sludge age, organic loading, pH, and temperature.

Nitrification warrants treatment system adjustments and can occasionally be a difficult process to achieve reliably. The aerobic autotrophs responsible for nitrification are more sensitive to ambient conditions, toxics, and inhibitors than the competing carbonaceous heterotrophs (Sharma and Ahlert 1977; Sutton, Murphy, and Jank 1974). Additionally, some confusion exists over the biological kinetics during typical wastewater treatment conditions (Charley, Hooper, and McLee 1980).

As shown in the stoichiometry, nitrification requires more oxygen than just the amount required by carboneous heterotrophs for carbon oxidation (BOD removal). A longer sludge age must be maintained due to the slower growth rate of autotrophic nitrifiers. Also, if sufficient alkalinity is not present in wastewater, the pH can drop substantially. This low pH limits the extent of nitrification.

Table 8.4.1 shows typical process parameters for single-stage nitrifying systems.

Denitrification

Biological denitrification reduces nitrate (NO_3^-) to nitrogen gas (N_2), nitrous oxide (N_2O) or nitric oxide (NO). This nitrogen removal process is the one most widely used in municipal wastewater treatment (Water Pollution Control Federation 1983). Denitrifying organisms are primarily facultative aerobic heterotrophs that can use nitrate in the absence of DO. Many genera of bacteria are capable of denitrification: *Achromobacter, Baclus, Brevibacterium, Enterobacter, Micrococcus, Pseudomonas,* and *Spirillum* (Davies 1971; Prescott, Harley, and Klein 1990). Several conditions enhance the amount of biological denitrification: nitrate, a readily available carbon source, and a low DO concentration. Low DO is the most critical condition since denitrification is simply several modifications of the aerobic pathway used for BOD oxidation (U.S. EPA 1975).

The stoichiometric reaction describing this biological reaction depends on the carbon source involved as follows (Randall, Barnard, and Stensel 1992):

TABLE 8.4.1 TYPICAL PROCESS PARAMETERS FOR SINGLE-STAGE NITRIFYING SYSTEMS

Parameter	*Range*
MLSS, mg/l	2500–5000
Hydraulic retention time, hr	6–15
Sludge age, days	5–15
Residue DO, mg/l	1.0–2.0
pH	6.5–8.0
Temperature, C	10–30

Carbon Source	Theoretical Stoichiometry
methanol	$NO_3- + 0.83\ CH_3OH \longrightarrow 0.5\ N_2 + 0.83\ CO_2 + H_2O + OH^-$ 8.4(4)
acetic acid	$NO_3- + 0.63\ CH_3COOH \longrightarrow 0.5\ N_2 + 1.3\ CO_2 + 0.75\ H_2O + OH^-$ 8.4(5)
sewage	$NO_3- + 0.1\ C_{10}H_{19}O_3N \longrightarrow 0.5\ N_2 + CO_2 + 0.3\ H_2O + 0.1\ NH_3 + OH^-$ 8.4(6)
methane	$NO_3- + 0.63\ CH_4 \longrightarrow 0.5\ N_2 + 0.63\ CO_2 + 0.75\ H_2O + OH^-$ 8.4(7)

The denitrification rate is a function of, among others, the temperature of the wastewater and the type of carbon source used for electron transfer. However, determining which carbon substrate is most suitable for wastewater treatment has been difficult (Tam, Wong, and Leung, 1992).

The following equation describes overall synthesis reaction using methanol as the carbon source and nitrate as the nitrogen source (U.S. EPA 1975):

$$NO_3- + 1.08\ CH_3OH + 0.24\ H_2CO_3 \longrightarrow 0.056\ C_5H_7O_2N + 0.47\ N_2 + 1.68\ H_2O + HCO_3- \quad 8.4(8)$$

Theoretically, 2.47 mg of methanol are required to reduce 1 mg of nitrate. Also, the equation predicts that 3.57 mg of alkalinity (as $CaCO_3$)/mg nitrate-nitrogen are produced. However, studies show that this estimate may be aggressive and that 3.0 mg of alkalinity are more likely (U.S. EPA 1993; Wanielista and Eckenfelder 1978). The alkalinity production capability of denitrification enables a combined nitrification–denitrification system to maintain a more stable pH. Additionally, a system that uses influent BOD as the carbon source and has the anoxic zone located before the aerated portion of the process will require less energy for aeration. This energy savings is because a portion of the waste is consumed by denitrification (anoxic stabilization).

ION EXCHANGE

FERTILIZER PLANT WASTEWATER

500 mg/l nitrogen concentration, 50% in ammonium and 50% in nitrate form at about 1 mgd total wastewater flow rate

TREATED EFFLUENT QUALITY

2 to 3 ppm NH_3 and 7 to 11 ppm NO_3

CONCENTRATED BY-PRODUCT STREAM

20% solids

CHEMICAL REGENERANT CONSUMPTION (PERCENT OF STOICHIOMETRIC)

135 to 150% acid and 110% base

NITROGEN REMOVAL EFFICIENCY

99%

Industrial wastewater discharges, especially from the ammonium nitrate fertilizer industry, typically contain concentrations of 500 mg/l of nitrogen, equally divided between ammonium and nitrate forms. Environmental engineers use the ion-exchange process to successfully remove such high concentrations of nitrogenous compounds. Ion exchange can purify the wastewater to a quality that complies with zero-pollutant-discharge criteria or that permits complete recycling of the wastewater. The ion-exchange process can also completely recover plant products lost into the waste stream and can efficiently recycle the recovered products into the plant processes.

Ion-exchange has been used successfully to treat the wastewater from a large nitrogen fertilizer plant producing 140,000 tn of prilled ammonium nitrate and 190,000 tn of nitrogen solutions annually. A captive ammonia plant producing 175,000 tn annually and a captive nitric acid plant producing 195,000 tn annually provide the feed stock for primary products. The wastewater from this typical plant contained a concentration of about 500 mg/l nitrogen, equally divided between ammonium and nitrate forms, in about 900,000 gal of effluent per day.

The fertilizer plant chose ion exchange as the treatment process at the conclusion of a two-year study which showed that no other process had the capability of approaching the zero-pollutant-discharge goal or permitted total water reuse and recycling. As a treatment process, ion exchange can provide effluent water of adequate quality for reuse or discharge, provided that the collected contaminant ions can be properly disposed, preferably as a recovered product.

Because of the nature of the manufacturing processes, the principal pollutant in nitrogen fertilizer plant effluent is ammonium nitrate. This salt is satisfactorily recovered from solution by ion exchange. Also, by regenerating cationic resin with nitric acid, and anionic resin with aqua ammonia, the ion-exchange process can provide for complete product recovery since the excess regenerants combine to form more principal product. Because the regenerants are inplant products, they are obtained at minimum cost.

The other ions resulting from resin backwash are principally those from water hardness. The ion-exchange process readily recycles recovered principal product, ammonium nitrate, and other recovered water impurities into the nitrogen solution production of typical nitrogen fertilizer plant. The resulting deionized water can also be reused.

The fertilizer plant selected continuous, countercurrent, moving-bed-type of ion exchange for this application since the system provides the low leakage of exchanged ions and high concentration of regenerant backwash favoring product recovery.

Table 8.4.2 shows an analysis of a typical nitrogen fertilizer plant wastewater as influent to the ion exchange treatment system together with analysis of the deionized effluent stream. Table 8.4.3 shows an analysis of the contaminants recovered from this wastewater.

In addition to the principal material, ammonium nitrate, the recovered ions constitute micronutrients for plants. However, a high concentration of chlorides makes extensive evaporation of the solution hazardous because of the probability of explosion. Consequently, the most feasible disposal of the recovered material is recycling into the production of nitrogen solutions.

Table 8.4.4 shows the materials recovered daily from 900,000 gal of nitrogen fertilizer plant wastewater. Table

TABLE 8.4.2 NITROGEN FERTILIZER PLANT WASTE AND TREATED WATER ANALYSIS

	Influent	*Effluent*
Component	*ppm*	*ppm*
NH_3	340	2–3
Mg	5	—
Ca	60	—
Na	0	—
NO_3	1240	7–11
Cl	53	—
SO_4	72	—
OH	—	—
pH	5–9	5.9–6.4
SiO_2	15	15

Note: The ammonium nitrate removal is 99.4%.

TABLE 8.4.3 AMMONIUM NITRATE PRODUCT STREAM ANALYSIS

Component	%
NH_4NO_3	17.83
Ca	0.260
Mg	0.020
NH_3	4.070
NO_3	14.880
Cl	0.220
SO_4	0.3120
Total Solids	19.762
Water	80.238

TABLE 8.4.4 DAILY QUANTITY OF MATERIAL RECOVERED (FROM 900,000 GAL WASTEWATER)

	Cation lb/d	*Anion lb/d*	*Blended lb/d*
$Ca(NO_3)_2$	1848	—	1848
$Mg(NO_3)_2$	222	—	222
NH_4Cl	—	600	600
$(NH_4)_2SO_4$	—	699	699
NH_4NO_3	12,021	12,021	31,211
Water	82,255	56,780	139,035
HNO_3	5585	—	—
NH_3	1290[a]	294	—

Total Product Ammonium Nitrate Solution 173,615
Note: [a]Adding NH_3 neutralizes excess HNO_3.

8.4.2 shows that 15 ppm of silica, SiO_2, is in the wastewater. The ion-exchange treatment system does not remove silica since the anion resin used is a weak-base type. Consequently, the deionized water produced by the ion-exchanger is not entirely satisfactory for use as cooling-tower makeup water or boiler feed water because of the silica content.

The plant can add silica polisher (a standard fixed-bed exchanger using a strong-base anion resin) to the system to remove silica and produce deionized water satisfactory for all inplant use. This addition permits complete wastewater recycling and reuse in a closed-loop system.

Phosphorus Removal Processes

Phosphorus removal processes were used in full-scale applications in the 1970s. Since phosphorus is typically the limiting nutrient for algae growth, removal from point sources can potentially halt eutrophication. The phosphorus content of typical wastewater process bacteria ranges between 1.5–3% (on a dry-weight basis). However, the bacteria responsible for EBPR contain a larger amount (4–6% on a dry-weight basis). This section reviews the major biological phosphorus removal processes and discusses the advantages, disadvantages, and flow schematics.

ANAEROBIC/OXIC (A/O) PROCESS

Part a in Figure 8.4.4 is a schematic representation of the A/O process. The unique feature of this aerobic (oxic) activated-sludge process is an anaerobic (both oxygen- and nitrate deficient) zone at the influent end of this process. The anaerobic and aerobic zones are each divided into several equally sized compartments. Influent wastewater and return activated sludge (RAS) are fed to the first compartment of the anaerobic zone. Typically, either centrally mounted or submersible mixers provide gentle mixing in the anaerobic zone. Wastewater treatment facilities can use various aeration methods, such as fine-bubble diffusers, surface mechanical aerators, and oxygen aeration to meet the oxygen demands in the oxic zone.

The A/O process is a high-rate process characterized by low hydraulic detention times (2.5–3.5 hr) and high F/M ratios (0.5–0.9 1/day). Wastewater treatment facilities can adapt this process for simultaneous phosphorus removal and nitrification by simply adjusting for the aerobic sludge age. Depending on climatic conditions and influent characteristics, especially influent TSS, the hydraulic detention time in the oxic zone can range from 4.0 to 8.0 hr, while the anaerobic detention time remains approximately 1 hr. More than sixty plants in the United States have used this process configuration. Figure 8.4.5 shows the operating and performance data of the North Plant at Titusville, Florida (no nitrification required), while Figure 8.4.6 shows the operating data at a plant in Lancaster, Pennsylvania (nitrification using pure oxygen aeration).

Even though simultaneous phosphorus and nitrogen removal has been successful, excessive nitrate concentrations in the RAS stream are a concern in the anaerobic zone. Therefore, environmental engineers recommend a modified basin configuration (see Part b in Figure 8.4.4). This process removes the nitrate in the return sludge before it is mixed in the anaerobic zone.

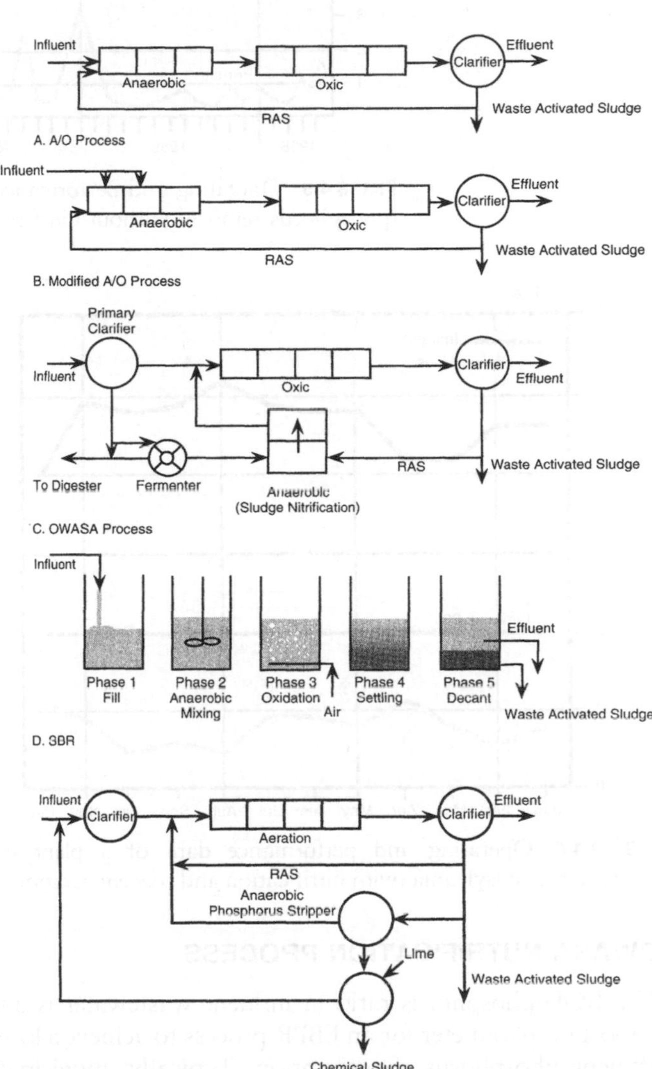

FIG. 8.4.4 Schematic diagrams of the phosphorous removal processes. (Reprinted, with permission, from S.N. Hong, 1982, A biological wastewater treatment system for nutrient removal, presented at the EPA Workshop on Biological Phosphorus Removal, Annapolis, Md., 1982); C.S. Block and S.N. Hong, 1984, *Treatment of wastewater containing phosphorus compounds,* U.S. patent 4,488,967; K. Kalb et al., 1990, Nutrified sludge—An innovative process for removing nutrients from wastewater, presented at the 63rd WPCF Conference, Washington, D.C. 1990; J.F. Manning and R.L. Irvine, 1985, The biological removal of phosphorus in sequencing batch reactors, *J. WPCF* 57:87; and R.T. Irvine et al., 1982, Summary report—workshop on Biological Phosphorus Removal in Municipal Wastewater Treatment, Annapolis, Md.)

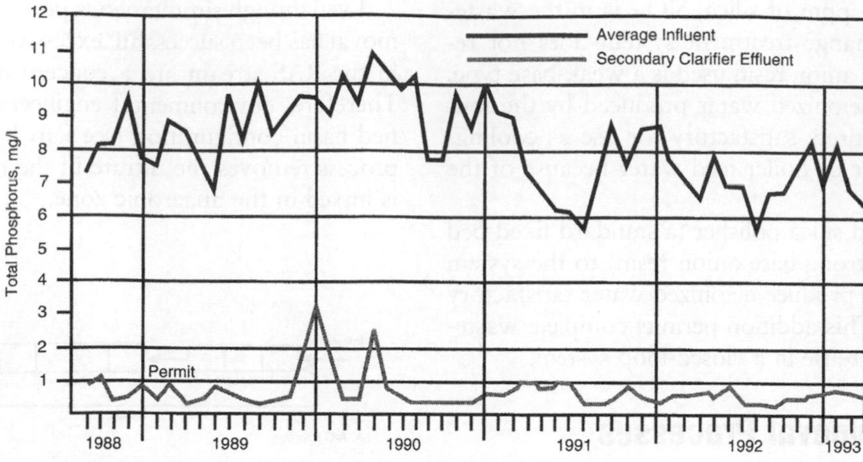

FIG. 8.4.5 Operating and performance data of the North Plant in Titusville, Florida (phosphorus removal without nitrification).

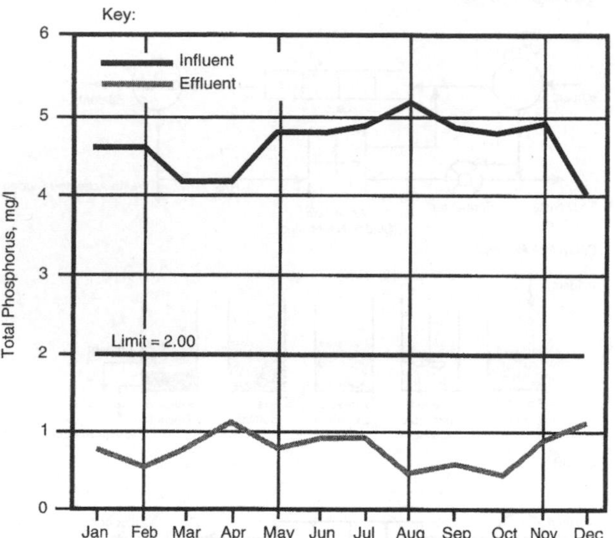

FIG. 8.4.6 Operating and performance data of a plant in Lancaster, Pennsylvania (with nitrification and oxygen aeration).

OWASA NUTRIFICATION PROCESS

The BOD/phosphorus ratio in influent wastewater is an important parameter for an EBPR process to achieve a low effluent phosphorus concentration. Typically, municipal wastewater contains a sufficient BOD/phosphorus ratio for biological phosphorus removal. However, in cases where excessive infiltration by rainwater into sewer lines or substrate consumption in pretreatment processes exist, the influent BOD concentration for biological phosphorus removal can be insufficient. Therefore, separate unit processes for side-stream production of BOD, mainly VFAs, supplement BOD requirements. Different methods for producing VFAs have been reported (Rabinowitz et al. 1987).

The OWASA process (see Part c in Figure 8.4.4) uses anaerobic fermentation of primary sludge to produce VFAs. This process mixes the VFA-enriched supernatant stream with the RAS in an anaerobic zone. The trickling-filter-treated influent wastewater flows directly into the oxic zone together with anaerobically conditioned RAS. Since the required BOD is supplemented by fermentation, a sufficient BOD/phosphorus ratio is maintained. Phosphorus removal performance is reliable and consistent with an effluent phosphorus concentration of 0.4 mg/l or less. Wastewater treatment facilities can also use this process for simultaneous biological phosphorus and nitrogen removal by increasing detention time in the oxic zone. Currently, one U.S. plant is in operation, located in Carrboro, North Carolina.

SEQUENCING BATCH REACTORS

All laboratory studies of the A/O process were performed with automatic-batch-fill-drawing (ABFD) apparatus with sequential operations of anaerobic mixing, aeration, and clarification in the same reactor. This concept has been used for full-scale operation and is an SBR. Part d in Figure 8.4.4 shows the sequence of operations in an SBR. An SBR can accomplish biological phosphorus removal alone or with nitrification (Manning and Irvine 1985). However, the application of SBRs is limited to low flow rates, typically 5 mgd or less.

PHOSTRIP PROCESS

Part e in Figure 8.4.4 shows a schematic configuration of the PhoStrip process. This process combines both biological and chemical phosphorus removal. Similar to other EBPR processes, this process subjects the biomass to alternating anaerobic/oxic environments to cultivate phosphorus-accumulating species. However, the process diverts a side-stream of phosphorus-enriched, activated sludge into a stripper where anaerobic conditions promote the release of intracellular phosphate. Then, the process treats the phosphorus-enriched, supernatant stream with lime to chemically precipitate phosphorus. The activated sludge, stripped of phosphorus, returns to the aeration basin for further uptake of phosphorus into the biomass under oxic conditions.

This process was used in several installations during the late 1970s and early 1980s (Peirano 1977; Northrop and Smith 1983). These installations reported good operating data but also had long periods of operation and maintenance problems. In some cases, the traditional chemical precipitation of phosphorus had lower operating costs than the PhoStrip process (Peirano 1977). These higher costs resulted in many installations terminating the PhoStrip process.

CHEMICAL PRECIPITATION

TYPES OF PRECIPITATING CHEMICALS

a) Alum
b) iron: (b1) ferrous, (b2) ferric
c) lime

PHOSPHORUS DISCHARGES PER CAPITA PER YR

In the form of human waste, 1.2 lb; in the form of detergents, 2.3 lb

RANGE OF ALUM CONCENTRATION USED

50 to 300 mg/l

IDEAL pH RANGE FOR PRECIPITATION

4 to 6 (a), 4.5 to 5 (b2), 7 to 8 (b1), and 10 to 11 (c)

WEIGHT RATIO OF ALUM REQUIRED TO PHOSPHORUS REMOVED

22:1 for 95% removal; 16:1 for 85% removal

ALUM REQUIREMENT TO REMOVE 85% OF A 10 MG/L PHOSPHORUS CONTENT

160 mg/l or 1335 lb per million gal

Wastewater treatment facilities can effectively use metal salts such as alum, ferrous sulfate, and ferric chloride to precipitate phosphorus in wastewater in conventional activated-sludge processes. Alum and ferric chloride are typically added into mixed liquor at the end of the aeration basin. Ferrous sulfate is added toward the front, allowing oxidation of ferrous into ferric ions. The molar ratio of metal ions to phosphorus depends on the required effluent phosphorus concentration and wastewater pH. Typically, the lower the effluent phosphorus concentration, the higher the molar ratio.

Environmental engineers often incorporate chemical precipitation into the design of EBPR processes as a standby or emergency measure to supplement biological processes in overcoming upset situations or augment phosphorus removal beyond the capacity of the biological process in meeting low effluent limits.

Many ionic forms effectively precipitate phosphorus from solution. The most notable are aluminium, calcium, and iron due to their low cost and general availability. Table 8.4.5 shows that all three of these ionic materials (multivalent metallic cations) form insoluble precipitates with phosphorus. In general, the degree of phosphorus removal by chemical precipitation is a function of the following factors:

- initial phosphorus concentration
- precipitating cation concentration
- concentration of other anions competing with phosphorus for precipitating cations
- wastewater pH

The tendency of aluminum and iron to hydrolyze in aqueous solution creates competition between the hydroxide and phosphate ions for the precipitating metal ions. Thus, the efficiency of phosphorus removal depends on the relative concentrations of these two anions in solution and is consequently pH dependent. A decrease in pH or hydroxide favors phosphate precipitation with metallic cations. When calcium is the precipitant, competition for calcium is predominantly between the phosphate and carbonate anions. As Table 8.4.5 shows, hydroxylapatite, $Ca_{10}(OH)_2(PO_4)_6$, is the most stable calcium phosphate solid phase.

Nitrifying/Nitrogen Removal Systems

The demand and applications for nitrogen removal from wastewater have steadily increased. For example, wastewater treatment plants discharging directly to aquifers are required to remove nitrates to limit drinking water contamination. A large amount of full-scale experience is available in nitrifying/denitrifying systems. This section reviews nitrogen removal systems in two categories: systems that do and do not use internal recycling streams. It also provides flow schemes.

A common problem for plants that cannot completely nitrify is their inability to maintain a sufficient aerobic (oxic) sludge retention time (SRT). Aerobic SRT is the average amount of time that a single microorganism spends

TABLE 8.4.5 EQUILIBRIUM CONSTANTS RELATED TO PHOSPHATE PRECIPITATION WITH METAL IONS

Reaction	*Log equilibrium constant,*[a] *25°C*
$Fe^{3+} + PO_4^{3-} = FePO_4(s)$	23
$3Fe^{2+} + 2PO_4^{3-} = Fe_3(PO_4)_2(s)$	30
$Al^{3+} + PO_4^{3-} = AlPO_4(s)$	21
$Ca^{2+} + 2H_2PO^-_4 = Ca(H_2PO_4)_2(s)$	1
$Ca^{2+} + HPO^-_4 = CaHPO_4(s)$	6
$10Ca^{2+} + 6PO_4^{3-} + 2OH^- = Ca_{10}(OH)_2(PO_4)_6(s)$	90

Note: [a]The higher the equilibrium constant, the more stable the precipitate formed.

in the aerated portion of the process. To increase this parameter, wastewater treatment facilities can make several operational adjustments including increasing the operating mixed liquor suspended solids concentration (MLSS); or increasing the influent detention time (IDT) in the aerobic section of the treatment process.

Most activated-sludge treatment processes, can realize nitrification when a sufficient sludge age is maintained, DO is available, and no inhibitors are present in the influent wastewater (such as specific compounds or pH extremities). Additionally, nitrification and phosphorus removal can be realized in the same process; the A/O, OWASA, SBRs, and PhoStrip processes can all achieve nitrification.

SYSTEMS WITHOUT INTERNAL RECYCLING

The Wuhrman process, Ludzack–Ettinger process, and oxidation ditches are nitrogen removal systems without internal recycling.

Wuhrmann Process

The Wuhrmann process, or post-denitrification, achieves nitrification and carbonaceous oxidation before the wastewater enters the anoxic zone for denitrification (see Part a in Figure 8.4.7). Endogeneous respiration provides the required carbon source since all available extracellular carbon has been removed.

Although the efforts of Wuhrmann helped to develop other single sludge nitrification/denitrification systems, this process has never been used in full scale. Operational problems include high turbidity levels of the clarified effluent, ammonia release from cell lysis in the anoxic zone, and high nitrate levels due to low denitrification rates (U.S. EPA 1993).

Ludzack–Ettinger Process

Part b in Figure 8.4.7 shows the Ludzack–Ettinger process. Because influent wastewater is directed first into an anoxic zone followed by an aerobic zone, this process is called pre-denitrification. Since nitrification occurs after the anoxic zone, the RAS stream recycles the nitrates. As such, this process typically operates with a high RAS return rate (75–150% Q). The influent wastewater serves as the carbon source for denitrification and thus has a higher denitrification rate than the Wuhrmann process.

Oxidation Ditches

Oxidation ditches successfully control total nitrogen effluent concentrations either by encouraging simultaneous biological nitrification/denitrification (SBND) in the same reactor or alternating biological processes between reactors.

Wastewater treatment facilities can achieve SBND by strategically locating the aeration equipment in the process reactors. By doing so, they can create alternating aerobic and anoxic zones. Phased isolation ditch technology, used extensively in Europe, alternates biological nitrification and denitrification in separate reactors. However, this technology differs from SBRs since wastewater is continuously discharged from the system. The BioDenitro process, developed by Krüger A/S of Denmark, has demonstrated applicability and potential in the United States (Tetreault et al. 1987). Part c in Figure 8.4.7 is a schematic diagram of the BioDenitro process.

SYSTEMS USING INTERNAL RECYCLING STREAMS

The modified Ludzack–Ettinger (MLE) process and the four-stage Bardenpho process remove nitrogen using internal recycling streams.

MLE Process

The MLE process is based on the premise that an insufficient amount of nitrate is returned in the RAS stream. By adding an internal recycle (typically 100–400% Q) from the end of the aerobic zone, this process increases the amount of nitrate returned to the anoxic zone for subsequent denitrification. As such, the MLE process realizes a greater degree of total nitrogen (TN) reduction. Part d in Figure 8.4.7 is a flow schematic diagram of the MLE process.

Four-Stage Bardenpho Process

The Bardenpho process provides a TN removal capability that cannot be obtained in the MLE process. The four-stage Bardenpho process subjects the nitrate that was not recycled from the primary aerobic zone to anoxic conditions in a secondary anoxic zone (see Part e in Figure 8.4.7).

Although biological denitrification is occurring in both anoxic reactors, the carbon source is different; the carbon source of the primary anoxic zones is supplied by influent wastewater, whereas endogenous respiration is responsible for any denitrification in the secondary anoxic zone. A small secondary aerobic zone prior to secondary clarification strips away any nitrogen gas entrained in the solids and nitrifies any ammonia released from cell lysis.

Simultaneous Phosphorus and Nitrogen Removal Processes

Several versions of biological processes for simultaneous phosphorus removal and nitrification/denitrification are installed and operated throughout the world. Table 8.4.6 summarizes the capability of each process, and Figure 8.4.8 summarizes their configurations. The following paragraphs describe each process.

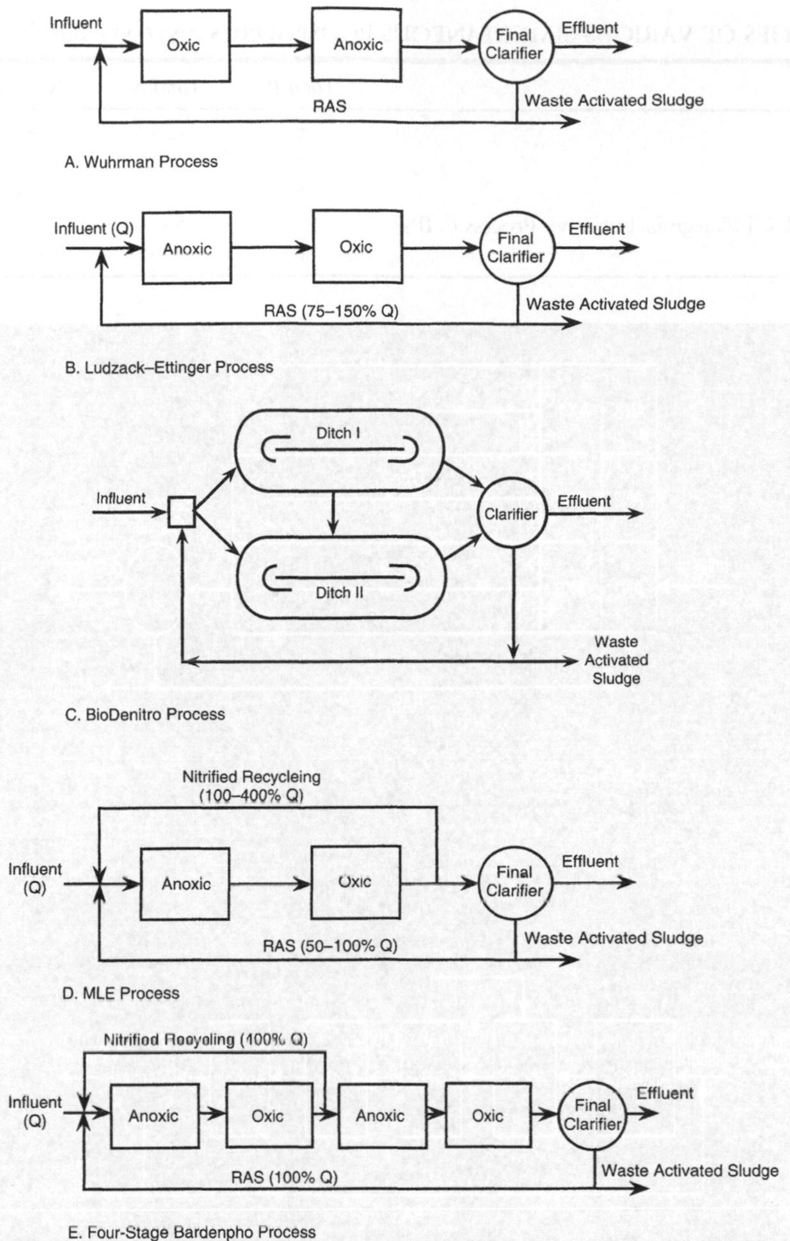

FIG. 8.4.7 Schematic diagrams of the nitrogen removal processes without internal recycling.

A^2/O PROCESS

Part a in Figure 8.4.8 is a typical schematic representation of the A^2/O process. The process consists of three zones in series: anaerobic, anoxic, and oxic. Each zone is further divided into several compartments. The mechanisms of phosphorus removal and nitrification are described in the preceding sections. This process recycles the nitrified mixed liquor containing nitrate and the nitrite at the end of the oxic zone. This internal recycled stream goes to the anoxic zone for denitrification at a rate of 100 to 200% of the influent wastewater flow rate. The process uses the organics in the influent wastewater as the carbon source for denitrification.

Influent detention times for the anaerobic and anoxic zones are generally 1 to 2 hr, while the oxic zone is 4 to 8 hr, depending on the influent wastewater characteristics. This process typically uses submersible mixers for solids mixing in the anaerobic and anoxic zones. Different types of aeration devices are used to satisfy oxygen demands in the oxic zone.

The A^2/O process can achieve an effluent quality with 1 mg/l or less in total phosphorus and ammonia. However, effluent NO_x–N concentration is typically limited to 6 to 10 mg/l and depends on the NH_3–N concentration in the influent wastewater as well as the internal recycling rate. This process is used in more than twenty installations in

TABLE 8.4.6 CAPABILITIES OF VARIOUS SIMULTANEOUS PHOSPHORUS AND NITROGEN REMOVAL PROCESSES

Process	*Total P*	*Total N*	*NH_3–N*	*Years in Operation, yr*
A^2/O	1	6–10	<1	>15
Improved A^2/O	1	3–5	<1	1
Modified Bardenpho	1	3	<1	>15
University of Cape Town (UCT)/Virginia Initiative Process (VIP)	1–2	8–10	<1	3
BioDenipho	1	3–5	<1	>15

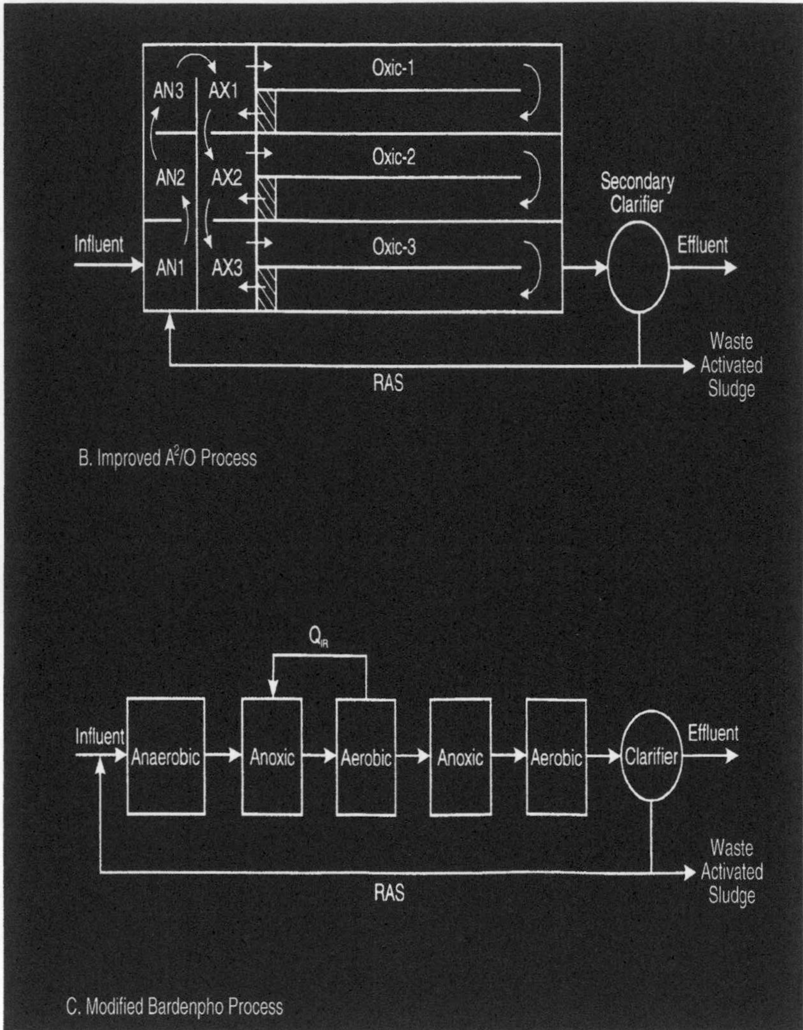

FIG. 8.4.8 Schematic diagrams of simultaneous phosphorus and nitrogen removal processes.

the United States. Table 8.4.7 shows the design parameters of some of these A^2/O systems.

IMPROVED A2/O PROCESS

This process further optimizes the NO_x–N removal in the A^2/O process. It recycles nitrified mixed liquor from each oxic stage to an anoxic stage for denitrification. Compared to the A^2/O process, this process maintains higher BOD concentrations in the first two anoxic stages and realizes higher denitrification rates. Furthermore, this process has an exhauster with 15 to 30 min detention time installed at the end of each oxic stage to deplete the DO in the mixed liquor prior to recycling to the anoxic stage. This scheme further improves the denitrification rate.

This process can achieve an effluent NO_x–N of 3 to 6 mg/l. Part b in Figure 8.4.8 is a schematic diagram of this process.

MODIFIED BARDENPHO PROCESS

This process, also known as the Phoredox process, is composed of an anaerobic zone preceding the Bardenpho process to achieve biological phosphorus removal. Part c

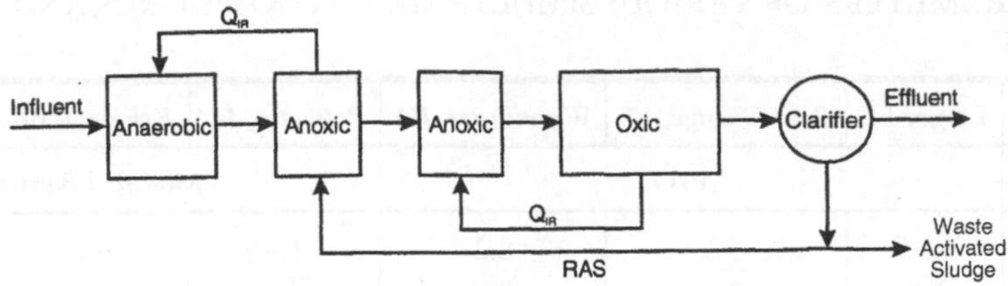

D. UCT Process

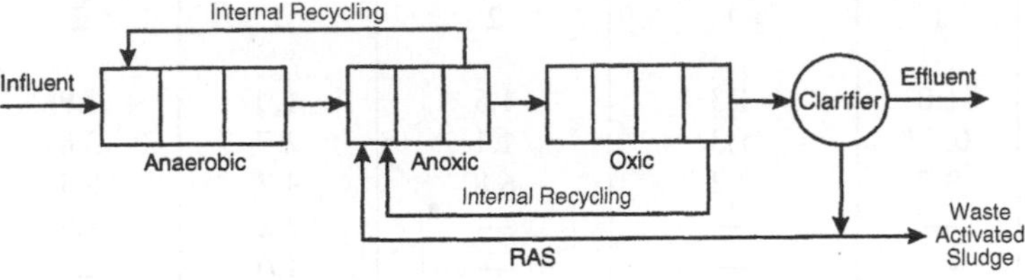

E. VIP Process

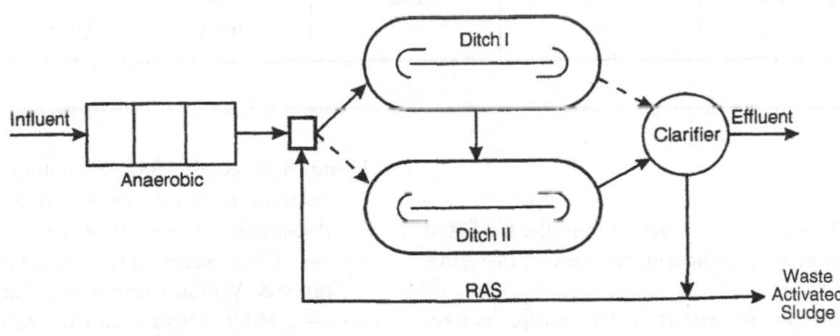

F. BioDenipho Process

FIG. 8.4.8 *Continued*

in Figure 8.4.8 is a schematic diagram of the process. It is similar to the Bardenpho process in that the design includes an influent detention time of 18 to 24 hr, an internal recycling rate of 200 to 400% of the influent flow rates, and no partitioning of each zone.

This process can achieve an effluent NO_x–N of 2 mg/l or less with proper detention time in the primary and secondary anoxic zones. However, the inherently low F/M of the process requires chemical polishing to achieve an effluent total phosphorus concentration of 1 mg/l. This process is used by more than 10 installations in the United States. Table 8.4.7 shows the design parameters.

UCT/VIP PROCESSES

Parts d and e in Figure 8.4.8 are schematic diagrams of the UCT and VIP processes. The flow arrangement of the two processes is identical except that the VIP process is partitioned in the anaerobic, anoxic, and oxic zones. Both processes discharge RAS into the anoxic zone instead of the anaerobic zone to avoid the detrimental effects of recycling NO_x–N on biological phosphorus removal. An additional return stream recycles MLSS from the end of the anoxic zone to the anaerobic zone. Currently, only one installation in the United States uses these processes.

BIODENIPHO PROCESS

Phased isolation-ditch technologies are also used for simultaneous phosphorus and nitrogen removal. The BioDenipho process includes an anaerobic zone that precedes the BioDenitro process to achieve enhanced biological phosphorus removal (EBPR). Part f in Figure 8.4.8 is a schematic diagram of the BioDenipho process. This process typically operates at a low F/M with a long hydraulic detention time and can achieve low effluent TN requirements.

—*Sun-Nan Hong*
R. David Holbrook

TABLE 8.4.7 DESIGN PARAMETERS OF VARIOUS SIMULTANEOUS PHOSPHORUS AND NITROGEN REMOVAL PROCESSES

	Largo, FL	*Port Orange, FL*	*Warminster, PA*	*Palmetto, FL*	*Kelowna, BC*	*Orange Co., FL*	*Louisburg, NC*
Process	*A^2/O*			*Modified Bardenpho*			*BioDenipho*
Plant Capacity, mgd	12.5	12	8.2	1.4	6.0	7.5	1.5
Primary Clarification	Yes	No	Yes	Yes	Yes	No	No
Wastewater Temp, °C	21–30	21–30	9–22	21–30	8–20	21–30	10–20
Discharge Limitation							
NO_x–N	—	10	10 (Summer)	—	—	—	—
Total N	8	—	—	3	6	3	Future limit
Total P	1	1	2	1	2	1	2
Influent Detention Time, hr							
Anaerobic	1.0	1.3	1.5	1.0	1.9	1.9	1.5
Anoxic 1	0.67	1.3	1.1	2.7	3.8	3.4	—
Oxic 1	3.7	7.2	6.8	4.7	8.5	10.7	22.5
Anoxic 2	—	—	—	2.2	1.9	1.9	—
Oxic 2	—	—	—	1.0	2.8	0.3	—
Total	5.37	9.8	9.4	11.6	18.9	18.2	24[1]
Internal Recycle							
First Anoxic	1:1	1:1	1:1	4:1	4 ~ 6:1	4 ~ 6:1	—
Anaerobic	—	—	—	—	—	—	—
Chemical Addition	No	No	No	Yes	Yes	Yes	No
Type	—	—	—	Alum	Alum	Alum	—

Note: [1]State mandated.

References

Aruw, V., et al. 1988. Biological mechanism of acetate uptake mediated by carbohydrate consumption in excess phosphorus removal systems. *Water Research* 22, no. 5:565.

Bargman, R.D., et al. 1971. *Nitrogen–phosphate relationships and removals obtained by treatment processes at the Hyperion Treatment Plant.* Pergamon Press Ltd.

Berber, A., and C.T. Winter. 1984. The influence of extended anaerobic retention time on the performance of phoredox nutrient removal plant. *Water Science Technology* 17, no. 81.

Borchardt, J.A., and H.A. Azad. 1968. Biological extraction of nutrients. *J. WPCF* 40, no. 10:1739.

Bordacs, K., and S.C. Chiesa. 1988. Carbon flow patterns in enhanced biological phosphorus accumulating activated sludge cultures. IAW-PRC Conference, Brighton, England.

Charley, R.C., D.G. Hooper, and A.G. McLee. 1980. Nitrification kinetics in activated sludge at various temperatures and dissolved oxygen concentrations. *Water Research,* 14:1387–1396.

Claete, T.E., and P.L. Steyn. 1988. The role of acinetobactor as a phosphorus removing agent in activated sludge. *Water Research* 22, no. 8.

Connell, C.H., and D. Vacker. 1967. Parameters of phosphate removal by activated sludge. *Proceedings 7th Industrial Water and Waste Conference, University of Texas, Austin, Texas.* II-28–37.

Davelaar, D., et al. 1978. The significance of an anaerobic zone for the biological removal of phosphate from wastewaters. *Water SA* 4, no. 2:54.

Davies, T. 1971. Population description of a denitrifying microbial system. *Water Research,* 5:553.

DeFoe, R.W., et al. 1993. Large scale demonstration of enhanced biological phosphorus removal with nitrification at a major metropolitan wastewater treatment plant. WEF Conference, Anaheim, CA.

Fuchs, G.W., and M. Chen. 1975. Microbial basis of phosphate removal in the activated sludge process for the treatment of wastewater. *Microbial Ecology,* no. 119.

Hong, S.N. et al. 1982. A biological wastewater treatment system for nutrient removal. Presented at the EPA workshop on Biological Phosphorus Removal, Annapolis, Maryland.

———. 1983. Recent advances on biological nutrient control by the A/O process. WPCF Conference, Atlanta, GA.

———. 1989. Design and operation of a full-scale biological phosphorus removal system. The 52nd WPCF Conference, Houston, TX.

Hong, S.N., and K.L. Andersen. 1993. Converting a single sludge oxygen activated sludge system for nutrient removal. The 66th WEF Conference, Anaheim, CA.

Joint Task Force of the Water Environment Federation and the American Society of Civil Engineers (ASCE). 1991.

Kang, S.J., et al. 1985. Full-scale biological phosphorus removal using A/O process in a cold climate. In *Management strategies for phosphorus in the environment.* London: Selper Ltd.

Krichten, D.J., et al. 1985. Applied biological phosphorus removal technology for municipal wastewater treatment by the A/O Process. In *Management strategies for phosphorus in the environment.* London: Selper Ltd.

Levin, G.V., and J. Shapiro. 1965. Metabolic uptake of phosphorus by wastewater organisms. *J. WPCF* 37, no. 6:800.

Manning, J.F., and R.L. Irvine. 1985. The biological removal of phosphorus in sequencing batch reactors. *J. WPCF* 57, no. 87.

Marais, G.V.R., and G.A. Ekema. 1976. The activated sludge process: Steady state behavior. *Water S.A.* 2:162–300.

McLaren, et al. 1976. Effective phosphorus removal from sewage by biological means. *Water SA,* no. 1.

Metcalf and Eddy, Inc. 1991. *Wastewater engineering: Treatment, disposal and reuse.* 3d ed. McGraw-Hill.

Nicholls, H.A., and P.W. Osborn. 1979. Bacterial stress: Prerequisite for biological removal of phosphorus. *J. WPCF* 51, no. 4:557.

Northrop, J., and D.A. Smith. 1983. Cost and process evaluation of phostrip at Amherst, NY. WPCF Conference, Atlanta, GA.

Peavy, M.S., D.R. Rowe, and G. Tchobanoglous. 1985. *Environmental Engineering.* McGraw-Hill.

Peirano, L.E. 1977. Low-cost phosphorus removal at Reno-Sparks, Nevada. *J. WPCF* 49:568.

Prescott, L.M., J.P. Harley, and D.P. Klein. 1990. *Microbiology.* Wm. C. Brown Publishers.

Rabinowtiz, B., et al. 1987. A novel operational model for a primary sludge fermenter for use with the enhanced biological phosphorus removal process. IAWPRC Specialized Conference, Rome, Italy.

Randall, C.W., J.L. Barnard, and M.D. Stensel. 1992. *Design and retrofit of wastewater treatment plants for biological nutrient removal.* Vol. 5. Technomic Publishing Co.

Sell, R.L., et al. 1981. Low temperature biological phosphorus removal. The 54th WPCF Conference, Detroit, MI.

Sharma, B., and A.C. Ahlert. 1977. Nitrification and nitrogen removal. *Water Research* 11:897–925.

Spector, M.L. 1977. Production of non-bulking activated sludge. U.S. Patent 4,056,465.

Sutton, P.M., K.L. Murphy, and B.C. Jank. 1974. Biological nitrogen removal—The efficiency of the nitrification step. 47th Annual WPCF Conference, Denver, Colorado, October, 1974.

Tam, N.F.Y., Y.S. Wong, and G. Leung. 1992. Effects of exogenous carbon sources on removal of inorganic nutrient by the nitrification–denitrification process. *Water Research* 26:9, no. 9:1229–1236.

Tetreault, M.S., B. Rusten, A.M. Benedict, and J.F. Kreissel. 1987. Assessment of phased isolation ditch technologies. *J. WPCF* 59, no. 9:833–840.

Tracy, K.D., and A. Flammino. 1987. *Biochemistry and energetics of biological phosphorus removal.* Pergamon Press.

U.S. Environmental Protection Agency (EPA). 1975. *Process design manual for nitrogen control.*

———. 1987. *Summary report—The causes and control of activated sludge bulking and foaming.* EPA 625-8-87-012.

———. 1993. *Nitrogen control manual.* EPA 625-12-93-010.

Van Haandel, A.C., Ekema, G.A. and G.V.R. Marais. 1981. The activated sludge process—3. *Water Res.* 15:1135–1152.

Wanielista, M.P., and W.W. Eckenfelder, Jr. 1978. *Advances in water and wastewater treatment—Biological nutrient removal.* Ann Arbor Science.

Water Pollution Control Federation. 1983. *Nutrient control—Manual of practice No. FD-7, Facilities Design.*

Wells, N.W. 1969. Differences in phosphate uptake rates exhibited by activated sludges. *J. WPCF* 41, no. 5:765.

Wentzel, W.C., et al. 1985. Kinetics of biological phosphorus release. *Water Sci. Tech.* 17, no. 57.

———. 1988. Enhanced polyphosphate organism cultures in activated sludge systems—Part I: Enhanced culture development. *Water SA*, 14 no. 81.

9

Chemical Treatment

Donald B. Aulenbach | Béla G. Lipták | Thomas J. Myron, Jr.

9.1 NEUTRALIZATION AGENTS AND PROCESSES

NEUTRALIZING REAGENTS

a) Acidic: a1) Sulfuric acid, a2) Hydrochloric acid, a3) Carbon dioxide, a4) Sulfur dioxide, a5) Nitric acid

b) Basic: b1) Caustic soda, b2) Ammonia, b3) Soda ash, b4) Hydrated lime, b41) Dolomitic hydrated quicklime, b42) High-calcium, hydrated quicklime, b5) Limestone

SOLUBILITY AND pH OF LIME SOLUTIONS

For a concentration range of 0.1 to 1.0 gm of CaO per liter, the corresponding pH rises from 11.5 to 12.5 at 25°C. Saturation occurs at about 1.2 gm/l

LIME AVAILABILITY

Dry forms with several particle-size distributions or liquid forms usually as a slurry with +20 wt % solids

CAUSTIC AVAILABILITY

In dry form as 75% Na_2O or as a 50% NaOH solution, precipitates formed are usually soluble.

TOXICITY OF REAGENTS

See Tables 9.2.7 and 9.2.8.

REAGENT REQUIREMENTS FOR NEUTRALIZATION

See Figure 9.2.5.

Process Description

Neutralization is the restoration of the hydrogen (H^+) or hydroxyl (OH^-) ion balance in solution so that the ionic concentrations of each are equal. Conventionally, the notation pH (puissance d'hydrogen) describes the hydrogen ion concentration or activity present in a solution as follows:

$$\mathrm{pH} = -\log_{10} [\mathrm{H}^+] \qquad 9.1(1)$$

$$\mathrm{pH} = -\log_{10} a_{\mathrm{H}^+} \qquad 9.1(2)$$

where:

$[H^+]$ = the hydrogen ion concentration, gmol per liter
a_{H^+} = hydrogen ion activity

For a dilute solution of strong acids, i.e., acids considered completely dissociate (ionized in solution), the following equation applies:

$$a_{\mathrm{H}^+} = [\mathrm{H}^+] \qquad 9.1(3)$$

At neutrality, the concentration of hydrogen and hydroxyl ions is equal. The product of their ion concentration (K_w) at 25°C is as follows:

$$(-\log_{10} [\mathrm{H}^+])(-\log_{10} [\mathrm{OH}^-]) = -\log_{10}K_w$$

$$K_w = 1.008 \times 10^{-14} \qquad 9.1(4)$$

At neutrality, the following equations apply:

$$\mathrm{pH} = \frac{-\log_{10}K_w}{2} = 7.0 = \mathrm{pOH} \qquad 9.1(5)$$

$$\mathrm{pH} + \mathrm{pOH} = 14 \qquad 9.1(6)$$

Thus, if a solution has a pH = 2.0 at 25°C, hydrogen ion concentration is 1×10^{-2} moles H^+ per liter, pOH = 12, and hydroxyl ion concentration is 1×10^{-12} moles OH^- per l. The ion product of water depends highly on temperature, changing approximately two orders of magnitude over a 60°C span (see Figure 9.1.1).

The pH notation as a means of expressing the hydrogen ion concentration is logarithmic. A pH change from 2.0 to 1.0 does not mean that the ion concentration has doubled; a change of one pH unit is an order of magnitude change. Thus, if an acid influent changes by three pH units, the $[H^+]$ is changing by a factor of one thousand. This logarithmic nature becomes an important consideration when reagent delivery systems are sized because if the ion load to be neutralized changes by a factor of 1000, the reagent delivery system must have the same turndown (rangeability).

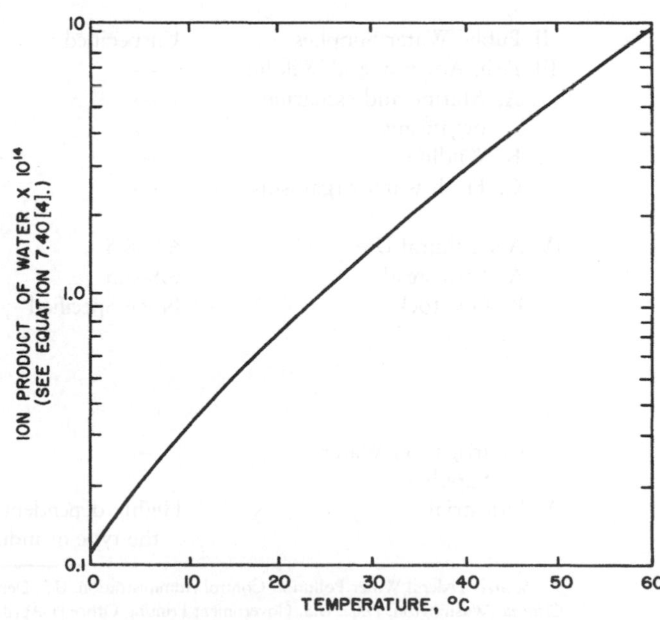

FIG. 9.1.1 Ion product of water as a function of temperature.

The need to neutralize, or at least place limits on, the pH variation of environmental waters has resulted in the promulgation of water quality standards legislation in virtually every state. The physical well-being of all life forms depends not only on the absolute value of the pH but also on the frequency of pH variation. Thus, for example, the lacrimal fluid of the human eye has a nominal pH of 7.4 and a high buffering capacity, i.e., it resists changes in pH. Variations of the lacrimal fluid as low as 0.1 pH unit can result in eye irritation. The Federal Water Pollution Control Administration has published a report that details not only the pH requirements for water of a designated end use but also the requirements for twenty other ions, as well as organic chemical limitations and physical and microbiological properties. Table 9.1.1 summarizes the preferred or acceptable pH ranges for various water quality categories.

Common Neutralization Reagents

Wastewater treatment facilities must counter the hydrogen or hydroxyl ion imbalance in a waste effluent by adding a material that restores the ion balance. Thus, if the waste effluent is acidic, i.e., pH < 7.0, they must blend a reagent having basic characteristics with the waste to achieve neutrality. Conversely, if the waste effluent is basic, i.e., pH > 7.0, they must use a reagent having acid characteristics. Table 9.1.2 lists common neutralization reagents.

In addition to the reagents listed in Table 9.1.2, waste acids and bases can also serve as neutralizing reagents. In somc cases, particularly in ion-exchange resin regeneration, in which the resin bed is treated first with a caustic solution and then with an acid solution, wastewater treatment facilities can store these solutions and then blend them to achieve a neutral solution rather than discharge them to the sewer immediately after use.

Four widely used reagents are sulfuric acid, caustic soda, hydrated chemical lime, and (to a limited degree) limestone. The main reasons for their popularity are economy and ease of handling.

LIME

Chemical lime is produced by the calcination of high-quality limestone which produces either high-calcium quicklime or dolomitic quicklime. Further treatment of high-calcium quicklime and dolomitic quicklime produces their hydrated counterparts. Figure 9.1.2 shows the chemical lime production process.

Table 9.1.3 shows typical analyses of the two types of quicklimes. The difference in chemical composition be-

TABLE 9.1.1 pH RANGES FOR VARIOUS WATER QUALITY CATEGORIES

Water Quality Category	*pH Range* *Preferred*	*pH Range* *Acceptable*	*Restrictions and Comments*
I Recreation and Aesthetics	6.5–8.3	5.0–9.0	8.3 < pH < 6.5 discharges to have limited buffering capacity
II Public Water Supplies	Unspecified	6.0–8.5	
III Fish, Aquatic, and Wildlife	—	6.0–9.0	
A. Marine and estuarine organisms	—	6.7–8.5	Bulk location[a] change ≤ 0.1 pH
B. Wildlife	—	7.0–9.2	
C. Fresh water organisms	—	6.0–9.0	Alkalinity ≥ 20 mg $CaCO_3$ per liter
IV Agricultural Uses	6.0–8.5	5.5–9.0	
A. Farmstead	6.8–8.5	—	
B. Livestock	None Specified	Fish indicator ponds encouraged at terminal watershed locations	
C. Irrigation Water Supplies	—	4.5–9.0	
V Industrial	Highly dependent on the type of industry		

Source: Federal Water Pollution Control Administration, U.S. Department of Interior, 1968, *Report of the Committee on Water Quality Criteria* (Washington, D.C.: U.S. Government Printing Office [1 April]).

Note:[a] The entire water body receiving the water should not change its pH by more than 0.1 as a result of wastewater pH variations.

TABLE 9.1.2 COMMON NEUTRALIZATION REAGENTS

Acid Reagents	*Base Reagents*
Concentrated (66°Bé) sulfuric acid	Caustic soda (NaOH)
Concentrated (20 or 22°Be) hydrochloric acid	Ammonia
Carbon dioxide	Soda ash (Na_2CO_3)
Sulfur dioxide	Hydrated chemical lime ($Ca[OH]_2$)
Nitric acid	Limestone ($CaCO_3$)

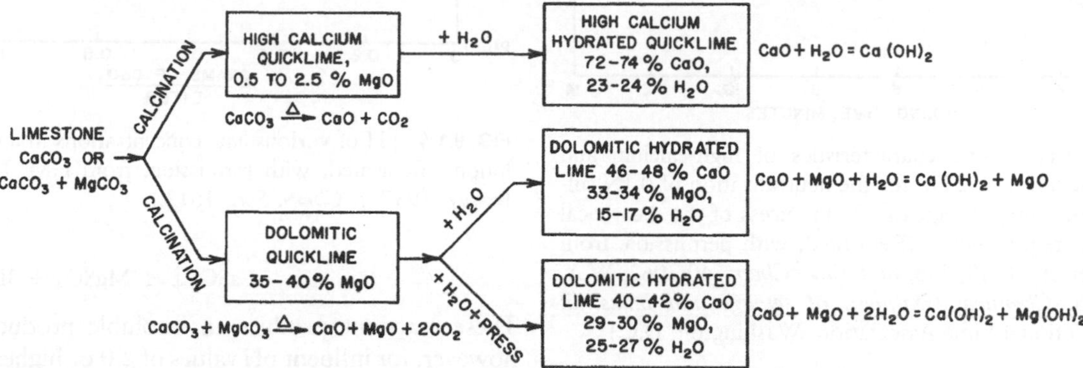

FIG. 9.1.2 Chemical lime production process. (Reprinted from National Lime Association, *Chemical lime facts,* Bulletin 214, Washington, D.C.)

TABLE 9.1.3 TYPICAL ANALYSES OF COMMERCIAL QUICKLIMES

Component	*High-Calcium Quicklime, %*	*Dolomitic Quicklime, %*
CaO	93.25–98.00	55.50–57.50
MgO	0.30–2.50	37.60–40.80
SiO_2	0.20–1.50	0.10–1.50
Fe_2O_3	0.10–0.40	0.05–0.40
Al_2O_3	0.10–0.50	0.05–0.50
H_2O	0.10–0.90	0.10–0.90
CO_2	0.40–1.50	0.40–1.50

Source: National Lime Association, *Chemical lime facts,* Bulletin 214 (Washington, D.C.).
Note: The range of values are not necessarily minimum and maximum percentages.

tween high-calcium and dolomitic quicklimes results in varying reaction rates and reactivities when these materials treat the same wastewater. Figure 9.1.3 demonstrates the reaction rate characteristics of high-calcium and dolomitic lime (2% excess) at 25°C when reacted with 0.1N H_2SO_4.

The reaction rate curve for high-calcium lime in the figure was obtained with 2% excess lime. Lesser amounts of excess lime result in longer reaction times. This consideration becomes important when environmental engineers size neutralization vessels. Due to the continuing reaction of lime, pH measured at the neutralization tank discharge may not be the final effluent pH downstream of the discharge, if the nominal time constant of the neutralization vessel is too short.

The solubility of lime (CaO) in aqueous solutions decreases with increasing temperature, e.g., 0.131 gm per 100 cc at 10°C and 0.07 gm per 100 cc at 80°C. Figure 9.1.4 plots the pH range of lime solutions up to saturation at 25°C. Lime as a neutralizing reagent is normally supplied in slurries or as a dry feed along with water to a small, agitated, holding container that overflows to the neutralization process. Normally, 20% or less lime slurries are used. The lower the lime concentration, the easier the handling in terms of abrasion and clogging. Quicklime is available in bulk or bag form in reasonably standard sizes (see Table 9.1.4).

In addition to the dry form, a hydrated lime produced

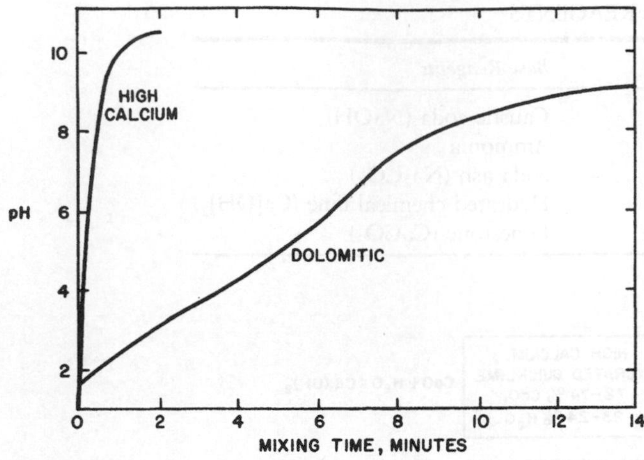

FIG. 9.1.3 Reaction rate characteristics of high-calcium and dolomitic quicklimes at 25°C for the neutralization of 0.1N sulfuric acid with a lime dosage of 2% in excess of the theoretical stoichiometric requirements. (Reprinted, with permission, from R.D. Hoak et al., 1948, *Ind. and Eng. Chem.* 40:2062; W.A. Parson, 1965, *Chemical treatment of sewage and industrial wastes,* 58, National Lime Association, Washington, D.C.)

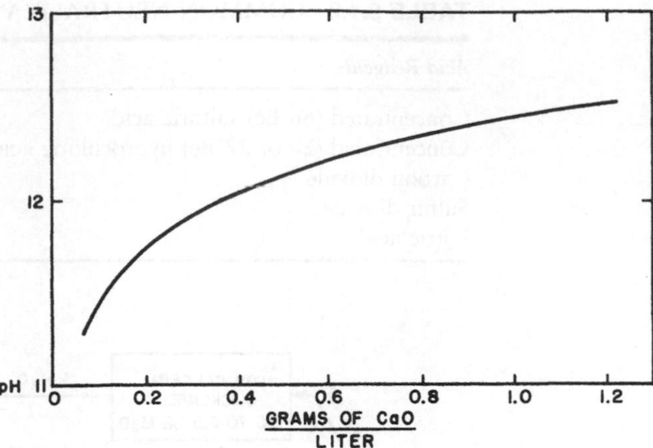

FIG. 9.1.4 pH of various lime concentrations in an aqueous solution. (Reprinted, with permission, from F.M. Lea and G.E. Bessey, 1937, *J. Chem. Soc.,* 1612.)

as a by-product in the manufacture of acetylene is available in slurry form having a nominal 35 wt % solids content. This lime slurry is delivered in tankwagons having a nominal capacity of 4500 gal. Table 9.1.5 gives a typical chemical analysis—particle size distribution—of this lime slurry. The cost of this lime form depends on the transportation cost.

Chemical reactions associated with hydrated limes depend on the type of acid(s) being neutralized. The reaction between sulfuric acid and hydrated, high-calcium and normal, dolomitic quicklimes shown in the following equations is typical:

$$H_2SO_4 + Ca(OH)_2 \longrightarrow CaSO_4\downarrow + 2H_2O \qquad 9.1(7)$$

$$2H_2SO_4 + Ca(OH)_2 + MgO \longrightarrow CaSO_4\downarrow + MgSO_4 + 3H_2O \qquad 9.1(8)$$

These reactions produce an insoluble product ($CaSO_4$); however, for influent pH values of 2.0 or higher, the quantity of $CaSO_4$ produced is insufficient to cause precipitation. Hydrofluoric acid reacted with lime produces a reaction product that is about two orders of magnitude less soluble than $CaSO_4$.

LIMESTONE

Limestone ($CaCO_3$) is an effective means of neutralizing waste acids. The process involves flowing the waste acids (mixtures of HCl and H_2SO_4) through a bed of limestone granules. The bed can be 3 ft deep at waste flows of from 1.3 to 2.2 gpm/sq ft of bed area. Recycling the treated effluent (from one to three times) and aeration can also improve performance.

This approach to waste neutralization appears to be financially attractive but has limited reliability. The main

TABLE 9.1.4 AVAILABLE QUICKLIME SUPPLIES

	Percentage Passing Through the Noted Mesh-Size Screen					
	Mesh Size and Opening					
Name	*No. 8 (0.093 in)*	*No. 20 (0.0328 in)*	*No. 100 (0.0058 in)*	*No. 200 (0.0029 in)*	*No. 325 (0.0017 in)*	*Remarks*
Lump Lime	—	—	—	—	—	Up to 8 in diameter
Crushed or pebble lime	—	—	—	—	—	¼ to 2½ in diameter
Ground lime	100	60–100	40–60	—	—	—
Pulverized lime	100	100	85–90	—	—	—
Dried, hydrated lime	100	100	100	95	—	—
Air-classified lime	100	100	100	100	99.5	—

TABLE 9.1.5 CHEMICAL ANALYSIS AND PARTICLE-SIZE DISTRIBUTION OF HYDRATED CHEMICAL LIME FROM ACETYLENE PRODUCTION

Chemical Analysis	*Particle-Size Distribution* — *Percentage of Particles That Pass the Noted Mesh-Size Screens*			
	No. 20	*No. 48*	*No. 100*	*No. 325*
95% $Ca(OH)_2$				
1.5% $CaCO_3$				
0.25% MgO	99.9	99.2	97	85
1.6% Fe_2O_3 and Al_2O_3				
1.1% Insolubles				

Source: Chemline Corporation.

reason for decreased reliability is the contamination of limestone by materials such as oil and grease with the subsequent loss of bed activity.

SODIUM HYDROXIDE (CAUSTIC)

The use of sodium hydroxide (caustic) is about equal to lime as a neutralizing agent. Although caustic is more expensive than lime, its reaction characteristics (virtually instantaneous) and handling convenience are factors behind its widespread use. Researchers evaluated the reactivities of various basic reagents on the same chemical system by mixing them with pickle liquor (60 gm iron and 20 gm sulfate per liter). They measured the effectiveness of reagent reactions with pickle liquor in terms of iron gm remaining in solution after 6 hr. These experiments were conducted at room temperature and at 60°C with and without aeration (agitation). Table 9.1.6 shows the results of these experiments.

Concerning the speed of reaction of NaOH (see Table 9.1.6) and the reaction rate curve for quick and dolomitic limes (see Figure 9.1.3), the data show that when reaction time is important, the efficacy of NaOH, particularly in solution form, is virtually instantaneous. The ease of delivery to the neutralization process is another advantage of NaOH solutions. However, NaOH solutions are corrosive, and safety showers located in the process area are suggested. Personnel working with this material should use eye and skin protective devices.

Sodium hydroxide is available in solid (75% Na_2O) or solution (50% NaOH) form. In concentrated solution form, heating containers and lines may be required for transfer operations during cold weather. Most large suppliers of inorganic chemicals can furnish solutions of any strength in tank trucks and pump them directly to the user's reagent storage tank. Users should equip reagent storage tanks located outdoors with heaters and appropriate control equipment to protect against freezing during winter. As with lime, neutralization reactions depend on the acid or acids being neutralized as follows:

$$2NaOH + H_2SO_4 \longrightarrow Na_2SO_4 + 2H_2O \qquad 9.1(9)$$

$$NaOH + HCl \longrightarrow NaCl + H_2O \qquad 9.1(10)$$

$$NaOH + HNO_3 \longrightarrow NaNO_3 + H_2O \qquad 9.1(11)$$

For the reactions described by Equations 9.1(9), (10), and (11), all products are highly soluble, and unlike lime, at least for high H_2SO_4 concentrations, precipitate accumulation is not a problem.

SODA ASH

Soda ash (Na_2CO_3) is not widely used for neutralization. It does, however, have widespread use in applications requiring minor pH adjustment (one unit or less) and in wa-

TABLE 9.1.6 IRON CONCENTRATION (gr/L) REMAINING AFTER SIX HOURS OR LESS

Neutralizing Agent \ *Test Conditions*	*Room Temperature and No Aeration*	*60°C and No Aeration*	*Room Temperature with Aeration*	*60°C with Aeration*
NaOH	Reaction Is Practically Instantaneous			
Na_2CO_3	0 in 0.75 hr	0 in 0.75 hr	0 in 0.75 hr	0 in 0.5 hr
CaO	0 in 0.25 hr	0 in 0.25 hr	0 in 5 min	0 in 5 min
CaO · MgO	1.88	3.14	1.04	0.30
$Ca(OH)_2$	0 in 0.5 hr	0 in 0.5 hr	0 in 0.5 hr	0 in 0.5 hr
$Ca(OH)_2$ · MgO	1.23	1.53	0.55	0 in 3.5 hr
Acetylene Sludge[a]	1.66	1.04	0 in 3.5 hr	0 in 3.5 hr
$CaCO_3$[b]	20.40	18.80	2.95	0.03

Source: R.D. Hoak, 1950, *Sewage and Ind. Wastes* 22:212.
Notes: [a]$Ca(OH)_2$ waste product. From $CaC_2 + 2H_2O \longrightarrow C_2H_2\uparrow + Ca(OH)_2$.
[b]Pulverized limestone.

TABLE 9.1.7 TOXIC RATING CODE

Toxic Rating	*Description of Rating*
0(a)	*None:* No harm under any conditions
0(b)	*None:* Harmful only under unusual conditions or in overwhelming dosages
1	*Slight:* Causes readily reversible changes that disappear after end of exposure
2	*Moderate:* Can involve both irreversible and reversible changes; not severe enough to cause death or permanent injury
3	*High:* Can cause death or permanent injury after very short exposure to small quantities
U	*Unknown:* Available information with respect to man is considered invalid.

Note: Calcium compounds (chemical quicklime, etc.) have a toxic rating of 1 in all local categories and U in the systemic category; sodium hydroxide has a rating of 3, 3, 2 in all local categories and U in the systemic category; sodium carbonate has a rating of 2, 2, 2 in all local categories and U in the systemic category.

TABLE 9.1.8 TOXIC RATING OF ACID REAGENTS

	Acute Local			
Acids	*Irritant*	*Ingestion*	*Inhalation*	*Acute Systemic*
H_2SO_4	3	3	3	U
HCl	3	3	3	U
HNO_3	3	3	3	3
SO_2	3	3	3	U
CO_2	0	0	0	1, inhalation

Source: N.I. Sax, *Dangerous properties of industrial materials,* 2d ed. (New York: Rhinhold.)

ter softening. Ordinarily, it is supplied in solid form, which means solid feeders or solution makeup equipment.

In addition to being about five times as expensive as hydrated chemical lime, soda ash produces carbon dioxide in reaction for applications where strong acids are treated. Carbon dioxide can cause frothing, particularly in agitated neutralization vessels. Like lime, minimal safety hazards are connected with handling this material.

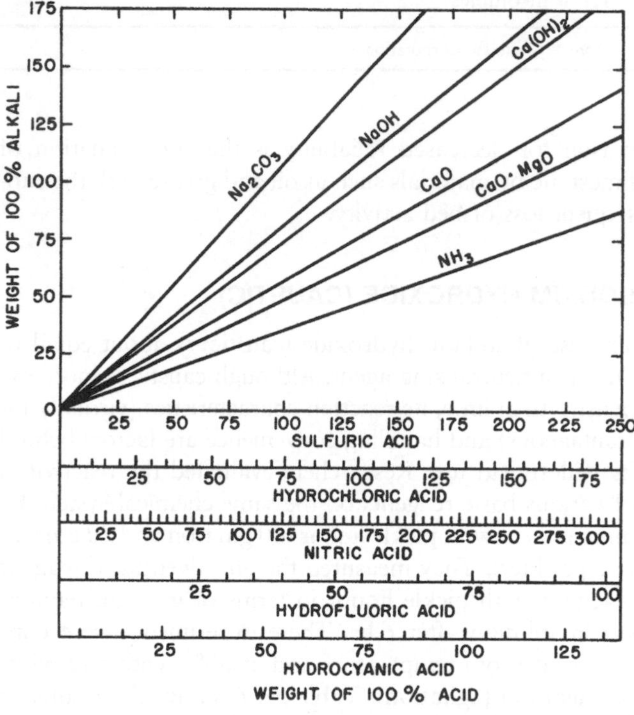

FIG. 9.1.5 Alkali neutralization graph. (Reprinted from W.A. Parson, 1965, *Chemical treatment of sewage and industrial wastes,* 58, Washington, D.C.: National Lime Association.)

AMMONIA

Ammonia, like soda ash, is not a prime neutralizing reagent. It is about twice as expensive as lime, and is the most toxic of all the alkaline neutralization reagents discussed, having a toxic hazard rating of 3 in all three acute local categories, i.e., irritant, ingestion, and inhalation, and U in the acute systemic category. Table 9.1.7 outlines these toxic hazard rating codes.

Ammonia is widely used in the petroleum industry where it is added to crude oil to neutralize acid constituents. The most serious deterrent for ammonia use is that it produces ammonium salts that supply nutrients (nitrogen) for algal growth.

The most important alkaline reagents are lime and sodium hydroxide. The choice between them depends largely on the economics of the total neutralization or treatment facility. When large volumes of waste are treated, requiring large amounts of reagents, wastewater treatment facilities chose lime because the cost is a significant portion of the total treatment cost. Since lime usually requires a substantial investment in slakers, tanks, pumps, and additional agitators, the small wastewater treatment facility can not afford such equipment, and a single tank filled (on a scheduled basis) with caustic solution by a local chemical supplier is the logical selection.

ACIDIC REAGENTS

Table 9.1.8 outlines the toxic hazard associated with acid reagents. Although five acids are listed as reagents, sulfuric acid, especially the 66°Bé (93.2% H_2SO_4), is by far the most widely used. It is the least expensive in highly concentrated solutions (66°Bé), it is noncorrosive (eliminating the need for special construction materials), and efficient in highly concentrated solutions. This last reason can cause problems if waste flows are small because the availability of small reagent delivery equipment is limited and small valves, pumps, or feeders can be plugged by small amounts of scale accumulating in storage equipment.

Hydrochloric and nitric acids are more costly than sulfuric acid and are highly corrosive requiring special construction materials. Carbon dioxide and sulfur dioxide, besides being expensive, must be dissolved in wastewater to produce carbonic acid and sulfurous acid.

To obtain efficient gas dispersion in large vessels requires concentric draft tubes and turbine-type blades, while a single draft tube (if any) and axial flow blades are used to efficiently blend waste and reagent for neutralization.

Figure 9.1.5 shows the alkali requirements for neutralizing various acids and the acid requirements for neutralizing various alkalis.

—*T.J. Myron*

9.2 pH CONTROL SYSTEMS

Final Element Rangeability
Requirement is extraordinary and depends upon titration curve and influent flow variability. Metering pumps are capable of 20:1 to 200:1, valves with positioners are capable of 50:1 or more, and a pair of split-ranged valves are capable of 1000:1 or more. pH swing that valves can handle can be determined as the base 10 logarithm of the rangeability for unbuffered titrations. If rangeability is 1000:1, the controllable pH swing is 3; for 100,000:1, it is 5

Final Element Precision and Characteristics
Requirement is exceptional and depends upon titration curve and desired control band. Electronically set metering pump and valves with positioners have repeatability from 0.1 to 2.0%. Linear valve characteristics are generally preferred.

Control Loop Dynamics
Effect of extreme nonlinearity and sensitivity of a pH process is diminished by a reduction in loop dead time. Ratios of loop dead time to time constant less than 0.02 are needed for setpoints on the steep portion of the titration curve to dampen oscillations. Reagent delivery delay is often the largest source of loop dead time.

Design Considerations
For proper mixing, process requires either an in-line mixer with 0.2 minute measurement filter and upstream and downstream volumes or a vertical, well-mixed vessel for attenuation of oscillations. Close-coupled control valve with ram valve or check valve to injection point and reagent dilution is needed to reduce reagent delivery delay. Additional stage of neutralization is needed for an inlet pH more than 2 pH units away from a small control band or whenever the final element rangeability or sensitivity requirement is excessive.

Reaction Tanks
Liquid depth should equal diameter, retention time should not be less than 5 minutes (10 to 30 minutes for lime), and dead time should be less than 1/20th of the retention time. For strong acid–strong base neutralizations, one, two, or three tanks are recommended for influent pH limits of 4 to 10, 2 to 12, and 0 to 14, respectively. Influent should enter at top, and effluent should leave at the bottom.

Agitator Choices
Propeller (under 1000 gallon [3780 1] tanks) or axial-flow impellers (over 1000 gallons [3780 1]) are preferred. Flat-bladed radial flow impellers should be avoided. Acceptable impeller-to-tank diameter ratio is from 0.25 to 0.4. Peripheral speeds of 12 fps (3.6 m/s) for large tanks and 25 fps (7.5 m/s) for tanks with volumes less than 1000 gallons (3780 1) are acceptable

Mixing Equipment
In-line mixer residence time should be less than 10 seconds and a well-mixed vessel residence time should be greater than 5 minutes. The vessel agitator should provide both a pumping rate greater than 20 times the throughput flow and a vessel turnover time less than 1 minute. Solid and gas reagents require a residence time greater than 20 times the batch dissolution time. Gas reagent injection needs a sparger designed to reduce bubble size and improve bubble distribution.

Setpoint Location
A setpoint on the flat portion of a titration curve reduces pH process oscillation and sensitivity and the control valve precision requirement.

pH Sensor Location
Insertion assemblies in pumped recirculation lines are preferred for increased speed of response, decreased coating, improved accessibility, and auto on-line washing and calibration. Insertion in recirculated lines is preferred by some.

Continuous Control Techniques
Flow feedforward for high loop dead time or rapid flow upsets. pH feedforward is only effective for influent pH on a relatively steep portion of the titration curve. A head start of reagent flow is needed for first stage when flow feedforward is not used. Signal linearization of measurement is beneficial for a constant titration curve. Self-tuning is helpful

Section 2.2 treats the subject of pH measurement. This section begins with an explanation of the difficult nature of the pH process; next, the process equipment used in pH control systems is described, including such topics as the selection of reagent delivery systems, mixing equipment, tank sizing, and other considerations. A discussion of the pH controller and its tuning rounds out the first half of the section, and the second half describes the various pH control applications.

Nature of the pH Process

The difficulty of pH control stems from the exceptionally wide range of the pH measurement, which for a 0 to 14 pH range covers 14 orders of magnitude of hydrogen ion concentration (Figure 9.2.1). It is commonly relied upon to detect changes as small as 10^{-7} in hydrogen ion concentration at mid-range. This incredible rangeability and sensitivity is the result of the nonlinear logarithmic relationship of pH to hydrogen ion activity as defined in Section 2.2. The process control implications are most severe for a process with only strong acids and based because the hydrogen ion concentration is proportional to the manipulated acid or base flow. The titration curve for such a system at 25°C is illustrated in Figure 9.2.2. The ordinate is the controlled variable (pH) and the abscissa is the ratio of the manipulated variable (reagent flow) to influent flow. Since the acids and bases are strong (completely ionized), the abscissa is also the hydrogen ion concentration.

As shown in Figure 9.2.2, change in pH for a change in reagent flow is 10^7 times larger at 7 pH than at 0 pH. The slope and hence the process gain changes by a factor of 10 for each pH unit deviation from the equivalence point at 7 pH. An expanded view of the apparently straight steep portion of the titration curve reveals another S-shaped curve (see Figs. 9.2.5 and 9.2.6). The controller gain for stability must be set inversely proportional to this process gain [as is shown in Equations 9.2(17) through 9.2(21)]. Therefore, changes in the operating pH require drastic changes in controller tuning. Even if a controller has a low enough gain to provide stability on the steepest portion of the titration curve, its response to upsets elsewhere will be so sluggish that the controller will only be able to handle disturbances that last over days. Such a controller response approaches the behavior of an integral-only mode and can be viewed more as an optimizer rather than a regulator, which should be the first line of defense against disturbances.

Seemingly insignificant disturbances are magnified by the steep portion of the titration curve, as shown by a small oscillation in the abscissa resulting in a large oscillation in the ordinate of Figure 9.2.2. The oscillations in the abscissa could be caused by an upset in the influent flow, influent concentration, reagent pressure, reagent concentration, or control valve dead band, or by the controller's reaction to noise. Even if the influent conditions were truly at steady state, just the commissioning of a pH loop can cause unacceptable fluctuations in pH if the setpoint is on the steep portion of the titration curve.

For an influent wastewater which is received with a pH between 0 and 6, a valve or other final element with a rangeability of 10,000,000:1 and with a precision of better than 0.00005% is required to control the neutralization process within 1 pH of setpoint for the titration curve of Figure 9.2.2. Because a single valve can not provide this control, it is necessary to have three stages of neutralization with split-ranged valves.

If the total loop dead time was zero, which also implies zero valve dead band, and if the control valve trim characteristics and positioning were perfect, and if the measurement error and noise was zero, control using a single valve would be possible. Perfect control in general is possible only in a loop having no dead time and no instrument error. Such a loop could immediately see and correct for any disturbance and would never stray from setpoint. While such perfect control is not possible, it does

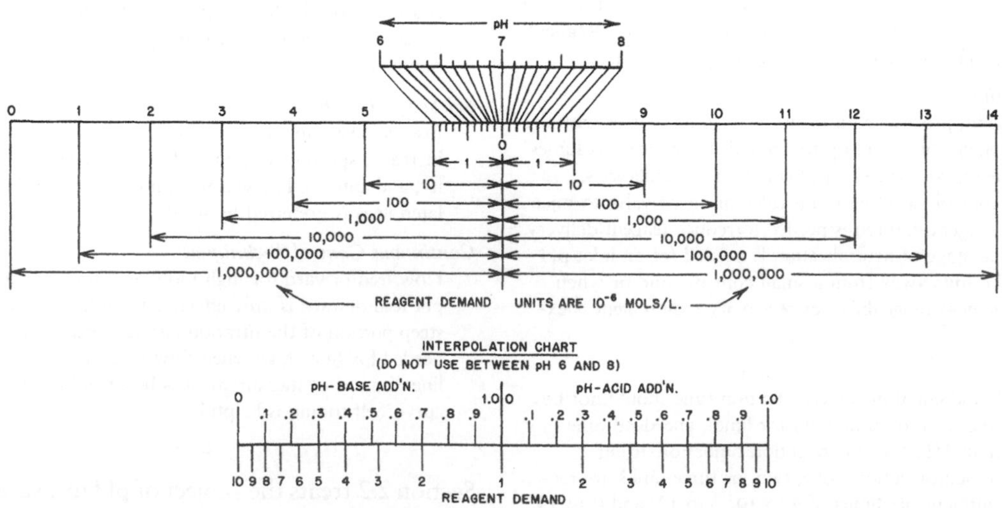

FIG. 9.2.1 pH versus reagent demand: strong acid–strong base

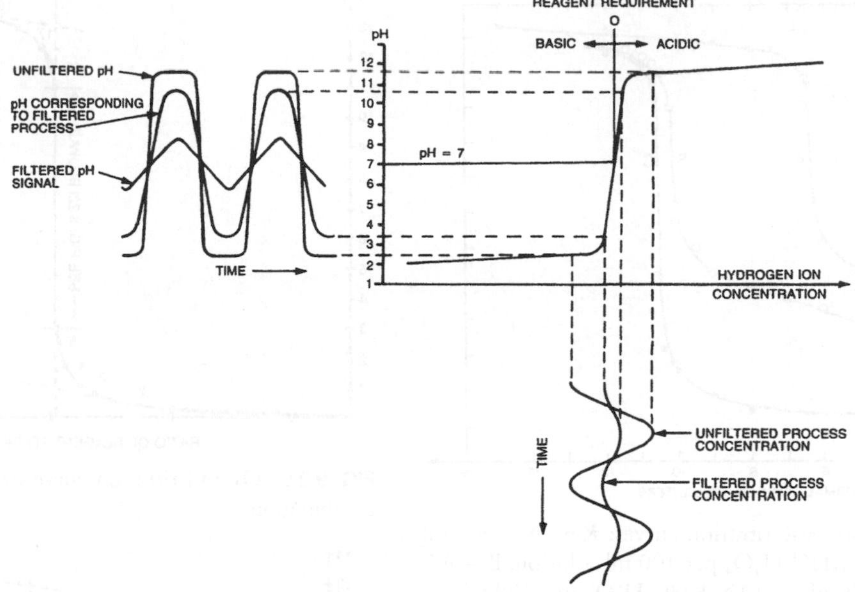

FIG. 9.2.2 The process gain of the neutralization process drops by a factor of 10 for each unit of pH from neutrality.

demonstrate that the goal for extremely tough loops such as the pH loop shown in Figure 9.2.2 should be to reduce dead time and instrument error as much as possible. As the dead time approaches zero, the detrimental effects of high process sensitivity and nonlinearity are also greatly reduced.

Nonlinear Controllers

Special controllers have been developed to compensate for the nonlinearity of most pH neutralization processes. These nonlinear controllers change their gain characteristics proportionally to the ion load (pH) of the process. The characteristics of the controller are as shown in Figure 9.2.3. The diagonal line represents the error-output relationship for the controller (in response to an error, a corrective signal is generated—the output—which eliminates the deviation from setpoint) with a 100% proportional band (gain = 1.0) without the nonlinear adjustments available with this controller. The first available adjustment is a slope adjustment that allows the proportional band to be increased (gain reduced) about the zero deviation point by a factor of 50. This means that when the gain setting of the controller is 1.0 (100% proportional band), the effective proportional band is 5000%, or a gain of 0.02 (insensitive controller) at the zero deviation point. The slope can be adjusted manually or by an external signal. The second adjustment is the error deviation range, over which the slope adjustment is operative; this is referred to as dead band.

The dead band is adjustable from 0 to ±30% error (deviation from setpoint). This latter feature allows the gain of the control loop to be adapted proportionally to the ion load. If the process to be controlled resembles Figure 9.2.4, a reagent flow rate or valve position signal can automatically adjust the dead band. At high ion loadings (curve A) the controller gain will be low, a desirable condition when the process gains are high. At lower ion loadings (curve B) the dead band can be reduced, thereby increasing the gain of the controller, a condition that is desirable when the

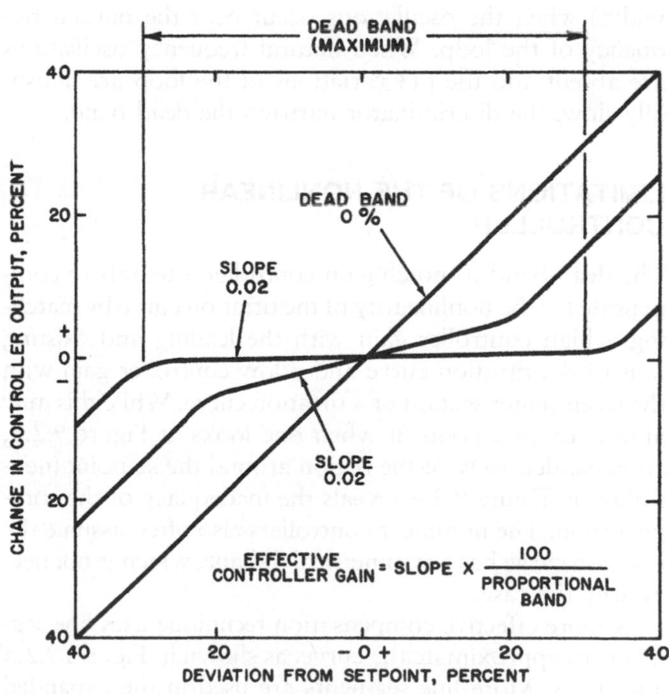

FIG. 9.2.3 Nonlinear controller characteristics.

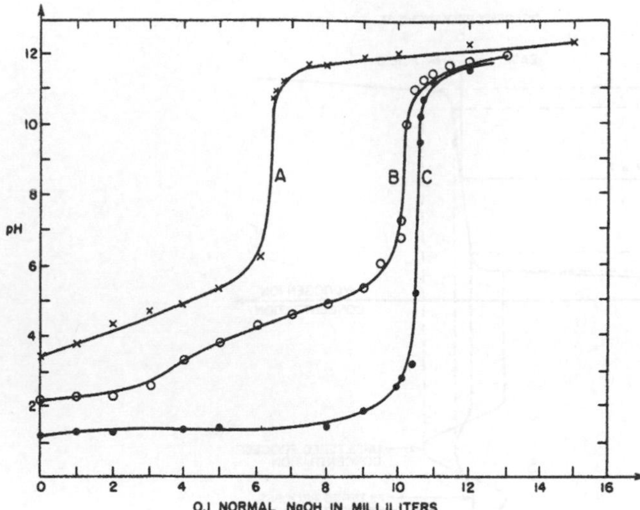

FIG. 9.2.4 Typical acid–base titration curves: *Key:* A = 9.9 ml HCl + 50 ml of 0.1N $KHC_8H_4O_4$ per 100 ml solution; B = 6.7 ml of 0.1N HCl + 50 ml of 0.1N $KHC_8H_4O_4$ per 100 ml solution; C = 100 ml of 0.1N HCl.

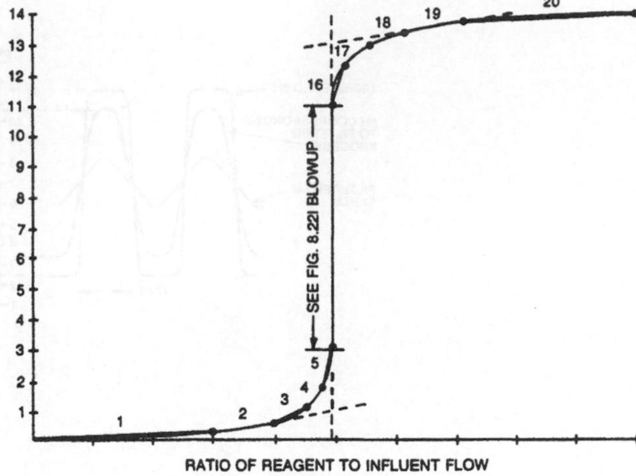

FIG. 9.2.5 Overall titration curve of a strong acid strong base combination.

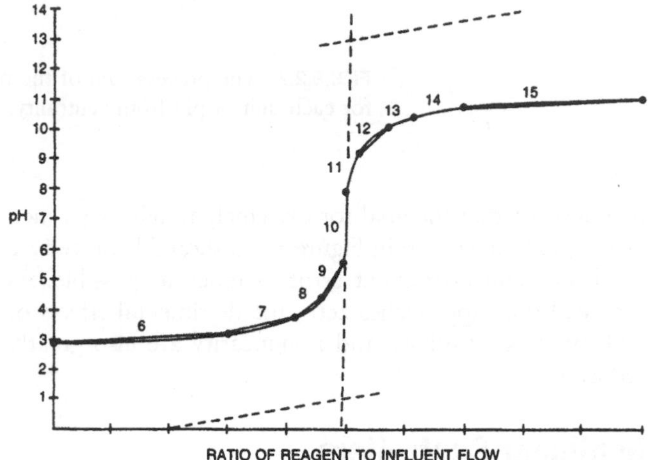

FIG. 9.2.6 Expanded view of the titration curve in Figure 9.2.5.

process and valve gains are low. The effectiveness of this type of controller and the benefits achieved by adapting the control loop characteristics to those of the process have been demonstrated on operating installations.

Adaptive controllers are also available for the automatic adjustment of the dead-band width, based on the condition of the pH loop. When the adaptive controller notices that the pH is cycling, it slowly widens (integral action only) the dead band and thereby extinguishes the oscillations after a few cycles. To do this the controller uses a discriminator, which increases its output (the dead-band width) when the oscillations occur near the natural frequency of the loop. When natural frequency oscillations are absent and the pH variations of the loop are unusually slow, the discriminator narrows the dead band.

LIMITATIONS OF THE NONLINEAR CONTROLLER

The dead-band or notch-gain controller attempts to compensate for the nonlinearity of the titration curve by matching a high controller gain with the leading and existing tails of the titration curve and a low controller gain with the steep center section of a titration curve. While this may appear to be a good fit when one looks at Figure 9.2.5, an expanded view of the region around the setpoint (neutrality) in Figure 9.2.6 reveals the inadequacy of the compensation. The nonlinear controllers also often assume the titration curve has a symmetrical S shape, which is not necessarily the case.

A more effective compensation technique uses line segments to approximate the curve, as shown in Figures 9.2.5 and 9.2.6. More line segments are used in the expanded view near the setpoint. Line segments are preferred to polynomials because smooth transitions from one polynomial to another are difficult to obtain and high order polynomials often have small humps and gain reversals which prevent the use of high controller gains or the use of rate action. The line segments are used to compute the abscissa (the ratio of reagent flow to influent flow) from the ordinate (pH). The abscissa is used as the controlled variable. If the curve is constant and accurate, the process gain becomes 1. Combined with a linear final element and a constant feed, the result is a nearly linear loop. The curve can be adjusted for temperature effects. The pH setpoint also goes through the same line-segment calculation. The display for the operator can be kept in units of pH, by the use of an output tracking strategy which provides a dummy controller for the operator interface.

Process Equipment and Reagent Delivery

Composition processes, whether pH or "pIon" (such as pCl and pAg), should be recognized as having two distinct aspects, one chemical and the other physical.

Several physical or process design considerations are associated with composition (pH or pIon) control applications. The most important is the primary device used to mix the reagent with the process stream. This can be as simple as reagent addition upstream of a pump or in-line mixer with a downstream measurement point or as complex as two mixed reaction vessels followed by an attenuation vessel.

Where reaction vessels are required, design decisions must be made to determine (1) size and number, (2) baffling, (3) agitation (how much and what type), (4) measurement probe location(s), and (5) reagent addition point location.

The other and equally important physical aspect of pH control is the design of the reagent delivery system. Both of these aspects will be discussed here.

The ease or difficulty of most industrial control applications is closely related to a property of the process referred to as dead time. Analogous terms, such as "transport time," "pure delay," and "distance velocity lag," describe the same effect. Dead time is defined as the interval between the introduction of an input disturbance to a process and when a measuring device *first* corrects for the effect of that disturbance. Qualitatively, the relationship between dead time and controllability is simple: The more dead time, the more difficult the problem of control. The presence of dead time in pH or pIon processes is extremely detrimental to controllability. The major reason is the severe sensitivity of the measurement of interest at the control point. One of the major goals of system design is to eliminate the dead time or to reduce it to an absolute minimum.

REAGENT DELIVERY SYSTEMS

Reagent addition requirements can be handled in diverse ways, depending on the process loads (flow of material to be neutralized) into the neutralization facility and the variation of the hydrogen or hydroxyl ion concentration, or both, in that flow. It should be recognized at the outset that because of the logarithmic nature of the pH measurement, a pH change of one unit can cause a tenfold change in load, whereas a 100 to 300 GPM (378 to 1134 l/m) change in flow (assuming no change in pH) is only a threefold change. Thus, the consequences of flow variations in waste streams can be relatively minor in comparison with ion concentration variations.

The equipment used to deliver reagents to the process under automatic control includes a metering device or a control valve. Metering pumps as a choice for reagent delivery are very accurate; however, delivery rangeability capability is limited to approximately 20:1 if speed is manipulated. Both speed and stroke can be manipulated to yield 200:1 rangeability, but the resulting relationship is squared and may require characterization. This means that where speed alone is manipulated, pH variations for a strong acid–strong base reaction greater than ±0.65 will result in cyclic or inadequate control. (A pH change of 1.3 means a 20-fold change in reagent requirement.) When pH load variations are minor (0.4 or less) and flow variations are less than 4:1, the choice of a metering pump with speed control is sufficient.

Control valves, like the metering pump, have limited rangeability. In this category two types of internal plug forms are usually considered for throttling service. They are the linear and the equal-percentage throttling characteristics. The term "equal percentage" means that the valve will produce a change in flow rate corresponding to a unit change in lift (valve plug movement), which is a fixed percentage of the flow rate at that point.

Both valves are available with minimum turndown of 50:1, and some have recently been developed (mainly in the smaller sizes) with claimed rangeabilities as high as 500 to 1.

Digital valves that can furnish rangeabilities of 2000:1 are also available. Cost, size, complexity, and materials of construction limit the application of these devices.

It is desired that the installed characteristic of the final element be linear so as not to introduce another nonlinearity into the pH loop. For control valves, the pressure drop available typically doesn't change much and is large compared to the system drop due to the low reagent flow rates normally associated with pH control. The result is an installed characteristic close to the inherent characteristic. Consequently, linear trim is preferred over equal-percentage trim for most pH applications to provide a more constant gain. The actual gain deviates from the theoretical constant gain, particularly at valve openings exceeding 80%. Low reagent flow (C_v less than 2.0) can cause valve sizing problems since the flow may not be completely turbulent. When the flow is viscous, the fully turbulent C_v for a control valve should be multiplied by a dimensionless coefficient Fr, which is a function of Reynolds number for the valve and which reduces valve capacity when viscosity is high. Since the hydraulic friction losses in the valve external to the trim can be assumed to be negligible, the flow is fixed by conditions within the trim and by the geometry of the plug and seat. A Mikroseal packless valve (available from H.D. Baumann Assoc., Inc.) having maximum C_vs between 0.0006 and 0.7, forces laminar flow. Such valve designs provide high rangeability but also nonlinearity, since flow is proportional to the third power of valve position. An accurate current-to-air (I/P) converter and output signal characterization within a microprocessor-based controller is needed to linearize the valve characteristic.

Valve Specification

The maximum valve capacity, rangeability, and precision requirements can be found by determining the parameters A and B from the titration curve, shown in Figure 9.2.7 for a setpoint on the flat portion and in Figure 9.2.8 for

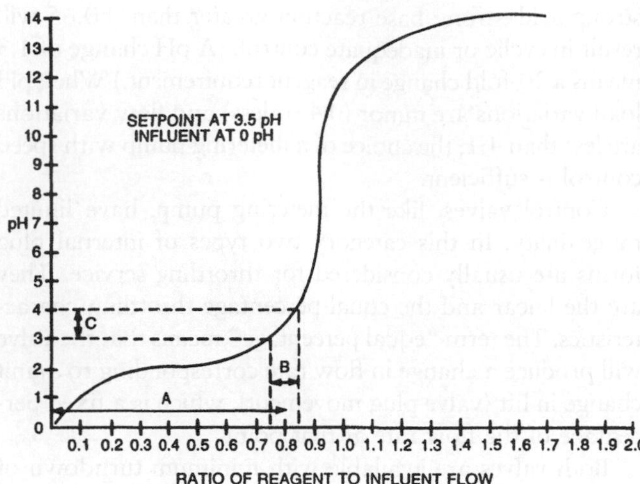

FIG. 9.2.7 Data required for reagent control valve specification when the pH setpoint is in a flat portion of the titration curve.

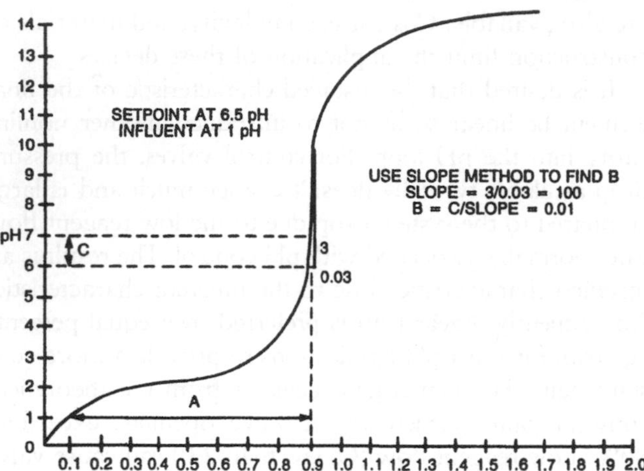

FIG. 9.2.8 Control valve specification data for application with pH setpoint in the steep portion of the titration curve.

a setpoint on the steep portion of the titration curve and by inserting them into Equations 9.2.1, 9.2.2, and 9.2.3. Parameter A is the distance along the x axis from the influent pH to the setpoint. It is multiplied by the maximum influent flow for a given operating condition. The $(A_{max})(F_{imax})$ combination that yields the largest product (increased by 25% to improve the valve gain uniformity) is used to set the maximum capacity for sizing the control valve. The maximum rangeability of reagent flow (R_{rmax}) is obtained by dividing the maximum product pair by the minimum product pair. The precision of the control valve in terms of its repeatability and dead band requirements is proportional to the ratio of the minimum B value to the maximum A value. The 80% value in Equation 9.2.3 corresponds to the 25% extra capacity used for sizing. For steep curves, B may be too small to estimate unless the titration curve is expanded in the control region (Figure 9.2.8).

$$F_{rmax} = (1.25)(A_{max})(F_{imax}) \quad 9.2(1)$$

$$R_{rmax} = (1.25)\frac{(A_{max})(F_{imax})}{(A_{min})(F_{imin})} \quad 9.2(2)$$

$$E_{rmin} = (80\%)\left(\frac{B_{min}}{A_{max}}\right) \quad 9.2(3)$$

where:

A_{max} = maximum abscissa distance influent pH → setpoint (ratio)
A_{min} = minimum abscissa distance influent pH → setpoint (ratio)
B_{min} = minimum control band translated to abscissa (ratio)
E_{rmin} = minimum control repeatability and dead band (%)
F_{imax} = maximum influent flow (gpm)
F_{imin} = minimum influent flow (gpm)
F_{rmax} = maximum reagent valve capacity for sizing (gpm)

Figure 9.2.9 illustrates the determination of the tolerable reagent flow variations. A strong acid has been added to water to achieve a pH of 6 and a pH of 2 (curves A and B). The reagent flow (10% NaOH) requirement for each solution is plotted on separate scales (lower and upper abscissas). Assuming a control specification of pH 7.0 ± 0.5, a reagent flow of ±28% variation can be tolerated when the pH of the inlet material is 6.0 (Curve B). When the pH of the inlet material is 2.0, the tolerable reagent flow variation is ±0.0028% (Curve A) and the problem is 10,000 times more difficult.

Valve Linearization

The drawback of using equal-percentage valves is a high gain contribution to the overall loop gain, especially at high reagent flows (sensitivity if a function of valve opening). One approach to countering this variable gain is by a characterizer with an input-output characteristic which is opposite that of the equal-percentage valve. This approach is illustrated in Figure 9.2.10. The resultant valve characteristic is approximately linear, which is highly desirable from an automatic control point of view, since the variable gain nature of the process makes the control problem difficult enough.

Using Multiple Valves

When the rangeability exceeds what can be obtained from a single control valve, a small and a large control valve can be split-ranged and sequenced so the transition point is in the usable throttling range of each valve. The split-range or switchover point should be chosen to keep the final element gain relatively constant. Equation 9.2(4) cal-

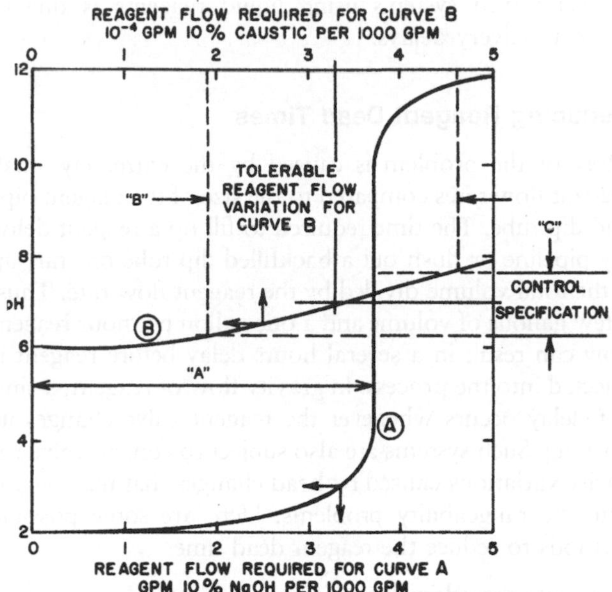

FIG. 9.2.9 Relationship of accuracy and rangeability to ion loading.

culates the switchover point for the more general case when the split-ranged valves control different types or concentrations of reagents. The split-range computation should be done in the microprocessor-based controller for greater accuracy, flexibility, and standardization of positioner calibrations. This way, separate outputs are created for each valve and displayed on the operator interface.

$$S = 100/\left(1 + \frac{(F_{2max})(C_2)}{(F_{1max})(C_1)}\right) \qquad 9.2(4)$$

where:

F_{1max} = maximum flow of reagent valve 1
F_{2max} = maximum flow of reagent valve 2
C_1 = concentration of reagent 1 (normality)
C_2 = concentration of reagent 2 (normality)
S = best split-range point (% of controller output signal)

Conventional Valve Sequencing

Conventional analog hardware can also be used for control valve sequencing. Sequencing of a pair of equal-percentage valves can achieve an overall rangeability approaching the product of the individual valves rangeabilities, e.g., 50 × 50 = 2500. The loss of rangeability is mainly caused by the amount of overlap between valves. A plot of the performance characteristics of this pair of valves is shown in Figure 9.2.11. The valve positioner of the smaller sequenced valve is calibrated for full stroke over 0 to 52% controller output signal (closed at 0%; fully open at 52%). The positioner of the larger sequenced valve is calibrated for full stroke over the range

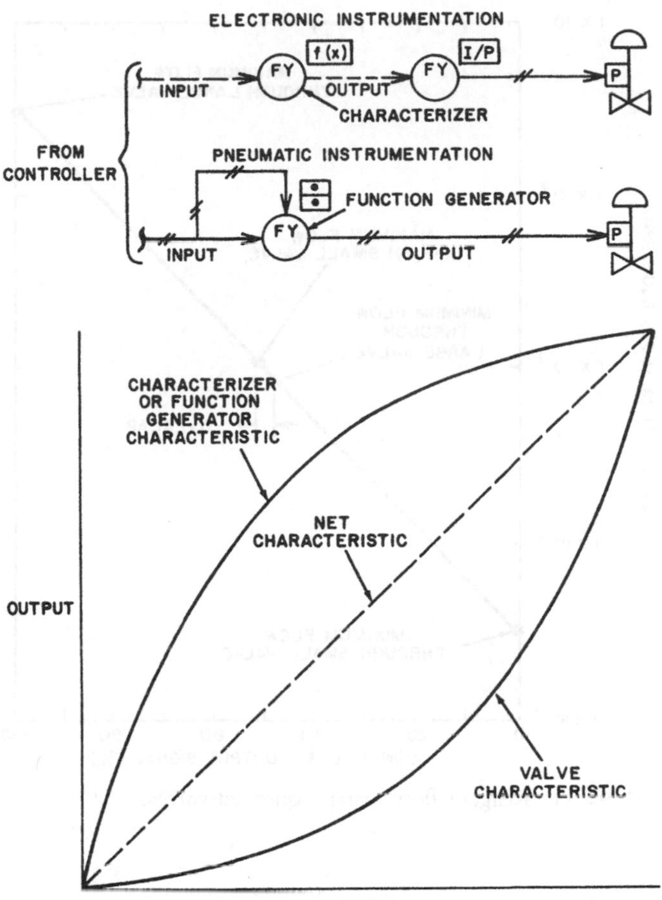

FIG. 9.2.10 Linearization of equal-percentage valves.

of 48 to 100% of controller output. Transfer between the valves (as the controller output changes) can be implemented with either pneumatic or electronic control elements. Since only one valve at a time is operating while the other valve is closed, the characteristic of the pair is equal percentage, as the semilog plot of Figure 9.2.11 illustrates. If the smaller valve, for example, were permitted to remain open when the larger valve came into service, the valve characteristic curve would have a discontinuity at the transfer point that could result in an unstable control system. There is a small and temporary flow transient at the transfer point, but the characteristic curve is maintained.

Sequencing of linear valves to provide an overall linear characteristic with wide rangeability is generally not satisfactory, because the transfer point occurs at 10% or less of the controller output signal. This means that the larger valve is essentially doing all the work. When using linear valves a high gain relay is also required to expand the controller output signal to operate the smaller valve. On the other hand, sequenced pairs of equal-percentage valves can be linearized the same way as can individual ones (Figure 9.2.10).

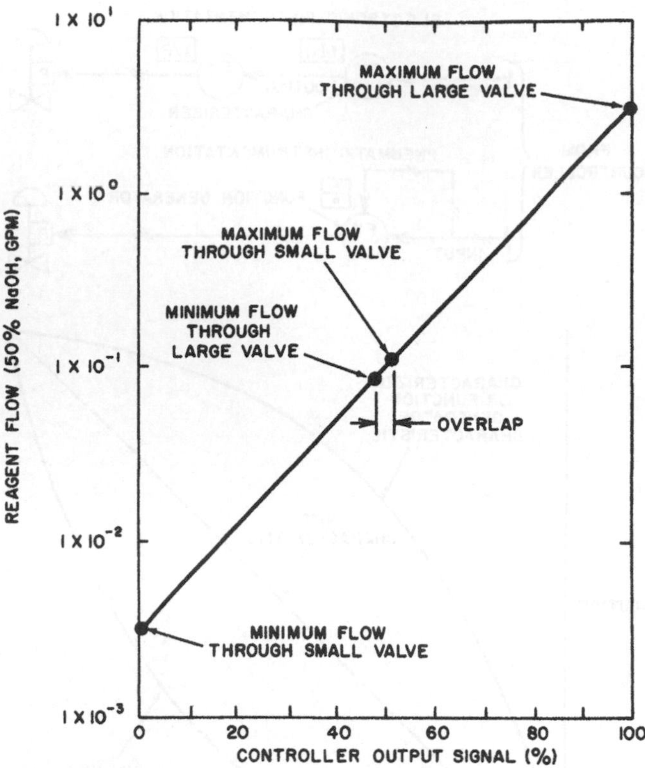

FIG. 9.2.11 Reagent flow using sequenced valves.

Linear valves with rangeabilities of 200:1 or greater can be sequenced successfully using conventional split-range techniques. The wide rangeability minimizes the step change in reagent delivery at the transfer point where the small valve remains open as the large valve begins to open. To counteract the large gain change at the transfer point, a characterizer must be used to ensure constant control loop gain.

Metering Pumps

The specifications for metering pumps (Section 2.1) should be carefully checked for rangeability and precision. The pump speed and stroke should be electronically set rather than pneumatically. The mechanism should be chosen that requires the least maintenance and meets the precision and rangeability requirements. Many flexible diaphragm pumps have a repeatability of only 2% and a turndown of 20:1. The pump should be located as close as possible to the injection point. Since this is usually more difficult for pumps than for control valves, greater reagent delivery delays (dead time), especially on start-up, are experienced with metering pumps. Also, it is necessary to maintain a backpressure on the pump in order to sustain accuracy. Consequently, pumps with a free discharge (reagent dumped on the liquid surface in an atmospheric vessel) should be provided with a backpressure regulator on the pump discharge. The biggest source of dead time in well-mixed systems using liquid reagents is due to reagent delivery delays.

Reducing Reagent Dead Times

Most of the problem is caused by the extremely small reagent flow rates compared to the size of the reagent pipe and dip tube. The time required to fill up a reagent delivery pipeline or flush out a backfilled dip tube on start-up is the total volume divided by the reagent flow rate. Thus, a few gallons of volume and a one gallon per hour reagent flow can result in a several hours delay before reagent is injected into the process. In gravity flow of reagent, a similar delay occurs whenever the reagent valve changes its opening. Such systems are also subject to control valve capacity variations caused by head changes that may aggravate the rangeability problems. Here are some possible methods to reduce the reagent dead time:

1. Locate throttle valve at injection point.
2. Mount on-off valve (preferably ram type) at injection point.
3. Reduce diameter and length of injector or dip tube.
4. Add a reliable check valve to injector or dip-tube tip.
5. Dilute the reagent upstream.
6. Inject the reagent into vessel side just past baffles.
7. Inject reagent into recirculation line at vessel entry point.
8. Inject reagent into influent line at vessel entry point.

Reagents such as ammonium hydroxide, calcium hydroxide, and magnesium hydroxide are weak bases and have pKa values close enough to many control bands to provide some flattening of the titration curve and some increased controllability from the standpoint of reduced sensitivity to disturbances. However, ammonium hydroxide can flash and create gas bubbles that escape from vessels or choke in-line systems or travel downstream undissolved. Calcium hydroxide (lime) and magnesium hydroxide in solid form may take 15 to 30 minutes to dissolve and in slurry form 5 to 10 minutes to dissolve. This slowness of reagent response adds tremendous dead time to the system. Also, the residence time must be 20 times greater than the dissolution time to insure that less than 1% remains undissolved in the effluent. Consequently, large volumes must be used for pH control, which further increase the loop dead time.

Reagent Delivery Hysteresis

Consider a reagent delivery device such as a control valve, a metering pump, or a dry feeder. The smallest incremental change that these devices can make is approximately 1%. Converted to the logarithmic pH scale and using an influent pH of 14 and a setpoint of 7, 1% excess acid produces a pH of 2, and 1% too little yields pH 12. These values were derived from Figure 9.2.1, where 1×10^6

reagent units are required to neutralize pH 14 to 7. One percent of this total is 10,000 reagent units, which correspond to pH 2 and 12.

In a similar fashion, the effect of the same error can be estimated for any other setpoint. Using a setpoint of 12, for example, $1 \times 10^6 - 10{,}000$, or 990,000, reagent units are required for neutralization from a 14 pH influent. One percent excess acid (9900 reagent units) corresponds to 10,000 − 9900, or 100 reagent units, which is pH 10. One percent too little acid corresponds to 19,900 reagent units, or approximately 12.3 pH using the interpolation chart on Figure 9.2.1.

The same procedure is used to illustrate the sensitivity of the process to hysteresis. Increasing the error to 1.5% for a setpoint of 12 gives a low pH of 3.13 and a high pH of 12.39 (10,000 − 14,850 = 4850 on the acid side and 10,000 + 14,850 on the caustic side).

Methods of reducing valve hysteresis, such as pulse interval control and the uses of digital valves, have been proposed. Although these techniques add cost and complexity to the control system, they should be investigated as alternatives to the installation of stirred tanks.

If hysteresis cannot be eliminated, it can profoundly influence pH loop performance unless some other element can be introduced to smooth out this incremental response or, in effect, to reduce the gain of the control loop.

MIXING AND AGITATION

The two types of mixing that are important to the control system are intermixing and backmixing. The reagent must be intermixed with the process stream to furnish complete elimination of the areas of unreacted reagent or untreated influent. Adequate intermixing between influent and reagent can be readily achieved by adding the reagent at a point of small cross-sectional area where there is some turbulence. Figure 9.2.12 illustrates the reagent being added in the pipeline before the influent enters the treatment facility. This is a desirable practice because it eliminates poor intermixing, which can cause a noisy signal to be observed in the effluent pH. A loop seal arrangement, particularly when long reagent transfer lines are required, allows the reagent line to remain full up to the point of introduction to the process and thus eliminates a potential source of process dead time.

Backmixing is more important than intermixing for close pH control. The treated stream must be held in a vessel sufficiently long for the reagent to react and be backmixed. In general, the degree of backmixing can be defined in terms of the pumping capacity of an agitator with respect to the flow and volume of the neutralization vessel. In practice, however, this definition has limited usefulness because of variables such as agitator construction and blade pitch, baffling of the neutralization vessel, and placement of inlet and outlet measuring electrodes. Experience shows that the best way to define backmixing for control purposes is by the ratio of the system dead time to retention time of the neutralization vessel. The retention time is the volume of the vessel divided by the flow through the vessel. A ratio of dead time to retention time equal to 0.05 is adequate for good control.

Suitable baffles or agitator positioning should be used in mixed neutralization vessels to avoid a whirlpool effect. The power supplied by the impeller must be used to turn the contents of the vessel over, not to whirl them about. With these effects in mind, a propeller or an axial-flow impeller should be selected to direct the flow of the vessel contents toward the bottom of the tank. The flat bladed radial-flow impeller should be avoided, since it generally tends to divide the vessel into two sections and increases system dead time.

Figure 9.2.13 is a plot of tank size against agitator pumping capacity per unit tank volume on logarithmic coordinates. The family of curves shown for various dead

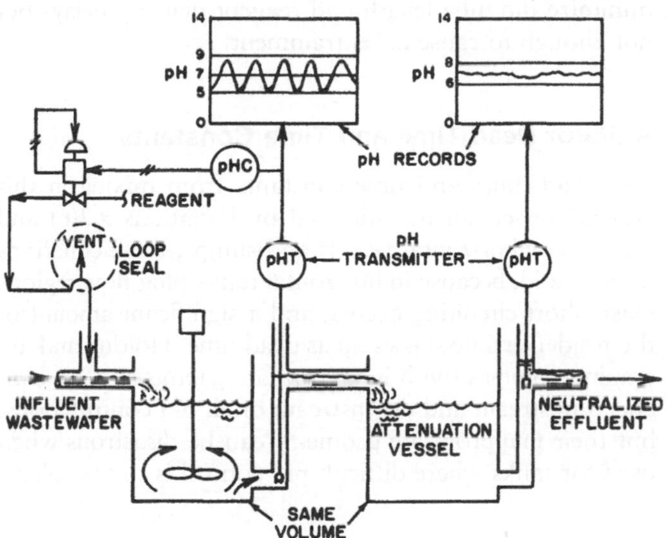

FIG. 9.2.12 Effect of attenuation vessel.

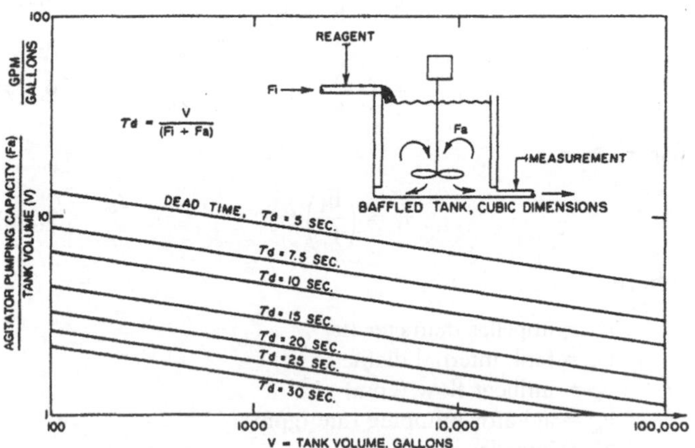

FIG. 9.2.13 Dead time (τ_d) as a function of mixing intensity.

times was developed from empirical data in tanks with capacities of 200, 1000, 10,000, and 18,000 gallons (756, 3780, 37,800, and 68,040 1). They apply to baffled tanks of cubic dimensions with the inlet at the surface and the outlet at the bottom on the opposite side of the tank. The ratio of impeller diameter to tank diameter varies from 0.25 to 0.4. Square pitch propellers at an average peripheral speed of 25 fps were used in up to 1000-gallon capacity tanks. Axial-flow turbine impellers at an average peripheral speed of 12 fps (3.6 m/s) were used in the larger tanks.

To be classified as a well-mixed vertical tank by the standards of pH control applications, the liquid height should be between 100% and 150% of the vessel width or diameter. The vessel walls should have baffles to prevent liquid rotation, the agitation pattern should be axial, and the agitator pumping rate calculated by Equation 9.2(5) should be at least 20 times the influent flow rate. The agitation should be great enough to break the surface and pull down the reagent (injected near the surface to minimize dip-tube length and reagent delivery delay) but not enough to cause air entrainment.

Agitator Dead Time and Time Constants

The dead times and time constants from mixing in this type of vessel can be estimated by Equations 9.2(6) and 9.2(7). For horizontal vessels and sumps, these equations do not hold, because in horizontal tanks plug flow regions exist, short circuiting occurs, and a significant amount of the residence time shows up as dead time. Holdup and averaging volumes (with inappropriate geometries) are beneficial upstream and downstream of all pH control loops, but their inappropriate geometry can be disastrous when used for tanks where difficult pH control is to take place.

$$F_a = (7.48)\frac{0.4}{\left(\frac{D_i}{D_t}\right)^{0.55}}(N_i)(D_i^3) \qquad 9.2(5)$$

$$\tau_d = \frac{V}{F_i + F_a} \qquad 9.2(6)$$

$$\tau_1 = \frac{V}{F_i} - \tau_d \qquad 9.2(7)$$

for $\tau_d \gg \tau_i$

$$\tau_1 = \left(\frac{F_a}{F_i}\right)(\tau_d) \qquad 9.2(8)$$

where:

D_i = impeller diameter (ft)
D_t = tank internal diameter (ft)
F_i = influent flow (gpm)
F_a = agitator pumping rate (gpm)
N_i = impeller speed (rpm)
τ_1 = mixing time constant (minutes)
τ_2 = mixing dead time (minutes)
V = vessel liquid volume (gallons)

Unfortunately, for steep titration curves most of the mixing time constant is lost due to acceleration of the pH measurement. Figure 9.2.14 shows how a 19-minute time constant is reduced to 0.04 minutes for a strong acid and strong base system by translating the points of 63% and 100% of the pH change for an upset to the abscissa. By the translation of the controlled variable from pH to the abscissa of the titration curve, linearization of the pH signal can restore the time constant to its original value.

CONTROL DYNAMICS

The performance of a stirred tank to periodic disturbances can be evaluated by consideration of the dead time and time constant properties of the tank.

For example, if the total system dead time is τ_{dt}, it can be defined as:

$$\tau_{dt} = \tau_{d1} + \tau_{d2} \qquad 9.2(9)$$

where:

τ_{d1} = tank dead time, inlet to outlet
τ_{d2} = remaining loop dead time (sampling system and control valve motor)

Given

$$\tau_{dt} = 0.05\ V/F \qquad 9.2(10)$$

where:

V = vessel volume
F = flow through vessel

The time constant (τ_1) for an agitated vessel with dead time (τ_{d1}) can be expressed as:

$$\tau_1 = V/F - \tau_{d1} \qquad 9.2(11)$$

Assuming that the stirred tank has the minimum 3.0-minute time constant previously mentioned and that the total dead time is divided 80% to (τ_{d1}) and 20% to τ_{d2}, Equation 9.2(11) can be restated:

$$\tau_1 = 0.96\ V/F \qquad 9.2(12)$$

Expressing τ_1 in terms of dead time by combining Equations 9.2.10 and 9.2.12:

$$\tau_1 = 19.2\ \tau_{dt} \qquad 9.2(13)$$

The dynamic gain of a stirred tank to periodic disturbances is given by Equation 9.2(14):

$$G_d = \frac{\tau_0}{2\pi\tau_1} \qquad 9.2(14)$$

where:

G_d = dynamic gain of the stirred tank 5
$= \frac{\text{percent change in output}}{\text{percent change in input}}$

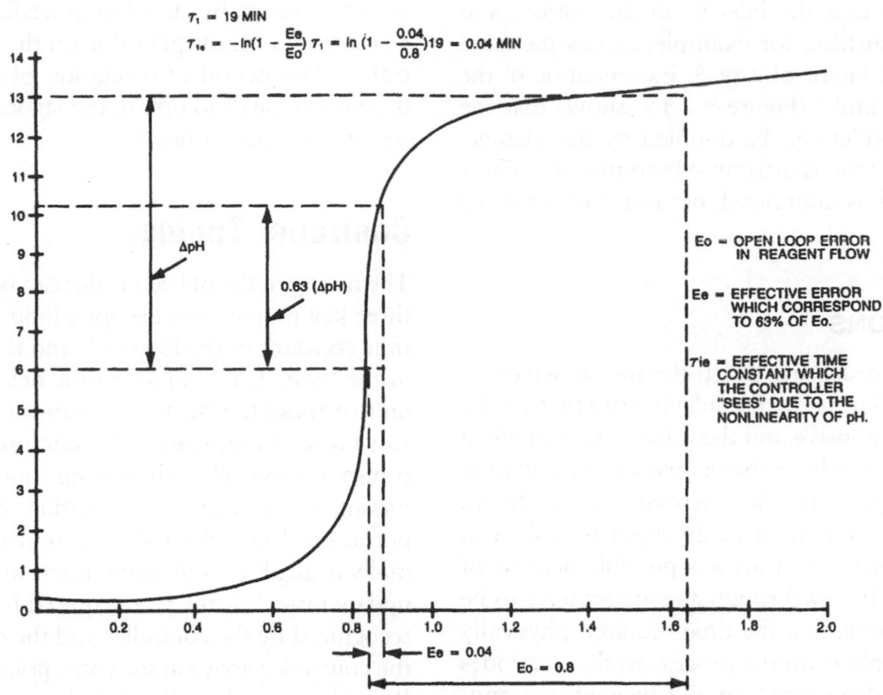

FIG. 9.2.14 When the titration curve is steep, the mixing time constant is much reduced (from $\tau_1 = 19$ min to $\tau_{1e} = 0.04$ min) due to the acceleration of the pH measurement. Linearization (Figures 9.2.5 and 9.2.6) restores the time constant.

τ_0 = period of oscillation of the disturbance
τ_1 = first-order time constant of the tank; approximately equal to (tank volume/flow through the tank system dead time)

To visualize the effect of dynamic gain, consider a flowing stream whose pH falls from 7 to 4 and returns to 7 in one minute. If the stream flowed through a tank with one minute retention time (volume/flow), the spike in pH would pass through virtually unchanged, and the effluent pH would closely track the influent pH. If, however, the stream flowed through a tank with 60 minutes retention time, practically no upset would be observed in the effluent pH because of the capacity effect of the large volume.

The period of oscillation, τ_0, of a typical composition process under closed-loop control with an optimally tuned (controller settings adjusted to match the process it controls) three-mode controller can be approximated as a function of the system dead time.

$$\tau_0 \simeq 4\tau_{dt} \qquad 9.2(15)$$

Substituting for τ_1 from Equation 9.2(13) and τ_0 from Equation 9.2(15) into Equation 9.2(14):

$$G_d = \frac{4\tau_{dt}}{2\pi(19.2\tau_{dt})} = 0.033 \qquad 9.2(16)$$

In this example the stirred tank has reduced the overall process gain by a factor of 30 (1/0.033). Two tanks used in series reduce the process gain (slow the process down) by the product of their individual gains. Assuming a second tank identical to the first, two tanks in series would reduce the process gain by a factor of 30^2, or 900. With the stirred tank, therefore, it is possible to reduce the process gain to a controllable level. An added benefit of an increased tank capacity is to smooth out high-frequency errors in reagent delivery caused by measurement noise.

This example is readily related to Figure 9.2.12, in which the output of the reaction vessel is the input disturbance in the attenuation vessel. If the frequency or period (τ_0) of the input disturbance can be kept short (on the order of seconds)by virtue of a *tight* control loop around the reaction vessel, then the dynamic gain number of the attenuation vessel will be very low (0.033 for the example), thereby increasing its attenuation capability. This results in a stable effluent pH that averages the input disturbance.

TANK CONNECTION LOCATIONS

The inlet and outlet in the treatment vessel should be located at opposite sides—one high and one low—with respect to the bottom of the tank. Generally, it is most convenient to introduce the influent stream on the surface of the tank and to locate the outlet at the bottom of the vessel.

Variations in the location of the inlet and outlet can considerably change the dead time. Reversing the flow through the tank so that the inlet is on the bottom and the outlet is at the surface, for example, causes the dead time to increase by a factor of 2 or 3. Examination of the flow patterns in the tanks (Figure 9.2.15) shows that the path from inlet to outlet can be doubled by this change. The additional dead time is attributable to the swirl effect of the agitator, which is minimized, but not eliminated, by baffling.

SENSOR LOCATIONS

The location of the measuring electrodes also deserves serious consideration. The general guidelines are that the locations should be responsive and the information supplied by them should be timely. Submersible or recirculation pipeline insertion type electrode assemblies are preferred when the measurement is used as an input to a control system. This preference is not always possible because of physical constraints. If flowthrough assemblies have to be used, the sampling time, i.e., the time required physically to transport the sample from the process to the electrodes (which is essentially dead time), should be kept to a minimum. Figure 9.2.12 shows a submersible-type assembly on the reaction vessel located as close as possible to the vessel exit. Location within the tank proper increases the measurement noise, principally because of concentration gradients. The requirements of the monitoring electrodes shown on the attenuation vessel are not as severe. Either flowthrough or submersible detectors can be used. The information supplied by these electrodes provides a clean record for any regulatory agencies involved.

EQUALIZATION TANKS

Upstream of a stirred neutralization vessel, a lagoon or a holding tank can be very useful because it serves to smooth out upsets in influent pH and flow, thus allowing the use of a simple feedback system rather than a more costly feedforward control system. A lagoon can also be used to store the material that is bypassed around the neutralization process in case of failure, a very important consideration if off-specification effluent causes a plant shutdown. The one thing that a lagoon cannot do is replace a mixed vessel as part of a control system. Any attempt to control the pH of a lagoon by closed-loop feedback control can only result in an effluent pH value on the opposite side of neutrality. The period of oscillation of such pH swings will depend on the dead time of the lagoon, but typically it will be on the order of hours.

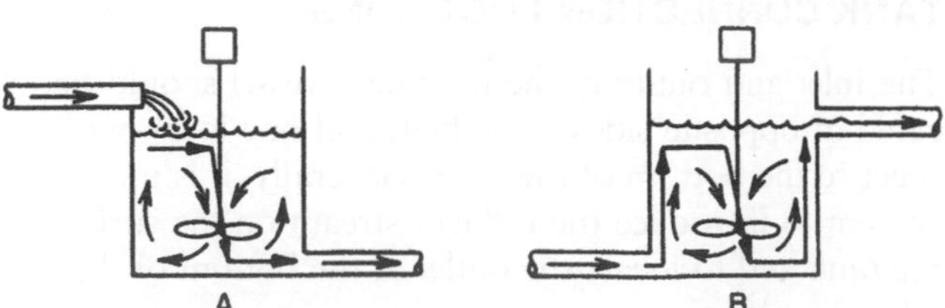

FIG. 9.2.15 Flow patterns in stirred tanks. *A.* Recommended flow path. *B.* Undesirable flow path.

Controller Tuning

The tuning of the pH controller can be approximated from three key parameters: the open-loop gain (Ko), the largest time constant of the loop (τ_1), and the total dead time (τ_d) in the loop. The total loop time delay is the most important of these terms. It is the sum of the dead times from valve dead band, reagent dissolution time, reagent piping transportation delay, the mixing equipment turnover time, mixing equipment transportation delay, sample transportation delay, electrode lag, transmitter damping (normally negligible), and digital filters and digital system scan update time. It is the time required for a disturbance to be recognized by the controller and the corrective reaction by the controller arrive at the entry point of that disturbance. Regardless of where the disturbance enters, the total loop time delay is the time it takes the disturbance effect to traverse the loop in Figure 9.2.16. The controller integral time and derivative time settings depend upon the loop dead time as shown in Equations 9.2(18), (19), and (21). The largest time constant slows down the excursion and gives the controller time to compensate for the upset. The largest time constant in a well-designed installation which is in excellent working condition and is provided with substantial back-mixed volumes is the process time constant (τ_1). The controller gain is proportional to the ratio of this time constant to the loop dead time (τ_d) multiplied by the open-loop gain (Ko) per Equation 9.2(17), if the dead time (τ_d) is less than the time constant (τ_1). Kc is proportional to the open-loop gain (Ko) per Equation 9.2(20) for systems where the dead time is greater than the time constant.

$$Kc = (\tau_1/\tau_d)/Ko \qquad 9.2(17)$$

$$Ti = 2\ \tau_d \qquad 9.2(18)$$

$$Td = 0.5\ \tau_d \qquad 9.2(19)$$

For $\tau_d > \tau_1$ and when using a PI controller:

$$Kc = 0.3\ Ko \qquad 9.2(20)$$

$$Ti = 1.0\ \tau_d \qquad 9.2(21)$$

where:

Kc = controller gain
Ko = the open-loop steady-state gain (dimensionless)
τ_1 = largest time constant with titration curve effect (minutes)
τ_d = total loop dead time (minutes)

Td = derivative time setting (minutes)
Ti = integral time setting (minutes/repeat)

DEAD TIME EXCEEDING TIME CONSTANT

The loop dead time can become larger than the largest time constant for in-line mixer installations, because these units provide mostly axial mixing instead of back-mixing. τ_d can also exceed τ_1 in several situations: in poorly mixed tanks; when the reagent dip tubes are poorly designed; in systems where transportation time exceeds turnover time; and when electrodes are improperly located or severely fouled. τ_d can exceed τ_1 in seemingly well-designed and well-mixed vessels also if the setpoint falls on a particularly steep section of the titration curve because (as was illustrated in Figure 9.2.14), most of the time constant is lost due to rapid movement of pH. Under such conditions, the measurement can actually accelerate and the process can appear to be non-self-regulating to the controller. For this case, dead-time dominance causes the window of allowable gains to close, and loop instability occurs regardless of tuning. Also, for non-self-regulating processes, it is important to maximize derivative or rate action and minimize the use of integral or reset action. Neither of these steps is possible when dead time exceeds the time constant in the loop.

PID CONTROLLER TUNING ($\tau_d < \tau_1$)

Tuning of the controller must be done at the normal setpoint and under the normal operating conditions due to the potentially severe nonlinearity of the process, as illustrated by the titration curve. Time constants predicted from equipment volume and agitator pumping rate must be corrected for the effect of the titration curve as shown in Figure 9.2.14 per the discussion in the section on mixing equipment. Open-loop tests will show the effect. Closed-loop methods of tuning, such as the Ziegler-Nichols ultimate oscillation method, must be done carefully to distinguish the gain where oscillations first start, because the loop will show nearly equal amplitude oscillations for a wide range of gains as it bounces back and forth between the flat portions of the titration curve. Increases in amplitude are difficult to detect once the oscillation moves outside the steep slope region of the titration curve.

The peak error for continuous pH control is proportional to the ratio of dead time to time constant multiplied by the open-loop error (the error with the loop in manual), if the dead time is less than the time constant. If τ_d exceeds τ_1, the peak error is proportional to the full open-loop error. The integrated error is proportional to this peak error multiplied by the dead time. It is critical to mark these errors along the abscissa of the titration curve and translate them to the ordinate (pH axis). Peak errors will be much larger than expected for excursions along the steep slope of the curve.

BATCH CONTROLLER TUNING

For batch control, the offset from setpoint can be made smaller than the peak error for continuous pH control if the sequential requirements of batch pH control are recognized and addressed.

Batch pH control is analogous to the titration done in chemistry lab. If the student has enough patience to use sequentially smaller doses and to wait longer as the pH approaches the endpoint, the final pH can end up within the measurement error of the endpoint. The increased difficulty of continuous pH control could be simulated by cutting a hole in the side of the beaker and adding a sample of variable flow and concentration.

If three vessels are provided, which are sequenced to fill, treat, and drain influent, plus if sufficient processing time and a strategy for variable dosing which duplicates the lab titration procedure is provided, the results will give good batch control of pH. The processing time must be long enough and the reagent dose sizes must be small enough to provide several doses even when the target set-

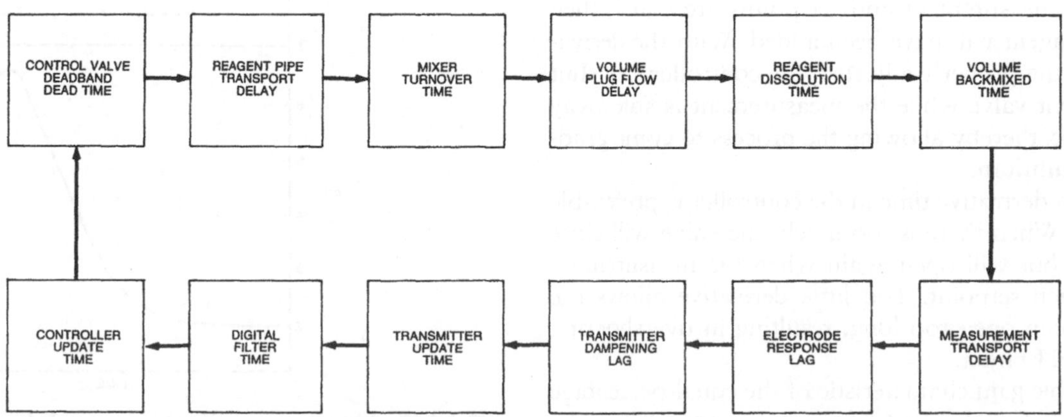

FIG. 9.2.16 The total loop time delay (τ_1) is the sum of all dead times and time delays (lags) in the loop.

point is on a steep section of the titration curve and the wait time between charging reagent doses exceeds the loop dead time as the pH approaches the setpoint. The use of integral action (PI or PID control) for reagent addition to a batch volume will cause overshoot and will necessitate cross-neutralization of acidic and basic reagents.

When the flow rate of material to be treated is reasonably small (perhaps less than 100 GPM, or 378 l/m), batch treatment may be a cost-effective pH control approach. As the flow rate increases, the tankage required rapidly shifts the economics in favor of a continuous pH control arrangement. Two unique characteristics of the pH batch process are:

1. The measurement (actual pH) and the setpoint (desired pH) are away from each other most of the time.
2. When the measurement and setpoint are equal (endpoint), the load on the process (reagent requirement) and, hence, the controller output are zero.

The controller characteristics for the batch pH control application should be proportional plus derivative. Reset must not be used, since reset windup will result in overshoot of the controlled variable. In a proportional controller, the corrective action generated is proportional to the size of the error; in a reset controller, to the area under the error curve; and in a rate controller, to the rate at which the error is changing. Once the measurement goes past the setpoint there is no way for the control system to bring it back to the setpoint, unless, of course, two controllers and two reagent supplies are used. In the absence of the reset control mode (proportional-only), a controller is usually supplied with a 50% bias so that the controller output is 50% when the measurement and the setpoint are equal. For the batch application with a proportional plus derivative controller, the bias must be 0% so that when measurement and setpoints are equal, the controller output is 0%.

The effect of secondary lags in the valve, process vessel, and measurement are compensated for by the derivative action of the controller. If, for example, reagent is added but its effect has not yet been seen by the pH electrode, when measurement and set points are equal, then too much reagent will have been added. With the derivative-time setting properly adjusted, the controller will shut off the reagent valve while the measurement is still away from setpoint, thereby allowing the process to come gradually to equilibrium.

Too much derivative time in the controller is preferable to too little. When there is too much, the valve will close prematurely but will open again when the measurement does not reach setpoint. Too little derivative allows the valve to remain open too long, resulting in overshooting the desired pH target.

The variable gain characteristic of the equal-percentage valve is an asset to this type of control system. When the measurement is far away from setpoint, the valve will be wide open, permitting essentially unrestricted reagent flow to the process. As the measurement approaches setpoint and the valve closes, the decreasing gain of the valve counters the increasing gain of the process. Figure 9.2.17 illustrates the measurement-valve behavior of the batch process.

Although the installation and process design considerations for the batch process are not as severe or demanding as the continuous operation, care should be taken to ensure that (1) adequate mixing is provided, (2) tank geometry precludes the existence of stagnant areas, (3) reagent delivery piping between valve and process is as short as possible, and (4) electrodes are placed in responsive locations.

Control Applications

The classical method of continuous pH control is a vertical, well-mixed tank for each stage of neutralization, as depicted in Figure 9.2.18. Each control loop should have a time-constant-to-dead-time ratio of at least 20:1 and a total loop dead time of less than a minute. This 20:1 ratio can be achieved by a ratio of agitator pumping rate to throughput flow of 20:1 [per Equation 9.2(8), developed earlier]. In other words, the sole source of dead time is assumed to be mixing. However, practical experience indicates that there are many other sources of dead time, as depicted in Figure 9.2.16, and that even vessels that are

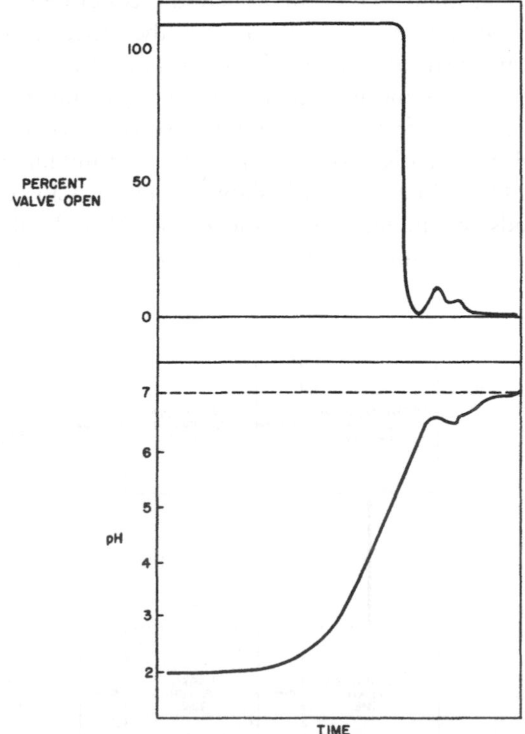

FIG. 9.2.17 Measurement and valve opening behavior for a batch process.

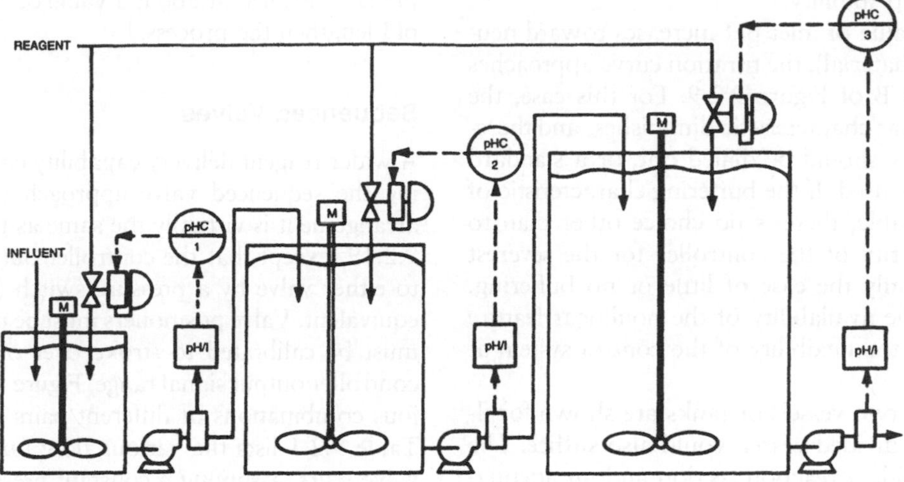

FIG. 9.2.18 Classical three-stage pH control system.

well-designed often have excessive dead time from a mixing standpoint. Thus, the 20:1 or better ratio can only be achieved by proper attention to final element selection, reagent piping design, vessel geometry, agitation patterns, electrode location, electrode cleaning, and scanning or update times. The vessels used in each stage typically are of different sizes to protect against having equal periods of oscillation of the loops. For a control loop the toughest disturbance to handle is one with the same frequency, because oscillations which are in phase can get magnified. If the ratio of agitator pumping rate to throughput flow is kept constant, the dead time is proportional to vessel volume, per Equation 9.2(6). This would suggest a requirement that the vessels not be the same size, assuming that all the loop dead time comes from mixing turnover time. It is a good practice to use different tank volumes; however, special efforts must be made to ensure that the other sources of dead time are kept smaller for the smaller volumes.

In Figure 9.2.18 the first stage should have the least dead time and be the fastest loop so that it can react quickly and compensate for disturbances before they affect the downstream stages. This requirement is similar to the principle that the inner loop should be the fastest for cascade control. This way the integrated error and volumes of off-spec material from the first stage for a given upset are significantly reduced as they are proportional to the dead time squared. However, rapid fluctuations of the influent above and below the setpoint are more effectively averaged out in a large volume with considerable savings in reagent usage. For these situations it is particularly advantageous to introduce an attenuation volume upstream of the pH control systems. If this is not possible, it may be best to use the largest vessel for the first stage and the smallest vessel for the second stage.

The setpoint of the first stage in Figure 9.2.18 is selected to be sufficiently far from the incoming pH along should have the smallest final element under normal operation and a larger one to deal with failures or bypasses of the preceding stages. A conservative method of estimation of the number of stages needed is to require one stage for every two pH units outside the control band. Influent at 1 pH and a control band from 7 to 8 pH places the influent 6 pH units away; therefore, the estimated requirement is three stages of neutralization. Feedforward control or signal linearization combined with valves of sufficient rangeability and precision can sometimes eliminate one stage.

FEEDBACK CONTROL SYSTEMS

Feedback control can be used very effectively in wastewater neutralization, provided the process is not subjected to dramatic or frequent load variations, or both. Maintained step changes in either load or setpoint can be handled effectively. Figure 9.2.19 illustrates a feedback control system in which the reagent flow rangeability requirements are not severe and can be handled by a single valve having linear characteristics. This system can accommodate (for a strong acid–strong base) inlet pH variations of approximately ±0.9 units around some normal value. If a linear valve is unavailable because of material or size limitations, an equal-percentage valve can be used but should be characterized to provide linear reagent delivery as shown in Figures 9.2.10 and 9.2.19. A valve positioner is required to eliminate valve hysteresis (difference in opening and closing characteristics) and to provide responsive valve movement.

The feedback controller in Figure 9.2.19 is a nonlinear controller with the characteristics shown in Figure 9.2.3. The overall loop stability depends on the characteristics of the treatable process material. For a process with a titration curve like that shown in part A of Figure 9.2.9, there is no question that the nonlinear characteristic will be helpful in achieving loop stability.

As the normal value of inlet pH increases toward neutrality (for an acid material), the titration curve approaches that shown in part B of Figure 9.2.9. For this case, the value of the nonlinear characteristic diminishes, and therefore the nonlinearity should be dialed out, or a standard controller should be used. If the buffering characteristic of the material is variable, there is no choice other than to adjust the nonlinearity of the controller for the severest case—which is usually the case of little or no buffering. The point is that the availability of the nonlinear feature markedly increases the flexibility of the control system at a moderate cost.

In Figure 9.2.19 two vessels or tanks are shown for illustration. A single divided vessel would also suffice. The objective is to provide a reaction section and an attenuation section. The former should be as small as possible but should provide efficient backmixing, thus permitting a tight control loop for the reaction portion of the process. If the accuracy capability of the valves is less than required, a noisy measurement will result, and the attenuation portion will provide a smoothing effect.

For those control conditions in which the setpoint is low or high, say 2 or 12, the process gain is very low, i.e., it takes a large change in reagent flow to cause a small change in measured pH, and a linear controller with a high gain (high sensitivity, narrow proportional band) suffices. In fact, on-off control (reagent valve is either fully open or closed) may be adequate. Low values of pH setpoint are used for the destruction of hexavalent chromium (Figures 9.4.8 and 9). The destruction proceeds rapidly when the pH is controlled at about a value of 2.5. Higher values of pH lengthen the process.

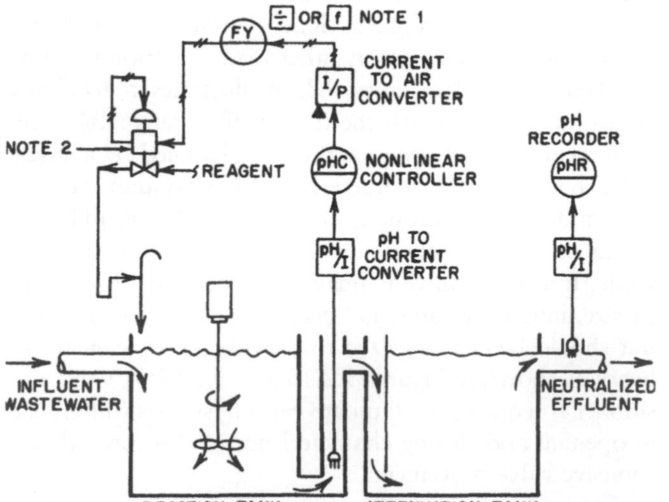

FIG. 9.2.19 Feedback control of pH. *Note 1:* For linearization of equal-percentage characteristic, commercially available divider or function generator may be used here. *Note 2:* Characteristics linear or equal percentage, depending on reagent delivery requirements. Positioner recommended for either choice.

Sequenced Valves

A wider reagent delivery capability can be obtained by using the sequenced valve approach (Figure 9.2.20). The arrangement is virtually the same as that shown in Figure 9.2.19, except that the controller output can be switched to either valve by a pressure switch (PS) or its electronic equivalent. Valve positioners must be used, since each valve must be calibrated to stroke over only a portion of the controller output signal range. Figure 9.2.21 illustrates various combinations of different pairs of sequenced valves. Table 9.2.1 lists the various flow rangeabilities for some valve pairs, assuming a constant pressure drop across the valves (equivalent to 9.5 feet, or 2.85 m, of 66° Bé sulfuric acid) and assuming an individual valve rangeability of 35:1. The valve size coefficients (C_vs) are 1.13, 0.14, 0.08, and 0.04, respectively, for CV-1, 2, 3, and 4.

The overlap between each valve pair becomes smaller as the rangeability increases. The pressure switch to transfer the valves can be set anywhere in the overlap region, because in this region the process loads can be satisfied by either valve.

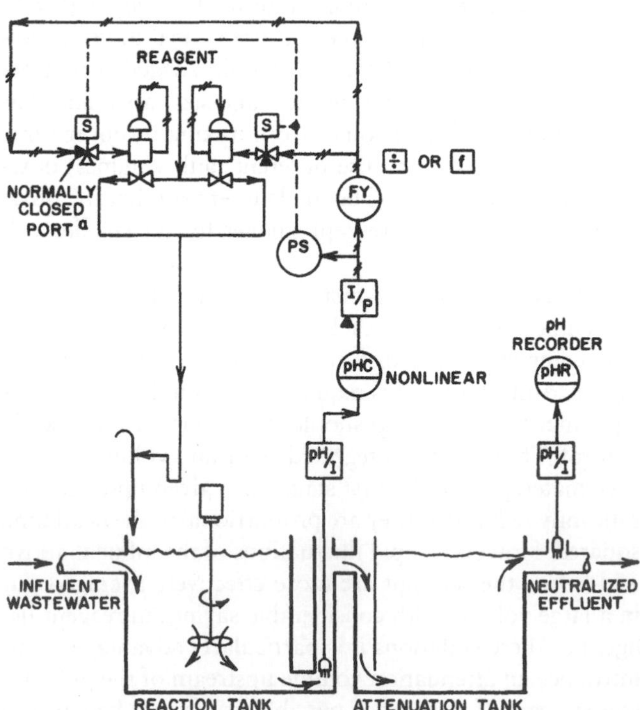

FIG. 9.2.20 Wide-range feedback control of pH. [a]This port is closed when the coil is de-energized because the pressure switch (PS) did not close the electric circuit to supply current to the solenoid coil.

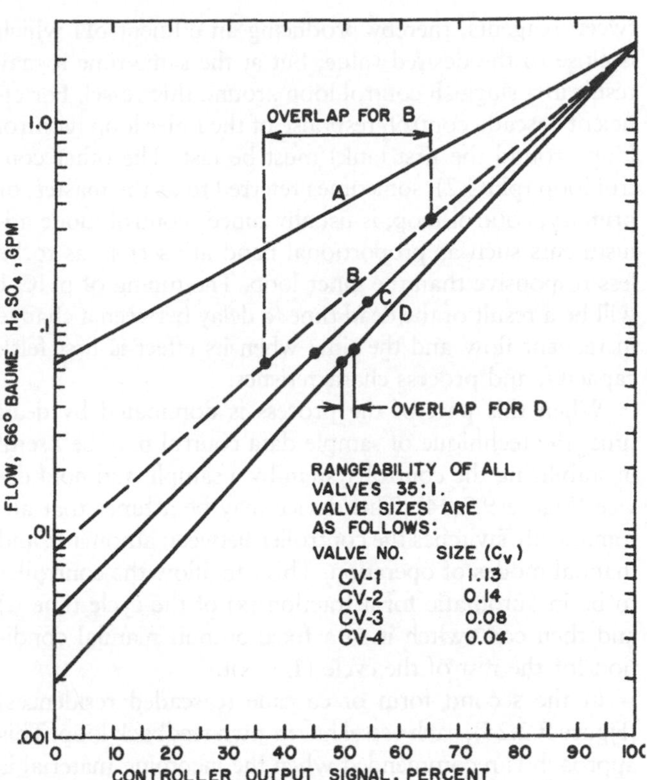

FIG. 9.2.21 Delivery capability for various valve pairs. *Key:* A = CV-1 alone; B = CV-1 + CV-2; C = CV-1 + CV-3; D = CV-1 + CV-4.

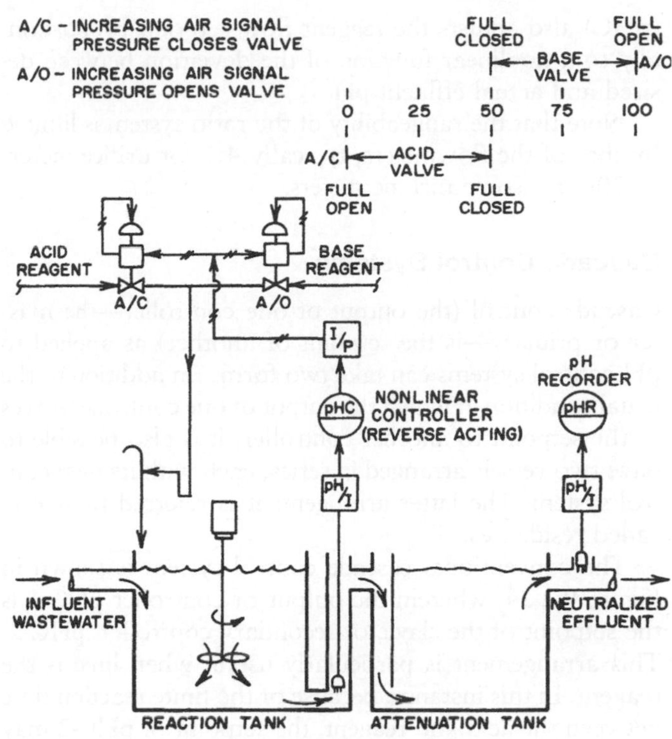

FIG. 9.2.22 Two-sided feedback control of pH.

Two Reagent Systems

Situations may arise wherein the influent may enter the system on either side of neutrality. Figure 9.2.22 illustrates the two-sided feedback control system. Although only one valve for each side is shown, it would be possible to have a sequenced pair for one side of neutrality and a single valve for the other, or a sequenced pair for both sides. Since this is a feedback control system, load changes cannot be frequent or severe in order for this system to give acceptable performance. For those applications in which load changes are frequent and severe, a combination feedforward-feedback should be considered. If sequencing is used, the reagent delivery system will have a high gain characteristic, since the stroking of the pair (moving from closed to open) is accomplished with only half the controller output signal, thereby doubling the gain (making it twice as sensitive). The valve gain will vary with the turndown, and a characterizer will be required for each set of sequenced valves to provide uniform loop gain.

Ratio Control Systems

Ratio control of pH can be extremely effective when the process flow rate is the major load variable, and the objective is to meet increased flow with a corresponding increase in reagent. Since flow measurements may be in error and reagent concentration may vary, a means for on-line ratio adjustment must be provided. Figure 9.2.23 illustrates a ratio control system in which the reagent set point is changed proportionally to changes in process flow. A feedback signal supplied by the feedback controller

TABLE 9.2.1 REAGENT DELIVERY TURNDOWN (RANGEABILITY) FOR SEQUENCED PAIRS OF EQUAL-PERCENTAGE VALVES

Valve Pair	*Line on Figure 8.40.21*	*Turndown*	*Log Towndown**	*Valve Positioner Calibration(s) (%)*
CV-1 (alone)	A	35:1	1.54	0–100
CV-1 + CV-2	B	275:1	2.44	0–63; 37–110
CV-1 + CV-3	C	570:1	2.76	0–58; 44–100
CV-1 + CV-4	D	1150:1	3.06	0–51; 50–100

*Signifies the approximate pH swing that valves will accommodate.

(pHC) also adjusts the reagent flow setpoint proportionally to a nonlinear function of the deviation between desired and actual effluent pH.

Note that the rangeability of the ratio system is limited by that of the flowmeters, typically 4:1 for orifice meters to 30:1 for some turbine meters.

Cascade Control Systems

Cascade control (the output of one controller—the master or primary—is the setpoint of another) as applied to pH control systems can take two forms. In addition to the usual condition in which the output of one controller serves as the setpoint to another controller, it is also possible to have two vessels arranged in series, each with its own control system. The latter arrangement is referred to as cascaded residences.

The conventional cascade control system is shown in Figure 9.2.24, wherein the output of controller pHC-1 is the setpoint of the slave, or secondary controller, pHC-2. This arrangement is particularly useful when lime is the reagent. In this instance, because of the finite reaction time between the acid and reagent, the setpoint of pHC-2 may have to be lower than the desired pH of the final effluent because the materials are still reacting with each other after they have left the first tank. If the setpoint pHC-2 is too high, the pH of the final stream will be greater than desired. When flocculation is to be carried out downstream of the pH treatment facility, stable pH values can be extremely important.

A delicate balance must be struck in this type of system with respect to the size of the first vessel. A long residence time in the first tank ensures long contact time between reagents, thereby producing an effluent pH which is close to the desired value, but at the same time it may result in a sluggish control loop around this vessel. For efficient cascade control, response of the inner loop (control loop around the first tank) must be fast. The other control loop (pHC-2), sometimes referred to as the master, or primary, control loop, is usually tuned (control mode adjustments such as proportional band are set) so as to be less responsive than the inner loop. The tuning of pHC-1 will be a result of the dead time (a delay between a change in reagent flow and the time when its effect is first felt), capacity, and process characteristics.

When this part of the process is dominated by dead time, the technique of sample data control may be useful in stabilizing the control system by a sample and hold device (Figure 9.2.24). This device may be a timer that automatically switches the controller between automatic and manual modes of operation. This can allow the controller to be in automatic for a fraction (x) of the cycle time (t) and then can switch it to a fixed-output, manual condition for the rest of the cycle $(1 - x)t$.

In the second form of cascade (cascaded residences) (Figure 9.2.25), each vessel has its own feedback loop. This approach is recommended when the incoming material is very strongly acidic or basic (pH values less than 1 or greater than 13). The first stage controls the effluent at a pH of approximately 4 or 10, and the second stage brings the effluent to its final value, near 7. The choice of pH setpoint for the first stage depends on the characteristics of the material. For example, material having a titration characteristic like that of A in Figure 9.2.9 may have a setpoint of approximately 3.5. The purpose is to make the control problem as simple as possible by staying on as linear a portion of the titration curve as possible.

This approach is logical when one considers the process gain characteristic as well as the accuracy limitation of a reagent delivery system. The remainder of the neutralization control problem is then similar to that illustrated in part B of Figure 9.2.9. In this manner, the control system does not have to cope with the entire nonlinear charac-

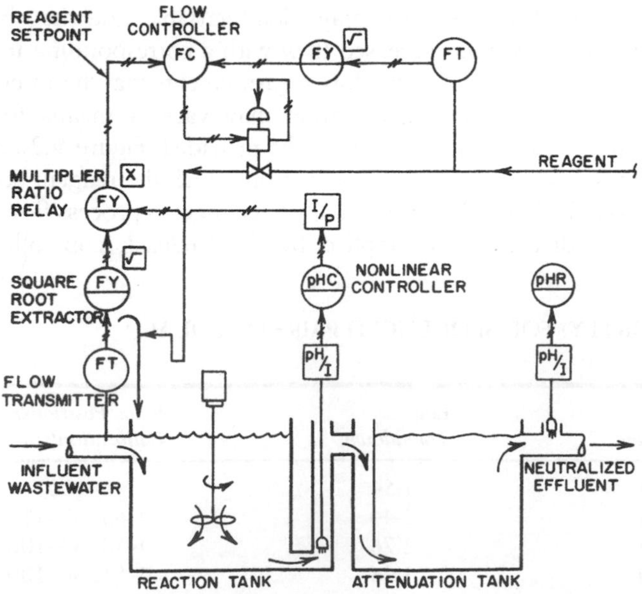

FIG. 9.2.23 Ratio control of pH.

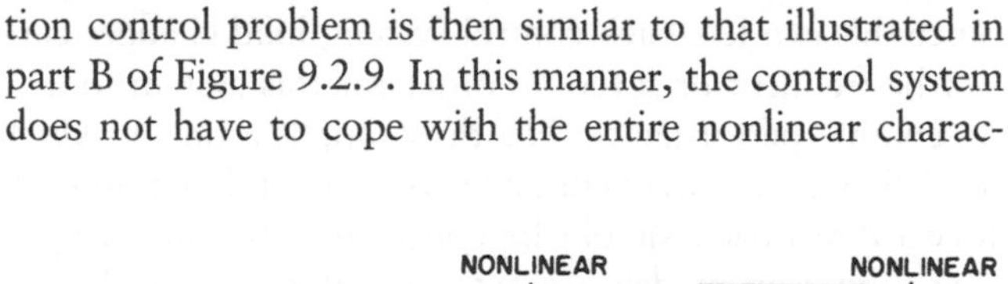

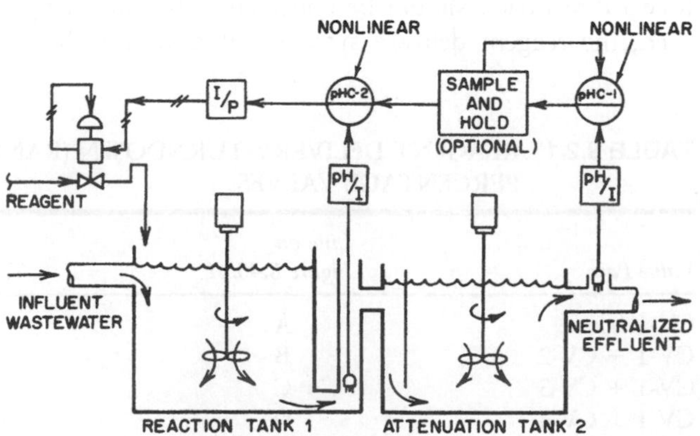

FIG. 9.2.24 Cascade control of pH.

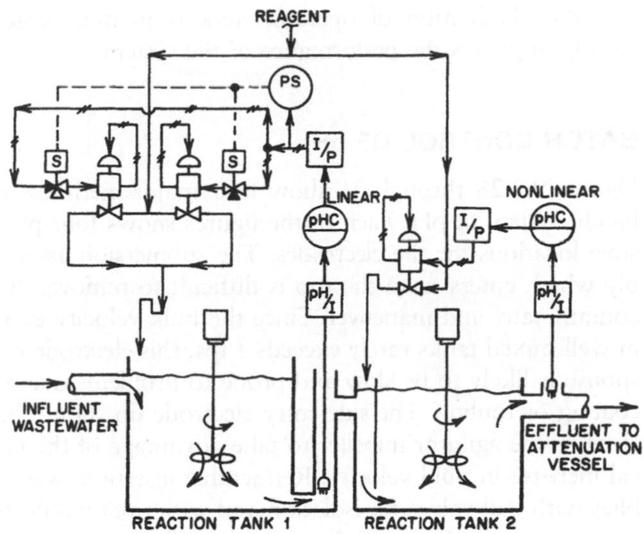

FIG. 9.2.25 Cascade residence control of pH.

teristic of the process all at once. A sequenced pair of valves is shown in conjunction with the first stage in order to handle the pH load variations. A single valve would probably suffice for the second stage, because its influent is pH-controlled. Depending on the valve sizes and on the individual valve rangeabilities, this three-valve arrangement has a maximum possible reagent flow turndown of approximately 125,000:1 (50 × 50 × 50).

FEEDFORWARD CONTROL SYSTEMS

A feedforward control (the consequences of upsets are anticipated and counteracted before they can influence the process) system is dedicated to initiating corrective action as soon as changes occur in process load. The corrective action is implemented using a control system that is essentially a mathematical model of the process. Ordinarily, the inclusion in the model of each and every load to which the process is subjected is neither possible nor economically justifiable. This means that a feedback control loop (usually containing the nonlinear controller for pH applications) is required in conjunction with the feedforward system. The function of the feedback controller is to trim and correct for minor inaccuracies in the feedforward model.

FEEDBACK-FEEDFORWARD COMBINATION CONTROL

The fractional control signal (X_B) to the valves is a function of the characterized flow signal f(F′). Therefore, the total output signal to the split-ranged valves (X_B) becomes:

$$X_B = f(F') + \text{feedforward output} - 2(\text{feedback output}) \quad 9.2(22)$$

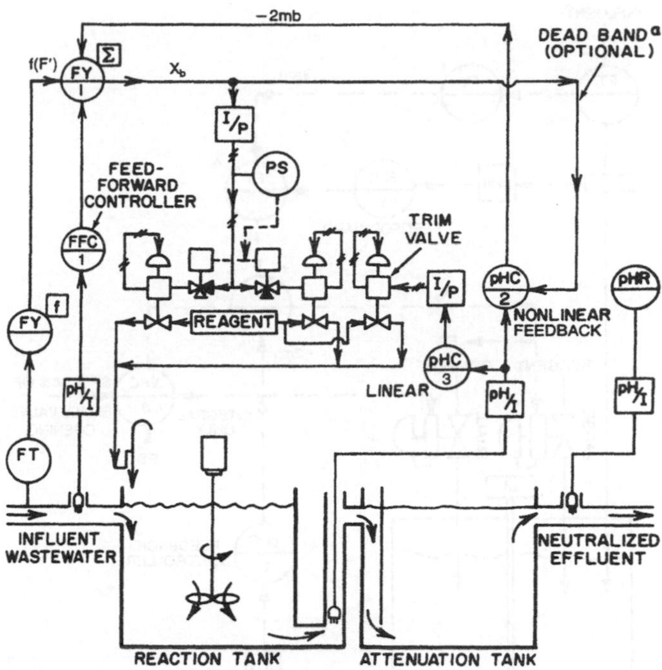

FIG. 9.2.26 Three-valve feedforward pH control system. [a]If pHC-2 is provided with a dead band, it will be inactive when the trimming controller (pHC-1) alone is operating.

Figure 9.2.26 illustrates a feedforward control system arrangement in which the flow characterization of the influent wastewater flow signal and the use of the dead-band adjustment feature previously discussed (Figure 9.2.3) are shown. A system as outlined in Figure 9.2.26 has demonstrated the need for the combination of feedforward-feedback control, because each type of control was tested individually and was found to be unsatisfactory.

A combination of feedforward-feedback on one side of neutrality and conventional feedback on the other is also possible. The nature and characteristics of the problem to be solved will indicate the nature of the solution.

Figure 9.2.27 shows how pH feedforward is used to rapidly position a large valve for major upsets. In this configuration an integral-only valve position controller (VPC) slowly optimizes the large valve position to keep the trim valve in the middle of its throttling range. Since the feedback manipulation of the big valve must be slow to prevent interaction between the two valves, the pH feedforward action provides a performance edge of rapid action for major upsets. Other methods to coordinate the movement of a large valve and a trim valve involve the use of output tracking strategies between two pH controllers, dedicated and tuned for each valve, so that both controllers are not in service at the same time.

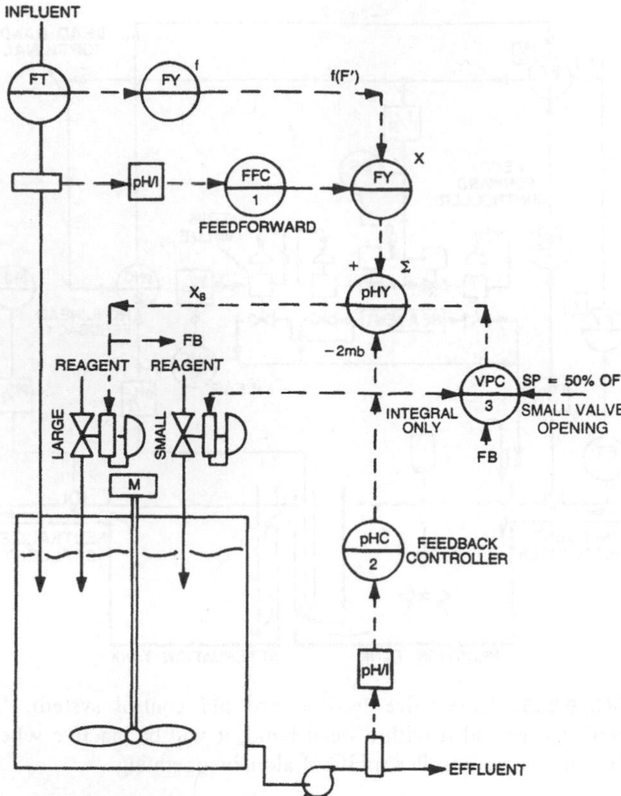

FIG. 9.2.27 Feedback-feedforward control system which keeps the small reagent valve near 50% open.

OPTIMIZATION AND AUTO-START-UP

The existence of many volumes and the presence of numerous reagent addition points provides some potentials for optimization. The goal can be the minimization of reagent usage and of salt production (from cross-neutralization of reagents) while keeping the pH within acceptable limits for the process materials of construction. Fuzzy logic can be useful in such optimization schemes as follows: the magnitude and direction of changes in the pH of various tank volumes without pH control loops and the magnitude and direction of reagent consumption in volumes with pH control loops can be used to turn on or off the addition of upstream reagent. For example, if a downstream volume has either a high or increasing rate of base addition, the existence of an intervening volume of low or decreasing pH would result in shutting off the upstream reagent flow.

The automated start-up and shutdown of pH loops is feasible if redundant sensors configured into voting or median selector systems protect against measurement failure. These operations can be smooth and reliable enough to eliminate the need for operator attention. Most pH loops are too sensitive and nonlinear for manual manipulation, and the elimination of operator actions in most cases greatly improves the performance of the system.

BATCH CONTROL OF pH

Figures 9.2.28 through 31 show four major methods of batch control of pH. Each of the figures shows four possible locations for the electrodes. The submersion assembly which enters from the top is difficult to remove, decontaminate, and maneuver. Since the bulk velocity even in well-mixed tanks rarely exceeds 1 fps, this electrode response is likely to be slow and prone to problems due to coating or fouling. The side-entry electrode tip should be close to the agitator impeller to take advantage of the local increase in fluid velocity. Retractable insertion assemblies with ball valves for isolation and with optional flush connections (see Section 2.2) are used to allow withdrawal while the vessel is full. Locating the electrodes in a recirculation pipeline is the best from the standpoint of probe response, self-cleaning action, and ease of access, but this approach is generally more expensive due to the need for block, drain, and bypass valves (not shown) for retractable installations or because of the cost of the flow assemblies when the pipeline can be emptied for removal of the electrodes. Electrodes installed on the discharge side of the pump have a slightly larger transportation delay than do those on the pump suction but are less likely to be damaged, because the pump strainer catches and the pump impeller breaks up clumps of material in the process fluid. When the reagent is added to the pump suction, the electrode must be in the discharge location for feedback measurement. The electrodes should be about 10 feet from the pump discharge and have double junction references to prevent contamination from the high-frequency pressure pulsations.

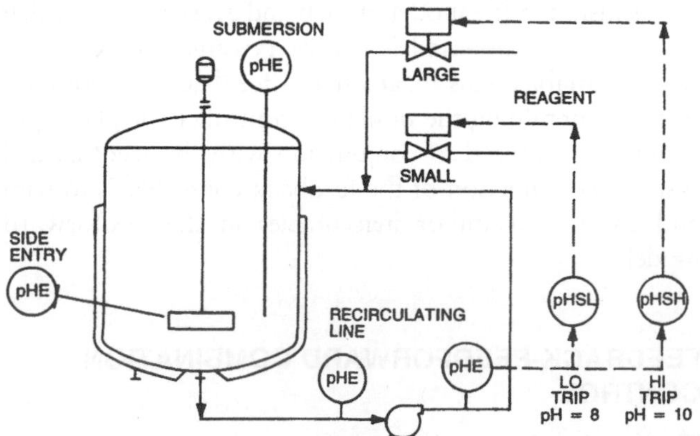

FIG. 9.2.28 Batch neutralization can be controlled by two on-off valves. If the batch is basic the large valve might close at pH = 10 while the small one closes at pH = 8.

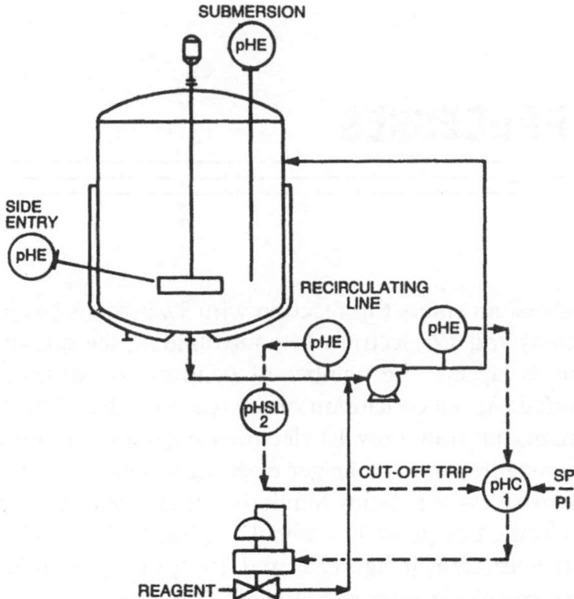

FIG. 9.2.29 Batch pH control can be configured with a proportional and integral in-line controller (pHC-1) and a safety trip cut-off (pHSI-2) guaranteeing that the batch pH does not drop too low.

The first method (Figure 9.2.28) is the simplest and works well only when the titration curve is flat, the influent flow is stable, and the control specifications are loose. Control is provided by two pH switches and two automatic on-off reagent valves. The large reagent valve is used at the beginning of batch treatment while the pH is in the flat portion of the titration curve. The small valve is used for the final adjustment, which usually occurs in the steep portion of the curve. Both valves are closed before the pH reaches the desired endpoint because the pH will continue to coast for the duration of the system dead time. If a basic batch is being neutralized the large valve might close at a pH of 10 while the small valve closes at pH of 8. The second method (Figure 9.2.29) uses in-line sensors. The in-line controller (PHC-1) setpoint is biased to kept the reagent valve open beyond its low output limit and thereby to minimize batch treatment time. A suction or vessel pH switch (pHSL-2) terminates the addition of reagent as the pH approaches the endpoint. While the reagent valve is temporarily shut, a vessel pH reading is obtained from the electrodes at the pump discharge. When needed, the in-line system can be momentarily restarted if the pH coasts to a value which is short of the endpoint. Variations of this strategy are used to operate continuous pH systems in a semi-batch mode. This can be done when sufficient vessel capacity is available so that the vessel discharge valve can be closed while the pH of the vessel contents is being adjusted during start-up or after a big upset.

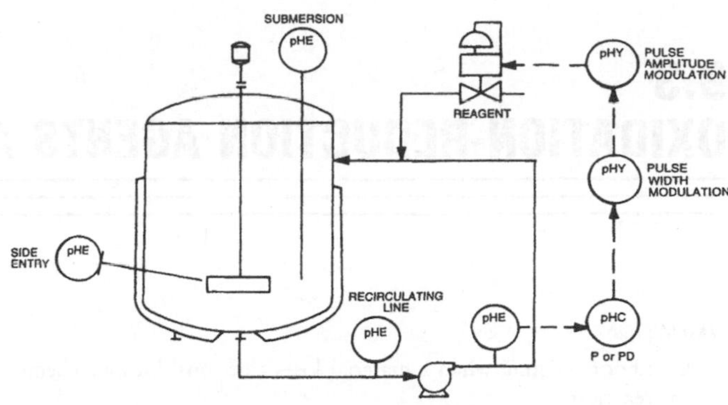

FIG. 9.2.30 Batch pH control with pulse width and amplitude modulation of the controller output.

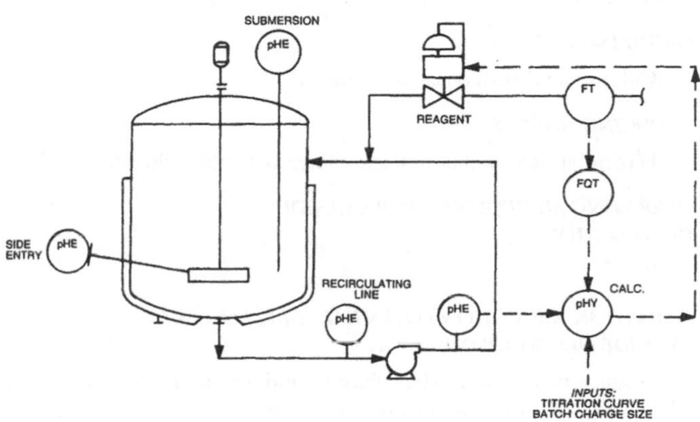

FIG. 9.2.31 Batch pH control can be based on the size of the batch and the required cut-off of reagent based on the titration curve.

The third method (Figure 9.2.30) uses a proportional-only or proportional-plus-derivative controller (Fig. 9.2.17) with pulse width and amplitude modulation of the controller output to mimic the laboratory titration process. The further away the pH is from the endpoint, the bigger and longer are the reagent flow pulses. Manual mode outputs bypass the pulsation algorithms to facilitate manual stroking of the valve. While a variety of customized strategies can be developed to achieve the same result, this scheme has the advantage of a typical controller interface and tuning adjustments for operations and maintenance.

The last method (Figure 9.2.31) uses the titration curve and the vessel volume to predict required charge of reagent and the setpoint for a totalizer. It depends heavily upon the accuracy of the curve and may best be implemented by partitioning the curve into segments and by using multiple charges. The titration curve should be corrected for temperature and for composition variations. Titration of samples taken just prior to the batch being charged could be used to verify the curve.

9.3 OXIDATION-REDUCTION AGENTS AND PROCESSES

OXIDATION

Condition existing when a material loses electrons during a chemical reaction

REDUCTION

Condition existing when a material gains electrons during a chemical reaction

OXIDIZING AGENTS

Chlorine and peroxygen compounds

REDUCING AGENTS

Ferrous sulfate, sodium metabisulfite, and sulfur dioxide

OPERATING pH REQUIRED FOR CYANIDE DESTRUCTION

9

OXIDATION-REDUCTION POTENTIAL (ORP) LEVELS IN CYANIDE DESTRUCTION

Cyanide is destroyed when chlorine addition results in a +400 millivolts (mV) solution potential, and cyanate destruction is accomplished at +600 mV.

REAGENT REQUIREMENTS OF CYANIDE DESTRUCTION

Each part of cyanide requires 9.56 parts of chlorine to oxidize it to carbon dioxide, and each part of chlorine requires 1.125 parts of sodium hydroxide to neutralize it. The actual requirements can exceed the stated stoichiometric values by a factor of 2 or 3.

OPERATING pH REQUIRED FOR HEXAVALENT CHROME REDUCTION

Between 2 and 3 pH

ORP LEVELS IN CHROME REDUCTION

See Table 9.3.5.

Process Description

Oxidation-reduction (OR) refers to a class of chemical reactions in which one of the reacting species gives up electrons (oxidation), while another species in the reaction accepts electrons (reduction). At one time, the term oxidation was restricted to reactions involving oxygen; similarly, the term reduction was restricted to reactions involving hydrogen. Current chemical technology has broadened the scope of these terms to include all reactions in which electrons are given up and assumed by reacting species; in fact, electron donating and accepting must take place simultaneously. Thus, magnesium can burn in chlorine as well as in oxygen as shown in the following equation:

$$Mg + Cl_2 \xrightarrow{\Delta} MgCl_2 + \text{heat and light} \qquad 9.3(1)$$

Magnesium enters this reaction with 12 protons (+) in its nucleus and 12 electrons (−) surrounding the nucleus in various layers. The number of neutrons is purposefully omitted. At the conclusion of the reaction, the 12 protons remain, but now only 10 electrons surround the nucleus. The magnesium is no longer electrically neutral because it has an excess of protons. Similarly, the chlorine enters with an electric charge of 17+ and 17−, but at the conclusion of the reaction, it has 17+ and 19−; in equation form, these chemicals react as follows:

$$Mg^{\circ} \xrightarrow{-2e} Mg^{+2} \qquad 9.3(2)$$

$$Cl^{\circ}_2 \xrightarrow{+2e} 2\ Cl^{-1} \qquad 9.3(3)$$

Magnesium gives up electrons and is thereby oxidized; chlorine assumes electrons and is thereby reduced. An additive that can take on (accept) electrons is an oxidizing agent (OA) and one that donates electrons is a reducing agent (RA).

In chemically treating noxious inorganic or organic waste to produce harmless or less harmful waste, wastewater treatment facilities can use the OR principle to monitor the presence or absence of adverse chemical species. The OR principle can also indicate the suitability of an environment for a type of treatment.

Wastewater treatment facilities perform monitoring by measuring the electrical potential of the chemical system with respect to a known reference before and after treatment and holding the electrical potential constant by adding a suitable reagent. They keep the electrical potential at a value that indicates that no adverse species is present or that it has been destroyed. The measurement is a voltage (emf) usually referred to as the ORP.

The emf measurement for a system has no specificity, i.e., it indicates neither the presence nor absence of a particular ion. The emf measurement indicates the activity ratio of the oxidizing species present to that of the reducing species present. The pH electrode is an example of a measurement that specifies the activity of a particular ion, i.e., ionized hydrogen in solution. The ORP electrode in conjunction with a reference electrode, develops an emf value as given by the Nernst equation as follows:

$$E_{meter} = E_o + \frac{0.0591}{n} \log \frac{[Ox.]}{[ReD.]} \qquad 9.3(4)$$

where:

E_o = a constant dependent on the choice of the reference electrode and half-cell potential of the reaction
n = number of electrons in the OR reaction
[Ox.] and [ReD.] = activities of the oxidized and reduced species, respectively

Application of the Nernst Equation

The following equations show the application of the Nernst equation with a simple reaction system, such as the conversion of ferrous to ferric ions by the addition of a solution containing ceric ions:

$$Fe^{2+} \longrightarrow Fe^{3+} + e \qquad 9.3(5)$$

$$Ce^{4+} + e \longrightarrow Ce^{3+} \qquad 9.3(6)$$

$$Fe^{2+} + Ce^{4+} \longrightarrow Fe^{3+} + Ce^{3+} \qquad 9.3(7)$$

Equations 9.3(5) and (6) are half-cell reactions. Equation 9.3(5) is written in the oxidized form, i.e., the charge(s) appear to the right of the arrow, and Equation 9.3(6) is written in the reduced form, i.e., the charge(s) appear to the left of the arrow. Each half-cell reaction has a standard or half-cell electrode potential.

The terms *standard electrode potential* and *half-cell potential* are closely related; the major distinction between them is conventional. If the half-cell reaction is written in the reduced form, as in Equation 9.3(6), the standard electrode potential and the half-cell potential are the same. If the reaction is written in the oxidized form (Equation 9.3[5]), the sign of the reported potential must be changed for the half-cell potential. Butler provides more information on the sign convention of OR reactions. Table 9.3.1 gives the standard electrode potentials for Equations 9.3(5) and (6).

The Nernst equation for the two reactions is as follows:

$$E_{meter\ Fe} = 0.771 + \frac{0.0591}{n}\log\frac{[Fe^{3+}]}{[Fe^{2+}]} \qquad 9.3(8)$$

$$E_{meter\ Ce} = 1.61 + \frac{0.0591}{n}\log\frac{[Ce^{4+}]}{[Ce^{3+}]} \qquad 9.3(9)$$

TABLE 9.3.1 ELECTRODE POTENTIALS FOR EQUATIONS 9.3(5) AND (6) WITH RESPECT TO THE STANDARD HYDROGEN ELECTRODE

Half-Cell Reaction	*Standard Potential (V),* E_n°
$Fe^{3+} + e \longrightarrow Fe^{2+}$	+0.771
$Ce^{4+} + e \longrightarrow Ce^{3+}$	+1.61

Source: The Chemical Rubber Company, Handbook of physics and chemistry, 45th ed. (Cleveland, Ohio).

TABLE 9.3.2 POTENTIALS OF STANDARD REFERENCE ELECTRODES RELATIVE TO THE STANDARD HYDROGEN ELECTRODE AT 25°C

Half-Cell Reaction	*Standard Potential (V),* 3°_{Ref}
$AgCl + e \longrightarrow Ag^\circ + Cl^-$ (1M KCl)	+0.235
$AgCl + e \longrightarrow Ag^\circ + Cl^-$ (4M KCl)	+0.199

where n = 1 and the standard hydrogen electrode is the reference electrode. For process applications, the standard hydrogen electrode is not used; instead the silver–silver chloride electrode in 1 M or 4 M KCl solution is the reference electrode. Table 9.3.2 lists the half-cell potentials of these standard reference electrodes.

The following equation gives the value of the Nernst E_o term:

$$E_o = E_n^\circ - E_{ref.}^\circ \qquad 9.3(10)$$

where:

E°_n = the standard potential of the reaction written in reduced form. Subscript *n* is the chemical symbol for the atom, molecule, or radical being considered.

Rewriting Equations 9.3(8) and (9) for use with the 1 M KCl electrode gives the following equations:

$$(E_{meter})_{Fe} = (0.771 - 0.235) + 0.0591\log\frac{[Fe^{3+}]}{[Fe^{2+}]}$$

$$(E_{meter})_{Fe} = 0.536 + 0.0591\log\frac{[Fe^{3+}]}{[Fe^{2+}]} \qquad 9.3(11)$$

$$(E_{meter})_{Ce} = (1.61 - 0.235) + 0.0591\log\frac{[Ce^{4+}]}{[Ce^{3+}]}$$

$$(E_{meter})_{Ce} = 0.375 + 0.0591\log\frac{[Ce^{4+}]}{[Ce^{3+}]} \qquad 9.3(12)$$

Note that $(E_{meter})_{Fe}$ and $(E_{meter})_{Ce}$ equal their respective Nernst E_0 values (0.536 and 0.375 V) when the ratio of [Ox.] to [ReD.] = 1. This ratio occurs at the half-titration point.

The following equation gives the ORP value at the equivalence or endpoint of the reaction:

$$E = \frac{n_1E_{Fe}^\circ + n_2E^\circ}{n_1 + n_2} \qquad 9.3(13)$$

where:

E = ORP, V
n_1 and n_2 = number of electrons in the oxidizing and reducing reactions, respectively

For the preceding example where $n_1 = n_2 = 1$, E is as follows:

$$E = \tfrac{1}{2}(E°_{Fe} + E°_{Ce}) = \tfrac{1}{2}(0.536 + 0.375)$$
$$E = +0.456 \text{ V}$$

Titration Curve

Figure 9.3.1 shows the approximate titration curve for the Fe^{2+}/Ce^{4+} system in two sections. Section A proceeds from the top left to the equivalence point, and section B proceeds from the equivalence point down to the right.

In the following example, Fe^{2+} is a noxious chemical effluent of a plant, and Ce^{4+} is a reagent added to produce the less noxious or inert Fe^{3+} form. A set of ORP electrodes monitors the reaction. What value of ORP indicates that the reaction is complete? If the process operates to the end or equivalence point, the least amount of titrating reagent is used; however, a slight deficiency in the amount of Ce^{4+} reagent signifies that Fe^{2+} is still present. If the process operates to the E°_{Ce} point, a large excess of Ce^{4+} reagent is required, which is costly and can require adding a substance to the treated wastewater that is just as adverse as the original waste.

The best operating potential for this process is about +0.400 V. Only a slight excess of reagent is required, but all of the Fe^{2+} is oxidized. For ORP applications, operation away from the equivalence point is advisable, insuring that the adverse species is completely eliminated.

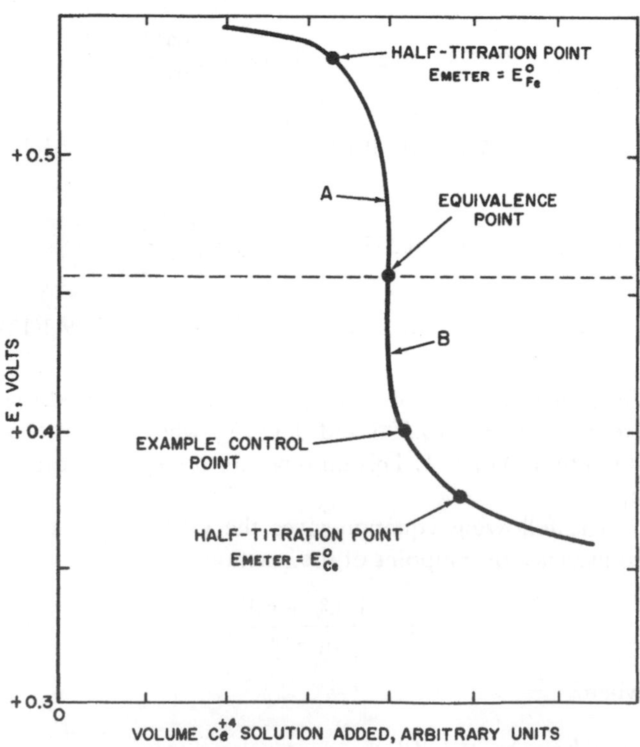

FIG. 9.3.1 Approximate titration curve for the Fe^{2+}/Ce^{4+} system.

The chemistry of this Fe^{2+}/Ce^{4+} reaction has been simplified to demonstrate the handling of the Nernst equation and the applicable sign convention. In an actual Fe^{2+}/Ce^{4+} system, the Fe^{2+} is usually in a highly acid solution.

Oxidation Processes—Cyanide Destruction

The destruction of cyanide waste by alkaline chlorination is an example of an oxidation process. The process is a two-stage operation, i.e., cyanide to cyanate and cyanate to carbon dioxide (usually as sodium bicarbonate) and nitrogen.

FIRST-STAGE REACTION

The first-stage reaction equation, assuming a sodium cyanide waste and liquid chlorine oxidizing reagent, is as follows:

$$NaCN + 2NaOH + Cl_2 \longrightarrow NaCNO + 2NaCl + H_2O \qquad 9.3(14)$$

The pH level of the reaction must be around 9 to preclude the formation of toxic gases such as cyanogen chloride (COCl) and hydrogen cyanide (HCN). When cyanide is in excess, which occurs when the untreated wastewater leaves the plant, the following reduction potential applies:

$$\underset{\text{oxidized}}{CNO^- + H_2O + 2e} \longrightarrow \underset{\text{reduced}}{CN^- + 2OH^-} \qquad 9.3(15)$$

This reaction has a standard potential of −0.97 V with respect to the standard hydrogen electrode. Assuming a Ag–AgCl (4M KCl) reference electrode (see Table 9.3.2) gives the following:

$$E_0 = -0.970 - 0.199 = -1.169 \text{ V or } -1169 \text{ mV}$$

The ratio of oxidants to reductants can be written using Equation 9.3(15) as follows:

$$\frac{[Ox.]}{[ReD.]} = \frac{[CNO^-]}{[CN^-][OH^-]^2} \qquad 9.3(16)$$

Substituting the value of E_0 already calculated and Equation 9.3(16) into the Nernst equation gives the following:

$$(E_{meter})_{CN} = -1.169 + \frac{0.0591}{n}\log\frac{[CNO^-]}{[CN^-][OH^-]^2} \qquad 9.3(17)$$

where n = 2. As stated in Equation 9.1(4) that $[OH^-][H^+] = K_w = 1 \times 10^{-14}$ at 25°C and $-\log[H^+] = pH$, Equation 9.3(17) can be written as follows:

$$(E_{meter})_{CN} = -1.169 + \frac{0.0591}{2}\log\frac{[CNO^-]}{[CN^-]} + 0.0591(14 - pH) \qquad 9.3(18)$$

Equation 9.3(18) shows the importance of the pH in the ORP process. If the pH is unknown and controlled, the meter reading is useless. When the process is controlled at pH = 9; $(E_{meter})_{CN}$ at the half-titration point can then be calculated as follows:

$$(E_{meter})_{CN} = -1.169 + 0 + 0.0591\ (14-9)$$

$$(E_{meter})_{CN} = -0.873\ V \qquad 9.3(19)$$

SECOND-STAGE REACTION

The second-stage treatment is a continuation of the alkali–chlorination process in which the cyanate is destroyed as follows:

$$2NaCNO + 4NaOH + 3Cl_2 \longrightarrow 6NaCl + 2CO_2 + N_2 + 2H_2O \qquad 9.3(20)$$

The reaction occurs at a controlled pH level, usually between 8 and 9.5, for a reaction time of 30 min or less. Table 9.3.3 shows the retention time relationship to pH requirements for the second-stage reaction.

The retention time requirement should not be confused with the vessel hold time constant; the latter is the ratio of volume divided by flow (V/F). The V/F ratio gives a nominal time constant. To determine the variance about the nominal value, environmental engineers should use a statistical technique to insure that all of the material treated has spent the required amount of time in the reaction vessel.

This second-stage reaction can be treated in a manner similar to the first-stage treatment. The reduction potential is estimated as +0.4 V as follows:

$$2CO_2 + N_2 + 2H_2O + 6e \longrightarrow 2CNO^- + 4OH^- \qquad 9.3(21)$$

$$E_m = E_o + \frac{0.0591}{6}\log\frac{[Ox.]}{[ReD.]}$$

$$E_o = 0.4 - 0.199 = +0.201\ V$$

$$\log\frac{[Ox.]}{[ReD.]} = \log\frac{[Ox.]}{[CNO^-]^2[OH^-]^4}$$

$$= \log\frac{[Ox.]}{[CNO^-]^2} - 4\log[OH^-]$$

$$= \log\frac{[Ox.]}{[CNO^-]^2} + 4(14 - pH)$$

$$(E_{meter})_{CNO} = 0.201 + \frac{0.0591}{6}\log\frac{[Ox.]}{[CNO^-]^2} + 4(14 - pH)\frac{0.0591}{6} \qquad 9.3(22)$$

When the pH is controlled at 9.0, the half-titration voltage $(E_m)_{CNO}$ can be estimated as follows:

$$(E_{meter})_{CNO} = 0.201 + 0 + 4(14-9)\frac{0.0591}{6} = 0.201 + 0.197$$

$$(E_{meter})_{CNO} = +0.398\ V$$

TABLE 9.3.3 RETENTION TIME REQUIREMENTS AS A FUNCTION OF pH LEVEL IN CYANATE DESTRUCTION

pH	*Retention Time (min)*
8.0	8
8.5	11
9.0	13
9.5	27
9.9	80
10	Infinite

The voltage at the equivalence point resulting from the half-cell reactions described by Equations 9.3(15) and (21) can now be computed. Since this system is unsymmetrical, i.e., n_1 electrons are transferred in one half-cell reaction and n_2 electrons are transferred by the other half-cell reaction, the following equation gives the voltage at the equivalence or endpoint:

$$E = \frac{n_1(E_{meter})_{CN} + n_2(E_{meter})_{CNO}}{n_1 + n_2}$$

$$E = \frac{2(-0.873) + 6(+0.398)}{6 + 2} = +0.0805\ V \qquad 9.3(23)$$

A typical value of the ORP control setpoint is +0.400 V, which is beyond the equivalence point value indicating the elimination of CN^-. Once the cyanate is undergoing destruction, the following equation gives the operative half-cell reaction:

$$ClO^- + H_2O + 2e \longrightarrow Cl^- + 2OH^- \qquad 9.3(24)$$

which has a standard potential of +0.89 V. When the reaction is controlled at pH = 9, the half-titration voltage of reaction in Equation 9.3(24) can be estimated as follows:

$$(E_{meter})_{Cl} = (0.89 - 0.199) + \frac{0.0591}{2}\log\frac{[ClO^-]}{[Cl^-]} + 0.0591(14 - pH)$$

$$= 0.691 + 0 + 0.296$$

$$(E_{meter})_{Cl} = +0.987\ V$$

The equivalence voltage of the half-cell reaction pair, Equations 9.3(21) and (24), is estimated from the half-titration voltages as follows:

$$E = \frac{6(E_{meter})_{CNO} + 2(E_{meter})_{Cl}}{2 + 6} = \frac{6 \times 0.398 + 2 \times 0.987}{8} = +0.545\ V$$

For cyanate destruction, the normal control point is about +0.600 V.

Figure 9.3.2 shows the reactions from cyanide to cyanate to cyanate destruction in titration form. The re-

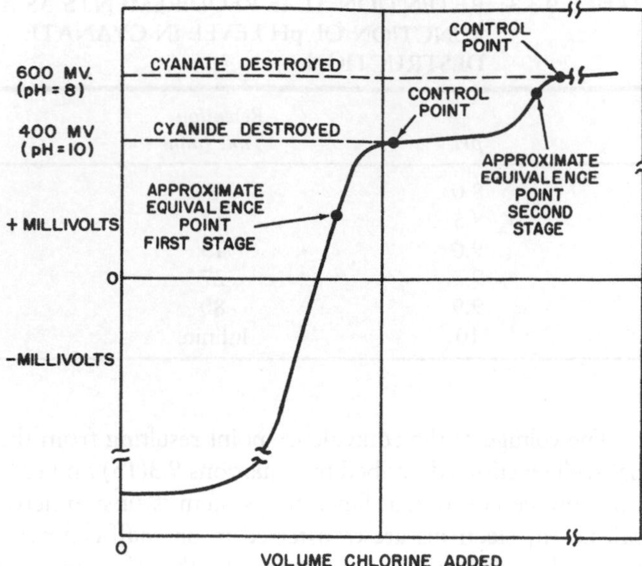

FIG. 9.3.2 Approximate path for cyanide destruction.

actions for the alkali chlorination of sodium cyanide are as follows:

1st Stage

$$Cl_2 + 2NaOH \longrightarrow NaOCl + NaCl + H_2O$$
$$NaOCl + NaCN + H_2O \longrightarrow CNCl + 2NaOH$$
$$CNCl + 2NaOH \longrightarrow NaCNO + NaCl + H_2O$$
$$Cl_2 + NaCN + 2NaOH \longrightarrow NaCNO + 2NaCl + H_2O$$

2nd Stage

$$3Cl_2 + 6NaOH \longrightarrow 3NaOCl + 3NaCl + 3H_2O$$
$$3NaOCl + 2NaCNO + H_2O \longrightarrow 2NaHCO_3 + N_2 + 3NaCl$$
$$3Cl_2 + 2NaCNO + 6NaOH \longrightarrow 2NaHCO_3 + N_2 + 2H_2O + 6NaCl$$

OXIDIZING REAGENTS

Other chlorine compounds are also oxidizing reagents. They are HOCl, NaOCL, and NH_2Cl_2. Liquid chlorine is used most often because of its convenient form (pure liquid shipped in cylinders) and the availability of chlorination equipment. As an economic measure, wastewater treatment facilities should consider the use of high-calcium lime as an alternate alkali for pH control when treating cyanide waste with greater than 200 ppm concentration.

Another process for cyanide destruction in zinc and cadmium electroplating operations uses a DuPont proprietary peroxygen compound in the presence of formalin to oxidize cyanide to cyanate. The endpoint for this process is determined by the cyanide ion electrode rather than by ORP measurement.

The quantity of alkaline material required to initially adjust the pH depends on the chemical and physical characteristics of each waste. No practical way is available for calculating this quantity, but it can be easily established by laboratory tests. Each part of cyanide requires 2.73 parts of chlorine to convert it to cyanate and 6.83 parts to oxidize it to carbon dioxide (as sodium bicarbonate) and nitrogen. In addition, each part of chlorine requires 1.125 parts of sodium hydroxide to neutralize the chlorine produced.

The actual chlorine quantities can be two or three times the theoretical requirements due to the chlorine demand on organic compounds, wetting agents, and so forth. However, either sodium or calcium hypochlorite is used instead of chlorine, the acids formed by hydrolysis are already neutralized, and little or no additional caustic is required.

Chlorine is also effective in oxidizing slaughterhouse waste, in which the treatment endpoint indication is visual rather than by instrument and is followed by coagulation. Phenolic wastes have also been successfully oxidized with chlorine. Chlorine dioxide, ozone, and potassium permanganate are alternate choices, but chlorine is the reagent of choice mainly because its cost per unit of oxidizing equivalent is lower than that of the alternates. Ammonia can increase chlorine consumption since the chlorine preferentially reacts with the ammonia before reacting with the phenol.

Reduction Processes—Hexavalent Chrome Removal

Reduction can be illustrated by the reaction in which toxic hexavalent chrome (Cr^{6+}) is reduced to the trivalent form (Cr^{3+}). In the latter, toxicity is reduced by a factor of about 100. The Cr^{3+} is then precipitated as the hydroxide and removed as a sludge. The following equation gives the half-cell reaction in reduced form:

$$Cr_2O_7^- + 14H^+ + 6e \longrightarrow 2Cr^{3+} + 7H_2O \qquad 9.3(25)$$

This reaction has a reduction potential of roughly 1.33 V with respect to the standard hydrogen electrode. The conversion time of Cr^{6+} + Cr^{3+} is pH dependent, as shown in Figure 9.3.3. To keep the reaction time under 30 min

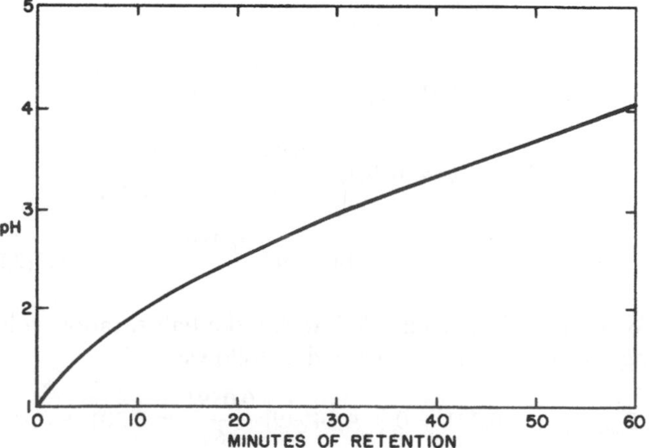

FIG. 9.3.3 Time–pH dependence of Cr^{6+} to Cr^{3+} conversion.

TABLE 9.3.4 CHROME REDUCTION AND PRECIPITATION REACTIONS

Ferrous sulfate ($FeSO_4$)	$2H_2CrO_4 + 6FeSO_4 + 6H_2SO_4 \longrightarrow Cr_2(SO_4)_3 + 3Fe_2(SO_4)_3 + 8H_2O$
	$Cr_2(SO_4)_3 + 3Ca(OH)_2 \longrightarrow 2Cr(OH)_3 + 3CaSO_4$
Sodium metabisulfite ($Na_2S_2O_5$)	$Na_2S_2O_5 + H_2O \longrightarrow 2NaHSO_3$
	$2H_2CrO_4 + 3NaHSO_3 + 3H_2SO_4 \longrightarrow Cr_2(SO_4)_3 + 3NaHSO_4 + 5H_2O$
	$Cr_2(SO_4)_3 + 3Ca(OH)_2 \longrightarrow 2Cr(OH)_3 + 3CaSO_4$
Sulfur dioxide (SO_2)	$SO_2 + H_2O \longrightarrow H_2SO_3$
	$2H_2CrO_4 + 3H_2SO_3 \longrightarrow Cr_2(SO_4)_3 + 5H_2O$
	$Cr_2(SO_4)_3 + 3Ca(OH)_2 \longrightarrow 2Cr(OH)_3 + 3CaSO_4$

most wastewater treatment facilities conduct Cr^{6+} reduction at pH levels less than 3.0; however, if the tank capacity is available, higher operating pH levels can reduce the consumption of acid reagents.

The chromium in the treated waste is generally in the form of chromic acid, chromate, or dichromate. Table 9.3.4 shows the reduction from the Cr^{6+} to the Cr^{3+} form with ferrous sulfate, sodium metabisulfite, and sulfur dioxide. The table also shows precipitation of the Cr^{3+} form is with $Ca(OH)_2$.

Of the three reagents in Table 9.3.4, sulfur dioxide has economic and handling advantages. The first two reagents require auxiliary acid to hold the pH at the proper level for a low reduction time. If auxiliary acid is not used, an excess of reagent chemicals is usually required to complete the reaction. For ferrous sulfate, about 250% excess is required; for sodium metabisulfate about 75% excess is required. The sulfurous acid formed when sulfur dioxide is hydrolized is normally sufficient to maintain the process at a low pH value, thereby precluding the need for excess acid.

Additionally, sulfur dioxide is easier to handle and feed due to its availability in bulk or large cylinders and standard feed-regulating equipment. Ferrous sulfate and sodium metabisulfite are dry powders that require feeders and mix tanks. Wastewater treatment facilities can require from 10 to 200% excess reducing reagent to overcome the quasi-buffering effect of oxidants in the process stream, especially if the stream is well aerated.

Figures 9.3.4 and 9.3.5 show titration curves for the Cr^{6+} reduction process in which the reducing reagents are sodium sulfite and sodium bisulfite, respectively. The reaction associated with the sodium sulfite reagent is as follows:

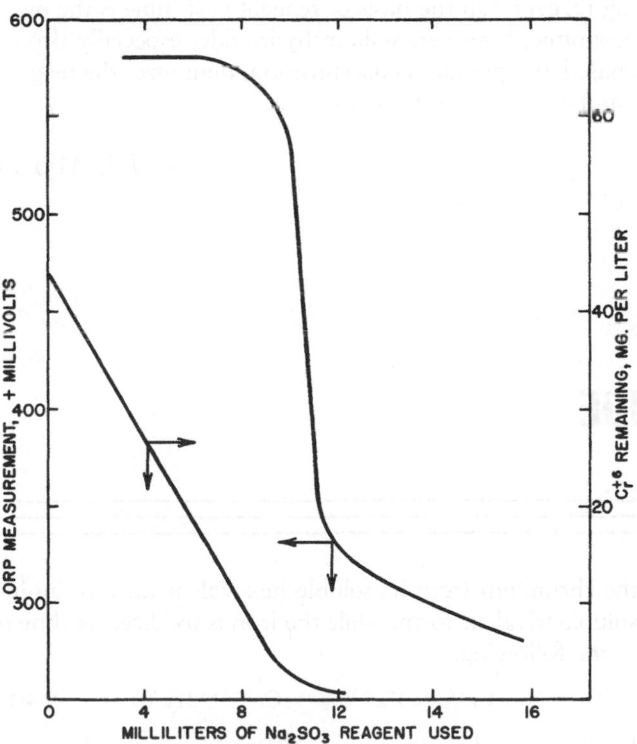

FIG. 9.3.4 Approximate path for Cr^{6+} to Cr^{3+} reduction. (Reprinted, with permission, from G.B. Hill, 1969, Complete removal of chronic acid waste with the aid of instrumentation, *Plating* 172 [February].)

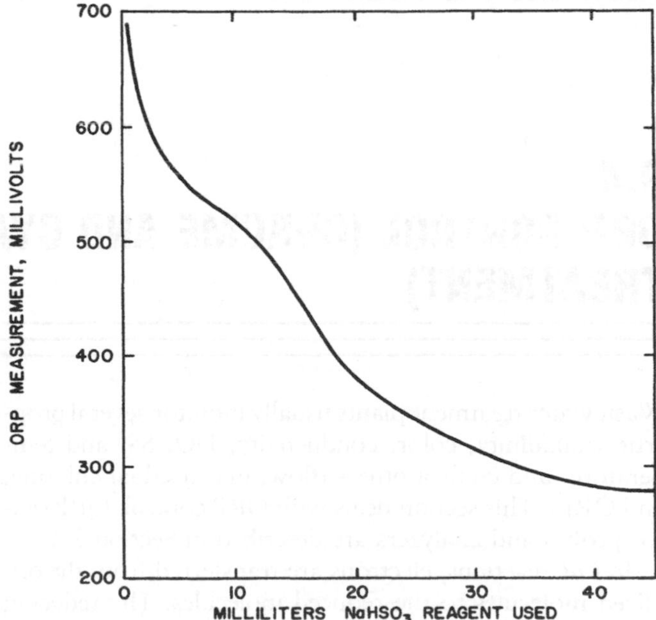

FIG. 9.3.5 Typical titration curve for Cr^{6+} to Cr^{3+} reduction at 2.5 pH. (Reprinted, with permission, from Hill, 1969.)

TABLE 9.3.5 CHROME REDUCTION ORP ENDPOINT VALUES FOR THREE REDUCING AGENTS

Reagent	*pH*	*ORP Value (mV) Standard Hydrogen Electrode*	*ORP Value (mV) AgAgCl (IM KCl) Reference Electrode*
Fe_2SO_4	2.0	500	265
$Na_2S_2O_5$	2.5	380	145
SO_2	2.9	165	−70

$$3Na_2SO_3 + 3H_2SO_4 + 2H_2CrO_4 \longrightarrow Cr_2(SO_4)_3 + 3Na_2SO_4 + 5H_2O \quad 9.3(26)$$

Like the cyanide oxidation process, the control point of the chrome reduction reaction is also monitored by an OR electrode combined with a reference electrode. Environmental engineers should carefully review published control point values for this process to ascertain the reference electrode used. If a standard hydrogen electrode is the basis rather than an industrial-type electrode, such as the AgAgCl (1M KCl) electrode, the reported values must be corrected by about −235 mV. Table 9.3.5 lists the approximate endpoint ORP values at the pH levels shown. The endpoint (equivalence point) and control point are not usually the same as shown by the Fe^{2+}/Ce^{4+} and cyanide examples (see Figure 9.3.1).

Wastewater treatment facilities should titrate each waste stream individually with the reagent to establish the correct control point value and verify the results by chemical analysis to make sure that the hexavalent chrome has been reduced to the required level.

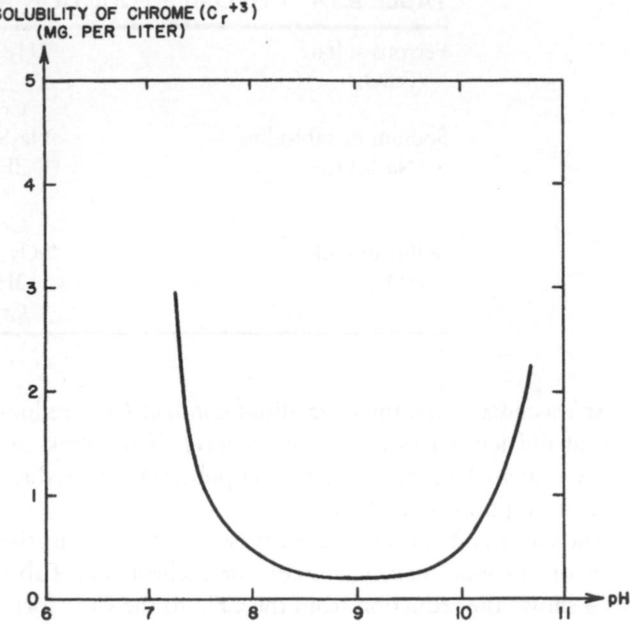

FIG. 9.3.6 pH dependence of Cr^{3+} solubility.

The precipitation reactions in Table 9.3.4 show lime, $Ca(OH)_2$, as the hydroxide source. Figure 9.3.6 shows the pH dependence of this reaction. The trivalent chromium concentration in solution is at a minimum at a pH of 9. Sodium hydroxide is equally acceptable as the precipitating reagent. On the basis of reagent cost, lime is the most economic; however, sodium hydroxide, especially if purchased at a prepared concentration, minimizes the reagent handling equipment requirement.

—*T.J. Myron*

9.4 ORP CONTROL (CHROME AND CYANIDE TREATMENT)

Wastewater treatment plants usually monitor several properties (alkalinity, color, conductivity, DO, SS, and temperature) and control others (flow, pH, residual chlorine, and ORP). This section deals with ORP control. ORP sensor probes and analyzers are described in Section 2.2.

In OR reactions, electrons are transferred from the oxidized molecules to the reduced molecules. The reducing agent donates the electrons and is oxidized in the process. The donated electrons are accepted by the reduced chemical. An example of such a process is the reduction of chromium by ferrous ions. Here, the ferrous ions reduce the chromium from its soluble hexavalent form to an insoluble trivalent form, while the iron is oxidized as shown in the following:

$$Cr^{+6} + 3Fe^{+2} \longrightarrow Cr^{+3} + 3Fe^{+3} \quad 9.4(1)$$

Environmental engineers use ORP to monitor and control biological or chemical reactions by quantitatively determining the oxidizing or reducing properties of solutions and the amount of ions present. The most widely used applications for ORP control in wastewater treatment are chrome reduction and cyanide destruction. This section discusses both continuous and batch versions of these processes. Other ORP applications include digestion and other biological processes, spa and swimming pool water treatment, and paper bleaching.

ORP Measurement

During the chemical reaction shown in Equation 9.4(1), an inert metal electrode placed in contact with the solution detects the solution's ability to accept or donate electrons. The resulting ORP redox potential is directly related to reaction progress. A reducing ion (ferrous) provides electrons and makes the electrode reading more negative. An oxidizing ion (Cr^{+6}) accepts electrons and makes the electrode reading more positive. The resulting net electrode potential is related to the ratio of concentrations of oxidizing and reducing ions in solution.

ORP is sensitive in measuring the degree of treatment provided by the reaction. However, it cannot be related to a definite concentration (only the ratio) and, therefore, it cannot be used as a monitor of final effluent concentration.

Wastewater treatment facilities can theoretically calculate the exact potential to ensure complete treatment, but in practice, this potential is subject to variations in reference electrode potential, pH, the presence of other waste stream contaminants, temperature, the purity of reagents, and so forth. Therefore, they usually determine it empirically by testing the treated wastewater for trace levels of the material to be eliminated. The optimum control point (ORP reading) occurs when just enough reagent is added to complete the reaction. Suggested control points are given later in this section; however, those control points are approximate and should be verified online by sample testing.

ORP instruments are calibrated like voltmeters, measuring absolute mV, although a standardized (zero) adjustment is often available on instruments designed for pH measurement also.

To verify the operation of electrodes, wastewater treatment facilities should have a known ORP solution composition using quinhydrone and pH buffer solutions. These solutions must be made up fresh to prevent air oxidation and deterioration. A more stable ORP reference solution consists of 0.1 M ferrous ammonium sulfate, 0.1 M ammonium sulfate, and 1.0 M sulfuric acid. Its ORP is +476 mV when measured with a silver–silver chloride, saturated potassium chloride reference electrode.

ORP electrodes must have a very clean metal surface. Routine cleaning of electrodes with a soft cloth, dilute acids, or cleaning agents promotes a fast response.

ORP Control

In designing ORP control (like pH control), the environmental engineer must recognize the chemistry of the process. In addition to vessel size, vessel geometry, agitation requirements (needed to guarantee uniform composition), and reagent delivery systems, they must also consider solid removal problems.

In some cases where one or both of the half-reactions (redox reaction) involve hydrogen ions, ORP measurements also become pH-dependent. The potential changes measured by the ORP electrode continue to vary with the redox ratio, but the absolute potential also varies with pH. Therefore, the wastewater treatment facility must determine the control point experimentally and use both pH and ORP measurements to control the process.

As with pH, reliable ORP control requires vigorous mixing to ensure uniform composition throughout the reaction tank. For continuous control, the tank should provide adequate retention time (process flow rate divided by filled tank volume), typically 10 min or more.

Complete treatment requires a slight excess of reagent and a control point slightly beyond the steep portion of the titration curve. Control in this plateau area, where process gain is low, is often provided by simple on–off control. Reagent feeders are typically metering pumps or solenoid valves. Wastewater treatment facilities can use a needle valve in series with a solenoid valve to set reagent flow more accurately and improve on–off control.

Chemical Oxidation

Water and wastewater are treated by chemical oxidation in specific cases when the contaminant can be destroyed, its chemical properties altered, or its physical form changed. Examples of chemicals that can be destroyed are cyanides and phenol. Sulfides can be oxidized to sulfates, thus changing their characteristics completely.

Iron and manganese can be oxidized from the soluble ferrous or manganous state to the insoluble ferric or manganic state, respectively, permitting their removal by sedimentation. Strong oxidants, such as chlorine, chlorine dioxide, ozone, and potassium permanganate are used. Chlorine is preferred because it is the least expensive and is readily available. Ozone is a strong second choice and is favored by some facilities because its excess converts to oxygen, while excess chlorine can react with industrial waste to produce cancer-causing substances.

All these chemical reactions are pH-dependent in relation to the reaction time required to proceed to completion for the required end products. Wastewater treatment facilities use residual oxidant or ORP measurement to control the process.

Cyanide Waste Treatment

Metal-plating and metal-treating industries produce the largest amounts of cyanide waste; however, other industries also use cyanide compounds as intermediates. Cyanide solutions are used in plating baths for brass, copper, silver, gold, and zinc. The toxic rinse waters and dumps from these operations require cyanide destruction before discharge.

The most frequently used technique for cyanide destruction is a one- or two-stage chemical treatment process. The first stage raises the pH and oxidizes cyanide to less toxic cyanate. When required, the second stage neutralizes and further oxidizes cyanate to harmless bicarbonate and nitrogen. Neutralization also allows the metals to be precipitated and separated from the effluent.

OR implies a reversible reaction. Since these reactions are carried to completion and are not reversible, the term is misleading. In practice, control is by electrode potential readings. An illustration is the oxidation of cyanide into cyanate with chlorine, according to the following reaction:

$2Cl_2$ Chlorine
$+$
$4NaOH$ Sodium hydroxide
$+$
$2NaCN$ Sodium cyanide
$\downarrow$
$2NaCNO$ Sodium cyanate
$+$
$4NaCl$ Sodium chloride
$+$
$2H_2O$ Water 9.4(2)

The electrode potential of the cyanide waste solution is -200 to -400 mV. After sufficient chlorine is applied to complete the reaction according to Equation 9.4(2), the electrode potential is $+200$ to $+450$ mV. The potential value does not increase until all cyanide is oxidized. Control of the pH is essential, with the minimum being 8.5. The reaction rate is faster at higher values.

Complete oxidation (destruction of cyanide) is a two-step reaction. The first step is oxidation to the cyanate level as described in Equation 9.4(2). The endpoint of the second-stage reaction is again detected by electrode potential readings and is $+600$ to $+800$ mV. The overall reaction is as follows:

$5Cl_2$ Chlorine
$+$
$10NaOH$ Sodium hydroxide
$+$
$2NaCN$ Sodium cyanide
$\downarrow$
$2NaHCO_3$ Sodium bicarbonate
$+$
N_2 Nitrogen
$+$
$4H_2O$ Water 9.4(3)

BATCH TREATMENT OF CYANIDE

Wastewater treatment facilities can perform the two-step cyanide destruction process in either batch or continuous processes. Because of the toxicity of cyanide and the rigid regulations on cyanide effluent discharges, some facilities prefer batch treatment, which guarantees completion of treatment before discharge.

Figure 9.4.1 describes the batch oxidation of cyanide. This process charges chlorine at a constant rate under flow control, while a pH control loop maintains the batch at a pH of around 9.5 by adding caustic. As the cyanide is oxidized into cyanate, the ORP probe senses a rise in millivolts from about -400 to over $+400$. At that point, all cyanide is destroyed, and the second stage of the reaction, cyanate destruction, begins.

At a millivolt reading of between $+600$ and $+750$, all cyanate is also destroyed, and the batch is done. At this point, the ORP switch actuates a 30-min timer. If the ORP has dropped at the end of that period, indicating that further reaction has taken place, the cycle is repeated. Otherwise, the batch is ready to be discharged. While batch treatment is in progress, a separate collection tank stores the untreated cyanide waste.

Figure 9.4.2 shows another arrangement for batch treatment with one pH and one ORP controller. This arrangement sequences the steps, changing the pH and ORP setpoints to obtain the required treatment while ensuring that treatment is complete before the next step begins.

First caustic is added to raise the pH to 11. Then hypochlorite to raise the ORP to approximately $+450$ mV, while simultaneously adding more caustic, as required, to maintain a pH of 11. An interlock prevents the addition of acid before the oxidation of all cyanide to cyanate is complete. Then, adding acid neutralizes the batch, and further hypochlorite oxidation completes cyanate-to-bicarbonate conversion.

This system can include a settling period to remove solids, or the batch can be pumped to another tank or pond for settling.

CHLORINATOR CONTROLS

Figure 9.4.3 is a schematic typical of systems with variable quality and flow rate. The chlorinator has two operators: one controlled by feedforward, the other by a feed-

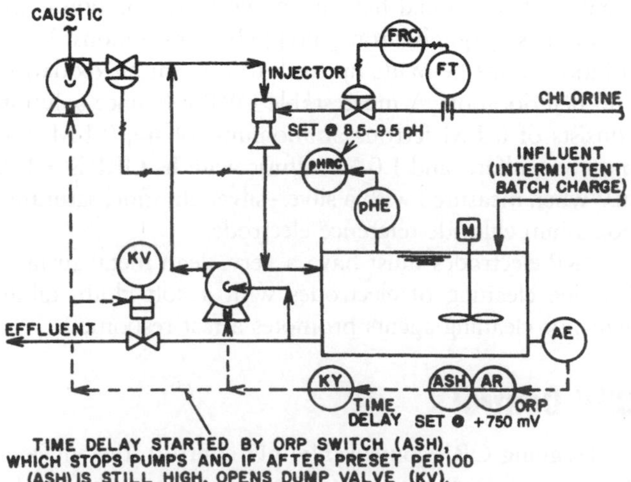

FIG. 9.4.1 Batch oxidation of cyanide waste with chlorine.

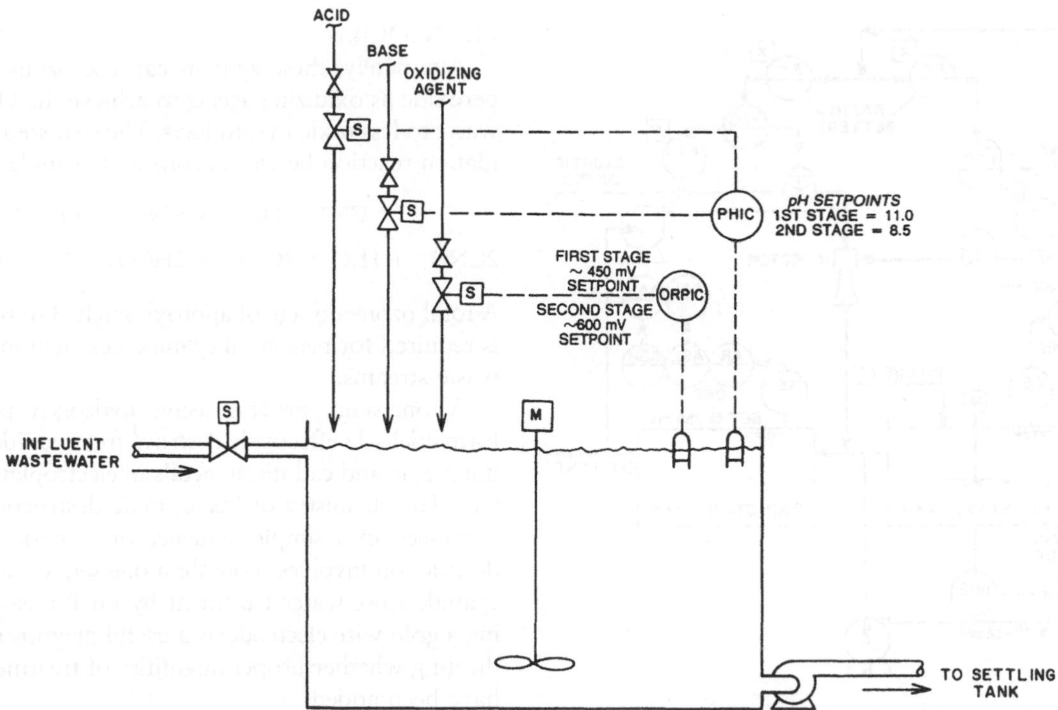

FIG. 9.4.2 Batch cyanide treatment.

back loop.

Most reactions are completed within 5 min. Except for cyanide treatment, most other chemical oxidation operations occur simultaneously with other unit operations, such as coagulation and precipitation, which govern the pH value. Although pH value affects the rate of reaction, it is seldom controlled solely for the oxidation process.

CONTINUOUS CYANIDE DESTRUCTION

Continuous flow-through systems have the advantage of reduced space requirements but require additional process equipment. In the system shown in Figure 9.4.4, the two reaction steps are separated.

In the first step, the ORP controller setpoint is approximately +300 mV. It controls the addition of chlorine to oxidize the cyanide into cyanate. The pH is maintained at approximately 10. The reaction time is approximately 5 min.

Since the second step (oxidation of the cyanates) requires an additional amount of chlorine charged at nearly the same rate as in the first step, the wastewater treatment facility obtains the chlorine flow rate signal by measuring chlorine feed rate to the first step and multiplying it by a constant to control the chlorine feed rate in the second step. The caustic requirement depends solely on the chlorine rate (pH control is not necessary); therefore, the facility can use the same signal to adjust caustic feed. The ORP instrument sampling the final effluent can signal process failure if the potential level drops below approximately +750 mV.

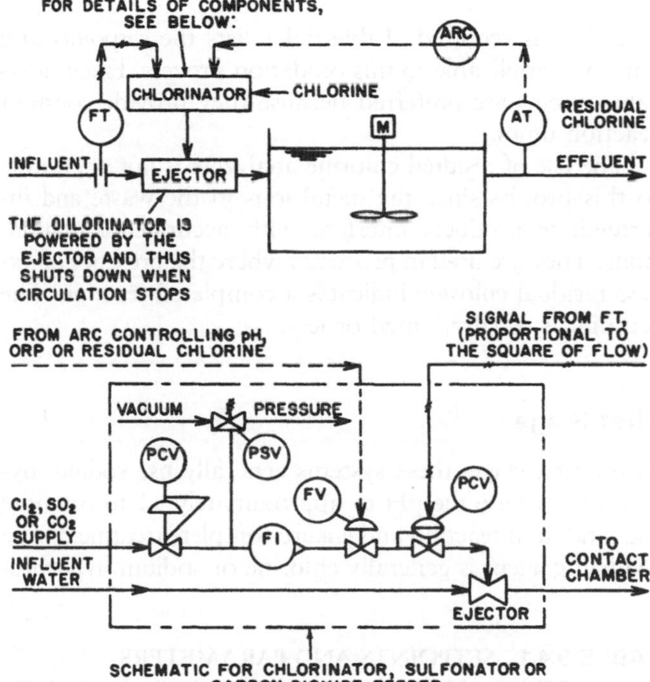

FIG. 9.4.3 Variable quality and flow rate of waste oxidized by chlorine.

Although a feedback loop from the second ORP analyzer to the second chlorinator seems beneficial, practice has not shown this loop to be necessary because the ratioing accuracy between the first-stage chlorine rate and secondary addition (approximately 1:1) is sufficiently high. At worst, the system as shown applies a little more chlo-

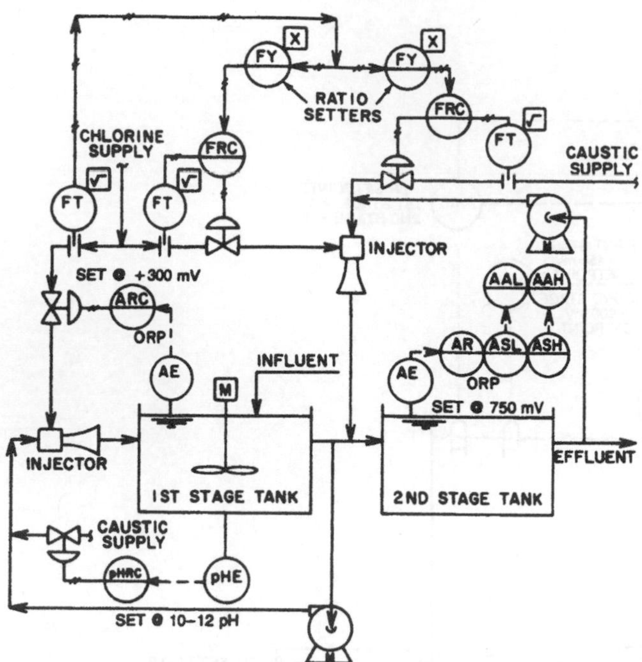

FIG. 9.4.4 Continuous oxidation of cyanide waste with chlorine. The influent has a continuous constant flow rate and variable quality.

rine than is required. Table 9.4.1 lists the setpoints and variables applicable to this oxidation process. Fixed-flow-rate systems are preferred because they provide constant reaction times.

The use of residual chlorine analyzers is not applicable to this process since the metal ions in the waste and intermediate products interfere with accurate determinations. They are used in processes where the presence of excess residual chlorine indicates a completed reaction. The setpoint is usually 1 mg/l or less.

First Stage

In the first stage, these systems generally use sodium hydroxide to raise the pH to approximately 11 to promote the oxidation reaction and ensure complete treatment. The oxidizing agent is generally chlorine or sodium hypochlorite (NaOCl).

TABLE 9.4.1 SETPOINTS AND PARAMETERS

	Process Steps	
Parameters	*Cyanide to Cyanate*	*Destruction of Cyanate*
pH	10–12	8.5–9.5
Reaction time (min)	5	45
ORP setpoint (mV)	+300	+750
Maximum concentration of cyanide (cyanate) that can be treated (mg/l)	1000	1000

Alternately, these systems can use ozone or hydrogen peroxide as oxidizing agents to achieve the OR of cyanide waste to less toxic by-products. The two-step chemical oxidation reaction between ozone and cyanide is as follows:

$$CN^- + O_3 \longrightarrow CNO^- + O_2 \qquad 9.4(4)$$

$$2CNO^- + H_2O + 3O_3 \longrightarrow 2HCO_3 + N_2 + 3O_2 \qquad 9.4(5)$$

A total ozone dosage of approximately 3 to 6 O_3/ppm CN is required for near-total cyanide destruction in industrial waste streams.

A one-stage process using hydrogen peroxide and formaldehyde effectively destroys free cyanide and precipitates zinc and cadmium metals in electroplating rinse waters. The chemistry of free cyanide destruction cannot be expressed in a simple sequence of reactions because the destruction involves more than one sequence. Monitoring cyanide rinse-water treatment by ORP measurement (using a gold wire electrode) is a useful diagnostic tool for indicating whether proper quantities of treatment chemicals have been added.

The following equation gives the overall reaction for the first stage using sodium hypochlorite (NaOCl), with cyanide expressed in ionic form (CN^-) and the result expressed as sodium cyanate (NaCNO) and chloride ion (Cl^-):

$$NaOCl + CN^- \longrightarrow NaCNO + Cl^- \qquad 9.4(6)$$

For cases when the oxidizing agent is chlorine, refer to Equations 9.4(2) and (3).

As shown in Figure 9.4.5, the first-stage reduction is monitored and controlled by independent control loops: base addition by pH control and oxidizing agent addition by ORP control. The pH controller adds base whenever the pH falls below 11. The ORP controller adds oxidizing agent whenever the ORP falls below approximately +450 mV.

The ORP titration curve (see Figure 9.4.6), shows the mV range covered when cyanide is treated in batches. Continuous treatment maintains operation in the oxidized, positive region of the curve near the +450 mV setpoint. Wastewater treatment facilities can determine the exact setpoint empirically by measuring the potential when all cyanide is oxidized but no excess reagent is present. They can verify this point with a sensitive colorimetric test.

In this reaction, the pH has a strong inverse effect on the ORP. Thus, wastewater treatment facilities must closely control the pH to achieve consistent ORP control, especially if they use hypochlorite as the oxidizing agent. Adding hypochlorite raises the pH, which, if unchecked, lowers the ORP, calling for additional hypochlorite. Controlling the pH at a setting above the pH level where hypochlorite has an influence and separating the ORP electrodes from the hypochlorite addition point can prevent this situation.

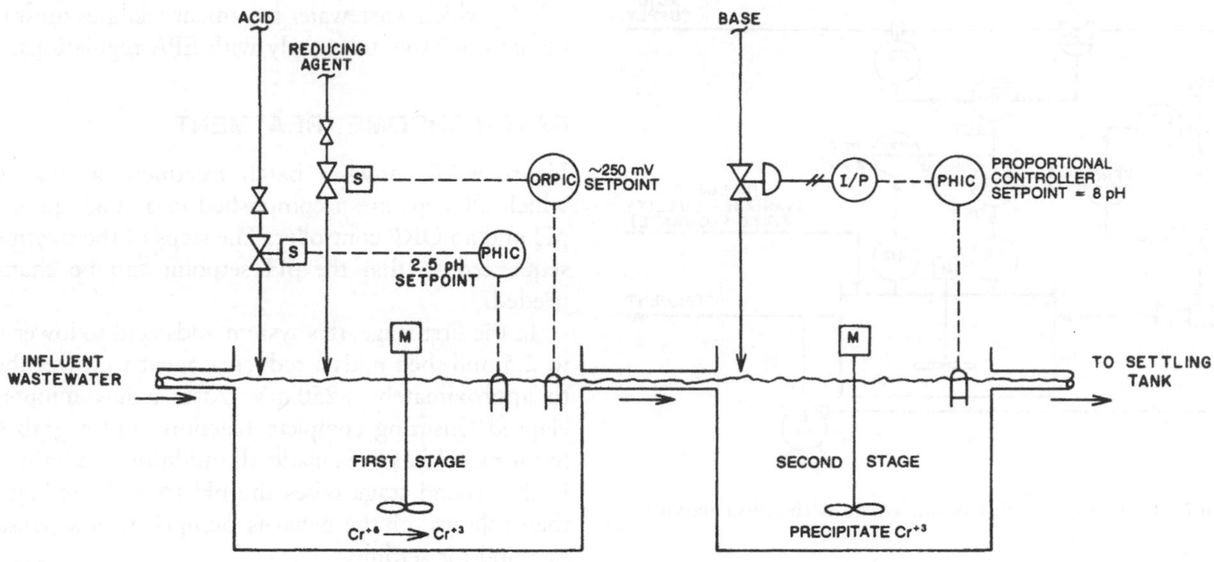

FIG. 9.4.5 Continuous cyanide treatment.

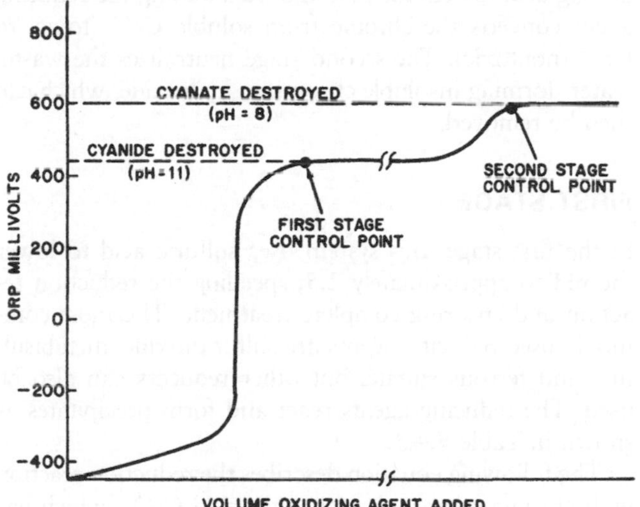

FIG. 9.4.6 Cyanide oxidation titration curve.

Gold ORP electrodes give more reliable measurement than platinum for this application. Platinum can catalyze some additional reactions at its surface and is more subject to coating than gold. The solubility of gold in cyanide solutions is not a problem since it contacts primarily with cyanate. Any loss of gold acutally keeps the electrode clean.

Second Stage

In this stage, the system neutralizes wastewater to promote additional oxidation as well as to meet the discharge pH limits. Sulfuric acid is typically used to lower pH to approximately 8.5, where the second oxidation occurs more rapidly. Acid addition must have a fail-safe design because below neutrality (pH = 7), highly toxic hydrogen cyanide can be generated if first-stage oxidation has not been completed.

The system adds hypochlorite either in proportion to that added in the first stage or by separate ORP control to complete the oxidation to sodium bicarbonate ($NaHCO_3$) as follows:

$$2NaCNO + 3NaOCl + H_2O \longrightarrow 2NaHCO_3 + N_2 + 3NaCl \quad 9.4(7)$$

ORP control in the second stage is similar to that in the first except that the control point is near +600 mV. In the second stage, pH control is more difficult than in the first because the control point is closer to the sensitive region of neutrality. The pH controller can be proportional only.

A subsequent settling tank or filter can remove suspended metal hydroxides although further treatment may be required.

Chemical Reduction

Chemical reduction is similar to chemical oxidation except that reducing reactions are involved. Commonly used reductants are sulfur dioxide and its sodium salts, such as sulfite, bisulfite, and metabisulfite. Ferrous iron salts are infrequently used. Typical examples are the reduction of hexavalent chromium, dechlorination, and deoxygenation.

Figure 9.4.7 is a schematic diagram of a typical system for the reduction of highly toxic hexavalent chromium to the innocuous trivalent form according to the following reaction:

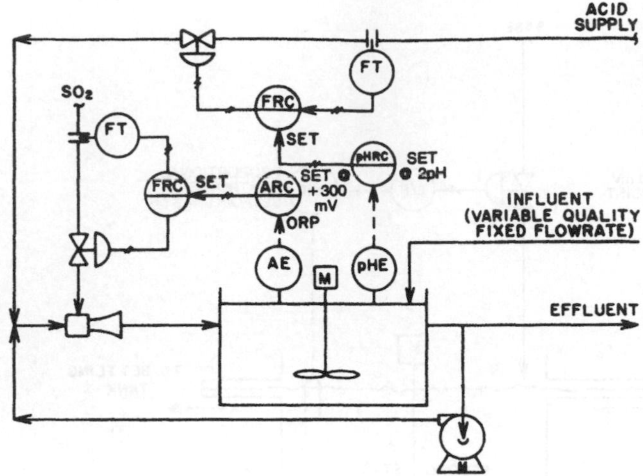

FIG. 9.4.7 Reduction of chromium waste with sulfur dioxide.

$3SO_2$	Sulfur dioxide
+	
$2H_2CrO_4$	Chromic acid
↓	
$Cr_2(SO_4)_3$	Chromic sulfate
+	
$2H_2O$	Water 9.4(8)

Most hexavalent chrome wastes are acid, but the reaction rate is much faster at low pH values. For this reason, pH control is essential. Sulfuric acid is preferred because it is less expensive than other mineral acids. The setpoint of the pH controller is approximately 2.

As in the treatment of cyanide, the chemical reaction is not reversible, and the control of sulfur dioxide addition is by electrode potential level, using ORP instrumentation. The potential level of hexavalent chromium is +700 to +1000 mV, whereas that of the reduced trivalent chrome is +200 to +400 mV. The setpoint on the ORP controller is approximately +300 mV.

The control system consists of feedback loops for both pH and ORP. The common and preferred design is for fixed-flow systems.

The system removes the trivalent chromic sulfate from solution by subsequently raising the pH to 8.5, at which point it precipitates as chromic hydroxide. The control system for this step can be identical with the one used in Figure 9.4.11. Table 9.4.2 summarizes the critical process control factors.

Chrome Waste Treatment

Chromates are used as corrosion inhibitors in cooling towers and various metal finishing operations, including bright dip, conversion coating, and chrome plating. The resulting wastewater from rinse tanks, dumps, or cooling tower blowdown contains the toxic and soluble chromium ion (Cr^{+6}), which wastewater treatment facilities must remove before discharge to comply with EPA regulations.

BATCH CHROME TREATMENT

Figure 9.4.8 shows a batch treatment arrangement in which all steps are accomplished in a single tank with a pH and an ORP controller. The steps of the treatment are sequenced so that the pH setpoint can be changed as needed.

In the first stage, this system adds acid to lower the pH to 2.5 and then adds a reducing agent to lower the ORP to approximately +250 mV. After a few minutes have elapsed (ensuring complete reaction) and a grab sample test for Cr^{+6} has been made, the addition of a basic reagent in the second stage raises the pH to 8. A settling period then follows, or the batch is pumped to a separate tank or pond for settling.

The most frequently used technique for chrome removal is a two-stage chemical treatment process. In the first stage, adding acid lowers the pH, and then adding the reducing agent converts the chrome from soluble Cr^{+6} (toxic) to Cr^{+3} (nontoxic). The second stage neutralizes the wastewater, forming insoluble chromium hydroxide, which can then be removed.

FIRST STAGE

In the first stage, this system uses sulfuric acid to lower the pH to approximately 2.5, speeding the reduction reaction and ensuring complete treatment. The most commonly used reducing agents are sulfur dioxide, metabisulfite, and ferrous sulfate, but other reducers can also be used. The reducing agents react and form precipitates as shown in Table 9.4.3.

The following equation describes the reduction reaction with chrome expressed as chromic acid CrO_3, which has a +6 charge on the chromium. The reducing agent is expressed as sulfurous acid (H_2SO_3), generated by sulfites at a low pH. The result is chromium sulfate, $Cr_2(SO_4)_3$, which has a +3 charge on the chromium, as follows:

$$2CrO_3 + 3H_2SO_3 \longrightarrow Cr_2(SO_4)_3 + 3H_2O \qquad 9.4(9)$$

Equation 9.4(9) describes the reaction when sulfur dioxide is the reducing agent.

TABLE 9.4.2 CRITICAL PROCESS CONTROL FACTORS

Variable	*Setpoints and Parameters Value*
pH	2.0
ORP setpoint (mV)	+300
Reaction time (min)	10 at pH 2.0 5 at pH 1.5

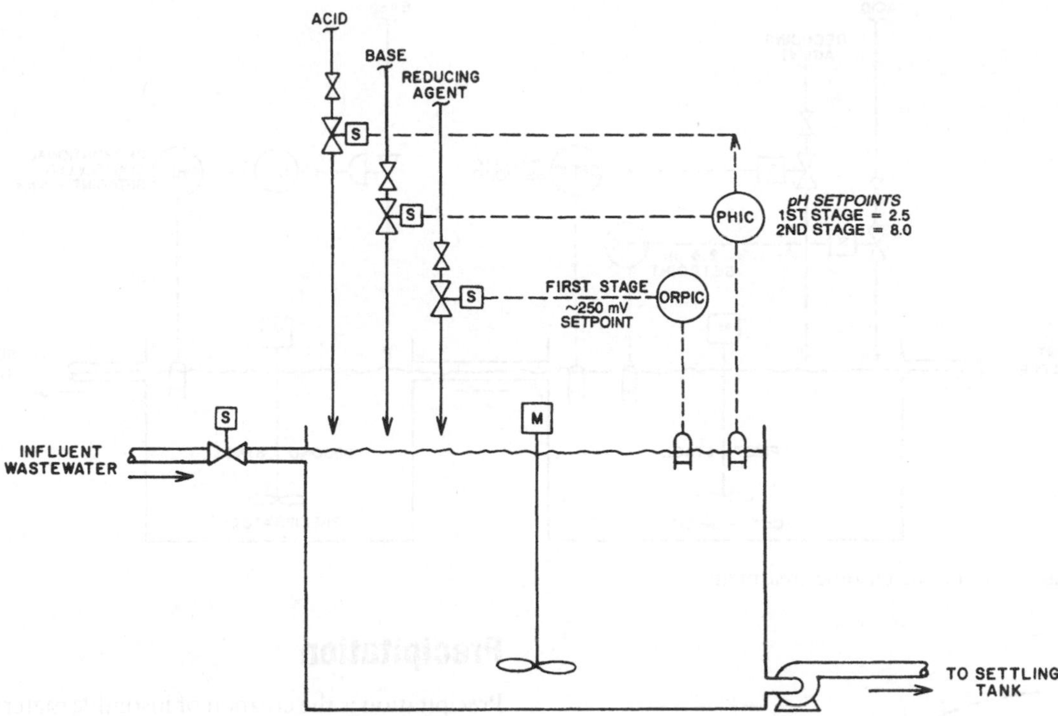

FIG. 9.4.8 Batch chrome treatment.

TABLE 9.4.3 CHROME REDUCTION AND PRECIPITATION REACTIONS

Reducing Agent	*Reaction*
Ferrous sulfate ($FeSO_4$)	$2H_2CrO_4 + 6FeSO_4 + 6H_2SO_4 \longrightarrow Cr_2(SO_4)_3 + 3Fe_2(SO_4)_3 + 8H_2O$ $Cr_2(SO_4)_3 + 3Ca(OH)_2 \longrightarrow 2Cr(OH)_3 + 3CaSO_4$
Sodium metabisulfite ($Na_2S_2O_5$)	$Na_2S_2O_5 + H_2O \longrightarrow 2NaHSO_3$ $2H_2CrO_4 + 3NaHSO_3 + 3H_2SO_4 \longrightarrow Cr_2(SO_4)_3 + 3NaHSO_4 + 5H_2O$ $Cr_2(SO_4)_3 + 3Ca(OH)_2 \longrightarrow 2Cr(OH)_3 + 3CaSO_4$
Sulfur dioxide (SO_2)	$SO_2 + H_2O \longrightarrow H_2SO_3$ $2H_2CrO_4 + 3H_2SO_3 \longrightarrow Cr_2(SO_4)_3 + 5H_2O$ $Cr_2(SO_4)_3 + 3Ca(OH)_2 \longrightarrow 2Cr(OH)_3 + 3CaSO_4$

As shown in Figure 9.4.9, the system monitors and controls the first-stage reaction with independent control loops: acid addition by pH control and reducing agent addition by ORP control. The system adds acid under pH control whenever the pH rises above 2.5; it adds the reducing agent under ORP control whenever the ORP rises above approximately +250 mV.

The ORP titration curve (see Figure 9.4.10) shows the mV range covered when Cr^{+6} chrome is treated in batches. With continuous treatment, however, operation is maintained in the completely reduced portion of the curve near the nominal +250 mV control point. The exact setpoint for an installation should be at a potential where all the Cr^{+6} is reduced but without excess sulfite consumption, which is accompanied by sulfur dioxide odor.

Chrome reduction is slow enough that 10 to 15 min can be required to complete the reaction. The reaction time increases if the pH is controlled at higher levels. Variations in pH also affect measured ORP readings; thus, the pH must be stable for consistent ORP control.

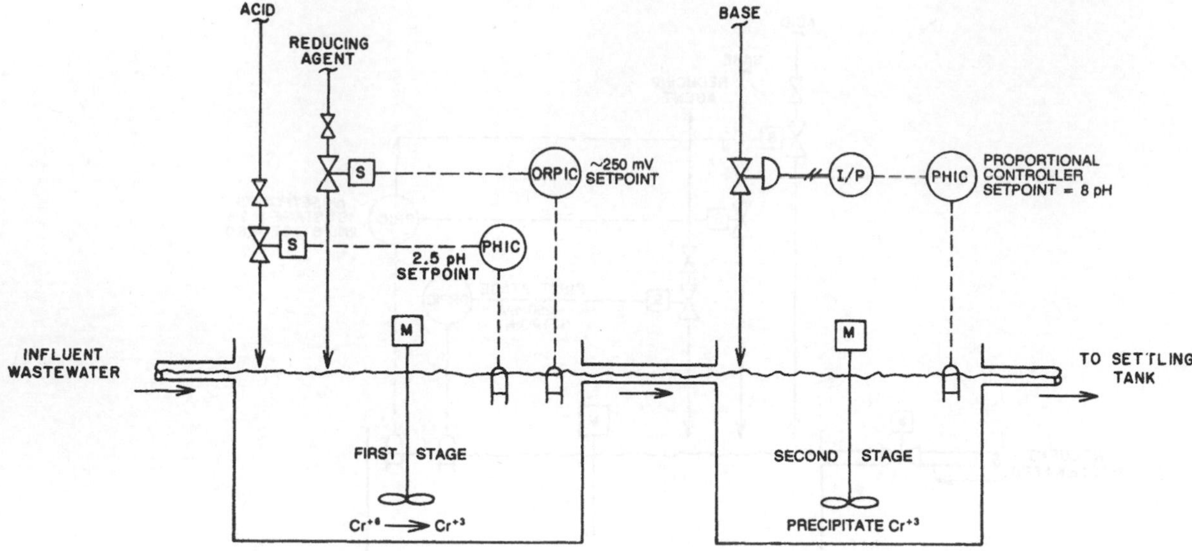

FIG. 9.4.9 Continuous chrome treatment.

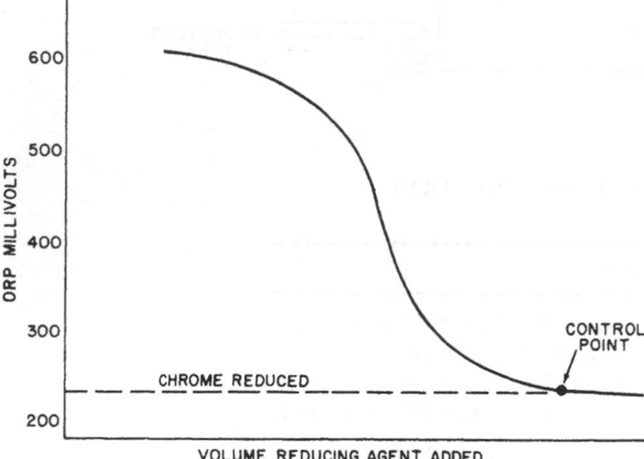

FIG. 9.4.10 Chrome reduction titration curve.

SECOND STAGE

In the second stage, the system neutralizes the wastewater to precipitate Cr^{+3} as insoluble chromium hydroxide, $Cr(OH)_3$, and to meet the discharge pH limits. The system uses sodium hydroxide or lime [$Ca(OH)_2$] to raise the pH to 7.5 to 8.5, as shown in the following reaction:

$$Cr_2(SO_4)_3 + 6NaOH \longrightarrow 3Na_2SO_4 + 2Cr(OH)_3 \qquad 9.4(10)$$

In the second stage, providing pH control is more difficult than in the first because the control point is closer to the sensitive region near neutrality. Although the second-stage reaction is fast, a retention time of at least 10 min is usually needed for continuous treatment processes to achieve stable operation. The pH controller is proportional only in this stage.

A subsequent settling tank or filter removes the suspended chromium hydroxide. Flocculating agents can assist in this separation.

Precipitation

Precipitation is the creation of insoluble materials by chemical reactions that provides treatment through subsequent liquid–solids separation. The removal of sulfates, removal of trivalent chromium, and softening of water with lime are typical precipitation operations. A variation of this process following the treatment discussed previously, in connection with the chemical oxidation process, removes iron and manganese. Lime softening is a common process and is used as an example.

The following reaction is involved:

$Ca(HCO_3)_2$	Calcium bicarbonate
+	
$Ca(OH)_2$	Calcium hydroxide
↓	
$2CaCO_3$	Calcium carbonate
+	
$2H_2O$	Water 9.4(11)

Calcium carbonate formed by this reaction is insoluble and can be removed by gravity separation (settling). The typical settling time is 30 min or less, but most systems are designed for continuous operation with typical detention times of 1 hr.

Water treatment with this process is called *excess-lime softening* because the application of lime is in excess of that required for the reaction described in Equation 9.4(11). Control consists of adding sufficient calcium hydroxide to maintain an excess hydroxide alkalinity of 10 to 50 mg/l, as shown in the following equation:

$$2P = MO + 10 \text{ to } 50 \qquad 9.4(12)$$

where P is the phenolphthalein alkalinity and MO is the methyl orange alkalinity. This addition results in a pH value of 10 to 11, but pH control is not satisfactory for economical operations. In this example, suitable analytical instrumentation is not available for continuous system

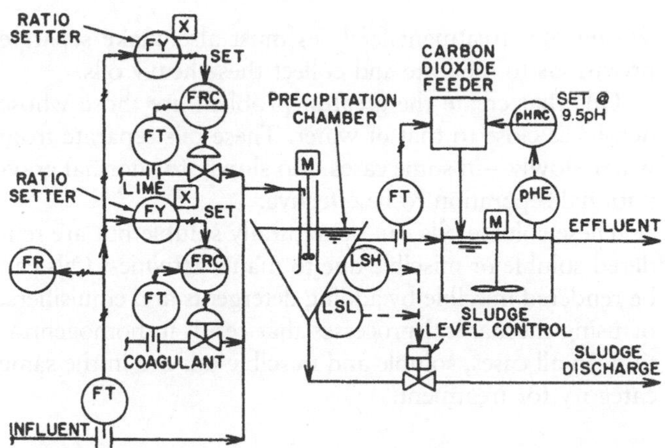

FIG. 9.4.11 Calcium carbonate precipitation control system. Details of the carbon dioxide feeder are shown in Figure 9.4.3.

control. If the quality of the untreated water varies, operator control of the lime dosage is essential. Wastewater treatment facilities usually pace the manual dosage by feedforward control from the flow rate.

A factor in calcium carbonate precipitation is a chemical phenomenon known as *crystal seeding*. This factor involves the acceleration of carbonate crystal formation by the presence of previously precipitated crystals. Wastewater treatment facilities accomplish crystal seeding in practice by passing the water being treated through a sludge blanket in an upflow treatment unit, shown schematically in Figure 9.4.11. The resulting crystals of calcium carbonate are hard, dense, and discrete, and they separate readily.

When colloidal suspended material is also to be removed, which occurs when surface waters are softened, water treatment facilities also add a coagulant of aluminum, iron salts, or polymers to precipitate the colloids. The dosage varies depending on the quantity of suspended material. The application of both coagulant and calcium hydroxide is controlled by flow–ratio modulation.

The resulting sludge, consisting of calcium carbonate, aluminum, or iron hydroxides, and the precipitated colloidal material are discharged to waste continuously. As previously noted, the presence of some precipitated carbonate is beneficial to remove all sludge. The use of an automatic level control system (see Figure 9.4.11) controls the sludge level at an optimum.

Water softened by excess-lime treatment is saturated with calcium carbonate and is therefore unstable. Water treatment facilities can achieve stability by adding carbon dioxide to convert a portion of the carbonates into bicarbonate, according to the following equation:

$$\begin{array}{cl} CO_2 & \text{Carbon dioxide} \\ + & \\ CaCO_3 & \text{Calcium carbonate} \\ + & \\ H_2O & \text{Water} \\ \downarrow & \\ Ca(HCO_3)_2 & \text{Calcium bicarbonate} \end{array} \qquad 9.4(13)$$

In contrast with the softening reaction, this process is suited to automatic pH control. Figure 9.4.11 shows this control system. The carbon dioxide feeder has two operators: one controlled by feedforward on the influent flow and the other by feedback on the effluent pH. The setpoint is about 9.5 pH.

Electrode fouling occurs as a result of the precipitation of crystallized calcium carbonate. Daily maintenance is required unless automated cleaners are used. The farther downstream (from the point of carbon dioxide application) the electrodes are placed, consistent with acceptable loop time delays, the less maintenance required.

—Béla G. Lipták

9.5 OIL SEPARATION AND REMOVAL

OIL SOLUBILITY

Decreased by either high base or high acid conditions

CHEMICALS USED TO ADJUST THE pH OF OIL SOLUTIONS

Sulfuric acid, lime, and soda ash

CHEMICALS USED TO COAGULATE AND MAKE OILS INSOLUBLE

Alum at a pH of 8 to 9 and ferrous sulfate at a pH of 8 to 10

PHYSICAL OIL SEPARATION TECHNIQUES

Settling, filtration, centrifuging, and flotation

Oil Sources

The two principal sites of oil waste are at the oil field and oil refinery. At the former, numerous problems are associated with well drilling. Drilling techniques frequently use water under pressure to bring material from the drill hole to the surface. In some cases, the drilling process passes through a significant amount of low-yield, oil-bearing strata prior to reaching the high oil-bearing region, resulting in oil being carried to the surface with the drilling mud. Also, on reaching the high oil-bearing stratum, the drilling process can regurgitate a small amount of highly concentrated oil.

Where there is naturally occurring or artificially induced high pressure on the oil-bearing areas, the oil can be forced up and out of the well. This situation represents a high concentration of oil that must be properly disposed. Brines are also associated with oil in the ground, and large amounts of brine contaminated with oil can reach the surface. The oil must be separated before the brine can be disposed in a satisfactory manner.

At the refinery, nearly all sources of oil waste result from spills. Although accidents can occur anywhere in the plant, the prime sources of spills are at loading and unloading sites. These areas should be curbed and separately sewered with provisions for separating oil from surface water runoff. Spills within the refinery present specific problems as a function of the oil type being processed at that point in the refinery. Treatment must be based on the type of oil involved.

Many other sources of oil reach the environment due to man's activities. Many of these are related to oil transport. Another widely dispersed source of oil spills is the automotive transportation system, with the local gasoline service station the focus of the problem. Due to federal taxation, reprocessing used oil for reuse is no longer economical.

Scavengers are reluctant to collect oil, and frequently oil reaches the local sewer system. The machine tool industry uses considerable volumes of oil for lubrication and cutting. All mechanized industries require oil for lubrication. In the steel industry, fabricated metals are frequently dipped in oil to prevent rust during storage. Runoff from highways and parking lots contains measurable amounts of oil leaked from motor vehicles. Another source of oil is the common two-cycle engine used frequently for lawn mowers and outboard motors. Tests show that at low speeds, two-cycle outboard motors bypass as much as one-third of the total fuel–oil mixture.

Numerous natural oil sources also exist. These sources occur frequently in marshy areas, primarily near natural oil-bearing strata. Natural oil seeps also exist in areas along the Pacific Coast. Generally, these natural sources are insignificant as pollution problems. However, if man or earthquakes disrupt the fissures, they can become enlarged to the point where a problem is created.

Oil Properties

Most oils are insoluble in water, aiding in their separation. Furthermore, most insoluble oils are lighter than water, and therefore they float on its surface. A few oils are heavier than water, and these settle to the bottom of water. Wastewater treatment facilities must also make separate provisions to separate and collect these heavy oils.

Oils that create the greatest problem are those whose density is close to that of water. These oils separate from water slowly—in some cases, too slowly for normal gravitational separation to be effective.

These soluble oils can be naturally soluble but are rendered soluble or miscible due to man's activities. Oils can be rendered miscible by adding detergents and emulsifiers, or using mechanical processes that result in homogenization. In all cases, soluble and miscible oils are in the same category for treatment.

Treatment

The removal of oil from water involves the separation by chemical or physical means and the ultimate disposal of oil.

CHEMICAL TREATMENT

Adding chemicals can improve the separation of oils whose density is close to that of water or oils that are slightly soluble. Generally, oils are less soluble under extreme acid and extreme base conditions. Commonly, sulfuric acid is used; however, lime at a pH of 8.4 is also effective.

To a degree, chemical dispersants are broken down by adding an acid or a base, but specific dispersants require specific chemicals. Truly soluble oils are difficult to render insoluble with acids or bases alone, and chemical coagulation at an appropriate pH is usually required.

Studies show that alum in the pH range of 8.0 to 9.0 and ferrous sulfate in the pH range of 8.0 to 10.0 are effective in coagulating soluble oil. Because controlling the pH in this range with lime is more difficult, wastewater treatment facilities generally use soda ash (Na_2CO_3) to raise the pH to the required level. If the facility uses chemical coagulation at an elevated pH, prior treatment with acid to improve separation is not recommended since this treatment requires larger amounts of alkali for final pH adjustment. Polyelectrolytes with the other coagulants can improve separation.

A recycling system has been devised to recover both added chemicals and separated oil. This system adds ferric chloride, lime, ferric sulfate, and a polymer to the oily liquid waste. It treats the precipitated sludge containing oils with sulfuric acid at 90°F (32.2°C) which releases the oil from the sludge. This oil is then separated and reused, and the ferric sulfate is also reclaimed from the sludge and reused.

PHYSICAL TREATMENT

Insoluble oils lighter than water can be easily separated in a settling tank with an adjustable skimming weir. These oils readily float to the surface, and the depth of the weir is adjusted according to the amount of oil in the water.

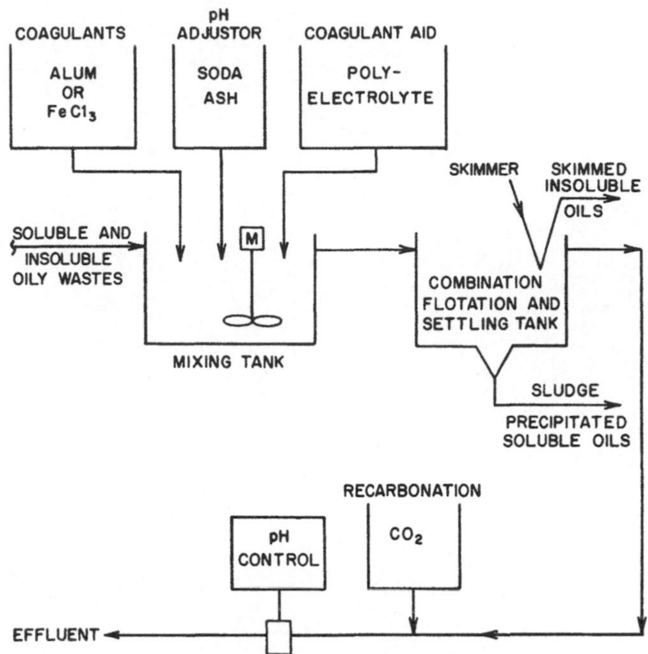

FIG. 9.5.1 Combined soluble and insoluble oil removal.

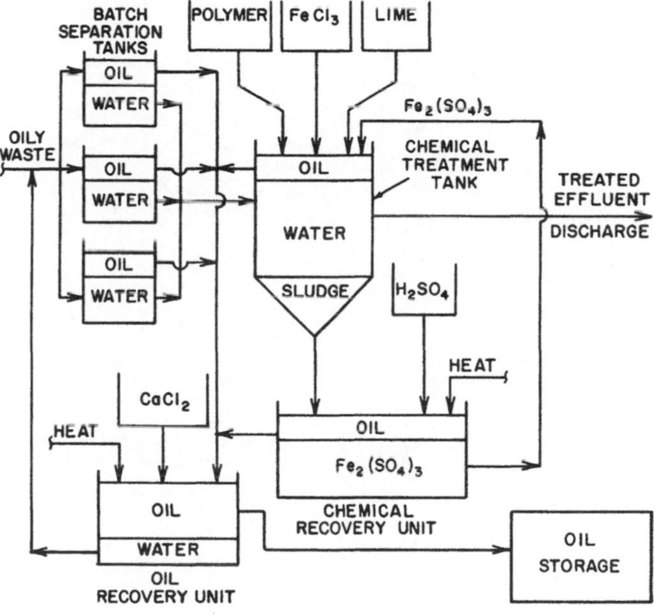

FIG. 9.5.2 Oily waste treatment system that uses chemical reuse and oil recovery.

Insoluble oils heavier than water can be recovered in a settling basin with a bottom sludge separator. The withdrawal rate of the sludge from the bottom must be proportional to the amount of heavy oil separated from these liquids. All coagulation techniques use the settling principle; the oil is rendered insoluble and heavier than the water so that it can be removed by settling with sludge removal.

For coagulated solids, wastewater treatment facilities use filtration through various media. Sand provides a satisfactory filter medium, but diatomaceous earth is more commonly used due to the disposability of the diatomaceous earth filter cake. Continuous centrifuging also separates oils from water, but this technique is expensive. Also, for varying densities and volumes of oil in waste, wastewater treatment facilities must continuously adjust the centrifuge system.

Dispersed air flotation is also effective in separating oil from water. This technique has been practiced with and without prior chemical coagulation. Sometimes breaking down the foam after it has carried the oil from the water is difficult. Dispersants can break down the foam so that the oil can be properly disposed of.

ULTIMATE DISPOSAL

The ultimate disposal of oil frequently creates a problem. Occasionally, reclaimed oil can be reprocessed in a refinery. In other cases, oils are mixtures that are not favorable for reclamation. In other cases, reclamation is not economical.

Some wastewater treatment facilities dispose oil on land, but they must take proper precautions to prevent the oil from contaminating the groundwater. To a limit, soil microorganisms can break down oil and render it inoffensive. Facilities for burning waste oil are also available. Problems arise from the differing volatilities of the oils collected. Reclamation of the heat produced is beneficial but not always practical due to variations in the amount of oil combusted.

TYPICAL TREATMENT SYSTEMS

In a typical combination treatment system for soluble and insoluble oil removal (see Figure 9.5.1), the soluble oils are precipitated by coagulants, elevated pH, and coagulant aids. Both soluble and insoluble oils are separated in the combination flotation and settling tank. The effluent may require recarbonation with appropriate pH control before being discharged to a receiving stream.

Figure 9.5.2 shows a system for chemical treatment with separation and recovery of oil and ferric sulfate. The proper application of chemical treatment methods can result in satisfactory separation and removal of oil from liquid waste.

—D.B. Aulenbach

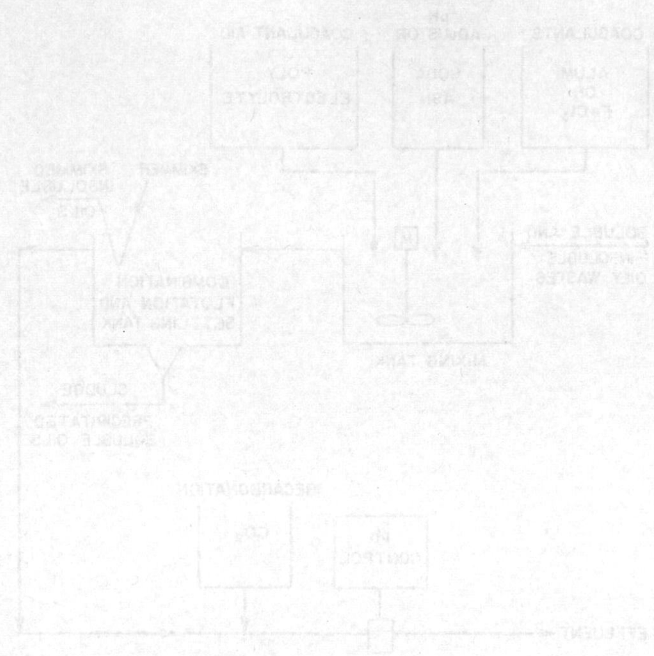

FIG. 9.8.1 Combined soluble and insoluble oil removal.

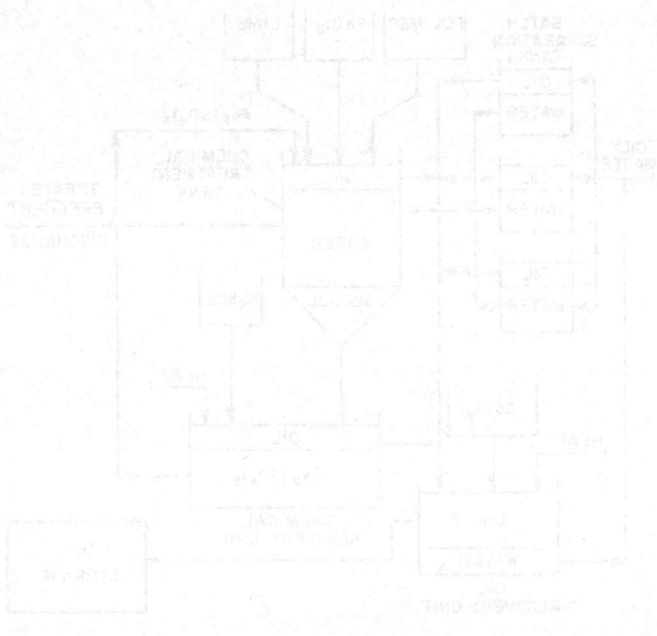

FIG. 9.8.2 Oily waste treatment system that uses chemical recovery and oil recovery.

Insoluble oils heavier than water can be recovered in a settling basin with a bottom sludge separator. The withdrawal rate of the sludge from the bottom must be proportional to the amount of heavy oil separated from these liquids. Oil coagulation techniques use the settling principle; the oil is rendered insoluble and heavier than the water so that it can be removed by settling with sludge removal.

For coagulated solids, wastewater treatment facilities use filtration through various media. Sand provides a satisfactory filter medium, but diatomaceous earth is more commonly used due to the disposability of the diatomaceous-earth filter cake. Continuous centrifuging also separates oils from water, but this technique is expensive. Also, for varying densities and volumes of oil in wastewater, treatment facilities must continuously adjust the centrifuge system.

Dispersed air flotation is also effective in separating oil from water. This technique has been practiced with and without prior chemical coagulation. Sometimes breaking down the foam after it has carried the oil from the water is difficult. Dispersants can break down the foam so that the oil can be properly disposed of.

ULTIMATE DISPOSAL

The ultimate disposal of oil frequently creates a problem. Occasionally, reclaimed oil can be reprocessed in a refinery. In other cases, oils are mixtures that are not favorable for reclamation. In other cases, reclamation is not economical.

Some wastewater treatment facilities dispose oil on land, but they must take proper precautions to prevent the oil from contaminating the groundwater. To a limit, soil microorganisms can break down oil and render it nontoxic. Facilities for burning waste oil are also available. Problems arise from the differing volatilities of the oils collected. Reclamation of the heat produced is beneficial but not always practical due to variations in the amount of oil produced.

TYPICAL TREATMENT SYSTEMS

In a typical combination treatment system for soluble and insoluble oil separation (see Figure 9.8.1), the soluble oils are precipitated by coagulants [illegible], and coagulant aids. Both solids and insoluble oils are separated in the combination flotation and settling tank. The effluent may require recarbonation with appropriate pH control before being discharged to a receiving stream.

Figure 9.8.2 shows a system for chemical treatment with separation and recovery of oil and ferric sulfate. The proper application of chemical treatment methods can effect satisfactory separation and removal of oil from liquid waste.

—D.B. Aulenbach

10

Sludge Stabilization and Dewatering

Carl E. Adams, Jr. | Frank W. Dittman | Brian L. Goodman | Frederick W. Keith, Jr. | Paul L. Stavenger | Michael S. Switzenbaum

10.1 STABILIZATION: AEROBIC DIGESTION

TYPES OF SLUDGES: a) Excess activated; b) Primary with activated or trickling-filter sludge

TOTAL QUANTITY OF WET SLUDGE: From an average treatment plant, between 1 and 2% of the sewage volume (influent) treated

VOLATILE SOLIDS REDUCTION EFFICIENCY: 40 to 60%

REDUCTION EFFICIENCY OF DEGRADABLE VSS: 85%

EXCESS SLUDGE PRODUCED: 0.2 lb per lb of BOD_5 removed

TYPICAL INFLUENT CONCENTRATION: Over 10,000 mg/l SS

DETENTION TIMES (DAYS): 10 to 15 (a); over 20 (b)

ORGANIC SOLIDS LOADINGS RATE (LB/1000 FT³): Below 100 (b); generally in the range of 40 to 120

AIR REQUIREMENTS (CFM/1000 FT³): 15 to 20 (a); 25 to 30 (b)

DIGESTOR TANK DESIGN: Similar to activated-sludge tanks; they are usually not covered or insulated

AERATION METHODS: Same as in conventional aeration tanks; mechanical surface aerators are superior to diffused aeration systems.

TYPICAL APPLICATIONS: Industrial and small municipal waste treatment plants in which mostly excess activated sludge is treated; cannot be used in cold climates

The aerobic stabilization of biological sludges, generated from wastewater treatment, is the basis for modifications in the activated-sludge process known as total oxidation and extended aeration. In many treatment plants, separate aerobic digestors stabilize mixtures of excess activated and primary sludge. The major objective of aerobic digestion is to produce a biologically stable end product suitable for disposal or subsequent treatment in a variety of processes. Aerobic digestion is generally more suited to the treatment of excess biological sludge than to primary sludge.

Primary sludge settles from raw waste prior to biological treatment. Secondary, biological sludge consists primarily of flocculated microorganisms and suspended organic material trapped or biosorbed to the floc. Secondary sludge is either excess activated sludge or trickling-filter humus. The mechanism of microbial degradation is different for various discrete mixtures of sludge. The degree of stabilization of volatile solids also varies with the sludge. This section discusses aerobic digestion as a separate treatment after solids–liquid separation is complete.

Theory and Mechanisms

Aerobic digestion is a process in which microorganisms obtain energy by endogenous or auto-oxidation of their cellular protoplasm. The biologically degradable constituents of cellular material are slowly oxidized to carbon dioxide, water, and ammonia, with ammonia being further converted to nitrates during the process.

The mechanism by which wastewater sludges is aerobically stabilized depends on the type of sludge being treated. For primary sludge, aerobic digestion follows a series of steps similar to anaerobic digestion. The suspended organic material must be enzymically converted to soluble constituents that can be degraded by the microbes for energy and nutrient supply. A simplified summary of the aerobic conversion of organic material into cellular material plus carbon dioxide and water is shown in Figure 10.1.1 and the following equation:

$$\underset{\text{(bacterial cell)}}{C_xH_yO_z} + NH_3 + O_2 + \text{Bacteria} \longrightarrow$$

$$C_5H_7NO_2 + CO_2 + H_2O(\text{cell synthesis}) \qquad 10.1(1)$$

DIGESTION OF PRIMARY SLUDGE

In the aerobic digestion of primary sludge, the primary sludge acts as a food supply for microorganisms. During the initial stages of aerobic stabilization, assuming an unlimited food supply with sufficient nutrients, the bacterial growth is limited only by the microbial rate of reproduction (log growth phase). Figure 10.1.2 shows this phenomenon. The oxygen uptake rate continually increases due to the increasing population of new bacteria.

As organic matter oxidation continues, the microorganisms enter a declining growth stage due to the limited food supply. The oxygen uptake rate also declines during this period. As the food supply becomes depleted, the organisms are forced to depend on internal storage products as a source of energy, and endogenous metabolism or respiration becomes prevalent. If endogenous bacterial cells

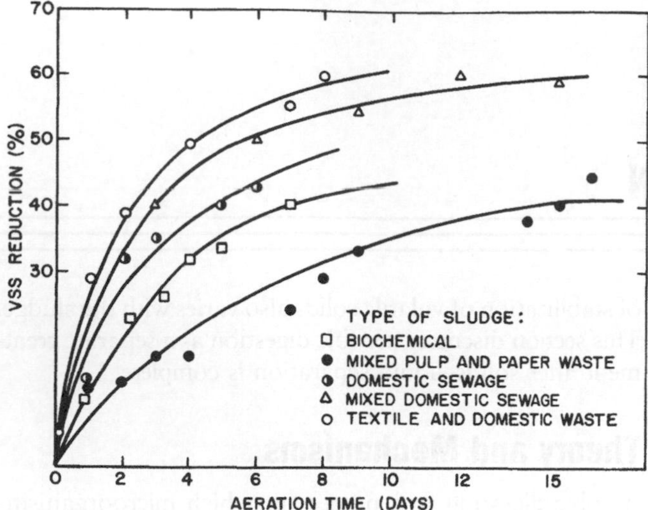

FIG. 10.1.1 Typical VSS reduction by aerobic digestion.

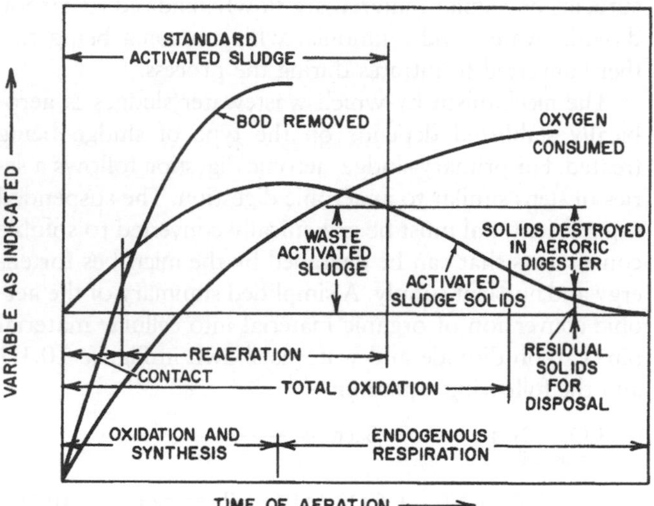

FIG. 10.1.2 Aerobic biological treatment.

are represented by the chemical formula $C_5H_7NO_2$, then the following equation gives the cellular destruction through aerobic digestion:

$$C_5H_7NO_2 + 5O_2 \longrightarrow 2H_2O + NH_3 + 5CO_2 \qquad 10.1(2)$$

As the surrounding food supply in the aerobic system becomes depleted and cannot supply sufficient organic matter for synthesis and energy, the rate of cellular destruction exceeds the organism growth rate. Although this phase of the metabolic cycle is generally referred to as the endogenous phase, it cannot be regarded simply as a period for individual cells using internal cellular carbon sources. The reason for this phenomenon is because a heterogeneous population of microorganisms is present, representing a complex ecosystem in which various microbial species serve as food for other members of the population. Eventually some organisms undergo cellular lysis, releasing protoplasm into the environment; the protoplasm is then used by other bacteria.

Aerobic stabilization of primary sludge results in a high F/M ratio in the aeration basin. Therefore, organic material in the primary sludge converts to bacterial cell material by synthesis, and the resulting change in the total VSS concentration is minimal. Hence, destruction of the bacterial cellular material requires long detention times.

DIGESTION OF SECONDARY SLUDGE

The aerobic digestion of excess secondary sludge is a continuation of the activated-sludge process. The F/M ratio is low, and little cell synthesis occurs. The major reaction is the oxidation and destruction of cellular constituents by lysis and auto-oxidation.

The cellular wall is composed of a polysaccharide-like material that is resistant to decomposition, resulting in a residual VSS concentration from the aerobic digestion process. However, this residual volatile portion is stable and does not produce problems in subsequent sludge handling operations or land disposal.

GENERAL BACKGROUND CONSIDERATIONS

The following parameters affect the aerobic digestion process:

1. Nature and characteristics of the sludge
2. Rate of sludge oxidation
3. Sludge age
4. Sludge loading rate
5. Temperature
6. Oxygen requirements

Generally, the majority of VSS digestion in the aerobic process occurs during the first 10 to 15 days of aeration. Figure 10.1.1 shows both the VSS reductions for a variety of sludges and the major portion of the reduction that occurs during the initial 10 days. The mixed pulp and paper waste sludge is most resistant to digestion (typical of paper mill sludge due to their high content of lignins and cellulose material).

Figures 10.1.3 and 10.1.4 show that maximum stabilization occurs within the 15 days of digestion. Figure 10.1.4 also reflects the increased reduction of volatile matter with increasing temperature.

The sludge age is defined as the ratio of the weight of VSS in the digestor to the weight of VSS added daily. The maximum sludge age is when no significant reduction occurs in the concentration of VSS. Laboratory studies indicate that VSS removal efficiency correlates with sludge age. One such study, summarized in Figure 10.1.5, was conducted at organic loadings between 42 and 112 lb per 1000 ft^3 and detention times of 15 to 30 days. The waste was a mixture of activated sludge and primary sludge. The following equation describes the relationship between the sludge age and VSS removal efficiency:

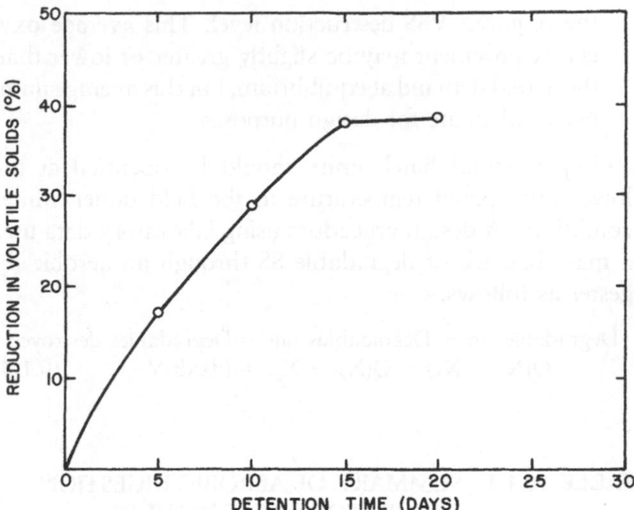

FIG. 10.1.3 Effect of detention time on aerobic digestion of activated sludge.

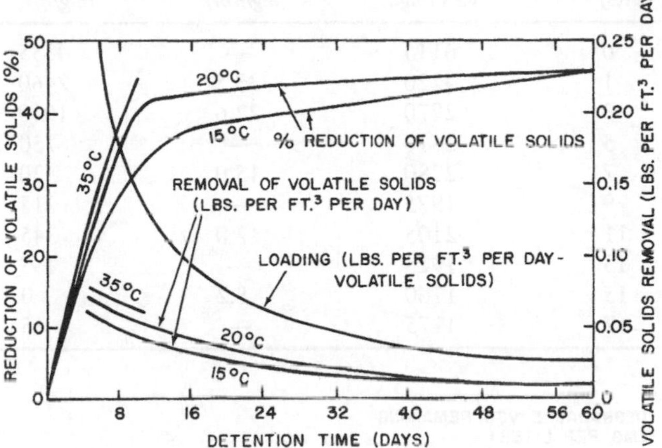

FIG. 10.1.4 Temperature effects on aerobic digestion.

$$\%\text{VSS Reduction} = 2.84 + 35.07 \log_{10}(\text{sludge age}) \quad 10.1(3)$$

Specific oxygen uptake rates (gm O_2 used per gram VSS per day) vary with detention time. Figure 10.1.6 shows the data reported for excess activated sludge.

DESIGN CONSIDERATIONS

Aerobic digestion is usually applied to extended-aeration or contact-stabilization, activated-sludge plants. However, the process is also suitable for many industrial and municipal biological sludges, including trickling-filter humus and excess activated sludge. Information pertaining to design criteria is not abundant. However, the principal design considerations are as follows:

1. Estimated daily quantity of sludge entering the digester
2. Specific oxygen requirements supplied by diffused or mechanical surface aerators
3. Digester detention time

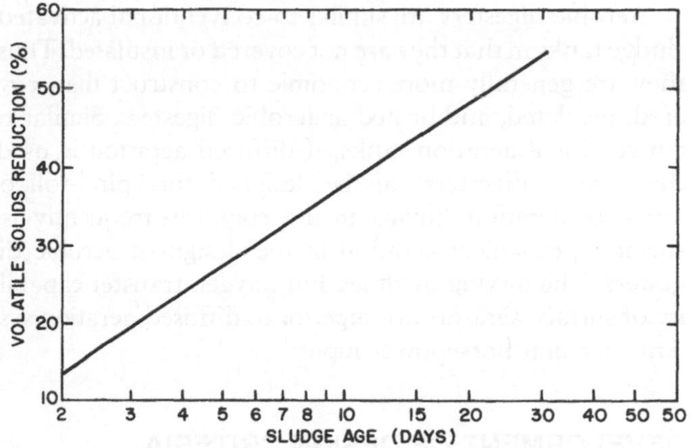

FIG. 10.1.5 VSS reductions as a function of sludge age.

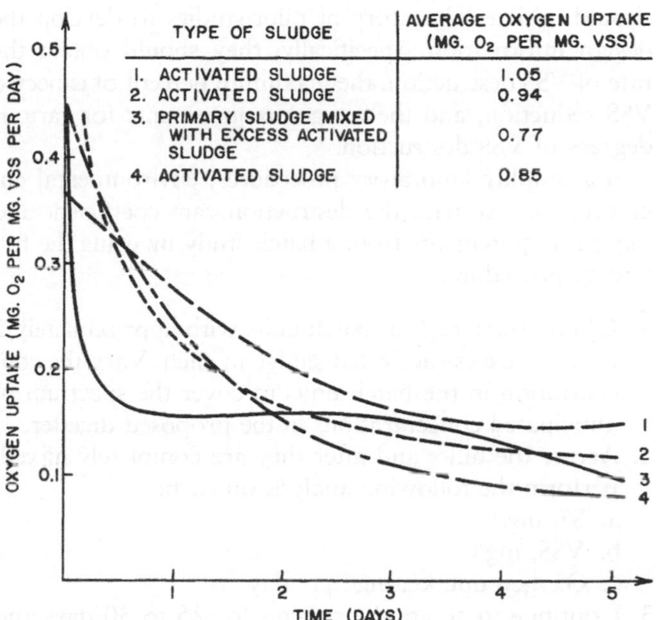

FIG. 10.1.6 Typical oxygen requirements for the aerobic digestion of sludge.

4. Efficiency of VSS reduction required
5. Solids loading rate

Recommended loadings for aerobically treated mixtures of primary and activated sludge or primary and trickling-filter sludge are less than 100 lb per 1000 ft^3 with a minimum recommended detention time of 20 days. The suggested minimum detention time for excess activated sludge is 10 days and preferably 15 days. If the temperature in the digestion basin is less than 60°F, additional capacity should be provided.

Recommended oxygen (air) requirements are 15 to 20 cfm per 1000 ft^3 of tank capacity; however, if only primary sludge is treated or waste-activated sludge is withdrawn directly from the final clarifier, the air supply should be 25 to 30 cfm per 1000 ft^3 of tank capacity.

Aerobic digesters are similar to conventional activated-sludge tanks in that they are not covered or insulated. Thus, they are generally more economic to construct than covered, insulated, and heated anaerobic digesters. Similar to conventional aeration tanks, if diffused aeration is used, the aerobic digesters can be designed for spiral-roll or cross-roll aeration. Environmental engineers frequently use surface mechanical aeration in the design of aerobic digesters. The mixing qualities and oxygen transfer capability of surface aerators are superior to diffused-aeration systems per unit horsepower input.

DEVELOPMENT OF DESIGN CRITERIA

Because of the varying nature of sludge and the lack of sufficient design information, environmental engineers should initiate laboratory or pilot studies to develop the design information. Specifically, they should obtain the rate of VSS destruction, the maximum percent of expected VSS reduction, and the oxygen requirements for various degrees of VSS destruction.

To simplify laboratory procedures, environmental engineers can estimate the destruction rate coefficient and oxygen requirements from a batch study by using the following procedure:

1. Obtain three or four batch units with approximately 2 to 4 l of excess activated sludge in each. Vary the concentration in the batch units to cover the spectrum of anticipated concentrations in the proposed digester.
2. Aerate the units and after they are completely mixed, perform the following analysis on each:
   a. SS, mg/l
   b. VSS, mg/l
   c. Oxygen uptake, mg/l per day
3. Continue to aerate the systems for 25 to 30 days and perform the preceding analyses every 3 days on each unit (see Table 10.1.1).
4. Plot the VSS and SS concentration remaining versus the sludge age (aeration time). From this plot, approximate the oxidizable or degradable fraction of the solids, i.e., the residual VSS remaining after 25 or 30 days of digestion can be taken as the nondegradable portion of the volatile matter (1775 mg/l in Table 10.1.1). The volatile solids destroyed during this aeration are the maximum degradable portion of the volatile matter.
5. For each sampling period, recalculate the remaining degradable VSS. Plot the degradable VSS remaining as a function of detention time (see Figure 10.1.7), and calculate the reaction rate coefficient (k). Note the effect of initial VSS on the destruction rate, and indicate this relationship, if noticeable.
6. Record the oxygen uptake rate as a function of sludge age or aeration time (see Figure 10.1.8). When sizing aeration equipment, estimate the oxygen utilization value as the average value exerted during aeration for the required VSS destruction level. This average oxygen requirement may be slightly greater or lower than the actual demand at equilibrium, but this average number is adequate for design purposes.

Experimental batch units should be operated at the lower anticipated temperature in the field under winter conditions. A design procedure using laboratory data uses a mass balance of degradable SS through an aerobic digester as follows:

Degradables in − Degradables out = Degradables destroyed

$$Q(X_1 - X_n) - Q(X_2 - X_n) = (dx/dt)V \qquad 10.1(4)$$

TABLE 10.1.1 SUMMARY OF AEROBIC DIGESTION LABORATORY TEST RESULTS

Time of Aeration (days)	*VSS (mg/l)*	*O_2 Uptake (mg/l/hr)*	*Oxidizable VSS Remaining (mg/l)*
0	6115	—	4355
1	4220	33.5	2460
3	2770	22.6	1010
5	2510	—	750
7	2280	15.0	520
9	1975	—	215
11	2105	12.0	345
13	1925	—	165
15	1760	8.2	0
17	1775	—	15

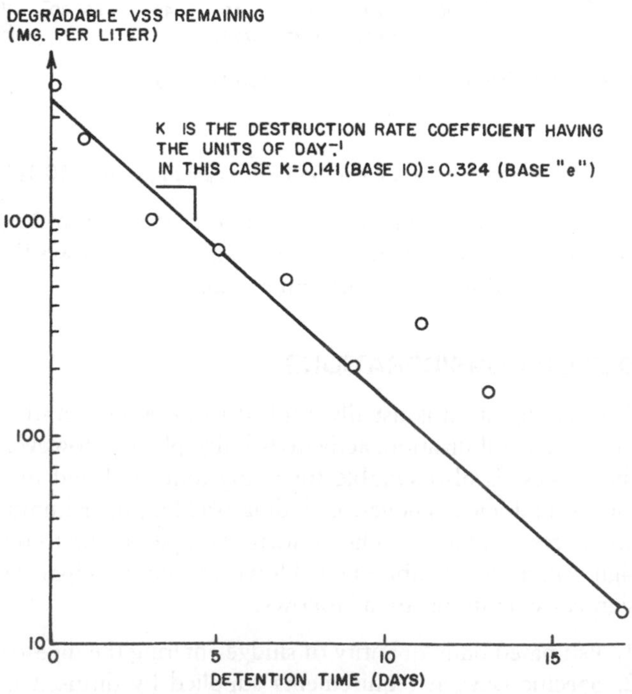

FIG. 10.1.7 Determination of batch rate coefficients in an aerobic digester.

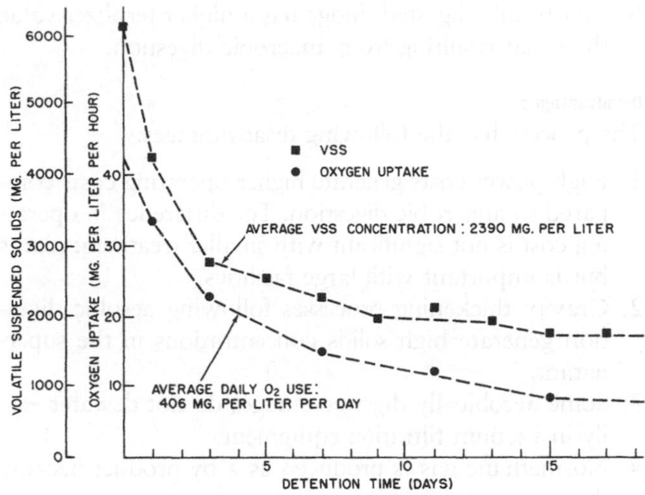

FIG. 10.1.8 Chronological data from a batch aerobic digester.

where:

X_1 = total VSS in influent (mg/l)
X_2 = total VSS in effluent (mg/l)
X_n = total nondegradable VSS, assumed identical for influent and effluent (mg/l)
Q = hydraulic flow (mgd)
V = reactor volume (mg)
$\frac{dx}{dt}$ = rate of destruction of degradable influent VSS ($dx/dt = k_b(X_2 - X_n)$)
k_b = destruction rate coefficient of degradables, base e (day^{-1})

Thus:

$$(X_1 - X_n) - (X_2 - X_n) = k_b(X_2 - X_n)V/Q = k_bt(X_2 - X_n) \quad 10.1(5)$$

or

$$t = \frac{X_1 - X_2}{k_b(X_2 - X_n)} \quad 10.1(6)$$

DESIGN EXAMPLE

In this example, the task is to design an aerobic digester for excess activated sludge from a secondary clarifier. The influent SS to the digester contain 12,000 mg/l SS, of which 83% are volatile. The design flow to the digester is 50,000 gpd, and the goal is 85% reduction of the degradable VSS content. The procedure is as follows:

1. Prepare batch reactors and correlate the data as shown in Table 10.1.1 and Figure 10.1.7. From the latter, the rate coefficient (k) is 0.141 (base 10), corrected to 0.324 for the logarithmic base e.
2. Calculate the nondegradable residue and the degradable concentration in the effluent.

   From Table 10.1.1, nondegradables: $(X_n) = 1775/6115 = 29\%$ of X_1

   Influent VSS (X_1): $= 0.83\ (12{,}000) = 10{,}000$ mg/l

   thus: $X_n = 0.29\ (10{,}000) = 2900$ mg/l

   Influent degradables:

   $(X_1 - X_n) = 10{,}000 - 2900 = 7100$ mg/l

   Effluent degradables:

   $(X_2 - X_n) = 0.15\ (7100) = 1070$ mg/l

   Total effluent VSS:

   $X_2 = 1070 + 2900 = 3970$ mg/l

3. Calculate the required detention time:

$$t = \frac{X_1 - X_2}{k(X_2 - X_n)} = \frac{10{,}000 - 3970}{0.324(1070)} = 17.4 \text{ days. Use 18 days.}$$

   Required volume = 0.05 mgd (18) = 0.9 mg = 120,000 ft^3

4. Calculate the oxygen requirements. Compute the average oxygen requirement by calculating the area under the curve in Figure 10.1.8 and dividing it by the design detention time.

   Area under the oxygen curve = 7310 mg/l O_2
   Average daily use = 7310/18 = 406 mg/l day

   Similarly, estimate the average VSS during a test period to determine the specific oxygen requirements.

   Area under curve = 43,040 day-mg/l
   Average VSS concentration = 2390 mg/l
   Specific oxygen utilization = 406/2390 = 0.17 mg/mg/day

   The total oxygen requirements based on an average VSS concentration X_2 in the digester effluent of 3970 mg/l are then:

$$\left(0.17 \frac{\text{mg } O_2}{\text{mg VSS-day}} \times 3970 \frac{\text{mg VSS}}{1}\right)\left(\frac{8.34 \text{ lb}}{\text{MG-mg/l}}\right)(0.9 \text{ mg}) = 5050 \text{ lb } O_2/\text{day} = 210 \text{ lb } O_2/\text{hour}$$

   Check the power requirements for mixing to see if mixing or oxygen requirements control the design.

Summary

Aerobic digestion is practiced in many industrial and a few municipal treatment facilities. However, because specific technical data are lacking, the inference is that aerobic digestion is a new, unproven technique for solids handling. However, the aerobic digestion process has many potential benefits and warrants attention by design engineers.

Advantages

Aerobic digestion has the following advantages:

1. A biologically stable end product is produced.
2. The stable end product is relatively noxious; hence, land disposal by holding lagoons or spray irrigation is recommended.
3. Due to simplicity of construction, capital costs for the aerobic system are low compared with anaerobic digestion and other solid handling schemes.
4. Aerobically digested sludge generally possesses efficient dewatering characteristics. It drains well when placed on a sand bed and resists rewetting during rainfall.
5. Volatile solids reduction equivalent to the anaerobic digestion results is possible with aerobic systems in treating secondary sludges.
6. Supernatant (floating) liquors from aerobic digestion possess a lower BOD content than those from anaerobic digestion. The aerobic supernatant commonly possesses a BOD of less than 100 mg/l. This advantage is significant because many conventional biological treatment plants are overloaded due to recycling of high-BOD supernatant liquors from anaerobic digestors.
7. Fewer operation problems occur in aerobic digestion processes than in the more complex, anaerobic process due to the higher stability of the aerobic system. Therefore, lower maintenance costs and less skillful labor are needed with an aerobic facility.
8. Aerobically digested sludge has a higher fertilizer value than that resulting from anaerobic digestion.

Disadvantages

The process has the following disadvantages:

1. High power costs generate higher operating costs compared to anaerobic digestion. The difference in operating cost is not significant with smaller treatment plants but is important with large facilities.
2. Gravity thickening processes following aerobic digestion generate high solids concentrations in the supernatant.
3. Some aerobically digested sludges do not dewater easily in vacuum filtration equipment.
4. No methane gas is produced as a by-product because the process is aerobic.
5. The solids reduction efficiency of the aerobic digester can vary with extreme changes in ambient temperature, which subsequently affects the aeration basin temperature.

The aerobic digestion process is well suited for industrial sludge treatment and small, municipal, activated-sludge plants. The industrial community favors aerobic digestion because of the low capital investment and simple operation. Industry often uses mechanical aerators in inexpensive open tanks followed by holding or disposal lagoons. Although a difference in emphasis exists at municipal waste treatment plants with regard to economics, logically, aerobic digestion should be evaluated, particularly for activated-sludge facilities.

—*C.E. Adams*

10.2 STABILIZATION: ANAEROBIC DIGESTION

TYPES OF DESIGNS

a) Conventional (unmixed and unheated) or low rate
a1) Conventional primary (digester for most VSS reduction)
a2) Conventional secondary (digester for solids–liquid separation)
b) High-rate (completely mixed and heated)
b1) High-rate primary (digester for most VSS reduction)
b2) High-rate secondary (digester for solids–liquid separation)

CONDITIONING OPERATIONS APPLIED BETWEEN PRIMARY AND SECONDARY DIGESTERS

Elutriation and chemical additive treatment

VOLATILE SOLIDS REDUCTION EFFICIENCY

50 to 60% average

TYPICAL INFLUENT CONCENTRATION

65 to 70% VSS

DETENTION TIME (DAYS)

10 to 20 (b); 30 to 60 (a); under 15 (b1); and up to 30 (a2)

ORGANIC SOLIDS (VSS) LOADING RATES (LBS/1000 FT3/DAY)

30 to 70 (a); 100 to 400 (b)

OPERATING TEMPERATURE

Heated to 85 to 95°F (a1, b1)

OPERATION

Intermittent (a,b); continuous (b)

DIGESTER DESIGNS

Primary digester with fixed; secondary with floating covers

GAS PRODUCTION

15 ft^3 per lb VSS destroyed or 1 ft^3 per capita per day

DIGESTER GAS HEATING VALUE

640 to 700 Btu/ft^3

DIGESTER HEAT LOSS

1.3 to 2.6 Btu/hr/ft^3 as a function of geographic location

Solids removed from wastewater by treatment plants and those accumulating from the conversion of degradable organic compounds to microbial mass contain substantial amounts of biologically degradable matter. Wastewater treatment facilities frequently use anaerobic digestion to reduce this organic material to carbon dioxide, methane, and other inert end products. In addition to the stabilization of degradable organic matter, anaerobic digestion also substantially decreases the weight of solids that require subsequent processing or disposal. In addition, anaerobic digestion produces a combustible gas that can be used for heating and incineration.

Conventional and High-Rate Digesters

Wastewater treatment facilities use both conventional and high-rate anaerobic digesters to reduce volatile solids in wastewater treatment sludge (see Figure 10.2.1).

Conventional digesters are loaded at a rate of 30 to 100 lb VSS per 1000 ft^3 per day (commonly 30 to 70 lb per 1000 ft^3 per day). Both feeding and sludge withdrawal are intermittent. Detention times of 30 to 60 days are common. Conventional primary digesters are heated to 85° to 95°F. Primary digesters are often followed by one or more secondary, unheated digesters, where primary digester detention time can be shorter. Although primary digesters are frequently equipped with fixed covers, secondary digester covers are often the floating type. Mixing is not used in primary digesters unless the digesters are followed by secondary, unmixed digester units.

High-rate digesters are loaded at a rate of 100 to 500 lb VSS per 1000 ft^3 per day (commonly 100 to 400 lb per 1000 ft^3 per day). Feeding and sludge withdrawal can be intermittent or continuous. High-rate primary digesters are heated to 85° to 95°F. High-rate units can operate in series, or they can be followed by conventional, unheated digesters, sludge holding and separation tanks, or other means of solids–liquid separation. Mixing by either gas recirculation or mechanical means is used with high rate units.

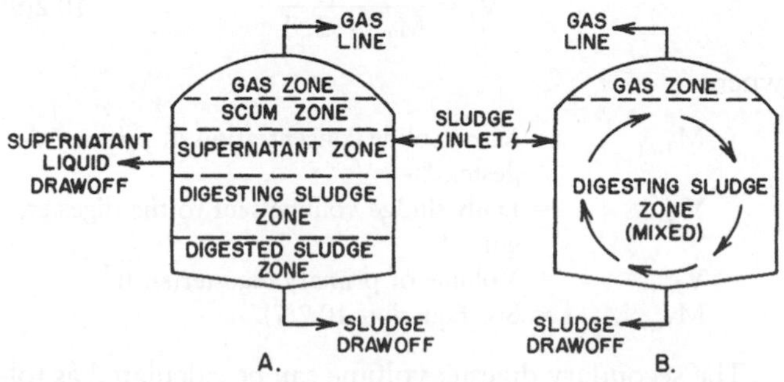

FIG. 10.2.1 Types of anaerobic digesters. **A,** Conventional; **B,** High rate.

Physical Design

The high-rate design is best for primary digesters with fixed covers, water heaters, external heat exchangers, and gas recirculation or mechanical mixing means. Secondary digesters should have floating covers. Interconnecting primary and secondary gas piping is recommended. Accessories, such as gas scrubbers for hydrogen sulfide removal, waste-gas burners, gas meters, gas compressors, gas holders, condensate–sediment traps, and regulators, should also be provided.

The detention time for a primary digester (high-rate) is usually 15 days or less. The maximum detention time for a secondary digester is 30 days.

DIGESTER DIMENSIONS AND ACCESSORIES

Many wastewater treatment facilities use a 3 in 12 (3 ft vertical, 12 ft horizontal) digester bottom slope (14°). The digester design can be based on a 15-ft minimum side wall height with 2 ft of freeboard. Medium and large wastewater treatment facilities should consider duplicate digestion units. A means for supernatant liquid drawoff should be provided for secondary digesters only if the primary digesters are the high-rate type. The digesters should be equipped with flame arresters and vacuum relief devices.

The gas holder (usually a sphere) is designed on the basis of the following relationship:

$$C = V\left(\frac{P - P'}{14.7}\right)\left(\frac{520}{T + 460}\right) \qquad 10.2(1)$$

where:

C = Capacity of the gas holding sphere (ft^3 of gas at 14.7 psia and 60°F)
V = Volume of sphere, ft^3
P = Maximum operating pressure, psig
P′ = Minimum operating pressure, psig
T = Temperature of stored gas, °F

GAS PRODUCTION AND HEATING REQUIREMENTS

The storage capacity of the gas holder should provide for a detention time of 0.25 to 0.50 days. The normal gas production is 0.9 to 1.0 ft^3 per capita of population per day when mixed primary and activated sludges are digested. Gas production is also estimated as 15 ft^3 per lb VSS destroyed. The Btu value of digester gases is generally 640 to 703 Btu/ft^3.

The following equation gives the digester heat loss:

$$Q = U(A)(T_2 - T_1) \quad 10.2(2)$$

where:

Q = heat loss, Btu/hr
U = Overall coefficient of heat transfer, Btu/hr/ft^2/°F (see Table 10.2.1).
A = Area (ft^2) normal to direction of heat flow
T_2 = Digestion temperature, °F
T_1 = Average external temperature for coldest two-week period, °F

Alternatively, the heat loss is 1°F per day for the entire digester contents, or 2600 Btu/hr/1000 ft^3 of tank contents in the northern United States. For the southern United States, half of this rate can be assumed.

The following equation gives the sludge heat requirement:

$$Q = WC(T_2 - T_1) \quad 10.2(3)$$

where:

Q = Heat required to raise sludge to the digestion temperature, Btu per day
W = Sludge added to the digester daily, lb
C = Mean specific heat of sludge, ≈1.0, Btu/lbm-°F
T_2 = Digestion temperature, °F
T_1 = Sludge temperature, coldest two-week period, °F

In the absence of other information, the sludge temperature can be assumed to be 50°F in the southern United States, 45°F in the central parts of the United States, and 40°F in the northern United States.

The following equation gives the length of the external, jacketed, sludge pipe, heat exchanger:

$$L = \frac{A}{\pi \frac{D}{12}} = \frac{H}{\pi U\, \Delta T_{LM} D/12} \quad 10.2(4)$$

where:

H = Total heat requirement, Btu/hr
U = Overall heat transfer coefficient (a value of 140 is used in the absence of other information) Btu/hr/°F/ft^2
A = Heat exchange surface area, ft^2
D = Diameter of sludge pipe, in
ΔT_{LM} = Logarithmic mean of the temperature difference $\left(\frac{\Delta T_1 - \Delta T_2}{\log_e \Delta T_1/\Delta T_2}\right)$

TABLE 10.2.1 OVERALL HEAT TRANSFER COEFFICIENTS FOR ANAEROBIC DIGESTERS

Digester Type	*Overall Coefficient of Heat Transfer (Btu/hr/ft²/°F)*
Concrete roof	0.5
Floating cover	0.24
Concrete wall air space	0.35
Concrete wall in wet earth	0.25
Concrete wall in dry earth	0.18
Floor	0.12

Digester Sizing

The following equations predict the VSS destruction:

Conventional digesters:

$$V_d = 30 + \frac{t}{2} \quad 10.2(5)$$

High-rate digesters:

$$V_d = 13.7 \operatorname{Log}_e (t) + 18.94 \quad 10.2(6)$$

where:

V_d = VSS destroyed, %
t = Time of digestion, days

These equations specifically apply to sludge with an initial VSS content of 65 to 70%. The quantity of solids remaining in the primary digester at the end of the digestion period can be calculated as follows:

$$M_{T_t} = M_{T_w}(t_1)\left(1 - \frac{\%V(V_d)}{2}\right) \quad 10.2(7)$$

where:

M_{T_t} = Solids remaining in digester, lb
M_{T_w} = Daily solid waste input into digester, lb
t_1 = Digestion period, days
$\%V$ = VSS in the sludge received by the digester, %

The following equations calculate the required primary digester volume:

$$M_{T_{m1}} = \frac{[M_{T_w}(t) + M_{T_t}]/2}{t(V_W)(8.34)} \quad 10.2(8)$$

$$V_1 = \frac{M_{T_t}}{M_{T_{m1}}(62.5)} \quad 10.2(9)$$

where:

$M_{T_{m1}}$ = Mean solids concentration in primary digester, %
V_W = Daily sludge volume sent to the digester, gal
V_1 = Volume of primary digester(s), ft^3
M_{T_w}, M_{T_t} t = See Equation 10.2(7).

The secondary digester volume can be calculated as follows:

$$V_2 = \frac{t_2\left(\frac{M_{T_t}}{t_1}\right)}{M_{T_{m2}}(7.5)(8.34)} \quad 10.2(10)$$

where:

V_2 = Volume of secondary digester(s), ft^3
t_1, t_2 = Detention time in primary and secondary digesters, days
$M_{T_{m2}}$ = Mean solids concentration in secondary digesters, %

The following equation calculates digester diameters:

$$r = \sqrt{\frac{V/n}{h\pi}} \quad 10.2(11)$$

where:

r = Radius, ft
V = Total volume required, ft^3
n = Number of digesters
π = 3.1416
h = Mean liquid depth, ft

ANAEROBIC DIGESTER DESIGN EXAMPLE

In this example, high-rate, primary digesters (heated and completely mixed) followed by secondary digesters (unheated and unmixed) provide anaerobic digestion of primary and excess activated sludge from a 10 mgd wastewater treatment plant. Secondary digesters provide a maximum 30-day detention time and are intended mainly for solids–liquid separation, sludge thickening, and sludge storage. The daily loading of solids to the primary digesters is as follows:

1. Primary (raw) sludge: 12,010 lb per day, (65% VSS)
2. Excess activated sludge: 8162 lb per day, (77% VSS)

Total solids = 20,172 lb per day, (70% VSS)
Volatile solids = 14,091 lb per day

The daily loading in terms of sludge volume is as follows:

1. Primary (raw sludge): 20,000 gpd (7.2% solids)
2. Excess activated sludge: 15,495 gpd (6.3% solids)

Total sludge volume = 35,495 gpd (6.8% solids)

VSS reduction in the primary digesters with a 15-day digestion period can be calculated in accordance with Equation 10.2(6) as follows:

$$V_d = 13.7 \text{ Log}_e (15) + 18.94 = 56\%$$

The quantity of solids remaining in the primary digesters at the end of the 15-day digestion period can be determined with Equation 10.2(7) as follows:

$$M_{T_t} = 20{,}172(15)\left(1 - \frac{[.70][0.56]}{2}\right)$$
$$= 243{,}274 \text{ lb}$$

Equations 10.2(8) and (9) determine the required volume V_1 of the primary digesters without the specific gravity correction as follows:

Mean solids concentration =

$$\frac{([15][20{,}172] + 243{,}274)/2}{(15)(35{,}495)(8.34)} = 6.15\%$$

$$V_1 = \frac{243{,}274}{0.0615(62.5)} = 63.291 \text{ ft}^3$$

The digester loading is therefore as follows:

$$\frac{14{,}091 \text{ lb. VSS per day}}{63{,}291 \text{ ft.}^3} = 220 \text{ lb VSS per 1000 ft}^3 \text{ per day}$$

The amount of solids withdrawn to the secondary digesters on a daily basis is as follows:

$$\frac{243{,}274 \text{ lb.}}{15 \text{ days}} = 16{,}218 \text{ lb per day}$$

If the average sludge solids concentration in the secondary digesters (achieved through supernatant decantation) is 8%, Equation 10.2(10) calculates the required digester volume V_2 as follows:

$$V_2 = \frac{(30)(243{,}274)/15}{0.08(7.5)(8.34)} = 97{,}230 \text{ ft}^3$$

This calculation assumes no digestion in the secondary digesters.

DIGESTER DIMENSIONS

With two primary and two secondary digesters, a mean water depth of 17.5 ft, a 15-ft sidewall depth, and 2 ft of freeboard, Equation 10.2(11) determines the required individual digester sizes as follows:

Primary digesters each:

$$r = \sqrt{\frac{63{,}291/2}{(3.1416)(17.5)}} \cong 24 \text{ ft}$$

Secondary digesters each:

$$r = \sqrt{\frac{97{,}230/2}{(3.1416)(17.5)}} \cong 30 \text{ ft}$$

GAS HOLDER SIZING

The gas holder can be designed as a sphere with sufficient volume to provide 0.5-day storage capacity. A maximum gas compression of 30 psig and a minimum pressure of 10 psig at a stored gas temperature of 80°F can be the design basis. The gas production rate is assumed to be 15 ft^3 per lb of VSS destroyed as follows:

$$14{,}091(0.56)(15) = 118{,}364 \text{ scfd}$$

Solving Equation 10.2(1) gives the required gas sphere capacity as follows:

$$0.5(118{,}364) = V\left(\frac{30 - 10}{14.7}\right)\left(\frac{520}{80 + 460}\right)$$

$$V = 45{,}190 \text{ ft}^3$$

The sphere diameter, therefore, is as follows:

$$D = 2r = 2\sqrt[3]{\frac{V}{(4\pi)/3}} = 2\sqrt[3]{\frac{45{,}190}{4.189}} \cong 45 \text{ ft}$$

The normal range of gas production is ≈ 0.9 – 1.0 ft^3 per capita per day when mixed primary and activated sludges are digested. In this example, the production is as follows:

$$\frac{118{,}364 \text{ ft}^3 \text{ per day}}{100{,}000 \text{ population}} = 1.2 \text{ ft}^3 \text{ per capita per day}$$

PRIMARY DIGESTER HEATING REQUIREMENTS

For the purposes of this example, a digestion temperature of 95°F and a plant location in the northern U.S. where the lowest (two-week mean) daily temperature is 10°F are used. All digester sidewall areas are earth-covered, with 75% of the areas covered by dry earth and the remaining 25%, as well as the floor area, in groundwater. The mean winter ground temperature is 25°F and the groundwater temperature is 40°F. Equation 10.2(2) calculates the heat losses for both primary digesters as follows:

1. Fixed cover (concrete): $Q = (\pi)(r^2)(2)(U)(T_2 - T_1)$

$$Q = 3.1416(24)^2(2)(0.5)(95 - 10)$$
$$= 153{,}813 \text{ Btu/hr}$$

2. Dry-earth-covered sidewall area:

$$Q = 0.75(3.1416[48][15][2][0.18][95 - 25])$$
$$= 42{,}751 \text{ Btu/hr}$$

3. Wet sidewall area:

$$Q = 0.25(3.1416[48][15][2][0.25][95 - 40])$$
$$= 15{,}551 \text{ Btu/hr}$$

4. Floor area:

$$Q = 3.1416(24)^2(2)(0.12)(95 - 40) = 23{,}886 \text{ Btu/hr}$$

5. Total primary digester area heat loss:

153,813
42,751
15,551
+23,886
236,001 Btu/hr

Alternatively, the approximate digester heat loss in the northern United States can be computed as 2600 Btu/hr/1000 ft^3 of digester volume as follows:

$$2600(63.3) = 164{,}580 \text{ Btu/hr}$$

This estimate, however, must be modified for the prevailing conditions, actual winter temperatures, or the presence of groundwater, e.g., at the plant site.

6. Sludge heating requirement using Equation 10.2(3): (The sludge temperature as it enters the primary digester is 50°F for this example.)

$$Q = \frac{20{,}172}{0.068}(1)(95 - 50)$$
$$= 13{,}349{,}118 \text{ Btu/day}$$
$$= 556{,}213 \text{ Btu/hr}$$

7. Total primary digester heating requirement:

Tank	236,001
Sludge	+556,213
Total	792,214 Btu/hr

For digester heating, a gas-fired heat exchanger can use the digester gas as fuel. The digester gas is assumed to have a heating value of 650 Btu per ft^3. Therefore, the excess heat can be calculated as:

1. 118,364 ft^3/day (650 Btu/ft^3) = 76,936,600 Btu/day = 3,205,692 Btu/hour

   3,205,692 – 792,214 = 2,413,478 Btu/hr excess heat is available for sludge incineration, building heat, and so forth.
2. Heat exchanger length: In this example, an external heat exchanger with 4-in sludge tubes jacketed by 6-in water tubes is used. Water enters the exchanger at 145°F and exits at 100°F. The sludge enters at 80°F and exits at 95°F. An overall heat transfer coefficient of 140 can be assumed. Equation 10.2(4) calculates the required heat exchanger length is calculated as follows:

$$\Delta T_1 = 145 - 80 = 65°F$$
$$\Delta T_2 = 100 - 95 = 5°F$$

$$\Delta T_{LM} = \frac{65 - 5}{\text{Log}_e\, 65/5} = 23.4°F$$

$$A = \frac{H}{U\,\Delta T_{LM}} = \frac{792{,}214}{(140)(23.4)} = 242 \text{ ft}^2$$

$$L = \frac{242}{(3.1416)(4/12)} = 231 \text{ ft}$$

—B.L. Goodman

10.3 SLUDGE THICKENING

TYPE OF THICKENER

a) Gravity settling
b) Pressure flotation

TYPICAL DEGREE OF SLUDGE THICKENING ACHIEVED

By a factor of 3 (a); or 4 to 6% solids in thickened stream (a,b)

TYPICAL SOLIDS FLUX RATE (LB/FT²/HR)

0.5 (a)

TYPICAL OVERFLOW RATE (GPD/FT²)

200 (a); and 2500 to 5000 (b)

HYDRAULIC DETENTION TIME (HRS)

0.3 (b); and from 6 and up (a)

Excess activated sludge and trickling-filter humus are almost always withdrawn at a solids concentration of less than 1% by weight. The underflow solids concentrations of primary clarifiers are frequently less than 5% by weight. To decrease the hydraulic loading on sludge digestion and dewatering units and thus increase their efficiency in terms of the weight of solids processed per unit area or volume per unit time, wastewater treatment facilities commonly use thickeners.

Gravity Thickening

Gravity-type thickeners (see Figure 10.3.1), in which waste sludge is continuously fed and thickened sludge is continuously withdrawn, are commonly used in wastewater treatment systems. The required thickener area depends on the rate of solids arrival at the thickener bottom, which depends on the settling velocity of the sludge solids and the concentration of the thickened sludge removed from the unit. Solids move downward through the thickener at the combined flux rate given in the following equation:

$$G = C_{in}(V_s + V_u) \qquad 10.3(1)$$

where:

G = Solids flux rate, lb/ft²/hr
C_{in} = Solids concentration in thickener influent, lb/ft³
V_s = Solids settling velocity due to gravity, ft/hr
V_u = Solids downward transport velocity due to underflow solids removal, ft/hr

The following equation expresses the downward transport velocity (V_u) due to concentrated solids withdrawal:

$$V_u = \frac{G}{C_{out}} \qquad 10.3(2)$$

where:

C_{out} = Concentration of thickened solids removed, lb/ft³

Thus, a limiting solids flux rate G_L exists and can be determined for any combination of values for the pertinent variables. The following equation gives the required thickener area A_T (in ft²):

$$A_T = \frac{M_{Tw}}{G_L} \qquad 10.3(3)$$

where:

M_{Tw} = hourly solid waste input into thickener, lb/hr

Environmental engineers can derive the required design data from settling tests conducted in cylinders with diameters of at least 3 to 6 in, and preferably greater, having a depth of from 3 ft to full-scale depth. These cylinders should be equipped with a stirring device having a tip speed of about 40 ft/hr. Figure 10.3.2 plots the data derived from these tests.

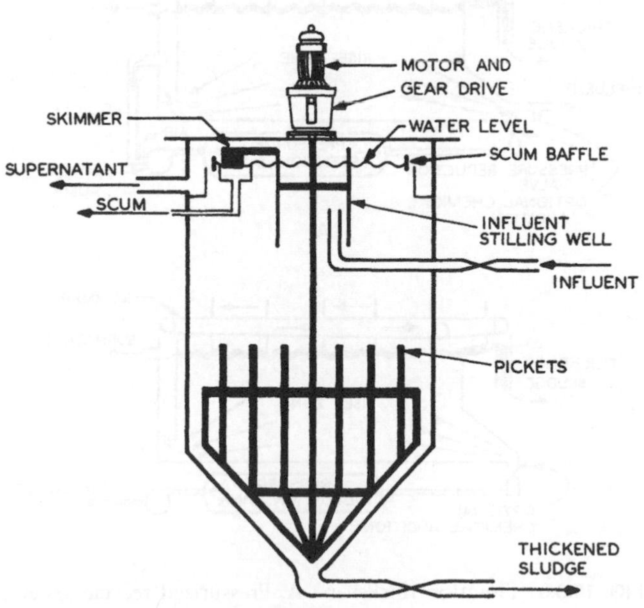

FIG. 10.3.1 Gravity thickener.

GRAVITY THICKENER DESIGN EXAMPLE

For the purposes of this example, the solids to be thickened are excess activated sludge with an initial solids concentration of 7 gm/l (0.434 lb/ft³). This sludge continuously enters the thickener, and the thickened sludge is continuously withdrawn at a concentration of 21 gm/l

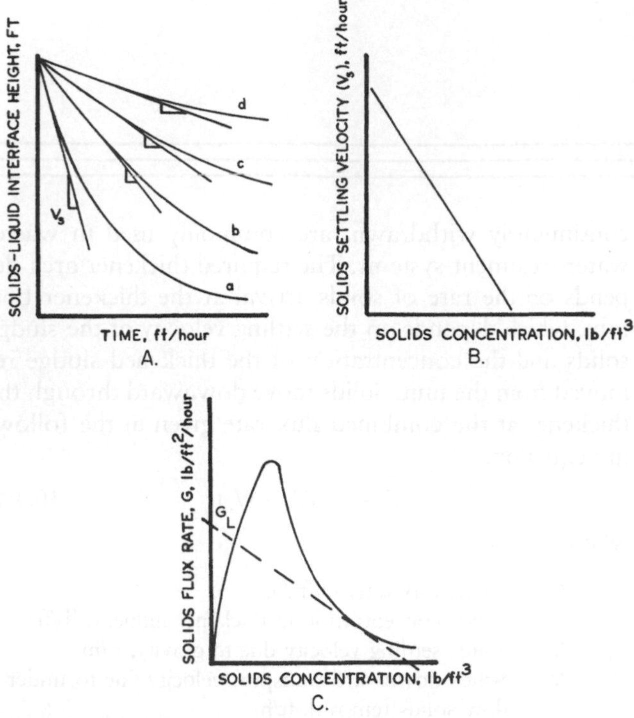

FIG. 10.3.2 Gravity thickener design data plots. Data plot procedure: 1. Conduct settling tests at various initial solids concentrations (a through d in panel A), covering the range of expected thickener influent solids concentrations; 2. Determine the settling velocity V_s for each concentration as the slope of a line tangent to the initial lineal settling rate portion of each curve; 3. Plot V_s values versus solids concentration, and construct a line of best fit; 4. Compute values of the flux rate G for a variety of solids concentrations (G = solids concentration, lb/ft³ × settling velocity, ft/hr), and plot as in panel C; and 5. Determine the value of the limiting flux rate G_L as the intercept on the flux rate axis of a line tangent to the flux curve drawn from the desired underflow solids concentration.

(1.302 lb/ft³). The G_L is determined from data plots (see Figure 10.3.2) 0.52 lb/ft²/hr. The volume of sludge to be concentrated is 5200 gph (693.3 ft³/hr). The required thickener area can be determined by using Equation 10.3(3) as follows:

$$A_T = \frac{0.434(693.3)}{0.52} = 578.6 \text{ ft}^2 = \frac{D^2\pi}{4}$$

$$\text{Diameter (D)} = \sqrt{\frac{578.6}{0.7854}} = 27.14 \text{ ft}$$

A good selection for this application is to provide two 27- to 30-ft diameter units equipped with vertical pickets. With one unit in service, the unit overflow rate (rate of effluent discharge) is as follows:

$$\frac{124{,}800 \text{ gal/day}}{578.6 \text{ ft}^2} = 215.7 \text{ gal/ft}^2\text{/day}$$

Flotation Thickening

A variety of pressurized, air flotation units are commercially available as complete packages. The basic units are either rectangular or round, with rectangular units more common in wastewater treatment systems (see Figure 10.3.3). Both direct pressurization and pressurized recycling systems are available. Wastewater treatment facilities use the later when the liquid contains flocculant or other fragile particles that disperse or suffer attrition when subjected to the high shear forces in the pressurization process.

The performance of pressure flotation units is related to the ratio of the weight of the air used to the weight of the solids supplied to the unit, A/S. However, this approach neglects both the number of bubbles into which the air is divided after the pressure release and the number of particles into which the weight of incoming solids is divided.

No ready means exists to assess the potential impact of these variables. The degree of flocculation of an activated sludge can vary in response to changes in influent waste characteristics or process management. Therefore, wastewater treatment facilities should use conservative designs. Alternatively, they can add flocculating chemicals to the unit to insure a constant number of particles per unit weight of applied solids. Polymers are usually selected as the flocculating chemicals. Facilities can substantially increase the particle rise rate with low polymer dose rates.

Wastewater treatment facilities can use the following equation to select pressurized recycling flotation units:

$$\frac{A}{S} = \frac{C(s_a)(f')(p - 1)R}{C_{in}(Q_w)} \qquad 10.3(4)$$

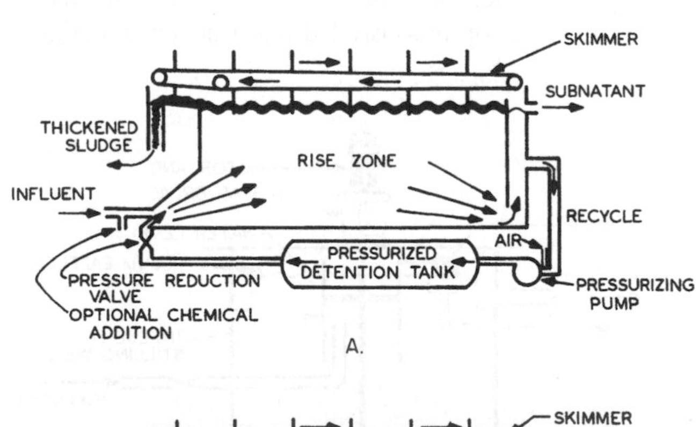

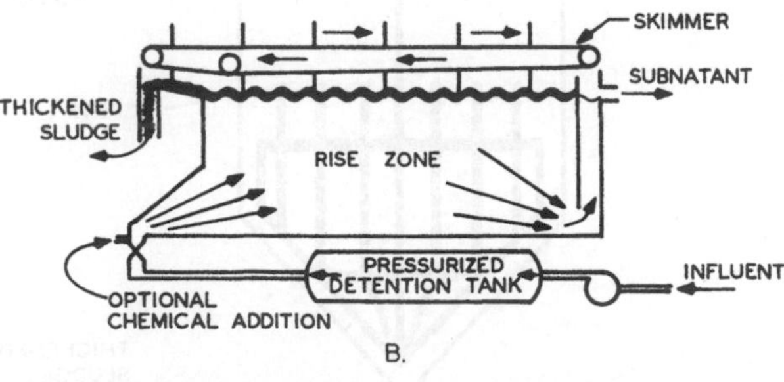

FIG. 10.3.3 Flotation thickeners. A, Pressurized recycle system; B, Direct pressurization system.

where:

A/S = ratio of lb of air to lb of incoming solids
C = A conversion factor, 0.834
s_a = Solubility of air in water, 18.68 cc/l at 20°C
f′ = Percent saturation achieved (decimal fraction)
p = Operating pressure, atmospheres
R = Pressurized recycling rate, mgd
C_{in} = Solids concentration in thickener influent, mg/l
Q_w = Waste sludge influent flow rate, mgd

Wastewater treatment facilities can use the following equation to select direct pressurization flotation units:

$$\frac{A}{S} = \frac{C(s_a)(f')(p - 1)}{C_{in}} \qquad 10.3(5)$$

Environmental engineers usually select flotation units that provide a surface overflow rate of 2500 to 5000 gal/ft²/day and a hydraulic detention time of about 20 min. They should also make provisions for dual units, bypassing, dewatering, and flocculant addition.

Environmental engineers develop design data by pressurizing either the waste influent (direct pressurization) or a simulated recycled sample, such as settled or filtered waste or sludge supernatant, and releasing it into an open graduated cylinder. For the pressurized recycling system, they release various amounts of recycled liquid into the required amount of waste or sludge to achieve the required recycling ratio (see Figure 10.3.4). Both the time of pressurization and the pressure are based on the latitude of design permitted by the commercial equipment. For custom designs, environmental engineers should study a range of values. The effect of temperature should also be studied, especially if the flotation units are not housed or influent temperatures vary.

Pressurized Recycling Design Example

For the purposes of this example, the solids to be thickened are excess activated sludge with an initial solids concentration of 7000 mg/l. The operating temperature is 20°C and the operating pressure selected is 60 psig (5.1 atm). A 60% air saturation has been achieved at a pressurized detention time of one min. The A/S ratio value selected is 0.03. The volume of incoming excess activated sludge to be thickened is 124,800 gal/day with an overflow rate of 2500 gal/ft²/day. Applying Equation 10.3(4)

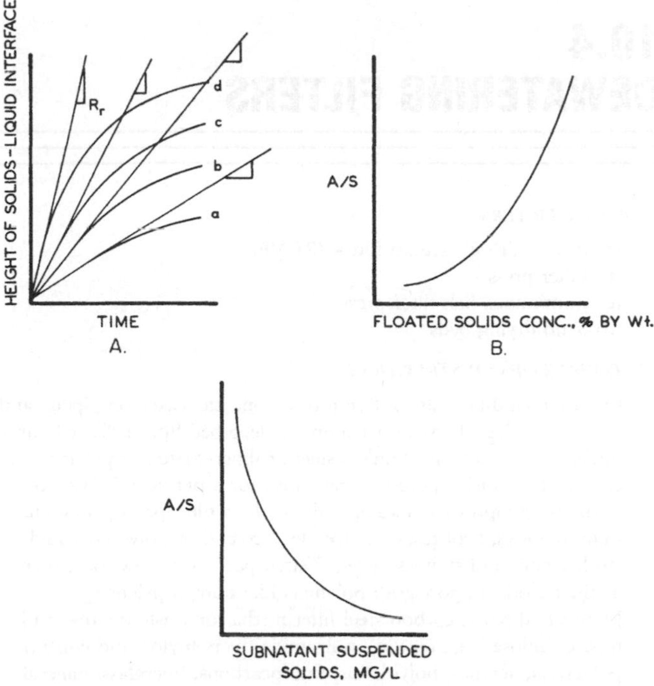

FIG. 10.3.4 Pressure flotation design data plots. Data plot procedure: 1. Conduct rise rate tests at various initial solids concentrations (a through d in panel A) covering the range of expected unit influent solids concentrations; 2. Determine the initial rise rate R_r for each concentration as the slope of a line tangent to the initial lineal rise rate portion of each curve; 3. Plot the concentration of floated solids (% by weight) versus the A/S for each influent solids level; 4. Plot the subnatant SS concentration (unit effluent SS) versus the A/S for each influent solids level; and 5. Select the design A/S ratio to achieve the solids capture or concentration objective, or alternatively, optimize for maximum concentration and capture.

yields the following:

$$0.03 = \frac{(0.834)(18.68)(0.60)(5.1 - 1)R}{7000(0.1248)}$$

$$R = 0.684 \text{ mgd}$$
$$= 475 \text{ gpm}$$

The required unit surface area is as follows:

$$A = \frac{124{,}800 \text{ gal/day}}{2500 \text{ gal/ft}^2\text{/day}} \cong 50 \text{ ft.}^2$$

—*B.L. Goodman*

10.4 DEWATERING FILTERS

TYPES OF FILTERS

(a) Rotary drum vacuum filters (RDVF)
(b) Filter press
(c) Continuous belt filter press
(d) Sand drying beds

MATERIALS OF CONSTRUCTION

Outdoor sand-bed filters: Reinforced concrete, cast-iron piping, and cast steel valves; Pressure vacuum, or deep-bed filters: Wetted parts under tensile loading: stainless steels; rubber-coated or polymer-coated steel; and copper-bearing alloys for a pH near 7; Wetted parts for compressive loading only, such as filter press plates: aluminum alloys, bronzes, cast steel, lead-coated or polymer-coated steel, wood, and stainless steel; Wetted parts, very low load, such as filter cloth support grid: polymers like polypropylene; Nonwetted parts: carbon steel filter media for waste treatment filters: Cellulose (e.g., cotton), polyamide fibers (nylon and nomex), polyesters, acrylics, polyolefins, fluorocarbons, fiberglass, mineral fibers, and woven metals

TOTAL COSTS

About $20 per ton of dry solids

OPERATING COSTS

$10 or more per ton of dry solids

FILTER SIZES AVAILABLE

Up to 550 ft^2 (a,d); up to 3500 ft^2 (d)

NUMBER OF DRIVE MECHANISMS

2(c,d); 6(a,b)

SLUDGE CONCENTRATIONS

Primary sludge: 6 to 11% solids; excess activated sludge: 2000 to 4000 mg/l VSS; anaerobic digester effluent: 4.5 to 7% VSS

FILTER CAKE SOLIDS CONTENT

20 to 30%

VACUUM FILTER FILTRATION RATES (LB/HR/FT²)

On primary sludge, 5 to 7; on excess activated sludge after heat or chemical treatment, 1 to 4; on lime precipitated chemical sludge, 2 to 6; on aerobically digested sludge, 1 to 2

PARTIAL LIST OF SUPPLIERS

Andritz-Ruther (b,c); Arlat Inc. (b); Ashbrook-Simon-Hartley (b,c); Eimco Process Equipment (a,c); Dorr-Oliver Inc. (a,c); Humbold Deacanter (b); Komline-Sanderson (a,b,c); Westech Engineering Inc. (a)

Table 10.4.1 provides an orientation on filter selection and application. Screens are discussed in Section 4.2. This section discusses those filters frequently used in waste treatment plants.

Domestic Wastewater

The distinguishing characteristic of nearly all domestic liquid waste entering municipal treatment plants is the low concentration of dissolved and suspended contaminants and the large volume of wastewater that must be treated. This characteristic means that domestic and industrial water is readily available and inexpensive. In the U.S., the per capita domestic water consumption is about 100 gallons per day; the solids concentration of generated wastes is 0.1 wt% or less.

Environmental engineers have proposed processes to reduce daily water consumption without jeopardizing public health. Such measures may eventually be required by law and would increase the solids concentration in waste streams. Meanwhile, wastewater treatment facilities must treat these dilute waste streams to protect public health, the water supply, and the environment.

Incoming domestic waste always contains large solid objects that are difficult or impossible to treat. Such tramp solids include bottle caps, cigarette filter tips, clumps or cloth or lint, and rubber goods, and they must be removed initially by coarse screens (see Section 4.2). Table 10.4.2 lists the typical composition of raw domestic waste after

TABLE 10.4.1 FILTRATION METHODS FOR WASTE TREATMENT APPLICATIONS

Filtration Application	*Filter Design Selected*
Separation of bacterial slurry, activated sludge, or submicron inorganic slurry, 0.5 to 5% solids, into unwashed solids plus cloudy filtrate	Continuous rotary vacuum belt filter with special cake discharge method; continuous rotary pressure filter
Separation of digested primary sludge into wet solids plus cloudy filtrate	Coil filter, belt filter
Separation of valuable chemical liquid from waste solids, with extensive cake washing	Batch pressure filters with or without automatic cake discharge
Separation of slurry into clean liquid plus solids mixed with filter aid	Rotary drum or batch-type pre-coat filter
Separation of slurry into clear liquid plus heavy slurry of separated solids	Deep-bed filter with backwash

TABLE 10.4.2 COMPOSITION OF RAW DOMESTIC SEWAGE AND MUNICIPAL SEWAGE

Constituent	Domestic Sewage[2] (mg/l)	Municipal Sewage (mg/l)			Combined Industrial and Municipal Sewage (mg/l)	
		Maximum	Mean	Minimum	Plant A	Plant B
TDS	200	396	305	237	—	—
Total SS	300	264	148	85	320	92
TS	500	640	453	322	—	—
TOC	120	—	—	—	—	—
BOD (dissolved and suspended)	200	276	147	75	429	208
COD (dissolved and suspended)	350	436	288	159	—	—
BOD/SS ratio	0.67	1.04	1.95	1.87	1.34	2.26

initial screening. This table also gives the composition of municipal sewage based on long term measurements.

FILTRATION VERSUS CLARIFICATION OF PRIMARY DOMESTIC WASTE

Filtration of dilute waste (sewage) is not feasible because of the high investment. In addition, the filter cake is impossible to form. Primary clarifiers usually separate solids from most of the liquid and produce more concentrated solids for filtration and additional processing (see Figure 10.4.1). Other low-power devices, such as sand-bed filters, are also used for this purpose. Solids-concentrating or solids-separating devices requiring high velocities, pressures, and power consumption, such as centrifuges, hydrocyclones, filter presses, or leaf filters, are ordinarily uneconomical for primary municipal waste treatment but are sometimes used for industrial pretreatment on a smaller scale.

The mechanical separation (filtration or clarification) of primary sludge is only partially effective as a treatment because 30 to 40% of BOD and COD are water soluble and cannot be so removed.

PRIMARY SLUDGE

The SS content of the primary clarifier or thickener underflow (primary sludge) varies depending on the detention time. Researchers found volatile solids of 8.2 to 10.7% weight in undigested primary sludge going from thickeners to high-rate anaerobic digesters (Jeris 1968). They measured material with a solids content of 6 to 10% weight in the same stream when vacuum filtration equipment was fed from thickeners. This sludge is usually hydrophilic and nondewaterable. One of the purposes of the anaerobic digester is to convert it into a readily dewaterable form.

Researchers reported a filtration rate of 6.9 lb/hr/ft^2 for settled primary sludge on an RDVF (Eckenfelder 1970). Other results include a filtration rate of 5 lb/hr/ft^2 for settled primary sludge on the vacuum filter (dry basis), with 32% solids in the filter cake. Undigested raw sludge should not be filtered because of odors, cleaning problems, and potential health hazards.

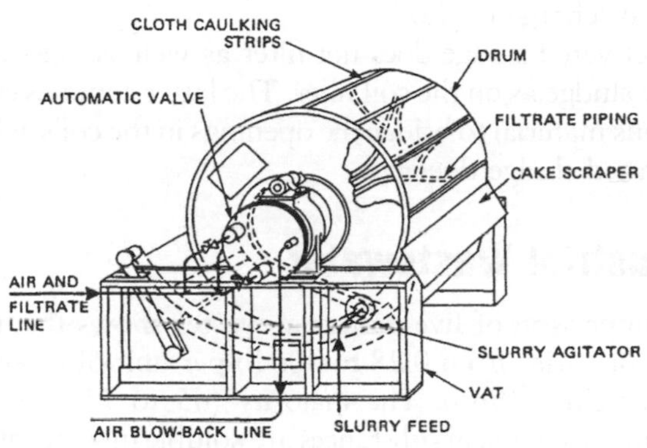

A. CUTAWAY VIEW OF A DRUM OR SCRAPER-TYPE RDVF

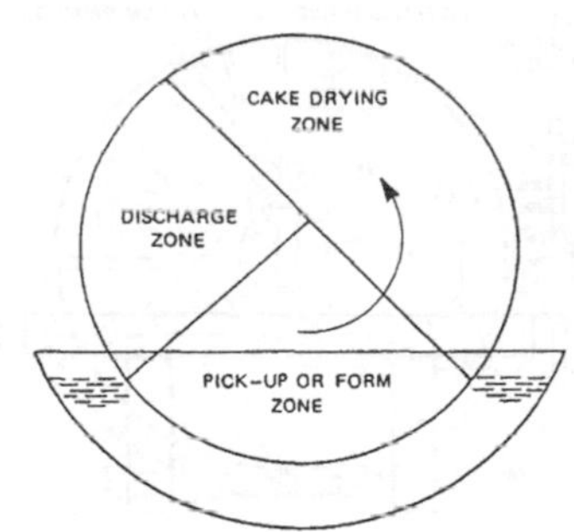

B. OPERATING ZONES OF A RDVF

FIG. 10.4.1 Schematic diagram of vacuum filter dewatering equipment.

ACTIVATED AND DIGESTED SLUDGE

Derived solid sludge from anaerobic digestion of settled raw sludge is readily dewaterable by filtration. The SS concentration in the digester product is lower (4.6 to 6.7% VSS) than that in the feed because of the anaerobic conversion of some organic material to methane. This sludge can be successfully filtered on coil-type RDVFs (see Figure 10.4.2).

Sludge from secondary treatment (activated-sludge process) consists of bacteria, dead cells, and associated material formed under oxidizing conditions. Suspended bacterial solids in the aeration tank are usually in the range of 2500 to 4000 mg/l. These solids can be removed on a deep-bed filter (see Figure 10.4.3) which can take the full flow of the plant.

Ordinarily, however, the aeration tank suspension is continuously circulated to a clarifier, whose underflow (bottom discharge sludge) contains 1 to 3% dry solids concentration or higher by weight depending on the settling time and properties. When excess activated sludge is conditioned with ferric chloride (10 lb per 100 lb dry solids), filtration usually proceeds well on vacuum belt filters at solids rates of 1 to 4 lb/hr/ft^2, or filtrate rates of 15 to 40 gph/ft^2 (see Table 10.4.3). Figure 10.4.4 shows the special cake discharge methods.

Activated sludge does not filter as well as digested primary sludge as on the coil filter. The latter contains enough fibrous material to bridge the openings in the coils, whereas activated sludge does not.

Industrial Wastewater

A comparison of five industrial wastes shows that the SS content varies from 0.28 to 400 lb per unit of production (Eckenfelder 1966). The majority (65 to 75%) of nonpolymeric chemical substances are soluble in water at 25°C

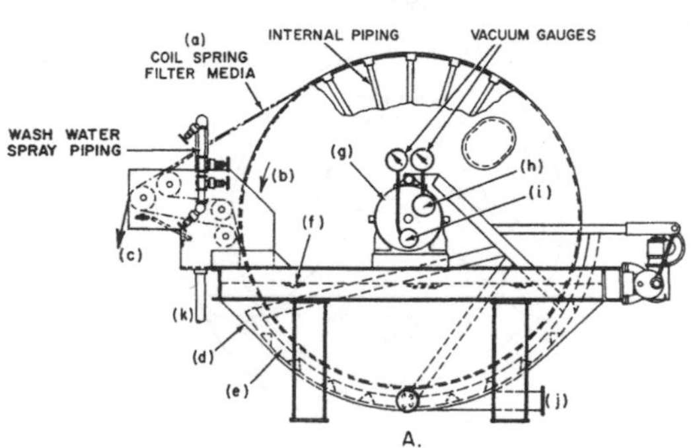

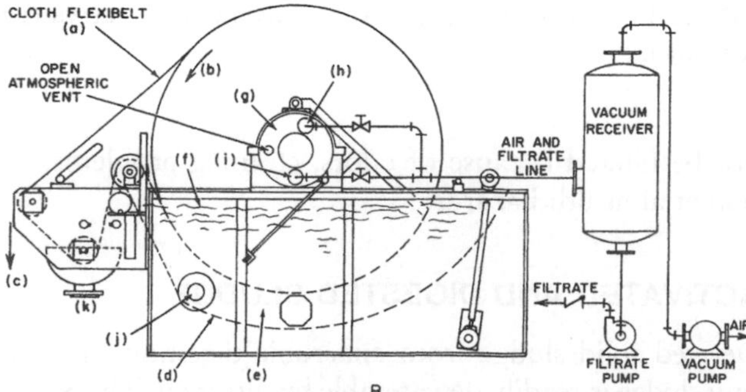

FIG. 10.4.2 Continuous RDVF. **A**, Coil filter; **B**, Belt filter. Sizes range to 550 ft^2 per unit. *Legend:* a, filter medium (coil or cloth); b, filter drum; c, cake discharge; d, slurry vat; e, agitator; f, slurry level; g, filter valve; h, drying air outlet; i, filtrate outlet; j, vat drain; k, belt wash drain.

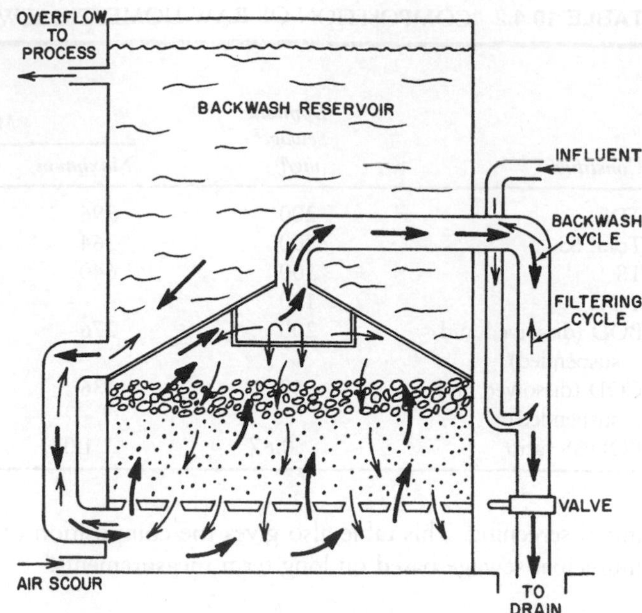

FIG. 10.4.3 Deep-bed filter with intermittent backwash. Water supply for backwashing occupies the upper part of the filter tank. When the drop across the filter reaches 4 ft of water, an air scour blows through the bed, carrying collected matter from the media. Then, the stored water flows through the bed in reverse, leaving the sand and anthracite clean.

TABLE 10.4.3 CONTINUOUS RDVF DATA FOR ACTIVATED SLUDGE

Parameter	*Range of Values*
Solids concentrations in slurry, % dry basis weight	0.64 to 3.43
Filtration operating vacuum level, " Hg	17.7 to 25.0
Slurry temperature, °C	11.5 to 15.5
Filtrate rates, gph/ft^2	18.2 to 27.4
Solids rates, dry basis; lb/hr/ft^2	0.47 to 3.47[a]
Cake thickness, in	$\frac{1}{64}$ to $\frac{3}{32}$
Drum diameter, ft	3.0
Drum width, ft	1.0
Drum speed, rpm	0.40 to 0.909

Note: [a]This rate is lower than the solids rate obtained from slurry flow because of solids passing through cloth.

and a pH of 7. Therefore, as the proportion of industrial waste in a system increases, the ratio of total BOD (dissolved and suspended) to SS increases. High polymers in industrial waste forms colloidal suspensions or solutions.

FILTER AIDS

In addition to inert chemical substances, raw industrial waste can contain microorganisms—bacteria, molds, and yeasts. For this group, the settling rates are often low, and the filtration resistance is high; yet frequently, the concentrations (1 to 10% by weight) make them unaccept-

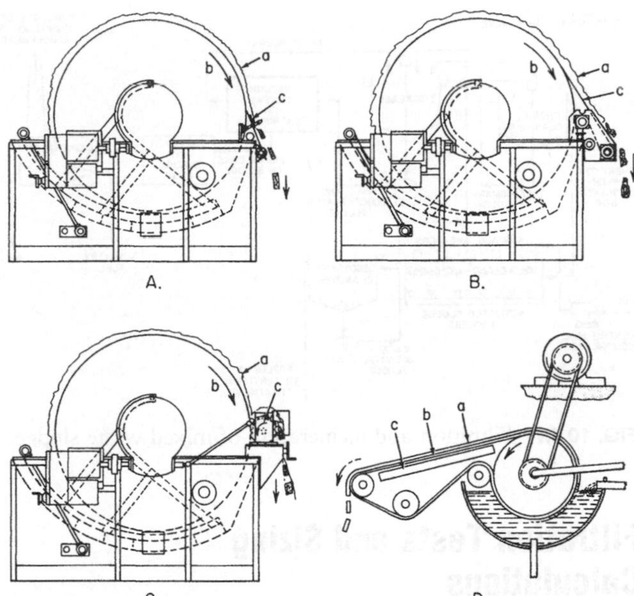

FIG. 10.4.4 RDVFs—alternate types of cake discharging mechanisms. **A,** Scraper discharge: filter cake (a) is removed from cloth on drum (b) by doctor blade (c), and air blow (2 to 4 psig) aids in dislodging cake; **B,** String discharge: filter cake (a) is lifted from cloth on drum (b) by parallel strings (c) tied completely around drum, about $\frac{1}{2}$ in apart; **C,** Roll discharge: filter cake (a), consisting of cohesive materials like clay or pigment, is lifted from cloth on drum (b) by adhesion to discharge roll (c); **D,** Heated belt discharge: filter cake (a), which adheres to cloth belt (b) because of adhesive effect of filtrate, is separated from cloth by heater (c), which dries the cloth; cake does not need drying.

able in public collection systems. Many industries solve this problem by filtration with generous amounts ($\frac{1}{2}$ to 1 lb per pound dry solids) of filter aid, usually diatomaceous earth. However, this solution creates a solids disposal problem.

Industries can minimize their use of filter aids by using the heated-belt discharge method on the rotary vacuum belt filter. Activated sludge (a derived solid consisting entirely of bacteria) can be successfully filtered this way without any filter aid when it is coagulated with ferric chloride. Consequently, attempts should also be made to coagulate other bacterial slurries.

Any coagulant used in the filtration of disposable waste requires Food and Drug Administration (FDA) approval, and ferric chloride is not acceptable. Nevertheless, an acceptable coagulant should be sought from the approved list to lower pretreatment costs.

Unless an industrial plant's effluent contains less than the maximum allowable solids content established by the regulatory agency, filtration or other pretreatment can be required. When an industry pretreats an effluent to meet other requirements, such as pH, the pretreatment can result in the precipitation of additional solids, a step that adds to the filtration or separation load.

SOLIDS CONCENTRATED BY PHYSICAL METHODS

Processes for physically concentrating solids from industrial waste include coagulation and aeration, which achieve either faster settling or flotation rates. Pretreatment by filtration applies only to particulate solids that settle in water. Because settleable solids vary widely in particle size, properties, and composition, filtration tests are needed. In selecting and designing filtration equipment, environmental engineers should also consider the toxicity, health, and fire hazards of the materials to be filtered.

SOLIDS DERIVED FROM CHEMICAL REACTIONS

Chemical reactions during pretreatment produce additional SS in industrial wastewater. The most common precipitating agent is lime slurry (10% calcium hydroxide plus H_2O). It precipitates a variety of calcium compounds, both organic and inorganic in water. Dolomitic lime ($Ca[OH]_2 \cdot Mg[OH]_2$) is a superior precipitating agent for some waste material. These two forms of lime represent the best combination of availability, economy, and versatility.

Compounds precipitated by lime usually settle well and can be filtered at a rate of 2 to 6 lb/hr/ft^2 on an RDVF (see Figure 10.4.5). Lime adds little to the dissolved solids content, and most calcium compounds have low solubility. Regardless of the alkali used, environmental engineers must perform filtration testing of the precipitates for design.

Frequently, the submicron industrial sludge formed by the reaction with lime should not be preconcentrated be-

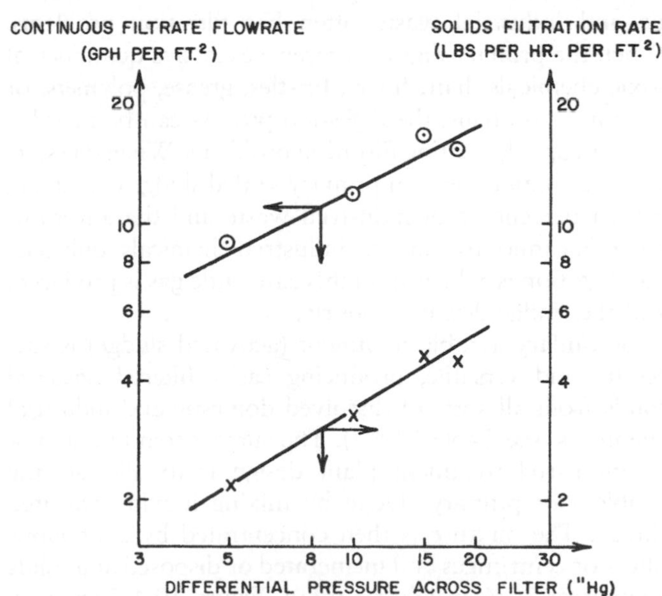

FIG. 10.4.5 Continuous RDVF of acid neutralization slurry ($CaSO_4$ + $Fe[OH]_3$ in H_2O).

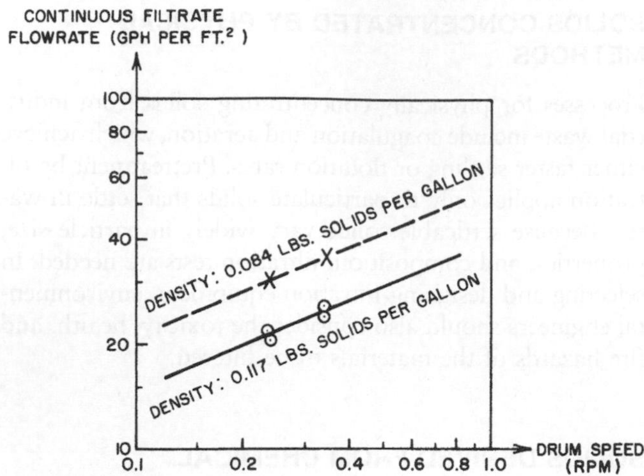

FIG. 10.4.6 Effect of drum speed and slurry density on filtrate flow rate.

cause filtration proceeds better on a dilute slurry (see Figure 10.4.6).

Combined Domestic and Industrial Wastewaters

In combined waste from many sources, the SS content is still too low (0.1%) to be filtered. The most common problem in solids removal from municipal plus industrial waste is an increased proportion of dissolved impurities. Table 10.4.2 shows typical figures (yearly averages) for two currently operating plants (A and B) handling high proportions of industrial waste. Little experience is available in filtration of this type of settled primary sludge.

Anaerobic digested sludge derived only from domestic waste is easily dewatered on the coil or belt filter (see Figure 10.4.2), but the same sludge derived from mixed domestic and industrial waste often has filtration problems. When the primary digesters receive excessive quantities of toxic chemicals, hair, fibers, bristles, grease, polymers, or gelatinous proteins, the digestion process can be retarded or interrupted, creating filtration problems. When digesters operate continuously on primary settled sludge containing a high percentage of industrial waste and the anaerobic bacteria cannot tolerate the industrial chemicals, only partial digestion is achieved. In this case, little gas is produced, and the sludge dewaters poorly.

Secondary aerobic treatment (activated sludge) is successful and versatile, producing easily filtered bacterial solids from all sorts of dissolved domestic and industrial chemicals (see Table 10.4.3). The current trend in domestic–industrial treatment plant design is to alleviate the problem of primary sludge by mixing it with activated sludge. The mixture is then concentrated by continuous filters or centrifuges and incinerated or disposed at a solids content of 20% weight or more. Figure 10.4.7 shows a flow diagram of this type.

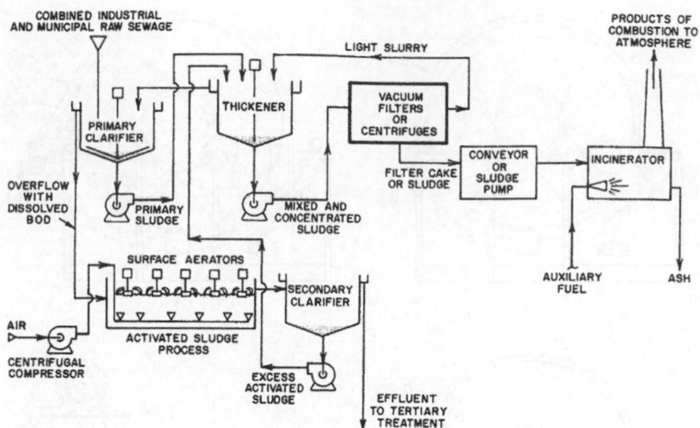

FIG. 10.4.7 Filtration and incineration of mixed waste sludge.

Filtration Tests and Sizing Calculations

While Buchner funnel tests often determine SS filtration characteristics, numerous improvements have been made in the method (see Figure 10.4.8). The filter cloth and its support grid should be the same as those on the full-scale or pilot-scale filter. To simulate an RDVF, this test arrangement holds the leaf downward in the slurry and gently moves it about during filtration. To produce a thicker cake, as on a filter press, wastewater treatment facilities can convert the leaf into a cylindrical funnel by adding a section of PVC pipe 4 in long.

These tests are batch types performed at a constant pressure. The following equation models these tests:

$$\frac{dV}{AdT} = \frac{\Delta P}{K_1V + K_2} \qquad 10.4(1)$$

where:

V = filtrate volume
T = time (measured from the beginning of test)
A = filter leaf area
ΔP = pressure differential, maintained constant across the leaf

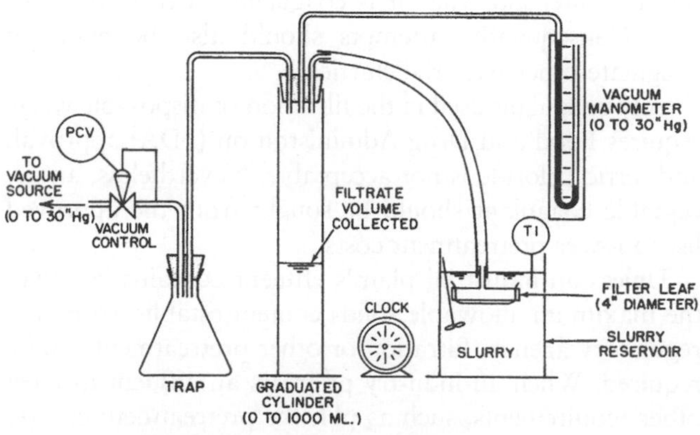

FIG. 10.4.8 Preferred leaf test arrangement.

K_1 = resistance coefficient of the filter cake
K_2 = resistance of the filter membrane

When the filtrate volume collected at uniform time intervals is read and the values of $(\Delta T/\Delta V)$ are calculated, Equation 10.4(1) can be rewritten as follows:

$$\frac{A\Delta P(\Delta T)}{(\Delta V)} = K_1V + K_2 \qquad 10.4(2)$$

Plotting the left side of Equation 10.4(2) versus V gives a straight line of a slope K_1 and an intercept K_2.

The following equation calculates the specific cake resistance for batch filtration:

$$\text{Let } K_1 = \alpha\mu W/A \qquad 10.4(3)$$

where:

α = specific resistance of the cake (to be calculated)
μ = viscosity of the filtrate (from liquid properties)
W = mass of dry solids/volume of slurry (from drying procedure)

Equation 10.4(3) neglects the filtrate content of the filter cake and calculates the value of α for the cake from A, μ, W, and K_1.

CONTINUOUS ROTARY FILTERS

For continuous rotary filters, environmental engineers must calculate the volumetric filtrate rate, test with mini-filters, and determine the solids retention and cake thickness.

Volumetric Filtrate Rate

If $K_2 \cong 0$, the following equations apply:

$$\frac{dV}{AdT} = \frac{\Delta P}{K_1V} = \frac{\Delta P}{\left(\frac{\alpha\mu W}{A}\right)V} \qquad 10.4(4)$$

and

$$\frac{VdV}{A^2} = \left(\frac{\Delta P}{\alpha\mu W}\right)dT$$

Integrating from 0 to V_f and from 0 to T_f gives the following equations:

$$\frac{V_f^2}{2A^2} = \left(\frac{\Delta P}{\alpha\mu W}\right)T_f$$

or

$$\frac{V_f}{A} = \sqrt{\frac{2\Delta PT_f}{\alpha\mu W}} \qquad 10.4(5)$$

where V_f and T_f are the filtrate volume and time, respectively, for one batch of a batch filter.

For one cycle of a small area A on a continuous filter, if n = cycles per minute and T_c = minutes per cycle, then $n = 1/T_c$. Also, $T_f = BT_c$ when $0 < B < 1.00$. B is the fraction of total area that is filtering at any given time. The cycles per hour = 60n.

On a continuously rotating drum, the following equation gives the continuous filtrate volume per unit area per hour:

$$Z_c = 60n\left(\frac{V_f}{A}\right) = \left(\frac{7200(\Delta P)Bn}{\alpha\mu W}\right)^{1/2} = Z_c, \frac{\text{gph filtrate}}{\text{ft}^2 \text{ total area}} \qquad 10.4(6)$$

The exponent $\frac{1}{2}$ on this equation holds for many solids as shown in Figures 10.4.5 and 10.4.6. For activated sludge, the exponent on the group can be different than $\frac{1}{2}$; occasionally, the individual variables in the group have different exponents. Nevertheless, the equation is a valuable guide in data correlation.

In principle, a value of α from a leaf test allows the calculation of total full-scale filter area A required to achieve a total filtrate rate V′ in gph, i.e., $A = V'/Z_c$. In practice, leaf tests give only approximate values of α, and the accuracy is usually not enough for precise design. This inaccuracy is partially because the filter cake does not pack and compress the same way on a test leaf as on a continuous filter. The wall effects in the filter funnel are also important.

Mini-filters

For reliable filter design, the best guides are previous experience with the same slurry and tests with miniature, continuous rotary filters (mini-filters) of the belt or coil type. These filters are available on a rental basis.

Solids Retention and Cake Thickness

On a continuous rotary filter, if W = the solids content of feed slurry (lb dry solids per gallon filtrate), the following equation gives the continuous solids rate in the feed slurry:

$$Z_cW = \sqrt{\frac{7200B\Delta PnW}{\alpha\mu}}\ \frac{\text{lb}}{\text{hr} \times \text{ft}^2} \qquad 10.4(7)$$

If the solids content of the filtrate = W_f (lb per gallon), then Z_cW_f represents the solids that pass through the cloth. If ρ_c is the cake density, the following equation calculates the cake thickness as discharged (L_c, ft):

$$L_c = \frac{\text{cake volume, ft}^3}{\text{hr} \times \text{drum area} \times \text{drum speed, revolutions per hr}}$$

$$= \frac{\text{cake mass rate per cake density}}{\text{hr} \times \text{drum area} \times \text{drum speed, revolutions per hr}}$$

$$= \frac{Z_c(W - W_f)}{\rho_c(60n)} = \frac{1}{60\rho_cn}\left(\sqrt{\frac{7200B(\Delta P)nW}{\alpha\mu}} - Z_cW_f\right) \qquad 10.4(8)$$

A comparison of Equations 10.4(7) and (8) indicates that as the drum rotation speed increases, the solids rate also increases, but the filter cake becomes thinner. The passage of fine solids through the filter cloth is often unavoidable in continuous filtration and can require recycling the filtrate to the clarifier. Cake washing is of minor importance in waste treatment filters, but when it is required, batch filters are usually superior to continuous ones.

FILTER PRESSES AND OTHER BATCH PRESSURE FILTERS

With a set of operating data on a continuous rotary filter, environmental engineers can calculate the specific resistance α using Equation 10.4(6), in which ΔP is the differential pressure on a continuous filter.

Based on time in minutes, the following equation applies:

$$Z_c/\,60 = nV_f/A = \sqrt{\frac{2(\Delta P)Bn}{\alpha\mu W}}$$

$$= K_3B^{1/2}(\Delta P_c)^{1/2}n^{1/2}, \text{ gpm per sq foot} \qquad 10.4(9)$$

where:

$$K_3 = \sqrt{\frac{2}{\alpha\mu W}} = \sqrt{\frac{2}{K_1A}} \qquad 10.4(10)$$

Equation 10.4(7) gives the continuous dry solids rate (assuming complete retention of solids by the filter cloth).

For nearly incompressible filter cakes, the following equation calculates the approximate required area of a batch filter to replace a continuous filter for the same slurry at the same average rate. The filtrate liquid per unit area per batch is as follows:

$$\frac{V_{fb}}{A} = \sqrt{\frac{2\Delta P_bT_{fb}}{\alpha\mu W}} = K_3\Delta P_b^{1/2}(T_{fb})^{1/2} \qquad 10.4(11)$$

For constant-pressure, batch filtration, V_{fb} is related to T_{fb}. If the following conditions exist,

$$T_{cb} = \text{total batch cycle time, min}$$
$$= \text{filter cake form time + discharge time}$$
$$= T_{fb} + T_{db}$$

then the average batch filtrate rate is as follows:

$$Z_b = \frac{V_f}{T_{cb}A} = \frac{K_3(\Delta P_b)^{1/2}(T_{fb})^{1/2}}{(T_{fb} + T_{db})}, \frac{\text{gpm}}{\text{ft}^2} \qquad 10.4(12)$$

The following equation calculates the relative rates on batch and continuous filters for the same slurry at the same temperature:

$$\frac{Z_c}{Z_b} = \frac{\text{average rate, continuous filter}}{\text{average rate, batch filter}}$$

$$= \frac{\Delta P_c^{1/2}(Bn)_c^{1/2}}{\left[\frac{\Delta P_b^{1/2}(T_{fb})^{1/2}}{T_{fb} + T_{db}}\right]} \qquad 10.4(13)$$

For quick-cleaning batch filters $T_{db} \ll T_{fb}$, then $T_{cb} \simeq T_{fb}$, and Equation 10.4(13) simplifies to the following equations:

$$\frac{Z_c}{Z_b} = \left(\frac{\Delta P_c}{\Delta P_b}\right)^{1/2} (Bn)_c^{1/2}(T_c)_b^{1/2}$$

$$= \left(\frac{\Delta P_c}{\Delta P_b}\right)^{1/2}\left(\frac{B}{T_c}\right)_c^{1/2}(T_{cb})^{1/2}$$

or

$$\frac{Z_c}{Z_b} = \left(\frac{\Delta P_c}{\Delta P_b}\right)^{1/2}\left[\frac{(T_c)_b}{(T_c)_c}\right]^{1/2} B^{1/2} \qquad 10.4(14)$$

FILTER SIZING EXAMPLE

In this example, the following operating data are for a continuous rotary vacuum filter that separates a solid from suspension at 22 in Hg vacuum (10.8 psi) with speed = $\frac{1}{2}$ rpm, drum area = 100 ft^2, B = 0.33 ($B = T_f/T_c$), and continuous cycle time = $(T_c)_c$ = 2 min = $\frac{1}{30}$ hr.

Environmental engineers can calculate the size of filter press or other batch pressure filter required to filter the same slurry at the same average rate when the filter cake is assumed to be incompressible. This example assumes that a multistage centrifugal pump supplies the filter press with slurry at an average differential pressure of 150 psid.

To fit the working schedule, this example uses cycle times of 2 hr, 4 hr, and 8 hr for the filter press. Due to automatic cleaning equipment (see Figures 10.4.9 and 10.4.10), the cleaning time is negligible compared to the cycle time. The equipment and piping resistances on the filter press are also negligible.

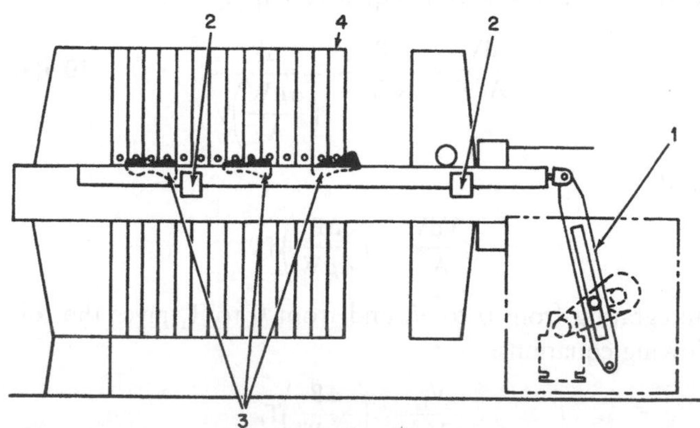

FIG. 10.4.9 Filter press with automatic cleaning. Sizes are available up to 3500 ft^2 per unit. At the end of the cycle, the opening and closing gear moves the end of the press to the right and starts the plate moving cycle. The reciprocating mechanism (1) drives the slide bars mounted on bearing blocks (2) to the right. One pair of pawls (3) mounted in the slide bars picks up only the first plate (4) and moves it to the right. The filter cake falls out. The mechanism reverses and moves the next plate. No separate plates or frames exist. Each plate is recessed on each side to provide cake space. During filtration, the slurry enters and filtrate exits through an internal passage (not shown).

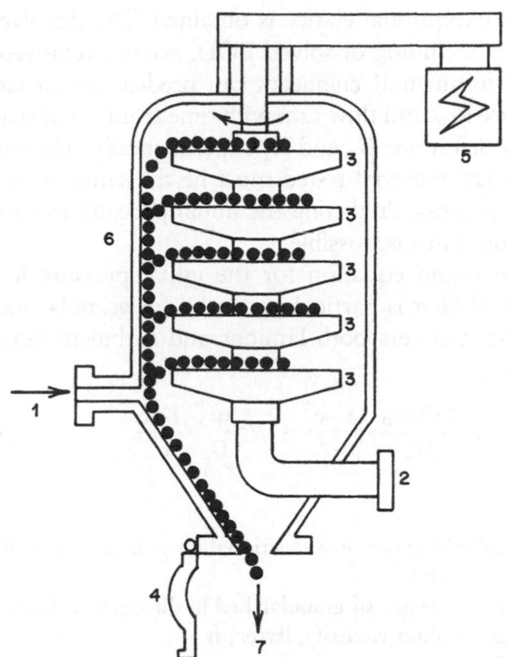

FIG. 10.4.10 Rotating-leaf filter with automatic cleaning. During filtration, the slurry enters at (1), the filtrate exists at (2), and solids are retained on leaves (3) and covered with a filter cloth. Upon completion of filtration, the washing and drying bottom closure (4) opens. The drive motor (5) starts and rotates the stack of filter leaves. Centrifugal force causes the solids to move off the filter leaves, strike the inside wall of the tank (6), and flow down to the solids exit (7). Sizes are available up to 540 ft^2 per unit.

The following equation calculates the 2-hr cycle on the filter press:

$$\frac{Z_c}{Z_b} = \left(\frac{10.8}{150}\right)^{1/2}\left(\frac{2 \times 60}{1/30}\right)^{1/2}(0.33)^{1/2}$$

$$= (.0721)^{1/2}(3600)^{1/2}(0.33)^{1/2}$$

$$= (.269)(60)0.57$$

$$= 9.21 = A_b/A_c$$

Filter press area = 921 ft^2

The following equation calculates the 4-hr cycle on the filter press:

$$A_b = 921\sqrt{2} = 1300 \text{ ft}^2$$

The following equation calculates the 8-hr cycle on the filter press:

$$A_b = 921\sqrt{4} = 1842 \text{ ft}^2$$

This procedure is only approximate if the pressure varies during the batch, and graphic integration using the pump characteristic curve provides better results.

The results are not valid for batch pressure filters if the cake is appreciably compressible. For either batch or continuous filtration of compressible solids, environmental engineers should conduct tests at various average pressures.

For batch filters, in addition to adequate filtration area the design must provide sufficient cake space to contain the total volume of the wet filter cake deposited during the proposed cycle time; otherwise the cycle can come to a premature end. This specification requires knowledge of the wet cake density and solids content, which should be measured in the laboratory and not assumed.

If T_{db} is not negligible compared to T_{fb}, environmental engineers can include its value (the batch cake discharge time) in the batch equations, and the calculation is more precise. However, high precision should not be expected in filtration calculations. As in the rotary filter case, environmental engineers obtain the best results for a batch filter in laboratory measurements using a mini-filter of the same general design as the full-scale filter.

This example shows that a batch filter requires a larger area than a continuous filter for the same slurry and average filtrate rate. The batch filter, however, has only two electrical drives—the feed pump and the discharge mechanism. The continuous rotary drum filter usually has six drives, including the feed pump, pan mixer, drum drive, belt discharge drive, vacuum pump, and filtrate pump.

Batch versus Continuous Filters

In municipal waste treatment plants, batch filters have not been used. The large scale of the operation, the compressibility of the sludge, and the lack of automatic cleaning devices are probably the reasons for this general lack of use.

For pretreatment units in an individual plant where the waste volume is small and the sludge is incompressible, plant management should compare the use of batch filters with automatic cleaning (see Figures 10.4.9 and 10.4.10) with continuous rotary filters.

CAKE WASHING

Cake washing, which is unimportant in large municipal–industrial treatment plants, can be beneficial in industrial pretreatment when a valuable material dissolved in the filtrate is separated from waste solids. Better washing can be obtained on batch filters than on continuous filters.

A thorough theoretical and experimental study of cake washing concludes that removing 80% of a solute dissolved in the filtrate left in the void spaces of the filter cake requires at least six void volumes of wash liquid and sometimes more. This statement explains why batch filters are superior to continuous rotary filters when extensive washing is required.

BATCH TESTS IN PRESSURE FILTERS WITH VARIABLE PRESSURE

The previous calculation of K_1 and K_2 requires a constant controlled pressure. An actual pressure filter, pilot-scale or full-scale, is usually supplied with slurry by a centrifugal pump at a varying pressure. If the pump suction and filter discharge are both at atmospheric pressure and the pipe and valve losses are negligible, the pump discharge pressure always equals the batch filter differential pressure (ΔP). The filtrate volume (V), pressure (ΔP_b), and time (T) data can then be plotted on companion plots (see Figure 10.4.11).

The true mean ΔP_b and true mean flow rate can be obtained by graphic integration. The required filtration area for any total mean flow rate can then be calculated.

DEEP-BED FILTER DESIGNS

Filters with deep beds of sand, diatomaceous earth, coke, charcoal, and other inexpensive packing materials have been successfully used in the filtration of potable water and can also be used in the small-scale treatment of dilute wastewater (see Figure 10.4.3). Their best applications are in polishing the filtrate from a continuous filter or the overflow from a primary or secondary settling tank. Without preseparation, the bed becomes loaded quickly. When the particles and bacteria in sizes smaller than the interstices of the bed, plus suspended BOD, are removed from the liquid, exceptional clarity is obtained. The dissolved substances, including dissolved BOD, are not removed.

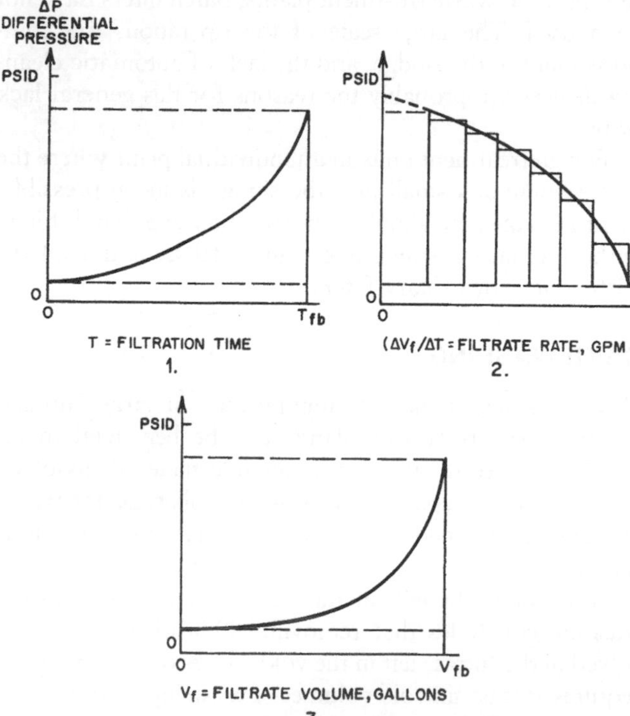

FIG. 10.4.11 Batch pressure filtration with variable pressure. Graphs 1 and 3 are generalized graphs of the original data taken during a run; graph 2 can be calculated from 1 and 3 or obtained from a flowmeter. The curved line on 2 is on or close to the characteristic curve of the pump. At any time, the following equation calculates the specific resistance α:

$$\alpha = \frac{(\text{area})^2(\text{psid})}{\mu(W)(V)(\Delta V/\Delta T)}$$

The variability or constancy of α indicates the compressibility or lack of it in the filter cake.

Environmental engineers can predict the variation of pressure loss and flow rate with time from small-scale tests if they determine K_1 and K_2 experimentally. The sand bed and waste material tested must be the same as that used in the process. Predicting the initial pressure-loss for various flow rates is possible.

The Ergun equation for the initial pressure loss in a deep-bed filter is particularly useful for granular materials because it covers both laminar and turbulent regimes as follows:

$$\left(\frac{\Delta P}{L}\right)g_c = \frac{150\mu u_0}{D_p^2}\cdot\frac{1-e^2}{e^3} + 1.75\frac{r\mu_0^2}{D_p}\cdot\frac{1-e}{e^3} \qquad 10.4(15)$$

where:

ΔP = pressure loss (frictional only) in direction of flow, lb/ft^2
L = length of granular bed in direction of flow, ft
μ = fluid viscosity, lb/sec, ft
D_p = mean particle diameter, ft
= void fraction of bed, dimensionless
ρ = fluid density, lb/ft^3
u_0 = fluid velocity based on empty cross section, ft/sec
g_c = gravitational acceleration

The general shapes of curves constructed from Equation 10.4(15) agree with experimental values; however, the coefficients vary appreciably for different fluid–solid systems. Therefore, environmental engineers should supplement the equation with tests when possible. Impurities deposited in the bed during batch filtration must be removed periodically, whenever the pressure loss rises and the flow rate decreases to the limiting acceptable values. Backwashing the bed to remove impurities creates a by-product of contaminated liquid, which must be disposed.

PRECOAT FILTERS

Environmental engineers must use variations of these deep-bed design methods wherever batch leaf precoat filters and rotary vacuum precoat filters are used. These precoats usually consist of diatomaceous earth or asbestos fibers preapplied to the filter medium. Since these materials do not have measurable particle diameters, D_p is unknown and the Ergun equation is not applicable. Therefore, test data are required even for initial pressure loss.

On a batch leaf precoat filter, the differential pressure increases continuously to the maximum allowable value. The precoat and impurities are then washed off, and a new precoat is applied. These filters can be operated without a vacuum pump when the filtrate is continuously removed with a centrifugal pump capable of operating at a low NPSH.

On a continuous rotary precoat filter (similar to the one in Figure 10.4.4), a slowly advancing knife continuously shaves off a few thousandths of an inch of the precoat plus the accumulated impurities as the drum rotates. As a result, the pressure loss is kept fairly constant throughout a run, and the run can be extended to one week when a 2- to 4-in precoat of processed diatomaceous earth is used. Because replacing the precoat only requires a few hours, the operation is virtually continuous. These filters are successful in removing SS and oil from slightly polluted water (0 to 200 ppm impurities). They are not used on sewage.

OUTDOOR SAND-BED FILTERS

Large, inexpensive, deep-bed filters for sewage treatment, constructed at ground level from naturally occurring sand, can be used on raw sewage, but presettling of solids is advisable. Isolated locations are recommended for these filters because some odors are inevitable and large land areas are required.

The filtered sludge leaves a mat of solids on the sand bed, which must be removed periodically. Operation and maintenance are simple, and operating costs are low. These units are suitable for small communities and isolated institutions, if approval from the authorities can be obtained. Table 10.4.4 shows data on the deep-bed sand filter.

TANK-TYPE, DEEP-BED FILTERS

Deep-bed filters within closed tanks are commercially available. They can be filled with sand or any other inert granular material appropriate for the filtration process.

TABLE 10.4.4 MAXIMUM ALLOWABLE LOADING OF INTERMITTENT SAND FILTER

Type of Sewage	*Gallons per day per acre*
Raw	20,000–80,000
Settled (overflow)	50,000–125,000
Biologically treated	Up to 500,000

Backwashing is carried out in batches and continuously. The operator selects a backwashing velocity that fluidizes the granular bed without transporting it out of the equipment while the backwash removes the waste solids. The bed packing particles must have a higher settling velocity and usually a higher density than the particles or flocs of solid waste being removed.

Because of the variability of finely divided SS, buying filters cannot be done in the same way as pumps, motors, and instruments, i.e., on the basis of design specifications only. The buyer should participate in filtration tests long enough to obtain a statistical probability that the filter will work. Table 10.4.1 summarizes the types of filtration tasks and the equipment that best performs each one.

—*F.W. Dittman*

10.5 DEWATERING: CENTRIFUGATION

CENTRIFUGE DESIGN TYPE

a) Imperforate bowl knife
b) Solid bowl conveyor
c) Disc centrifuge with nozzle discharge

CENTRIFUGE SELECTION AND APPLICATION

See Table 10.5.1

HYDRAULIC CAPACITY

For a, up to 60 gpm for feed solids ≤1% decreasing to 40 gpm at 3 to 5% solids; for b, to 400 gpm on lime sludges, as low as 75 gpm on some sewage sludges for the largest unit; for c, from 20 to 300 gpm normal and 400 gpm maximum

SOLIDS HANDLING CAPACITY

For a, to 12 tn/hr wet cake on paper mill waste; seldom limited by solids loading in waste treatment applications

CAKE HOLDING CAPACITY

For a, practical maximum about 14 ft^3 for centrifugal acceleration > 1200 × gravity

SOLIDS DISCHARGE

For a, intermittent; solids showing plastic flow by skimmer at full-bowl speed; stiffer cakes by plowing knife at reduced speed. For b, continuous by helical conveyor operating at 10 to 30 rpm differential speed from bowl. For c, continuous up to 6% concentration by weight on secondary biological sludge.

POWER REQUIRED

For a, windage and friction plus about $\frac{1}{3}$ hp per gpm, totaling to 45 hp at maximum rates. For b, to 250 hp, but varies directly with liquid and solids loading. For c, from 3 gpm per hp at high rates to $1\frac{1}{2}$ gpm per hp at low rates.

MATERIALS OF CONSTRUCTION

For a, normally stainless steel on waste treatment; carbon steel, special alloys, and various coatings available. For b, commonly stainless steel in waste treatment; carbon steel available; monel or titanium for special corrosion problems. For c, bowls always stainless steel, usually type 316; covers usually stainless steel, but carbon or cast steel sometimes used.

TABLE 10.5.1 ORIENTATION TABLE FOR CENTRIFUGATION

	Centrifuge Characteristics									
	Suitability of Centrifuge by Type°				*Capacity Range*			*Discharged Cake Concentration*		
Type of Sludge or Waste°°	*Good*	*Fair*	*Poor*	*No*	*High*	*Medium*	*Low*	*High*	*Medium*	*Low*
Sewage°°, primary raw	B		A	C	B		A	B	A	
Sewage, primary anaerobic digested	B	A		C	B		A		B	A
Sewage, primary aerobic digested	A	B		C		A,B			A,B	
Sewage, primary raw limed	B		A	C	B		A	A,B		
Sewage, secondary biological (activated and humus)	A,C		B		C	A	B	A	B, C	
Sewage, secondary biological (activated) with alum	A, C	B			C	A,B		A	B, C	
Sewage, primary + secondary biological, cosettled	A	B		C		A, B			B	A
Sewage, primary + secondary biological, anaerobic digested	B	A		C		B	A	B		A
Sewage, whole, modified biological (activated)	A	B	C		C	A	B		A, B	C
Sewage, whole, aerobic digested	A	B	C		C		A, B		A, B	C
Sewage, primary, heat-treated raw	A, B			C	B	A		B	A	
Sewage, secondary biological, heat-treated raw	A	B		C		A, B			A, B	
Sewage, primary + secondary biological, heat-treated raw	A, B			C	B	A		B	A	
Sewage, tertiary with lime (for phosphate)	B		A	C		B	A	A, B		
Sewage, teritary with alum (for phosphate)	A	B, C			C	A, B		A	B	C
Industrial, coarse solids	B	A		C	B		A	B	A	
Industrial, clean biological	A, C		B		A,C	B			A, B	C
Industrial, hydrous or flocculant solids	A	B, C			A, C	B			A, B	C
Industrial, oil–water emulsion	C	A		B		C	A			
Industrial, fine solids	A, C	B			A	C	B		A, B	C
Water treatment, lime softening	B		A	C	B		A	A, B		
Water treatment, alum	A, C	B			C	A, B			A, B	C
Classification of lime and hydrous solids	A, B			C	A, B			B	A	
Classification of digested sewage (aerobic or anaerobic)	A, B			C	B	A		B	A	

Notes: °Type of Centrifuge: A = Imperforate bowl with knife and skimmer discharge; B = Solid bowl with conveyor discharge; C = Disk bowl with nozzle discharge

°°In each of these waste streams, the feed to the centrifuge is the sludge produced by the operation. For example, the first waste listed in the table is a raw sludge taken from the underflow of a primary clarifier.

DESIGN PRESSURE

Normally atmospheric for a, b, and c. Design b is also available at 150 psig rating in vertical construction.

DESIGN TEMPERATURE

Up to 200° to 225°F without special design for a, b, and c.

The three types of centrifuges applied in waste treatment processes rely on settling to separate solids and differ only in discharging the settled solids. Environmental engineers base their centrifuge selection on the particle size, concentration, feed rate, performance required, disposal methods, and costs. Table 10.5.1 lists three criteria (1) suitability based on the anticipated type of solids and minimal coagulant requirements, (2) capacity range based on the largest commercial units at the same recovery of SS, and (3) discharged cake concentration relative to other dewatering equipment at the same recovery level.

The conveyor centrifuge is used for coarse and heavy loadings of solids and normally requires a coagulant for reasonable recoveries. The disc centrifuge with nozzle discharge is restricted to sludges containing no coarse particles and usually having low concentrations and large volumes. The imperforate bowl with knife and skimmer can handle fairly low feed rates at high recoveries, usually without coagulant (see Figure 10.5.1).

Imperforate Bowl (Basket) Knife Centrifuge

This centrifuge, the simplest type in general use, is suited for soft or fine solids that are difficult to filter and waste that varies in concentration and solids characteristics.

That design can be either a top-driven suspended bowl or an underdriven bowl with three-point casing suspension; the drive can be a hydraulic or an electric motor. A 12- or 14-in diameter bowl is often used for test work, whereas 30 to 48 in is the range of commercial units. Older designs operate below 1000 G (G is the ratio of centrifugal acceleration to the acceleration of gravity), whereas current units use 1300 G. Cake concentration and clarification both improve with increased acceleration.

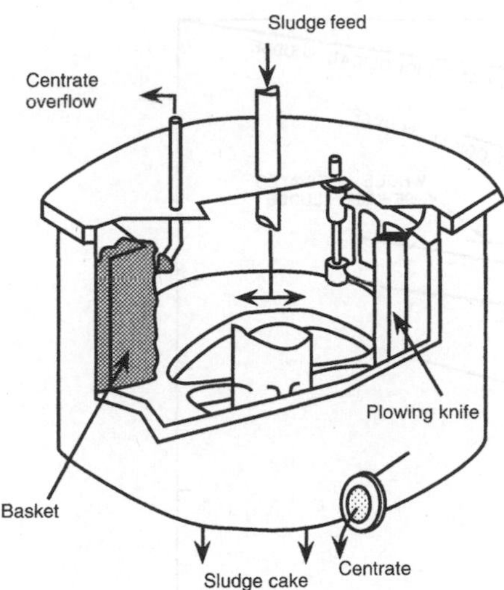

FIG. 10.5.1 Basket centrifuge. (Reprinted from U.S. Environmental Protection Agency [EPA], 1987, *Design manual: Dewatering municipal wastewater sludges,* EPA/625/1-87/014 [September].)

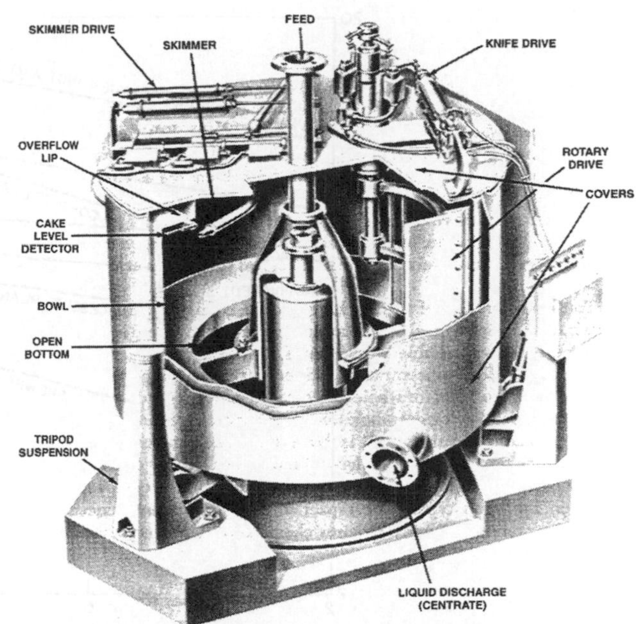

FIG. 10.5.2 Imperforate bowl knife centrifuge with skimming tube.

A fully automated knife centrifuge for waste treatment (see Figure 10.5.2) includes a rotating bowl, covers to collect clarified effluent, a feed distributor, an overflow lip to maintain an annular layer of liquid in the bowl, a cake-level detector, a skimming tube with drive, a plowing knife with axial or rotary drive, and an open-bottom bowl through which the plowed cake can drop. Feed entering the top and introduced near the bottom of the bowl flows axially, with the clarified effluent discharged at the top. The SS settle on the bowl wall and eventually impede clarification. This centrifuge then interrupts the feed for $1\frac{1}{2}$ to $2\frac{1}{2}$ min to discharge the cake. The total centrifugation cycle usually takes 6 to 30 min.

Instrumentation automatically initiates the discharge cycle. During the cycle, the skimmer moves outward to remove supernatant liquid and may continue to remove soft solids. Coarser or fibrous solids are then plowed out with the knife while the bowl is held at a reduced speed. A concentration gradient exists across the cake from the softest material at the inside to the heaviest and coarsest at the bowl wall. The cake can be removed as a whole, or it can be classified into concentration fractions for separate handling when the actions of skimmer and knife are preset.

APPLICATIONS

Since clarification occurs in a quiescent zone undisturbed by moving elements such as a conveyor, settling is effective, and recovery is good. Because the bowl speed is low, shearing of flocculant material is minimal. These centrifuges can efficiently capture (90%) of even difficult biological or alum sludge solids without coagulants.

Applications for these centrifuges include metal hydroxide waste, aerobic sewage sludge, and water treatment alum sludge. Figure 10.5.3 gives the typical performance curves for four biological sludges showing the recovery of SS as a function of the feed rate averaged over a full cycle. Recovery refers to the proportion of the entering solids that are retained in the bowl for discharge as cake.

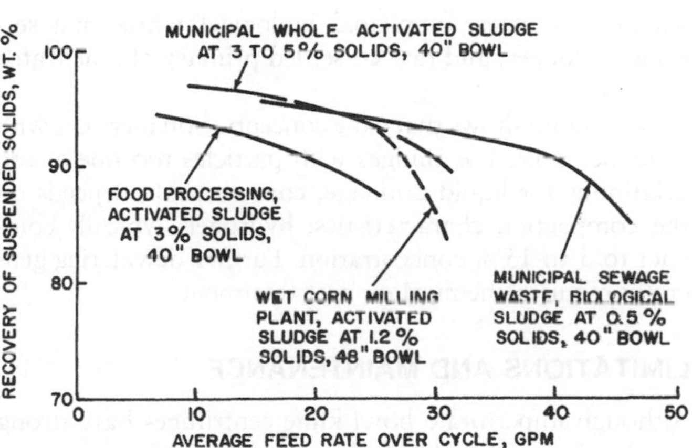

FIG. 10.5.3 Clarification of biological sludges in imperforate bowl knife centrifuge.

As with all centrifuge designs, increasing the feed rate decreases the residence time and recovery. These centrifuges produce recoveries of 90% or better over a range of feed rates. Industrial activated sludge is often more difficult to clarify than the consistent biological sludge generated in municipal sewage plants.

Concentration of discharged cake is a function of the sludge but is also influenced by the residence time in the bowl and the G level. Residence time, controllable within limits, is a function of centrifuge size, throughput rate, feed concentration, and recovery. Figure 10.5.4 gives typical performance data for whole cakes from several industrial

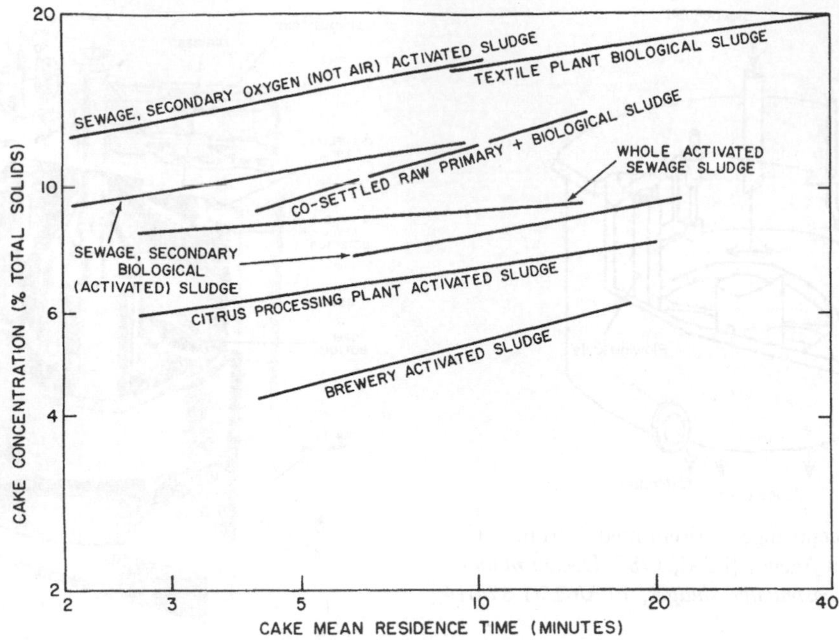

FIG. 10.5.4 Whole cake concentration from imperforate bowl knife centrifuge applied to various sludges.

activated sludges, municipal biological whole and secondary sludges, and raw co-settled primary plus activated sludges.

The figure shows that cake concentration increases with residence time. For sludges with particles too fine or too gelatinous for liquid drainage, concentration depends on the compaction characteristics; hydrogels typically compact to 5 to 15% concentration. Further dewatering generally requires chemical or heat treatment.

LIMITATIONS AND MAINTENANCE

Although imperforate bowl knife centrifuges have strong capabilities on difficult sludges, the hydraulic parameters or the frequency of cake discharge limits these units to low concentration feeds at about 50 gpm maximum.

These centrifuges can handle wastes that are incapable of gravity-thickening to more than a few percent and that are subject to natural flocculation. These centrifuges adapt readily to changes in the feed rate, concentration, and solids characteristics. The complementary action of the skimmer and knife during cake discharge classifies the cake.

Because the operation is at low speeds compared to other centrifuges, the maintenance is low, with the bearing life in the range of 1 to 3 years. Lubrication is simple, power consumption is low, and little replacement of parts is needed.

Solid Bowl Conveyor Centrifuge

A conveyor centrifuge is used when the SS are coarse, high concentration cakes are required, solids loading is high, or classified solids recovery is required. The unit is flexible, easy to operate over a range of feed rates and concentrations, and requires little attention. The combination cylindrical–conical bowl ranges from a 6-in diameter for testing to 14 to 36 in for commercial units. Long bowls at more than a 4-to-1 length-to-diameter ratio have higher capacities and are commonly applied on waste sludge at 1000 to 2700 G acceleration.

The rotating assembly of a conveyor centrifuge consists of a bowl shell and a conveyor supported between two sets of bearings. This bowl and conveyor are linked through a planetary gear system designed to rotate the bowl and conveyor at slightly different speeds. The bowls are belt-driven. The rotating assembly is covered by a stationary casing for safety, odor, and noise control.

In the usual countercurrent design (see Part A in Figure 10.5.5), the feed enters through a pipe in the hollow shaft of the conveyor, accelerates through ports, and enters the annular pond of liquid at the wall of the bowl close to the conical section. Liquid flows to the far end of the bowl and discharges as clarified effluent (centrate) over adjustable weirs that regulate the depth of the pond. The conveyor moves the settled solids to the conical section where they drain on the unsubmerged beach before discharge.

In the cocurrent design (see Part B in Figure 10.5.5), the feed is introduced at the end of the bowl. The liquid and settled solids move together toward the conical beach. An adjustable skimmer removes the supernatant liquid, while the solids drain on the beach before discharge.

The abrasive nature of waste sludges ranges from extremely abrasive pulp mill waste to clean biological sludge. Paper mill and primary sewage sludges have necessitated

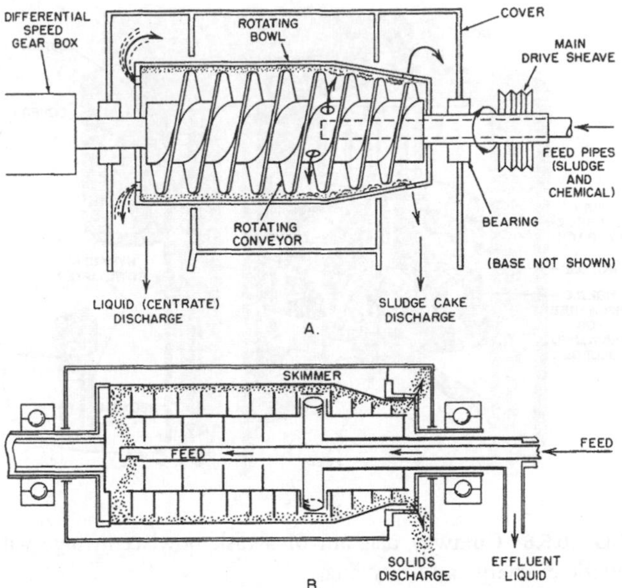

FIG. 10.5.5 Solid bowl conveyor centrifuge. A, Countercurrent; B, Cocurrent.

the use of abrasion-resistant, hard-surfacing materials for direct application or as replaceable inserts.

OPERATING VARIABLES

Design variables affect clarification and cake dryness. The design and application determine the bowl diameter, length, beach angle, and conveyor configuration. The bowl capacity is proportional to the length at a given diameter; the angle of the conical beach, normally 5 to 10°, can affect the discharge of solids.

Wastewater treatment facilities can adjust the bowl speed, relative conveyor speed, and pond depth to change the performance after installation. Relative conveyor speed is basically set by gearbox design but is also controlled by an external drive or braking assembly attached to the pinion shaft of the box. The lowest differential speed compatible with adequate solids removal is the optimum choice because it decreases turbulence, increases drainage time, and decreases the linear velocity between the conveyor, bowl, and solids and, consequently, the associated abrasion. The pond depth adjustment is a compromise between better clarification in a deeper bond and a dryer cake from a shallower pond.

The process variables are primarily the feed rate and the addition of chemicals. Increasing the feed rate characteristically reduces recovery (see Figure 10.5.6) and increases the cake dryness due to the selective loss of finer material.

Adding polymeric or other coagulants in 0.1 to 0.2% solution internally to the pond or before the centrifuge improves recovery but at a cost (see Figure 10.5.7). Cationic polymers are usually best for sewage unless alum or lime addition requires a shift to anionic. Since the feed rate per centrifuge determines the number and size of the units to meet the plant capacity, unnecessarily high recoveries are costly in capital expenses and/or polymer usage.

APPLICATIONS

Municipal sewage plants can use the conveyor centrifuge for raw primary and raw mixed or cosettled primary plus biological sludges, anaerobically digested primary or mixed sludges, and heat-treated or limed chemical sludges. It can be applied, (sometimes at high coagulant costs) to dewatering whole, mixed, or secondary waste biological (excess activated) sludges, aerobic digested sludges, and alum or ferric chemical sludges. In water treatment plants, it is excellent on water-softening lime sludges but less effective on alum sludges. In industry, applications include

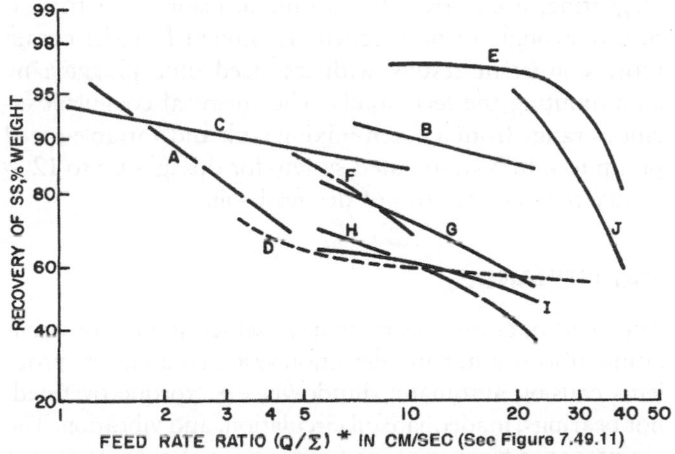

FIG. 10.5.6 Recovery–capacity performance curves for conveyor centrifuge. *Sludge Description Legend: A,* converter paper mill, coated paper; *B,* kraft mill, boxboard and coated paper; *C,* cold rolling mill, clarifier sludge; *D,* sewage, raw limed primary clarifier, classified without polymer; *E,* sewage, raw limed primary clarifier, treated with 1.5 lbs. anionic polymers added per ton dry feed solids; *F,* raw tannery waste; *G,* sewage, raw primary clarifier; *H,* sewage, raw primary plus secondary biological sludge; *I,* sewage, anaerobic digested primary plus secondary biological; *J,* basic oxygen furnace scrubber liquor.

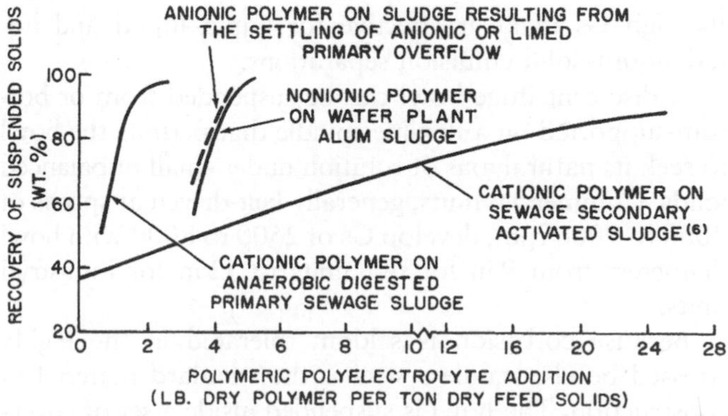

FIG. 10.5.7 Recovery response to polyelectrolyte addition in a conveyor centrifuge.

clarifying scrubber liquors, recovering solids from packing-house and metals-treating waste, roughing out solids upstream of disc centrifuges on refinery sludges, and dewatering potash and mining tailings. Controlled settling allows the recovery of calcium carbonate from limed sludge.

Figure 10.5.6 shows the expected solids recovery against feed rate ratios for a number of wastes. On municipal sewage sludges, comparing the G and H curves shows the effect of adding secondary biological sludge to primary sludge. Anaerobic digestion of the latter does not change the clarification characteristics markedly (I). The cake concentration depends largely on the sludge and can vary (see Table 10.5.4).

AUXILIARY EQUIPMENT

Degritting, even down to 150 mesh, prior to centrifugation is strongly recommended. Treatment facilities obtain more consistent results with less feed-zone plugging by comminuting the feed solids. The chemical coagulant facilities range from a batch-mixing tank and variable-speed pump to a fully automated system for dosages up to 12 lb of dry polymer per ton of dry feed solids.

MAINTENANCE

The gearboxes and bearings have self-contained or circulating lubrication. Fault detection systems can locate problems causing automatic shutdown, e.g., torque overload, hot bearings, inadequate oil circulation, and vibration. The maintenance frequency is a function of the abrasiveness of the solids; an initial inspection is suggested after 2000 hr of operation with subsequent inspections as required.

Disc Centrifuge with Nozzle Discharge

This centrifuge is specifically suited to the thickening of excess activated sludge or alum sludge that are free of coarse solids. It can handle high feed rates with effective clarification for very fine particles at low concentrations. Its high Gs make it effective for liquid–liquid and liquid–liquid–solid emulsion separations.

A disc centrifuge bowl can be suspended from or bottom-supported on a flexible spindle that permits the bowl to seek its natural axis of rotation under small unbalanced loads. Commercial units, generally belt-driven at speeds of 3000 to 9000 rpm, develop Gs of 2500 to 8000 with bowl diameters from 9 in for test units to 32 in for industrial units.

Because corrosion is seldom tolerated in the highly stressed bowls, stainless steel is the standard material of construction. The bowl is suspended inside a set of covers (see Figure 10.5.8) containing two decks that separately receive the clarified effluent and thickened sludge. For an emulsion, a third deck collects the second liquid phase.

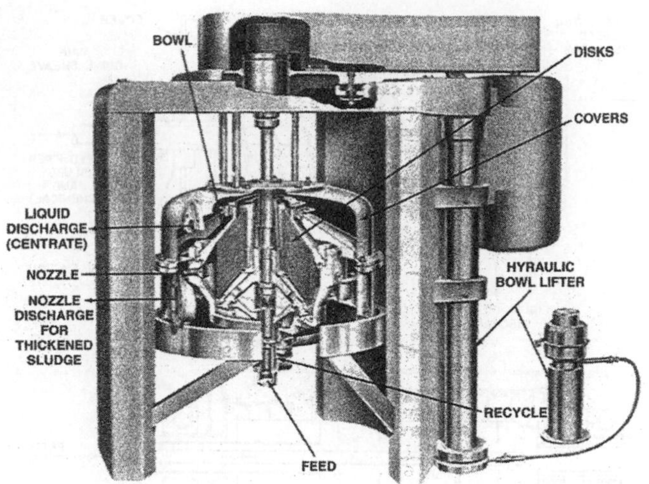

FIG. 10.5.8 Cutaway diagram of a disk bowl centrifuge with nozzle discharge and recycling.

The distinguishing feature of the bowl is a stack of cones or discs arranged so that the feed stream is divided among them, reducing the particle settling distance. The angle of the cone causes aggregated solids to slide and settle on the inner walls of the bowl that slope toward a peripheral zone containing orifices, called nozzles. The number of nozzles varies with the bowl size, but the size of the nozzle orifice is usually 0.030 to 0.080 in.

For efficient performance, the solids should not be fibrous or pack into a hard cake, and the incoming feed must be screened. The solids concentration in the nozzle discharge can reach ten to twenty times that in the feed. Recycling some of the nozzle discharge stream back into the bowl can control underflow concentrations (thickened sludge).

APPLICATIONS

The clarification of biological sludge (excess activated) from feed concentrations of 0.3 to 1.0% SS is the most important application of these centrifuges. Figure 10.5.9 shows the capacity range of a large disc centrifuge used at six municipal plants for excess activated-sludge thickening to 5½% solids concentration without the use of coagulants.

Contrary to some reports for midrange feed rates of disc centrifuges, varying the SVI from 40 to 137 or altering the centrifuge underflow concentration below 6½% has little effect on the solids recovery (see Figure 10.5.10). Figure 10.5.9 also shows the direct effect of the feed rate on the quality of clarification of oil separated from an aged petroleum refinery waste emulsion.

LIMITATIONS AND MAINTENANCE

The disc centrifuge should only be used for continuous extended running because the bowl must be cleared of solids before the centrifuge is restarted after a shutdown.

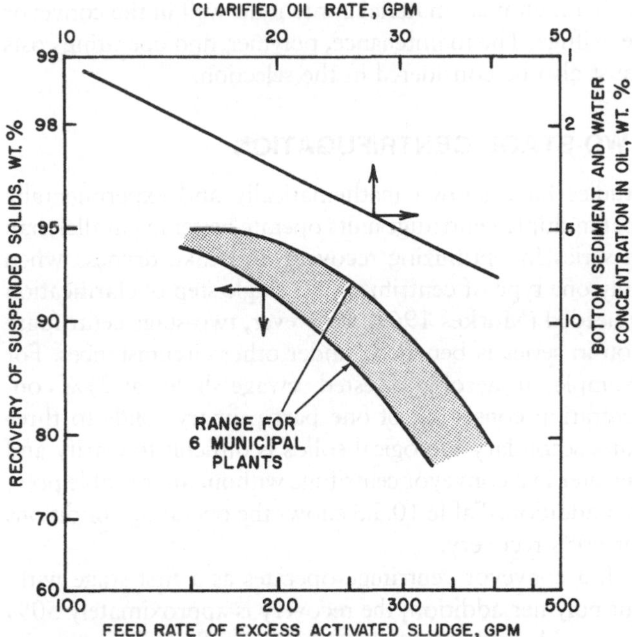

FIG. 10.5.9 Recovery–capacity performance curves for large disk centrifuges.

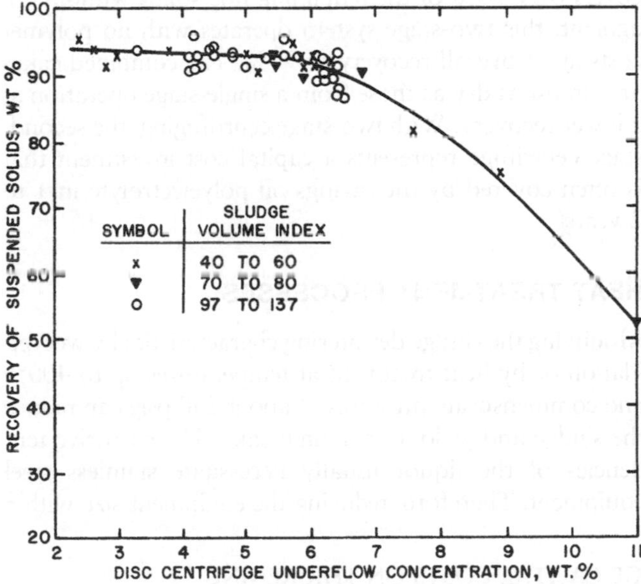

FIG. 10.5.10 Effect of the SVI and underflow concentration on recovery for mid-capacity flows of municipal excess activated sludge in disk centrifuges.

Maintenance is nominal since bearings in smaller units are grease-packed, whereas larger units use circulating oil or spray-mist systems. On abrasive applications, some areas require hard-surfacing; nozzle bushings, on the other hand, are replaceable.

AUXILIARY EQUIPMENT

Adequate feed screening is vital to keep these units online as long as 8 weeks without cleaning. Inline self-cleaning strainers are generally satisfactory. Due to changes in the feed, control is recommended for consistency of thickened sludge. This control can be manual or fully automated, which uses a viscosity sensing loop to detect the sludge discharge consistency and modulate the recycling rate.

PERFORMANCE CORRELATION

Theoretically, the Σ concept allows performance comparison between similar centrifuges operating on the same feed material. Since Σ reflects a fixed maximum bowl capacity, for any flow rate (Q), the Q/Σ ratio represents the hydraulic loading.

In theory, any two similar centrifuges treating the same sludge should show the same settling performance at the same value of Q/Σ, and differences reflect nontheoretical factors. Since these efficiency factors are nearly identical in hydrodynamically similar centrifuges, i.e., in two sizes of disc centrifuge bowls of identical dimensional ratios, the clarification performance in one bowl can be directly related by the Σ ratio to the anticipated performance of a different size bowl.

When the centrifuges are not similar, as in a conveyor bowl compared to a disc bowl, the Q/Σ plots indicate relative performance, but environmental engineers cannot use the performance curve of one type for scaleup to the other type. Nevertheless, a Q/Σ plot showing flow ratios against solids recovery represents a generalized correlation and is a standard method of evaluating centrifuge performance.

Figure 10.5.11 is a generalized plot for a municipal excess activated sludge in disc, conveyor, and knifing centrifuges. The low Q/Σ value for the disc centrifuge results from its high shear with consequent feed degradation. The conveyor and basket centrifuges cause less floc degradation; the former has a scrolling inefficiency on a soft cake. Even considerable polymer addition cannot entirely overcome this effect.

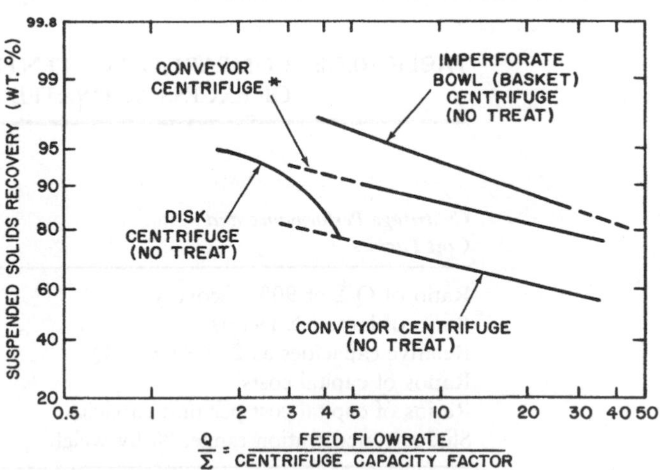

FIG. 10.5.11 Relative performance of centrifuges on municipal excess activated sewage sludge. °12 pounds of polymer added per ton of dry solids in feed.

LABORATORY TESTING

The only valid method for testing a limited amount of sludge for centrifuge performance is to run the material in a pilot unit. These tests generally require feed rates up to 10 gpm but also permit the evaluation of chemical acids. The clarification, capacity, and cake dryness scaleup predictions are good. For many municipal sewage sludges, sufficient information is available for prediction. For industrial wastes, testing is recommended.

SELECTION OF CENTRIFUGE

The operating curves in Figure 10.5.11 show one aspect of centrifuge selection. For example, if a wastewater treatment facility thickens municipal excess activated sludge at 90% recovery to limit the return of solids with the recirculated centrifuge liquid discharge (centrate), the feed flow rate must be large enough to require one or more of the largest units of any centrifuge type.

The conveyor centrifuge without polymer addition cannot reach 90% recovery and cannot be considered in this example. For the other three curves in Figure 10.5.11, the ratios of Q/Σ at 90% recovery are listed in Table 10.5.2 together with the ratios of Σ for the respective large units. The relative capacities of these units are determined by the product $\Sigma \times Q/\Sigma$.

The disc centrifuge shows a 4-to-1 advantage over the other units on this basis despite its lower Q/S value. The ratios of the unit capital costs do not differ markedly even when auxiliaries are included. The installation costs are not listed but are similar for the three types. The relationship of capital cost per unit capacity shows the disc centrifuge has more than a 3-to-1 advantage. The table also lists the solids contents of the thickened sludges.

Environmental engineers must consider other selection factors. The method and cost of disposal determine the required sludge concentration level. For smaller flows, the selection may favor discontinuous operation using the knifing (basket) centrifuge. Some industrial biological sludges with a high inert material content do well in the conveyor centrifuge. The maintenance, polymer, and operating costs must also be considered in the selection.

TWO-STAGE CENTRIFUGATION

Studies have shown mathematically and experimentally that multiple-centrifuge units operate better in parallel than in series in optimizing recovery and cake dryness when only one type of centrifuge or a single step of clarification is needed (Murkes 1969). However, two-stage centrifugation in series is beneficial under other circumstances. For example, an aerobic digested sewage sludge at $2\frac{1}{2}$% concentration consisting of one part primary solids to three parts secondary biological solids is difficult to clarify and dewater in a conveyor centrifuge without appreciable polymer addition. Table 10.5.3 shows the operating conditions for 90% recovery.

If a conveyor centrifuge operates as a first stage without polymer addition, the recovery is approximately 50% because the coarser particles and some flocculent material are removed. If the effluent is then directed to an imperforate bowl knifing centrifuge (basket), which readily recovers over 90% of the remaining fine solids without coagulant, this two-stage system operates with no polymer costs at an overall recovery of 97%. The combined cakes are almost as dry as those from a single-stage operation at a lower recovery. With two-stage centrifuging, the second-stage centrifuge represents a capital cost investment that is often covered by the savings on polyelectrolyte in 1 to 2 years.

HEAT TREATMENT PROCESSES

Modifying the sludge dewatering characteristics by wet oxidation or by heat treatment at temperatures up to 400°F and commensurate pressures of about 250 psig can reduce the sludge and yield a drier final cake. The corrosive tendencies of the liquor usually necessitate stainless steel equipment. Therefore, reducing the equipment size with a

TABLE 10.5.2 COMPARISON OF CENTRIFUGE CAPITAL COSTS ON THICKENING OF EXCESS ACTIVATED SLUDGE

Centrifuge Performance and Cost Factors	*Type of Centrifuge*		
	Disc	*Conveyor°*	*Basket (skimmer and knife)*
Ratio of Q/Σ at 90% recovery	1.0	2.0	5.5
Ratio of largest Σ factors	22.8	2.3	1.0
Relative capacities as $\Sigma \times Q/\Sigma = Q$	22.8	4.6	5.5
Ratios of capital costs	1.3†	1.2‡	1.0
Ratios of capital cost per unit capacity	1.0	4.6	3.2
Sludge concentration range, % by weight	5–6	4–12	9–11

Notes: °With polymer addition; 90% recovery is impossible without this additive.
†Includes cost of prescreening and concentration control equipment.
‡Does not include feed comminution, polymer preparation equipment, or polymer costs.

TABLE 10.5.3 COMPARISON OF ONE-STAGE AND TWO-STAGE CENTRIFUGING OF AEROBIC DIGESTED SEWAGE SLUDGE

Performance Factors	Centrifuge Operation: Single Stage	Two Stage
Single or First Stage		
Usage of polymer, lb/tn of dry solids	8–12	0
Solids recovery, %	90	50
Solids in cake, %	12–13	13–14
Second Stage		
Usage of polymer, lb/tn of dry solids		0
Overall solids recovery, %		97
Solids in combined (first and second stage) cakes, %		10–11

preconcentration of the feed at 6 or 7% solids is economically advantageous.

Wastewater treatment facilities can preconcentrate the feed by combining gravity-settled primary sludge with disc-centrifuged secondary sludge concentrate or by partially dewatering cosettled primary and secondary sludges in a conveyor or knife (basket) centrifuge depending on the flow rates. Sludge concentrations higher than 7% are difficult to pump and require an increase in the heat transfer surface. Conveyor centrifuges and RDVFs can dewater the treated sludge to cakes containing 35 to 45% solids without chemical addition.

RECOVERY OF CHEMICAL ADDITIVES

Figure 10.5.6 shows the performance of centrifuging for the selective recovery of calcium carbonate solids from a limed sewage sludge. When the economics of plant size and location warrant calcining and recycling, such recovery is readily made from limed sewage sludges, tertiary limed sludges for phosphate removal, water-softening lime sludges, and limed industrial sludges, frequently at efficiencies of 70 to 85% of the concentration of the calcium carbonate in the feed.

The calcium carbonate content is often 40 to 70% of the SS. The SS content of flocculated phosphate complexes and magnesium hydroxide varies from almost zero to 30%. Figure 10.5.12 shows the relative recoveries of magnesium hydroxide and calcium carbonate for two limed sludges. Calcium hydroxyapatite recovery approximates that of the hydroxide. Selective recovery of calcium carbonate also assures a reduced content of unfavorable solids in the recycled material.

The cake is deposited in an imperforate bowl knifing centrifuge (basket) with a high concentration gradient due to both differences in residence time and the relative densities of the collected solids. Whole sewage from a contact stabilization plant can produce a cake with a 5% solids content at the inner edge and 30% at the bowl wall, with a nonlinear concentration gradient. Because the lighter and lower concentration material is largely organic, returning to stabilization by selective removal with the skimming

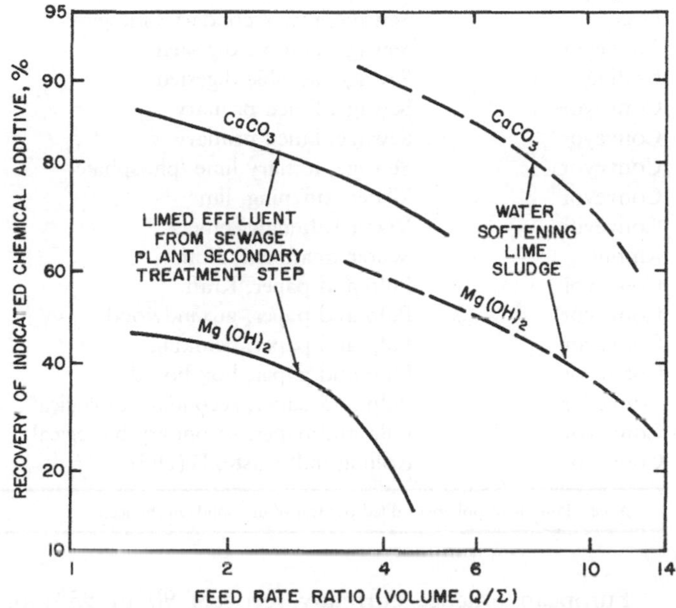

FIG. 10.5.12 Relative recoveries of magnesium hydroxide and calcium carbonate from limed sludges.

tube is sometimes feasible. The knifed-out residue then has a higher concentration and is more suitable for incineration. Adjusting the cut point between the skimmed and knifed material can accommodate changes in the feed and required dryness of the heavier discharge.

Selected Applications and Trends

Trends to improve performance in municipal and industrial waste treatment plants lead to new processes and more complex combinations of sludge. Primary and secondary sludges can be dewatered separately or mixed. The former often results in a higher initial cost but also gives a dryer combined cake and lower operating cost. Table 10.5.4 shows the effect on dewatering of adding secondary to primary sludge. Many smaller plants using only aerobic contact stabilization have low sludge rates that are handled at somewhat higher capital costs in knifing centrifuges without coagulant use.

TABLE 10.5.4 TYPICAL CENTRIFUGE PERFORMANCES ON WASTE SLUDGE

Type of Centrifuge	*Type of Waste Sludge*	*Solids Recovery, %*	*Solids in Cake, %*	*Chemical Addition, lb/tn*°
Conveyor	Sewage, raw primary	65–80	30–35	0
Conveyor	Sewage, raw primary	80–95	25–35	2–8
Conveyor	Sewage, digested primary	75–85	30–35	0
Conveyor	Sewage, digested primary	80–95	22–30	2–8
Conveyor	Sewage, raw primary and biological	50–70	18–22	0
Conveyor	Sewage, raw primary and biological	80–95	15–20	4–12
Knifing	Sewage, raw primary and biological	90–97	11–14	0
Conveyor	Sewage, digested primary and biological	55–70	23–30	0
Conveyor	Sewage, digested primary and biological	80–95	20–25	4–10
Conveyor	Sewage, raw secondary biological	85–95	5–15	8–15
Knifing	Sewage, raw secondary biological	90–95	9–11	0
Disk	Sewage, raw secondary biological	85–90	5–7	0
Conveyor	Sewage, aerobic digested	80–90	10–18	6–20
Knifing	Sewage, aerobic digested	90–95	9–12	0
Conveyor	Sewage, limed primary	55–70	40–60	0
Conveyor	Sewage, limed primary	80–95	15–30	1–5
Conveyor	Sewage, tertiary lime (phosphate)	35–70	55–70	0
Conveyor	Water-softening, lime	60–90	40–60	0
Conveyor	Water-softening, lime	90–100	35–50	2–8
Knifing	Water treatment, alum	90–95	8–20	0
Conveyor	Pulp and paper, Kraft	80–85	25–40	0
Conveyor	Pulp and paper, groundwood	85–90	15–20	0
Conveyor	Pulp and paper, deinking	85–90	15–25	0
Conveyor	Pulp and paper, box board	85–95	25–35	0
Conveyor	Pulp and paper, secondary biological	60–70	12–14	0
Conveyor	Pulp and paper, secondary biological	90–95	10–13	2–10
Conveyor	Rolling mill waste, $Fe(OH)_3$	95	12–18	1

Note: °Pounds of polymer added per ton of dry solids in the feed.

European practice currently recovers 90 to 95% of solids during dewatering. Because the need for such high recoveries has not been fully proven, the United States practice cannot justify the increased equipment and polymer costs and the resulting wetter cakes. In the United States, the usual requirements are in the range of 80 to 90% recovery, with a trend to lower recoveries, particularly in connection with wholly aerobic plants in which the return sludge quantities are large.

Many waste treatment plants are providing for chemical treatment to remove phosphate. Lime is frequently favored in many western states, and alum or iron salts is favored in the eastern states. The addition of lime at any point in the process produces sludge that is best treated in conveyor centrifuges either for high recovery with anionic polymers or for classification. Alum is commonly added before the secondary biological sludge clarifier but has little effect on centrifugal dewatering except for the shift to an anionic polymer. Alum (tertiary) treatment of secondary biological effluent produces sludge similar to water-treatment alum sludge.

Sludge containing appreciable quantities of iron salts often is amenable to clarification and dewatering in a conveyor centrifuge. Lime sludge from water softening is readily dewatered in conveyor centrifuges producing dry cakes (see Table 10.5.4). In simple water treatment by alum, resultant sludge quality and cake concentration obtained in an imperforate bowl knife centrifuge seem to be a function of the turbidity of the raw water.

ALTERNATE DEWATERING METHODS

Other methods of sludge dewatering in addition to centrifugation are available. Because the power requirements for centrifuges are a direct function of volumetric throughput, treatment facilities preconcentrate sludge, usually by gravity to the point of hindered settling (see Figure 10.5.4) and compaction, to reduce operating costs.

Flotation can preconcentrate large waste streams. Coagulant and labor costs are appreciable for concentrations of 4% or more. Centrifugation produces higher concentrations. RDVFs, when used to dewater biological sludge, have higher labor, maintenance, and coagulant costs. Their power requirements per unit throughput are lower, but their space and building requirements are higher. A vacuum filter produces a greater clarity filtrate but a slightly wetter cake.

Filter presses, recently automated, are used widely in Europe, but their capital costs frequently exceed those for centrifugation. No coagulant is used, but flyash or an equivalent filter aid is normally needed at ratios from 1:1 to 5:1 relative to the sludge solids. Filter presses reduce the moisture load on the incinerator but require additional cake-handling facilities. Continuous-gravity or pressure-screening devices are often more economical than centrifuges for a population of 10,000 or less due to their low operating speed and low maintenance and power costs. Their high coagulant use does not alter their overall economy.

—*F.W. Keith*

10.6 HEAT TREATMENT AND THERMAL DRYERS

Porteous Heat Treatment

OPERATING PRESSURE: 180 psig

OPERATING TEMPERATURE: 360°F

UNIT CAPACITIES: Up to 100 gpm

COOKING DETENTION TIME: 30 min

IMPROVEMENT IN DEWATERING BY HEAT TREATMENT: 200%

SOLIDS CONTENT OF THICKENED, HEAT-TREATED SLUDGE: 7 to 11%

CLARIFIED EFFLUENT BOD: 4500 mg/l or more

COD REDUCTION BY HEAT TREATMENT: 22%

VOLATILE SOLIDS REDUCTION BY HEAT TREATMENT: 28%

MATERIALS OF CONSTRUCTION FOR EQUIPMENT: Carbon steel

Heat treatment followed by filtration is economical for dewatering sludge without using chemicals. Heating sludge at elevated temperatures and pressures causes the cells to rupture, releasing bound water from the sludge particle and the water of hydration. The gel system structure is irreversibly destroyed, and the solids lose affinity for water and are readily separated from the liquid by decantation. The technical term for this process is *heat syneresis*.

The exact nature of mechanisms that release bound water and water of hydration from the sludge is not fully understood. It *is* known that coalescence of particles occurs, and that the high-velocity particle collisions (as a function of the energy level at high pressure [180 psig] and temperature [360°F]) disturb the molecules and assist in particle structure breakdown.

PROCESS

The continuous, automated process is sized on the basis of the flow rate. Figure 10.6.1 shows a basic flow sheet and the components of the heat treatment process. This process draws the sludge (any type) from the storage tank through a grinder, reducing particle size and fiber lengths. Then, a feed pump pressurizes the sludge to approximately 250 psig and passes it through a water-to-sludge heat exchanger that elevates its temperature to approximately 300°F before it enters the reactor.

The process simultaneously injects steam from a boiler into the reactor with the incoming preheated sludge, raising sludge temperature to approximately 360°F. The sludge cooks in the reactor for about 30 min at the corresponding pressure of 180 to 200 psig and is then discharged through the cooling exchanger discharge valve to the decant thickener tank. The cooled sludge temperature is approximately 110°F.

For high pressure pumps, the triplex ram (a three-cylinder, single-acting displacement pump without rings) or progressive cavity-type (a precision screw conveyor pump,

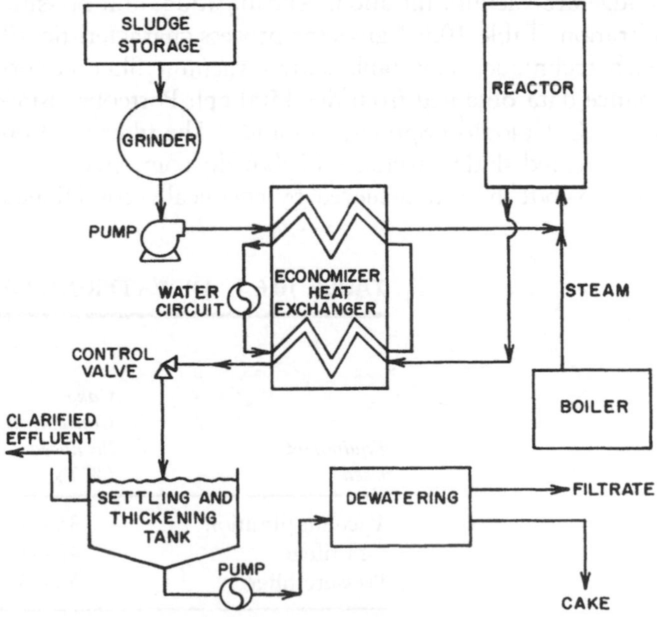

FIG. 10.6.1 Diagram of the Porteous process.

such as Moyno) pumps are commonly used. The heat exchanger has removable end caps and water-to-sludge circuitry to eliminate tube fouling and maintenance. The process removes grit from the influent sludge to minimize tube plugging and protect pumps, valves, and pipes from erosion.

Sludge flows both to and from the reactor through the heat exchanger inner tube. The exchanger is a concentric tube-in-tube design. A closed water circuit system with pressure tank and pump transfers heat from the reactor effluent to the incoming feed sludge. Water flows through the tube jacket, and the heat exchanger efficiency in this design is above 80%.

The reactor vessel is designed in accordance with the ASME unfired pressure vessel code requirements. Radiation or probe-type level sensors control the reactor sludge level. The process cools and liquifies gases released from the reactor to avoid odor formation. Also, the enclosed settling tank prevents odors from leaving the tank, and a small afterburner destroys odoriferous vapors.

The settling tank is a conventional, picket-type thickener where solids and liquids separate and readily settle to produce a sludge with 8 to 14% solids content. The thickened sludge from the thickener is pumped to the dewatering device.

Separate sludge mixing tanks should be provided when treated sludge is stored for extended periods because supersettling to a 40% solids content can occur, creating pumping problems. Vacuum filters, centrifuges, filter presses, and horizontal vacuum extractors have all been successfully applied to dewater and decanted sludge.

Dewatering Results

Three field-proven techniques for dewatering heat-treated sludge are vacuum filtration, centrifugation, and pressure filtration. Table 10.6.1 gives the process characteristics of each technique. This table shows vacuum filter performance data obtained from the 4500 gph Porteous installation in Colorado Springs, Colorado. The filter yield on heat-treated sludge averages 12 lb/ft^2/hr compared to the 3- to 5-lb/ft^2/hr rate achieved by chemically conditioned sludge. The moisture content of the heat-treated sludge filter influent is 50 to 60% compared to 80% for the chemical sludge, and the volume of filter cake also decreased by 50%.

The filter feed solids concentration from the thickener varies from 7 to 11%. Figure 10.6.2 shows the effects on cake yield. The filter rate also varies as a function of the ratio of primary sludge to secondary sludge. Filtration rates of 4 to 10 lb/ft^2/hr can be obtained on activated sludge and 12 to 26 lb/ft^2/hr on primary sludge. The filtration performance on mixed raw primary and activated sludge is proportional to the quantity, fiber content, and individual rate of each sludge.

Wastewater treatment facilities have achieved centrifugation results of 50 to 65% moisture in cake concentrations and an efficiency of 70 to 80% in solids capture. With polymer additions in test work, the Aire Plant in Geneva, Switzerland, reported a 95% capture of solids. Filter pressing is used extensively in England and continental Europe. The pressing time is directly related to the cooking temperature (see Figure 10.6.3).

Performance Factors

Variables that determine the dewatering properties of heat-treated sludge include process variables (temperature, pressure, steam flow, and detention time) and sludge variables (feed solids concentration, sludge type, volatiles-in-feed sludge, chemical content of sludge, and fiber content of sludge). Heat treatment improves the dewatering characteristics of all sludges, and environmental engineers should use laboratory studies to determine the optimal operating conditions.

The relationship of time, pressure, and temperature controls the degree of sludge solubilization and the BOD and COD of the effluent liquor and filtrate.

Liquor Treatment

The clarified effluent liquor and dewatered filtrate are returned to the treatment plant (see Figure 10.4.7) and produce a biological load on secondary oxidation treatment

TABLE 10.6.1 DEWATERING OF HEAT-TREATED SLUDGE

Equipment Used	*Cake Concentration Produced (% TS)*	*Efficiency of Solids Removal*° *(%)*	*Dewatering Capacity Increase Relative to Nonheat-treated Sludge (%)*
Vacuum filtration	35–45	95	200–300
Centrifuge	40–55	80–90	200–300
Pressure filter	50–65	100	200–300

Note: °Assuming that sludge does not include wash water solids.

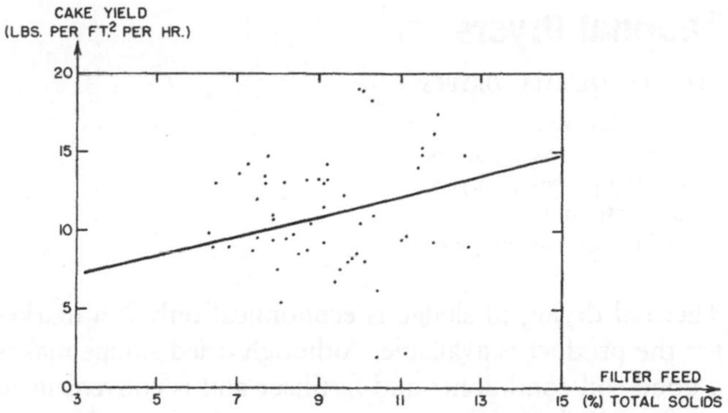

Fig. 10.6.2 Relationship between filter feed concentration and cake yield.

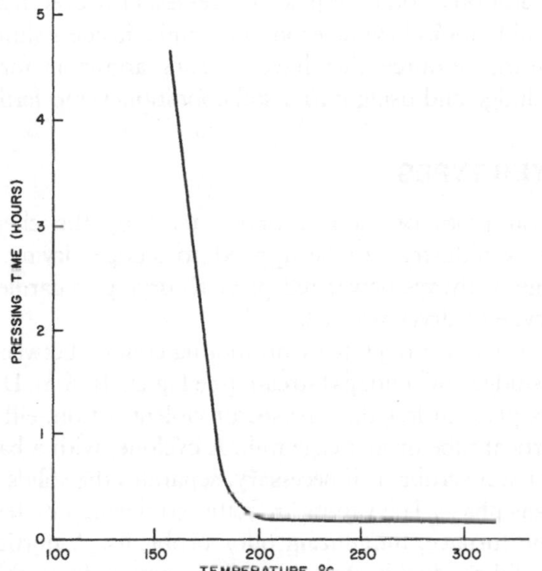

FIG. 10.6.3 Relationship of filter pressing time to cooking temperature.

processes. Table 10.6.2 shows typical liquor by-product compositions utilizing data of European plants. Figure 10.6.4 shows the BOD and COD values of the liquors produced at the first installation at Colorado Springs as a function of feed (to the grinder) solids concentration.

From both sources, the typical clarified effluent BOD concentration is approximately 4500 mg/l. These plants usually return this liquor to the sewage influent daily over several hours to minimize the effects of high BOD concentrations. They achieved BOD reductions of 78 to 97% in activated-sludge, pilot-plant tests.

The heat-treatment process accomplishes a COD reduction of 22%, which closely agrees with the volatile solids reduction (about 28%). The COD reduction is almost entirely due to the oxidation of the solids rather than the liquor.

The sewage plant's secondary treatment must handle the total increase in BOD load because most of the BOD load in the liquor is soluble and is not removed in the clarification step. Each pound of BOD recycled to the treatment plant produces roughly 0.5 to 0.6 lb of sludge. Plastic and rock trickling filters used in English plants as pretreatment units prior to the heat-treatment step reduce the BOD level by 52%. The pH of the clarified effluent liquor is about 0.5 pH units less than the feed pH. Wet-combustion systems generally reduce the feed sludge pH by 3 to 5 units.

Materials of Construction

All components of the heat-treatment system, except for the stainless steel balls in the valves and the reactor pressure control line, are constructed of carbon steel. Carbon steel materials have been used at the Halifax, England installation since 1935 without corrosion problems. Tube

TABLE 10.6.2 ANALYSES OF SOME LIQUOR BY-PRODUCTS FROM SLUDGE DEWATERING PROCESSES IN EUROPEAN PLANTS (RESULTS IN MG/L)

	Type of Liquor				
Constituent	*Heat-Treatment Liquor, Halifax (Secondary sludge)*	*Heat Treatment Liquor, Luton (All sludge)*	*Press Liquor, Bradford Sewage Department*	*Press Liquor, Halifax (Primary sludge)*	*Supernatant Liquor, Halifax Experimental Digestion Plant*
TS	10,820	7500	10,430	10,440	8610
Total ash	1460	1700	4240	6300	2750
Oxygen absorbed in 4 hr	2610	1500	698	288	1600
5-day BOD	4620	4500	4170	3780	1480
Organic nitrogen	1110	410	504	150	—
Ammoniacal nitrogen (NH_3 ion)	418	830	190	220	—
Albuminoid nitrogen (protein)	506	—	135	17.3	—

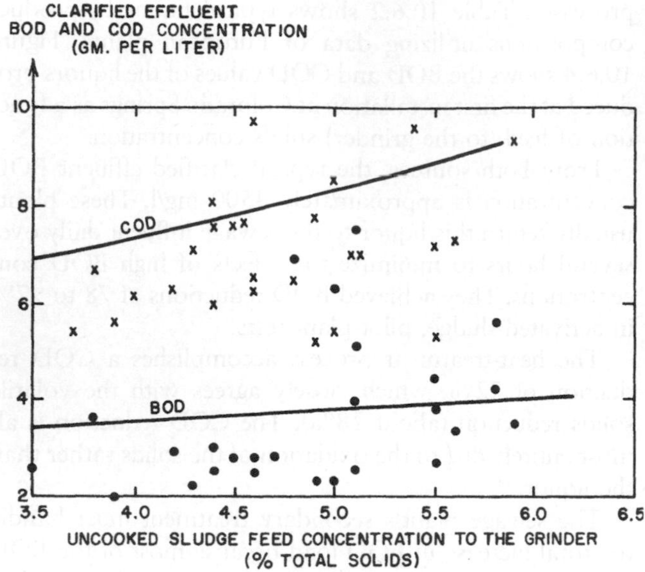

FIG. 10.6.4 Clarified effluent BOD and COD concentration as a function of solids content of heat treated sludge into the thickener.

turns and sludge piping are fabricated of heavy thickness due to the erosive forces of grit and sand in the feed sludge.

Wet-combustion systems are fabricated from stainless steel.

Table 10.6.3 shows operating data for the Porteous process.

As previously mentioned, this process is more economic than the chemical conditioning of sludges to improve their dewatering characteristics. Heat-treatment fuel costs are further reduced if the dewatered cake is incinerated and the waste heat is recovered. The final sludge cake is ultimately disposed as a low-grade soil conditioner, land fill material, or fuel. The high temperature used in the process sterilizes the sludge from pathogenic germs and weed seeds.

Thermal Dryers

TYPES OF THERMAL DRYERS

a) Flash dryers
b) Screw conveyor dryers
c) Multiple-hearth dryers
d) Rotary dryers
e) Atomized or spray dryers

Thermal drying of sludge is economical only if a market for the product is available. Although dried sludge makes a good soil conditioner and fertilizer and is convenient to use, especially for the home gardener, the cost of preparing and packaging it is seldom recouped from the profits. Accordingly, sludge drying is seldom used in the United States and does not represent an economical alternative to incineration or other disposal processes. However, if a municipality looks beyond economics and considers sludge as a natural resource, they have a strong argument for drying sludge and using it as a soil conditioner and fertilizer.

DRYER TYPES

Several types of thermal dryers used by the chemical process industry can be applied to sludge drying. The sludge is always dewatered prior to drying, regardless of the type of dryer selected.

Flash dryers operate by promoting contact between the wet sludge and a hot gas stream (see Figure 10.6.5). Drying takes place in less than 10 sec of violent action, either in a vertical tube or in a cage mill. A cyclone, with a bag filter or wet scrubber, if necessary, separates the solids from the gas phase. The vapors are returned through preheaters to the furnace, minimizing odor problems. A portion of the solid product is often returned to precondition the wet sludge.

The *screw conveyor dryer* uses a hollow shaft and blades through which hot gas or water is pumped (see

TABLE 10.6.3 PORTEOUS PROCESS DATA

Porteous System: gph	*1000*	*2000*	*3000*	*4500*	*6000*	*7500*	*9000*
Total dry solids feed, lb per day	10,000	20,000	30,000	45,000	60,000	75,000	90,000
Percent dry solids	5	5	5	5	5	5	5
Total wet sludge feed, gpd	24,000	48,000	72,000	108,000	144,000	180,000	216,000
Operation period, hr per week	168	168	168	168	168	168	168
Porteous system rate, gph	1000	2000	3000	4500	6000	7500	9000
System horsepower	12	20	35	48	64	70	80
Power demand, Kwh per day (hp)(0.60)(24)	173	280	505	692	922	1008	1152
Fuel required, million Btu per day (gal/day)(800 Btu)(10^{-6})	19.2	38.4	57.6	86.4	115.2	144.0	176.0
Boiler feed water, gpd (0.08 gal)(gpd)	1920	3840	5760	8640	11,520	14,400	17,600
Flushing water, gpd (3 hr/day)(gph)	3000	6000	9000	13,500	18,000	22,500	27,000

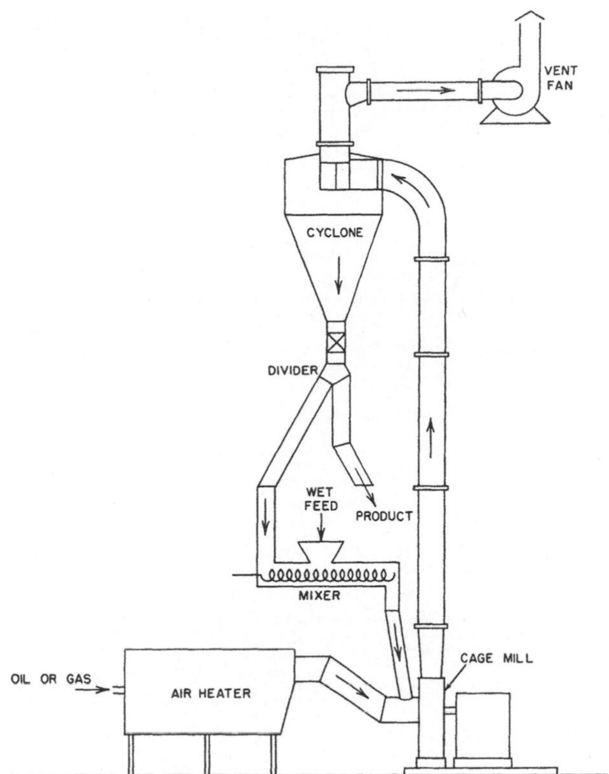

FIG. 10.6.6 Flash dryer.

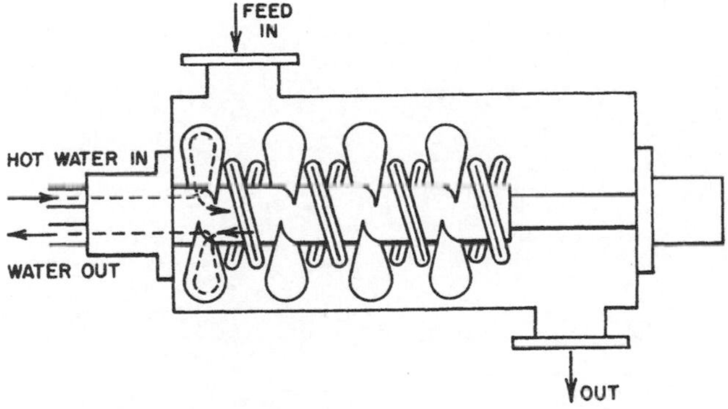

FIG. 10.6.6 Screw dryer.

Figure 10.6.6). The heat is transferred to the sludge as it is conveyed through the dryer.

Multiple-hearth dryers are converted multiple-hearth furnaces. The wet sludge can be mixed with dry product as it descends through the furnace. Fuel burners are located both on top and bottom.

Rotary dryers consist of a rotating cylinder through which the sludge moves (see Figure 10.6.7). Various types of blades or flights are installed in the dryers depending on the type of material being dried. Drying takes place by direct contact with heated air.

The chemical process industry has used *atomized or spray dryers* for many years (see Figure 10.6.8). For example, they make detergents by spraying the wet slurry

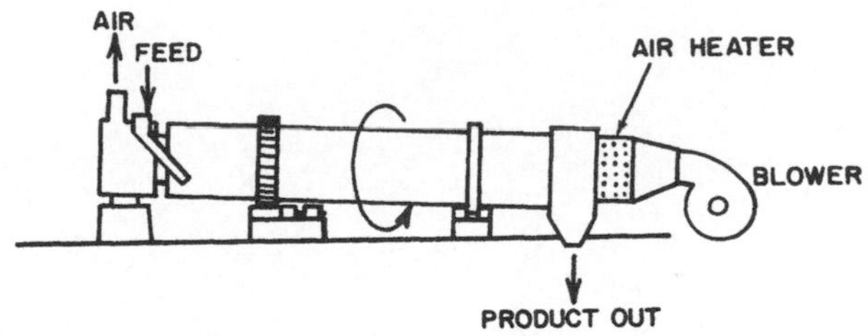

FIG. 10.6.7 Rotary dryer.

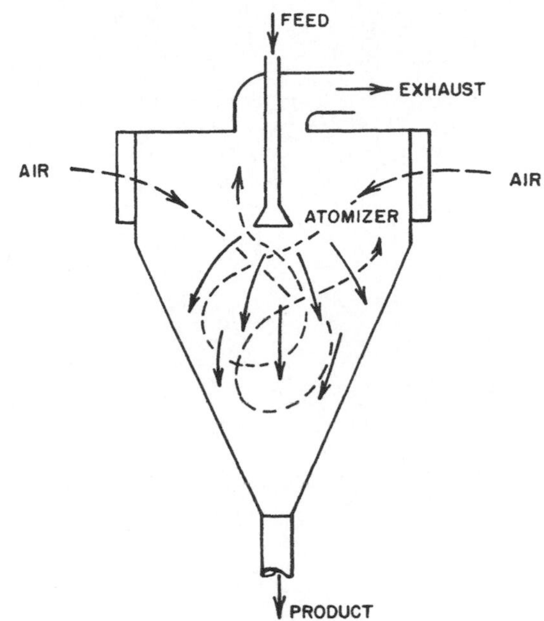

FIG. 10.6.8 Atomized or spray dryer.

into a heated tower. Spraying solids countercurrently into a downward draft of hot gas dries them although concurrent spray dryers are also used in the chemical industry.

OPERATION AND ECONOMICS

Experience shows that purchasers of dried sludge reduce their offers once a plant is in operation. Lack of competition for the product is usually the reason. Waste heat from refuse incineration can be used for drying sludge. Having an incinerator allows a plant to burn the dry sludge if no market exists for it. This burning eliminates the need for stockpiling.

Operational problems with heat drying (flash and rotary types used on sludge) include explosions due to grease accumulation, storage problems due to the absorption of moisture, and the fine-powder form of the product which is difficult to handle and apply. Pelletizing is a method of alleviating the last problem.

Air pollution, in the form of dust and odor, is another problem. Expensive control equipment is often required to meet local air quality standards.

11

Sludge Disposal

T.F. Brown | Wayne F. Echelberger, Jr. | Louis C. Gilde, Jr. | Béla G. Lipták | Joseph G. Rabosky | Frank P. Sebastian | John R. Snell | Michael S. Switzenbaum

11.1 SLUDGE INCINERATION

Multiple-Hearth Radiant Incineration

PER CAPITA DRY SLUDGE PRODUCTION

0.2 lb per day

SLUDGE FEED CONCENTRATION

Up to 75% moisture

CAPACITY IN UNITS OF WET FEEDRATE

13 in operating diameter (OD)–16 lb/hr; 30 in OD–300 lb/hr; 22 ft 3 in OD–30,000 lb/hr; 28 ft 3 in OD–over 50,000 lb/hr

INCINERATOR ASH PRODUCTION

10% of feed sludge volume

LOADING RATE

7 to 12 lb/ft^2/hr

EXHAUST GAS TEMPERATURE

700° to 800°F (without afterburner)

Since its initial use in 1934, the multiple-hearth incineration system has become the most widely used means of ultimate sewage sludge disposal, with more than 200 installations in the United States accounting for a capacity of over 50,000 wet tons of sludge per day. Multiple-hearth process development has kept pace with the increasing concern over air pollution, odor control, and resource reuse. New developments include the use of multiple-purpose furnaces for sludge incineration and chemical reclamation.

Following the various dewatering processes, waste solids still laden with water remain for disposal. More stringent laws and codes have reduced the choices for disposing such waste solids or sludge.

PROCESS DESCRIPTION

Multiple-hearth furnaces were originally developed in 1889 to roast pyrites for the manufacture of sulfuric acid. Modern multiple-hearth systems have some 120 proven uses, including:

- Burning (or drying) raw sludge, digested sludge, and sewage greases
- Recalcining lime sludge (the burning of $CaCO_3$ into CaO) and waste pond lime from sugar manufacturing
- Pyrolysis: the manufacture and regeneration of activated-granular carbon and the regeneration of diatomaceous earth; pyrolyzing fruit, nut, and lumber waste (peach pits, walnut shells, almond shells, sawdust, and bark) for charcoal briquettes
- Reclaiming oily chips from boring machines to metal briquettes and reclaiming cryolite in aluminum smelting operations
- Other roasting uses like mercury, molybdenum sulfide, carbon, magnesium oxide, uranium yellow cake, and nickel

The multiple-hearth furnace is a simple piece of equipment, consisting primarily of a steel shell lined with refractory on the inside. The refractory can be either castable or brick, depending on the size of the furnace.

The interior is divided by horizontal brick arches into separate compartments called hearths. Alternate hearths have holes at the periphery to allow the feed solids to drop onto the hearth below. The center shaft, driven by a variable-speed motor, rotates the rabble arms situated on each hearth (see Figure 11.1.1). The rabble teeth on these arms are at an angle so that the material is moved inward and then outward on alternate hearths. The shaft and rabble arms are cooled by air introduced at the bottom; this air is recycled as required by the thermal process.

The sludge is fed through the furnace roof by a screw feeder or a belt and flapgate. The rotating rabble arms and rabble teeth push the sludge across the hearth to drop holes where it falls to the next hearth and continues downward until the sterile phosphate-laden ash is discharged at the bottom.

The multiple-hearth system has the following three distinct operating zones:

1. The top hearths where the feed is partially dried
2. The incineration and deodorization zone where temperatures of 1400° to 1800°F (760° to 982°C) are maintained
3. The cooling zone where the hot ash gives up heat to the incoming combustion air.

The warmed air rises to the combustion zone in counterflow to the sludge flow, and the hot combustion gases

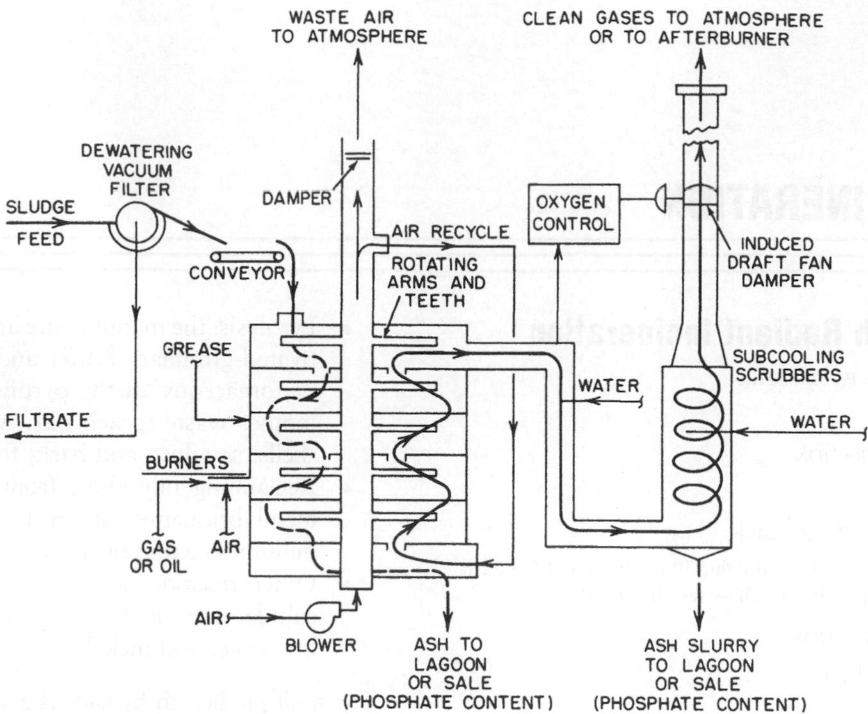

FIG. 11.1.1 Multiple-hearth sludge incineration.

sweep over the cold incoming sludge, evaporating the sludge moisture to about 48% At this moisture content, a phenomenon called *thermal jump* can occur in the combustion zone. This beneficial energy exchange allows the generation of odor-free exhaust gas at temperatures of 500° to 1100°F (260° to 593°C). Table 11.1.1 gives the typical temperature profile across the sludge furnace.

CONSTRUCTION

The steel shell of the incinerator is either of welded or bolted construction. Furnaces with diameters less than 10 ft are shop-welded and shipped in one piece. The larger sizes require field assembly and are usually bolted.

AUXILIARY EQUIPMENT

Belt conveyors, ribbon screws, and bulk-flow conveyors are commonly used. They discharge to a hopper with a counterweighted flapgate that intermittently drops the material into the furnace. Screw feeders and bulk-flow conveyors often discharge directly into the furnace.

The burners can handle all common liquid and gaseous fuels, including distillate and residual oils, natural gas, sewage digestion gas, propane, or combinations of these fuels. The design of burner tiles and boxes is important because slagging can occur within the tiles.

TABLE 11.1.1 SLUDGE FURNACE TEMPERATURE PROFILE

Hearth No.	*At Approximately Half Capacity (°F)*	*Nominal Design Capacity (°F)*
1	670	800
2	1380	1200
3	1560	1650
4	1450	1450
5	1200	1200
6	325	300

EQUIPMENT SELECTION AND PROCESS DESIGN

At one time, multiple-hearth furnaces were sized by a series of slide-rule and chart computations. Now a computer routine normally performs sizing. The size of the multiple-heart furnace is determined by the processing rate of the wet feed per square foot of furnace area, i.e., the loading rate equals 7 to 12 lb/ft^2/hr. Table 11.1.2 lists some standard sizes.

EXHAUST GASES–AIR POLLUTION

Exhaust gases at most installations pass from the incinerator furnaces through refractory-lined flues and enter three-stage, impingement-type scrubbers. Frequently, making the stack discharges essentially invisible is favorable. Accordingly, following the scrubber, some systems subcool the gases (below saturation temperature) to 110°F to condense the water vapor and reduce the acid formation in wet gases.

In advanced waste treatment systems that use lime, the use of CO_2 gases to recarbonate the effluent also reduces

TABLE 11.1.2 STANDARD MULTIPLE-HEARTH FURNACE SIZES

Unit Size	OD for Wall Thickness of: 6″	9″	13½″	Column Height	*Square Feet of Effective Hearth Area and Normal Shell Height* — Hearths 1	2	3	4	5	6	7	8	9	10	11	12
13″	° 18″ °@2½″			1′	1	2¼ 1′8″	2¼	3 3′4″								
30″	° 3′ °@3″	44″ °@7″		1½′	4	8 2′2″	12	16 3′10″	20	24 6′	28	32 7′8″	36	40 9′4″		
39″	4′3″	4′9″	5′6″	2½′	7	14 3′6″	19	28 6′	32	37 8′6″	42	48 11′0″	54	61 13′6″		
54″	5′6″	6′	6′9″	4′	15	31 4′2″	42	63 7′4″	74	85 10′6″	98	112 13′8″	126	140 16′10″		
5½′	6′6″	7′	7′9″	4′	24	47 4′8″	63	94 8′3″	110	125 11′10″	145	166 15′5″	187	208 19′0″		
7′	8′0″	8′6″	9′3″	5′	32	65 6′3″	96	130 10′10″	161	193 15′5″	225	256 20′0″	288	319 24′7″	351	383
8½′	9′6″	10′	10′9″	6½′	47	94 6′3″	138	188 10′8″	235	276 15′1″	323	364 19′5″	411	452 23′10″	510	560
12′	13′0″	13′6″	14′3″	6½′	97	195 6′9″	287	390 11′8″	487	575 16′7″	672	760 21′5″	857	944 26′4″	1041	1128 31′2″
14½′	15′6″	16′	16′9″	7′	143	286 8′0″	422	573 13′2″	716	845 18′7″	988	1117 24′1″	1260	1400 29′6″	1540	1675 35′0″
16½′	17′6″	18′	18′9″	7′	181	363 8′4″	534	727 14′3″	908	1068 20′2″	1249	1410 26′0″	1591	1752 31′11″	1933	2090 37′9″
18′	19′0″	19′6″	21′3″	8′	215	431 8′4″	634	863 14′4″	1078	1268 20′2″	1483	1660 26′1″	1875	2060 31′11″	2275	2464 37′10″
20′	21′0″	21′6″	23′3″	8′	269	538 9′6″	790	1077 16′1″	1346	1580 22′9″	1849	2084 29′6″	2350	2600 36′2″	2860	3120 42′11″
23½′	24′6″	25	25′9″	8′	382	764 11′4″	1145	1528 18′9″	1909	2292 26′2″	2674	3056 33′7″	3438	3818 41′	4200	4584 48′5″
26′	27′0″	27′6″	28′3″	8′	463	926 12′7″	1389	1852 20′9″	2315	2778 28′10″	3241	3704 36′11″	4167	4630 45′	5093	5556 53′1″

Note: °OD corresponding to noted wall thickness.

the amount of air pollution caused by the incinerator. In these systems, the higher degrees of wastewater treatment also reduce air pollution. Thus, cleaner water and air go with the lime PCT process and the hydrolysis adsorption process.

ODOR CONTROL

The operating and exhaust temperature of fluidized-bed incinerators is usually high enough to destroy odorous substances.

The thermal jump helps the sewage sludge to bypass the temperature zone where offensive odors are distilled and sometimes enables the multiple-hearth furnace to also effectively burn sewage solids without producing odors even though the gas outlet temperatures are as low as 500°F.

The malodorous volatile organic acids, butyric and caproic, exist in low concentrations in ordinary sludge; they exist as nonvolatile salts in raw sludge conditioned with lime and ferric chloride because of the high pH. Consequently, they are not released during ordinary drying operations but are carried into the burning zone where they are destroyed.

While the possibility of odor production always exists, wastewater treatment facilities can usually accomplish odor-free incineration. As a precautionary measure, they can add standby afterburners on the top hearth.

ASH RESIDUE

The final output of the multiple-hearth sludge incineration system is sterile, inert, and free of putrescible material and obnoxious odors. The ash volume amounts to approximately 10% of the furnace feed, based on a sludge cake containing 75% moisture with 70% of the sludge solids volatile. The ash is dry and contains less than 1% combustible matter, which is normally fixed carbon. This ash can be transferred hydraulically, mechanically, or pneumatically and can be used for landfill or roadfill. It is also being used experimentally to make bricks and concrete blocks.

NEW DEVELOPMENTS

For advanced waste treatment processes using lime, wastewater treatment facilities can enhance the economics of onsite lime recalcination by using the organic sludge as a partial fuel source for lime and recalcining in a single, multipurpose furnace.

CHARACTERISTICS OF MULTIPLE-HEARTH SYSTEMS

The following statements summarize the characteristics of multiple-hearth systems.

1. They generally handle sludges containing up to 75% moisture without auxiliary fuel.
2. They can incinerate or pyrolyze a variety of sludge materials, individually or in combination, including difficult-to-handle materials such as scum, grease, and ground refuse.
3. They can be used (or retrofitted) as a reclamation furnace for chemicals such as lime in combination with sludge or separately.
4. They can operate continuously or intermittently over a range of sludge feed capacities because the excess air for combustion and exit-gas temperatures can vary to suit conditions.
5. They have a long life and low maintenance costs over decades of operation.
6. They have a single feed point without a pressure feed requirement.

Reuse of Incinerator Ash

Various forms of slag tiles, bricks, and concrete blocks are made from ash residues. The city of Tokyo sells a large tonnage of its multiple-hearth furnace ash to C. Itoh Fertilizer Sales Company, Ltd.

The source of the ash is Tokyo's Odai treatment plant. This plant services a drainage area of 11,248 acres, with a planned treatment capacity of 111 mgd. The Odai plant extends over an area of 23.4 acres and uses a 100-tn/day, multiple-hearth furnace. An increasing portion of the city of Tokyo's total sewage sludge is processed by incineration.

The ash from the Odai plant is marketed under the trade name Vitalin (the Japanese word "lin" means phosphorus). A bag of Vitalin has the following percentage composition:

- Silica oxide 30.00
- Magnesium oxide 3.30
- Calcium oxide 30.00
- Phosphoric oxide 6.20
- Ferric oxide 18.20
- Potassium 1.00
- Nitrogen 0.20
- Manganese 0.06
- Copper 0.61
- Boron 200.00 ppm

Vitalin is sold under the special fertilizer category because material containing less than 12% phosphate cannot be classed as fertilizer. (Some states also require that the material have a nitrogen content of 6% or a total NPK range of 20 to 25%.)

In addition to Odai sewage sludge ash, the city of Nagoya has sold sludge ash from a multiple-hearth furnace under the name of Hormolin.

The phosphate in incinerator ash can be used in plant metabolism even though the P_2O_5 is insoluble above a pH of 3. In acidic soils, silicate and lime increase the pH of the soil. For such purposes, Japanese farmers use mixtures of organic SiO_2 and CaO. Thus, the components in the sludge ash are valuable to the soil for this purpose even if the phosphate content has limitations.

Table 11.1.3 contains an analysis of the ash from the South Lake Tahoe Water Reclamation plant. The Tahoe ash has more phosphate (7 to 10%) than the Japanese products. The lime used in the tertiary (phosphate removal) phase is removed with the sludge stream and is present in the ash in concentrations of 30 to 35%.

The city of Osaka has also used ash as a base material for roads around the Nakahama Sewage Treatment plant, but the ash is not used commercially in that manner.

With increased interest in resource recycling, the need for alternatives to ocean dumping of sludge, and a projected U.S. phosphate supply of only 80 years, the prospect of using the phosphate contained in sewage merits further investigation. Sludge or sludge ash containing appreciable quantities of metals such as zinc and chromium can damage crops (not grass or cereals) by heavy and repeated applications. However, the toxic effects are manifested only on acid soils, and sludge containing lime can probably offset some of the harmful effects.

The cumulative effect of boron (200 ppm in the Odai ash) requires further investigation, and monitoring and controlling the toxic material content in all sludge streams is advisable. Phosphate-rich sludge ash provides opportunities for recycling materials and recapturing value to defray the cost of advanced sewage treatment plant operations.

Fluidized-Bed Incineration

TYPES OF PROCESSES

a) Combustion
b) Pyrolysis

TABLE 11.1.3 TYPICAL ANALYSIS OF ASH FROM TERTIARY-QUALITY, ADVANCED WASTE TREATMENT SYSTEMS

	Percent of Total			
Content	*Sample 1 Lake Tahoe*	*Sample 2 Lake Tahoe*	*Minneapolis–St. Paul*	*Cleveland*
Silica (SiO_2)	23.85	23.72	24.87	28.85
Alumina (Al_2O_3)	16.34	22.10	13.48	10.20
Iron oxide (Fe_2O_3)	3.44	2.65	10.81	14.37
Magnesium oxide (MgO)	2.12	2.17	2.61	2.13
Total calcium oxide (CaO)	29.76	24.47	33.35	27.37
Available (free) calcium oxide (CaO)	1.16	1.37	1.06	0.29
Sodium (Na)	0.73	0.35	0.26	0.18
Potassium (K)	0.14	0.11	0.12	0.25
Boron (B)	0.02	0.02	0.006	0.01
Phosphorus pentoxide (P_2O_5)	6.87	15.35	9.88	9.22
Sulfate ion (SO_4)	2.79	2.84	2.71	5.04
Loss on ignition	2.59	2.24	1.62	1.94

FLUIDIZED-BED TEMPERATURE

1300° to 1500°F is normal; 2000°F is maximum.

TEMPERATURE VARIATIONS THROUGHOUT THE BED

5° to 10°F

HEAT BALANCE

Combustion releases; pyrolysis consumes heat energy.

MAXIMUM SLUDGE MOISTURE CONTENT WITHOUT NEED OF AUXILIARY FUEL

65% without air preheater, 73% with preheater

Although incineration using fluidized beds is relatively new, the fluidization of solid particle media has been used for some time in the chemical and petrochemical industries. The petroleum industry has used fluidized-bed reactors for catalytic cracking of complex hydrocarbons to produce simple, molecular hydrocarbon structures. If the fluidized medium consists of catalyst particles (a substance that enhances a chemical reaction without being consumed), chemical reactions are promoted within the medium.

Incineration in a fluidized bed consists of injecting substances, namely solid waste or liquid sludge and air, into a catalytic bed that causes a chemical reaction to occur. In this reaction, as in the cracking process, simpler products are formed. Frequently, the catalyst for this reaction is a bed of heated, fluidized sand.

PROCESS DESCRIPTION

The fluidized-bed incinerator is a vertical cylinder with an air distribution plate containing many small openings near the bottom (see Figure 11.1.2). This plate, which allows air to pass into the media, also supports the sand or other bed media. An external air source forces the air into the bottom of the cylinder. Once the air is distributed, the bed expands, i.e., it becomes fluidized (see Figure 11.1.3).

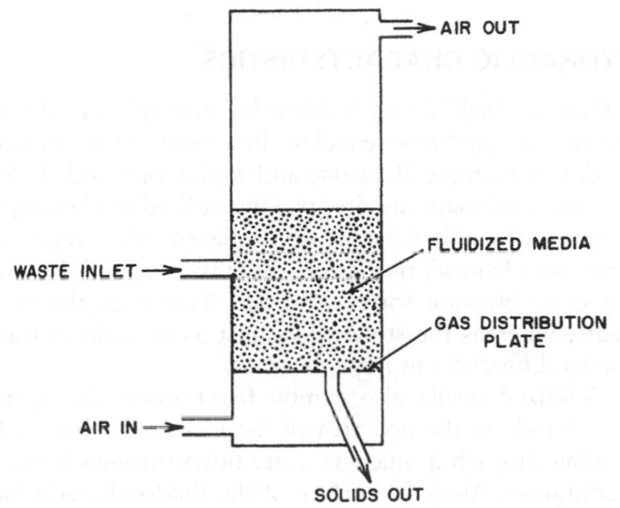

FIG. 11.1.2 Fluidized-bed incinerator.

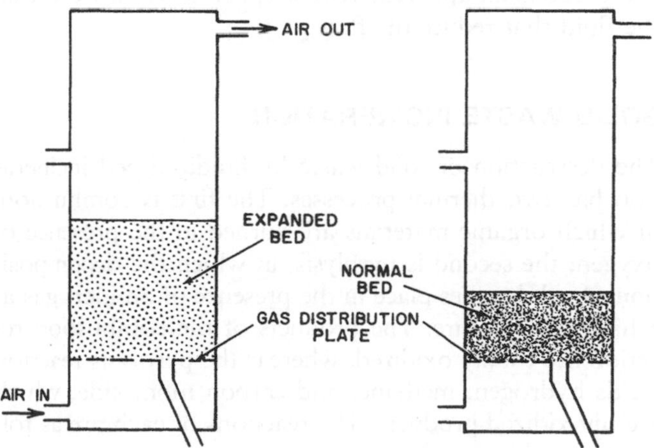

FIG. 11.1.3 Comparison of expanded and normal bed.

The volume of air that can pass through the bed is limited. If air is admitted at a low rate, it travels through the tortuous channels among the particles and escapes from the top of the bed without causing the individual particles of the bed to move, and the particles remain in their original packed configuration. This condition exists until the force exerted by the air overcomes the weight of the particles.

As the airflow increases, the bed expands. Initially, the expansion is such that the particles remain in contact with each other, however, an additional increase in airflow results in sufficient bed expansion so that the catalyst particles no longer remain in contact. When this expansion occurs, each particle is surrounded by air, and true fluidization begins. If the airflow or velocity continues to increase, the bed expands further. Additional velocity increases result in entrainment of the media particles in the discharging air stream until eventually all are carried out of the bed.

HYDRAULIC CHARACTERISTICS

A fluidized-bed follows Archimedes' principle, i.e., the bed exerts a buoyant force equal to the weight of the displaced fluid. For example, if a stone and a piece of wood of identical size and shape are dropped into a fluidized bed, equal volumes of catalyst media are displaced. Since these volumes are identical, they are equal in weight, and therefore the same buoyant force is exerted. However, the wood floats, whereas the stone sinks, just as they do in water due to differences in mass densities.

Fluidized media also exhibit free-flowing characteristics. A hole in the bed area of the incinerator has media flowing through it much as water flows through a hole in a container. Also, the surface of the fluidized media parallels that of the earth; if its container is tilted, the bed surface remains parallel with the ground. Lastly, the fluidized bed exhibits an apparent viscosity, i.e., forces arise within the fluid that reduce the flow.

SOLID WASTE INCINERATION

The destruction of solid waste by fluidized-bed incineration has two thermal processes. The first is combustion, in which organic materials are burned in the presence of oxygen; the second is pyrolysis, in which the decomposition of solids takes place in the presence of an inert gas at a high temperature. The products of the combustion reaction are totally oxidized, whereas the pyrolysis reaction yields hydrogen, methane, and carbon monoxide, which are unoxidized products. The reactions of each are as follows:

$$\text{Solid waste} + \text{Oxygen} \xrightarrow{\text{Combustion}} CO_2 + H_2O + \text{Ash} + \text{Heat} \quad \mathbf{11.1}(1)$$

$$\text{Solid waste} + \text{Inert gas} \xrightarrow{\text{Pyrolysis}} H_2 + CO + CO_2 + \underset{\text{(methane)}}{CH_4} + \text{Charcoal heat} \quad \mathbf{11.1}(2)$$

Since these reactions take place in fluidized beds at about 1400°F, the heat produced from the combustion reaction helps maintain the bed temperature, while pyrolysis requires heat.

The advantages of using fluidized beds for the destruction of solid waste materials include the following:

1. The heated particles store large quantities of readily available heat.
2. Particle movement throughout the bed prevents the formation of hot spots or temperature zones.
3. High heat-transfer rates result in rapid combustion.
4. Bed agitation prevents solids stratification.
5. Unfavorable gases and products can undergo total combustion, eliminating the need for expensive air pollution control equipment.
6. Few moving parts are located within the bed, reducing maintenance.
7. Temperature variations are minimal (less than 5° to 10°F) throughout the bed.

The disadvantages of using fluidized beds for solid waste incineration are as follows:

1. The maximum temperatures cannot exceed 2000°F when sand is used as a bed medium because sand softens at this temperature.
2. The power costs are high.
3. Equipment is necessary to recover fine solids because the catalyst media become entrained.
4. Auxiliary fuel is usually necessary because the composition and heating value of solid waste vary.

SLUDGE INCINERATION

Fluidized-bed incinerators can be used for the combustion of both organic and inorganic sludge. Sewage sludge with a high concentration of organics has been incinerated, and sludge with high inorganic content, e.g., salts of sodium and calcium, can be disposed in this manner. The pulp and paper industry destroys waste liquors using the fluidized-bed technique.

Two important factors that control sludge destruction in the bed are airflow and operating temperature. Gas passes upward through a bed of sand that contains solid sludge particles less than $\frac{1}{4}$ in in size. Vigorous mixing within the bed optimizes the reaction. Since air velocity is the source of bed agitation, treatment facilities can increase or decrease the reaction rates by controlling the air supply. Temperatures in these beds normally range from 1300° to 1500°F, and the amount of excess air required is 25%. These two factors insure complete combustion so that odorous materials are completely destroyed.

Previously dewatered sludge, which sometimes contains as much as 35% total solids, is normally introduced into the freeboard area of the furnace above the fluidized-bed material (see Figure 11.1.4). Hot gases from the bed evaporate the water in the sludge while the sludge solids enter the fluidized medium.

Once in the bed, the organic material oxidizes to carbon dioxide and water vapor, which exit from the reactor as exhaust gas. Inorganic material either deposits on the bed particles (which is called the agglomerative operation) or leaves the bed in the exhaust gas, which is the nonagglomerative system. Then, dust collection equipment removes the particles in the exhaust gas. For the agglomerative system, continuously or intermittently withdrawing excess bed material maintains a constant bed volume.

The advantages and disadvantages of liquid incineration using fluidized beds are similar to those listed for solid waste disposal. Waste incineration is a new application for fluidized beds. The total combustion of the waste does not produce obnoxious gaseous products. The fluidized-bed method is applicable for both solid and liquid waste disposal.

Wet Oxidation

ACCEPTABLE FEED COD RANGE

25 to 150 gm/l

MINIMUM WASTE HEATING VALUE

To keep the reactor thermally self-sustaining requires 1500 Btu/gal, which corresponds to about 7% solids or 80 gm/l COD. With lower heating values, external heat is needed.

COD REDUCTION

Increases with temperature, reaction time, and feed concentration; ranges from 5 to 80%

VSS REDUCTION

Ranges from 30 to 98%

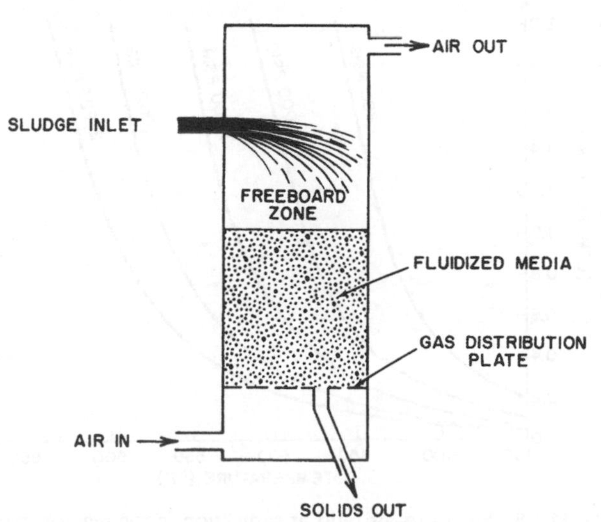

FIG. 11.1.4 Fluidized-bed sludge incinerator.

MATERIALS OF CONSTRUCTION

Stainless steel

OPERATING TEMPERATURE

Cannot exceed 705°F; usually between 300° and 600°F

OPERATING PRESSURE

Usually between 300 and 2000 psig

The patented Zimmermann process involves flameless or wet combustion in aqueous solutions or dispersions. In aqueous dispersions, this process can oxidize a range of organic and industrial wastes to carbon dioxide and water by adding air or oxygen. The wet-air oxidation process (this term preferred by the patentees) does not require dewatering prior to combustion and oxidizes the combustible matter in the liquid phase by applying heat and pressure.

The advantages of this process are that it creates no air pollution and generates sterile, easily filtered, and biodegradable end products. The inherent cleanliness of the fully enclosed system and the potential to generate or recover steam, power, and chemicals are other advantages.

The disadvantages include the need to use stainless steel construction materials and the complex equipment that requires high capital investment and well-trained operators.

PROCESS DESCRIPTION

This process mixes the waste liquor with air and preheats it by steam during startup and by the reactor effluent during operation to about 300° to 400°F (see Figure 11.1.5). At this reactor inlet temperature, oxidation starts with the heat release which further increases the temperature as the liquor–air mixture moves through the reactor. The higher the operating temperature, the greater the COD reduction for the same contact time period (see Figure 11.1.6). The operating temperature cannot exceed the critical temperature of water (705°F) because the continuous presence of a liquid water phase is essential.

The operating temperature in the reactor is the temperature of the saturated steam at the partial pressure of

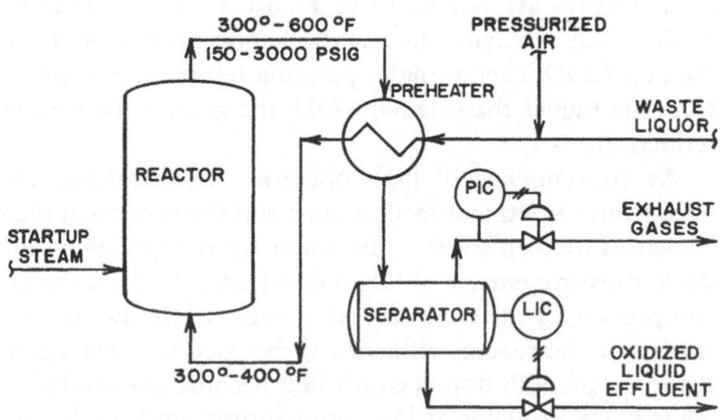

FIG. 11.1.5 Wet oxidation process.

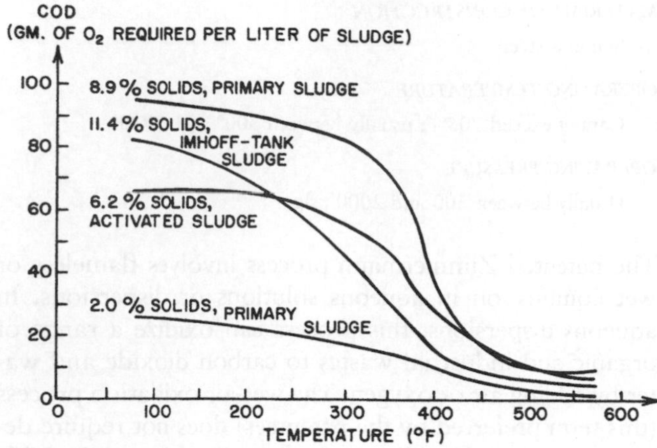

FIG. 11.1.6 COD reduction from sludge exposed to excess air for 1 hr at various temperatures.

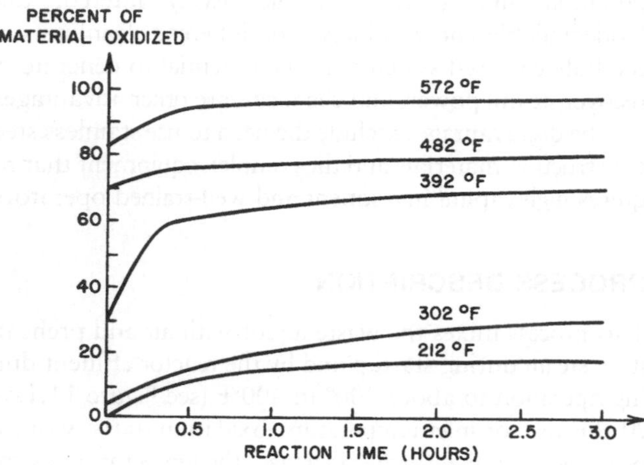

FIG. 11.1.7 High COD reduction and low reaction time with high operating temperatures.

steam in the air–steam mixture. The air–steam mixture is a gas with an increasing CO_2 and a decreasing O_2 content as the material moves through the reactor.

The wastewater treatment facility selects the operating temperature according to the required COD reduction and reaction time. When the reaction time must be short and the reduction in COD must be substantial, high operating temperatures are required (see Figure 11.1.7). The COD of the effluent leaving the reactor is independent of the incoming COD level at high operating temperatures; therefore, the higher the original COD, the greater the percent reduction.

As consequence of high operating temperatures, the wastewater treatment facility must run the process at high pressures to keep some of the water from vaporizing. The static pressure energy of the exhaust gas can drive the air compressor or generate electric power, while the thermal energy of the reactor effluent can be used for steam generation. Air pollution is controlled because the oxidation takes place in water at low temperatures and no fly ash, dust, sulfur dioxide, or nitrogen oxide is formed.

LIMIT ON AIR USAGE

Figure 11.1.8 shows the weight ratio of steam to air at saturation in the reactor vapor space for various operating pressures and temperatures. These curves are based on the following equation:

$$S/A = (53.3\ T)/144\ (Pt\text{-}Ps)Vs \qquad 11.1(3)$$

where:

S = steam (lbm)
A = air (lbm)
T = temperature (°R = °F + 460)
Pt = total pressure (psia)
P_s = saturated steam pressure (psia)
V_s = specific volume of saturated steam (ft^3/lbm)

The curves in Figure 11.1.8 show the maximum amount of air that can be added per gallon of waste liquor without elimination of the liquid phase in the reactor. The steam-to-air ratio is 2 lbm per lbm at 2000 psig and 595°F, 1500 psig and 553°F, or 1000 psig and 510°F. At this ratio, if each gallon of waste liquor contains 8 lb of water, the addition of 4 lb air per gallon of waste vaporizes all water. Therefore, the wastewater treatment facility must select a lower ratio, such as 3.5 lbm of air per gallon of waste, to maintain some water in the liquid phase.

SEWAGE SLUDGE APPLICATIONS

The wet combustion process has been used in sewage sludge treatment since the early 1960s, and in this period, both continuous and batch plants were installed for either high-pressure or low-pressure operation. Table 11.1.4 describes the capabilities of this process based on data provided by the patentee. The tabulation assumes that the waste has 3.5% solids and a COD of 43 gm O_2/l before

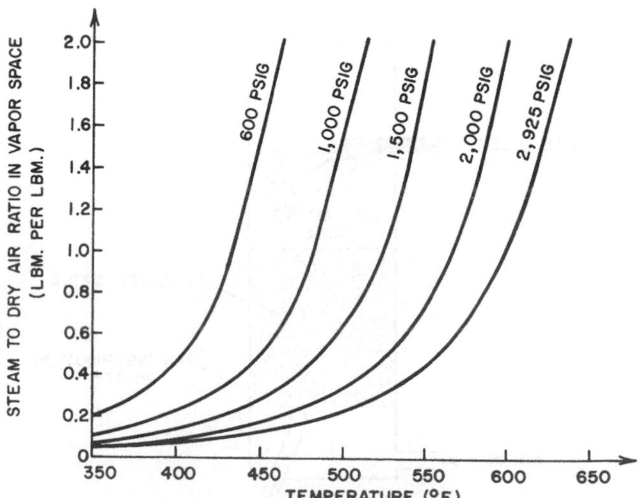

FIG. 11.1.8 Steam-to-air ratio at saturation in the reactor vapor space for various operating temperatures and pressures.

TABLE 11.1.4 SEWAGE SLUDGE CHARACTERISTICS FOLLOWING DIVERSE LEVELS OF WET-AIR OXIDATION

	Before Oxidation	*After Low Oxidation*			*After Intermediate Oxidation*			*After High Oxidation*		
	Raw Sludge	*Oxidized Slurry*	*Filtrate*	*Filter Cake*	*Oxidized Slurry*	*Filtrate*	*Filter Cake*	*Oxidized Slurry*	*Filtrate*	*Filter Cake*
% COD Reduction	0	5–15	—	—	40–50	—	—	70–80	—	—
% Insoluble Volatile Solids Reduction	0	30–50	—	—	70–80	—	—	92–98	—	—
Filtration Resistance ($sec^2/gm \times 10^7$)	3500	6	—	—	6	—	—	10	—	—
Volume, l	1.00	°1.05	1.00	0.05	1.00	0.97	0.03	1.00	0.97	0.03
TS, gm/l	35.1	29.8	8.2	21.6	21.6	7.1	14.5	16.0	5.5	10.5
Volatile Solids, gm/l	23.4	18.1	5.2	12.9	9.9	4.1	5.8	4.3	3.1	1.2
Ash, gm/l	11.7	11.7	3.0	8.7	11.7	3.0	8.7	11.7	2.4	9.3
pH	6.0	5.2	—	—	4.9	—	—	6.2	—	—
Moisture in Filter Cake, %	—	—	—	64	—	—	58	—	—	50
Drained Cake Dry Weight (lb/ft^3) Packed	—	—	—	20	—	—	35	—	—	55
Settled Volume after 4 Hr (%)	100	37	—	—	15	—	—	7.5	—	—
Phosphorus as P°°, gm/l	0.61	0.61	0.18	0.43	0.61	0.06	0.55	0.61	0.02	0.59
Total Nitrogen, gm/l	1.49	1.49	1.15	0.34	1.49	1.42	0.07	1.49	1.44	0.05
Ammonia Nitrogen, gm/l	0.57	0.75	0.72	0.03	1.00	0.98	0.02	1.20	1.19	0.01
Total Sulfur, gm/l	0.30	0.30	0.20	0.10	0.30	0.25	0.05	0.30	0.28	0.02
SO_4 as S, gm/l	—	0.17	0.14	0.03	0.30	0.25	0.05	0.30	0.28	0.02

Source: Zimpro Division of the Sterling Drug Incorporated of New York.
Notes: °Increased volume due to steam injection.
°°× 3.065 = PO_4 content.

oxidation. The table lists three levels of oxidation corresponding to ranges of 5 to 15%, 40 to 50% and 70 to 80% reduction in COD.

The wet oxidation process can accommodate sludge concentrations between COD values of 25 and 150. It reduces the filtration resistance and improves the sludge draining characteristics so that it becomes compatible with vacuum filtration. Sulfur is oxidized to sulfate with no sulfur dioxide leaving with the gas exhaust. The process removes most phosphorus with the filter cake and can accomplish complete precipitation without lime addition. The filtrate liquid contains short-chain, water-soluble organic compounds that are biodegradable. Nitrogen is not oxidized to nitrite or nitrate, and no nitrogen dioxide (NO_2) leaves with the exhaust gas because the organic nitrogen is degraded to ammonia.

OTHER APPLICATIONS

Powdered, activated-carbon regeneration is practical using the Zimmermann process. The spent carbon slurry at 6 to 8% solids enters the wet oxidation process. The reaction temperature allows the adsorbed organic compounds to be oxidized without destruction of the activated carbon. The application of powdered carbon in wastewater treatment can benefit from the economic carbon regeneration.

Wet oxidation can render plastics, detergents, insecticides, and other nonbiodegradable materials compatible with conventional sewage treatment processes. When the waste contains both paper and plastic material, these need not be separated because both are decomposed at the same reaction temperature.

Wastes that are deficient in nitrogen require the addition of this element for satisfactory biological treatment. Wet oxidation of the sludge can reduce the cost of adding nitrogen because this process returns nitrogen to the biological treatment step as ammonia. Other potential applications of the Zimmermann process include the treatment of tannery, glue factory, plating, sulfide, phenol, paper, cyanide, textile mill, brewery, or photochemical waste and the recovery of chrome, magnesium, titanium, and silver.

Flash Drying or Incineration

FUEL REQUIREMENTS FOR SLUDGE DRYING

6500 Btus/lb sludge with 18% solids

The flash drying or incineration of sewage sludge was a new process in the mid 1930s. It allowed the plant operator to either incinerate the sludge or flash dry it using conventional fuels, such as coal, fuel oil, or natural gas. With this flexibility, the operator produced only the amount of dried sludge that the market could bear and incinerated the remainder.

The odor and dust problems with this process resulted in less use, and in some cases, plant expansions use alternate processes, such as the multihearth incinerator or the fluidized-bed incinerator. Both these methods, however, lack the flexibility of drying or incinerating or both.

The flash-drying system is less costly if no dried finished sludge is produced or the pollution controls are minimized. With new improvements and scrubbers and an increased demand for dried sludge, this process should have an increased demand.

PROCESS DESCRIPTION

The flash-drying method has become common for drying or incinerating sewage solids because of its low capital costs and flexibility. The pretreatment includes sludge thickening in some conventional manner such as by vacuum filters. Then, this method mixes this dewatered sludge with previously dried sludge to reduce its moisture content and its effective particle size. The preconditioned mixture is fed into the drying system where it moves at a velocity of several thousand ft/min in a stream of gas having a temperature of 1000° to 1200°F. The sludge passes through this high-temperature, turbulent zone in a few seconds during which time its moisture content reduces to 10% or less. Next, a cyclone separator separates the hot gases from the fine, fluffy, heat-dried sludge.

If incineration is used, the system introduces the dried sludge produced in the flash dryer into the furnace through special sludge burners and burns it at about 1400∞F. The heat from this burning process is recycled into the drying operation.

If the sludge is conserved and sold or sold in bulk or in bags, this method must burn significant quantities of auxiliary fuel in the furnace either separately or with a small quantity of dried sewage sludge. Environmental engineers estimate that this method required 0.4 lb coal plus 0.94 ft^3 of natural gas (6500 Btu) to produce a dry pound of sewage sludge when starting with a thick, liquid sludge of about 18% solids.

FLASH DRYING VERSUS OTHER PROCESSES

The multiple hearth process cannot dry sludge without incinerating it. This process is the main competitor to the flash-dryer system and dominates it in use. The fluidized-bed unit is essentially a high-temperature furnace with hot sand fluidized by air jets. Sludge enters the top of the chamber and falls into the hot bed and combusts during the violent mixing of sludge and hot sand. The use of this process

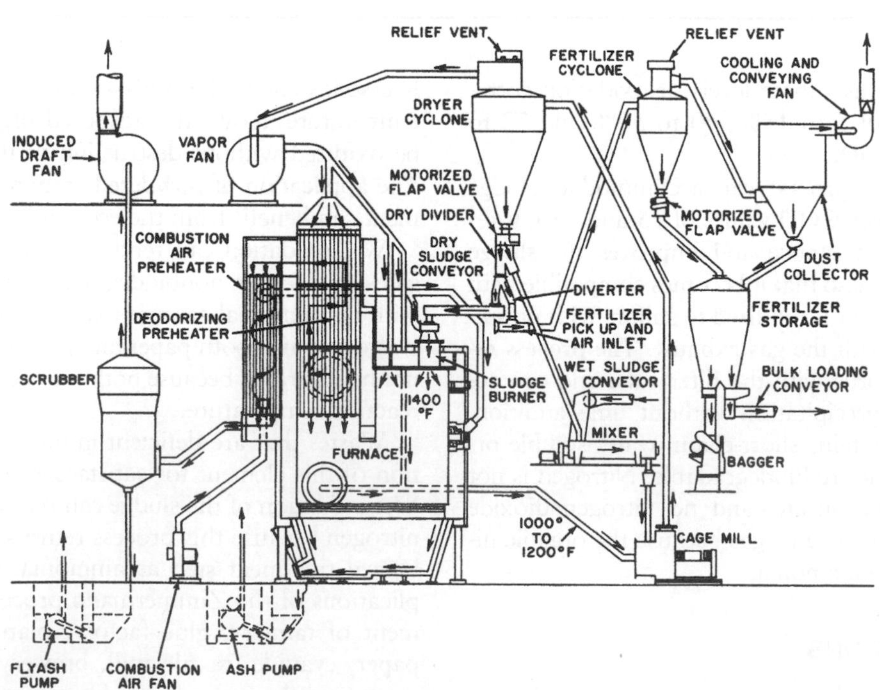

FIG. 11.1.9 Sludge drying and incineration using a deodorized flash-drying process.

is increasing over the flash-dried system although it does not have the flexibility of either drying or incinerating the sludge.

Composting sewage sludge is an innovation because no commercial installations exist in which sewage solids are composted alone. The process has potential both ecologically and economically but requires using shredded municipal refuse. When sludge is composted alone, without shredded refuse, successful treatment requires recycling the drier, already composted, sewage solids with the raw, wetter, dewatered solids, such as occurs in the second step of the flash-drying method. This recycling produces a drier, more porous sludge, which is necessary for good composting.

Although flash drying of sewage sludge is not common due to its previous odor and air pollution problems (which have since been corrected), it has the advantage of complete flexibility in incinerating or drying the sludge at a reasonable price. Thus, a sewage plant using this process can be flexible with one piece of equipment and dry only the amount of sludge that the market demands and incinerate the rest.

Figure 11.1.9 shows the operation of a flash-dryer incinerator system. Flash dryers are also used in the paper industry and to dry sewage sludge.

VALUE OF HEAT-DRIED SLUDGE

Heat-dried sludge compared to sludge dried on sand beds is free of pathogens and weed seeds and is therefore safer to use. However, under normal conditions, heat-dried sludge is more powdery and more difficult to spread and mix with soil than conventional sludge. It is initially repellent to water although once it becomes partially moist, it readily absorbs more water. Therefore, heat-dried sludge must be further treated before being sold as a fertilizer.

It sells in 50- to 65-lb sacks. Although Milorganite, due to its guaranteed 6% nitrogen content, (coming from Milwaukee's beer waste) is still in demand, a similar product made in Chicago (only $3\frac{1}{2}$ to 4% nitrogen) is not as popular, and only a small percentage of their current supply is being sold.

The destruction of pathogens during heat drying is such that only 2 coliform bacteria/gr remain in over 100 samples. Bacterial, parasitic, and viral enteric pathogens commonly found in sewage have the same order of heat sensitivity as coliforms.

—*F.P. Sebastian*
J.G. Rabosky
Béla G. Lipták

11.2 LAGOONS AND LANDFILLS

EFFECT OF WET SLUDGE BEING APPLIED TO LAND

Nitrogen Loss to Leachate

60 mg/l for 1-in layer per year, 1070 mg/l for 12-in layer per year

Other Pollutants

Bacteria can travel up to 100 ft through granular soil; heavy metals migrate only after the soil is saturated.

SLUDGE LAGOON LOADING RATES

400 to 1000 tn dry solids/acre/yr

The disposal of waste sludge through lagoons and landfills is an economical means of ultimate sludge disposal. The lagoons can receive undigested primary sludge, excess activated sludge, or digested sludge as either an interim process in the total sludge handling scheme or as a method of ultimate sludge disposal. Normally, landfills are the ultimate disposal locations for dried (dewatered) sludge, and this disposal method can be economical depending on the haul distance from the wastewater treatment plant to the landfill.

In considering the location, design, operation, and maintenance of sludge disposal lagoons and landfills, environmental engineers must consider the sludge loading criteria, possible health effects through groundwater pollution, the potential for heavy metal accumulation in the soil and groundwater, the possibility of fertilizer nutrients like nitrogen and phosphorus reaching the surface water, and general nuisance developments.

The land availability and climate are important considerations when lagooning is considered as a dewatering technique. Large land areas are generally required. Poor dewatering occurs in cold and rainy climates. Operation and maintenance costs are associated mainly with the removal of dried sludge. Lagoons should not be used as a final treatment for coagulant sludge if the ultimate disposal requires solids concentrations greater than 9%.

Design and Operation

Lagoons are typically earthen constructions equipped with an inlet control device and overflow structures (Montgomery 1985). Building lagoons involves enclosing a land surface with dikes or excavation (Masschelein 1992). Impermeable liners placed in the bottoms of lagoons minimize drainage (Borchardt et al. 1981). Drying occurs by removal of the supernatant and evaporation. Wastewater treatment facilities often use lagoons as a storage and sludge-thickening step prior to further dewatering and ultimate disposal (Westerhoff et al. 1978).

The volume requirements in lagoon design are a function of both the volume of water being treated as well as the degree of dewatering achieved within the lagoon. Figure 11.2.1 shows information on the volume requirements on the basis of these two parameters. For softening sludges, practical experience in several midwestern cities indicates that 0.45 to 0.65 acre–ft of lagoon volume are required per 1 mgd of water treated per 100 mg hardness (as $CaCO_3$) removed (Faber et al. 1969). This design estimate assumes an average sludge concentration within the lagoon of 50% dry solids.

Lagoons can operate either as continuous fill (permanent lagoons) or fill and dry (dewatering lagoons). However, operating lagoons as a thickening process as opposed to an ultimate disposal method is best (Montgomery 1985).

Continuous-Fill Lagoons

For a continuous-fill lagoon, a side water depth of 8 to 13 ft with a 3- to 5-yr capacity is recommended (Montgomery 1985; Faber et al. 1969). Multiple cells equipped with decanting devices are also preferred (Faber et al. 1969). In these lagoons, the sludge is applied in layers, and the supernatant is removed periodically for air drying. This process is repeated until the lagoon is filled with solids. The lagoon is then covered, and the land is reclaimed. The type of disposed sludge dictates any future use of the land.

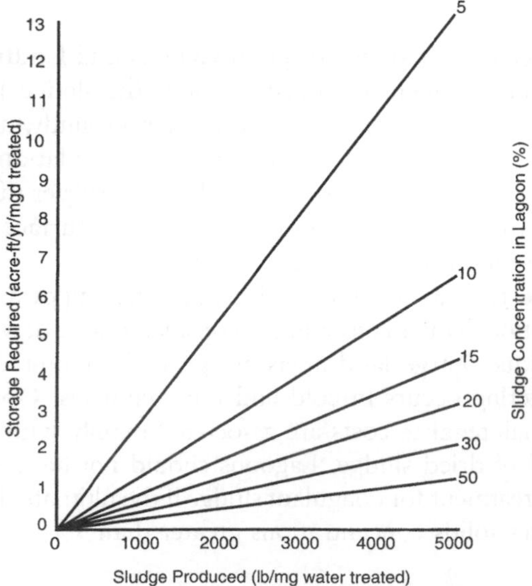

FIG. 11.2.1 Relationship between water volume treated, solids concentration, and required lagoon storage volume. (Reprinted, with permission, from L. Streicher et al., 1972, Disposal of water treatment plant wastes, *Jour. AWWA* [December].)

Lime sludge can reach 40% solids when settled through ponded water. Solids concentrations as low as 20 to 30% are also reported. If the supernatant is allowed to flow off the site or is removed with decanting equipment, up to 50% solids can be achieved (Faber et al. 1969).

Alum and iron coagulant sludge can reach 10 to 15% solids, with iron sludge dewatering faster than alum sludge. Neither is suitable for landfill at this dry-solids content. After 2 to 3 years, 30 to 40% solids can be reached (Masschelein 1992). The percent solids values are overall averages. The percent dry solids concentration varies with depth, being greater near the bottom of the lagoon.

Fill and Dry Lagoons

When wastewater treatment facilities use fill and dry lagoons, they require multiple lagoons to alternate filling, draining and drying the supernatant, and removing the dried sludge. The lagoons are sized based on variables such as sludge production and characteristics and average air temperature. Side water depths of approximately 3 to 6 ft contain sludge discharges from 1 to 3 ft.

The sludge is then allowed to settle. The wastewater treatment facility decants off the supernatant periodically to increase the sludge-to-air contact until the sludge is sufficiently dry. They repeat this process until the lagoon is filled with approximately 4 to 12% solids for coagulant sludge and 40 to 50% for lime-softening sludge.

Several months to more than a year can be required to achieve these solids concentrations (Montgomery 1985). The dried sludge is then removed with a dragline, clamshell, or front-end loader, and the lagoon is used again. Because of the low solids concentrations achieved with coagulant sludge, lagooning typically requires further dewatering for landfill disposal.

Freeze and Thaw

In climates with long periods of temperatures below freezing, freezing and thawing offers a method of further concentrating coagulant sludge (Krasauskas 1969). Freezing releases the hydration water from the aluminum hydroxide complex producing a volume as small as $\frac{1}{6}$ the original volume. Thawing produces small granular particles like coffee grounds that dry to a brown powder which is easily dewatered and disposed (Krasauskas and Streicher 1969).

At least two lagoons are necessary. One lagoon is left to dewater prior to winter and then allowed to freeze. Another lagoon receives the sludge. Wastewater treatment facilities can achieve a final solids concentration of 17.5% for alum sludge with the freeze–thaw method.

Health Considerations

Concerning the possible health effects associated with these methods of sludge disposal, the information to date indicates that the potential hazard of disease transmission by pathogenic organisms originating from sludge disposal on land is not significant. Bacteria normally do not travel distances greater than 100 ft through granular soils, and viruses do not pass through 2 ft of clean sand at moderate liquid application rates during 7 months.

The fate of heavy metal pollutants (such as iron, cobalt, nickel, copper, and zinc) in soils is not known, but their chemistry indicates that they generally form insoluble precipitates. The major mechanism for the retention of heavy metals in soil may be sorption on hydrous oxides of iron and manganese, thus significantly retarding the migration of these pollutants in the soil. As the capacity of soil to retain heavy metal elements is exceeded, a breakthrough to the groundwater occurs, and environmental engineers must consider this possibility when planning a permanent lagoon or landfill disposal facility.

The possibility of fertilizer nutrient buildup in groundwater and surface water from sludge lagoon and landfill leachates does exist. Therefore, environmental engineers should be concerned with this possibility because high concentrations of nitrates in drinking water can have toxic effects on humans and nitrogen and phosphorus contribute to eutrophication in surface water.

The nitrogen loss to leachate can be as high as 60 mg/l as N for 1 in of wet sludge applied to land per year and can increase to 620 and 1070 mg/l for 6- and 12-in application rates, respectively. Because sludge lagoons are ordinarily flooded, biological denitrification processes may have a minimizing effect on the nitrogen loss to lagoon leachate.

The potential phosphorus pollution of ground and surface water by leachate is not too serious since soils normally have a high capacity to retain phosphates. Only when the phosphate retention capacity of the soil is exceeded can this type of pollution create a problem; this capacity can eventually be exceeded in conjunction with permanent sludge lagooning operations. Soil erosion in lagoon and landfill areas can also be a source of nutrient pollution by supplying soil-adsorbed phosphorus to surface water.

Odors and troublesome insects are common complaints of the general public concerning the location and operation of sludge lagoons and landfills. Proper operation and maintenance of the disposal sites through restricted access, weed control, and effective landfill covering minimizing these adverse concerns.

Engineering Considerations

Sludge lagoons have been classified as 1) thickening, storage, and digesting lagoons; 2) drying lagoons; and 3) permanent lagoons. Class 1 is specified when the capacity of conventional sludge handling facilities is exceeded, but digestion in a lagoon is a long process and can become a nuisance. With this type of facility, the sludge eventually has to be removed and properly dried.

Ordinarily, drying lagoons are a substitute for sludge drying beds, and the wastewater treatment facility must remove the dried sludge prior to refilling the lagoon. This treatment can require multiple-lagoon units.

The permanent lagoon, from which the sludge is not removed, is the most inexpensive method of sludge disposal provided that adequate land is available close to the waste treatment plant. Facilities for the removal of the supernatant liquid are suggested for this type of lagoon operation.

The engineering layout and design of sludge lagoons should include provisions for uniform distribution of the applied sludge and a convenient method for removing the dry sludge, if necessary. A discharge system that restricts sludge travel to 200 ft has been suggested along with embankments at an exterior slope of 1:2, an interior slope of 1:3, and a top width adequate for maintenance vehicle travel. For the different sludge lagoon operations, the solids loading rates vary from 400 to 1000 tn of dry solids per acre per year.

The higher loading rate is for dewatering lagoons. Sludge lagoons are commonly located on the wastewater treatment plant grounds and can also be constructed in permeable soils when ground and surface water pollution by lagoon leachate is not a potential difficulty. If adequate land is not available on the treatment site, pumping the sludge to locations within 5 or 10 mi of the plant can be economic. Meteorological parameters such as temperature, precipitation, and evaporation influence the operation of sludge lagoons and should be considered in the location and design of these facilities.

Operation

The operation of sludge lagoons and landfills can be based on a 3-yr cycle. In such an operation, the lagoon is filled for a year and then allowed to dry for 18 months followed by cleaning. The supporting soil lies fallow for 6 months before the lagoon returns to operation.

For a dewatering lagoon, filling to a depth of 2.5 to 4 ft is suggested. Wastewater treatment facilities can do this filling by first adding a layer of 1 ft of sludge, then switching temporarily to a second lagoon to allow drying, and then adding the remaining layers to the first lagoon. With one wet year, a 4-ft depth can provide 2 to 3 years of capacity with this procedure.

Sludge lagooning for land reclamation does not require removal of the sludge. Landfilling with dewatered wastewater sludge mixed with municipal solid waste can improve the operation of sanitary landfills by accelerating degradation in the landfill and thereby shortening the time until the landfill can be used.

The principal advantages of lagooning and landfilling for ultimate waste sludge disposal are the low operating and maintenance costs. Among the disadvantages, large areas are required, nuisance difficulties may exist, and ground and surface water can be polluted by leachate. Landfill operations also require an adequate earth supply for covering the fill area.

—*W.F. Echelberger*

11.3 SPRAY IRRIGATION

TYPES OF SPRAY IRRIGATION SYSTEMS: a) Infiltration type; b) Overland type

NITROGEN AND PHOSPHORUS REDUCTION: Up to 90%

BOD REDUCTION: About 99%

MAXIMUM ACCEPTABLE SODIUM CONTENT IN WASTEWATER: 1000 ppm

REQUIRED WASTEWATER PUMPING PRESSURE: 60 to 100 psig

PRETREATMENT OF WASTEWATER: Screening and grease removal only

HYDRAULIC LOADING BY SPRAY IRRIGATION: 1 in/day (a), 3 in/week (b)

ORGANIC LOADING IN FOOD INDUSTRY: 100 to 250 lb BOD/acre/day is normal; 500 to 1000 is maximum.

TYPICAL INFLUENT AND EFFLUENT CONCENTRATIONS (MG/L): Wastewater BOD: 600 to 700; Underdrain effluent BOD: less than 10

Spray irrigation is a modification of the system used in agriculture for irrigating crops. However, the objective is the disposal of liquid waste rather than providing moisture and nutrients to harvestable crops.

The first operative spray irrigation system in the United States was located in Pennsylvania in 1947. Since 1947, spray irrigation systems have been used for the disposal of waste from paper mills, kraft and neutral sulfite semichemical pulp mills, vegetable and fruit canneries, strawboard mills, dairies, fine chemical fermentations, and milk bottling plants. The acceptance of this method of waste disposal is verified by its use in Indiana, Wisconsin, Michigan, Ohio, Illinois, Oregon, New Jersey, Texas, Ontario, Kentucky, Tennessee, Pennsylvania, Minnesota, and Iowa. The system is attractive because of its flexibility and total treatment of applied fluids.

Flexibility in expanding or contracting the capacity of the treatment facilities is especially beneficial with fermentation waste because of the changing quantity and quality of this waste. Total treatment is a solid asset when highly concentrated waste, such as fermentation residue, is handled. Typical fermentation waste with 65,000 mg/l BOD after 98% treatment in a conventional, complete-treatment, biological facility still contains 1300 mg/l BOD, which is usually unacceptable for discharge.

After pretreatment and grease removal, this system sprays wastewater through a sprinkler system onto land that is planted with special grasses. The wastewater infiltrates the ground where soil microorganisms convert the organic waste into inorganic nutrients. The purified water is collected by an underground perforated pipe and sent to a final polishing pond.

The removal of soluble organic wastes by spray irrigation is a highly efficient process for treating industrial waste. Until recently, most of these systems required good soil infiltration characteristics. The system required a site where large volumes of water—as much as 1 in per day—could be applied and where the hydrological characteristics of the soil permitted water transfer underground and laterally out of the area of application. In many cases where the infiltration was adequate, the lack of sufficient lateral movement either limited the rate of application or created flooding. Usually, overcoming these problems involved installing artificial drainage similar to farm tile drainage except for extra precautions in the spacing and the means of avoiding siltation of the collection system because of the high application rates.

Techniques were also developed that use impervious soils for purification by overland flow. These systems are used with impervious clay-type soils in which significant infiltration is not possible.

Physical and Biological Nature

Spray irrigation of wastewater should not be confused with farm irrigation. Water can be purified either as it flows overland or as it percolates through the soil. Purification, in either case, occurs biologically and depends on the biota and organic litter on and in the soil. Pure sand, without organic debris, provides only mechanical filtration without reducing the soluble organic matter.

Most spray irrigation systems rely on water percolating through the soil and flowing away from the irrigation area by an underground route. Ordinarily, this underground flow is natural drainage, but it can also be enhanced by artificial drainage. When hydrological characteristics limit the lateral movement of underground water, wastewater treatment facilities can substitute an overland flow technique.

The grasses planted on the treatment field are multifunctional; they protect the soil surface from erosion and compaction and retard the flow of water across the slope in overland flow systems. They also provide a protective habitat for microorganisms and a vast surface area for adsorption, mass biological activity, and treatment of the impurities in the water. When the grass is cut for hay, it is a valuable crop that can effectively reclaim the plant nutrients released to the soil during decomposition of the organic waste material. Almost any species of grass is satisfactory provided that it produces abundantly, is water tolerant, and forms a turf.

Of all the interacting phenomena in the natural filtration system, microbiological activities are the most important and are carried on by all molds, fungi, bacteria, earthworms, snails, and insects that feed directly or indirectly on the organic waste (see Figure 11.3.1). The microbial populations in the disposal field, although specific for the plant effluent, comprise a highly complex community.

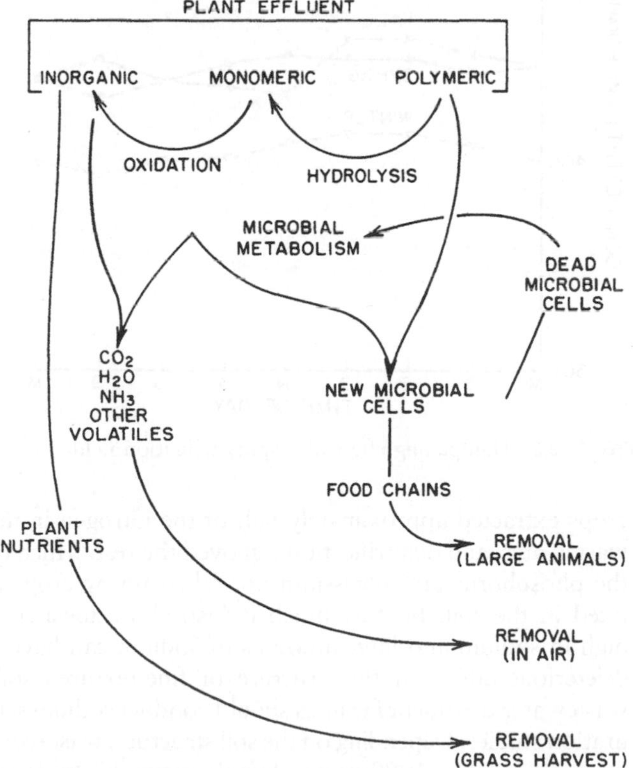

FIG. 11.3.1 Population succession cycle.

Organisms use the organic waste products from both carbohydrates and proteinaceous matter as nutrients. The carbon dioxide and water released by the degradation of carbohydrates escape into the air. The ammonia released upon decomposition of protein can be 1) released to the atmosphere, 2) used directly by microorganisms, and 3) converted into nitrite and nitrate. Microbial use of the organic effluent constituents converts a portion of this material into new forms of organic material that, if not removed, are used by different microbial populations. The process is repetitive, and a portion of the organic matter is converted into carbon dioxide, water, and ammonia at each cycle.

Tests show an evolutionary or seeding process whereby microorganisms specific for an effluent develop on the disposal site. The time required for the evolutionary process may be one reason for the greater capacity of older disposal systems. Also, the maturing of the system may be hastened when it is seeded with specific organisms.

Temperature and Shock Load Effects

A spray irrigation system continues to purify water when temperatures are near freezing. Since the respiration of microorganisms slows down as temperature decreases, researchers believed that the impurities were being adsorbed on the surface of the vegetation and held there until the weather grew warm again. However, biological studies show that as the temperature decreases, the number of organisms increases, thus maintaining a constant level of mass activity. Figure 11.3.2 shows this phenomenon.

Spray irrigation systems have the outstanding capability of handling shock loads as well as periods of long shutdowns and immediate startups, producing excellent results in either case. In addition, variations in effluent composition, such as the results of night cleanup, produce no adverse effects. The effluent pH of a spray irrigation system stays between 6.8 and 7.0 although the waste applied at night reaches a pH of 12 for approximately 1 hour and sometimes for as long as 3 hours.

Figure 11.3.3 provides evidence of this dampening effect, showing the diurnal variations of electrical conductivity of both the wastewater and the field effluent for each season. The higher conductivity in runoff during the summer months is due to the increase in evapotranspiration and the decrease in the runoff volume.

For an overland flow system, if a single terraced slope is accidentally overloaded due to mechanical failure, the effluent treatment continues in the other terraces and waterways before the effluent reaches the receiving stream. This capability makes the spray irrigation method a safe technique.

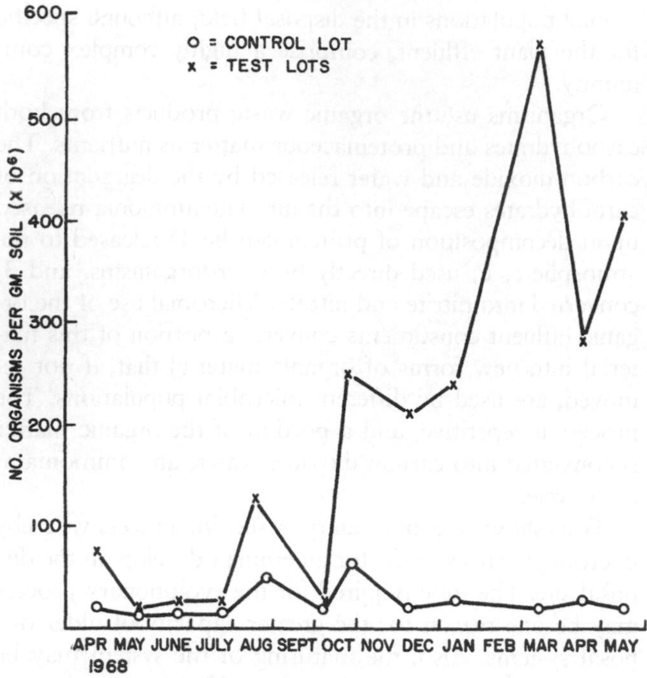

FIG. 11.3.2 Total microbial population on control and test lots.

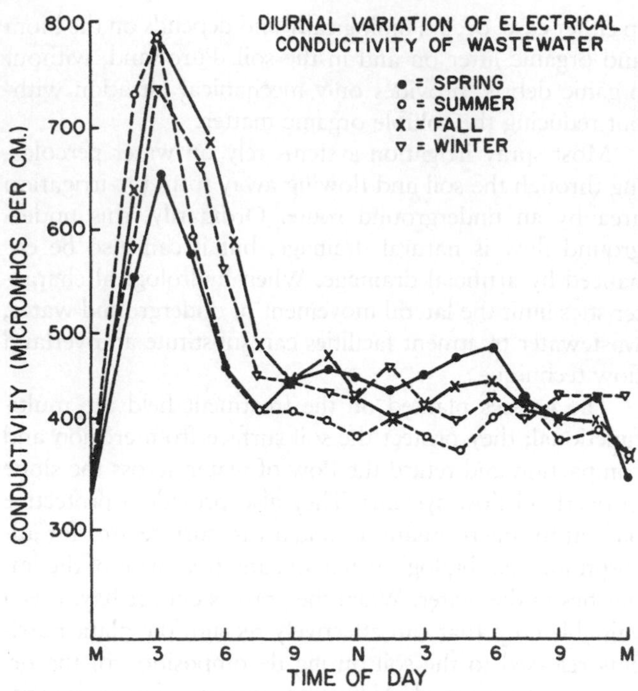

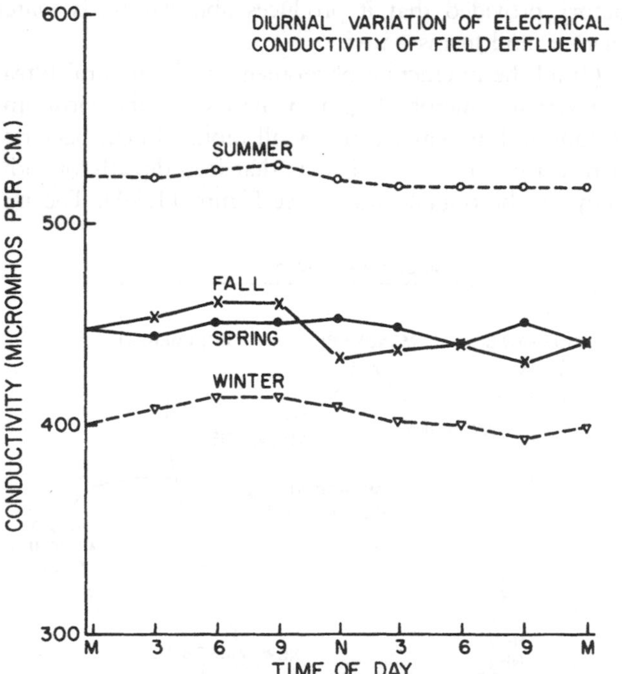

FIG. 11.3.3 Dampening effect of a spray irrigation field.

Wastewater Pretreatment

An advantage of spray irrigation systems is the small degree of pretreatment needed. For most industrial waste, this procedure consists of coarse screening and possible grease recovery. Screening can be achieved on screens with openings of 1 to 2 mm. Fine-mesh screening is not normally warranted since the sprinklers usually have openings of $\frac{3}{8}$-in or larger diameter. Grease separation by gravity is normally adequate for removing most floatable solids and reducing the problems caused by floating material in the pump reservoir. When the wastewater contains sand or fine grit, the use of a detritus separator is also advisable.

For most soluble organic waste, rapid handling and minimum detention time avoid the anaerobic decomposition of the organic matter that creates odors. Pumping is accomplished with high-head pumps, with most designs operating in the range of 60 to 100 psig. The minimum pressure at the sprinklers is 35 psig for efficient spray distribution and breakup.

Sprinkler systems are the most convenient method of uniformly applying waste to the ground without damaging the vegetative cover or soil structure. The organic loading varies depending on the type of wastewater applied. Most food processing waste applications operate at 100 to 250 lb BOD/acre/day although several systems operate in the 500- to 1000-lb/acre/day loading range.

Environmental engineers have conducted studies to determine the fate of nitrates, phosphates, potassium, and sodium added through effluent irrigation. In several cases, crops extracted approximately half of the nitrogen in the wastewater, and denitrification removed the rest. Much of the phosphorus and potassium not taken up by crops is fixed in the soil. Because much industrial wastewater is high in sodium and large amounts of sodium can have a deleterious effect on the structure of fine-textured soil, wastewater treatment facilities should conduct sodium saturation studies. Depending on the soil structure, waste containing more than 1000 ppm sodium is unsuitable for long-term irrigation of fine-textured soils.

System Description

In most cases, waste is pumped through lateral piping and sprayed through sprinklers or spray nozzles located at intervals (see Figures 11.3.4 and 11.3.5). The waste percolates through the soil, and the organic compounds undergo biological degradation. The liquid is either stored in the soil layer or discharged to the groundwater. Maintaining a cover crop, such as grass, vines, trees, or other vegetation, maintains porosity in the upper soil layers. As much as 10% of the waste flow evaporates or is absorbed by the roots and leaves of plants. Trees develop a high-porosity soil cover and yield high transpiration rates. A small elm tree can take up as much as 3000 gpd under arid conditions.

Normally, the waste is first pumped to a holding reservoir, which can be merely an earthen pit but serves to equalize fluctuations in daily operations. The waste is then pumped from the reservoir to the spray nozzles by a header and lateral piping system. The piping system can be permanent, using valves to direct the waste to various locations within the spray field (see Figure 11.3.6), or the piping can be temporary and can be physically moved for spraying different areas. The permanent piping method requires a greater initial investment; the temporary system entails a greater operating cost.

A combination of the two systems is a sound compromise. A permanent headerline from the waste reservoirs across the spray field combined with lightweight, movable laterals with quick-disconnect connections (see Figure 11.3.7) works well if the spray field is passable. A tough and durable steel pipe can be used for the headerline. Lightweight pipe, such as aluminum or plastic, is required for the laterals since they are usually moved daily or weekly.

If the pipe is laid above ground, it must be arched occasionally. Because spraying is done only during daylight, the header and laterals must be drained at night in subfreezing weather. Quick-disconnect connections make system draining easier. The spray nozzles should be durable

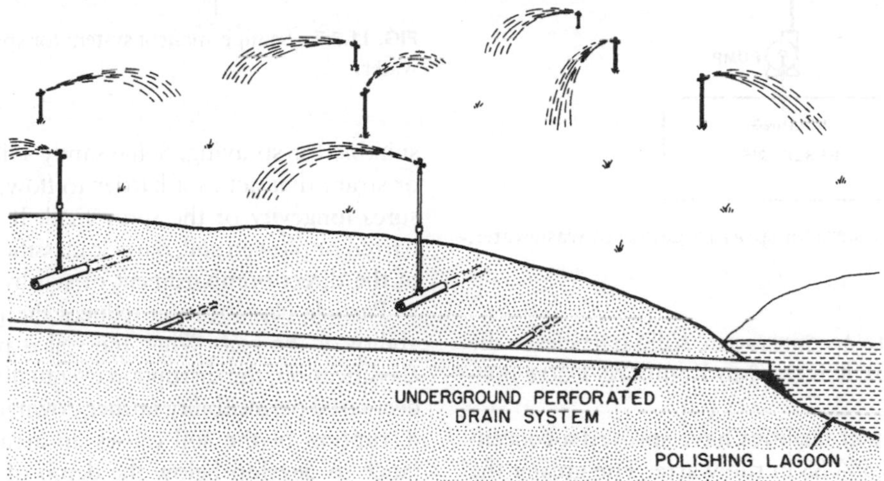

FIG. 11.3.4 Artificial underground drainage system design.

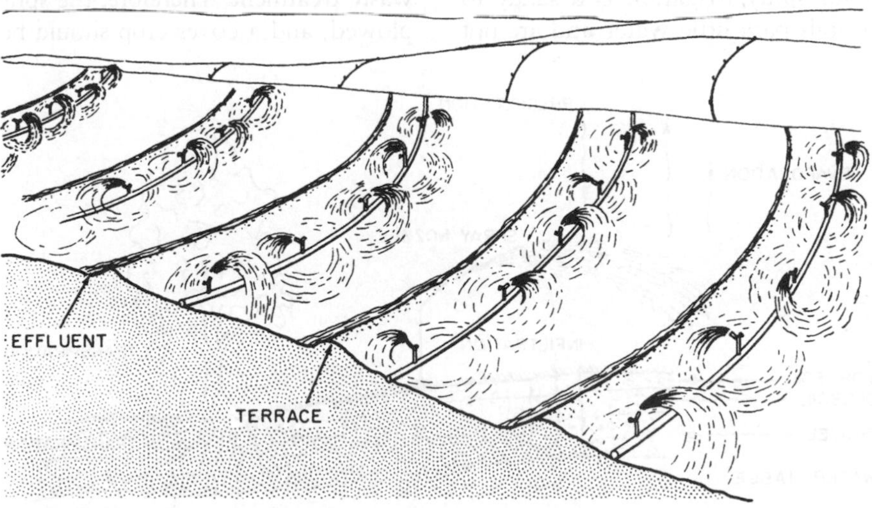

FIG. 11.3.5 Overland technique of spray irrigation.

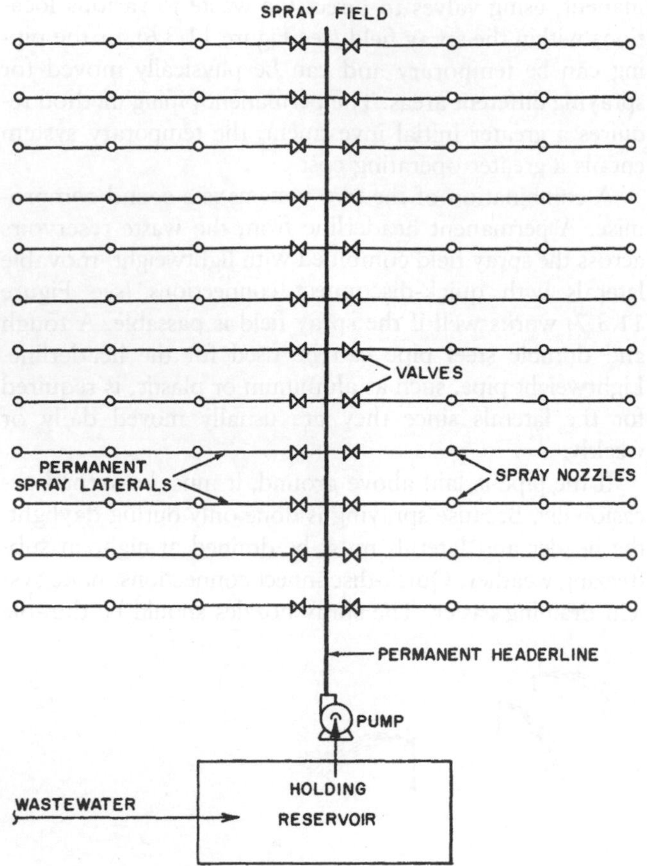

FIG. 11.3.6 Permanent system for spray irrigation of wastewater.

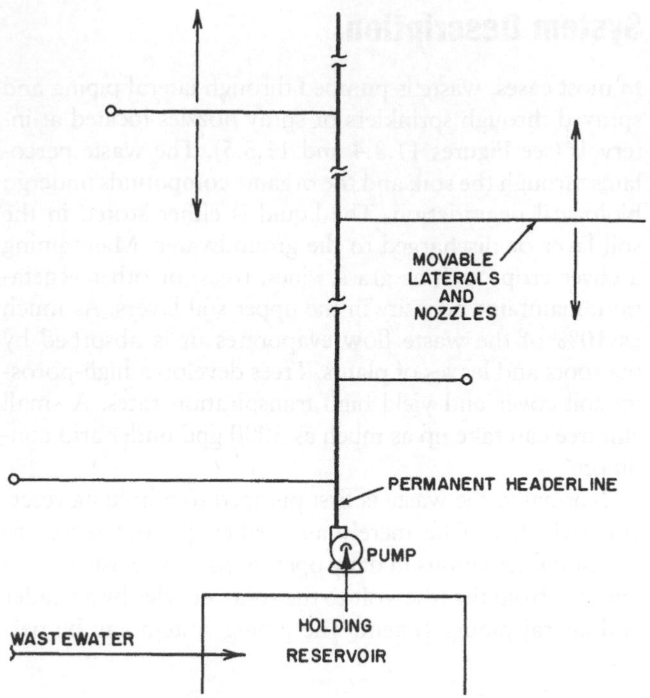

FIG. 11.3.7 Semipermanent system for spray irrigation of wastewater.

and sufficiently large to accommodate solid particles in the waste. An efficient nozzle is the $\frac{5}{8}$-in Rainburg 1-acre nozzle.

Wastewater treatment facilities have sprayed a concentration of as much as 5% SS in the waste. The greater the volume a nozzle can spray, the less chance the nozzle will freeze up.

The ideal soil type for spray irrigation is a sandy to sandy-loam type. Clay soils pass little water and are not suitable for spraying. Some sandy soils exhibit clay lenses or strata that act as a barrier to flow. A ground cover assures longevity of the spray field. If wastes is applied to barren or plowed soil, particle classification occurs because of the direct bombardment of soil by the liquid.

Particle classification reduces the porosity and permeability of the soil. When the soil is plowed after particle classification, the disruption eventually extends the area of decreased permeability farther down in the soil. Continued disruption of the soil causes an area of fluid-flow resistance to develop below the depth of plowing or subsoiling.

Once this condition has occurred, the area is useless for waste treatment. Therefore, the spray field should not be plowed, and a cover crop should be maintained. Experi-

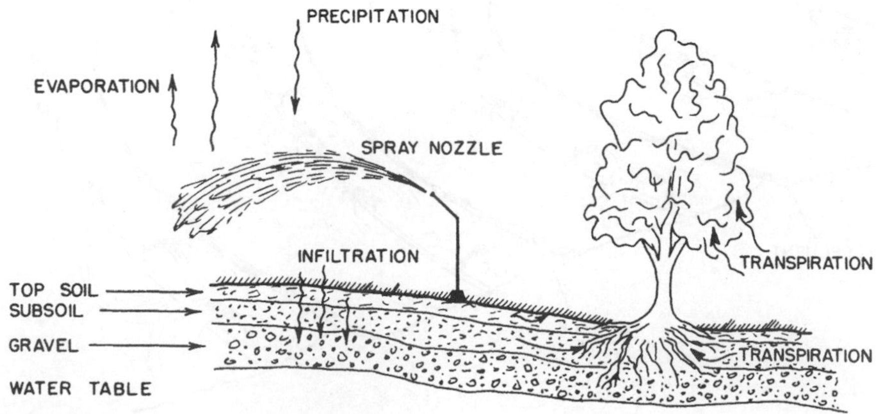

FIG. 11.3.8 Spray irrigation of wastewater.

ments using waste woodchips and bark from a papermill operation as a cover for the spray field are proving successful. The bark chips prevent direct bombardment of the soil and thus particle classification. Figure 11.3.8 shows the phenomenon used in a spray irrigation system.

Wastewater treatment facilities should also install a monitoring system to safeguard the groundwater in the area. Test wells located at various depths and locations can measure the increase and the spread of contamination. Usually, measuring one or two of the major contaminants in the waste (sulfate or nitrate) suffices for groundwater analysis. If those parameters do not increase, the wastewater treatment facility can assume that the groundwater is not being contaminated.

An excessively alkaline or acid waste is harmful to the cover crop and hampers operation. High salinity impairs the growth of a cover crop and causes sodium to replace calcium and magnesium by ion exchange in clay soils. This alteration causes soil dispersion and results in poor drainage and aeration in the soil. A maximum salinity of 0.15% can eliminate these problems.

Design Criteria and Parameters

The capacity of the soil to absorb liquid is proportional to the overall coefficient of permeability for the soil between the ground surface and the groundwater table as follows:

$$Q \ 5 \ 0.328 \ KS \qquad 11.3(1)$$

where:

Q = gallons per minute (gpm) per acre
K = overall coefficient of permeability, feet per minute (fpm)
S = degree of saturation of soil (near 1.0 for a steady-rate application)

The coefficient of permeability K depends on the soil characteristics (see Table 11.3.1).

The application rate depends on the soil structure, the land contour, the waste characteristics, the local evaporation–precipitation rate, and the supervision afforded the spray irrigation system. An efficient initial and design guide is the following schedule for the use of 1-acre spray plots:

1. Spray for 10 hr at 90 gpm; rest the plot for 2 weeks.
2. Spray for a second 10 hr; rest the plot for 2 weeks.
3. Spray for a third 10 hr, and discontinue use of plot after 30 hr dosage (162,000 gal) for approximately 90 days before restarting schedule.

The schedule is a guide; actual practice dictates which spray plots can receive more or less. Applying waste at too great a rate causes ponding and surface runoff. Ponding can result in anaerobic decomposition and thus cause odor problems. In addition, it renders the spray fields impassable for pipe moving; therefore, ponding should be avoided.

TABLE 11.3.1 COEFFICIENTS OF SOIL PERMEABILITY

Soil Type	*Permeability Coefficient K (fpm)*
Trace fine sand	1.0–0.2
Trace silt	0.8–0.04
Coarse and fine silt	0.012–0.002
Fissured clay soils and organic soils	0.0008–0.0004
Dominating clay soils	<0.0002

Environmental engineers should size the holding reservoir to contain at least 2 days of waste production. They should design pumping, piping, and nozzles in both number and size to apply at least one day's waste production during daylight.

Infiltration Techniques

To determine the soluble organic removal capability of a soil, environmental engineers must study the infiltration capacity and permeability of the soil. This capability is a function of the soil texture and depends on the nature of the vegetation, moisture content of the soil, and temperature. Infiltration rates vary from $\frac{1}{10}$ in/hr for low-organic-content clay soils to $\frac{1}{2}$ in/hr for sandy silt loam to more than 1 in/hr for deep sand.

The movement of treated water in soil is a function of the soil pore size, root structure, and evapotranspiration by plants. When the limitations of infiltration or lateral moisture movement are exceeded, ponding and (frequently) failure of the irrigation system occur. Therefore, environmental engineers must tailor the application rates and the hours during which the waste is applied daily to the soil drainage capacity at the site.

When the lateral transmissibility in the soil is not high enough to provide rapid drainage from the application area, an artificial underdrainage system can be installed. Figure 11.3.4 shows such a system. It consists of a bitumastic, impregnated-paper-fiber, underdrain pipe with $\frac{3}{8}$-in perforations at 9-in intervals. The pipe is 5 to 7 ft below the ground surface and is wrapped in a $\frac{1}{2}$-in fiberglass mat used as a filter guard to prevent siltation of the drain pipe by soil particles.

The field in Figure 11.3.4 has a high infiltration rate for the first 5 to 7 ft but is then underlain by dense substrata. Without an underdrainage system, the field can become a morass or marsh.

The spray irrigation system automatically applies approximately a 1-in layer of wastewater to the field per day. The sprinklers are spaced for complete overlap. The underground perforated pipe collects the purified water and

TABLE 11.3.2 TREATMENT EFFICIENCY OF OVERLOAD SPRAY IRRIGATION SYSTEM

	Mean Concentration mg/l		*Percent Removal*	
Parameter	*Wastewater*	*Section Effluent*	*Concentration Basis*	*Mass Basis*
TSS	263	16	93.5	98.2
TOC	264	23	90.8	—
BOD	616	9	98.5	99.1
Total phosphorus	7.6	4.3	42.5	61.5
Total nitrogen	17.4	2.8	83.9	91.5

sends it to the final polishing pond. Such a system that begins with raw waste sprayed on the field at an average 635 ppm BOD concentration results in an underdrain effluent concentration of less than 10 ppm BOD. The percent BOD reduction on a concentration basis is almost 99%.

Overland Techniques

The best-documented system of overland filtration is a food industry installation in Paris, Texas. The soil structure is gray clay underlaid by red clay at varying depths that are highly erodable. Little infiltration occurs in this system which is laid on a land so contoured that the waste flows in a thin sheet across the surface. The treatment depends on the microbiological activity of soil organisms to purify the water as it travels across the field. The treated water is collected in terraces, and four or five sprinkler lines are laid out on a hillside as shown in Figure 11.3.5. This installation applies the wastewater at approximately 0.6 in/day or 3 in/5-day week.

Studies indicate that 175 ft of downhill slope provide effective purification. Sprinklers normally blanket an area 100 ft in diameter, and the downslope requirement is 50 ft beyond the perimeter of the sprinklers. For maximum efficiency, the degree of slope should be between 2 and 6%. Flatter slopes encourage puddling and subsequent anaerobic conditions, whereas the retention time on a steep slope is insufficient for complete treatment at normal application rates.

The primary grass in this system is Reed Canary, which yields a large quantity of high-quality hay, containing up to 23% crude protein with twice the mineral content of other good-quality hay. In feeding tests, cattle preferred the hay grown on the disposal site to other types of hay.

By relating potential evapotranspiration to the quality and quantity of the hay crop environmental engineers can predict the time of year or stage of growth when the highest-value hay can be harvested. They can use the relationship between potential evapotranspiration and soil tractionability to plan a hay harvest that least disrupts the disposal system's normal operation. Poor soil tractionability can interfere with a planned harvest, whereas in other areas with lighter soils, an optimum harvest can result in highest-crop value and equipment utilization. Two or more harvests of Reed Canary grass per growing season may be feasible.

This installation has achieved removal rates of up to 90% phosphorus and nitrogen in the wastewater. The subsequent reclaiming of most of these nutrients by the hay crop extends the finite capacity of the soil to store nutrients.

While the soil concentration of TDS and sodium is increasing at this site (Paris, Texas), it has not reached the point that is injurious to plants. Some signs exist that the rate of increase is lessening and a state of equilibrium is being approached. Nitrates percolating through the groundwater reserve are not expected to build to a harmful level.

Table 11.3.2 summarizes the treatment performance of the Paris overland flow system. Early observations indicated that while BOD and nitrogen removal were high—99 and 90%, respectively—the phosphorus removal was low, about 45% (see Table 11.3.2). A change in the operating procedure that provided a longer rest period between applications, with no change in the total volume, increased the phosphorus removal to nearly 90% without affecting the BOD or nitrogen removal efficiency.

An analysis of groundwater samples showed that while mineral salts had increased over a 5-year period, the total accumulation was not critical and the rate of increase appeared to be dropping off. Based on the data accumulated to date, no significant disturbance to the soil structure is anticipated for 35 to 50 years, and an equilibrium stage (due to rainfall) will probably be reached sometime in the interim.

Conclusions

Industrial wastewater disposal by spray irrigation provides a low-cost method of waste treatment in many areas. Soils have an enormous capacity to absorb pollutants and convert them into plant nutrients or inorganic substances through microbiological activity. Placing soluble organic matter on soil oxidizes it to carbon dioxide or converts it to humus. Phosphates are held by the soil, and nitrates are taken up by the plants. Nitrates are also denitrified by soil microorganisms. Thus, a high-quality water effluent is produced.

The least-expensive systems are those on soil with high infiltration capacity and efficient hydrological characteristics. However, successful installations can be made on flat, poorly drained areas or gently sloping land.

—*T.F. Brown, Jr.*
L.C. Gilde, Jr.

11.4 OCEAN DUMPING

Dumping waste in the ocean, seas, estuaries, or inland lakes is regulated by emission standards for abating pollution of the oceans. The concept of an infinite ocean (a mile and a half deep on the average around the world) has given way to the reality that the ocean is a limited and valuable resource. This resource must be protected, otherwise it can become like Lake Erie and the Baltic Sea, and irreversible oceanic life systems may create an uninhabitable environment for people as well as for marine life. To quote the famous French underwater explorer–scientist Jacques Cousteau, "In 30 years of diving, I have seen this slow death everywhere underwater. In the past 20 years, life in our oceans has diminished 40%".

Some calculations suggest that the cycle time of oceans (the time required for an ocean's waters to evaporate, form clouds, return to the land in the form of rain, percolate down into the groundwater, and eventually return to the rivers and back to the ocean) is about 2000 years. Cousteau was reporting only the first visible consequences of ocean pollution. It will take 2000 years to learn the total impact of the pollution to date.

Most scientists say that more research is needed, but in the meantime, caution is essential. A research facility (New York Ocean Science Laboratory in Montauk, Long Island) is emphasizing the study of ocean pollution, including the effects of organic matter and heavy metals, particularly mercury.

Worldwide, losses from pollution are due to reduced eatable seafood despite the addition of organic matter and nutrients.

Effects

Some authorities argue that adding organic nutrients benefits the sea, citing statistics on the increased yield of fish. Increased fish yields in controlled environments such as nutrient-rich fish ponds support this view. Others argue that ocean dumping of sludge is safe because sludge contains only treated and stabilized biodegradable substances without any floatables.

Some also argue that as long as enough DO is in the water to support animal life and decompose organic waste, sludge dumping does not upset the ecological balance of the receiving water. With this logic, the ocean can be viewed as a great sink, capable of absorbing almost anything that is thrown into it. If this view were correct, San Francisco Bay could handle the waste of 200 million people since the tidal action in the bay replaces the water twice a day.

Some frequently argue that sludge, the end product of the sewage treatment process, is a benign substance. This statement is not completely true. Unfortunately, the sludge from a city like New York also contains toxic industrial waste because industrial plants frequently dump their waste in the municipal sewage system. The synergistic effects of the many synthetic chemicals, toxic substances, pesticides, PCBs, heavy metals, and medical wastes containing viruses and bacteria are not fully understood and are likely to be harmful to the ecology of the receiving water. Many experts feel that neither the DO level nor the organic-waste-assimilating capacity of oceans is a safe criterion for waste disposal. The effects and interrelationships are more complex, and the consequences are not understood well enough to accept such simplistic arguments.

Since dumping began at the 106 Mile Site, which receives New York's sludge, Rhode Island fishermen and lobstermen have reported diseased lobsters and crabs and a general drop in their catch of bottom-dwelling fish. Some controversy exists concerning the fishermenís reports that sludge is driving the fish away or that it causes shell disease (burn-spot disease) in lobsters and crabs. The National Oceanic and Atmospheric Administration (NOAA) indicated that the ocean dilution is sufficient to eliminate the harmful effects; however, it made that statement without studying the fish in the area.

Scientists have reported a proliferation of certain forms of sea life at the 106 Mile Site. Part of the reason why damage to the receiving water is reduced, if not eliminated, is due to the great depth and large area of this site. The heavy fraction of the sludge takes 3 to 4 days to sink to the bottom, while the lighter particles take up to a year. This delay allows more time for microorganisms to decompose the sludge. While scientists have reported that the ocean bottom at the dump site teems with sea life, they have not yet determined if this sea life is contaminated with bacteria, viruses, or heavy metals that might enter the food chain.

Biological studies on the Chesapeake Bay relate primarily to the disposal of bottom deposits dredged from the area and show the degree of pollution and its drift and harmful effect on the environment. For example, the total phosphates and nitrogen levels increased by a factor of 50 to 100 over normal levels as spoil material deposited over an area five times that designated for the disposal. The studies indicate that life was adversely affected, particularly that of the bottom organisms. A mathematical approach is useful to the understanding of benthic sludge decomposition and the degree and rate of purification of waste deposited on the sea bottom surrounding outfalls.

Regulation

Ocean dumping of sludge and other solid waste is widely practiced in Europe and Japan. For many decades in the United States, sewage sludge was barged to approved areas on the Gulf and Atlantic coasts. Toxic waste to be dumped in the ocean was usually put in containers and shipped to more remote locations. While many cities, such as San Diego and San Francisco, have banned ocean dumping, others continue to barge their sludge into the sea.

The marine disposal of radioactive waste was terminated in the United States in 1967. Yet in 1968, the yearly quantity of other waste dumped in the sea was still close to 50 million tn (see Table 11.4.1). Even though Congress banned ocean dumping of sewage sludge in 1992, this form of waste disposal will probably not stop completely until the turn of the century.

Unregulated disposal beyond the boundaries of the territorial sea imperils the waters, resources, and beaches of the maritime nations. Specific legislation is needed to give national and international authorities the responsibility for preventing ocean pollution and protecting ocean resources. Creating such authorities and enforcement methods is a slow, difficult process, and no effective policy for ocean management has evolved on either the national or international level.

ILLEGAL DUMPING

The incident of the *Khian Sea* shows the state of international controls on ocean dumping. In September 1986, the *Khian Sea,* owned by a Bahamian company, the Amalgamated Shipping Corporation, loaded in Philadelphia with 28 million lb of toxic incinerator ash. The ship attempted, unsuccessfully, to discharge its load in the Bahamas, the Dominican Republic, Honduras, Costa Rica, Guinea-Bissau, and the Cape Verde Islands. In February 1987, it discharged 4 million of its 28 million lb of ash in Haiti but then was ordered out of that country.

In September 1988, the *Khian Sea* was sighted in the Suez Canal with a new name, *Felicia,* and a new owner, Romo Shipping, Inc. In October 1988, Captain Abdel Hakim, vice-president of Romo Shipping, sent a message to Amalgamated Shipping (which has since gone out of business) indicating that the ash had been discharged; Hakim did not say where it had been discharged. The resolution of this case and who will answer for what in front of which legal authority is not clear. But this incident illustrates the chaotic state of international control over dumping in international water.

Illegal dumping is not limited to international water. In December 1988, the State Attorney General of New York accused the General Marine Transport Corporation of illegal dumping in the Raritan River, the Hudson Bay, and the coastal waters of New York City. The lawsuit also names four officers of the corporation and recommends placing environmental police on barges.

Some sludge haulers do not take their loads to designated areas but dump them closer inshore. To control this situation, the EPA now requires that each load of sludge be accompanied by a black box, which is dumped with the load. This requirement allows the EPA to protect against cheating. Another EPA requirement is that the barge must dump the load slowly to maximize dispersal.

Illegal dumping also includes medical waste, which has caused New York area beaches to close. Until this dumping occurred, no government agency was charged with tracking the safe disposal of hospital waste from the point of generation to the point of disposal. After the beach closings, Congress introduced bills requiring the EPA to create a paper trail to control the disposal of medical waste. New regulations in New York State require hazardous and infectious medical debris, including needles, to be placed in strong, moisture-resistant, red bags conspicuously labeled as infectious. Medical practitioners must either carry their infectious waste to an approved hospital incinerator or deliver it to a certified trucker. Regulations have also established an elaborate record-keeping system to track waste from the source to disposal.

Sludge Pipelines and Marine Fills

Some coastal cities have found that piping sewage sludge into the sea is less expensive than barging it. In the late 1960s, Los Angeles reduced its sludge-handling costs from \$14 to \$2/dry ton by constructing a 7-mi-long, 22-in-di-

TABLE 11.4.1 ESTIMATED MARINE DISPOSAL FOR 1968

	Pacific Coast		*Atlantic Coast*		*Gulf Coast*		*Total*		*Percent of Total*	
Type of Waste	*Annual Tonnage*	*Estimated Cost, $*	*Annual Tonnage*	*Estimated Cost, $*	*Annual Tonnage*	*Estimated Cost, $*	*Annual Tonnage*	*Estimated Cost, $*	*Tonnage*	*Cost*
Dredging spoils	7,320,000	3,175,000	15,808,000[a]	8,608,000	15,300,000	3,800,000	38,428,000	15,583,000	80%	53%
Industrial										
waste	981,000	991,000	3,011,000	5,406,000	690,000	1,592,000	4,682,000	7,989,000	10%	27%
containerized	300	16,000	2200	17,000	6000	171,000	8500	204,000	<1%	1%
Refuse[b]	26,000	392,000					26,000	392,000	<1%	<1%
Sludge[c]			4,477,000	4,433,000			4,477,000	4,433,000	9%	15%
Miscellaneous	200	3000					200	3000	<1%	<1%
Construction and demolition debris			574,000	430,000			574,000	430,000	1%	2%
Explosives			15,200	235,000			15,200	235,000	<1%	<1%
Total waste[d]	8,327,500	4,577,000	23,887,400	19,129,000	15,986,000	5,563,000	48,210,090	29,269,000	100%	100%

Notes: [a]Includes 200,000 tn of flyash
[b]At San Diego, 4700 tn of vessel garbage at $280,000 dumped in 1968 (discontinued in November 1968).
[c]Tonnage on wet basis. Assuming average 4.5% dry solids, this amount is about 200,000 tn/yr. of dry solids being barged to sea.
[d]Radioactive waste omitted because sea-disposal operations were terminated in 1967.

ameter pipe and discharging the sludge through the pipe at a depth of 320 ft on the edge of a submarine canyon. Later, Los Angeles mixed some of its sludge with sawdust and sold it as compost.

In designing outfalls in the ocean, environmental engineers try to achieve good mixing between the heavier saltwater and the lighter sewage effluent. Good mixing is essential to ensure that the waste does not surface like an oil slick and pollute the shoreline and beaches. Environmental engineers can use the following equation in designing an outfall:

$$N = (K)(Q^2)/(Y)(X^2) \qquad 11.4(1)$$

where:

N = The maximum tolerable shore pollution expressed as the arithmetic mean or 80 percentile
Q = The average sludge flow
Y = The depth at which the sludge is discharged into the ocean
X = The distance of the discharge point from the shore
K = A constant that varies from 5 to 10 million, when the units are in feet and gallons

Another method of disposing sludge and municipal solid waste is ocean landfills. Hong Kong has built marine fills for disposing its refuse in the estuaries of the bay. The marine fill is surrounded by a solid rock dyke, and refuse is loaded and compacted into it. Tidal action causes some leaching of the marine fill. After about 3 years, the refuse converts into compost and is used on nearby farms.

Disposal Methods

The most common ocean disposal method is to thicken the waste to a sludge or solid and barge it to the point of disposal. When the waste is toxic, it is usually put in containers and dropped in more remote places. So-called approved areas exist on the East, Gulf, and Atlantic coasts for waste disposal.

Detailed oceanographic studies indicate that inversion areas, in the water above the outfalls, of piped sludge limit the spread of solids and coliform bacteria to the surface although evidence exists that some digested sludge travels as much as 6 miles.

Wastewater treatment facilities can also dispose sludge at sea through a pipeline either by diluting digested sludge with the treated effluent from the plant or reducing the solids content and allowing the solids to be diffused into the ocean with the sewage. The advantage of removing the solids and digesting them prior to disposal is that this treatment results in an 80% reduction of volatile solids and more than a 99% removal of coliform and pathogenic bacteria.

The disposal of sludge and other solid waste in the ocean is more prevalent in other countries than in the United States (particularly in the industrialized areas of Japan and Europe).

Abstracts appearing regularly in the *Journal for Water Pollution Control Federation* stress the need for stricter regulations for ocean dumping as well as pollution abatement on a world scale. A low-cost alternative to ocean disposal is the disposal of digested sludge on croplands.

—J.R. Snell
Béla G. Lipták

11.5 AIR DRYING

LAND AREA REQUIRED

1 to 2 ft^2 per capita

LAYER THICKNESS ON SAND BEDS

7 to 8 in of digested primary sludge with 6 to 8% solids

MOISTURE CONTENT OF DRIED SLUDGE

60 to 70%

SALE PRICE OF DRIED SLUDGE

Usually free

For a small community, air drying digested sludge is the accepted, most common, and most economic process for sludge treatment and disposal. The advantages of simplicity and economy overshadow the disadvantages of potential nuisance, susceptibility to adverse weather, residual pathogens, weed seeds, and insect populations. Design criteria are well established for various parts of the United States. Wastewater treatment facilities have replaced most of the sand with wide strips of pavement to facilitate the mechanical removal of the sludge. Except in dry regions of the country, this pavement has reduced draining and greatly increased the drying time.

Various additives used to reduce the drying period (alum, lime, and polyelectrolytes) are not practical. Wastewater treatment facilities can improve drying in open sand beds by the following means:

1. Making the sand bed uncompacted and smooth prior to flooding
2. Providing 7 to 8 in of wet-sludge depth for optimum results in an uncovered dry bed
3. Providing 12 to 14 in of wet sludge in covered beds
4. Providing a prethickened sludge, which is better than a thin sludge

Except for odor control in developed areas or areas with an extremely cold climate, beds need not be covered. Final preparation of the dried sludge for public use, such as shredding, windrow composting, or heat drying, increases its value.

Sand-bed drying has many advantages for smaller communities. Over 70% of communities with a population less than 5000 use sludge drying beds. This use drops to 25% for municipalities with a population between 5000 and 25,000, and to 5% for cities with more than 25,000 inhabitants. The only economic alternative to sludge drying beds for a small community is a sludge lagoon or land disposal.

When communities use composting in digesting the organic part of municipal refuse, introducing a gravity-thickened, raw sludge is economically competitive. For larger coastal communities, ocean disposal has been used but is being scrutinized as polluting. Perhaps the strongest advantage of using the drying beds is their simplicity; no special skills are needed to operate them.

The greatest disadvantages of this technique are the large areas required, the potential nuisance from odors and insects, and the cost of labor to remove the sludge after drying. Open drying beds are susceptible to adverse weather, while covering them (except under unusual circumstances) is impractical. The weathering process of drying on an open sand bed causes some nitrogen loss. Pretreatment through anaerobic or aerobic digestion prior to dewatering is necessary to stabilize the sludge. Weed seeds and pathogens are not destroyed, and open sand drying beds attract insect populations.

Area Needed

The area required for open sand beds depends on the weather conditions (humidity, temperature, and rainfall). Tables 11.5.1 and 11.5.2 give the area needed for sludge drying beds under different circumstances.

Mechanical Sludge Removal

The minimum design for sludge removal should include two concrete strips leading into each sludge drying bed so that a pickup truck can back up close enough to all parts of the bed for an operator to cast the dried sludge directly onto the truck with a pitchfork. A depth of 9 to 18 in of well-washed sand should be used in the bed. A uniformity coefficient not over 4 and preferably under $3\frac{1}{2}$ should be used, and the effective size of the sand should be between 0.3 and 0.75 mm.

Underdrains should have 4-in diameter pipes spaced 8 to 10 ft apart so that they do not fill with sand. Backfilling with gravel around underdrains enhances their effectiveness. Partially covering the bed with asphalt surfaces sloping 1 to 2 in/ft allows front-end loaders to remove sludge.

Paved drying areas allow the use of mechanical equipment but decrease the drying rate compared to drainable, sand drying beds. An open-type, asphalt bed with a small layer of sand over it is a good compromise for maximizing drainage while supporting equipment.

CHEMICAL AND PHYSICAL CONDITIONING

Properly administered, chemical pretreatment of sludge before drying can reduce the drainage and drying time by 50%. Most operators do not use chemicals under normal

TABLE 11.5.1 SLUDGE DRYING BED AREA NEEDED FOR DEWATERING DIGESTED SLUDGE

	Area (ft²/capita)[a]	
Type of Sludge	*Open Beds*	*Covered Beds*
Primary	1.00	0.75
Trickling filters	1.50	1.25
Activated sludge	1.75	1.35
Chemical coagulation	2.00	1.50

Note: [a]South of 40° north latitude, these figures can be reduced by 25%; and north of 45° north latitude, they should be increased by 25%.

TABLE 11.5.2 SLUDGE DRYING BED AREA NEEDED FOR DEWATERING DIGESTED SLUDGE—RANGE OF VALUES

	Area (ft²/capita)[a]	
Type of Sludge	*Open Beds*	*Covered Beds*
Primary digested	1.0 to 1.5	0.75 to 1.0
Primary and humus digested	1.25 to 1.75	1.0 to 1.25
Primary and activated digested	1.75 to 2.5	1.25 to 1.5
Primary and chemically precipated digested	2.0 to 2.25	1.25 to 1.5

[a]With glass-covered beds, more sludge drawings per year are obtained because of protection against rain and snow; a combination of open and enclosed beds provides maximum use. Open beds can evaporate cake moisture faster than covered beds under favorable weather conditions.

circumstances; they use them only when extra quantities of sludge must be disposed on overcrowded facilities. The usual chemical choices are ferric chloride, chlorinated copperas, and alum. The dosage for alum is about 1 lb of commercial grade in 100 gal of digested sludge. It reacts not only with the hydroxide ion to produce a floc of $Al(OH)_3$ but also with carbonated salts in digested sludge to release carbon dioxide, causing the sludge particles to float and the liquor moiety to drain away more readily.

The best coagulant and optimum dose for a sludge are best determined in the laboratory and then tried on a pilot scale in the field. The sludge should be freshly drawn from the tank and properly mixed with the liquid coagulant just prior to the sludge running onto the sand drying bed.

Both inorganic coagulants and organic polyelectrolyte flocculants can be used. Polyelectrolytes save not only money but also floor space and improve safety and reduce operating time. The findings for polymer use in sludge preflocculating before vacuum filters also apply to preflocculating prior to sand bed drying.

The drying rate of well-digested sludge can be classified into two phases: (1) a constant-rate drying period and (2) the falling-rate drying period, which on the average is 5% greater than the evaporation rate of free water. Formulas for determining the drying time of well-digested sludge are believed to be applicable to both sludge drying beds and lagoons.

Fly ash can also be used in wastewater treatment and sludge conditioning. Since fly ash is generally a waste product from most power plants, it can be obtained for the cost of hauling. Fly ash alone or combined with lime can successfully be used a coagulant for either wastewater or sludge. Sawdust and coal have also been used with some success as an aid in sludge dewatering. Introducing these solid additives is more difficult than adding liquid ones.

FREEZING

The phenomenon of a rapid water release after sludge has been slowly frozen and thawed is well known. The problem is that freezing and thawing sludge on sand beds is difficult. It requires the installation of freezing tubes and a ready way of adding and removing insulation to and from the surface. Plant operators can frequently take advantage of freezing and thawing by placing sludge on the bed late in the fall and removing it early in the spring or during a winter thaw.

VACUUM UNDERDRAINS

Theoretically, evacuating the underdrains of sand drying beds speeds up the initial dewatering of sludge. As soon as water drains to the point at which air can enter, the vacuum does little or no good unless the surface of the sludge bed is covered with a plastic or membrane seal. The practicality of speeding dewatering by this method is arguable but should be further explored.

OPTIMUM DEPTH FOR VARIOUS SLUDGE

A depth of 7 to 8 in is optimum on open sand beds dosed with well-digested primary sludge of 6 to 8% solids. Greater depths can be used under certain conditions, and depths as high as 12 to 14 in are possible with covered sand beds. A thinner sludge can be placed in a deeper layer than a thickened sludge. The same may be true of sludge pretreated with chemicals or polymers. Trickling-filter or activated sludge alone or combined with primary sludge should not be placed in as thick a layer as primary sludge alone because of the slime present.

DEGREE OF DRYNESS

The purpose of a sludge drying bed is to reduce the moisture content of digested sludge from 4 to 8% solids to 20 to 40% solids. Figure 11.5.1 shows the relationship between the percent solids of sludge applied to the bed and the expected pounds of solids that can be obtained from each square foot of bed annually. Generally, mechanical equipment can remove sludge with as much as 70 to 80% moisture. Hand removal with a pitchfork with tines about 1 in apart is the most common way of removing dried sludge in smaller plants. The moisture content should be between 60 and 70% for best results. At this moisture content, the sludge begins to peal away from the sand but does not yet crumble.

The more moisture the sludge contains, the more weight must be removed. If sludge is shredded and stored for use by the citizens, the moisture content at the time of removal is critical. If time permits and ample sand bed capacity exists, the operator can sometimes reduce the moisture content to as low as 50%. If heat drying is used prior to bagging, the sludge should be dried to at least 50% moisture. The same is true if composting is used as a pretreatment to bagging or use by citizens. However, if the sludge is taken to a landfill or farm, it can be removed with a moisture concentration as high as 70 to 80%.

COMPOSTING PARTIALLY DRIED SLUDGE

Equipment for composting animal manures should be applicable for composting sewage sludge.

MECHANIZED REMOVAL

In large cities, mechanical equipment removes the sewage sludge after drying on sand beds. Most larger modern installations use vacuum filters or centrifuges to dewater the sludge prior to ultimate disposal. Intermediate-size plants, where labor costs have increased the cost of hand sludge removal, have installed lightweight, front-end loaders and scrapers.

To prevent crushing underdrains or disturbing the sand, installations use pavement in strips or fractional areas. In other cases, flotation-type equipment is used directly over the sand. Occasionally, specially designed equipment spans the sludge drying bed and is supported on rails on either side.

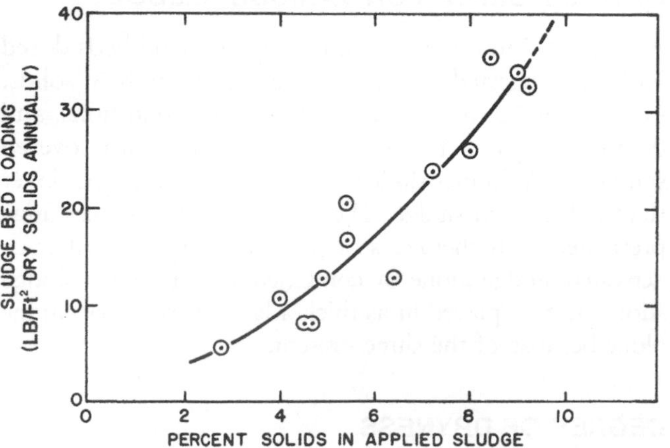

FIG. 11.5.1 Sludge volume reduction at a variable rate during drying, with the greatest volume reduction occurring at a 70 to 60% moisture content.

MAINTENANCE OF SAND BEDS

Installations should consider the following recommendations for good care of drying beds:

1. A coarse, 2.0-mm-effective-size sand should be used on the surface.
2. The sludge should be drawn slowly so that a hole is not pulled in the digester sludge blanket.
3. The sludge bed should be thoroughly cleaned, removing all small bits of dried sludge before the bed is reflooded.
4. The bed (preferably 8 in in depth) should be rotated if possible.

A report from Birmingham, England, showed a 9 in depth to be optimum after depths from 6 to 18 in were tried. These researchers also found that only 11% of the volume of sludge left the beds through the underdrains while the rest evaporated (Bowers 1957).

For tropical and semitropical areas, sludge drying beds should be substantially smaller than those in moderate or cooler climates. Because sand beds give only four loadings per year in Winnipeg, Canada and these loadings must be during the summer season, they are impractical. Drained lagoons in the same location generate no obnoxious odors and produced efficient drainage and drying in an 18-month period when poured to a depth of about 3 ft (Bubbis 1953). Without underdrains, 24 months are required. Drainage rates increase with coarser types of drainage media although at a cost of less solids in the filtrate.

COSTS AND SALES

No appreciable revenue can be expected from the sale of digested primary sludge dried on open sand beds if the quantities are small. The labor cost of removing and preparing this sludge for use is substantial. The purpose of encouraging the public to use it is more for sound public relations than to subsidize the treatment and disposal costs.

People generally like to have some humus on their gardens. They look to the city sewage plant as a place where it is readily available and are usually willing to go there for it but do not expect to pay for it. Because of the low quality of this type of sludge, it is generally not bagged and sold, as is heat-dried sludge.

—J.R. Snell

11.6 COMPOSTING

Composting is the aerobic, theromophilic, biological decomposition of organic material under controlled conditions. It is essentially the same process that is responsible for the decay of organic matter in nature except that it occurs under controlled conditions.

Over the past 20 years, legislative actions have imposed strict limits on the disposal of organic waste such as sewage sludge, municipal solid waste, and agricultural waste due to the potentially severe environmental problems associated with the management of such residuals. For example, 10 years ago, most sludge produced in the United States was disposed in oceans or landfills. Ocean disposal is now illegal, and landfills are rapidly closing. Also, increasing, stringent, air pollution regulations make incineration less attractive.

At the same time, increased amounts of organic wastes are being generated. This increase is especially true of sewage sludge because of the upgrading of wastewater treatment plants and the expansion of services (U.S. EPA 1993).

As a consequence, new practices are being encouraged that include the treatment of organic waste with resource recovery. Composting is one of these practices (Kuchenrither et al. 1987). Composting is a method of solid waste treatment in which the organic component of the solid waste stream is biologically decomposed under controlled aerobic conditions to a state in which it can be safely handled, stored, and applied to the land without adversely affecting the environment. It is a controlled, or engineered, biological system.

Composting can provide pathogen kill, volume reduction and stabilization, and resource recovery. Properly composed waste is aesthetically acceptable, essentially free of human pathogens, and easy to handle. Compost can improve a soil's structure, increase its water retention, and provide nutrients for plant growth. As Hoitink and Keener (1993) note "It is not surprising therefore that composting of wastes has resulted in a variety of beneficial effects in agriculture as the Western World progressed from a 'throw-away' mentality to a more environmentally friendly society."

Golueke (1986) points out that composting relies more on scientific principles as time passes. At the same time, advances have been made in the technology used for the process, such as the static pile (Epstein et al. 1976), and the development of in-vessel systems (U.S. EPA 1989). Coupling the need for waste diversion practices (from landfills) with the advances in the fundamental science and process technology has increased the use of composting for wastewater treatment. Approximately 159 sludge composting facilities operate in the United States (Goldstein, Riggle, and Steuteville 1992).

This section concentrates on the fundamentals of sludge composting. Several excellent sources provide more detail on composting principles (Hoitink and Keener 1993; Rynk 1992; Haug 1993).

Process Description

Numerous types of composting systems exist, but for the most part, composting systems can be divided into three categories: windrow, static pile, and in-vessel. Windrow systems are composed of long, narrow rows of sludge mixed with a bulking agent. The rows are typically trapezoidal in shape, 1 to 2 m high and 2 to 4.5 m wide at the base. The rows are usually uncovered but can be protected by simple roofs. The sludge mixture is aerated by convective air movement and diffusion. Wastewater treatment facilities periodically turn the rows using mechanical means to expose the sludge to ambient oxygen, dissipate heat, and refluff the rows to maintain good free air space. Windrows can also be aerated by induced aeration (Hay and Kuchenrither 1990). Windrows are space-intensive but mechanically simple.

Static pile systems are also composed of a sludge–amendment mixture but are aerated by forced-aeration systems installed under the piles (Epstein et al. 1976). The aeration can be either positive or negative. Finstein, Miller, and Strom (1986) stress the need for positive aeration for process control. Others note the advantage of negative (suction) aeration for better possibilities of capturing the process air for odor control. Currently, most facilities in the United States use the static pile method (Goldstein, Riggle, and Steuteville 1992).

In-vessel composting takes place in either partially or completely enclosed containers. A variety of schematics use various forced aeration and mechanical turning technologies (Tchobanoglous and Burton 1991). In-vessel composting is space efficient but more mechanically complex than the other two system categories. They offer excellent possibilities for process and odor control. Among the facilities commissioned within the past few years, a greater percentage are using in-vessel methods (Goldstein, Riggle, and Steuteville 1992).

Each of the system categories is capable of producing a good compost in a reliable and efficient manner. The choice of any given system depends on the site location, available space, and other local conditions.

Each system is composed of common basic steps. As shown in Figure 11.6.1, the basic steps of the composting process include the following:

1. Mixing dewatered sludge with a bulking agent
2. Aerating the composting pile by either the addition of air or mechanical turning
3. Further curing
4. Recovering the bulking agent
5. Final distribution

The first and second steps are critical to the process success. Recovering the bulking agent is an optional step that relates to system economy (reuse of the bulking agent) and product quality (the product compost with or without wood chips). The curing stage also relates to product quality because it influences compost stability. During this period, which can last as long as 30 to 60 days, further product stabilization with pathogen die-off and degassing occurs (Rynk 1992). Final disposal depends on the market for the product compost. The intended market for the compost dictates the need for bulking agent recovery as well as the length of the curing stage, and any other final operations (such as bagging).

While composting is a simple process, facilities must operate in a careful manner to ensure the production of a good-quality, stable compost while minimizing adverse environmental aspects, such as odor production. To ensure the production of a stable compost in a reliable and efficient manner while minimizing odor production, wastewater treatment facilities must operate any system to promote the growth of the microbial population and maintain these organisms under proper environmental conditions for a sufficient amount of time for the reactions (of stabilization) to occur.

The diagram proposed by Rynk (1992), as shown in Figure 11.6.2, shows the composting process. As described by Rynk (1992), the following conditions must be established and maintained:

- Organic materials appropriately mixed to provide the nutrients needed for microbial activity and growth, including a balanced supply of carbon and nitrogen (C:N ratio)
- Oxygen at levels that support aerobic organisms
- Enough moisture to permit biological activity without hindering aeration
- Temperatures that encourage vigorous microbial activity from thermophilic microorganisms

Process Fundamentals

The factors affecting the composting process include oxygen and aeration, nutrients (C:N ratio), moisture, porosity, structure, texture and particle size, pH, temperature, and time. These conditions are developed and maintained by process management. The following considerations are important for process management:

- Raw material selection and mixture
- Moisture management
- Aeration
- Time

These considerations are explained next.

RAW MATERIAL SELECTION AND MIXTURE

Wastewater treatment facilities add amendments to sludge to adjust the moisture and other characteristics (such as nutrient level) to improve composting. Bulking agents are added for structural support. The goal is to create a mixture of sludge and bulking agent and amendment with the proper characteristics to support aerobic digestion. The choice of material depends on the characteristics of the sludge, in particular the moisture content (which depends on the degree and type of dewatering process) and the nitrogen content of the sludge. The proper mixture has an appropriate C:N balance, proper porosity (to ensure aerobic conditions), and proper moisture content. Haug (1993) points out that the amount of free air space in the mixture is more crucial than the porosity, which is the amount of space not occupied by solids or water.

In terms of porosity and moisture content, the latter is usually the determining factor. A typical value for a mixed pile is about 60% moisture and 40% solids. High moisture levels lower the free air space of the pores and thus

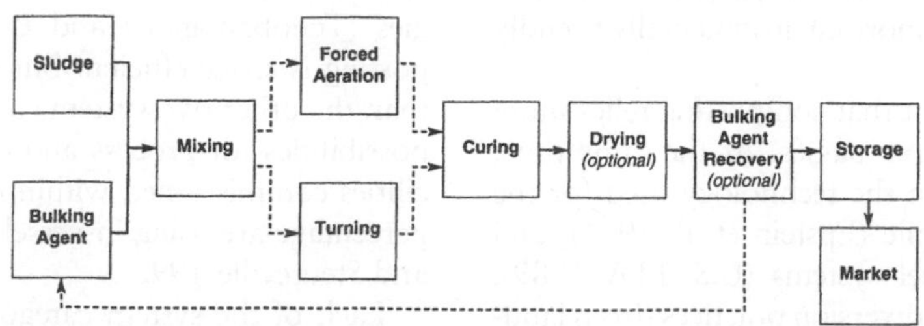

FIG. 11.6.1 Composting process flow diagram.

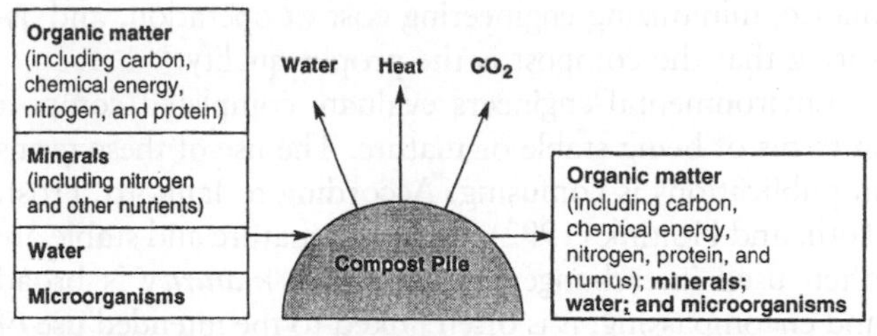

FIG. 11.6.2 Composting process.

inhibit aerobic activity, while low moisture levels do not support sufficient biological activity.

MOISTURE MANAGEMENT

Moisture management is an important part of composting. As stated previously, typically the initial moisture content of the sludge mixture is adjusted to about 60%. During the composting process, water is lost via evaporation. Water loss is driven by diffusion, air exchange, and heat generation. Some water can leach out of the mixture. Water is gained by precipitation (for uncovered systems) and as a product of respiration. In general, a net loss of water occurs. The final mixture has a moisture content of about 40%. As noted, both too high and too low levels are problems.

AERATION

Aeration serves three interdependent functions of composting. Aeration adds stoichiometric oxygen for respiration, removes water vapor, and dissipates heat. Finstein and Hogan (1993) note that heat removal determines the rate of aeration and stress that this removal is important for process control. For proper pathogen removal, the temperature must reach at least 55°C. However, allowing a composting system to reach temperatures of 70 to 80°C is self-limiting, results in poor operation, and leads to the production of unstable compost.

TIME

The length of time the composting process runs depends on the degree of stability of the compost being produced. In other words, the end point is variable. The degree of stability depends on the use of the compost end product. The time varies depending on the type of reactor and sludge mixture, but in general, the active composting time is 3 to 4 weeks. This time does not include additional curing time.

Design

Many factors must be considered in the design of a composting system. These factors are summarized by Tchobanoglous and Burton (1991) and are listed in Table 11.6.1.

TABLE 11.6.1 DESIGN CONSIDERATIONS FOR AEROBIC SLUDGE COMPOSTING PROCESSES

Item	*Comment*
Type of sludge	Both untreated and digested sludge can be composted successfully. Untreated sludge has a greater potential for odors, particularly for windrow systems. Untreated sludge has more energy available, degrades more readily, and has a higher oxygen demand.
Amendments and bulking agents	Amendment and bulking agent characteristics, such as moisture content, particle size, and available carbon, affect the process and quality of the product. Bulking agents should be readily available. Wood chips, sawdust, recycled compost, and straw can be used.
C:N ratio	The initial C:N ratio should be in the range of 25:1 to 35:1 by weight. Checking the carbon ensures that it is easily biodegradable.
Volatile solids	The volatile solids of the composting mix should be greater than 50%.
Air requirements	Air with at least 50% oxygen remaining should reach all parts of the composting material for optimum results, especially in mechanical systems.
Moisture content	The moisture content of the composting mixture should be not greater than 60% for static pile and windrow composting and not greater than 65% for in-vessel composting.
pH	The pH of the composting mixture should generally be in the range of 6 to 9.
Temperature	The optimum temperature for biological stabilization is between 45 and 55°C. For best results, the temperature should be maintained between 50 and 55°C for the first few days and between 55 and 60°C for the remainder of the composting period. If the temperature increases beyond 60°C for a significant period of time, the biological activity is reduced.
Mixing and turning	Mixing or turning the material being composted on a regular schedule or as required prevents drying, caking, and air channeling. The frequency of mixing or turning depends on the type of composting operation.
Heavy metals and trace organics	Heavy metals and trace organics in the sludge and finished compost should be monitored so that the concentrations do not exceed the applicable regulations for end use of the product.
Site constraints	Factors in selecting a site include the available area, access, proximity to the treatment plant and other land uses, climatic conditions, and availability of a buffer zone.

Numerous references provide additional information on the design of composting systems. In particular, Rynk (1992), Haug (1993), and U.S. EPA (1985) provide information for general design, and U.S. EPA (1989) provides information for the in-vessel system.

Special Considerations

With restrictions on the disposal of sludge, beneficial use of biosolids has become a significant trend. Composting is the leading beneficial reuse technology in terms of manufacturing a product for application to the land. Of course, the success of composting depends on marketing the final compost product. In other words, a use must exist for the compost generated from wastewater sludge.

In addition, the public must accept the composting process. Donovan (1992) notes that the most difficult challenge municipalities face in implementing sludge plans is facility siting. The general public is apprehensive concerning any waste handling facility, and specific concerns about odor, health, traffic, and land values have slowed or stopped many projects.

Composting is basically a simple process; it is quite robust and therefore a forgiving process. It can be managed in many cases (such as the backyard compost pile) with little or no technical knowledge. However, as composting applications increase and broaden in scope, the need exists for more sophistication in the design and operation of composting facilities. This need is underscored by the emphasis on aspects such as odor control and compost product quality. Such concerns demand a higher level of technology and management.

STABILITY AND PRODUCT QUALITY

In composting operations, the objective is decomposition rather than complete stabilization. The degree of decomposition, however, is not an absolute state since it depends on the final product use. In some cases, the degree is one where the material does not cause nuisances when stored even if it is wetted. If the final product is used on a plant system, the compost should not be phytotoxic.

Currently, many parameters can be used for composting process control and final product quality including the final drop in temperature; degree of self-heating capacity; amount of decomposable and resistant organic material; rise in redox potential; oxygen uptake and carbon dioxide evolution; starch test; color, odor, appearance, and texture; pathogen and indicator organisms; and inhibition of germination of cress seeds (Finstein, Miller, and Strom 1986; Inbar et al. 1990). This list covers many possibilities, but which are best for measuring the completion of composting is unclear. The optimal parameter or group of parameters is important for maximizing process performance, minimizing engineering cost of operation, and assuring that the compost is the proper quality.

Environmental engineers evaluate completed compost in terms of being stable or mature. The use of these terms in publications is confusing. According to Iannotti, Frost, Toth, and Hoitink (1992), the terms mature and stable are often used interchangeably. *Compost maturity* is broad and encompassing; it is often linked to the intended use of the compost and is therefore subjective. *Compost stability* is readily definable by its biological property of microbial activity. As such, Iannotti, Frost, Toth, and Hoitink (1992) propose a stability assay based on DO respirometry. Nonetheless, a simple, yet reliable, and universally acceptable analytical tool for evaluating compost stability does not exist.

In addition to stability, pathogen destruction is an important characteristic defining product quality. Other characteristics used for compost product specification include the concentration of specific constituents (e.g., metals and nutrients), particle size, texture, pH, moisture content, odor, weed seed inactivation, phytotoxicity, reduction of volatile solids, and product consistency (U.S. EPA 1989). The choice of characteristics depends on the compost use.

The major compost uses include large-scale landscaping (golf courses, public works projects, highway median strips), local nurseries, industries (as potting material), greenhouses, urban gardeners, land reclamation projects (strip mines), and landfill (daily and final cover).

Often the criteria used are legal regulations such as those for heavy metals and pathogens. Recently, federal regulations have been issued for the use and disposal of sewage sludge, including compost (U.S. EPA 1993). States are now in the process of adopting these regulations or formulating more stringent regulations.

PATHOGENS

Wastewater sludge is known to contain pathogens including bacteria, viruses, parasites, and helminths. Epstein and Donovan (1992) note that pathogens can be grouped under three major headings: primary pathogens, secondary or opportunistic pathogens, and endotoxins. They further note that the major concerns with pathogens related to composting wastewater sludge are product disinfection, worker health, and public health as impacted by facility location.

The U.S. EPA (1979; 1993) in the previous 40 *CFR* Part 257 regulations and the new 40 *CFR* Part 503 regulations is primarily concerned with product quality and safety of the compost. The possible presence of pathogens is a major concern. The previous regulation for pathogen control was technology based. Under 40 *CFR* Part 257, minimum standards were issued for processes to significantly reduce pathogens (PSRP). Compost that had been subject to PSRP could be used but was limited to certain restrictions. The previous regulations also defined processes to further reduce pathogens (PFRP). Fewer restrictions were placed on the use of PFRP compost.

Both PSRP and PFRP are based on a timeñtemperature requirement. For example, if a composting process reached at least 40∞C for at least 5 consecutive days, and 55∞C for at least 4 hr during that time, it met PSRP. If aerated, static piles and in-vessel systems maintained a temperature of at least 55°C for 3 consecutive days (in the coolest part of the pile), then that compost met PSRP. Such sludge was subjected to less restriction for distribution and marketing.

The new regulations regulate the product compost as well as the process. To obtain a compost that can be widely distributed or marketed (now called Class A), processors must use a PFRP time–temperature standard (or equivalent processing) and produce a product with less than or equal to 1000 fecal coliforms/g dry solids or less than or equal to 3 salmonella/g dry solids.

Numerous studies have been conducted for pathogen levels at composting facilities and in the final compost product (Epstein and Donovan 1992). Most of these studies focused on indicator organisms (fecal coliforms) and salmonellae. Yanko (1988) notes that composting is an effective method for the disinfection of sludge although considerable variability exists among the data related to the method of composting and system design and operation. In other words, proper disinfection requires careful system design and operation. In addition, the possibility exists of a repopulation of organisms in disinfected compost (Farrell 1993).

Several authors have reported the potential for the repopulation of salmonellae in composted wastewater sludge (Epstein and Wilson 1975; Brandon, Burge, and Enkiri 1977; Brandon and Neuhauser 1978; Burge et al. 1987). Brandon et al. (1977; 1978) relate the repopulation to the moisture content of the compost. Russ and Yanko (1981) evaluate the factors affecting salmonellae repopulation in sludge compost. They found that the moisture level, temperature, and nutrient content of the composted solids affect repopulation. They further report that repopulation is transient with population peaks occurring around 5 days followed by a subsequent die-off. Other authors note the importance of microbial competition for minimizing repopulation (Hussong, Burge, and Enkiri 1985; Yeager and Ward 1981).

The most important parameter for pathogen destruction is temperature. Adequate temperature can be reached, but it depends on proper facility design and operation. The most important considerations are preparing a good initial mix and maintaining aerobic conditions. These conditions are a function of the mix properties, moisture control, aeration, and the C:N ratio. As previously noted, these factors impact the composting process and affect pathogen destruction.

ODORS AND ODOR CONTROL

Odor control has become a major concern in the successful operation of any composting facility. Indeed, some operating facilities have been closed due to odor problems (Libby 1991). Numerous papers have been published identifying the causes of odors and management strategies to control odors (Hentz et al. 1992; Miller 1993; Goldstein 1993; Van Durme, McNamara, and McGinley 1992).

Many potential sources of odors exist at composting facilities. While the process air coming off a compost pile is most odorous, environmental engineers must evaluate all potential sources of odors. Therefore, a proper inventory of the potential sources of odors is necessary including liquid sludge and dewatered sludge facilities.

Haug (1990) states that odors are part of the composting process and cannot be eliminated, but they can be managed. Finstein et al. (1986; 1993) point out that controlling the composting process is crucial in minimizing odor production. This process control includes good aeration and maintaining the proper temperature, moisture, and structure of the piles.

A variety of compounds cause odors including fatty acids, amines, aromatics, inorganic and organic sulfur compounds, and terpenes (Miller 1993). They vary in their chemical properties. Some are acidic; some are basic. Some can be oxidized; others cannot. Differences exist in their solubility and adsorbability. Also, the compounds change over the course of time of operation. Therefore, any treatment system must have a broad spectrum of removal mechanisms.

Typical odor management (in addition to good process operation) involves containment of process gases, collection of gases, gas treatment, and proper dispersion. Other management possibilities include the dilution of odors with large volumes of air and the use of masking agents. Gas treatment options include oxidation processes, chemical scrubbers, and biofilters.

Summary and Conclusions

Composting is a cost effective and environmentally sound alternative for the stabilization and ultimate disposal of wastewater sludge. It produces compost—a stable, humus-like material which is a soil conditioner. Thus, the process can achieve waste treatment with resource recovery and represents a beneficial use of sludge.

Recent advances have been made in the basic fundamental science associated with composting along with the technology used for the process. These advances have increased the use of the process for wastewater sludge management.

While the composting process is simple in concept, it must be regarded as an engineered unit process. As such, it must be based on sound scientific principles, designed with good engineering, and operated with care by well-trained and motivated operators. With these practices, wastewater treatment facilities can produce a safe compost of consistently good quality in an environmentally sound manner.

—*Michael S. Switzenbaum*

References

Brandon, J.R., W.D. Burge, and N.K. Enkiri. 1977. Inactivation by ionizing radiation of *Salmonella enteritides* serotype montevideo grown in composted sewage sludge. *Applied and Environmental Microbiology* 33: 1011–1012.

Brandon, J.R., and K.S. Neuhauser. 1978. *Moisture effects on inactivation and growth of bacteria and fungi in sludges.* Pub. no SAND 78-1304. Albuquerque, N.M.: Sandia Lab.

Burge, W.D., P.D. Miller, N.K. Enkiri, and D. Hussong. 1987. *Regrowth of Salmonellae in composted sewage sludge.* EPA/600/S2-86/016.

Donovan, J.F. 1992. Developments in wastewater sludge management practices in the United States. Paper presented at the New Developments in Wastewater Policy, Management and Technology Conference, Sydney, Australia, May 18, 1992.

Epstein, E., and J.F. Donovan. 1992. Pathogens in compost and their fate. Paper presented at the WEF Seminar on Pathogens in Sludge, New Orleans, LA September 1992.

Epstein, E., and G.B. Wilson. 1975. Composting raw sludge. In *Municipal sludge management and disposal.* Rockville, Md.: Information Transfer Inc.

Epstein, E., G.B. Wilson, W.D. Burge, D.C. Mullen, and N.K. Enkiri. 1976. A forced aeration system for composting of wastewater sludge. *Journal Water Pollution Control Federation* 48: 688–694.

Farrell, J.B. 1993. Fecal pathogen control during composting. In *Science and engineering of composting,* edited by H.A.J. Hoitink and H.M. Keener. Hinesburg, Vt.: Upper Access Books.

Finstein, M.S., and J.A. Hogan. 1993. Integration of composting process microbiology, facility structure, and decision making. In *Science and engineering of composting,* edited by H.A.J. Hoitink and H.M. Keener. Hinesburg, Vt.: Upper Access Books.

Finstein, M.S., F.C. Miller, and P.F. Strom. 1986. Monitoring and evaluating composting process performance. *Journal Water Pollution Federation* 58: 272–278.

Goldstein, N. 1993. Odor control progress for sludge composting. *BioCycle* 34, no. 3: 56–59.

Goldstein, N., D. Riggle, and R. Steuteville. 1992. Sludge composting maintains growth. *BioCycle* 33, no. 12: 49–54.

Golueke, C.G. 1986. Compost research accomplishments and needs. *BioCycle* 27, no. 4: 40–43.

Haug, R.T. 1990. An essay on the elements of odor management. *BioCycle* 31, no. 10: 60–67.

———. 1993. *The practical handbook of compost engineering.* Boca Raton, Fla.: Lewis Publishers.

Hay, J.C., and R.D. Kuchenrither. 1990. Fundamentals and application of windrow composting. *Journal Environmental Engineering (ASCE)* 116: 746ñ763.

Hentz, L.H., Jr., C.M. Murray, J.L. Thompson, L.L. Gasner, and J.B. Dunson Jr. 1992. Odor control research at the Montgomery County Regional Composting Facility. *Water Environment Federation* 64: 13ñ18.

Hoitink, H.A.J., and H.M. Keener (eds). 1993. *Science and engineering of composting.* Hinesburg, Vt.: Upper Access Books.

Hussong, D., W.D. Burge, and N.K. Enkiri. 1985. Occurrence, growth, and suppression of Salmonellae in composted sewage sludge. *Applied and Environmental Microbiology* 50: 887–893.

Iannotti, D., Frost, B.L. Toth, and H.A.J. Hoitink. 1992. Compost stability. *BioCycle* 33: no. 11: 62–66.

Inbar, Y., Y. Chen, Y. Hadar, and H.A.J. Hoitink. 1990. New approaches to compost maturity. *BioCycle* 31, no. 12: 64–69.

Kuchenrither, R.D., D.M. Diemer, W.J. Martin, and F.J. Senske. 1987. Composting's role in sludge management: A national perspective. *Journal Water Pollution Control Federation* 59: 125–131.

Libby K. 1991. Lessons from a closed MSW composting plant. *BioCycle* 32, no. 12: 48–52.

Miller, F.C. 1993. Minimizing odor production. In *Science and engineering of composting,* edited by H.A.J. Hoitink and H.M. Keener. Hinesburg, Vt.: Upper Access Books.

Russ, C.F., and W.D. Yanko. 1981. Factors affecting Salmonellae repopulation in composted sludges. *Applied and Environmental Microbiology* 41: 597–602.

Rynk, R. ed. 1992. *On-farm composting handbook.* NRAES-54. Ithaca, N.Y.: Northeast Agricultural Engineering Service.

Tchobanoglous, G., and F.L. Burton. 1991. *Wastewater engineering,* 3d ed. New York, N.Y.: McGraw-Hill Inc.

U.S. Environmental Protection Agency (EPA). 1979. Criteria for classification of solid waste disposal facilities and practices. *Code of Federal Regulations.* Title 40, Part 257. *Federal Register* 179, (13 September 1979): 53438.

———. 1985. *Seminar publication on composting of municipal wastewater sludges.* EPA/625/4-85/014.

———. 1989. *In-vessel composting of municipal wastewater sludge.* EPA/625/8-89/016. Cincinnati, Ohio: CERI.

———. 1993. Standards for the use or disposal of sewage sludge. *Code of Federal Regulations.* Title 40, *Federal Register 58,* (19 February 1993); 9248.

Van Durme, G.P., B.F. McNamara, and C.M. McGinley. 1992. Bench-scale removal of odor and volatile organic compounds at a composting facility. *Water Environmental Research* 64: 19–27.

Yanko, W.A. 1988. *Occurence of pathogens in distribution and marketing municipal sludges.* EPA/600/1-87/014. Research Triangle Park, N.C.: HERL.

Yeager, J.G., and R.L. Ward. 1981. Effects of moisture content on long-term survival and regrowth of bacteria in wastewater sludge. *Applied and Environmental Microbiology* 41: 1117–1122.

Bibliography

Alkhatib, E.A., and L.T. Thiem. 1991. Wastewater oil removal evaluated. *Hydrocarbon Processing* (August).

American Society of Civil Engineers (ASCE). 1977. Wastewater treatment plant design. In *Water Pollution Control Federation Manual of Practice No. 8.* ASCE Manuals and Reports on Engineering Practice No. 36.

Bauman, E.R., and H.E. Babbit. 1953. *An investigation of the performance of six small septic tanks.* University of Illinois Engineer Experimentation Station, Bulletin Series No. 409.

Brown, J.D. and G.T. Shannon. 1963. *Design guide to refinery sewers.* Presented to the API Div. of Refining, Philadelphia, 14 May, 1963.

Buchanan, R.D. 1974. Pumps and pumping stations. In *Environmental engineering handbook,* edited by B.G. Liptak. Radnor, Pa.: Chilton Book Company.

Cotteral, J.A., and D.P. Norris. 1969. Septic tank systems. *J. Sanit. Engineer. Div. Amer. Soc. Civ. Engineer.* 95:715, August, 1969.

Eckenfelder, W.W. Jr. 1966. *Industrial water pollution control.* New York: McGraw-Hill.

Eckenfelder, W.W., J. Patoczka, and A.T. Watkin. 1985. Wastewater treatment. *Chem. Eng. Prog.* (September).

Fair, G.M., J.C. Geyer, and D.A. Okun. 1970. *Elements of water supply and wastewater disposal.* New York: John Wiley and Sons, Inc. (November).

La Grega, M.D. and J.D. Keenan. 1974. Effects of equalizing wastewater flows. Journal of Water Pollution Control Fed. 46, no. 1: 123.

Linsley, R.K., J.B. Franzini, D.L. Freyberg, and G. Tchobanoglous. 1992. *Water-resources engineering.* 4th ed. New York: McGraw-Hill, Inc.

Lipták, B.G., ed. 1974. *Environmental engineers' handbook. vol. 1, Water pollution.* Radnor, Pa.: Chilton.

Lipták, B.G. 1995. Pumps as control elements. In *Instrument engineers' handbook.* 3d ed. Radnor, Pa.: Chilton Book Company.

Lipták, J. 1974. Grit removal. Vol. I of *Environmental engineers' handbook,* edited by B.G. Lipták. Radnor, Pa.: Chilton Book Co.

Lipták, J. 1974. Screening devices and comminutors. In *Environmental engineers' handbook,* edited by B.G. Lipták. Radnor, Pa.: Chilton Book Co.

Metcalf & Eddy, Inc. 1981. *Wastewater engineering: Collection and pumping of wastewater.* New York: McGraw-Hill Book Company.

Metcalf & Eddy, Inc. 1991. *Wastewater engineering, treatment, disposal and reuse.* 3d ed. New York: McGraw-Hill Book Company.

Metcalf, L. and H.P. Eddy. 1935. *American sewerage practice (Vol. III). Disposal of sewage.* New York: McGraw-Hill.

Nemerow, N.L. 1962. *Theories and practices of industrial water treatment.* Reading, Mass.: Addison-Wesley.

Novotny, V., K.R. Imhoff, M. Olthof, and P.A. Krenkel 1989. *Karl Imhoff's handbook of urban drainage and wastewater disposal.* New York: John Wiley.

Phelps, E.B. 1944. *Stream sanitation.* New York: Wiley.

Qasim, S. 1985. *Wastewater treatment plant: Planning, design, and operation.* New York: Holt, Rinehart, Winston.

Rich, L.G. 1961. *Unit operations of sanitary engineering.* New York: Wiley.

Sawyer, C.N., and P.L. McCarty. 1978. *Chemistry for environmental engineering.* 3d ed. New York: McGraw-Hill.

Shieh, W.K., and C.Y. Chen. 1984. Biomass hold-up correlations for a fluidized bed biofilm reactor. *Chem. Eng. Des. Res.* 62, no. 133.

U.S. Environmental Protection Agency (EPA). 1975. Process design manual for suspended solids removal. EPA 625-I-75-003a. Washington, D.C.: U.S. EPA.

Yee, C.J. 1990. Effects of microcarriers on performance and kinetics of the anaerobic fluidized bed biofilm reactor. Ph.D. Dis., Department of Systems, University of Pennsylvania, Philadelphia.

Index

A

B

C

D

E

F

G

H

I

J

K

L

O

Q

R

S

T

U

V

W

Z

T - #0251 - 071024 - C0 - 279/216/25 - PB - 9780367399122 - Gloss Lamination